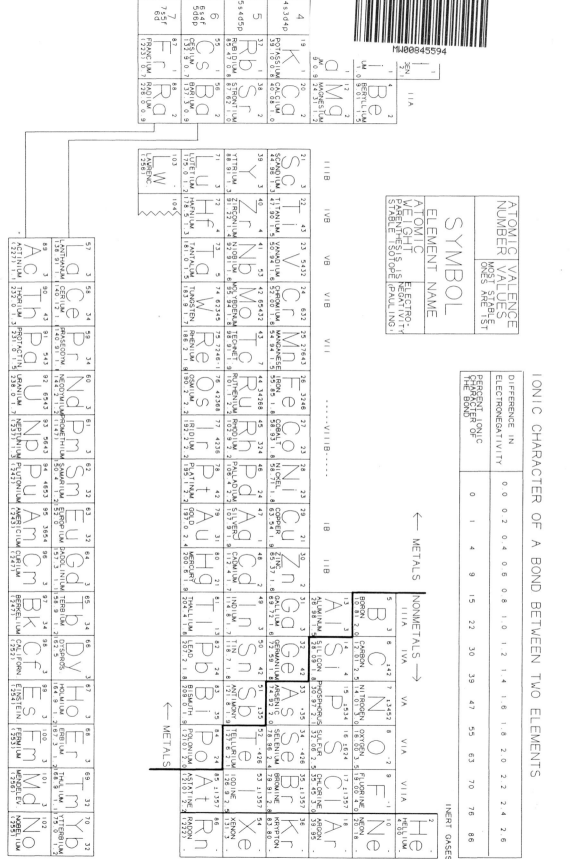

Handbook of
PULPING AND PAPERMAKING

SECOND EDITION

Handbook of
PULPING AND PAPERMAKING

SECOND EDITION

Christopher J. Biermann

DEPARTMENT OF FOREST PRODUCTS AND
CENTER FOR ADVANCED MATERIALS RESEARCH

OREGON STATE UNIVERSITY

CORVALLIS, OREGON

ACADEMIC PRESS
An Imprint of Elsevier

Find Us on the Web! http://www.apnet.com

This book is printed on acid-free paper. ⊗

Academic Press
An Imprint of Elsevier
525 B Street, Suite 1900, San Diego, California 92101-4495

United Kingdom Edition published by
Academic Press Limited
24-28 Oval Road, London NW1 7DX

Library of Congress Cataloging-in-Publication Data

Biermann, Christopher J.
    Handbook of pulping and papermaking / by Christopher J. Biermann,
 -- 2nd ed.
        p.   cm.
    Includes indexes.
    ISBN - 13: 978-0-12-097362-0
    ISBN - 10: 0-12-097362-6
    1. Wood-pulp.   2. Papermaking.   I. Title.
    TS1175.B54   1996   96-20451
    676--dc20   CIP

PRINTED IN THE UNITED STATES OF AMERICA
08 / 9 8 7

CONTENTS

PREFACE TO THE SECOND EDITION

The first edition of this book, under the title of *Essentials of Pulping and Papermaking*, was designed primarily as a teaching tool, but also found wide use by industry as a current source of information. The second edition of this work keeps the first 20 chapters with only minor alterations. In order to increase the dual use of this work, it has been expanded to include key areas of paper chemistry and optical properties, wood and fiber anatomy, paper use, and processing equipment, including a useful pump troubleshooting guide. The comprehensive nature of the work now merits the modified title of *Handbook of Pulping and Papermaking*.

The first eleven chapters of this book are introductory in nature for beginning students and workers in the field who desire some knowledge of the entire mill. The remainder of the book provides detailed information on topics central to the production and use of paper. Throughout, numerous references lead the reader to further information.

Several areas, such as pulp bleaching, environmental issues, and papermaking chemistry, are changing so rapidly that the reader will need to consult current journals to stay abreast of these fields. Nevertheless, this work provides the background and context for the improvements the industry is realizing in these areas.

I am grateful for the help of many people who provided information and figures and who reviewed portions of the manuscript. The Information on clay was supplied by F. Camp Bacon, Jr. of Evans Clay Co., Stuart Pittman of Thiele Kaolin Co., Robert F. Hurst of Huber Clay, and John A. Manasso of Dry Branch Kaolin Co.

Mankui Chen provided material on transportation safety. R.L. Krahmer and A.C. VanVliet provided most of the macroscopic cross section photographs of woods in Chapters 26 and 27.

Gary Vosler of Willamette Industries reviewed the section on corrugated containers. Fred Green and Carter P. Hydrick of Keystone International, Inc. provided information and figures on valves. Kelly Lacey provided the figure of a progressing cavity pump.

Todd Chase reviewed material for clarity. Reviews of selected material were provided by Kevin Hodgson and Richard Gustafson, both of the University of Washington, and Mahendra Doshi of Doshi and Associates, although the author is solely responsible for any deficiencies in the text.

PREFACE TO THE FIRST EDITION

It is said that "if you give a man a fish, he eats for a day; if you teach a man to fish, he eats for a lifetime." This adage represents the entire aim of science, that is, to characterize or explain as wide a range of phenomena or conditions as possible with the least number of rules. In short, science generalizes to as large a degree as possible. To memorize a formula that applies to only one set of conditions is no more learning science than giving a man a fish is teaching him how to fish. Yet many cultures, including our own, continue to teach by requiring memorization of rules or formulas. Students wind up learning formulas they do not understand with little idea of the underlying assumptions upon which they are based (and conditions to which they apply.) The result is students who show very little imagination and extension of thought to solve new problems, which is the very essence of science. The author has seen students with a great deal of tenacity, but not creativity, spend hours upon hours late into the night attempting to find a solution to a problem by searching dozens of books, rather than attempting to solve it by thinking it through, using their own internal resources.

This book stresses concepts and gives a series of problems to demonstrate understanding of the concepts rather than just the memorization of "facts"; the difference results in the molding of a mind into that of a scientist in preference to molding a mind into an insignificantly small, unreliable data base.

I hope instructors who use this book make the tests open book—after all, life in the mill is open book—there is no point in teaching any other way. Some of the problems in this book are deliberately irrelevant to anything one might encounter in reality to demonstrate the wide applicability of certain techniques and to discourage the student from trying to find a reference book to solve the problem. Frequent "curve balls" are included to keep one thinking in terms of the "big picture"; that is, when problem solving in a mill, one is not going to have all the necessary information delivered on a silver platter. It is important to figure out what information is required, what information is at hand, and where to obtain the remainder of the information to solve a problem. Often it is not what we do not know, but rather what we _think_ we know that is _not_ true, that causes us problems.

This book is designed as an educational tool for pulp and paper science courses and a reference book to those in industry. It is a scientific reference concentrating on principles of pulp and paper processes rather than the multitudinous types of equipment available to carry out these processes. TAPPI Standard methods are emphasized, but many of the TAPPI Standards give references to related methods such as those of ASTM, CPPA, SCAN, ISO, and APPITA.

Material is presented roughly in the same order as material flows through the pulp and paper mill. The first section of the book covers fundamentals, and later sections cover more advanced treatments. It is impossible to present so much information on so many diverse areas and supply all of the necessary background in each section without undue duplication. Also, one does not like to burden the novice with details only important to a few investigators. It is appropriate to use the extensive index as the ultimate, alphabetical guide to this volume to find all of the material available on a particular topic.

The author welcomes your comments regarding this book and suggestions for teaching aids.

ACKNOWLEDGMENTS

The author wishes to thank all of the people and companies who contributed information, photographs, and other help. Special thanks to the Pope and Talbot mill in Halsey, Oregon, the James River mill in Halsey, Oregon, the James River mill in Camas, Washington, and the Boise Cascade Corporate Research Center in Portland, Oregon for providing valuable help. The sketches by the late John Ainsworth were used with generous permission of the Thilmany Paper Division of International Paper Co.

Stephen Temple of Black Clawson was instrumental in providing information and figures of his company's equipment, as were Bonny Rinker of Andritz Sprout-Bauer, Inc.; Barbara Crave and Elaine Cowlin of Beloit Corp.; Henry Chen of James River Corp. in regards to Crown Zellerbach Corp. material; Robert Gill of Pfizer, Inc.; Manuel Delgado of Straw Pulping Technology; R. H. Collins of Kamyr, Inc.; Jody Estabrook of Fiberprep; Maria Krofta of Krofta Eng. Corp.; Steven Khail of Manitowoc Eng. Co.; Garth Russell of Sunds Defibrator; and many others who provided information for this book.

Alfred Soeldner took many of the excellent scanning electron micrographs that are shown in this volume. An extensive resource of notes in the OSU pulp and paper files was used in this work; contributions to these files were made by Walter Bublitz, James Frederick, Jr., Philip Humphrey, and others, although it was not always clear who supplied any given information. Alton Campbell, of the University of Idaho, contributed a copy of his class lecture notes, which were helpful to this work. Michael Haas, of Longview Fibre, contributed a set of the books prepared for the Pulp and Paper Manufacturing Technology Program offered through Lower Columbia College. Andre Caron, of the National Council of Air and Stream Improvement, supplied information concerning environmental regulations and control. Myoung-ku Lee prepared the index and assisted with many aspects of this work. Troy Townsend and Jinfeng Zhuang reviewed much of the manuscript.

Victor Hansen reviewed the paper machine wet end section and contributed much information on water removal on flat wire machines.

Special acknowledgment is made to the late Kyosti Sarkanen who inspired me to start this project, sent copies of his class lecture notes to me, and was to have collaborated on this work. The editorial and production staff at Academic Press contributed substantially to this work. One could not ask for a more reasonable publisher.

Additional recognition is given to Jerry Hull, my right-hand man who runs my laboratory when I am not to be found for days at a time, for numerous suggestions of research activities and comments on this work, and editorial help on the manuscript. I would especially like to thank my wife, Lora Jasman, M.D., for numerous suggestions and editorial help.

Many other people have been involved in the review process, but must remain publicly unthanked as they remain unknown to me.

This book was typeset from WordPerfect® 5.1. Many of the drawings were done with Generic CADD™ 5.0. Many of the chemical structures were drawn with Wimp™2001. Many graphs were drawn with Quattro®Pro 3.0. The original copy was printed on a Hewlett Packard LaserJet 4 with CG Times font.

A conscientious effort has been made to include trademarks when cited in the text. Any omission is unintentional, and I welcome notification of any oversights.

The inclusion of color plates has been made possible by the generous support of the following companies and organizations:

The Black Clawson Company
Contributors to Oregon State Univ.
Pacific Section Tappi
Rhône Poulenc
  Specialty Chemicals
  Basic Chemicals
Willamette Industries, Incorporated

ABBREVIATIONS

ABS	acrylonitrile-butadiene-styrene terpolymer
AC	alternating current
AFLA	another four letter acronym
AFPA	American Forest and Paper Assoc.
AKD	alkyl ketene dimer (size)
AOX	adsorbable organic halides
API	American Paper Institute (now AFPA)
APMP	alkaline peroxide mechanical pulping
APP	alkaline peroxide CTMP
ASA	alkenyl succinic anhydride (size)
ASB	aerated stabilization basins
ASTM	American Society for Testing and Materials
AQ	anthraquinone
BAT	best available technology economically achievable (the standard of the 1987 CWA; it is more stringent than BPT)
BACT	best available control technology
bbl	barrel
BCTMP	bleached CTMP
BDU	bone-dry unit
BLRB	black liquor recovery boiler
BOD	biochemical (biological) oxygen demand
BPT	best practicable technology (standard of the 1977 CWA)
Byu	British thermal unit
C1S	coated one side
C2S	coated two sides
CAA	Clean Air Act (1970, 1990)
CAAA	Clean Air Act Amendments of 1977
CD	cross (paper) machine direction
CDC	Centers for Disease Control
CED	cupriethylenediamine (viscosity test)
CEMS	continuous emission monitoring system
CFD	computational fluid dynamics
cfm	cubic feet per minute
CIE	international color scale
COD	chemical oxygen demand
CPPA	Canadian Pulp and Paper Association
CSF	Canadian Standard freeness
CTMP	chemi-thermomechanical pulp
CTR	cationic titration ratio
CWA	Clean Water Act of 1977. The February, 1989 amendment to the CWA is called the Water Quality Act of 1987

DADMAC	poly(diallyldimethylammonium chloride)
DARS	direct alkali recovery system
DC	direct current
DCS	distributive control system
DDA	dihydro-dihydroxy anthracene, see SAQ
DDJ	dynamic drainage (Britt) jar
DIN	Deutsche Industrie Normen (German Industry Standard)
DP	degree of polymerization or differential pressure (across dryer drums)
dpi	dots per inch
EBS	ethylenebisstearamide
ECF	elemental chlorine free (bleaching)
EDTA	ethylenediaminetetraacetic acid
EEC	European Economic Community
EMC	equilibrium moisture content
EO	ethylene oxide
EPA	Environmental Protection Agency
ESP	electrostatic precipitator; emergency shutdown procedure
FAS	formamidine sulfinic acid
FCG	fine ground carbonate
FDA	Food and Drug Administration (U.S.)
FIFO	first in—first out (rotation method)
FGR	flue gas recirculation
FPR	first pass retention
FPAR	first pass ash retention
FSP	fiber saturation point
GCC	ground calcium carbonate
GPC	gel permeation chromatography
gsm	grams per square meter
HLB	hydrophilic-lipophilic balance
HMWT	high molecular weight (polymers)
HPLC	high performance liquid chromatography
HST	Hercules size tester
HW	hardwood
IPST	Institute of Paper Science and Technology; formerly, the IPC
IR	infrared
ISE	ion-specific electrode
ISO	International Organization for Standardization
L	breaking length
LMWT	low molecular weight (polymers)

LTV	long tube vertical bodies		PTFE	teflon, polytetrafluorethylene
LWC	lightweight coated (groundwood)		PVC	polyvinylchloride
MC_{GR}	moisture content on wet weight basis		RCRA	Resource Conservation and Recovery Act of 1976 (amended 1984)
MC_{OD}	moisture content on oven-dry basis		RDH	rapid displacement heating
MCC	modified continuous cooking		RH	relative humidity
MD	(paper) machine direction		rpm	revolutions per minute
MEE	multiple-effect evaporators		S_n	a secondary cell wall layer
MF	melamine-formaldehyde		SAQ	soluble anthraquinone (see DDA)
MG	machine glazed		SARA	Superfund Amendments and Reauthorization Act of 1986
MSDS	material safety data sheet		SBR	styrene butadiene rubber
MT	metric ton		SEC	size exclusion chromatography
NCASI	National Council for Air and Stream Improvement		SEM	scanning electron microscopy
NCG	non-condensable gases (pulping)		SGW	stone groundwood (pulp)
NCR	no carbon required (carbonless) paper		SMA	styrene maleic anhydride
NIR	near infrared		SOP	standard operating procedure
NMR	nuclear magnetic resonance spectroscopy		SPC	statistical process control
NSPS	New Source Performance Standard		SQC	statistical quality control
NSSC	neutral sulfite semi-chemical (pulp)		SS	stainless steel
OCC	old corrugated containers		SW	softwood
OCS	outdoor chip storage		t	metric ton (ton written out is 2000 lb)
OMG	old magazines		TAPPI	Technical Association of the Pulp and Paper Industry
ONP	old newspapers			
ORP	oxidation-reduction potential		TCDD	2,3,7,8 tetrachlorodibenzo-p-dioxin
OSHA	Occupational Safety and Health Administration		TCF	totally chlorine free (bleaching)
			TIS	technical information sheet (TAPPI)
PAC	polyaluminum chloride		TLA	three letter acronym
PAM	polyacrylamide		TMDL	total maximum daily loads
PCC	precipitated calcium carbonate		TMP	thermomechanical pulp
PE	polyethylene		TOCl	total organically bound chloride
PEI	polyethylenimine		TOX	total organic halides
PEO	polyethyleneoxide		tpy	tons per year
PGW	pressure groundwood (pulp)		TQM	total quality control
pH	$-\log[H^+]$		TRS	total reduced sulfur
$p(x)$	$-\log(x)$ where x is the argument		TSS	total suspended solids
Pima	Paper Industry Management Association		UF	urea-formaldehyde
PLI	pounds per linear inch		UFGC	ultra fine grounf carbonates
PM	paper machine		UFGL	ultrafine ground limestone
PO	propylene oxide		UM	Useful Method (TAPPI)
PPI	*Pulp & Paper International*		UV	ultraviolet
ppm	parts per million		WG	water gravity (suction leg)
ppt	parts per trillion		WPCA	Water Pollution Control Act
ppq	parts per quadrillion		WQA	Water Quality Act of 1987
psi	pounds per square inch (pressure)		WRV	water retention value
psig	pounds per square inch gauge			

LIST OF COLOR PLATES

PLATE 1

PLATE 2

PLATE 3

PLATE 4

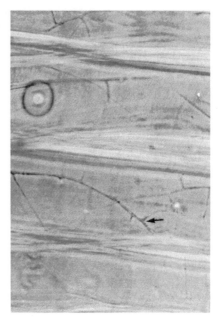

PLATE 5

PLATE 6

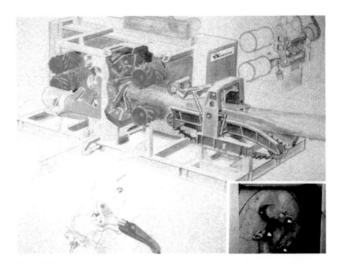

PLATE 7

PLATE 8

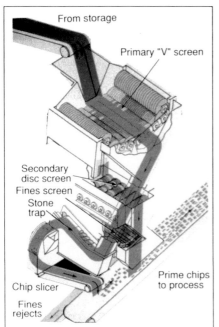

PLATE 9

PLATE 11A

PLATE 11B

PLATE 10

PLATE 11C

PLATE 12

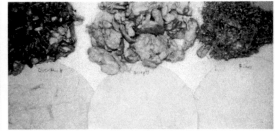

PLATE 13

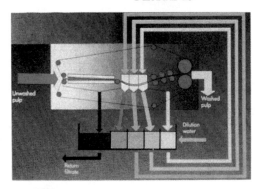

PLATE 15

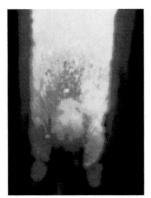

PLATE 17

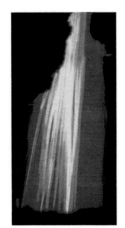

PLATE 18

PLATE 14

PLATE 16

PLATE 19

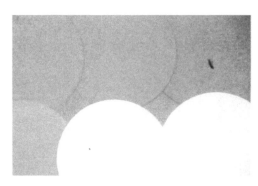

PLATE 20

PLATE 21

PLATE 22

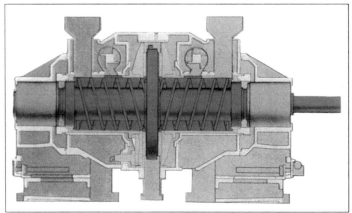

PLATE 23

PLATE 24

PLATE 27

PLATE 25

PLATE 28

PLATE 26

PLATE 29

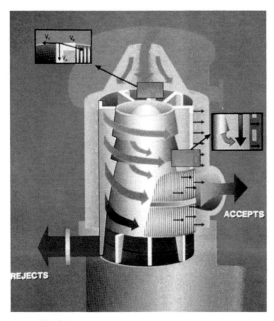

PLATE 30

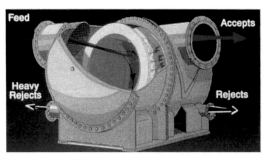

PLATE 31

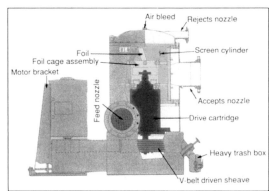

PLATE 32

PLATE 33

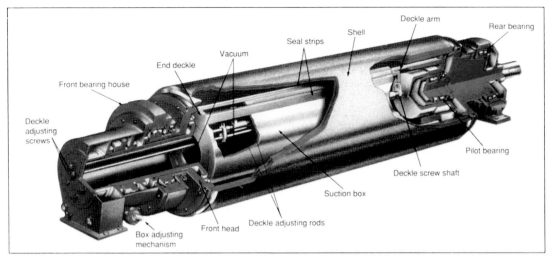

Labels: Front bearing house, Deckle adjusting screws, End deckle, Vacuum, Seal strips, Shell, Deckle arm, Rear bearing, Pilot bearing, Deckle screw shaft, Suction box, Box adjusting mechanism, Front head, Deckle adjusting rods

PLATE 34

PLATE 35

PLATE 36

PLATE 37

PLATE 38

PLATE 39

PLATE 40

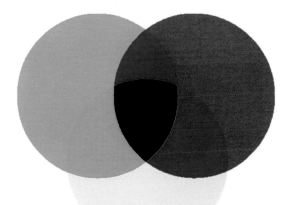

Subtractive Colors

PLATE 41

15°

0°

45°

75°

90°

PLATE 42

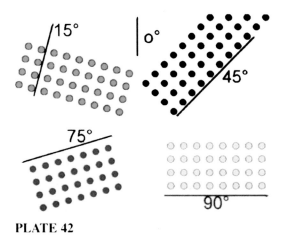

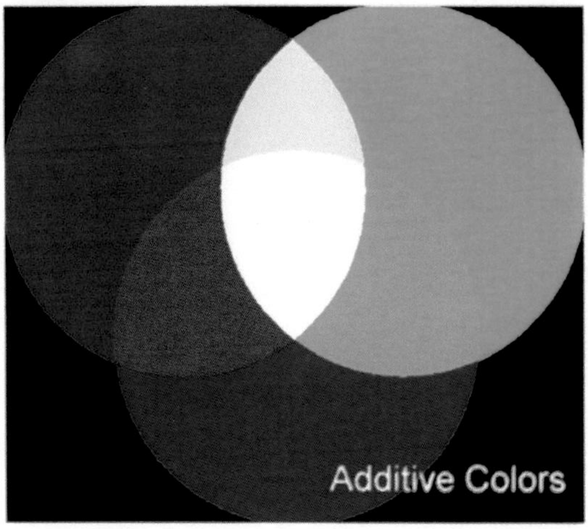

Additive Colors

PLATE 43

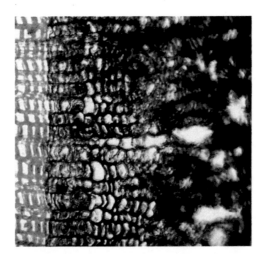

PLATE 44

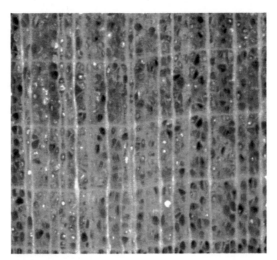

PLATE 45A

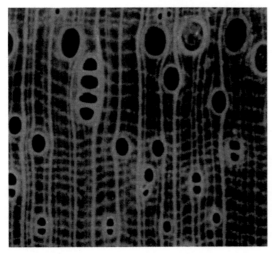

PLATE 45B

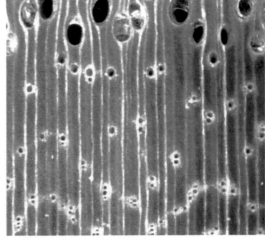

PLATE 45C

1

INTRODUCTION AND THE LITERATURE

1.1 INTRODUCTION TO PAPERMAKING

The availability of writing material has always gone hand in hand with the development of society. The earliest media for writing was the stone tablet. Egyptians invented papyrus around 3000 B.C. Papyrus is made by crisscrossing thin sections of the papyrus reed, which is ubiquitous in the marshy delta of the Nile River. Papyrus is held together by natural glues within the reed when pressure is applied. The papyrus was smoothed by rubbing with smooth stones. Around 200 B.C. parchment, the tanned skin of sheep or goats, was developed. Parchment is now a high quality grade of paper made from vegetable fibers.

The invention of paper is credited to Ts'ai Lun of China, who, in the year 105 A.D., was working on a papyrus substitute by decree of Emperor Hoti. Within 20 years he had paper from the bark of mulberry trees that was treated with lime, bamboo, and cloth. The Chinese considered paper a key invention and kept this a closely guarded secret for over five centuries until the technology slowly made its way westward. The Arabs captured a Chinese city containing a paper mill in the early 700s and from this started their own papermaking industry. Paper was first made in England in 1496. The first U.S. mill was built in 1690, the Rittenhouse mill, Germantown, Pennsylvania.

The development of the paper machine is the most important milestone of the industry. Louis Robert, working at the paper mill owned by Ledger Didot, made his first model of the continuous paper machine in 1796 near Paris. The first model was very small and made strips of paper about as wide as tape. He received a French patent for his machine in 1799, at the age of 37, based on drawings he had submitted. Shortly after being awarded the patent he began construction based on these drawings.

Development of the paper machine was led by the two Fourdrinier brothers, Henry and Sealy, who, in 1801, bought one-third interest in the British patent rights of Robert's machine. They hired Bryan Donkin who took three years to develop the first practical paper machine, which was in operation at Two Waters Mill, Hertz, England, in 1804. A sketch of this machine is shown in Fig. 1-1. Donkin's company continued to manufacture and improve the fourdrinier machine for many years. His company supplied most of the early fourdrinier machines throughout the world.

At about the same time John Dickinson, a colleague and friend of Donkin, was working on his cylinder machine, which was refined by 1809. In fact, both Dickinson and Donkin contributed important ideas to each of these machines. For example, in 1817, John Dickinson became the first person to mention steam-heated drying cylinders. He used cast iron cylinders complete with stationary siphon tubes. In 1820, Thomas Crompton added dryer felts using single tier

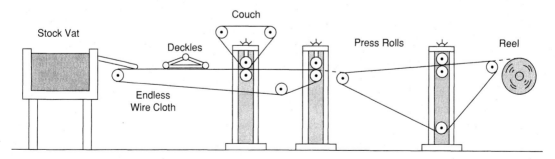

Fig. 1-1. Diagram of Bryan Donkin's paper machine, circa 1804. Redrawn from J. Ainsworth, *Papermaking,* **©1957 Thilmany Paper Co., with permission.**

Table 1-1. The production of paper and machine speed with time.[1]

Year	US paper use (10^6 tons)	Record Paper Machine Speed (ft/min)	Record PM wire width (in.)
1810	0.2	40	60
1860	0.2	80	60
1880	0.5	200	60
1900	2.3	500	160
1920	5.5	1000	205
1940	11	1600	320
1965	45	3000	360
1990	80	6800	380

[1]These are approximate values to show a trend. Maximum speeds apply to tissue machines, which tend to much faster than other machines.

Fig. 1-2. Pulp logs leaving the forest. Reprinted from *History and Description of Paper Making*, ©1939 Crown Zellerbach Corp., with permission.

drying. Double tier driers soon followed. Crompton studied textile drying, which may have provided the thought for using drier felts. Drying cylinders and cylinder arrangements have not changed appreciably in concept, since this time.

The dandy roll was patented in 1825 by the two brothers John and Christopher Phipps of Kent county. They made rolls from 4 to 12 inches (10-30 cm) in diameter. Some people claim that John Marshall may have been the first person to make a "riding roller", and he began manufacturing them in 1826.

M. Canson of Annonay, France put a suction box under the wire of his fourdrinier machine in 1826, as had already been done on the cylinder machine, but kept this a secret.

In 1828, George Dickinson received a patent for a machine that, among many other things, had the forerunner of the suction roll. It was, however, about 80 years later, with the invention of the cycloidal vacuum pulp, that suction rolls were used in the wet end or press section. In 1829, John Dickinson used a reverse press with

double felting to decrease the two-sidedness of paper. In fact, many of the important inventions of the paper machine were discovered within the first three decades of Robert's invention.

With the invention of the paper machine, the amount of paper that could be produced was soon limited by the fiber supply, since cotton was the main constituent in paper. During the mid-nineteenth century the technology for converting wood to pulp suitable for paper was developed. With a plentiful supply of pulp available, the amount of paper production was then closely related to improvements in paper machine speeds and widths as shown in Table 1-1. The production of paper in the U.S. increased by a factor of 10 in the 19th century and almost 50 in the 20th century. In 1950, the U.S. made 60% of the world's paper, by 1990 this was less than 30%.

Recently, paper production in developed countries seems to be limited once again by fiber supply. Gone are the days when huge logs went directly to pulp mills as shown in Fig. 1-2. Much roundwood is obtained from precommercial thinnings using skidders to drag them to the roadside as shown in Fig. 1-3. In the Northwest U.S. wood prices are high enough that some mills are now pulping construction wastes, pallets, and other materials that are chipped. Even chemical pulping of old corrugated boxes is practiced in parts of the world to stretch fiber resources. Fig. 1-4 gives the steps in making paper for a present-day mill.

Regardless of the method of making paper, whether by machine or by hand, there are some common steps that are shown in Fig. 1-5. The raw material (in this case the inner bark of cedar) is pulped to allow separation of the individual vegetable fibers. The pulp is then washed to remove pulping chemicals. The pulp may then be bleached if it is to be used in white papers. The pulp is treated with beating or refining to separate the fibers from each other and to roughen the surface of the fiber. Most papers do not use adhesives, but rely on hydrogen bonding between fibers. Paper is then formed by running a dilute slurry of pulp through a screen, and then most of the water is allowed to drain out by gravity. Additional water is then pressed from the sheet. The last water is removed by evaporation.

1.2 INTRODUCTION TO THE LITERATURE

It is very important to be able to access articles in the scientific literature in particular subject areas. Information is one of the most underused resources in the pulp and paper industry. The information in original research papers and other references can be used to solve many problems and questions that arise in pulp and paper mills. If you are not aware of a particular journal or reference book mentioned in this chapter or how it is used, be sure to ask your librarian for assistance. Many libraries can obtain copies of articles or books, which they do not have in the library, through interlibrary loan programs. The lists in this chapter are by no means exhaustive; however, some of the more routinely used references are included.

The literature can be divided into the primary literature and the secondary literature. The primary literature consists of patents, journal articles, and proceedings of meeting presentations. This is the most current information and the most difficult to access. One can keep abreast of only a few journals. To fully use this resource, one should periodically conduct library searches with the abstract indices in areas of critical importance to a mill. For example, if a mill is considering conversion to alkaline papermaking, one would want to obtain as much current information as possible on the subject to save downtime during the conversion. The primary literature is never obsolete. Indeed, it is often important to go back to the primary literature to check the source of equations, tables, etc. used in the secondary literature. Often, on investigation of an original article, one learns that information that has become widespread is being used under inappropriate conditions.

The secondary literature is information that is summarized in books, monographs, reference books, and so forth. It tends to be older and may

Fig. 1-3. Skidders drag small trees from the forest so the remaining trees can grow faster. Reprinted from *Making Pulp and Paper*, ©1967 Crown Zellerbach Corp., with permission.

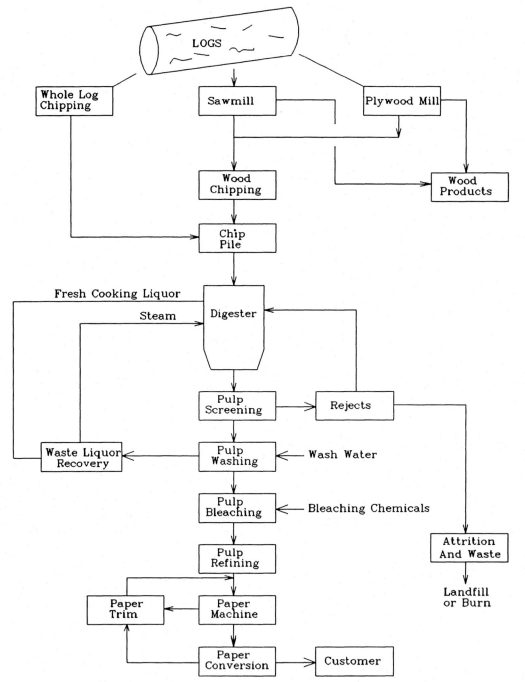

Fig. 1-4. An overview of chemical pulping and papermaking processes. Courtesy of W. Bublitz.

be outdated; however, it is usually the best source to get background information in a particular field.

If one considers articles (and other references) in specialized fields with applicability to pulp and paper, such as corrosion, process control, simulation, and so forth, to be useful, then the importance of abstract indices to access the literature becomes obvious. One should not rely on merely the pulp and paper journals; many

Fig. 1-5. The basic steps to papermaking. The fiber source is pulped (and bleached for white papers); the pulp is washed; the pulp is refined to insure the fibers are separated from each other and to roughen the surface for better fiber bonding; a paper web is formed from a dilute pulp slurry letting most of the water drain by gravity; additional water is pressed from the web; and the remaining free water is removed by evaporation to give the product.

important advances to pulp and paper technology are merely modifications of technologies that are well established in other disciplines.

On-line computer searches

Literature searches using computer terminals is quite common and convenient. Virtually any computer with a modem can be used from any location once an account is set up with a vendor. Most libraries have facilities for doing literature searches from computers and would probably know what sources to use for the pulp and paper industry. This information is also available from other sources (Steelhammer and Wortley, 1992).

I recommend that some time be spent with thorough manual searching of one or two years with abstract indices when searching a new topic, especially a topic outside of ones normal activities. This is helpful to learn what key words are most effective so that important work will not be overlooked. One should also be familiar with the pulp and paper literature before getting overly accustomed to using computer-aided searches of the literature.

1.3 ABSTRACT INDICES

Abstract Bulletin of the Institute of Paper Science and Technology

The most complete abstracting service of research directly related to pulp and paper is the *Abstract Bulletin, Institute of Paper Science and Technology* that is published monthly, in addition to an annual index. It is available on line for computer searching. Before 1989 it was the *Abstract Bulletin of the Institute of Paper Chemistry*.

Chemical Abstract Index

The *Chemical Abstract Index* is a comprehensive abstract bulletin covering all areas of chemistry and chemical engineering. It is extremely useful for getting current information on all aspects of chemical problems affecting the pulp and paper industry. Author, subject, chemical, and chemical formula indices are available to find abstracts. Indices cover a six-month period; collective indices cover ten volumes, corresponding to five years.

Scientific Citations Index

The principal use of the *Scientific Citations Index* is to keep track of the articles cited in the primary scientific literature. For example, suppose Dr. I. M. First published an article in 1988, and the citation for that article appeared in the *Literature Cited* section of an article published in *Tappi J.* in January, 1990 by Ms. U. R. Second. Under First, I. M. in the 1990 *Citations Index* (not the *Source Index*), one would find that U. R. Second cited the article of I. M. First. This is a useful form of feedback to determine the significance of a particular scientific publication. In some fields there are a few key authors whose work is usually cited. By determining who cited these authors, one can find some of the most recent work.

1.4 TECHNICAL AND TRADE JOURNALS

Nordic Pulp and Paper Research Journal

This journal publishes about 40 high quality, fundamental, original research papers per year. (Arbor Pub. AB, Stockholm, Sweden, appears quarterly.) This journal started to publish research papers at about the same time (in 1986) the long-standing *Svensk Papperstidning*, ceased publishing research papers.

Journal of Pulp and Paper Science

This journal also publishes about 40 high quality, fundamental, original research papers per year. (Technical Section, Canadian Pulp and Paper Association, Montreal, Quebec, appears bimonthly.)

Tappi Journal

This journal publishes about 300 high quality, original research papers per year. It also includes industry news, summaries of TAPPI conferences, advertisements, and other topics. (TAPPI, Norcross, Georgia, appears monthly.)

Because of the sheer number of articles published, this is often a useful source to collect some of the current information on a technical subject. By using the annual index in the December issue, several years of this magazine can be rapidly checked for articles on a particular subject.

References cited in those articles will give further information, though not necessarily the most current information.

Other pulp and paper research journals

Pulp and Paper Canada publishes about 70 high quality, original research articles annually in addition to industry news, feature articles, advertisements, and so forth. (Southam Business Information and Communications Group Inc., Don Mills, Ontario, published monthly.)

Appita Journal publishes about 40 high quality, original research papers annually in addition to industry news, features, advertisements, and so forth. (Technical Association of the Australian and New Zealand Pulp and Paper Industry, Inc., Parkville, Victoria, appears bimonthly.)

Paperi ja Puu, Paper and Timber publishes about 60 original research, review, and feature articles annually along with departments, and other news items. (Toimitusjohaja, Helsinki, Finland, 10 issues annually.)

Progress in Paper Recycling appears quarterly since late 1991 and contains research articles and practical, useful articles in paper recycling. (Doshi & Associates, Appleton, Wisconsin.)

Other research journals

There are many other journals which have original research articles on pulping, fiber modification and other chemical subjects applicable to the pulp and paper industry. Some of these include *Journal of Wood Chemistry and Technology* (Marcel Dekker, Inc., appearing quarterly since 1981), *Wood Science and Technology* (Springer-Verlag, the journal of the International Academy of Wood Science, appears quarterly), *Wood and Fiber Science* (Society of Wood Science and Technology, Madison, Wisconsin, appears quarterly), *Holtzforschung* (Walter de Gruyter, appears bimonthly), *Mokuzai Gakkaishi* (Japan Wood Research Society, appears monthly), and the *Forest Products Journal*, (Forest Products Research Society, Madison, Wisconsin, 10 issues annually). These journals are essentially confined to research articles.

Trade journals or magazines

Other pulp and paper journals have industry news, feature articles, and advertisements, on a wide variety of issues in pulp and paper presented in a relatively non-technical fashion. Many are available without charge to industry personnel.

PaperAge is published monthly (Global Publications, Inc., Westwood, New Jersey). *Pima Magazine* is published monthly and is designed for mill management (Paper Industry Management Association, Arlington Heights, Illinois). *American Papermaker* (A/S/M Communications, Inc., Atlanta, Georgia) is published monthly. It was published in 4 editions by region of the U.S. from 1987-1992; prior to 1987 it was published as *Southern Pulp and Paper*. *International Papermaker* is published quarterly as of late 1992. Other magazines include *Pulp and Paper Magazine* (Stockholm, Sweden, published quarterly) and *Pulp & Paper Journal* (Maclean Hunter Ltd., Toronto, Ontario, 11 issues per year).

Pulp & Paper (the international version is *Pulp & Paper International, PPI*) is published monthly and is a very good source of current industry production figures for North America (Miller Freeman Publications, San Francisco, California). (In November, 1986 the *Paper Trade Journal* ceased publication and was incorporated into *Pulp & Paper*.) It has about six to ten articles in each issue that focus on one or two topics in addition to feature articles that might be on any topic. The focus of a given month does not change much from year to year, so this is often an **excellent source for general, up to date information** on a given topic. In 1995, the topics were:

Month	Focus
JAN	Outlook '95; Capital spending
FEB	CEO Profiles; Recycling
MAR	Process Control; Non-chlorine bleaching
APR	Papermakingchemicals;Forming/Pressing
MAY	Coating; Recycling; Environmental; Papermaking
JUN	Stock Preparation; Quality Management; Mechanical Pulping
JUL	Mill Maintenance; Woodyard
AUG	Process Support Equipment; Market Pulp
SEPT	Engineering; Anatomy of a Startup
OCT	Chemical pulping/bleaching; Finishing and converting
NOV	Power/recovery; transportation
DEC	Mill Manger's Report; Pulping/bleaching

1.5 REFERENCE BOOKS

PaperWorld

This highly recommended, five volume reference (in addition to the General Index and World Atlas) covers the international pulp and paper industry. Each volume is divided into five parts covering 1) Pulp and Paper Industry, 2) Exporters and Importers, 3) Suppliers (for example, machines and chemicals), 4) Associations, and 5) Trade Press. Names, addresses, and phone numbers are given in most cases. (Birkner® & Co., Hamburg, Germany.).

Lockwood-Post's Directory

The *Lockwood Post's Directory* lists pulp and paper mills, converting mills, and suppliers with information on personnel, types and amounts of products produced, water, wood and power usages, equipment, etc. by state or province for the United States and Canada, respectively. It is updated annually. (*1990 Lockwood-Post's Directory*, Miller Freeman Publications, Inc., New York, N.Y. 1990.)

Pulp & Paper Fact Book

This 426 page reference book lists production by grades, sales, prices, and historical trends of companies for each grade. It is updated annually. (*1990 North American Fact Book*, Miller Freeman Publications, Inc., San Francisco, CA 1990.)

Tappi Test Methods, Tappi Useful Methods

This two-volume set contains standardized test procedures for everything from wood chips to containers and coated paper including many raw materials used in these processes. Volume 1 covers fibrous materials, pulp testing, and paper and paperboard testing. Volume 2 covers nonfibrous materials, container and structural materials testing, and testing practices. These test methods are used throughout the industry to insure quality control and comparability of results. Other countries have similar sets of testing procedures that are mentioned later. (*Tappi Test Methods*, TAPPI, Atlanta, Georgia, 1991. In 1992 it was published as a single volume.) *Tappi Useful Methods* is a supplement of additional methods that have not been as thoroughly reviewed as the official test methods.

Tappi Technical Information Sheets

This three-volume set contains data sheets on densities of solutions, troubleshooting and other guides, inspection reports, tables, figures, glossaries, etc. (*Technical Information Sheets*, TAPPI, Atlanta, Georgia, 1989.)

Standard Testing Methods

This contains a series of standardized test methods and data sheets of use to the pulp and paper industry. It is similar to the two previously listed references put out by TAPPI Press, but contains tests of the CPPA. (*Standard Testing Methods*, Technical Section, Canadian Pulp and Paper Association, Montreal, Quebec.)

Pulpwoods of the United States and Canada

This two-volume set (Volume 1, conifers; Volume 2, hardwoods) lists tree species, their silvics, wood properties, and general pulping characteristics with references. It includes the commercial species of North America. This information is increasingly important as pulp mills look to alternate species as wood fiber supplies become more difficult to obtain. (*Pulpwoods of the United States and Canada*, 3rd edition, Institute of Paper Chemistry, Edited by Isenberg, I.H., revised by Harder, M.L., and L. Louden, Appleton, Wisconsin, 1981.)

Wood Handbook

This is a general handbook on the physical, mechanical, electrical, thermal and other properties of wood and wood composites. (*Wood Handbook: Wood as an Engineering Material*. Agriculture Handbook number 72, Washington, DC: U. S. Dept. of Agriculture; revised 1987.)

1.6 TEXTBOOKS

Handbook for Pulp and Paper Technologists

This resource of pulp and paper technology includes numerous illustrations. It is weak in the theoretical and practical sides of the field and slightly dated, but strong on the equipment aspect. (*Handbook for Pulp and Paper Technologists*, Smook, G.A., Joint Textbook Committee of the Paper Industry, CPPA, Montreal, Quebec or TAPPI, Atlanta, Georgia, 1982. 395 p.)

Pulp and Paper: Chemistry and Chemical Technology

This four-volume set gives a comprehensive look at the chemistry of pulp and paper technology. It is not heavily illustrated, nor does it give much detail on equipment descriptions, but neither of these detract from its purpose. It is a useful reference book, a good starting point for detailed information on many topics in pulp and paper. (*Pulp and Paper: Chemistry and Chemical Technology*, Casey, J.P., Ed., Wiley-Interscience, New York, N.Y. 1980, 1980, 1981, 1983 vol. 1-4, respectively. 2609 p. total) Topics, not book titles, are:

1. Wood chemistry, pulping and bleaching
2. Papermaking and environmental control
3. Chemical additives, paper properties and analysis
4. Coating and converting

Wood Chemistry: Fundamentals and Applications

This book is a fundamental volume on wood chemistry of very high relevance to the pulp and paper industry. It covers the chemistry of wood and bark components, pulping and bleaching chemistry, chemistry of cellulose plastics and other derivatives (derived from dissolving pulp), and carbohydrate chemistry. (*Wood Chemistry: Fundamentals and Applications*, Sjöström, E., Academic Press, New York, 1981. 223 p.; 2nd edition, San Diego, 1993.)

Textbook of Wood Technology

Volume 1--structure, identification, uses, and properties for the commercial woods of the United States and Canada--is very useful. Extensive micrographs and descriptions by wood species also make it an important reference book. Since the cost of wood is now about 20-40% of the final product (although quite variable) and can be expected to increase as a percentage with more competition for this resource, knowledge about wood becomes very important. (*Textbook of Wood Technology, Vol. 1*, 3rd edition, Panshin, A.J., and C. deZeeuw, McGraw-Hill Book Company, New York, N.Y., 1970. 705 p.)

Others

Other textbooks are still useful if somewhat dated. These include *Pulping Processes*, Rydholm, S.A., Interscience, New York, 1965 (corrected version, 1967) 1269 p.; *Handbook of Pulp and Paper Technology*, Britt, K.W., Reinhold, Publishing Corp., New York, 1964, 537 p.; the two-volume set *Pulp and Paper Science and Technology*, Libby, C.E., McGraw-Hill, New York, 1962, 436 and 415 p., respectively; *Paper and Paperboard*, Kline, J.E., Miller Freeman, San Francisco, 1982, 232 p. (2nd ed., 1991, 245 p. and, except for updated statistics, very similar to the first edition); *Pulp Technology and Treatment for Paper*, Clark, J.d'A., Miller Freeman, San Francisco, 1978, 751 p. (2nd ed., 1985, 878 p., which includes nominal updating of some fundamentals like mechanical pulping); and the three-volume set *Pulp and Paper Manufacture*, 2nd ed., MacDonald, R.G., Ed., McGraw-Hill, New York, 1969, (769, 542, and 655 p., respectively) listed below.

1. The Pulping of Wood
2. Control, Secondary Fiber, Structural Board, Coating
3. Papermaking and Paperboard Making

Work on the 3rd edition of the *Pulp and Paper Manufacture* series (with 7 × 10 in. format) began in 1983. While generally informative, they often lack continuity, making them difficult for newcomers to use. The titles are as follow:

1. Properties of Fibrous Raw Materials and Their Preparation for Pulping, 1983, 174 p.
2. Mechanical Pulping, 1987, 281 p.
3. Secondary Fiber and Nonwood Pulping, 1987, 266 p.
4. Sulfite Pulping and Technology, 1985, 352 p.
5. Alkaline Pulping, 1989, 637 p.
6. Stock Preparation and Nonfibrous Additives, 1992, 316 p.
7. Paper Machine Operations, 1991, 693 p.
8. Coating, Converting and Specialty Processes, 1990, 386 p.
9. Mill Control and Control Systems: Quality & Testing, Environmental, Corrosion, Electrical, 1992, 386 p.
10. Process Control & Information Systems, 1993.

1.7 CHEMISTRY REFERENCE BOOKS

Merck Index

The *Merck Index* contains data on over 10,000 chemicals, drugs, and biologicals with information on synonyms, formula, structure, elemental composition, commercial manufacture, toxicity, uses, physical properties, and solubilities in various solvents. It is updated approximately every eight years, with the 11th edition appearing in 1989, and available on-line for computer searches. There are also tables on subjects such as pH indicators and buffer solutions, conversion factors, abbreviations, and names of organic chemistry reactions. (*The Merck Index*, eleventh edition, Merck & Co., Inc. Rathway, N.J., 1989.)

Handbook of Chemistry and Physics

This is a widely used, extensive collection of information that is revised annually. It is divided into sections containing mathematical tables, properties of the elements, inorganic compounds, and organic compounds, general chemistry, and physical constants. (*Handbook of Chemistry and Physics*, 70th edition, 1989-1990, Weast, R.C., Ed., CRC Press, Boca Raton, Florida, 1989.)

Lange's Handbook of Chemistry

The *Handbook of Chemistry* contains tables of mathematical and statistics, fundamental constants, conversion factors, atomic and molecular structures, physical constants of inorganic and organic compounds (crystalline form, refractive index, melting and boiling points, density, and selected solubilities), analytical chemistry (activity coefficients, equilibrium constants, pH measurements, gravimetric factors, etc.), electrochemistry, spectroscopy (X-ray, NMR, IR), thermodynamic properties (enthalpies and Gibbs free energies of formation, entropies, and heat capacities of compounds and elements), and physical properties (solubilities, vapor pressures, melting and boiling points, viscosity, dipole moments, etc.). This handbook is revised approximately every six years. (*Lange's Handbook of Chemistry*, thirteenth edition, Dean, J.A., Ed., McGraw-Hill Book Co. New York, N.Y., 1985.)

Perry's Chemical Engineers' Handbook

Although most of the information covered in this extensive reference is related to engineering rather than chemistry, this is an extremely useful reference book for topics such as thermodynamics, corrosion, reactor design, distillation, process control, and evaporation in the area of chemical engineering. Information on virtually any unit operation used by the pulp and paper industry can be found in this book. (*Perry's Chemical Engineering Handbook*, sixth edition, Perry, R.H. and D.W. Green, editors, McGraw-Hill Book Co., New York, N.Y., 1984.)

1.8 OTHER REFERENCES

This list of references does not include the many books, bibliographies, conference proceedings, and other sources of information available in specific areas of pulp and paper. To locate these, one should obtain a *Publications Catalog* from TAPPI Press (Atlanta, Georgia), a *Bibliographic Series List* from the Institute of Paper Science and Technology [Atlanta, Georgia and formerly the Institute of Paper Chemistry (Appleton, Wisconsin)], a list of Pulp and Paper Technical Books available from the Technical Section of the Canadian Pulp and Paper Association, or similar resources.

There are many reports of the U.S. Department of Agriculture Forest Product Laboratory (FPL) in Madison, Wisconsin that are not well known. The early summary publication that includes bibliographies of FPL reports and journal publications of the FPL staff is useful to access much of the early work of this laboratory. The reference to this information is Report No. 444, List of publications on pulp and paper, August, 1960, 58 pp. Semiannual lists are available as well on the work done at the FPL. Some of these reports contain pulping and papermaking studies on wood species that may become commercial in the future as fiber becomes more and more difficult to obtain.

1.9 ANNOTATED BIBLIOGRAPHY

On-line literature searches using computers

1. Steelhammer, J.C. and B. Wortley, The computer connection, *PIMA Mag.* 75(5):40-

41(1992). This brief article lists several of the vendors of computer literature searches with their addresses and phone numbers.

History of the industry

2. Clapperton, R.H., *The Paper-making Machine, Its Invention, Evolution and Development*, Pergamon Press, New York, 1967, 365 p. (Out of print.) This work is easy reading and made interesting by the diagrams and photographs. A section includes biographies of early papermakers including L. Robert, L. Didot, J. Gamble, H. Fourdrinier, J. Hall, J. Dickinson, Escher Wyss and J. Voith. As I read this, I soon find myself vicariously looking over the shoulders of the great inventors as they first design and assemble their machines. This was an important source of information for the text above and is my favorite reference on the history of the paper machine.

3. Smith, D.C., *History of Papermaking in the United States (1691-1969)*, Lockwood Pub. Co., New York, 1970. 693 p.

4. Hunter, D., *Papermaking, The History and Technique of an Ancient Craft*, 2nd ed., Alfred A. Knope, New York, 1943. 611 p. with over 317 photographs and diagrams. Dard Hunter came from a long line of papermakers and printers. His writings on the history of papermaking are well known, and a papermaking museum, once housed at the Massachusetts Institute of Technology, but now at the Institute of Paper Science and Technology, is named after him.

5. Hunter, D., *Papermaking Through Eighteen Centuries*, William Edwin Rudge, New York, 1930. 358 p. This work deals with early methods of papermaking and includes numerous sketches and pictures of old techniques.

6. Rudlin, B., *Making Paper - A Look into the History of an Ancient Craft*, Vallingby, Sweden, 1990, 278 p. with 16 color plates (8 leaves) of contemporary, artistic, handmade

papers using a variety of vegetable fibers. This work covers both handmade paper and machine-made paper.

EXERCISES

Sources of information

1. In 1981 and 1982 some articles on carbohydrate analysis came out of the laboratory of G. D. McGinnis. Using the *Chemical Abstracts* find one or two citations for these articles.

2. Once again, using the *Chemical Abstracts*, find the references for some recent articles on silylation (trimethylsilylation) of carbohydrates (monosaccharides). How would you find some of the more important references in this field without looking through every single *Chemical Abstract* index over the last 30 years?

3. In 1985 an article in *Analytical Chemistry* cited the following article: C.J. Biermann, T.P. Schultz, and G.D. McGinnis, titled Rapid steam hydrolysis/extraction of mixed hardwoods as a biomass pretreatment. *J. Wood Chem. Tech.*, 4(1), 111-128 (1984). Using the *Science Citations Index* find the article in *Analytical Chemistry*.

4. Is anthraquinone considered toxic? How is it manufactured? (*Merck Index*)

5. What is the boiling point for H_2S? (*CRC Handbook Chem/Phys*)?

6. Name a journal that appears in the *Current Contents* (not previously discussed in this chapter) that might have articles of interest to a wood chemist.

7. The *American Men and Women of Science* is a handy reference for learning about many of the scientists with Ph.D. or M.D. degrees-- for example, someone who will interview you for a research position. Using this find out where some scientists you know or C. J. Biermann was educated.

8. Use the *Lockwood-Post's* directory to find out how many pulp and paper mills are operating in your state or province.

9. Use the *Abstract Bulletin of IPST* to find a recent reference to an article on one of the following current topics. After you have read the article, write *your own* one page summary of the article. Be sure to give the article citation in your summary so your reader may locate this article.

    A. The use of anthraquinone in pulping.
    B. Oxygen delignification.
    C. Use of chlorine dioxide in pulping.
    D. Dioxin in paper or mill effluent.

10. Use *Pulp and Paper* to find out last month's production by grade of paper for the United States.

11. Using *Lange's Handbook of Chemistry*, find the value for the first acid dissociation constant of aluminum ion in water. Try to calculate the pH of 0.01 M $AlCl_3$ in water.

Using the literature

12. Of course, the literature comes from people who are willing to write. This is an important skill that takes time and practice to develop. It is useful to write essays from one or two paragraphs to one or two pages throughout the course. Some suggested topics are as follows:

    A. Select a mill in your vicinity and write a two paragraph synopsis of pulping method(s) used, the grades of paper produced, the names of the mill manager and technical director, and other information necessary to brief your "supervisor" who is going to visit the mill next week.

    B. Write an essay describing the difference between heat and temperature.

    C. A one to two page essay on a topic of you own choosing.

13. Which would you use the *Abstract Bulletin of IPST* or the *Chemical Abstracts* to quickly find an article on dioxin in pulp mill effluent? Why? Which would you use to find the latest information on the long term toxicity of dioxin? Why?

14. You have an interview with Dr. John Paper for a research position. You want to find some information about him and his research work. Give two resources that might be suitable for this.

2
WOOD AND FIBER FUNDAMENTALS

2.1 WOOD AND BARK

Bark

Bark (Fig. 2-1) is the outermost layer of tree trunks and branches that protects the tree from its environment. It comprises about 10%--20% of the tree stem and has complex anatomy and chemistry. Bark is a contaminant in the wood supply used for making pulp because it decreases the quality of pulp proportional to its level of contamination. There is very little usable fiber in bark (mostly due to the very small size of the fibers in bark), and it consumes chemicals during pulping and bleaching stages; furthermore, it causes dark specks in the final paper product. The relatively high level of *nonprocess elements* (impurities) such as silica and calcium interfere with chemical recovery processes. For the pulp industry, typical bark tolerances in wood chips are 0.3 to 0.5%,

although the kraft process is more tolerant than the other pulping processes. Bark removed from wood is usually burned as a fuel. Whole tree chipping in the forest (a practice some argue will become important in the future as it gives a higher yield of wood chips) requires cleaning of the chips before pulping to remove bark, dirt, needles or leaves, twigs, etc.

Chemically, bark typically consists of about 10-30% extractives (depending on the species and extraction procedures), 15-45% cellulose, 15-40% lignin, and the remainder is other carbohydrates (termed polyoses that are similar to the hemicelluloses) and tannins (condensed polyphenolic compounds). Bark, especially the outer bark, may contain waxy materials called suberin and cutin, which are polyesters of dicarboxylic acids with C_{16} to C_{24} structures and bifunctional alcohols among other molecules.

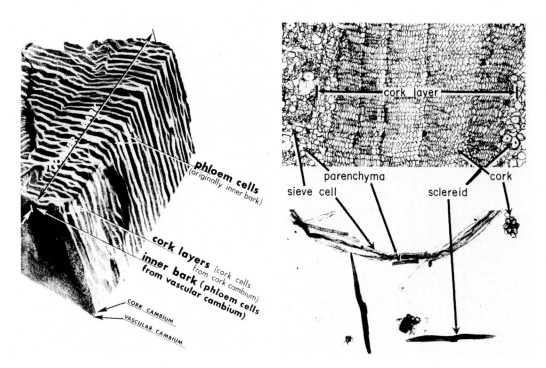

Fig. 2-1. The anatomy of Douglas-fir bark including three dimensional view of whole bark (left), cross section of bark (upper right), and isolated fibers (lower right). Courtesy of R. L. Krahmer.

The distinction between lignin and other polyphenolic compounds is not always possible. The bark of a few tree species is used as a source of tannins due to the high extractives and polyphenolic compounds. Tannins are used commercially for dyes, astringents, and leather tanning. The fuel value of bark is about 18.6-25.6 MJ/kg (8000-11000 Btu/lb) of oven-dry bark. The ash content of softwood bark is about 0.5-4%, while hardwood bark is about 1-6%. The ash is primarily calcium (80-94% of the cations) and potassium (2-7% of the cations) salts of oxalate, silicates, and phosphates. The effect of moisture content on the fuel value is described on page 21.

Physically, bark usually has a *basic specific gravity* between 0.40-0.65. There are a variety of bark layers, but mature bark is loosely divided into the inner bark (consisting of living phloem tissue) and outer bark (consisting of dead rhytidome tissue).

Wood

Technically, wood is xylem tissue, which arises from the cambium (inner bark) of trees and shrubs and consists of cellulose, hemicellulose, lignin, and extractives, hence a lignocellulosic material. Its function is support of the crown and conduction of water and minerals from the roots to the leaves of a tree. *Sapwood* is the outer part of the trunk and contains some living cells. *Heartwood* is found in the center of older trees, contains only dead cells, and is generally drier than sapwood. Each *annual growth ring* contains *earlywood* (sometimes called *springwood*), which is laid down in the early part of the growing season (spring and summer) and is characterized by large cells with thin cell walls, and *latewood* (sometimes called *summerwood*), which is characterized by small cells and thick cell walls. Fig. 2-2 shows the gross structure of wood, and Table 2-1 shows the function and characteristics of woody tissues.

The inner portion of wood in a tree eventually dies with a corresponding deposition of extractives, the process of *heartwood* formation. Heartwood is more difficult to pulp than sapwood. The moisture content of sapwood is high as the wood is normally saturated with water, but with heartwood formation, the moisture content decreases, with air replacing some of the water. Some hardwoods, such as the white oaks, but not the red oaks, form *tyloses* in the vessels during heartwood formation, which greatly reduces the permeability of the wood to fluid flow. For this reason, white oaks are impermeable and suitable for wine barrels, whereas red oaks are quite permeable and unsuitable for barrels. In many

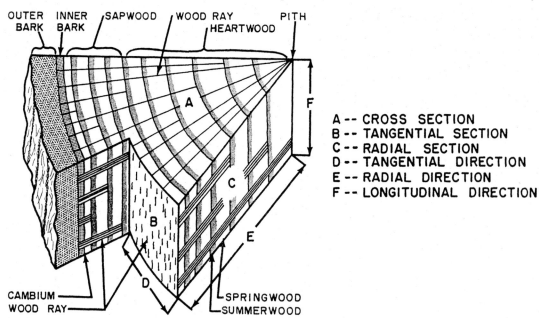

OUTER BARK INNER BARK SAPWOOD WOOD RAY HEARTWOOD PITH

A -- CROSS SECTION
B -- TANGENTIAL SECTION
C -- RADIAL SECTION
D -- TANGENTIAL DIRECTION
E -- RADIAL DIRECTION
F -- LONGITUDINAL DIRECTION

CAMBIUM
WOOD RAY
SPRINGWOOD
SUMMERWOOD

Fig. 2-2. Gross structure of a softwood stem cross section. Reprinted from W.I. West, Wood Pole Conference, Corvallis, Ore., Mar. 21, 1960, with permission.

Table 2-1. The function and characteristics of various tree stem tissues.

Tissue	Function
Outer bark	Physical and biological protection
Phloem (inner bark)	Conduction of food up and down the stem
Vascular cambium	Thin layer of cells giving rise to all the wood and inner bark fibers. The tree stem grows outward
Rays	Storage and lateral food movement from the phloem to the living cells of the cambium and sapwood
Pith	The center of the tree; from the apical meristem
Growth ring -Earlywood -Latewood	One year's growth of wood Low density wood designed for conduction of water High density wood for strength to support the tree
Sapwood	Conduction of sap (water, soil nutrients) up to the leaves.
Heartwood	Provides strength to support the crown. Often relatively low moisture content in softwoods.
Juvenile wood	The first 10 growth rings surrounding the pith. Usually low density and relatively short fibers.

species the heartwood is obviously darker than the sapwood; in other species the presence of starch in *sapwood* allows it to be distinguished from heartwood by a test with iodine.

Some of the important pulping variables of wood and wood chips are:

1. Moisture content--percentage of water reported relative to dry or wet wood weight.
2. Specific gravity--the density of wood material relative to the density of water.
3. Tension and compression strength properties.
4. Bark content.
5. Chemical composition--cellulose, hemicellulose, lignin, and extractives.
6. Length of storage--amount of decay and extractives content.
7. Chip dimensions.
8. Wood species.

The moisture content of wood is an important factor, since one pays for wood on an oven-dry basis, which represents the actual amount of wood material present. Often it is desirable to have a low moisture content to reduce the energy requirements to chemically pulp the wood and reduce transportation costs, but there is little that can be done to control the moisture content of wood sources. The wood density and wood chip bulk density (including air between the wood chips) are also important in determining the amount of wood material one purchases and in determining digester charge levels.

The properties of wood (or chips from wood) depends on growth factors such as the location of the tree from which it came and its location within the tree. For example, the average fiber length typically increases gradually as the first 50 inner growth rings are formed, and fibers are longest in wood 10-20 ft above the ground. For this reason there is some concern about the properties of wood harvested from short rotation wood plantations or forests that have trees with high amounts of *juvenile wood*, the wood made during the first 10 years after a stem forms. Trees growing on wet, warm, and sunny locations are fast growing and have coarse, stiff fibers, whereas trees growing on dry, cold, and less sunny locations are slow growing and have fine, dense fibers.

Fig. 2-3. Scanning electron micrograph (SEM) of a cypress heartwood cube.

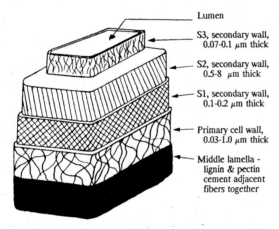

Lumen

S3, secondary wall, 0.07-0.1 μm thick

S2, secondary wall, 0.5-8 μm thick

S1, secondary wall, 0.1-0.2 μm thick

Primary cell wall, 0.03-1.0 μm thick

Middle lamella - lignin & pectin cement adjacent fibers together

Fig. 2-4. A mature softwood fiber. Adapted from U.S. For. Ser. Res. paper FPL-5, 1963.

Softwoods

Wood, or trees, from the Gymnosperms, a subdivision of the division Spermatophytes (plants with seeds) are known as softwoods. Gymnosperms are also called conifers or evergreens. These trees retain their needles (leaves) in winter.

Softwoods are characterized by relatively simple wood anatomy consisting of 90-95% longitudinal fiber tracheids 2.5-7 mm long and 25-60 μm wide, 5-10% ray cells, and 0.5-1.0% resin cells. Fig. 2-3 shows a wood cube from cypress.

All wood fibers have many similar structural and chemical features. Fig. 2-4 shows a typical softwood fiber. The S_2 layer is generally the thickest layer of the fiber and its effect dominates the overall properties of the fiber. In this layer, the cellulose microfibrils are oriented at about 10-30° from the main longitudinal axis of the fiber. This gives the fiber a higher tensile strength in this direction compared to the radial or tangential directions. Fibers do not shrink and swell appreciably in the longitudinal direction while there is very high shrinkage in the other two directions. The microfibril angle is discussed further in regards to fiber strength in Section 2.7.

Softwood fiber length and coarseness

Most softwood fibers average 3 to 3.6 mm in length. Some notable exceptions are three of the four southern pines--longleaf, shortleaf, and slash pines at 4.6-4.9 mm, sugar pine at 5.9 mm, western larch at 5.0 mm, sitka spruce at 5.9 mm, redwood at 7.0 mm, Douglas-fir at 3.9 mm, and baldcypress at 6.2 mm. Fiber coarseness (the mass per chain of fibers 100 meters long) is typically 18-30 mg/100 m. (Data from *Pulpwoods of the United States and Canada*, 1980.)

Hardwood

Wood, or trees, from the Angiosperms, a subdivision of the division Spermatophytes, are known as hardwoods. Hardwood trees are also called broadleafs or deciduous. These trees lose their leaves in winter. Hardwoods have complex structures including vessel elements, fiber tracheids, libriform fibers, rays cells, and parenchyma cells. The fibers of hardwoods are on the order of 0.9-1.5 mm long, leading to smoother paper of lower strength compared to softwood fibers. The morphology of hardwoods is much more complex than that of softwoods. The cellular composition is 36-70% fiber cells, 20-55% vessel elements, 6-20% ray cells, and about 2% parenchyma cells by volume. Fig. 2-5 shows the structure of red oak.

Pits

Chemical pulping of wood depends upon the ability of cooking liquor to flow through the wood. Fibers in wood have numerous openings between each other on their radial surfaces that are known as pits or pit pairs. In the living sapwood they allow for the movement of liquids up and down the length of the tree. (The swelling of wood above a pH of 10-12 during alkaline pulping increases the tangential and radial permeability drastically.) The type of pit will depend on the fiber type in which the pits are found. The most common type is the *bordered pit* which occurs between two adjacent tracheids of softwoods. The exact nature of a bordered pit depends on the species

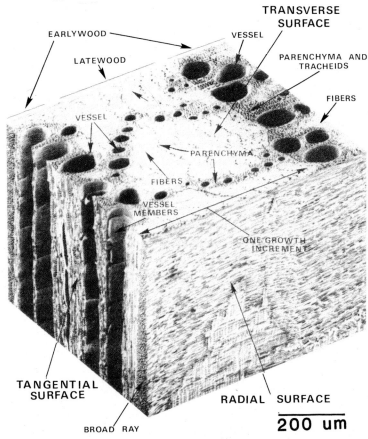

EARLYWOOD
LATEWOOD
TRANSVERSE SURFACE
VESSEL
PARENCHYMA AND TRACHEIDS
FIBERS
VESSEL
PARENCHYMA
FIBERS
VESSEL MEMBERS
ONE GROWTH INCREMENT
TANGENTIAL SURFACE
RADIAL SURFACE
BROAD RAY
200 um

Fig. 2-5. Red oak hardwood. From C.W. McMillin and F.G. Manwill, *The Wood and Bark of Southern Hardwoods.* **USDA Rep. SO-29, 1980.**

of wood. Pits are the only intercellular channels for conduction of liquids in softwood tracheids, which make them important in wood pulping and treatment with preservatives. In softwoods, the bordered pits become *aspirated* during heartwood formation; that is, the pit membrane is displaced to one side of the pit chamber allowing the torus to seal the aperture, which effectively stops the flow of liquids through the pit as shown below. This condition is known as *pit aspiration* and occurs whenever wood is air or kiln dried. This is one reason dried softwood does not pulp well.

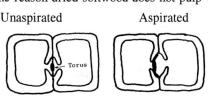

Unaspirated   Aspirated

Torus

Knots

A knot is the base of a branch that starts from the trunk or a large limb of the tree. The wood grows around the base of the branch and gives rise to several types of knots that influence the pulping and pulp characteristics of the material. For example, if the branch dies, the trunk will continue to form around the branch, encasing the entire branch stub, including the bark. Eventually there will be no trace of the knot from the exterior tangential surface. If the branch continues to live, the branch wood and trunk wood are then continuous. Knots are harder, more dense, often more resinous, and more difficult to pulp than the normal wood. Knots often remain unpulped after chemical pulping operations and decrease the screened pulp yield.

Reaction wood

Trees under physical stresses such as the weight of heavy branches or leaning trunks produce wood that *reacts* to these stresses and is known as *reaction wood*. In practice, there is little one can do about the presence (or absence) of reaction wood, but it is usually not present in large amounts.

Angiosperms (hardwoods) produce *tension wood*, which is usually located on the upper side of branches or leaning trunks where the wood is under a tension force. Tension wood has fewer vessels, and those are of smaller diameter, a thick *gelatinous layer* in the cell wall on the lumen side, and a higher cellulose content than normal wood. Tension wood produces pulp with higher yield, but with lower strength, than normal wood, which makes it well suited for dissolving pulps.

Gymnosperms (softwoods) form *compression wood* on the lower side of branches or leaning trunks, where the wood is under a compression force. Compression wood tends to have a higher proportion of latewood, higher lignin content, higher density, and higher hardness compared to normal wood, making it less suited for pulping. TAPPI Standard T 267 has photographs and methods for compression wood identification.

Moisture content (wet or green basis), MC_{GR}

The moisture content based on the wet weight of material is used in pulp and paper mills for wood as well as most other raw materials such as pulp, paper, and fillers. It represents the amount of water in wood as a fraction of the *wet* weight of wood. If no subscript is used one can generally assume the moisture content is on a wet basis in the pulp and paper literature. The weight of water is determined by weighing the wood before and after drying at 105°C. The green-basis moisture content of freshly cut wood is typically 50%, but can vary from about 30-60%.

$$MC_{GR} = \frac{\text{mass } H_2O \text{ in wood}}{\text{wet wood mass}} \times 100\%$$

Moisture content (dry basis), MC_{OD}

A measure of the moisture content of wood based on the oven-dry (an obsolete term is bone dry) weight of wood is used by wood scientists and foresters. It represents the amount of water in a wood sample divided by the *oven-dry* weight of wood material. The oven-dry weight is obtained by drying the wood to constant weight at 103-105°C (217-221°F). Freshly cut wood has an oven-dry basis moisture content on the order of 100%, although it varies from about 45-150%.

$$MC_{OD} = \frac{\text{mass } H_2O \text{ in wood}}{\text{oven-dry wood mass}} \times 100\%$$

Solids content

The solids content is a measure of the solid material in wet samples such as wood, pulp, and paper. The term *consistency* is used instead of solids content in pulp slurries.

$$\text{solids content} = 100\% - MC_{GR}$$

$$\text{solids content} = \frac{\text{oven-dry sample mass}}{\text{wet sample mass}} \times 100\%$$

Solid wood density

Solid wood density is a measure of the dry weight of wood per unit volume of green wood. Since wood contracts about 8-15% on a volume basis as it is dried below 30% moisture, it is important to specify the moisture content at which the volume was measured. Typical units are lb/ft³, g/cm³, or kg/m³.

Specific gravity

Specific gravity is the (unitless) ratio of the *solid wood density* to the density of water at the same temperature. The solid wood density may be determined using the green volume, the oven-dry volume, or intermediate volumes. This is notable as wood shrinks about 8-15% as it dries. The *basic specific gravity* always uses the green volume. [The density of water at 20°C (68°F) is 62.4 lb per cubic foot, 1 gram per cc, or 1 metric ton per cubic meter.] Softwoods have typical specific gravities of 0.35-0.50 g/cc on a green volume basis, but can vary from 0.29-0.60 among North American commercial species; hardwoods have typical specific gravities of 0.35-0.60 on a green volume basis, but can vary from 0.30-0.90 among the North American, commercial species. Balsa wood, used in model-building, has a basic specific gravity of 0.16, whereas ironwood is 1.05. The specific gravity of the cell wall material is 1.50. Note that (using appropriate units):

Specific gravity of wood × Density of water = Solid wood density

Wood decay and other deterioration

At ambient temperatures and suitable pH conditions, wood can decay in the presence of moisture and oxygen. This means wood undergoes decay, especially when above 20% MC_{OD} and exposed to air. In temperate environments, decay leads to much more loss of structural wood than is caused by termites. Wood in the form of logs kept under a water sprinkler (Fig. 2-6) or immersed in water (Fig. 2-7) tends to decay slowly. While water may seem to promote decay, under these conditions available oxygen is low in concentration due to air displacement by water, limiting decay. Keeping wood in any form below 20% moisture content prevents decay, but it is generally impractical to dry wood for pulping, and dry wood usually does not pulp well since liquor penetration is more difficult.

Wood chip piles, as a rule of thumb, lose 1% per month due to decay and chemical oxidation. Fig. 2-8 is a summary of wood deterioration in chip piles from Fuller (1980).

Wood deterioration begins by respiration of living parenchyma cells of the wood rays. This increases the temperature and accelerates bacterial and fungal decay. Heating above 45-55°C (115-130°F) is mostly due to chemical oxidation (especially of extractives), as most bacteria and fungi will not grow well at these temperatures. Temperatures above 55°C (130°F) lead to severe losses upon pulping. Degradation can be limited by keeping piles below 15-18 m (50-60 ft) high. Conditions that aggravate decay include the use of whole tree chips, high fines fraction that decreases air circulation, pile compaction, high piles, and the storage of hardwoods, which tend to have high starch contents. Loss of terpenes (important if collected at a mill) may be over 50% in the first month.

For these reasons, it is important to rotate chip inventories on a first in--first out (FIFO) basis. Mills recovering extractives often use FIFO in combination with pulping some chips directly as they enter the mill to obtain high terpene recoveries. Chip storage is sometimes advantageous to reduce pitch problems during sulfite pulping. Western U.S. kraft mills typically operate with a three-month supply of chips in storage, whereas mechanical pulp mills are limited to a one-week supply as chip darkening leads to inalterable pulp

Fig. 2-6. Logs stored under a water sprinkler to keep oxygen out and thereby prevent decay. The photo was taken at Mary's River Lumber in Philomath, Oregon and is courtesy of Susan Smith.

Fig. 2-7. Logs for export on the Columbia River in the northwest U.S. Keeping wood in water decreases decay and makes for easy handling and transportation in rafts.

darkening. (Chemical pulping and bleaching of chemical pulps remove most of the dark color.)

There are two principal decay fungi groups and their observable action is shown in Plate 4.

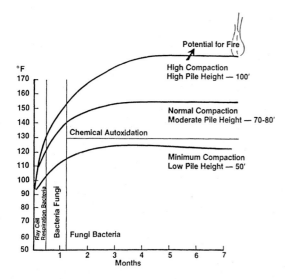

Fig. 2-8. Deterioration in wood chip piles. ©1985 TAPPI. Reprinted from Fuller (1985) with permission.

Actual fungi hyphae are shown in Plate 5. *Brown rot fungi* attack the carbohydrates leaving a brown chip. Small losses in weight lead to very large decreases in cellulose viscosity and, therefore, pulp strength. *White rot fungi* attack both the lignin and carbohydrates leaving a whitish wood. Small loses of wood material do not lead to an appreciable decrease in cellulose viscosity or paper strength; indeed paper strength often increases slightly. Treating wood with white rot fungi or enzymes derived from decay fungi is the basis of the experimental biopulping and some biobleaching methods. *Sapstains*, such as bluestain, darken woods but do not degrade their strength properties.

Fuel value of wood

The heating value of wood is about 21 MJ/kg (9000 Btu/lb) for oven-dry softwood and 19.8 MJ/kg (8500 Btu/lb) for oven-dry hardwood. The higher value for softwood is due to the higher lignin content. (Lignin has a much lower oxygen content than the carbohydrates it displaces.) Actual heating values depend on species, growing conditions, age, etc. The actual fuel value of wet wood or bark is calculated on the basis that 1 kg water takes 2.5 MJ to evaporate (1 lb of water takes 1100 Btu to evaporate). For example, 1 lb

of softwood at 50% moisture content is ½ lb of wood with 4500 Btu fuel value, but 550 Btu would be needed to evaporate the other ½ lb of water. This wet wood would have an effective fuel value of 3950 Btu/lb (wet basis).

2.2 WOOD CHIPS AND SAWDUST

Sawdust

Sawdust is the residue generated by saw teeth when wood is cut into lumber. In the past, it has had some limited use by the pulp and paper industry. It gives a pulp with short fibers that is suitable as part of the furnish for tissue and writing papers. Since the 1970s, saw blades have become thinner with more teeth, which, in many cases, makes the sawdust too small to be used as a fiber source for pulp.

Chips

Wood chips are mechanically disintegrated wood, traditionally in pieces 12-25 mm (1/2 to 1 in.) along the grain, variable in width, and 3-6 mm (1/8 to 1/4 in.) thick. Uniform chip size is very important in chemical pulping because large chips (particularly overthick chips in kraft cooking) undercook, leaving large amounts of shives, while small chips clog the liquor circulation system, use large amounts of chemical, and give a low yield of weak pulp. Bark, dirt and other materials should always be kept to a minimum (0.5% or less), especially in mechanical pulps where they give a dark pulp that cannot be brightened, since lignin must be retained in these pulps.

Short chips will give paper that is slightly weaker due to fiber cutting. Softwood chips less than 12 mm (0.5 in.) long (since this is the axis parallel to fiber orientation) will have reduced average fiber lengths since many of the fibers will be cut. For example, 25 mm long chips from Douglas-fir will have an average fiber length of 3.5 mm; 12 mm long chips will have an average fiber length of 3.0 mm; and 6 mm long chips will have an average fiber length of only 2 mm.

In the western U.S., 80% of wood is received as chip waste from primary wood processors and 20% is chipped on site. In the eastern and southern U.S. the figures are reversed.

(However, in 1947, over 80% of the wood used in the western U.S. at pulp mills was in the form of roundwood; this decreased to about 40% roundwood by 1960 and 20% by 1980.) Whether wood chips are generated on the mill site or purchased as wood mill residues, quality is extremely important. One key factor in obtaining the highest quality pulp possible with the most efficient use of pulping and bleaching chemicals and the least environmental impact is to have as uniform a chip as possible. **This factor cannot be over stressed!**

Chip silos

For the sake of efficiency, after screening, wood chips must be metered into the digesters without constant supervision. One method of accomplishing this is with the use of chip silos that hold 50-300 tons of chips. A representative diagram of a chip silo is shown in Fig. 2-9, while actual silos and a turntable (rotor table) are shown in Fig. 2-10. The silos may be located below a pile of wood chips and not visible in the woodyard. A large supply of chips in the silo assures a constant wood supply to the digesters. The turntable can meter the chips to the conveyer that then takes the chips to the digester. Controlling the speed of the turntable controls the chip supply.

Chip sources

Whole tree chips are made from the entire stem of the tree, usually in the woods. Whole tree chipping gives more wood chips but they are contaminated with bark, dirt, twigs, etc.; therefore, they must go through complicated cleaning operations prior to use. Even so, residual contaminants mean these chips are only useful for courser grades of paper. *Residual chips* are made from wood residues generated by solid wood conversion as by sawmills and plywood mills. For example, edgings and cores are chipped. *Sawdust* is the residue generated by the teeth of a saw. It is chemically pulped in continuous digesters with short residence times such as the *M&D* or *Pandia* and is used as a filler to give smoothness to papers. Different saw types generate sawdust of different size distributions., some consist of particles that are too small to give useful pulp.

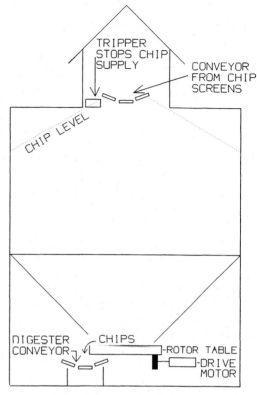

Fig. 2-9. Chip silo.

2.3 WOOD CHIP PREPARATION AND HANDLING AT THE PULP MILL

General considerations

In many mills, wood chips are made on site from roundwood. The discussion in this book focuses on typical operations at pulp mills producing their own chips or purchasing chips. Many mills rely on purchased chips that are generated by outside facilities such as saw mills, plywood mills, whole tree chips, etc. Note that the people generating such chips use other methods as well as the ones described here and often do not understand the importance of chip quality (or even what constitutes chip quality) from the point of view of pulp manufactures. For example, chip suppliers are not apt to use chip thickness screens, but oscillating, round-hole screens. The chip buyer for a given mill should work closely with chip suppliers *communicating with them* as to what the pulp mill requires and how the supplier might best meet these demands. This assumes that

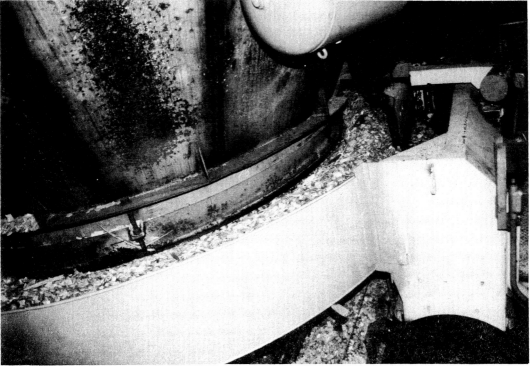

Fig. 2-10.   Chip silos (top) that each hold 320 units.   They are 44 feet in diameter and 80 feet high.

the pulp mill is monitoring each supplier for chip quality.

Slasher deck

The slasher is a deck where saws cut logs into shorter lengths, such as 2.5 m (8 ft), for easier handling. Because slashers require much maintenance, many mills are going to tree length handling of wood, although this is much more difficult with hardwoods because of their tendency to have several main trunks.

Barker

A barker (or debarker) is a device used to remove bark from wood before chipping. Removal of bark is necessary as it has negligible useful fiber, darkens pulp, requires extra chemical usage, and introduces contaminants such as calcium, silica and aluminum into the chemical recovery system. Bark adhesion is about 3-5 kg/cm² during the growing season and 2-3 times higher during the dormant season (winter).

One type of barker, the *drum barker*, is a large, rotating, steel drum 2.5-5.0 m (8-16 ft) in diameter by 8-25 m (30-90 ft) long (12 ft in diameter by 70 ft long is common), mounted with the exit lower than the entrance to promote the flow of logs (Fig. 2-11 and Plate 6). The drum rotates at about 5 revolutions per minute. A dam at the exit controls the log retention time, which is on the order of 20-30 minutes. Debarking occurs

by mechanical abrasion of the logs against each other. Typically the wood contains 0.5-1.0% bark after this method, although this depends on the type of wood, average log diameters, season of the year, and other factors. Logs that are not adequately debarked may be sent through the drum again. Some disadvantages of the drum barker are the relatively high power requirements, maintenance costs, and bruising of the logs.

The *cambial shear barker* works with a ring of knives that peel off the bark as the logs are fed individually as shown in Plate 7. Feeding the logs individually increases the operating costs due to the required supervision. This method does not work well on frozen logs. The *cutterhead* or *Rosserhead* barker works like a planar, where a turning wheel that is toothed or spiked is held against the log and removes the bark by abrasion (see the insert of Plate 7). It has low capacity since logs are fed individually, requires large amounts of supervision, and has high wood losses. It is used in low-production facilities such as small saw mills. It is suitable for logs that are difficult to debark such as frozen logs.

Hydraulic barkers operate by impinging high pressure water jets onto the wood. With *ring barkers*, logs 0.5-1.5 m (1-5 ft) in diameter travel at 1 m/s (3 ft/s) while 1500 gallons per minute of water at 1500 psi squirt out of four nozzles in a rotating ring (16 stationary nozzles in older designs) with a velocity of 400 ft/s. As much as

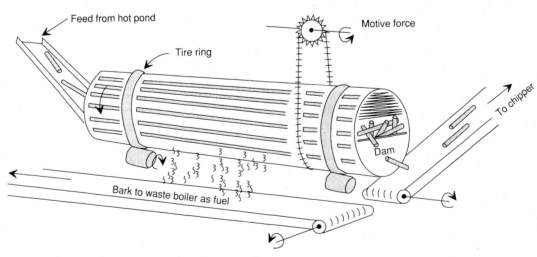

Fig. 2-11. Schematic diagram of a barking drum. Redrawn from J. Ainsworth, *Papermaking,* **©1957 Thilmany Paper Co., with permission**

Fig. 2-12. Bellingham hydraulic barker in action from *Making Pulp and Paper*, ©1967 Crown Zellerbach Corp., with permission.

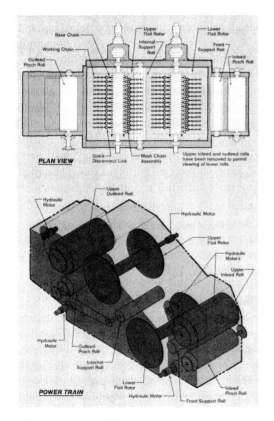

Fig. 2-13. A diagram of a flail debarker. Courtesy of Manitowoc Engineering Company.

2500 hp is required to drive the water pumps. The water must be very clean for ring barkers as grit and dirt quickly disintegrate the seal.

Single jet units such as the modified *Bellingham barker* (Fig. 2-12) use a traveling nozzle that moves up and down while the log is moved and rotated using airplane-type controls (joy sticks). Water is supplied by a six-stage impeller pump to produce 1200 gal/min at a nozzle pressure of 1400 psi. Numerous other designs for hydraulic barkers are used as well. The waste water from hydraulic barkers is high in BOD (a type of pollutant) and color and must be treated before release, a severe drawback. They were once common on the West Coast of the U.S. as they work well for debarking large logs.

Flail debarkers (Fig. 2-13) use a rotating cylinder with numerous chains hanging from it to delimb and debark small diameter material. It is useful for in-the-woods operations for processing precommercial thinnings that would otherwise be unusable. Without debarking, whole tree wood chips are less valuable. Plate 8 shows an in-the-woods operation using a flail delimber/debarker.

Chipper

Chippers are devices used for mechanically breaking down wood into chips (Fig. 2-14). Chipping takes approximately 7-14 kWh/t (0.4-0.8 HP-days/ton). Hardwoods generally are harder to chip and generate fewer fine chips but more large chips than softwoods. The traditional chipper for log lengths of 1.5-3 m (4-10 ft) is the *gravity feed* (or *drop-feed*) *disk chipper*, where the logs enter through a spout mounted on the top (or sometimes other) corner of the feed side. Mills that handle tree length wood must use *horizontal feed disk chippers* that have the feed at the top or the bottom of the feed side. Horizontal chippers use more knives than drop-feed chippers to avoid the production of more fine and pin chips. Disk chippers use *gravity discharge*, which increases the initial capital cost due to the higher elevation required for the chipper, or *blowing discharge*, which increases the pins and fines content due to the extra chip damage caused by the paddles forcing the chips from the chipper.

The best chipping of softwood logs leads to 85% accept chips, 4% overthick, 2% overlength

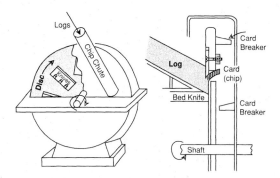

Fig. 2-14. Drop feed (gravity) wood chipper. Redrawn from J. Ainsworth, *Papermaking*, ©1957 Thilmany Paper Co., with permission

chips, 7% pin chips, and 2% fines (see below for these definitions). A worn blower (for blower discharge units) or excessive anvil gap leads to more pins and fines.

Drum and *double cone (V-drum) chippers* have very limited use for smaller sized wood. Because of their design and because they are processing small residues, the generation of pin chips (15-25%) and fines (5% or higher) is much higher than chipping softwood logs. Other chippers are designed for specific purposes such as *veneer chippers* that chip veneer residues from plywood plants and *core chippers* that chip the core of peeled logs used to make veneer.

Chip size sorting for production

Ideally all chips regardless of their source are sorted (screened) at the mill into several fractions according to their size to permit uniform pulping. In the past, most mills classified wood chips by size using oscillating, round-hole screens such as those shown in Fig. 2-15. During the 1970s it became apparent that for the kraft cooking process, chip thickness is of primary concern. Since 1980, almost all kraft mills have installed equipment that classifies the chips by thickness (Fig. 2-16 and Plate 9) to remove the overthick chips. Most mills use additional separations to remove fines. A few mills even separate pin chips going to the process and meter them back into the process. Most sulfite mills and sawmills continue to use chip classification by round-hole screens.

Laboratory chip screening

Chip classification is also done in the research and development laboratory (as opposed to production quality control) for experimental protocols and to determine the quality of chips from the various chip vendors, although the methods used are not designed for large numbers of samples on a routine basis. Laboratory classification was traditionally based on chip size using round-holed screens and is known as the *Williams classification* with pans containing 9/8, 7/8, 5/8, 3/8, and 3/16 in. holes. This method is now obsolete for most purposes. To more closely duplicate screening at kraft mills, a thickness screen is now included for laboratory screening. The exact definitions of the following chip fractions depend on the nature of the classification scheme. The laboratory screen is shown in Fig. 2-17.

The following definitions are based on typical laboratory screening. *Overs* are the oversized or overthick fraction of chips and are retained on a 45 mm (1.8 in.) diameter hole screen and are thicker than 10 mm for conifers or 8 mm for hardwoods. [For sawdust, overs are retained on a 12 mm (1/2 in.) diameter hole screen.] *Accepts* are the chip fraction of the ideal size distribution for pulping. These chips pass through an 8 or 10 mm slotted screen and are retained on a screen with holes 7 mm (0.276 in.) or 3/8 in. diameter. *Pin chips* are the chips that pass through a 7 mm screen but are retained on a 3 mm (0.118 in.) or 3/16 in. hole screen. *Fines (unders)* are the undersized fraction of chips or sawdust and are collected in the bottom pan. The definition of fines will vary with mill specifications, but fines generally consist of material passing through a 3 mm screen.

Wood chip quality control at the mill

Chip quality control uses devices that are designed to handle numerous samples quickly with a minimum of operator time. It is important to practice wood chip quality control for several reasons. The most basic reason is that the amount of dry wood must be determined for a truck, railcar, or barge load so that the supplier can be paid for the equivalent oven-dry wood. An equal-

Fig. 2-15. Vibratory, round-holed screens for chip size classification. The top shows oversized chip removal; the bottom shows accept and pin chip fractions each at a different mill.

ly important reason to determine chip quality from each load is so that the mill can work with each chip supplier to insure that high quality chips are supplied in terms of desired species, suitable chip size distribution, and small amounts of dirt, bark, and decayed material.

Fig. 2-16. Chip thickness screen for overthick chip removal. The insert shows a close-up of disks.

Like any method to insure quality, a representative sample from each truckload of incoming chips must be obtained. Some mills use continuous samplers to insure the sample represents the entire truckload, but most mills use a simple bucket sampler that is filled from the first 5% of the truckload. Fig. 2-18 shows a continuous sampler and a bucket sampler that grabs a sample from one section of the truck. Plate 10 shows a chip truck dumping in an arrangement that also uses a bucket-type sampler. (Stories circulate about unscrupulous suppliers that water-down or use inferior chips in the part of the truck that they know will not be sampled.) Many mills may also have a person collect a sample every hour of the chips going to the digester to see how the screening system is working. (Stories also circulate about the person who collects all eight samples for a given shift at one time and submits one sample every hour.)

Ideally, laboratory determinations are made for moisture content, chip size distribution, and bark, rot, and dirt contents. Determinations of wood species, extractives content, chip bulk density, and chip damage are made less often. Chip size distributions are determined in the laboratory using oscillating screen systems, systems with adjustable thicknesses between bars (Fig. 2-19), or other suitable systems. Collection of representative wood chip samples from rail cars or barges can be quite challenging (Fig. 2-20.)

Miscellaneous analyses

TAPPI Standard T 257 describes the sampling and preparation of wood for analysis whether logs, chips or sawdust. The basic density and moisture content of pulpwood is determined according to TAPPI Standard T 258. In this test, volume is measured by water displacement and moisture content by the difference in mass before and after drying at $105\,°C \pm 3°$. The overall weight-volume of stacked roundwood is determined according to TAPPI Standard T 268. TAPPI Standard T 265 is used to measure the natural (wood derived) dirt in wood chips. Dirt originates from the outer and inner barks, knots,

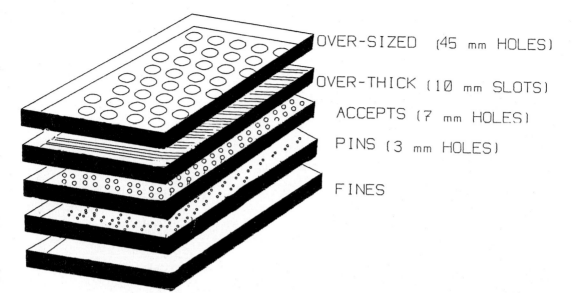

OVER-SIZED (45 mm HOLES)

OVER-THICK (10 mm SLOTS)

ACCEPTS (7 mm HOLES)

PINS (3 mm HOLES)

FINES

Fig. 2-17. Laboratory chip size distribution analysis. The top shows the pan configuration; the bottom shows the actual equipment.

stains, rot (decay), etc. Color photographs help with identification for the novice. The ash content (i.e., the mineral content including metals and their anions and silicates) of wood and pulp is determined by ignition in a muffle furnace at 575 $\pm$ 25°C as described by TAPPI Standard T 211.

TAPPI Standard T 263, with numerous diagrams and photomicrographs, covers identifi-

Fig. 2-18. Continuous sampling during chip truck emptying on the left; on the right a bucket that is filled when it swings out, obtaining a sample that represents only one section of the load.

cation of wood and fibers from conifers. One should consult the references listed in this method for additional information and high quality photomicrographs unless experienced in microscopy and, more particularly, wood anatomy.

The amount of hardwood and softwood in wood chips is easily determined with the Mäule test (Section 21.18), which gives a purple color for hardwoods while leaving softwoods uncolored.

2.4 SOLID WOOD MEASUREMENT

Wood is measured and sold on a variety of bases. The volume of solid wood in stacked roundwood can be determined by *scaling* (a labor-intensive method of sampling volume and log sizes and converting to solid wood volume with tables) or by *water displacement* methods. In practice, it is easier to weigh the wood (using truck scales)

and determine the bone dry weight using moisture contents of the wood. Table 2-2 gives the conversion factors for different units of solid wood and wood chips that are described below.

Board foot

A board foot is a volumetric measurement of solid wood. It is equal to 12 in. × 12 in. × 1 in. or 1/12 cubic foot of solid wood.

Cord

A cord is stacked roundwood occupying a total volume of 4 ft by 8 ft by 4 ft. Typically, a cord of stacked wood contains 80-90 cubic feet of solid wood, although this can vary widely, and will yield about 500 bd. ft. of lumber or 1.2 BDU of chips. (A face cord is used to sell firewood in some locations; it is 4 ft by 8 ft by the width of the wood pieces.)

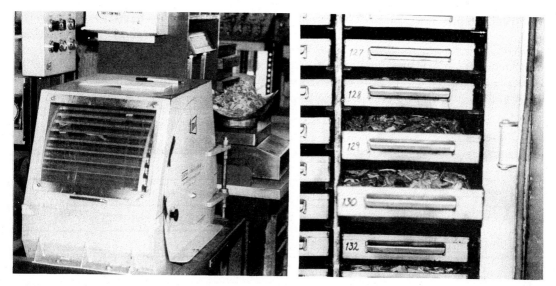

Fig. 2-19. Chip size classification (left) and moisture content determination (right) in a mill's wood quality control laboratory.

Cunit

A cunit is 100 cubic feet of solid wood in stacked roundwood. It is used to determine the wood content of pulp logs.

Fig. 2-20. Rail cars ready to be unloaded. They will be tipped back on the special platform to discharge their contents. Over six cars per hour can be unloaded by this method.

2.5 WOOD CHIP MEASUREMENT

Bulk density

The bulk density is the oven-dry weight of chips (or sawdust, or other wood residue) contained in a given volume of space. The bulk density of the chips depends on the specific gravity of the wood source, the chip geometry, and the chip size distribution. For example, Douglas-fir chips from roundwood are typically 192 kg/m^3 (12 lb/ft^3) (dry wood weights), while Douglas-fir chips made from veneer are 184 kg/m^3 (11.5 lb/ft^3). White fir and pine chips are about 168 kg/m^3 (10.5 lb/ft^3), and those of redwood are about 160 kg/m^3 (10 lb/ft^3). A rule of thumb is that 1 m^3 of wood yields about 2.6 m^3 of chips (1 ft^3 yields about 2.6 ft^3).

Bone dry unit, BDU

A bone dry unit is the equivalent of 2400 lb of oven-dry chips, sawdust, or other wood particles. A BDU of packed Douglas-fir chips occupies approximately 200 cu. ft.

Unit

A unit is 200 cu. ft. of wood chips, sawdust, or other wood particles. A 40 foot open-top chip truck carries about 18 units. One unit of Douglas-fir or western hemlock chips is about

Table 2-2. Approximate conversion factors of wood of 0.45 specific gravity in various forms (for 85 ft³ of solid wood per cord and 2.6 ft³ of chips per ft³ of solid wood). Entries with five significant digits are exact for any wood. Conversion factors for wood varying from these parameters should be calculated for individual situations. Board feet values do not include sawing loses.

Convert from↓	Convert to↓ by multiplication			
	m³ stacked wood	m³ of solid wood	MT of bone dry wood	m³ of wood chips
1000 board ft	3.55	2.3598	1.06	6.1
cord	3.6247	2.41	1.08	6.3
cunit	4.26	2.8318	1.27	7.4
BDU	3.64	2.42	1.0886	6.3
unit	3.28	2.18	0.98	5.6636
ft³ chips	0.0164	0.0109	0.0049	0.028318

0.85 cords of logs or 67 ft³ of solid wood. One unit of Douglas-fir or western hemlock sawdust is about 80 ft³ of solid wood.

2.6 WOOD CHEMISTRY

The composition of hardwoods and softwoods by the class of compounds is given in Table 2-3. The ultimate (elemental) analysis of wood is given in Table 2-4.

Lignin is more highly concentrated in the middle lamella and primary cell wall regions of the wood fiber than any other part of the cell wall. Most of the lignin, however, is actually in the secondary cell wall since the secondary cell wall accounts for most of the mass of the fiber. The concentration of the major components with varying cell wall position is shown in Fig. 2-21.

Cellulose

On the molecular level, cellulose is a linear polymer of anhydro-D-glucose connected by β-(1→4)-linkages as shown in Fig. 2-22. The *degree of polymerization* (DP), which is the number of units (glucose in this case) that make up the polymer, is above 10,000 in unaltered (so-called "native") wood, but less than 1000 in highly bleached kraft pulps.

Table 2-3. Typical compositions of North American woods (percent).

	Hardwoods	Softwoods
Cellulose	40-50	45-50
Hemicelluloses		
(Galacto)glucomannans	2- 5	20-25
Xylans	15-30	5-10
Lignin	18-25	25-35
Extractives	1-5	3-8
Ash	0.4-0.8	0.2-0.5

Table 2-4. The ultimate analysis of North American woods in percent.

Carbon	C	49.0-50.5	%
Oxygen	O	43.5-44.5	%
Hydrogen	H	5.8- 6.1	%
Nitrogen	N	0.2- 0.5	%

Physically, cellulose is a white solid material that may exist in crystalline or amorphous states. Cellulose in wood is about 50-70% crystalline and forms the "back-bone" structure of a wood fiber.

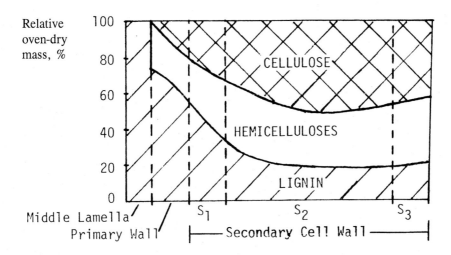

Fig. 2-21. **Typical composition of a conifer fiber across its cell wall.** **From Krahmer, R.L. and Van Vliet, A.C., Eds.,** *Wood Technology and Utilization*, **O.S.U. Bookstores, Corvallis, Oregon 1983.**

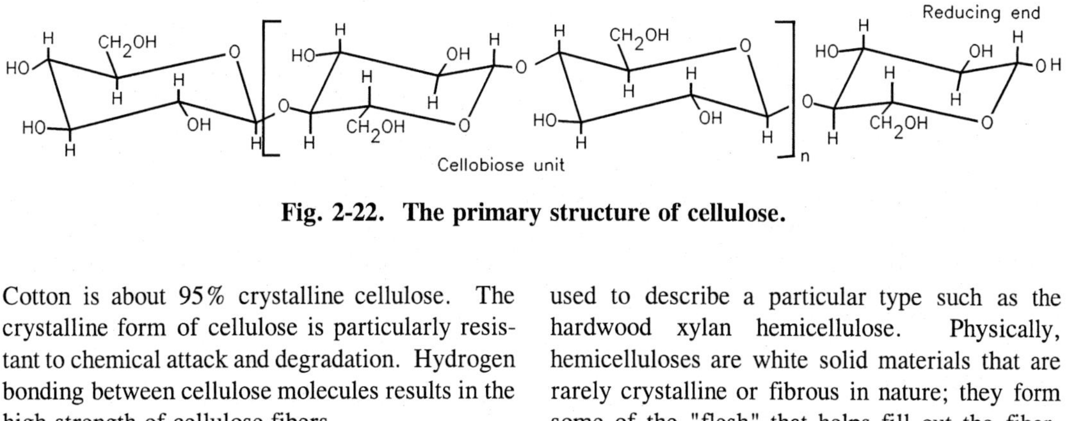

Fig. 2-22. **The primary structure of cellulose.**

Cotton is about 95% crystalline cellulose. The crystalline form of cellulose is particularly resistant to chemical attack and degradation. Hydrogen bonding between cellulose molecules results in the high strength of cellulose fibers.

Microfibrils are aggregations of cellulose molecules into thread-like structures approximately 3.5 nm in diameter, containing both crystalline and amorphous regions. Microfibrils occur in the secondary cell wall. Microfibrils are oriented in different directions in each of the three layers within the secondary cell wall; the fibril angle is measured from the longitudinal axis of the cell.

Hemicellulose(s)

Hemicellulose(s) are actually a class of materials. The plural form should be used to describe them generically, but the singular form should be used to describe a particular type such as the hardwood xylan hemicellulose. Physically, hemicelluloses are white solid materials that are rarely crystalline or fibrous in nature; they form some of the "flesh" that helps fill out the fiber. Hemicelluloses increase the strength of paper (especially tensile, burst, and fold) and the pulp yield, but are not desired in dissolving pulps. (Dissolving pulps are relatively pure forms of cellulose used to make cellulose-based plastics.) Starch is often added to pulp to increase the strength of paper and probably has a very similar mechanism of effect as the hemicelluloses.

Hemicelluloses chemically are a class of polymers of sugars including the six-carbon sugars mannose, galactose, glucose, and 4-*O*-methyl-D-glucuronic acid and the five-carbon sugars xylose and arabinose. The structures of these monosaccharides are shown in Fig. 2-23. (Pectin, a related compound, occurs to a small degree in the middle lamella, especially in the pith and young tissue, and consists of polygalacturonic acid.) Hemicelluloses are condensation polymers with a molecule of water removed with every linkage. All of the monosaccharides that make up the hemicelluloses have the D configuration and occur in the six-member pyranoside forms, except arabinose, which has the L configuration and occurs as a five-member furanoside. The number average DP is about 100-200 sugar units per hemicellulose molecule.

Hemicelluloses are much more soluble and labile, that is, susceptible to chemical degradation, than is cellulose. They are soluble in 18.5% NaOH (which is the basis of their measurement in Tappi Test Method T203). Low molecular weight hemicelluloses become soluble in dilute alkali at elevated temperatures, such as in kraft cooking. Hemicelluloses are essentially linear polymers except for single-sugar side chains and acetyl substituents. Hemicellulose chemistry is described below; representative hemicellulose structures are shown in Fig. 2-24.

Softwood hemicelluloses

Galactoglucomannans are polymers of glucose and mannose in the backbone linked by β-(1→4) bonds with galactose units as side chains connected by α-(1→6) bonds. Acetylation is also present. The ratio of glucose:mannose:galactose: acetyl groups is on the order of 3:1:1:1, respectively. Galactoglucomannans make up about 6% of the weight of softwoods.

Glucomannans have structures analogous to galactoglucomannans, but with about 90% of the galactose units replaced by mannose units. They make up 10-15% of the weight of softwoods.

Xylans or *arabinoglucuronoxylans* are found in all land based plants and have a backbone of poly-β-(1→4)xylose; corncobs are highly concentrated in xylans. In softwoods, side chains of α-(1→3) linked arabinose and (1→3) linked 4-O-methylglucuronic acid occur. The ratio of xylose to 4-O-methylglucuronic acid to arabinose is 4-7:1:>1. These xylans have a DP of 100-120, lack acetyl groups, and make up 5-10% of their mass.

Arabinogalactans are composed of poly-β-(1→3)-galactose with numerous (1→6) arabinose and galactose side chains (side chains of DP 2 or less are common) and an overall DP of 200. The ratio of galactose to arabinose is 6:1 in western larch. They occur at about 1% in most softwoods, but compose 5-30% of the weight of larch species. In larch they occur as two types: the first with a DP of 80-100 and the second with a DP of 500-600. Arabinoxylans are water-soluble, unlike other hemicelluloses, and for this reason they are occasionally classified with the extractives.

Hardwood hemicelluloses

Glucuronoxylans (xylans) are the principal hemicellulose of hardwoods. Side branches of 4-O-methylglucuronic acid are linked by α-(1→2) linkages. The ratio of xylose to 4-O-methylglucuronic acid is typically 7:1 but varies from 3-20:1. About 70-80% of the xylose units are acetylated at the C-2 or C-3 positions. These xylans have a DP of 40-200 (typically 180) and make up 15-30% of hardwoods.

Glucomannans have a backbone of β-(1→4) linked glucose and mannose groups in a 1:1 to 2:1 ratio with very small amounts of acetylation and a DP of 40-100. They occur at 2-5%.

Implications of hemicellulose chemistry

The carboxylic acid groups (RCOOH) of 4-*O*-methylglucuronic acid residues of xylans contribute to hemicellulose acidity, presumably contribute to hemicellulose solubility under alkaline conditions (by the formation of carboxylate salts), and, contribute to the ease of rosin sizing (hardwoods, containing more xylans, are easier to size with alum and rosin than softwoods). The 4-*O*-methyl-glucuronic acid groups tend to be preferentially removed during alkaline pulping either by selective cleavage or, if they are not evenly distributed among the xylans, by selective solubilization.

Acetyl groups are saponified (hydrolyzed to give free acetic acid) very quickly under alkaline conditions. The free acetic acid consumes a significant portion of the alkali used during kraft cooking. Typically softwoods have 1-2% acetyl groups, while hardwoods have 3-5%.

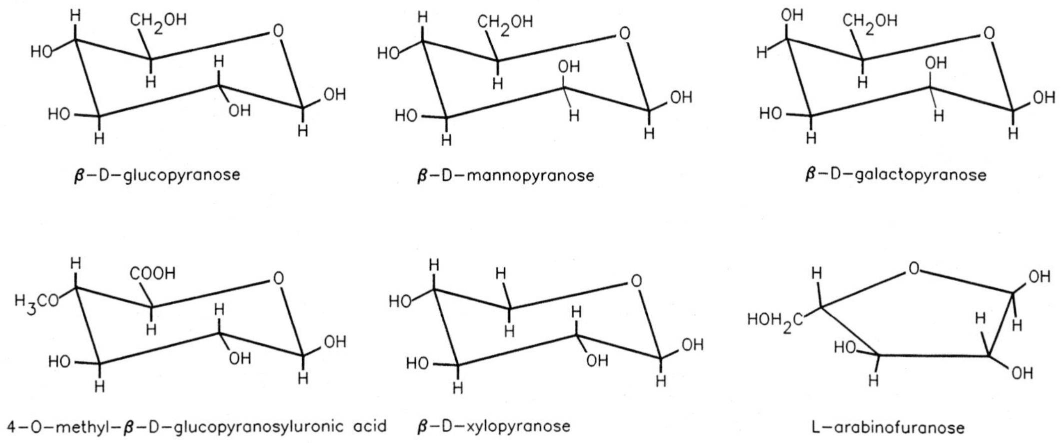

Fig. 2-23. The principal monosaccharides of wood hemicelluloses.

Softwood galactoglucomannans: O-Acetylgalactoglucomannans (p is 6 member pyranose ring)

-β-D-Glc$p$-(1→4)-β-D-Man$p$-(1→4)-β-D-Man$p$-(1→4)-β-D-Man$p$-(1→4)-β-D-Glc$p$-(1→4)-β-D-Man-
                6   2(3)
                ↑   ↑
          α-D-Gal$p$ (≈ 25%)   Acetyl (20% of backbone units)

Softwood xylans: Arabino-4-O-methylglucurononxylan (f is 5 member furanose ring)

-β-D-Xyl$p$-(1→4)-β-D-Xyl$p$-(1→4)-β-D-Xyl$p$-(1→4)-β-D-Xyl$p$-(1→4)-β-D-Xyl$p$-(1→4)-β-D-Xyl$p$-
                3   3
                ↑   ↑
     4-O-Me-α-D-GlcUp (15-25%)   α-L-Ara$f$ (≈ 10% of Xyl)

Hardwood glucomannans:

-β-D-Glc$p$-(1→4)-β-D-Man$p$-(1→4)-β-D-Glc$p$-(1→4)-β-D-Man$p$-(1→4)-β-D-Glc$p$-(1→4)-β-D-Glc$p$-
                                        2(3)
                                        ↑
                                   Acetyl (few)

Hardwood xylans: O-acetyl-4-O-methylglucuronoxylan

-β-D-Xyl$p$-(1→4)-β-D-Xyl$p$-(1→4)-β-D-Xyl$p$-(1→4)-β-D-Xyl$p$-(1→4)-β-D-Xyl$p$-(1→4)-β-D-Xyl$p$-
                2   2(3)
                ↑   ↑
     4-O-Me-α-D-GlcUp (≈ 15%)   Acetyl (70-80% of Xyl)

Fig. 2-24. Representative structures of the predominant hemicelluloses.

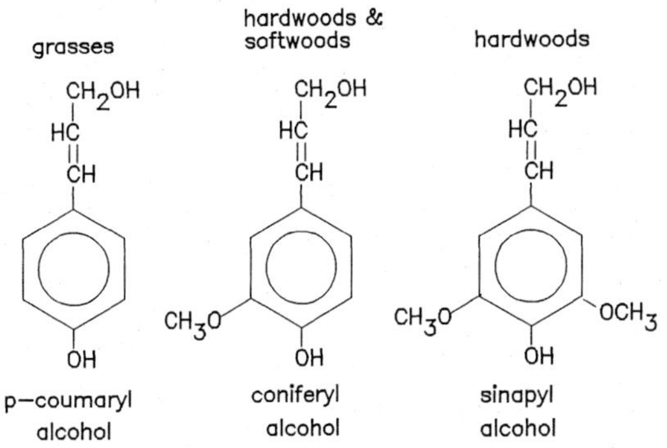

Fig. 2-25. Lignin precursors for plants. Softwoods have coniferyl alcohol, while hardwoods have coniferyl and sinapyl alcohols.

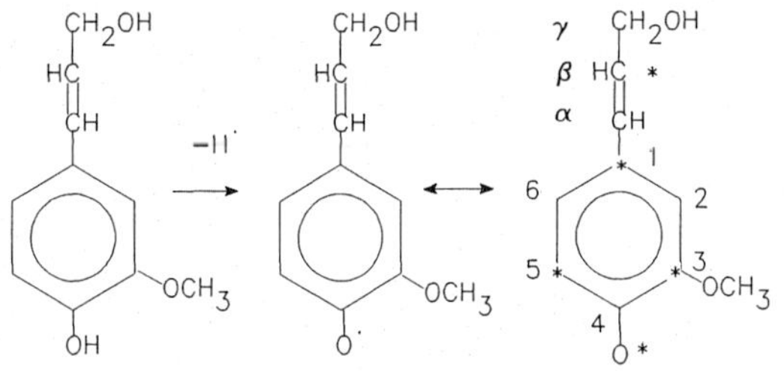

Fig. 2-26. Formation of free radicals from coniferyl alcohol. *Positions where the free radical occurs in resonance structures.

Lignin

Lignin is a complex polymer consisting of phenylpropane units and has an amorphous, three-dimensional structure. It is found in plants. Its molecular weight in wood is very high and not easily measured. Lignin is the adhesive or binder in wood that holds the fibers together. Lignin is highly concentrated in the middle lamella; during chemical pulping its removal allows the fibers to be separated easily. The glass transition temperature (softening temperature) is approximately 130-150°C (265-300°F). Moisture (steam) decreases the glass transition temperature slightly.

There are three basic lignin monomers that are found in lignins (Fig. 2-25.) Grasses and straws contain all three lignin monomers, hardwoods contain both coniferyl alcohol (50-75%) and sinapyl alcohol (25-50%), and softwoods contain only coniferyl alcohol.

Using coniferyl alcohol as an example, the first step of lignin polymerization in the plant cell wall involves formation of a free radical at the phenolic hydroxyl group (Fig. 2-26). This structure has five resonance structures with the free radical occurring at various atoms as shown on the right of Fig. 2-26. Carbon atoms C-1 and C-3 in softwoods and C-1, C-3 and C-5 in hardwoods do not form linkages due to steric hindrance (crowding). Carbon atoms of the propane unit are labelled from the aromatic ring outward as α, β, and γ, respectively. Carbon atoms of the aromatic ring are labelled from the propane group towards the methoxy group from 1 to 6, respectively. Some commonly occurring lignin linkages are also shown in Fig. 2-27. A "representative" lignin molecule is shown in Fig. 2-28.

Extractives

Extractives are compounds of diverse nature with low to moderately high molecular weights, which by definition are soluble (extracted) in organic solvents or water. They impart color, odor, taste, and, occasionally, decay resistance to wood. There are hundreds of compounds in the extractives of a single sample of wood. The composition of extractives varies widely from species to species and from heartwood to sapwood. Heartwood has many high molecular weight polyphenols and other aromatic compounds not found in sapwood, and these give the heartwood of many species (such as cedar and redwood) their dark color and resistance to decay. Some classes of extractives (Figs. 2-29 to 2-31) important to the pulp and paper industry are described with representative compounds.

Terpenes are a broad class of compounds appearing in relatively high quantities in the softwoods, where they collect in the resin ducts of those species with resin ducts. Species such as pines have large amounts of terpenes. Mills pulping highly resinous species with the kraft process collect the terpenes and sell them. Hardwoods have very small amounts of the terpenes.

Terpenes are made from phosphated isoprene units (Fig. 2-29) in the living wood cells. It is usually very easy to identify the individual isoprene building blocks of a terpene. Isoprene has the empirical formula of C_5H_8, *monoterpenes* have the empirical formula of $C_{10}H_{16}$, sesquiterpenes are $C_{15}H_{24}$, and the resin acids are oxygenated *diter-*

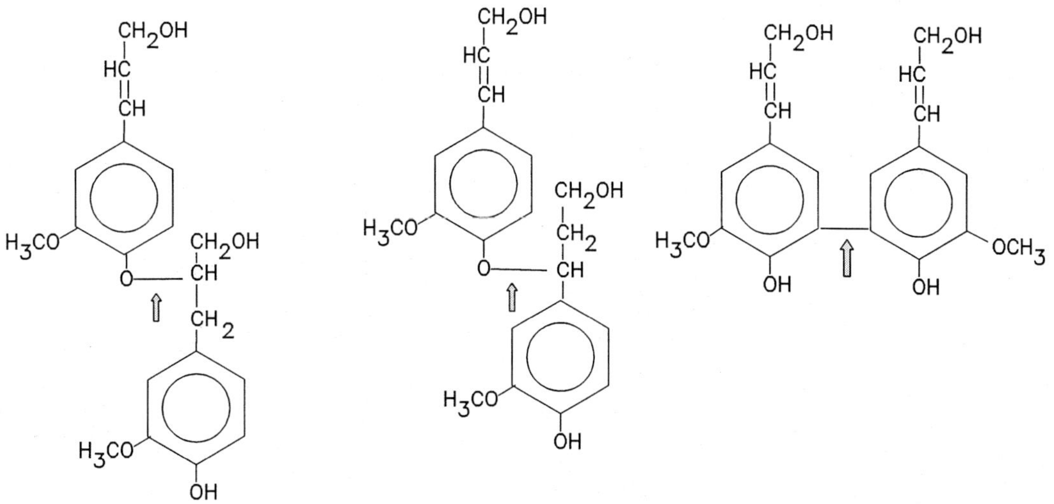

4-*O*-β-aryl ether linkage   **4-*O*-α-aryl ether linkage**   **C-C linkage.**

Fig. 2-27.   Example linkages between lignin monomers.

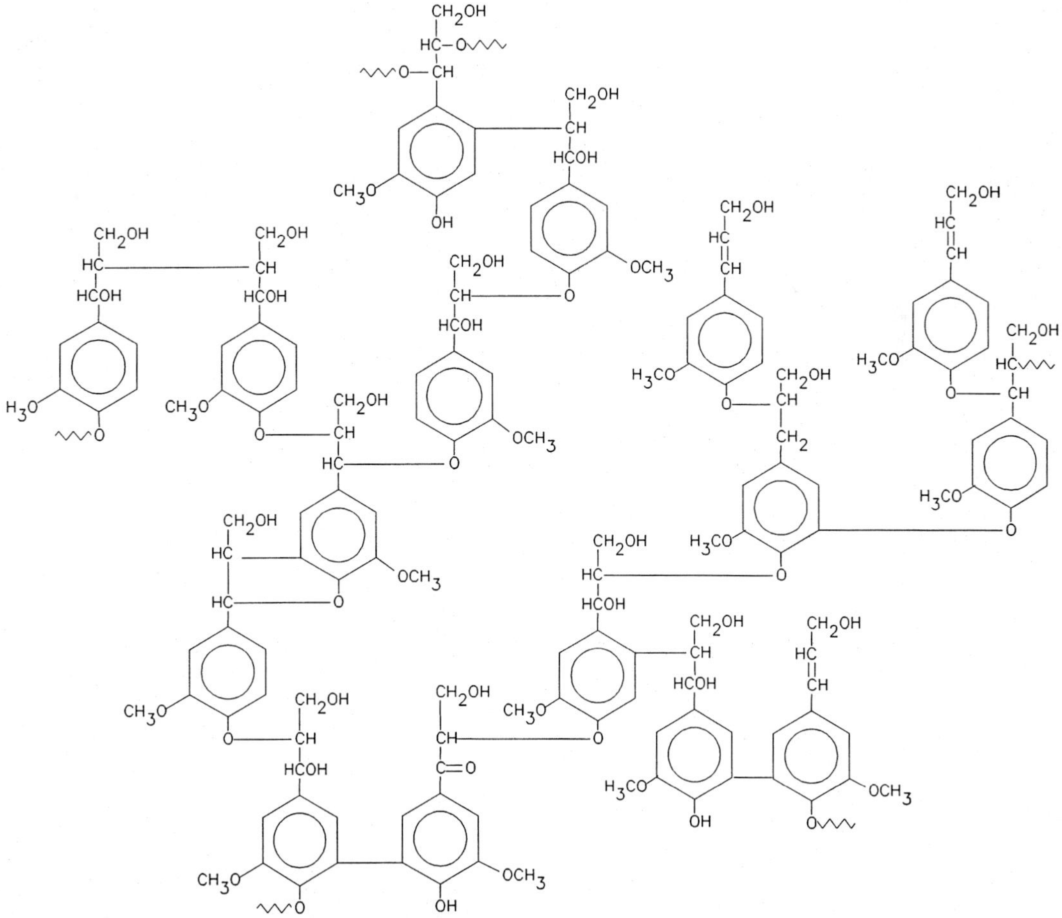

Fig. 2-28.   A hypothetical depiction of a portion of a softwood lignin molecule.

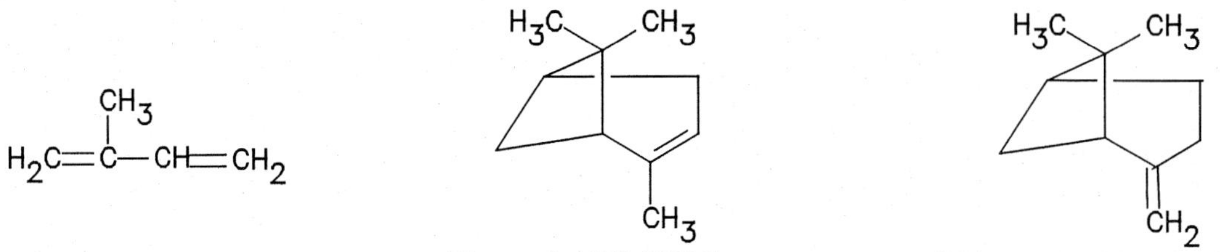

Fig. 2-29. Examples of monoterpenes each made from two isoprene units.

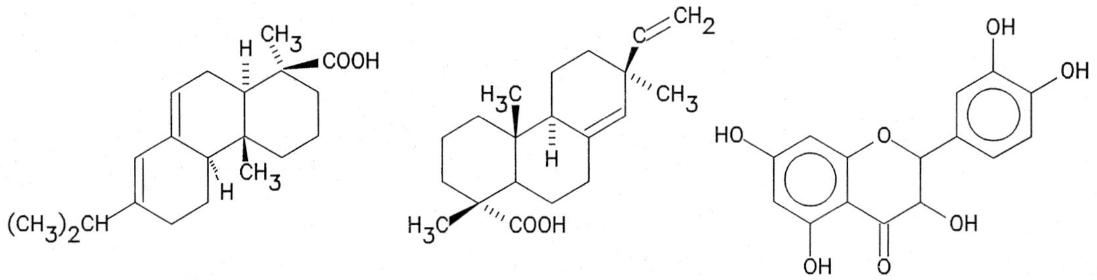

Isoprene

α-Pinene, bp 155-156°C

β-Pinene, bp 165-166°C

Abietic acid, mp 172-175°C

Pimaric acid, mp 219-220°C

Taxifolin

Fig. 2-30. Two resin acids (diterpenes) and taxifolin (dihydroquercetin).

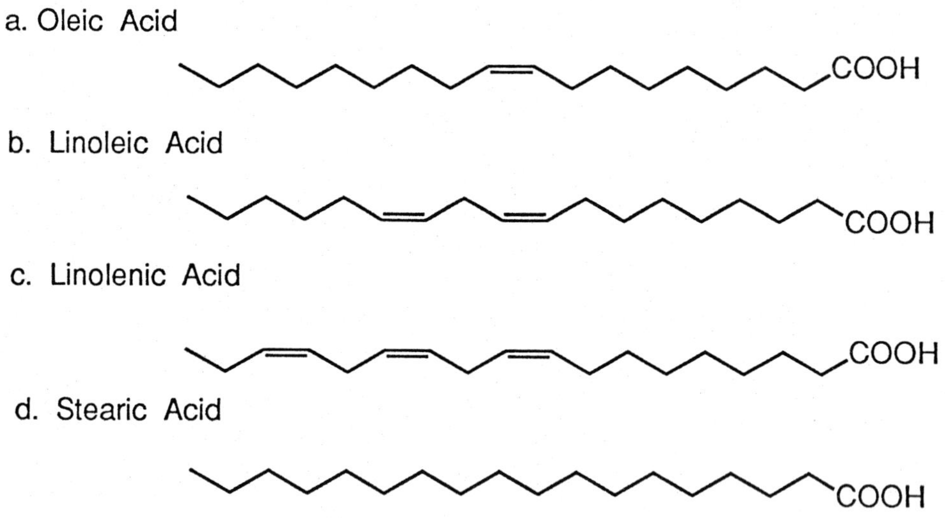

a. Oleic Acid

b. Linoleic Acid

c. Linolenic Acid

d. Stearic Acid

Fig. 2-31. Examples of fatty acids with 18 carbon atoms in wood.

penes and have the empirical formula of $C_{20}H_{32}O_2$. Higher terpenes are also found. Oxygenated terpenes with alcohol and ketone groups become prevalent with exposure to air, as in the case of pine stumps. *Turpentine* consists of the volatile oils, especially the *monoterpenes* such as α- or β- pinene; these are also used in household pine oil cleaners that act as mild disinfectants and have a pleasant aroma. (According to the *Merck Index*, α-pinene from North American woods is usually the dextrorotary type, whereas that of European woods is of the levorotatory type.) Because turpentine consists of volatile compounds, it is recovered from the vent gases given off while heating the digester. *Resin acids* such as abietic and pimaric acids, whose structures are shown in Fig. 2-30, are used in rosin size and are obtained in the *tall oil* fraction.

The *triglycerides* and their component *fatty acids* are another important class of extractives. Triglycerides are esters of glycerol (a trifuntional alcohol) and three fatty acids. Most fatty acids exist as triglycerides in the wood; however, triglycerides are *saponified* during kraft cooking to liberate the free fatty acids. (*Saponification* is the breaking of an ester bond by alkali-catalyzed hydrolysis to liberate the alcohol and free carboxylic acid. Saponification of triglycerides is how soap is made; this is how the reaction got its name. Sodium based soaps are liquids; potassium based soaps are solid.)

The principal components are the C-18 fatty acids with varying amounts of unsaturation, that is, the presence of carbon-carbon double bonds, whose structures are shown in Fig. 2-31. [Polyunsaturated fats, a term used to describe "healthy" food fats, are fatty acids (or triglycerides containing fatty acids) with two or three carbon-carbon double bonds like linoleic or linolenic acids.] Stearic acid is the saturated (with no double bonds) C-18 fatty acid. Other fatty acids, mostly with even numbers of carbon atoms, may be present as well depending on the species of wood.

Just as animal triglycerides (fats) contain small amounts of cholesterols, plant fats contain small amounts of *sterols* that are very similar to the cholesterols' structures. One example is β-sitosterol. *Fatty acids* and *resin acids* constitute the *tall oil* fraction recovered during black liquor evaporation by skimming the surface. The resin acids are separated by fractional distillation.

Phenolic compounds are more common in heartwood than sapwood and are major constituents in the bark of many wood species. In a few species these compounds can interfere with bisulfite pulping; for example, dihydroquercetin (Fig. 2-30) interferes with sulfite pulping of Douglas-fir. These compounds contain C_6 aromatic rings with varying amounts of hydroxyl groups. Some classes of these compounds are the *flavonoids*, which have a $C_6C_3C_6$ structure; the *tannins*, which

are water-soluble; polyflavonoids and other polyphenol compounds that are used to convert animal hides into leather; and the *lignans,* which have two phenyl propane units (C_6C_3-C_3C_6) connected between the β-carbon atoms.

Ash

Ash consists of the metallic ions of sodium, potassium, calcium, and the corresponding anions of carbonate, phosphate, silicate, sulfate, chloride, etc. remaining after the controlled combustion of wood. Wood ash is sufficiently alkaline so that when added to triglycerides it can be used to make soap; this was practiced by many cultures for centuries using animal fats.

Holocellulose

Holocellulose is a term for the entire carbohydrate fraction of wood, i.e., cellulose plus hemicelluloses.

Alpha cellulose

Alpha cellulose is a fraction of wood or pulp isolated by a caustic extraction procedure. While generally it is considered to be "pure" cellulose, actually it is about 96-98% cellulose.

Cellulose polymers and derivatives

Cellulose polymers (Fig. 2-32) are made from *dissolving pulp.* They include cellulose xanthate (a bright orange colored solution formed by reaction of alkali cellulose with carbon disulfide, which is an intermediate product that, upon acidification forms regenerated cellulose such as cellophane, rayon, and meat casings), cellulose acetate (a plastic used in films, eyeglass frames, cigarette filters, etc.), cellulose nitrate (smokeless powder, which replaces gunpowder in certain applications), carboxymethyl cellulose (a water-soluble thickener and dispersant), and methyl cellulose (a thickener and plastic).

Chemical analysis of wood

Wood is usually ground to 40 mesh (0.6 mm) before chemical analysis. Various chemical analyses of wood are covered in the TAPPI Standards. T 246 describes preparation of wood for chemical analysis including extraction with neutral solvents, such as ethanol and benzene, to remove the wood extractives. (**If one is doing wood extractions I highly recommend using toluene in place of**

Esterification:	Nitration:	$ROH + HNO_3/H_2SO_4 \rightarrow RONO_2 + H_2O$
	Acetylation:	$ROH + Ac_2O/HAc/H_2SO_4 \rightarrow ROCOCH_3 + H_2O$
Etherification:	Methylation:	$2\,ROH + (CH_3)_2SO_4/NaOH \rightarrow 2\,ROCH_3 + H_2SO_4$
	Carboxymethylation:	$ROH + ClCH_2COOH/NaOH \rightarrow ROCH_2COO^- Na^+$
Xanthation:	Formation:	$ROH + CS_2 + NaOH \rightarrow ROCSS^- Na^+ + H_2O$
	Regeneration:	$ROCSS^- Na^+ + H^+ \rightarrow ROH + CS_2 + Na^+$

Fig. 2-32. Commercial cellulose-based polymers.

benzene; benzene is harmful and toxic. **The difference in results will be negligible!**)

After acid hydrolysis of the cellulose and hemicellulose, the monosaccharides can be measured by chromatography. T 250 is an archaic method of monosaccharide analysis by paper chromatography. T 249 uses gas chromatography to separate and measure the monosaccharides, but much faster and better methods have been developed. (See Section 34.4 for the reference on carbohydrate analysis.) Pentosans in wood and pulp are measured by T 223; the pentoses are converted to furfural which is measured colorimetrically. The solubility of wood or pulp with 1% sodium hydroxide, T 212, is a measure of hemicellulose and cellulose degraded by decay, oxygen, chemicals, etc.

2.7 WOOD AND FIBER PHYSICS

Equilibrium moisture content

Because wood and paper are hygroscopic materials, when fully dried they adsorb water vapor from the atmosphere. The equilibrium moisture content of wood or wood pulp depends on the temperature and relative humidity of the atmosphere surrounding the specimen. Relative humidity is the partial pressure of water vapor divided by the maximum water vapor pressure of saturation at the same temperature.

Fiber saturation point, FSP

The fiber saturation point represents the moisture content of a lignocellulosic material such that additionally adsorbed water is not chemically absorbed to the wood. This occurs at about 30% MC_{OD} (at room temperature) in wood. For example, wood taken at room temperature and 99% relative humidity will have a moisture content approaching 30%. At lower humidities the equilibrium moisture content will be lower. Chemically adsorbed water requires additional energy to remove it from wood beyond the water's heat of vaporization.

Shrinkage

Wood shrinks and swells as a function of moisture content. Above the fiber saturation point there is no change in wood dimensions according to moisture content, but as wood dries below the FSP, it shrinks. Since the microfibrils are almost parallel to the longitudinal axis of the fiber in the thick S2 cell wall layer, and since water molecules do not increase the length of microfibrils but are added between them, there is very little shrinkage in the longitudinal direction, but about 4% shrinkage in the radial direction and 6% in the tangential direction. The difference in shrinkage between the radial and tangential directions occurs due to orientation of microfibrils around the cell wall pits and other factors.

Uneven grain orientation may cause severe warping or fracturing of lumber and furniture due to tremendous stresses that develop from uneven shrinkage as the wood dries.

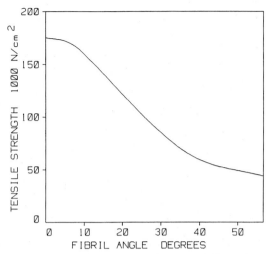

Fig. 2-33. Fiber strength versus S-2 fibril angle. (60% yield, kraft spruce fibers.) After Page, et al, 1972.

Fiber strength

Fiber strength is the strength of an individual fiber. Using small test devices, the strength of individual fibers can be measured. Paper strength can be viewed as a trade off between the strength of individual fibers and the strength of interfiber bonding (the strength of the bonds between fibers that hold them together). In unrefined pulp, the "weak link" in the paper strength is fiber-fiber bonding. In pulp which is overly refined, the weak link often becomes the strength of the individual fibers. Thus, with increased refining the properties influenced by interfiber bonding (tensile and burst) increase and the properties mostly influenced by fiber strength (tear) decrease.

The situation becomes more complicated if we consider the microfibril angle of the S-2 cell wall layer. Fig 2-33, after the work of Page, et al (1972), shows that the tensile strength of individual fibers in the longitudinal direction depends principally on the S-2 fibril angle (in the absence of fiber defects, etc.), and that the tensile strength decreases rapidly with increasing fibril angle. This work also showed that the fibril angle of latewood tends to be lower than that of earlywood in spruce, and the latewood fibers were stronger than the early wood fibers, which has been reported in the literature with other species as well. (Black spruce also tends to have a lower fibril

angle than white spruce; this indicates that the average fibril angle of wood species is as important as the average fiber length.) The authors concluded "fibres of the same fibril angle have similar strength, independent of species." The maximum fiber tensile strength of 17,000 N/cm^2 corresponds to a breaking length of over 100 km, or about 10 times stronger than paper!

Fiber bonding in paper, hydrogen bonding

In paper, fibers are held together by hydrogen bonding of the hydroxyl groups of cellulose and hemicelluloses. The carboxylic acid groups of hemicelluloses, also play an important role. Water interferes with hydrogen bonding between fibers; thus, paper losses much of its strength when wet. Hemicelluloses increase the strength of paper, while lignin on the surface of fibers, which is not able to form hydrogen bonds, decreases the strength of paper. Using water in the formation of paper greatly increases its strength as capillary action of water pulls the fibers together, and may partially solubilize the carbohydrates, so hydrogen bonding can occur. There are no methods available for dry formation of paper with appreciable strength, although such a process could make papermaking much more economical.

Hydrogen bonding holds lignocellulosic fibers together in paper. The strength of R-OH--R-OH hydrogen bonding is 3-4 kcal/mol. This is relatively weak compared to covalent bonds, which are on the order of 100 kcal/mol. However, the large number of potential hydrogen bonds along the length of the holocellulose molecules means that paper can be quite strong. Modification of hydroxyl groups by acetylation, methylation, etc. prevents hydrogen bonding and decreases the strength of paper dramatically. Also, paper made from a slurry in a solvent of low polarity is weak as the formation of hydrogen bonds is hindered.

Factors which increase the fiber surface area (or the area of fiber to fiber contact) increase interfiber bonding. Refining tends to allow the interfiber bonding to increase by fibrillation (increasing the surface area) and hydration of fibers (making them more flexible to mold around each other better), but the strength of the individual fibers decreases. More information on fiber physics is found in Sections 17.8-17.10.

2.8 PROPERTIES OF SELECTED WOOD SPECIES

This section presents wood properties of selected wood species. Much of the information presented here is from the reports of the U.S. Department of Agriculture Forest Product Laboratory (USDA-FPL) in Madison, Wisconsin. These reports are very useful to demonstrate wood properties and variability in individual species of wood. Remember that there is considerable variation in wood properties and, therefore, reported wood properties! Table 2-5 and Table 2-6 show the average fiber lengths, basic densities, basic specific gravities (oven dry weight divided by green volume), and chemical pulp yields for softwoods and hardwoods, respectively. The data are from "Pulp yields for various processes and wood species", USDA For. Ser., FPL-031 (Feb., 1964), which is a slight revision of "Density, fiber length, and yields of pulp for various species of wood", Tech. Note No. 191 (1923).

Table 2-7 and Table 2-8 show chemical compositions, basic specific gravities, and select-ed solubilities for softwoods and hardwoods, respectively. The data is from the USDA-FPL.

2.9 NONWOOD AND RECYCLED FIBER CONSIDERATIONS

Recycled or secondary fiber

Recycled fiber is fiber whose source is paper or paperboard arising outside of the mill. It is distinguished from *broke*, which is off-specification paper produced and reused within the mill. Recycled fiber is obtained from recycled paper.

It is very important to have "pure" sources of paper from which to make recycled fiber. Newspapers should not be contaminated with magazines and brown paper and boxes. Office papers should not be contaminated with newsprint or brown papers. For example, mixed waste paper is worth about $10-20/ton while clean paper clippings from an envelope factory are worth over $250/ton. In the U.S. almost 30% of the paper consumed in 1989 was recycled. This compares to a recovery rate of 50% in Japan, one of the highest rates. Other developed nations tend to have higher recovery rates than the U.S.

Table 2-5. Basic pulping properties of U.S. softwoods. From USDA FPL-031 (1923, 1964).

Species	Scientific name	Aver. fiber length (mm)	Dens. lb/ft³	Spec. Grav.	Pulp yield, %[1] Kraft	Sulfite
Baldcypress	*Taxodium distichum*	6.00	26	0.42	48	46
Cedar:						
Atlantic white	*Chamaecyparis thyoides*	2.10	19	0.30	45	
Eastern redcedar	*Juniperus virginiana*	2.80	27	0.43	45	
Incense	*Libocedrus decurrens*	2.00	22	0.35	45	40
Port-Orford	*Chamaecyparis lawsoniana*	2.60	25	0.40	45	45
Western redcedar	*Thuja plicata*	3.80	19	0.30	40	43
Douglas-fir, coastal	*Pseudotsuga menziesii*	4.50	28	0.45	48	48
Fir:						
Balsam	*Abies balsamea*	3.50	21	0.34	50	47
California red	*A. magnifica*	3.25	23	0.37	48	48
Grand	*A. grandis*	5.00	23	0.37	48	49
Noble	*A. procera*	4.00	22	0.35	47	48
Pacific silver	*A. amabilis*	3.55	22	0.35	49	49
Subalpine	*A. lasiocarpa*	3.15	21	0.34	48	48
White	*A. concolor*	3.50	22	0.35	48	48

[1]Screened yield for nonbleachable kraft (for bleachable subtract 2-3%) and bleachable sulfite.

Table 2-5. Basic pulping properties of U.S. softwoods. Continued.

Species	Scientific name	Aver. fiber length (mm)	Dens. lb/ft³	Spec. Grav.	Pulp yield, %[1]	
					Kraft	Sulfite
Hemlock:						
Carolina	*Tsuga caroliniana*	3.10	30	0.48	45	48
Eastern	*T. canadensis*	3.50	24	0.38	45	44
Mountain	*T. mertensiana*	3.70	26	0.42	45	
Western	*T. heterophylla*	4.00	24	0.38	47	46
Larch:						
Tamarack	*Larix laricina*	3.50	31	0.50	48	42
Western	*L. occidentalis*	5.00	32	0.51	48	42
Pine:						
Jack	*Pinus banksiana*	3.50	25	0.40	48	45
Loblolly	*P. taeda*	4.00	29	0.46	48	45
Lodgepole	*P. contorta*	3.50	24	0.38	48	45
Longleaf	*P. palustris*	4.00	34	0.54	48	45
Monterey	*P. radiata*	2.60	29	0.46	48	45
Ponderosa	*P. ponderosa*	3.60	45	0.72	48	45
Red	*P. resinosa*	3.70	26	0.42	48	45
Shortleaf	*P. echinata*	4.00	29	0.46	48	45
Slash	*P. elliottii*	4.00	35	0.56	48	45
Sugar	*P. lambertiana*	4.10	22	0.35	48	45
Virginia	*P. virginiana*	2.80	28	0.45	48	43
White, eastern	*P. strobus*	3.70	21	0.34	48	
White, western	*P. monticola*	4.40	23	0.37	48	45
Redwood	*Sequois sempervirens*	7.00	24	0.38	38	
Spruce:						
Black	*Picea mariana*	3.50	24	0.38	50	48
Blue	*P. pungens*	2.80	23	0.37	43	48
Englemann	*P. engelmannii*	3.00	20	0.33	47	48
Red	*P. rubens*	3.70	24	0.38	50	48
Sitka	*P. sitchensis*	5.50	23	0.37	47	48
White	*P. glauca*	3.50	23	0.37	50	48

[1]Screened yield of nonbleachable kraft (for bleachable subtract 2-3%) and bleachable sulfite.

Table 2-6. Basic pulping properties of U.S. hardwoods. From USDA FPL-031 (1923,1964).

Species	Scientific name	Aver. fiber length (mm)	Dens. lb/ft^3	Basic Spec. Grav.	Pulp yield, %[1] Kraft	Sulfite
Ailanthus	*Ailanthus, altissima*	1.20	23	0.37		
Alder, red	*Alnus, rubra*	1.20	25	0.40	50	
Ash, green	*Fraxinus pennsylvanica*	1.05	33	0.53	44	
Ash, white	*F. americana*	1.20	34	0.54		47
Basswood, American	*Tilia americana*	1.20	20	0.32		
Beech, American	*Fagus grandifolia*	1.20	35	0.56	49	44
Birch, paper	*Betula papyrifera*	1.20	30	0.48	50	46
Birch, yellow	*B. alleghaniensis*	1.50	34	0.54	53	45
Buckeye, Ohio	*Aesculus glabra*	0.90	21	0.34		47
Butternut	*Juglans cinerea*	1.20	22	0.35		47
Chestnut, American	*Castanea dentata*	1.00	25	0.40		42
Cucumber tree	*Magnolia acuminata*	1.30	27	0.43		45
Elm, American	*Ulmus americana*	1.50	29	0.46	46	
Elm, rock	*U. thomasii*	1.30	36	0.58		47
Elm, slippery	*U. rubra*	1.70	30	0.48		47
Hickory, mockernut	*Carya tomentosa*	1.40	40	0.64	48	40
Hickory, shagbark	*C. ovata*	1.35	40	0.64		
Mangrove	*Rhizophora mangle*	1.40	55	0.88		
Maple, red	*Acer rubrum*	1.00	31	0.50	43	45
Maple, silver	*A. saccharinum*	1.75	28	0.45		
Maple, sugar	*A. saccharum*	1.10	35	0.56	43	45
Oak, blackjack	*Quercus marilandica*	1.00	40	0.64	42	41
Oak, chestnut	*Q. prinus*	1.35	36	0.58	45	
Oak, northern red	*Q. rubra*	1.40	35	0.56	46	45
Oak, overcup	*Q. lyrata*	1.35	36	0.58	46	
Oak, post	*Q. stellata*	1.50	37	0.59		45
Poplar:						
aspen, quaking	*Populus tremuloides*	1.20	22	0.35	54	52
aspen, bigtooth	*P. grandidentata*	1.20	22		50	
balsam	*P. balsamifera*	1.00	19	0.30	50	49
cottonwood, eastern	*P. deltoides*	1.30	23	0.37	52	52
cottonwood, swamp	*P. heterophylla*	1.30	24	0.38	47	47
Sugarberry	*Celtis laevigata*	1.10	29	0.46	46	44
Sweetgum	*Liquidambar styraciflua*	1.60	29	0.46	50	46
Sycamore, American	*Platanus occidentalis*	1.70	29	0.46		47
Tupelo, black	*Nyssa sylvatica* var. *sylvatica*	1.70	29	0.46	48	48
Tupelo, swamp	*N. sylvatica* var. *biflora*	1.70	34	0.54	48	46
Tupelo, water	*N. aquatica*	1.60	29	0.46		47
Willow, black	*Salix nigra*	1.00	24	0.38	52	52
Yellow-poplar	*Liriodendron tulipifera*	1.80	24	0.38	47	47

[1]Screened yield of nonbleachable kraft (for bleachable subtract 2-3%) and bleachable sulfite.

Table 2-7. Chemical composition of softwoods.

Species, location, average diameter	Sample Ave. Age Years	Basic Spec. Grav.	Chemical composition, %				Solubility	
			Holo-cellul.	Alpha cellul.	Lignin	Total pento-sans	Hot, 1% NaOH	EtOH–Benzene
Cedar:								
Western red, res., west.US, 24.5"	205	0.306	48.7	38.0	31.8	9.0	21.0	14.1
Douglas–fir, Oregon, old growth, residues		0.439	67.0	50.4	27.2	6.8	15.1	4.5
Oakridge, Ore, second growth, residues		–	69.9	52.6	28.0	4.9	9.7	–
Wyoming, US, 7.8"	182	0.417	65.7	46.9	27.2	6.5	15.8	5.2
Fir:								
Balsam, Mich. US, 6.5"	48	0.330		42.2	28.5	10.8	11.1	3.5
Pacific silver, res., west.US, 19.6"	169	0.385	60.7	43.2	30.5	9.2	10.8	2.6
Subalpine, Mont, US, 6.4"	44	0.304	67.1	44.2	29.6	8.9	11.5	2.2
White, Calif. US		0.363	65.5	49.1	27.8	5.5	12.7	2.1
Hemlock:								
Eastern		0.38	68	43	32	10	13	
Western, Washington, US			66.5	50.0	29.9	8.8	18.1	5.2
Larch:								
Tamarack, Wisconsin, US, 4.7"	43	0.491	60.4	43.3	25.8	8.6	18.2	3.6
Western, Washington, US, 5.4"	53	0.514	66.5	50.0	26.6	7.8	13.4	1.4
Pine:								
Jack, Wisconsin, US		0.43	67.0	47.0	27.0	9.1	13.5	3.5
Loblolly, Arkansas, US		0.45		48.7	27.7	8.9	12.0	2.7
Lodgepole, Montana, U.S.								
Sound, 5.1", little decay	25	0.373	68.8	44.3	26.7	10.3	12.8	3.0
Sound, 7.4", little decay	135	0.415	71.6	47.3	25.9	10.9	11.6	2.8
Insect–killed, 8.9", some decay	109	0.400	65.1	44.1	26.5	9.2	14.9	4.2
" ", Down, 9.2", sig. decay	112	0.429	67.9	45.2	27.9	10.0	12.9	3.1
Ponderosa, 4.9", 1.9% heartwood	39	0.40	67.7	45.0	25.1	10.2	13.4	5.6
Red		0.43			24	11		
Shortleaf, Ark., US, 6.0"	27	0.49	69.3	48.5	27.6	8.9	12.2	3.3
Slash, thinnings, Louis., US, 3.9"	10	0.484	64.7	47.2	28.5	9.1	11.5	3.2
White, eastern, Maine, US, 8.6"	38	0.336	66.3	43.9	26.1	6.6	15.9	5.9
White, western, Idaho, US			73.8	50.4	25.4	7.8	11.3	2.9
Redwood, Calif. US, oldgrowth		0.358	55.1	42.6	33.4	7.2	18.8	9.9
" " second growth		0.344	60.5	45.7	33.1	7.2	13.9	0.4
Spruce:								
Black, Michigan, US, 6.5"	49	0.392		43.1	26.8	11.4	11.4	1.8
Englemann, Colorado, 13.6"	169	0.333	67.3	45.2	28.2	7.4	11.6	1.7
Red		0.38	73	43	27	12		
White		0.38	73	44	27	11	12	

Table 2-8. Chemical composition of hardwoods.

Species, location, average diameter	Sample Ave. Age Years	Basic Spec. Grav.	Chemical composition, % Holo-cellul.	Alpha cellul.	Lignin	Total pento-sans	Solubility Hot, 1% NaOH	EtOH–Benzene
Alder, red, Washington US, 7.8"	21	0.385	70.5	44.0	24.1	19.2	17.3	1.9
Ash, green, Ark., US, 10.2"	46	0.519		41.0	25.3	16.5	18.4	5.5
Ash, white		0.56						
Birch, paper, New York, US		0.49	78.0	46.9	20.5	21.8		
Birch, yellow		0.55						
Elm, American, Ark., US, 13.7"	56	0.470		46.9	24.3	18.1	16.7	2.6
Eucalyptus								
E. saligna, Brazil, 5", fibers 0.87 mm		0.546	72.3	49.7	25.3	14.7	13.3	1.7
E. kertoniana, ", 4.5", fibers 0.93 mm		0.513	74.3	50.3	28.1	15.0	13.6	1.5
E. robusta, Puerto Rico,		0.490	66.6	47.7	27.5	16.2	12.2	2.1
Hickory, mockernut, N.C.US, 7.2"	65	0.676	69.8	45.0	18.9	16.4	18.5	5.1
Hickory, shagbark, N.C. US, 4.4"	29	0.703	71.3	48.4	21.4	18.0	17.6	3.4
Maple, red, Mich. U.S.		0.45	77.6	48.6	20.6	18.3	15.9	3.0
Maple, sugar, Mich. U.S.				49.2	21.5	18.6	16.7	
Oak, blackjack, Ark., US, 5.6"	48	0.635		43.9	26.3	20.1	15.0	3.5
Oak, northern red, Mich. US, 5.9"	36	0.582	69.1	46.0	23.9	21.5	21.7	5.2
Oak, white, Virginia, US, 5.4"	41	0.616	62.6	46.1	27.7	18.4	19.8	3.0
Michigan, 7.3"	69	0.613	72.2	47.5	25.3	21.3	19.0	2.7
Poplar:								
aspen, quaking, Wisc. US		0.35	78.5	48.8	19.3	18.8	18.7	2.9
cottonwood, eastern		0.37		46	24	19	15	
Sugarberry, Ark., US, 9.9"	47	0.489		40.2	20.8	21.6	22.7	3.1
Tanoak, Calif. US, 11.3"	99	0.601	70.4	45.2	19.0	18.3	18.9	2.0
Tupelo, black, Miss. US, 6.3"		0.513	71.7	48.1	26.2	14.5	12.8	2.3
Willow, black, Ark., US, 15.5"	28	0.377		46.6	21.9	18.8	17.4	2.2
Yellow–poplar		0.40		45	20	19	17	

REFERENCES FOR TABLES 2–7 AND 2–8.

PP–110, Physical characteristics and chemical analysis of certain domestic hardwoods received at the Forest Products Laboratory for Pulping from October 1, 1948, to November, 1957. PP–112, Physical characteristics and chemical analysis of certain domestic pine woods received at the Forest Products Laboratory for Pulping from October 1, 1948, to September 4, 1956.   PP–114, Physical characteristics and chemical analysis of certain softwoods (other than pine) received at the Forest Products Laboratory from October 1, 1948, to August 7, 1947.

Details of some wood samples (with four digit FPL shipment numbers) were obtained from reports. Occasionally, missing numbers have been filled in from other FPL reports that were used to verify the plausibility of numbers, when available. Softwoods: Douglas–fir residues, old growth ship.

2655 and second growth ship. 2467, Rep. no. 1912 (Rev. July, 1956). Lodgepole pine, ship. 2414–2417, 2434, Rep. no. R1792 (June, 1951). Ponderosa pine sample was 20% w/w, 21% v/v, bark as received, Rep. no. R1909 (Oct., 1951). Pacific silver fir sawmill residues, ship. 2128, heartwood was 16.1" diameter, Rep. no. R1641 (Feb., 1947).   Western redcedar sawmill residues, ship. 2132, heartwood was 22.7" diameter, Rep. no. R1641 (Feb., 1947).   Eastern hemlock, red pine, red spruce, and white spruce, Rep. no. 1675 (Rev. Nov., 1955).

Hardwoods: Red alder, ship. 3050, Rep. no. 1912 (Revised, July, 1956). Black willow, American elm, sugarberry, green ash, and blackjack oak, ship. 1549, 1550, 1545, and 1508, respectively, Rep. no. R–1491 (Feb. 1944). American beech, eastern cottonwood, and yellow– poplar, Rep. no. 1675 (Rev. Nov. 1955). Eucalyptus, Rep. no. 2126 (Sept., 1958).

The product with the largest recovery by amount and percent in the U.S. is old corrugated containers (OCC). Over 50% of OCC is recovered in the U.S. One reason is that large amounts of OCC are generated at specific sites, such as grocery stores and other retail outlets. Newspapers and other *post-consumer wastes* are much more expensive to collect and tend to be highly contaminated with unusable papers and trash. Still, 33% of newsprint is recovered, but only one-third of this ends up in new newsprint, with the rest used in chipboard or exported. Several states, however, are in the process of enacting legislation which demands large amounts of recycled fiber (10-50%) in new newsprint. The recovery and reuse of newsprint will change rapidly over the next several years. On the order of 25% of U.S. recovered fiber is exported, and the remaining 75% is reused domestically.

Use of recycled paper

Generally freight costs limit the distance that waste paper may be transported. About 80% of all waste paper comes from one of three sources: corrugated boxes, newspapers, and office papers. Only about 10% of waste paper is deinked to be used in printing or tissue papers; mostly it is used in paperboards and roofing materials where color is not important. However, the percentage of deinked paper is expected to increase considerably over the next few years.

Reuse of discarded paper involves extensive systems for removal of foreign materials. This involves skimmers to remove floating items, removal of heavy items at the bottom of a repulper, and the removal of stringy items such as rope, wet strength papers, etc. The so-called non-attrition pulping method works like a giant blender to separate the fibers. Coarse screening is then used for further cleaning prior to use of fine screens. These processes are discussed in detail in Chapter 10.

Nonwood plant fibers

About 10% of the fiber used to make paper each year worldwide is from nonwood plant fibers, including cotton, straws, canes, grasses, and hemp. Non vegetable fibers such as polyethylene and glass fibers are also used. Fig. 2-34 shows electron micrographs of "paper" made from four nonwood fibers. In the U.S. paper contains only about 2% of nonwood fibers on average. Globally, however, the use of nonwood fiber is increasing faster than wood fiber. Nonwood fiber sources were used for hundreds of years before wood was used as a fiber source for papermaking.

Many factors influence the suitability of raw materials for use in paper. These include the ease of pulping and yield of usable pulp; the availability and dependability of supply; the cost of collection and transportation of the fiber source; the fiber morphology, composition, and strength including the fiber length, diameter, wall thickness, and fibril angle (primarily the thick S-2 layer); the presence of contaminants (silica, dirt, etc.); and the seasonability of the supply (storage to prevent decay is costly.)

Nonwood sources of plant fibers include straws such as wheat, rye, rice, and barley; grasses such as bamboo, esparto, and papyrus; canes and reeds such as bagasse (sugar cane), corn stalks, and kenaf; bast (rope material) such as flax (linen), hemp, jute, ramie, and mulberry; and seed hairs such as cotton. Tappi Standard T 259 describes the identification of nonwood plant materials that are used or have the potential to be used in the paper industry. Many photomicrographs and detailed information make this a particularly useful resource on nonwood fibers.

In the U.S. wood has replaced other fiber sources for a wide variety of reasons (that will be discussed and are later summarized in Table 2-9 for straw). For example, in the U.S. all corrugating medium was made from straw prior to the 1930s. Around this time the chestnut blight made a lot of hardwood available, which was effectively pulped by the new NSSC process to make corrugating board. By the end of the 1950s, most of the straw-using mills had closed or switched to hardwoods. Now almost all corrugating medium in the U.S. is made with hardwoods and/or recycled fiber.

In contrast to this, Europe's largest corrugating medium mill uses pulp from wheat, rye, oat and barley grain straws, along with secondary fiber. The mill is owned by Saica and operates in Zaragoza, Spain. Plate 11 shows some aspects of the process. The mill boasts a production of 1200

Fig. 2-34. Cotton paper (top left); kenaf newsprint press run [top right, see *Tappi J.* 70(11):81-83(1987)]; polyethylene nonwoven material (bottom left); and glass fiber battery separator.

t/day of medium containing 25% or 50% straw pulp, with the higher grade containing the larger amount of straw pulp. The balance of the fiber is from recycled paper. The mill is state-of-the-art with high speed paper machines using extended-nip presses. Saica plans to double its capacity and incorporate straw into linerboard using a paper machine with two extended-nip presses in the early 1990s. This mill produces 400 t/day of straw pulp using four continuous digesters operating at atmospheric pressure and about 100°C (212°F) with pulping times of 1 hour. The low temperature allows mild carbon steel to be used, as the sodium hydroxide cooking chemical is not so corrosive at these low temperatures. The mill recovers some of the organic material as fuel by anaerobic fermentation, while the sodium is not recovered. Straw Pulping Technology has sold several of these systems.

Most nonwood fiber is pulped in continuous digesters at temperatures around 130-160°C (265-320°F) for 10-30 min. The Pandia digester, which is a horizontal tube digester with screw feed, is a commonly used digester. However, in Denmark, Frederica Cellulose uses wheat and rye grain straw in batch digesters to make about 140 t/day of bleached straw market pulp. The mill did not have any liquor recovery or treatment before 1985, but is now the only mill in the world using direct alkali recovery (DARS, see Section 34.9 for a description of this process) to regenerate the sodium hydroxide from the spent cooking liquor. The DARS process is used since the mill is so small that a recovery boiler and lime kiln would be uneconomical.

Cotton

Natural cotton (from the seed pods) is a very pure form of cellulose. In its native form, cotton fibers are too long to make good paper, since poor formation (evenness of fiber distribution) in the paper would be the result. Also, cotton suitable for textiles is too expensive for use in paper. *Rag* (cotton remnants of the textiles industry) cotton has relatively short fibers and is much less expensive. Cotton linters (seed hair) have shorter fibers. Cotton rag fibers are typically 10-45 mm long, while cotton linters are 1-2 mm long; both are 0.02-0.04 mm in diameter. Because cotton is

almost pure cellulose, it has little hemicellulose and leads to low strength papers. Consequently cotton fibers are mixed with wood pulp to make a suitably strong paper. Fig. 2-34 shows paper made from cotton fibers.

Bast fibers

Hemp (used in ropes), *jute,* and *ramie* grow in subtropical areas, such as in the Philippines, and have long (20 mm), very strong fibers. These materials have specific applications in cigarette papers, tea bags, sack paper, and saturating papers. *Linen (flax)* is similar to *hemp*, but the fibers are shorter (2-5 mm by 0.02 mm diameter). These materials are usually processed by kraft or soda processes.

Straw

Straws, the stalks of grain crops obtained after threshing, may be processed into pulp. Straws from most edible grains are suitable. The most important types pulped are *wheat, rice, rye*, and *barley* with yields typically about 35% for bleached grades to 65% for high yield pulps suited for corrugating media or linerboard. Straw is low in lignin and is especially suited for fine papers. Because straw pulps have low drainage rates, it takes more water and large washers to wash them. The soda process is the most common pulping method for straw; anthraquinone is sometimes used with soda or kraft pulping of straw. Straw fiber lengths are typically 0.5-2.5 mm with diameters of 0.01-0.2 mm. Straw hemicellulose is mostly xylans. All of these factors make straw pulp similar to hardwood pulp. There are 75 million tons of straw available annually in the U.S. The advantages and disadvantages of straw are summarized in Table 2-9.

Chemical recovery of straw pulping liquors is complex and has not been practiced until environmental pressures in the 1980s forced some mills to begin this practice. Since large amounts of water are required to wash pulps, much energy is needed to concentrate the dilute liquor from the brown stock washers. However, about one-half of the liquor can be pressed out of the pulp and subjected to chemical recovery to avoid high evaporation costs. (Many mills that pulp wood by semi-chemical processes do this as well.) Many straws, espe-

Table 2-9. The advantages and disadvantages of straw as a fiber source.

Advantages	Disadvantages
Byproduct from agriculture	Transportation and storage problems
Often cheaper than wood	Straws are bulky and contain silica
Large annual crop - 1 to 10 tons per acre per year	Short harvest time of 1 to 2 months; thus heavy drain on capital
Needs little refining	Degrades very quickly - high losses
Makes excellent filler, good printing and smoothness	Low freeness (drainage) rates and thus low production rates.

cially rice straw, have large amounts of silica that are removed during alkaline pulping. The silica interferes with the chemical recovery process. There are about 10 countries which pulp significant amounts of rice despite its high silica content. New techniques for desilicanization of black liquor now allow chemical recovery.

Grasses

Grasses are pulped by the soda process and, like straw, are low in lignin. Grasses have long (up to 3 mm), thin (0.01-0.02 mm) fibers suitable for fine, high quality papers with good strength and opacity. Esparto grass grows in southern Europe and Northern Africa. Pulp from *bamboo* (grown in India, China and other Asian countries) is used in fine papers and the resulting paper is stronger than paper of straw pulp. Because of its high silica content, it is pulped using the kraft process. *Bamboo* pulp is somewhat similar to softwood sulfite pulp.

Canes and reeds

Bagasse (sugar cane residue) is pulped by the kraft or soda process. It must be depithed since the inner pith cells are useless for papermaking and decrease the pulp freeness considerably. It is used in fine papers. Improvements in processing bagasse have made it more popular around the world as a fiber source for pulp. Since bagasse is often used as a fuel at sugar processing plants, the plant must replace the bagasse with another fuel if it is to make it available for pulping.

Kenaf has been investigated in the U.S. by the USDA. It is a member of the *hibiscus* group of plants and is being investigated for use in news-

print. Fig. 2-34 shows a sample of newsprint made from kenaf from an actual press run.

Glass and polymers

Nonwovens are fiber mats made using synthetic fibers. They can be formed much like paper into a cloth material instead of using a weaving process, hence the term nonwoven. Paper receives some competition from nonwovens in products such as tote bags, envelopes, and computer diskette sleeves. The long fibers give these products very high tear strengths. Glass fibers (often with chemical additives) are used in products such as battery separators, glass fiber filter mats, and as reinforcement in a large variety of composite materials. Fig. 2-34 shows mats constructed with polyethylene and glass fibers.

2.10 ANNOTATED BIBLIOGRAPHY

Chipping

1. Robinson, M.E., Optimizing chip quality through understanding and controlling chipper design characteristics and other variables, *Proceedings, 1989 TAPPI Pulping Conference*, pp 325-338.

Here the importance of proper chipper maintenance is stressed. Some of the conclusions are: Using knives to chip more than 500 to 1000 tons (dry basis) of chips with dirt or gritty bark in the wood will result in poor chips due to nicks in the knives. Dry wood produces more oversized and overthick chips than green wood. High chipping velocity, post chipping damage to chips due to harsh handling, and frozen wood produces larger amounts of fines, especially in softwoods. Shortwood chips better in gravity feed machines than in horizontal chippers. Since chip thickness is directly proportional to chip length, chip length can be used as a means of controlling chip thickness. Reductions in the pins fraction can be gained by removing the card breakers, or slowing the chipper speed.

2. Hartler, N. Chipper design and operation for optimum chip quality, *Tappi J.*, 69(10):62-66(1986) (It is similar in *Proceedings, 1985 TAPPI Pulping Conference*, pp 263-271.)

Hartler is a respected expert on wood chip quality. A target cutting speed of 20-25 m/s for the knives is typical. A decrease in the spout angle results in a lower fines content in the wood chips, but has the disadvantages of increased damage to chips, decreased chip bulk density, and a decrease in the maximum diameter of wood that can be processed.

3. Twaddle, A.A. and W.F. Watson, Survey of disk chippers in the southeastern USA, and their effects of chip quality, *Proceedings, 1990 Tappi Pulping Conference*, pp 77-86. [Also in *Tappi J.* 75(10):135-140(1992). The second A of their equations 3 and 4 should be a B.]

The authors found from their survey of 101 chippers that about 40% of the chippers were either the 112 in.-15 knife (powered with 800-2500 hp) or 116 in.-12 knife (powered with 1250-3000 hp, but most with 2500 hp) combinations. Chippers typically ran at 300-450 rpm with almost 40% at 360 rpm. 43% of the chippers were manufactured by Carthage. 15% used disposable knives. About 50% ran with mid disk operating speeds below 25 m/s. 62% of the chippers used passive, gravity *bottom discharge*, while 37% used *blowing discharge*, induced by vanes mounted on the back of the disk.

Four correlation equations for pins and fines generation for hardwoods and softwoods each

were developed from the parameters of discharge type, chipper rpm, and chip set-up length. (They defined pins as less than 2 mm thick and retained on a 5 mm round hole pan, and fines as less than 2 mm thick passing through a 5 mm round hole pan, which was the classification used at 70% of the yards measuring fines.) The regression equations can be used to compare the performance of ones chipper with the average, although there can be large variations expected for various species of wood. The equations are as follows with r^2 of 0.27 for softwoods and 0.39 for hardwoods:

softwood pins (subtract 1.58% for bottom discharge units):

% = 6.34 + (0.0092 × rpm) + (-0.26 × chip setup length in mm)

softwood fines (subtract 1.16% for bottom discharge units):

% = 3.02 + (0.0062 × rpm) + (- 0.15 × chip setup length in mm)

hardwood pins (subtract 0.65% for bottom discharge units):

% = 3.64 + (0.0062 × rpm) + (-0.17 × chip setup length in mm)

hardwood fines (subtract 1.17% for bottom discharge units):

% = -0.28 + (0.0081 × rpm)

These equations show that an increase in speed of 50 rpm will lead to about 0.3-0.4% increase in fines and pins each. Analysis of chippers using softwoods and bottom discharge units showed that there was about a 50% increase in fines content (1% to 1.5%) as new knives aged to mid life, but no decrease from mid-life knives to old knives. Keep in mind that other factors besides the chip size distribution, such as chip bruising and geometry, are important in the pulping and papermaking properties of wood chips.

Chip quality, uniformity, and testing

4. Hatton, J.V. *Chip Quality Monograph, Pulp and Paper Series*, No. 5, Joint Textbook Committee of the Paper Ind., 1979, 323 p.

    This work is the classic on the topic of wood chip quality. Prior to Hatton's research in the 1970s on wood chip thickness screening for kraft mills, screening of wood chips was done with round-hole screens.

    The importance of uniform chip thickness and quality in kraft pulping cannot be over-stressed. Even under ideal cooking conditions, for a cook at a kappa number of 20 some of the fiber from pin chips will have a kappa as low as 10 and other fiber from over thick chips will have a kappa as high as 50, with the concomitant problems of each.

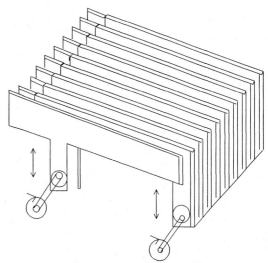

Fig. 2-35. Diagram of Dynagage™ bar screen with bars anchored on left hand side in lower position.

5. Christie, D., Chip screening for pulping uniformity, *Tappi J.* 70(4):113-117(1987). The article summarizes the importance of *chip thickness screening* in kraft pulping.

6. Luxardo, J. and S. Javid, New technology for chip thickness and fines screening, *Pulp Paper Can.* 93(3):39-46 (T56-T63)(1992). A similar paper appears as Smith, D.E. and S.R. Javid, Trends in chip thickness screening, *Tappi J.* 73(10):185-190(1990). This work also appears in at least three conference proceedings. Recent advances in Acrowood products are given here.

7. Nelson, S.L. and P. Bafile, Quinnesec woodyard focuses on chip thickness control at the chipper, *Tappi J.* 72(3):95-106(1989).

8. Thimons, T., Chip-thickness screening with an oscillating bar screen, *Tappi J.* 74(11):183-185 (1991) (*ibid.*, Chip thickness screening without rotating wear surfaces, *1991 TAPPI Pulping Conf. Proc.*, pp 553-555.)

    Chip thickness screens have saved mills millions of dollars. Most thickness screens use rotating metal disks that have high wear and must be properly maintained for good perfor-

mance. A new system has been developed that uses longitudinal distribution bars attached to two eccentric shafts with bars alternating as to which shaft they are connected. The system looks remarkably simple and effective. Overthick removal is very high while accept carry over is very low. The method is called Dynagage™ Bar Screen and is shown in Fig. 2-35.

9. Berlyn, R.W. and R.B. Simpson, Upgrading wood chips: the Paprifer process, *Tappi J.* 71(3):99-106(1988). A wide variety of chips including whole-tree chips, chips from partially decayed trees, and logging residues are claimed to be improved by the Paprifer process.

10. Marrs, G., Measuring chip moisture and its variations, *Tappi J.* 72(7):45-54(1989). Chip moisture meters, even though they have inherent imprecision, may give the best available estimates of chip moisture during processing.

Chip pile

11. Fuller, W.S., Chip pile storage--a review of practices to avoid deterioration and economic losses, *Tappi J.* 68(8):48-52(1985). This

article is a concise summary of wood chip pile management with 24 references.

Fiber physics
12. Page, D.H., F. El-Hosseiny, K. Winkler, and R. Bain, The mechanical properties of single wood-pulp fibres. Part I: A new Approach, *Pulp Paper Mag. Can.* 73(8):72-77(1972).

Nonwood fibers
13. Clark, T.F, Annual crop fibers and the bamboos, in *Pulp and Paper Manufacture*, Vol. 2, 2nd ed., MacDonald, R.G., Ed., McGraw-Hill, New York, 1969, pp 1-74. Processing of a variety of nonwood fibers is considered. It seems likely that the U.S. will start using nonwood fibers in brown papers or newsprint within the next two decades.

14. Atchison, J.E., World capacities for nonwood plant fiber pulping increasing faster than wood pulping capacities, *Tappi Proceedings, 1988 Pulping Conference*, pp 25-45. This is a good summary about who is pulping what around the world.

Wood handling
15. Lamarche, F.E., Preparation of pulpwood, in *Pulp and Paper Manufacture*, Vol. 1, 2nd ed., MacDonald, R.G., Ed., McGraw-Hill, New York, 1969, pp 73-147.

This is a thorough look at handling roundwood and wood chips at the mill, especially in regards to the equipment. It includes pulpwood measurement, log storage and handling, debarking, chipping and chip quality and handling. Chip screening by *thickness* is not mentioned as the article predates the widespread understanding of the importance of chip thickness screening for kraft pulping. It is well illustrated.

EXERCISES

Wood
1. True or false?   Bark in the wood chip supply is unimportant as it has good fiber for making paper. Why?

2. Circle the correct choice in each set of parentheses. (White *or* Brown) rot in wood chips affects the ultimate paper strength properties the most for a given weight loss. (Hardwood *or* Softwood) pulp is used to make strong paper while (hardwood *or* softwood) pulp is used to make smoother paper. (Hardwoods *or* Softwoods) have more lignin. Short storage time of wood chips is most important for (mechanical *or* kraft) pulping.

3. List two changes that take place in chips during storage in a chip pile.

4. Wood chip quality is very important in the final properties of paper. List five parameters for chip quality and their effect(s) on the final paper property.

5. What chip size fractions cause the most problems in kraft pulping, and what are the problems they cause?

Fraction	Problem
_____	_____
_____	_____

6. Approximately _____ % of the production cost of pulp is due to the cost of wood chips.

Wood chemistry
7. What are the three major components of wood?

8. Circle the correct choice in each set of parentheses. (Cellulose *or* Hemicelluloses) provide(s) individual wood fibers with most of their strength. (Cellulose *or* Hemicelluloses) is/are appreciably soluble in alkaline solutions at elevated temperatures.

9. What material holds the fibers together in wood?

10. How does softwood lignin differ from hardwood lignin?

11. What is turpentine? From what woods is it obtained?

12. Name three commercial materials made from dissolving pulp (cellulose). What are their uses?

13. Describe the process of making carboxymethyl cellulose and show the principal chemical reaction.

Wood and fiber physics

14. In the absence of fiber defects, what factor determines the longitudinal tensile strength of individual fibers? For a similar reason, wood fibers do not shrink nearly as much in their longitudinal direction as in their radial or tangential directions. Why is this so? See Fig. 2-36 to help see why this is so.

15. Wood and paper are hygroscopic materials. What does this mean?

16. What holds fibers together in paper?

Nonwood and recycled fiber

17. Is most recycled fiber deinked? Discuss.

18. What is (considered to be) the maximum percentage of fiber recovery that can be sustained. Why is this so? At what levels do the U.S., Japan, and W. Germany recycle?

19. The value of mixed paper waste is on the order of $_____/ton. The value of high quality trimmings from an envelope factory is on the order of $_____/ton. What concept do these facts demonstrate?

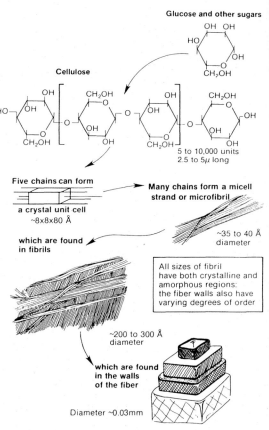

Fig. 2-36. Relative sizes of fibers and fibrils. ©**1982, 1991 James E. Kline. Reprinted from** *Paper and Paperboard* **with permission.**

20. Trees have replaced all other plants as a fiber source for paper around the world. Discuss this statement.

3

PULPING FUNDAMENTALS

3.1 INTRODUCTION TO PULPING

Pulp

Pulp consists of wood or other lignocellulosic materials that have been broken down physically and/or chemically such that (more or less) discrete fibers are liberated and can be dispersed in water and reformed into a web. Fig. 3-1 shows a brown paper (no bleaching) kraft mill process. According to *Pulp & Paper Week*, November 21, 1988, the price of delivered pulps in the 4th quarter of 1988 in the U.S. (per ton) were as follows:

bleached softwood kraft	$735-760
bleached hardwood kraft	$685-730
unbleached softwood kraft	$670-700
dissolving pulp	$800-810
bleached CTMP	$625

Pulping

There are four broad categories of pulping processes: *chemical, semi-chemical, chemi-mechanical*, and *mechanical pulping*. These are in order of increasing mechanical energy required to separate fibers (fiberation) and decreasing reliance on chemical action. Thus, chemical pulping methods rely on the effect of chemicals to separate fibers, whereas mechanical pulping methods rely completely on physical action. The more that chemicals are involved, the lower the yield and lignin content since chemical action degrades and soulblizes components of the wood, especially lignin and hemicelluloses. On the other hand, chemical pulping yields individual fibers that are not cut and give strong papers since the lignin, which interferes with hydrogen bonding of fibers, is largely removed. Fig. 3-2 shows electron micrographs of several pulp types that demonstrate this point. Details of these types of pulps will be considered below.

Table 3-1 summarizes important aspects of the most common classes of pulping processes. Table 3-2 shows the production of pulp by pulping process to show the relative commercial significance of the processes. The relative strength of kraft:sulfite:soda:stone groundwood pulps for a

given species of wood are roughly 100:70:40:30, although this depends on the species of wood, strength property, and pulping conditions. Table 3-3 gives some mechanical and physical properties of representative commercial pulps. While this may not mean much on the first reading through the book, it is useful information for understanding the reason why particular pulps are used in particular grades of paper.

Wood-free, free-sheet

Wood-free pulp or free-sheet paper contains no mechanical pulp or contains pulp subjected to a minimum of refining; consequently, during its manufacture the water drains very quickly from the pulp on the Fourdrinier wire.

Screening

Screening of pulp after pulping is a process whereby the pulp is separated from large shives, knots, dirt, and other debris. *Accepts* consist of the pulp that has passed through the screens. The *accept yield* is the *yield* of accepts. *Rejects* or *screenings* are the larger shives, knots, large dirt particles, and other debris removed by the screens after the pulping process.

Shives

Shives are small fiber bundles of fibers that have not been separated into individual fibers during the pulping process. They appear as "splinters" that are darker than the pulp.

Yield

Yield is a general term used in any phase of pulping, papermaking, chip screening, bleaching, etc. indicating the amount of material recovered after a certain process compared to the starting amount of material before the process. To have meaning, both samples must be compared on an oven-dry basis. In pulping operations the yield is the oven-dry pulp mass expressed as a percentage of the oven-dry wood mass. Mechanical pulp yields are typically 92-96% and bleached chemical pulp yields are typically 40-45%. For example,

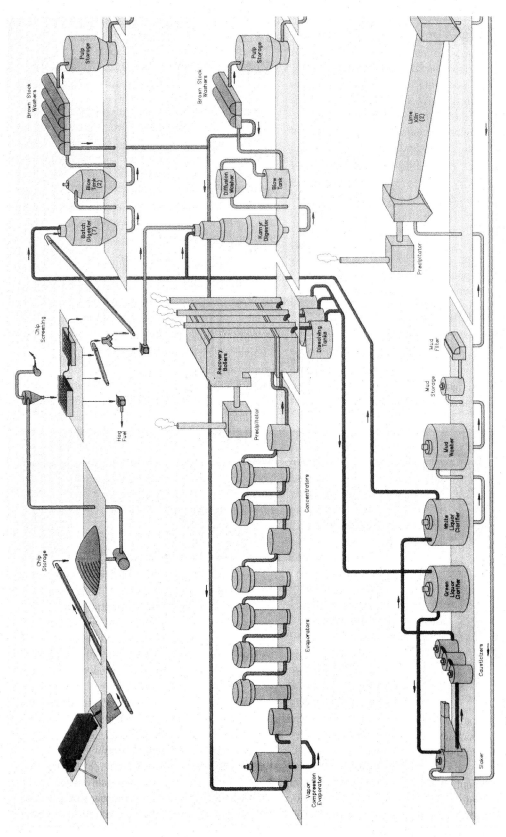

Fig. 3-1. Mill layout of a kraft linerboard mill (no bleaching). Courtesy of Weyerhaeuser Paper Co. (Figure continues linearly from the right-hand side of this page to the left-hand side of the facing page.)

56

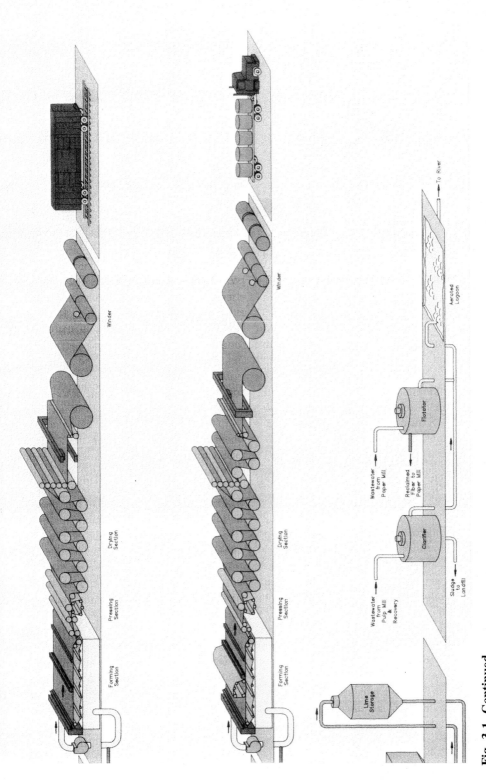

Fig. 3-1. Continued.

Fig. 3-2. (A) steam exploded hardwood. Papers of (B) SGW hardwood, (C) TMP softwoods, (D) newsprint, (E) NSSC hardwood/OCC corrugating medium, and (F) bleached kraft softwood fibers.

Table 3-1. Summary of pulping processes.[1]

Process	Chemicals	Species	Pulp Properties	Uses	Yield, %
Mechanical pulping	none; grindstones for logs; disk refiners for chips	Hardwoods like poplar or light-colored softwoods like spruce, balsam fir, hemlock, true firs	High opacity, softness, bulk. Low strength and brightness.	Newsprint, books, magazines.	92-96%
Chemi-mechanical pulping	CTMP; mild action; NaOH or NaHSO$_3$		Moderate strength		88-95%
Kraft process, pH 13-14	NaOH + Na$_2$S (15-25% on wood); unlined digester, high recovery of pulping chemicals, sulfur odor	All woods	High strength, brown pulps unless bleached	Bag, wrapping, linerboard, bleached pulps for white papers	65-70% for brown papers, 47-50% for bleachable pulp; 43-45% after bleaching
Sulfite, acid or bisulfite pH 1.5-5	H$_2$SO$_3$ + HSO$_3^-$ with Ca^{2+}, Mg^{2+}, Na$^+$, or NH$_4^+$ base; Ca^{2+} is traditional but outdated since no recovery process; lined digesters	hardwoods-poplar and birch and non-resinous softwoods; Douglas-fir is unsuitable	light brown pulp if unbleached, easily bleached to high brightness, weaker than kraft pulp, but higher yield	Fine paper, tissue, glassine, strength reinforcement in newsprint	48-51% for bleachable pulp; 46-48% after bleaching
	Mg^{2+} base	almost all species-spruce and true firs preferred	Same as above but lighter color and slightly stronger	Newsprint, fine papers, etc.	50-51% for bleachable pulp 48-50% after bleaching
Neutral sulfite semi-chemical (NSSC) pH 7-10	Na$_2$SO$_3$ + Na$_2$CO$_3$ about 50% of the chemical recovered as Na$_2$SO$_4$	Hardwoods (preferred) aspen, oak, alder elm, birch; softwoods Douglas-fir sawdust and chips	Good stiffness and moldability	Corrugating medium	70-80%

[1]Reprinted from Krahmer, R.L. and A.C. VanVliet, Ed., *Wood Technology and Utilization*, O.S.U. Bookstore, Corvallis, Ore., 1983 with permission.

Table 3-2. Approximate pulp production (thousands of MT) by pulping method.[1]

Category	Process		U.S. 1990	U.S. 1960	Canada 1989	Global 1986
Mechanical pulp	TMP		3400		7000	31,000 total
	CTMP		30		2800	
	SGW		3200	3292		
	PGW		600			
	RMP		342			
Chemical	Kraft	Total	50,000	14,590	10,850	88,000
		Bleached softwood	14,000		6868	29,000
		hardwood	13,000		1577	26,000
	Sulfite		1600	2578	1627	9000
	Dis- solv- ing	Kraft	800	1138 total	273 total	
		Sulfite	650			
	Soda (hardwoods)		300	420		
Semi- chemical	NSSC, etc.		5000	1991 total	432	8000
Chemi- mechanical	steam explosion/ Asplund		600			
	cold soda		200			
Total virgin wood pulp			62,700	25,316	23,700	136,000
Wastepaper, recycling			18,400	8000		56,000

[1]Data from *1990 Pulp and Paper Fact Book* as pulp productions or capacities, FAO, and Libby Volume I for U.S. 1960 data. There is significant rounding of values here as this table is meant only as an approximation. Different sources differ appreciably in amounts. In 1986 the world paper production was 202,000 thousand tons. The nonwood and non-secondary fiber portion of 10,000 tons consists of minerals for coating and fillers and other plant fibers such as bagasse, straw, and bamboo.

Table 3-3. Properties of commercial pulp samples ca. 1975 from manufacturer's specifications.[1]

Beating time (minutes)	CSF (ml)	Break length (km)	Tear factor (metric)	Double folds	Bulk (cc/g)	Air resist. (sec)	Opacity
IA. Softwood kraft pulp (Kamyr dig.) of long-fibered western hemlock, redcedar, and Doug.-fir. The alpha cellulose content is 82%, the pentosans content is 5.6%, and the lignin content is 2.7%							
0	662	4.7	222	70	1.88	4	
20	508	9.1	121	780	1.60	50	
45	205	11.3	104	1720	1.44	802	
IB. The same fiber mix as IA after fully bleaching by CEDED to 88+ brightness is 86.5% alpha cellulose and 4.5% pentosans. The cellulose viscosity is C_p 22, and the stretch is 3.2-3.5%							
0	661	5.0	242	120	1.83	3	73
15	494	9.9	122	900	1.56	34	67
30	249	11.0	114	1600	1.48	513	63
II. Sulfite softwood pulp of 94% brightness, 88% alpha cellulose, and 350 cps viscosity							
0	705	1.9	136	3	1.78	1	75
31	550	5.9	102	130	1.45	15	70
75	250	7.5	83	650	1.33	250	65
III. Hardwood kraft pulp of 91% brightness, 87.5% alpha cellulose, and 65 cps viscosity							
0	585	2.6	106	5	1.85	2	82
13	550	4.5	101	30	1.75	5	79
81	250	8.3	87	350	1.45	80	74
IV. Hardwood sulfite of 93% brightness, 88.5% alpha cellulose, and 160 cps viscosity							
0	625	1.5	54	0	1.85	1	82
15	550	2.8	67	5	1.67	5	80
74	250	5.5	68	40	1.35	65	86
V. Softwood thermomechanical pulp of 50% brightness (Determined in OSU lab. in 1991)							
0	396	1.3	5.4	1	3.49		97
120 (PFI)	123	3.4	7.8	8	2.65		97
240 (PFI)	102	3.8	6.7	9	2.53		96

[1]Note that testing conditions of 60 g/m² handsheets at 73°F at 50% relative humidity with refining by the valley beater (except TMP as noted). The relative mullen burst values are similar to those of the tensile breaking length. All tests are TAPPI Standard methods.

100 lb of dry wood yields about 40 to 45 lb of pulp for bleached printing paper.

$$\text{yield, \%} = \frac{\text{dry product mass out}}{\text{dry material mass in}} \times 100\%$$

Total yield, %

The total yield is equal to the amount of pulp removed during screening and the yield of pulp after the screens when all three are expressed as a percentage of the original wood put in the digester. When one speaks of pulp yield it is necessary to state whether it is the total yield or screened yield.

Total yield (%) = Screenings (%) + Screened yield (%)

Consistency

Consistency is a measure of the *solids content* as a percentage in a pulp slurry.

$$\text{consistency} = \frac{\text{dry solids mass}}{\text{slurry mass}} \times 100\%$$

3.2 MECHANICAL PULPING

Mechanical Pulp

Mechanical pulp is pulp produced by using only mechanical attrition to pulp lignocellulosic materials; no chemicals (other than water or steam) are used. Light colored, non-resinous softwoods and some hardwoods are often the fiber source. The total yield is about 90-98%. Lignin is retained in the pulp; therefore, high yields of pulp are obtained from wood. Mechanical pulps are characterized by high yield, high bulk, high stiffness, and low cost. They have low strength since the lignin interferes with hydrogen bonding between fibers when paper is made. The lignin also causes the pulp to turn yellow with exposure to air and light.

The use of mechanical pulps is confined mainly to non-permanent papers like newsprint and catalog paper. Mechanical pulps constitute 20-25% of the world production and this is increasing due to the high yield of the process and increasing competition for fiber resources. Furthermore, technological advances have made mechanical pulps increasingly desirable.

Table 3-4. Mechanical pulp capacity by region in thousands of metric tons[1].

Region	Year	TMP	CTMP	PGW
USA	1980	1619	0	0
	1988	3028	38	224
Canada	1980	996	110	0
	1988	4144	2421	0
Scandinavia	1980	1411	110	40
	1988	4596	685	743
Europe	1980	663	30	0
	1988	1559	155	307
Japan	1980	862	0	0
	1988	1161	310	150
Other	1980	320	0	0
	1988	1238	897	44

[1]Data from Kayserberg (1989). Europe values exclude Scandinavia.

Table 3-4 shows the production of various types of mechanical pulp in 1980 and 1988 for several areas of the world. During this time global mechanical pulp production increased from 6 to 15 million metric tons. Several important aspects of mechanical pulping are condensed from this article. In the last decade, TMP processes have largely replaced SGW. CTMP and bleached CTMP (BCTMP) are displacing small amounts of chemical pulps in certain grades of paper. Most CTMP is produced in Canada. PGW is important in Scandinavia, but has limited production outside this region.

Chip quality and cleanliness

Some aspects of chip quality are very important in mechanical pulping. Since mechanical pulps cannot be brightened very much by chemicals, chip quality (except for SGW, which uses wood logs) is of extreme importance. Generally chips less than two weeks old are used with stringent bark and dirt tolerances. Chips older than two weeks tend to be discolored too much by decay and air oxidation. Washing the chips (Fig. 3-3 shows one type of chip washing system) before pulping is a necessity, just as in any me-

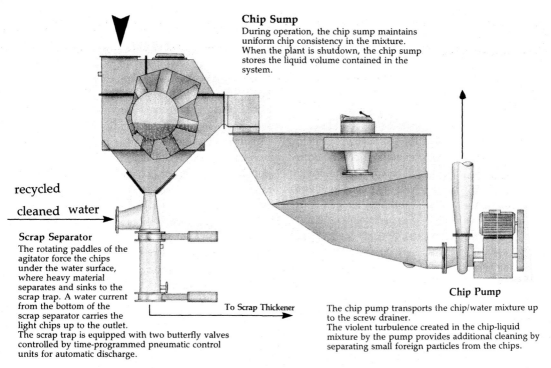

Chip Sump

During operation, the chip sump maintains uniform chip consistency in the mixture. When the plant is shutdown, the chip sump stores the liquid volume contained in the system.

recycled

cleaned water

Scrap Separator

The rotating paddles of the agitator force the chips under the water surface, where heavy material separates and sinks to the scrap trap. A water current from the bottom of the scrap separator carries the light chips up to the outlet.
The scrap trap is equipped with two butterfly valves controlled by time-programmed pneumatic control units for automatic discharge.

To Scrap Thickener

Chip Pump

The chip pump transports the chip/water mixture up to the screw drainer.
The violent turbulence created in the chip-liquid mixture by the pump provides additional cleaning by separating small foreign particles from the chips.

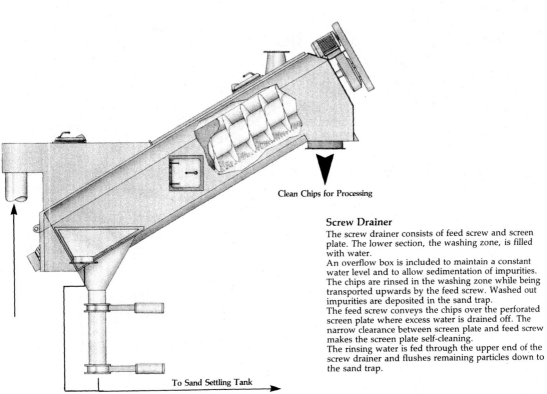

Clean Chips for Processing

Screw Drainer

The screw drainer consists of feed screw and screen plate. The lower section, the washing zone, is filled with water.
An overflow box is included to maintain a constant water level and to allow sedimentation of impurities. The chips are rinsed in the washing zone while being transported upwards by the feed screw. Washed out impurities are deposited in the sand trap.
The feed screw conveys the chips over the perforated screen plate where excess water is drained off. The narrow clearance between screen plate and feed screw makes the screen plate self-cleaning.
The rinsing water is fed through the upper end of the screw drainer and flushes remaining particles down to the sand trap.

To Sand Settling Tank

Fig. 3-3. Chip washing system that incorporates water reuse. Courtesy of Sunds Defibrator.

chanical or semi-chemical pulping operation where disk refining is ultimately required to break apart the wood chips. Sand, pebbles, tramp metal, and other gritty materials would otherwise cause undue damage to the refiners.

Stone groundwood (SGW)

Groundwood mechanical pulp is produced by grinding short logs, called *bolts*, with grindstones on the tangential and radial surfaces. Original grindstones were sandstone, but segmented stones with embedded silicon carbide or aluminum oxide are used now. The sandstone grindstones had to be treated carefully to prevent them from bursting due to thermal shocks; they were phased out in the early 1940s. SGW was first produced in the 1840s as a wood meal using technology from grain mills; the French naturalist DeReamur realized that paper could be made from wood when his study of wasp nests in 1719 showed that they were made from wood particles held together with proteins from wasp saliva. Early mechanical pulps were merely fillers to extend the supply of cotton and other fibers. However, advances in the process gave stronger pulps. The large amount of fines makes this pulp useful for increasing the opacity of some printing papers, although hardwood CTMP (discussed below) is a suitable substitute for this purpose. The energy use is about 1300 kWh/ton (50-80 hp-day/ton) with the higher amounts for high grade papers; yields are 93-98%.

The mechanism of grinding is not a cutting action, and stones with sharp edges are not desired. Instead, reasonably selective fiber-fiber separation is achieved by peeling caused by repeated shear stresses. The shear stresses begin when the fibers are about 3 to 5 fiber diameters away from the stone's surface. These are obtained by repeated compression and decompression of the wood generated by the groove pattern on the stone's surface. Woods of low density are particularly suited to groundwood production because they allow the most deformation due to large lumens that accommodate expansion and contraction of the fibers caused by impact with the grits on the surface of the stone. If the wood is not able to distribute the load by flexing, fibers tend to be cut rather than separated from each other, leading to pulp of high fines content and low strength.

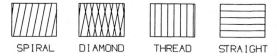

SPIRAL   DIAMOND   THREAD   STRAIGHT

Fig. 3-4. Burr patterns used for sharpening pulp stones for stone groundwood production.

The burr pattern is very important. The commonly used patterns are shown in Fig. 3-4. They include *spiral*, which is the most commonly used pattern type since it produces a high quality pulp; *diamond*, which is usually used to erase old patterns and resurface the stone; *thread*, which is not often used since it tends to generate both chunky fibers and fines (actually flour); and *straight*, which makes long, coarse pulp fibers suitable for building and insulation boards. A No. 9, 1 in. lead, spiral burr has 9 threads per inch with the thread rising 1 in. from the left to the right hand side. Burr patterns are cut by the metal burr which makes several passes over the stone until the burr pattern depth on the stone is about 1.6 mm. This is called stone sharpening and must be done every 50-150 hours.

Pulp strength properties are considerably lower right after sharpening of the stone, so stone sharpening must be staggered from grinder to grinder in a mill. Pulp burst and strength properties increase between stone sharpenings by 25-50% (advantageous). However, pulping specific energy increases 25-50%, production rate decreases 25-100%, and freeness decreases by a factor of 1 to 3, all three of which are disadvantages.

Water is used in the grinding process before and after the grinding area to control the temperature of the grindstone and to wash the pulp from the grindstone. The pulp leaves at 2-5% consistency. A 3,000-5,000 hp motor spinning a 1.5 m (typically 54-70 in.) diameter grindstone at 240-300 rpm (18-24 m/s or 3400-4500 ft/min at the grinding surface) will grind about 50 ton/day, corresponding to 30-40 hp/ft² grinding area. The grinding pressure is 40-80 psi. The water temperature in the pit is 140-180°F. Newsprint requires 55-70 hp-day/ton while book paper uses 60-85 hp-day/ton. The important grinding variables are:

1. Wood species and other wood variables.
2. Type of pulpstone.

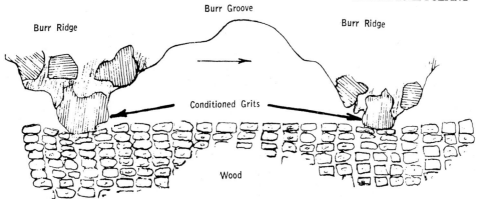

Fiberizing by Compression of the Wood by Conditioned Grits

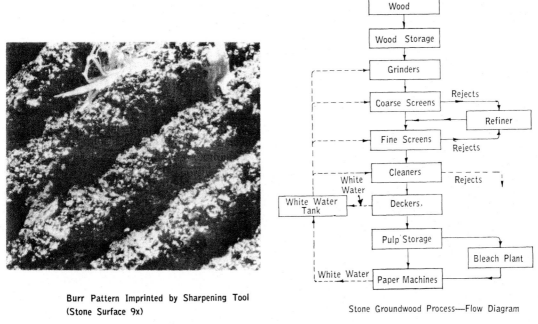

Burr Pattern Imprinted by Sharpening Tool
(Stone Surface 9x)

Stone Groundwood Process—Flow Diagram

Fig. 3-5. Grinding action and overall stone groundwood (SGW) process. Reprinted from *Making Pulp and Paper*, **©1967 Crown Zellerbach Corp., with permission.**

3. The use (or not) of a water-filled grinding pit.
4. Type of burr pattern on the stone.
5. Stone surface speed.
6. Hours on the stone since last burring.
7. Pressure of wood against the stone.
8. Temperature of grinding surface, 130-180°C (265-355°F).
9. Amount of water used (and, therefore, the final pulp consistency).

An example of SGW paper from hardwoods is shown in Fig 3-2. This paper was from an experimental press run of the March 26, 1959 *Savannah Evening News* where a mixture of seven southern U.S. hardwoods were used. While relatively intact fibers and vessel elements are observed in this figure, there are numerous fiber fragments as well.

The overall process and grinding action are shown in Fig. 3-5. Fig. 3-6 shows some of the

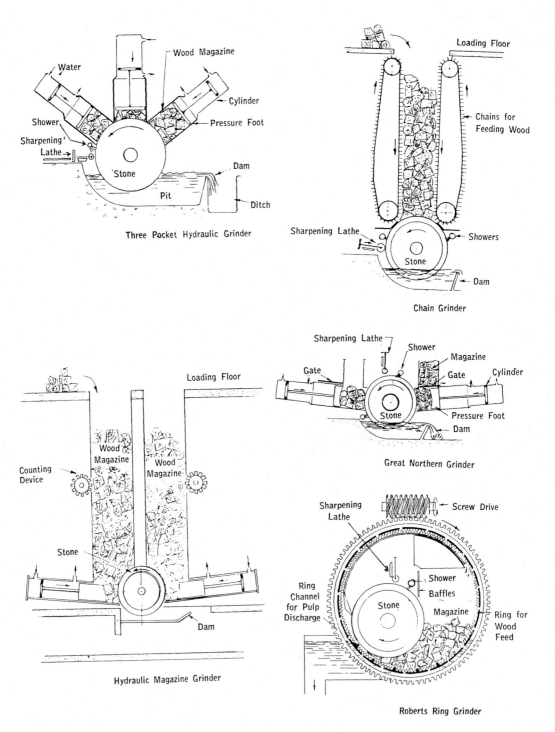

Fig. 3-6. Grinder configurations used in the stone groundwood process. Reprinted from *Making Pulp and Paper*, ©1967 Crown Zellerbach Corp., with permission.

many types of grinder configurations that have been used. Pocket grinders were once popular, but have been largely replaced by the various magazine grinders that have continuous feeding of wood, which greatly reduced the manpower required to load wood. The three pocket hydraulic grinder was introduced around 1900. When the wood in one of the pockets is consumed the door is opened after the pressure foot is retracted, wood is added manually, the door is closed, and the pressure foot is engaged with water pressure. The magazine grinder was introduced by Voith of Germany in 1908.

The Great Northern Grinder (and modifications such as that of Koehring-Waterous Ltd.) was installed at many mills making newsprint in the 1950s. Large units grind wood 1.6 m (5 ft) long at the rate of 80 tons/day using 5500 horsepower. The chain grinder is the most common of the continuous grinders introduced in the 1920s. Large units pulp wood 2 m (6.5 ft) long at the rate of 100 tons/day using 7000 horsepower. The continuous Roberts ring grinder was introduced in the 1940s.

Pressure groundwood (PGW)

By pressurizing the grinder with steam at temperatures of 105-125°C (220-255°F), the wood is heated and softened prior to the grinding process. This gives better separation of fibers with less cutting action and lower fines generation. This process yields a pulp that has higher tear strength and freeness and is brighter than SGW, yet has lower power requirements. About one-half of the world's PGW is produced in the Scandinavian countries. As one reads through this section it will become apparent that PGW is to SGW as TMP is to RMP.

Refiner mechanical pulp (RMP)

Refiner mechanical pulp is produced by disintegrating chips between revolving metal disks or plates with raised bars (Plate 23 and pages 144-146) at atmospheric pressure. This process was developed during 1948-1956 by Bauer Bros. and is patented by Eberhardt. Some of the steam generated in the process softens the incoming chips resulting in fibers that maintain more of their original integrity compared to SGW, because there

is less cutting action and increased separation of fibers at the middle lamella. Because of this action, there are less fines formed, and RMP is bulkier and stronger than SGW. Power requirements are 1600-1800 kWh/ton (90-100 hp·day/ton). Disk refiners are up to 1.5 m (60 in.) in diameter and rotate at 1800 rpm with 60 Hz power; this gives a velocity at the periphery of up to 140 m/s. The plates containing the metal bars must be replaced every 300-700 hours or low quality pulp is produced and energy use increases. Refining is usually carried out in two stages. The first is at 20-30% consistency to separate the fibers, while the second is at 10-20% consistency to alter the surface of the fibers for improved fiber bonding in the final paper. The pulping variables are:

1. Wood species and other wood variables.

2. Pulp consistency.

3. Sharpness and pattern of refiner plates.

4. Temperature of refining.

5. The gap between the refiner plates. (0.005 to 0.1 in.; 0.035 in. typically)

6. Rate of feed through the refiner.

7. Speed of the refiner plates.

The RMP process was an important development in the history of mechanical pulping. Now, however, TMP, a modification of the RMP process, and other mechanical pulping processes have made the RMP process almost obsolete.

Thermomechanical pulp (TMP)

TMP was developed about 10 years after RMP and has become the most important mechanical pulping method. For example, in 1975, less than 200,000 metric tons of TMP were produced in North America. In 1989 the U.S. capacity was just under 3 million metric tons, while Canada was just under 6 million metric tons.

The TMP process is very similar to the RMP process except that pulp is made in special refiners that are pressurized with steam in the first stage of refining. Fig. 3-7 shows a typical flow diagram for a TMP plant. TMP is usually carried out in two stages of refining. In the first stage, the refiners are at elevated temperature and pressure

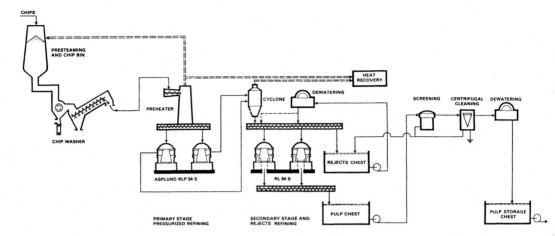

Fig. 3-7. Representative layout of a newsprint TMP plant. Courtesy of Sunds Defibrator.

to promote fiber liberation; in the second stage, the refiners are at ambient temperature to treat the fibers for papermaking. The higher temperature during refining in the first step, 110-130°C (230-265°F), softens the fibers and allows their recovery with minimal cutting and fines compared to SGW or even RMP. The refining is just under the glass transition temperature of lignin, which is approximately 140°C (285°F), so that separation of fibers occurs at the S-1 cell wall layer. This improves fibrillation (surface area) and access to hydroxyl groups for hydrogen bonding. A laboratory pressurized refiner for making TMP pulp is shown in Fig. 3-8.

The high strength of this pulp relative to the other mechanical pulps has made it the most important mechanical pulp. Energy requirements are 1900-2900 kWh/ton (100-150 hp-day/ton),

Fig. 3-8. Pressurized laboratory refiner for making TMP pulp.

slightly more than RMP, and significantly more than SGW. Over two-thirds of this is used in the primary pressurized refining step, and less than one-third is used in the secondary atmospheric pressure refining step. An even consistency of 20-30% is ideal for primary refining and is the most important operating variable; lower consistencies cause fiber damage from the refiner plate bars while higher consistencies cause the refiner to plug. TMP tends to be darker than SGW due to chemical reactions at the elevated temperatures and wood chip supplies often contain more bark, dirt, and other impurities than bolts used in SGW. The pulp yield is 91-95%.

Solubilization of wood components make BOD levels relatively high in mill effluents. If the steaming temperature is too high then the process has the problems of the Asplund process (Section 3.4), that is, the surface of the fibers becomes coated with lignin that interferes with hydrogen bonding. This results in weak, dark paper.

Fig. 3-2 shows two samples of TMP. The first is softwood TMP from the refiner chest of a newsprint mill. The second is a sample of a 1990 newsprint. These samples show much less fiber fragmentation than the SGW sample, but there are still more fiber fragments and fiber delamination than in the bleached kraft softwood fibers.

3.3 CHEMI-MECHANICAL PULPING

General aspects

What are now called chemi-mechanical processes were originally called chemi-groundwood processes since these chemical pretreatments were developed and used commercially at a time (early 1950s) when stone groundwood was the predominant mechanical process. (The original laboratory work on chemi-mechanical pulping and cold soda pulping of straw had been done much earlier.) However, these chemical pretreatments have application to all of the mechanical pulping processes.

ESPRA (Empire State Paper Research Associates), whose laboratories are in Syracuse, New York, applied the NSSC process (Section 3.5) in a mild form prior to groundwood pulping. In one operation, four-foot long hardwood logs were treated in large digesters (60 feet high by 10 feet

in diameter). A vacuum was first applied to remove much of the air from the logs and allow better liquor penetration. The liquor, containing aqueous Na_2SO_3 and Na_2CO_3, was then introduced and kept at 130-155°C (265-310°F) depending on the species of wood. The pressure was maintained at 150 psi. The pretreated logs were then pulped by the stone groundwood method. All of the advantages of the CTMP process (over TMP) were observed long before CTMP was thought of with the original chemi-groundwood method. The grinding requirements were about half that required without pretreatment and the CSF was 300-350 ml. At around the same time, the cold soda method was developed, although it has only had limited use.

The chemi-mechanical pulping process consists of two stages with yields of 85-95%. A particularly mild chemical treatment is followed by a drastic mechanical action, but not as drastic as without chemical pretreatment. The original lignin structure and content is preserved, but extractives and small amounts of hemicellulose are lost. When higher temperatures are used in the various steps, a darker pulp is usually obtained. One of several chemical pretreatments can be applied prior to SGW, PGW, RMP, or TMP. The most common chemi-mechanical process is now CTMP. The pretreatments are hot sulfite or cold soda and are particularly applicable to hardwoods that otherwise do not give mechanical pulps of high quality. Several mills have also begun to use an alkaline peroxide chip pretreatment.

Chemical pretreatments

In the *hot sulfite* process pressurized hot sulfite liquor is used to treat chips prior to fibration. This results in a pulp brighter than the cold soda process that is usable in book papers, catalogs, and newspapers. The pulp is usable in papers by itself in the case of softwood pulp, but 10-15% chemical pulp must be added to the hardwood pulp to achieve adequate strength. With hydrogen peroxide bleaching, brightness levels up to 82% may be obtained.

The *cold soda* process was first investigated by the U.S. Forest Products Laboratory in Madison in the early 1950s as a pretreatment prior to making RMP or stone groundwood. By 1960,

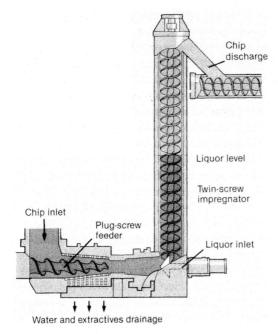

Chip discharge

Liquor level

Twin-screw impregnator

Chip inlet

Plug-screw feeder

Liquor inlet

↓ ↓ ↓
Water and extractives drainage

Fig. 3-9. Liquor impregnation using the pressure-expansion technique. Courtesy of Sunds Defibrator.

about 10 mills used the process, however by 1970, only a few mills used the process as other processes for pulping hardwoods had been developed.

The yield of the cold process is 80-95%. Chips are first soaked in 5-15% cold caustic soda (sodium hydroxide solution) as a swelling agent. Lignin is retained, but some hemicelluloses and all resins are lost, resulting in a pulp with good drainage. Softwoods resist this treatment, producing coarse fibers with unusually high power demands. When used prior to SGW (CSGW), this method results in a pulp which drains faster than groundwood pulp, meaning that faster paper machines can be used. Refining energy is 500-700 kWh/ton (28-40 hp-day/ton).

The *alkaline peroxide* pretreatment for mechanical pulping (APMP) was introduced by Sprout-Bauer in 1989. This process (McCready, 1992) is used to make 600 t/d of aspen pulp at the Millar Western mill in Saskatchewan, Canada, using a modification of Scott Paper's patented process. Four stages of chip compression (using a Hymac screw press with a 4:1 ratio) and chemical impregnation are used prior to refining. Much, if not all, of the bleaching is claimed to

occur here, often saving the cost of a separate bleach plant. The refiners are reported to require about 30-40% less energy than with sulfite pretreatment for BCTMP pulp. This pulp mill used a closed water loop by concentrating certain effluents to 70% solids for combustion. APMP does not require pressurized refiners. Increasing the amount of caustic in the pretreatment increases the strength of the pulp. Increasing the amount of peroxide increases the pulp brightness. This method is particularly profound because effluents with high BOD and inorganic chemical levels have prevented the use of CTMP in many locations.

Chemi-thermomechanical pulp (CTMP)

CTMP is a chemi-mechanical process that is similar to the TMP process except the chips are first pretreated with relatively small amounts (about 2% on dry wood) of sodium sulfite or sodium hydroxide (with hydrogen peroxide in the new alkaline peroxide method) under elevated temperature and pressure prior to refining. Liquor penetration is often achieved by a system that compresses the wood chips into a liquid tight plug that is then fed into the impregnator vessel where the chips expand and absorb the liquor (Fig. 3-9.) Unlike TMP, CTMP is effective with most hardwoods particularly with the cold soda process. CTMP processes produce effluents of high color and BOD which may be difficult to treat; environmental considerations have kept this process from being used in many locations.

3.4 RELATED PULPING METHODS

Asplund process

Interestingly, the Asplund process was developed about 30 years prior to TMP. It involves presteaming wood chips at temperatures above the glass transition temperature of lignin, 550-950 kPa (80-140 psi) steam pressure at 150-170°C (300-340°F), prior to refining between revolving metal disks or plates. The lignin is sufficiently soft that separation occurs at the middle lamella, and fibers are left with a hard lignin surface that is not amenable to fibrillation by refining or bonding by hydrogen bonding. Such fibers are very useful for hardboard materials where adhesives are used that react with the lignin and pressing at high temperatures will allow the lignin-coated fibers to adhere to each other.

Masonite process (steam explosion)

This process was invented by Mason and patented in 1926 as an attempt to make pulp for paper. Chips are steamed at 180-250°C (355-480°F), well above the glass transition temperature of lignin, for one to two minutes until the final pressure is on the order of 4-7 MPa (600-1000 psi). The contents of the reactor are suddenly blown to atmospheric pressure through a slotted plate and to a cyclone where the fibers are collected. The glassy lignin surface and dark color of the fibers preclude their use in paper, although they make a very useful, high density fiberboard product. In fiberboard the fibers are held together by resins such as phenol-formaldehyde, not by hydrogen bonding as in paper. The yield is 80-90%. However, a mild explosion pulping process is being used at one mill to process recycled paper.

Fig. 3-2 shows TMP fibers from mixed hardwoods of the southern U.S. Notice the fibers exist in fiber bundles. The color (not observable in the SEM photograph) is dark brown. The fibers have glassy surfaces high in lignin that prevent hydrogen bonding. This pulp is unsuitable for use in paper.

3.5 SEMI-CHEMICAL PULPING

Vapor-phase pulping

Vapor-phase pulping is a process where chips are impregnated with cooking liquor (Fig. 3-9), the free liquor is drained or otherwise removed, and cooking occurs in an atmosphere of steam.

Semi-chemical process

Also called high yield chemical pulping, semi-chemical pulping processes involve two steps with pulp yields of 60-80%. In the first step a mild chemical treatment is used which is followed by moderate mechanical refining. There is partial removal of both lignin and hemicellulose. The first step of the semi-chemical process may be similar to any of the commercial chemical pulping methods (to be described), except that the temperature, cooking time, or chemical charge is reduced. The NSSC and the kraft semi-chemical methods are the most common of this category.

For yields below 65%, the pulp is washed and chemical recovery continues as outlined in the sulfite or kraft section, except the pulp can be cleaned by dewatering presses to limit dilution of the liquor. Up to 75% of the liquor can be removed without dilution. Because the liquor is not easy to recover in small operations, the majority of semi-chemical mills are integrated with chemical mills so that the dilute spent liquors may be used as a source of makeup chemicals in the associated chemical process. This *cross recovery*, however, limits the size of the semi-chemical plant, because a large plant would produce excess chemical beyond that required for make-up in the kraft mill.

NSSC, Neutral Sulfite Semi-Chemical

The NSSC process is most frequently used (and is the most common method) for production of corrugating medium. This process was developed in the early 1940s by the U.S. Forest Products Laboratory in Madison as a means of using hardwood in the paper industry, especially the chestnut of the appalachian region, which was being devastated by the chestnut blight. In this process, high pulp yields are obtained (75-85%).

NSSC cooking liquors contain Na_2SO_3 plus Na_2CO_3 (10-15% of the chemical charge to act as a buffer); the liquor pH is 7-10. Cooking time is 0.5-2 h at 160-185°C (320-365°F). The residual lignin (15-20%) makes paper from this pulp very stiff, an important property for corrugating medium. Hardwood is usually the fiber source, and NSSC hardwood pulp is approximately as strong as NSSC softwood pulp and even stronger than kraft hardwood pulp. The low lignin removal makes chemical recovery difficult. Anthraquinone additives may be used to improve the pulp properties or yield, particularly from softwoods. The degree of cooking is controlled by the temperature, the chemical concentration, and the residence time in the digester. Subsequent refining energy of the pulp is 200-400 kWh/ton (10-20 hp-day/ton). Fig. 3-2 includes an electron micrograph of paper made with 50% hardwood NSSC pulp and 50% fiber recycled from old corrugated containers.

Kraft green liquor semi-chemical process

The green liquor semi-chemical pulping process for corrugating medium uses green liquor as the pulping liquor. The green liquor can be obtained from an associated kraft mill or from a recovery boiler specifically for the green liquor mill. In either case, the lime cycle is not required for liquor regeneration.

De-fiberator or hot stock refiner

The de-fiberator, also called the hot stock refiner, is an attrition mill designed to break apart the hard semi-chemical chips into pulp.

Cross recovery

Cross recovery is the use of the waste liquor of a semi-chemical mill as the make-up chemical in the kraft recovery plant. For example, the Na_2SO_3 in the waste liquor of an NSSC mill, when added to the kraft recovery boiler, is converted to Na_2S during burning of the liquor and furnishes the kraft mill with its make-up sulfur and sodium. Fresh make-up chemicals are used to generate the pulping liquor for the NSSC mill. Obviously, the size of the NSSC mill is limited by the make-up chemical requirements of the kraft mill. In a green liquor semi-chemical mill, fresh chemical is taken from the kraft green liquor storage while the spent pulping liquor is simply returned to the kraft recovery cycle.

3.6 GENERAL CHEMICAL PULPING

Delignification

Delignification is the process of breaking down the chemical structure of lignin and rendering it soluble in a liquid; the liquid is water, except for organosolv pulping.

Kappa number, permanganate number

The kappa number is a measure of the lignin content of pulp; higher kappa numbers indicate higher lignin content. A similar test is the permanganate number (or K number). The kappa number is used to monitor the amount of delignification of chemical pulps after pulping and between bleaching stages. For more detail on pulp tests, how they are carried out, and their significance, see Chapter 6.

Pulp viscosity

The pulp viscosity (see Section 6.3) is a measure of the average chain length (*degree of polymerization, DP*) of cellulose. It is determined after dissolving the pulp in a suitable solvent such as cupriethylenediamine solution. Higher viscosity indicates a higher average cellulose DP that, in turn, usually indicates stronger pulp and paper. Decreases in viscosity result from chemical pulping and bleaching operations and to a certain extent are unavoidable, but must be minimized by proper attention to important process parameters. Cellulose viscosity has little importance in mechanical pulping since the cellulose chains are not appreciably degraded by this operation.

Fiber liberation point

The fiber liberation point occurs when sufficient lignin has been removed during pulping that wood chips will be soft enough to break apart into fibers with little or no mechanical action.

Full chemical pulps, unbleached, bleached

A pulp produced by chemical methods only is known as a full chemical pulp. Most chips are at the fiber liberation point after cooking at 130-180°C (265-355°F) with appropriate pulping liquors. The total pulp yield is about 50 percent, and the pulp contains about 3-5% lignin. These pulps have high strength and high cost. The common methods are the kraft process (also called alkaline or sulfate process with a cooking pH above 12) and the various sulfite processes with a wide range of pH cooking conditions. An older method, enjoying only limited use these days on hardwoods and nonwood fiber such as straw, is the soda process that uses sodium hydroxide as the only active pulping ingredient. *Unbleached pulp* is a full chemical pulp as it comes from the pulping process. It is light to dark brown in color. *Bleached pulp* is a white pulp produced by bleaching full chemical pulps.

Dissolving pulp

Dissolving pulp is a low yield (30-35%) bleached chemical pulp that has a high cellulose content (95% or higher) suitable for use in cellulose derivatives such as rayon, cellulose acetate, and cellophane (see the end of Section 2.6 for a description of these materials.) Dissolving pulp is

Fig. 3-10. The top of a batch digester where digester loading occurs. In the background several additional digester tops are visible.

manufactured by the kraft process using an acid prehydrolysis step to remove hemicelluloses or by an acid sulfite process. Improved cellulose purity is achieved by a cold alkali extraction of the pulp.

Digester

The digester is a pressure vessel used for cooking chips into pulp. Digesters are designed to operate in either a continuous mode, if in a long, narrow tube-shape, or in a batch mode.

Batch digester

Batch digesters are large digesters, typically 70-350 m³ (2,500 to 12,500 ft³), that are filled with wood chips and cooking liquor. The top of a batch digester is shown in Fig. 3-10. A diagram of a batch digester is shown in Fig. 3-11. A laboratory batch digester is shown in Fig. 3-12.

Typically a mill has a bank of six to eight digesters so that while several are cooking, one is filling, one is blowing, one might be under repair, etc. Heating with steam may be *direct*, where steam is added directly to the digester which diles the cooking liquor, or *indirect*, where steam is passed through the inside of tubes within the digester which allows reuse of expensive steam condensate and gives more uniform heating. The *cooking time* is the length of time from initial steaming of chips to the start of digester blowing; the *time to temperature* is the length of time from initial steaming to the point where the desired cooking temperature is reached; and the *time at temperature* is the length of time from when the cooking temperature is reached until the digester blow starts.

A typical sequence of events for an entire cook is as follows: 1) The digester is first opened and filled with chips, white liquor, and black liquor. 2) After initial circulation of the liquor additional chips are added as the contents settle. 3) The digester is then sealed and heating with steam begins. The temperature rises for about 90 minutes until the cooking temperature is achieved. 4) The cooking temperature is maintained for about 20-45 minutes for the kraft pro-

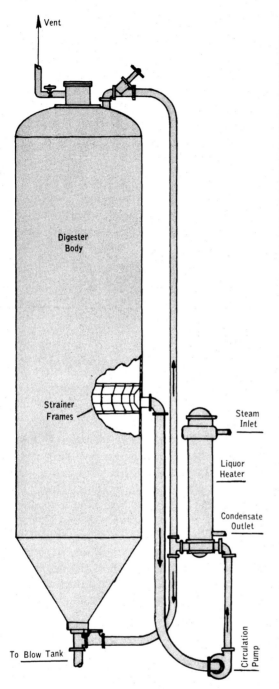

Fig. 3-11. A stationary batch digester with indirect heating of the liquor. Reprinted from *Making Pulp and Paper*, ©1967 **Crown Zellerbach Corp., with permission.**

Fig. 3-12. A laboratory batch digester with product (below).

cess. During the heating time, air and other noncondensable gases from the digester are vented. 5) When the cook is completed, as determined by the kappa of pulp from the digester, the contents of the digester are discharged to the blow tank. 6) The digester is opened and the sequence is repeated.

Continuous digester

A continuous digester is a tube-shaped digester where chips are moved through a course that may contain elements of presteaming, liquor impregnation, heating, cooking, and washing. Chips enter and exit the digester continuously. Continuous digesters tend to be more space efficient, easier to control giving increased yields and reduced chemical demand, labor-saving, and more energy efficient than batch digesters.

Since continuous digesters are always pressurized, special feeders must be used to allow chips at atmospheric pressure to enter the pressurized digester without allowing the contents of the digester to be lost. *Screw feeders* are used for materials like sawdust and straw. A moving plug of fiber is used to make the seal. *Rotary valves* (Fig. 3-16) work like revolving doors. A pocket is filled with wood chips or other fiber source at atmospheric pressure. When the valve is rotated it is sealed from the atmosphere and then opened into the digester where the contents are deposited.

Kamyr digesters (Figs. 3-13 to 3-15) are large, vertical digesters, where chips enter the top and exit the bottom continuously in a plug flow. In 1992, 390 Kamyr digesters were operational worldwide with a rated capacity of 77 million tons per year. The chips are exposed to various areas

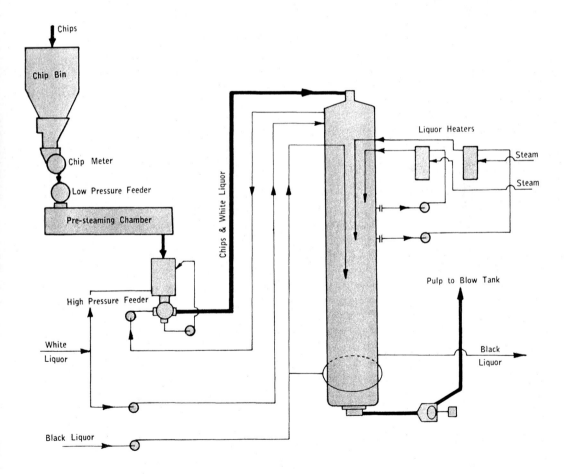

Fig. 3-13. The Kamyr continuous digester. Reprinted from *Making Pulp and Paper*, ©1967 Crown Zellerbach Corp., with permission.

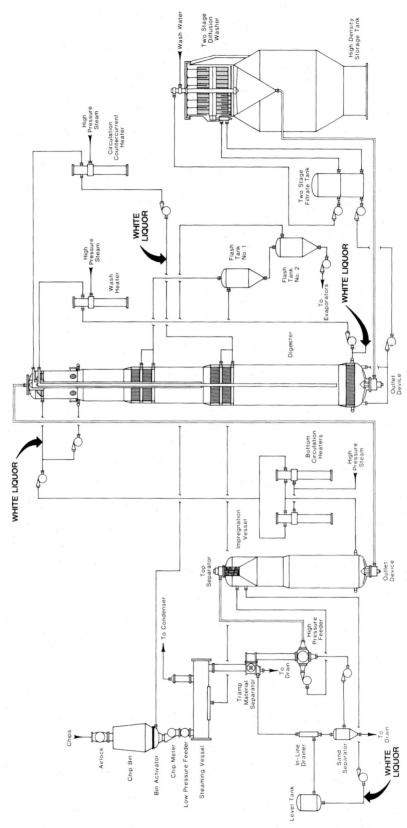

Fig. 3-14. Two vessel hydraulic digester with extended MCC™ and two stage diffuser washer. Courtesy of Kamyr, Inc.

Fig. 3-15. A two-vessel Kamyr digester.

of the digester where they are impregnated with liquor, heated to the cooking temperature, held at the cooking temperature to effect pulping, and partially washed. The pulp is then sent to the blow tank.

In the typical process chips from the bin are introduced to a low pressure pre-steaming chamber where the chips are transported by a screw feed. Here flash steam at atmospheric pressure preheats the chips and drives off air so that liquor penetration will be enhanced. The chips pass through the high pressure feeder where cooking liquor carries them to the digester. Most of the liquor is returned to the high pressure feeder. Impregnation occurs for about 45 minutes at 130°C so cooking

will be much more uniform. Liquor heating occurs in the zone formed by two external liquor heaters. The cooking zone merely supplies the appropriate retention time for cooking to take place. The pulping reaction is terminated with the addition of cold, wash liquor from the brown stock washers that displaces the hot, cooking liquor.

The *M & D* digester (Messing and Durkee) is a long digester inclined at a 45° angle. A diagram of this is shown in Fig. 3-16, and an actual digester is shown in Fig. 3-17. This digester is often used for kraft pulping of sawdust or semi-chemical pulping methods. It has some use for kraft pulping of wood chips. The chips or sawdust enter the top, go down one side of the digester, return back up the other side being propelled by a conveyer belt, and exit out the top. The cooking time is about 30 minutes. A midfeather plate separates the two sides of the digester. Since the size of these shop-fabricated digesters are limited to about 2.4 m (8 ft) diameter, they are used for relatively small production levels.

Another continuous digester is the *Pandia* digester (Figs. 3-18 and 3-19), which is horizontal, uses a screw feed, and is often used in kraft pulping of sawdust, semi-chemical pulping of chips with short cooking times, and pulping of nonwood fiber such as straw.

Digester charge, relief, and blowing

The *digester charge* includes the wood chips and cooking liquor in a digester. *Digester relief* occurs while the chips are heated to temperature and during the cooking process to relieve the pressure caused by the formation of volatile gases. These gases are released and condensed to improve the pulping process. This removes air from within and around the chips; increasing the liquor circulation and evenness of cooking; reduces the digester pressure, which reduces fiber damage during the blow; and allows collection of turpentine, if desired.

Digester blowing occurs at the end of a cook when the contents of a digester are cooled to about 100°C (212°F) and allowed to escape to atmospheric pressure. There is usually sufficient force in a full chemical pulp to cause fiber separation. If the contents are blown from temperatures near that of cooking, 170°C (340°F) for kraft cooking,

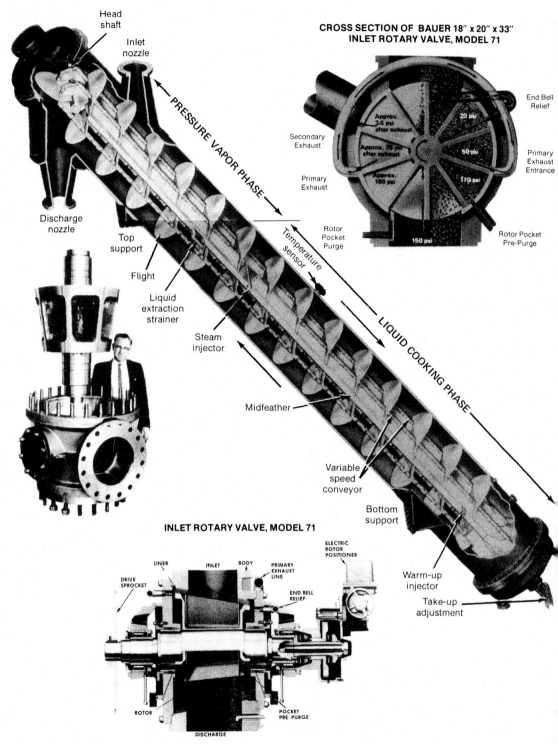

Fig. 3-16. The M&D continuous digester. The inserts show various aspects of the rotary valve. Courtesy of Andritz Sprout-Bauer, Inc.

Fig. 3-17. The M & D continuous digester. To the right is the blow tank. The top insert shows a hydraulically driven exit valve and the bottom insert shows an electric-motor driven exit valve.

the pulp often loses a significant (10-15%) amount of its strength.

Blow tank

Blow tanks are large, cylindrical vessels that receive the hot pulp from the digesters (Fig. 3-20). Agitators mix the pulp from the digester with dilute black liquor so the pulp slurry can be pumped and metered for the correct consistency. The heat of the hot gases from the blow tank are recovered by the *blow heat accumulator*, a large heat exchanger. Up to 1000 kg (2000 lb) of steam per ton of pulp is generated by batch digesters and some continuous digesters and must be condensed. Continuous digesters like the Kamyr are able to recover the heat more efficiently by liquor displacement methods, although this technology is being applied to batch digesters in the rapid displacement heating method.

Condensation of the blow gases also decreases pollution by recovering most of the volatile reduced sulfur compounds, organic compounds such

Fig. 3-18. A Pandia continuous digester with two tubes. The insert shows the feed valve.

as methanol, and related materials known together as *foul condensate*. The condensate is sometimes used for dilution water in applications such as lime mud dilution or brown stock washing where the foul compounds will be trapped prior to a combustion process. Non condensable gases are often

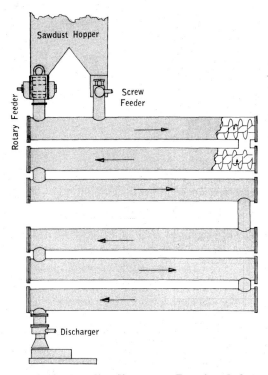

Fig. 3-19. Pandia digester. Reprinted from *Making Pulp and Paper*, ©1967 Crown Zellerbach Corp., with permission.

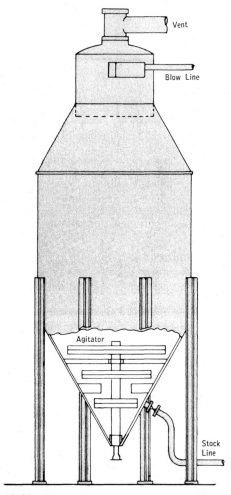

Fig. 3-20. Blow tank. Reprinted from *Making Pulp and Paper*, ©1967 Crown Zellerbach Corp., with permission.

diverted to the lime kiln for combustion. The pulp is then sent to be screened and cleaned in the brown stock washers.

Liquor

Cooking or pulping liquors are aqueous solutions of chemicals used for delignification of wood by pulping.

Chemical concentration

The chemical concentration is measure of the concentration of the pulping chemical in the liquor. For example, in sulfite pulping the liquor may be 6% SO_2, indicating 6 grams of sulfite chemical (SO_2 basis) per 100 ml of liquor. In this case if the liquor:wood ratio is 4:1, the percent chemical on wood is 24% as SO_2. The following is an important relationship, not the definition of chemical concentration.

chemical concentration in liquor =

$$\frac{\text{percent chemical on wood}}{\text{liquor to wood ratio}}$$

Chemical charge (to a process), percent chemical on wood (or pulp for bleaching)

The chemical charge or chemical on wood is another fundamental parameter of chemical pulping processes. It is the measure of the weight of chemical used to process a material relative to the weight of the material itself. This applies to any chemical process, such as pulping and bleaching. For example, typically, kraft pulping is carried out with 25% total alkali on wood. This would indicate 500 lb of alkali as sodium oxide (in the U.S. sodium-based chemicals are reported in terms of sodium oxide) for 2000 lb of dry wood. Chemicals in sulfite pulping are expressed on an SO_2 basis. Also, when bleaching mechanical pulp, one might use "0.5% sodium peroxide on pulp". This means that for every one ton of oven-dry pulp, 10 lb of sodium peroxide are used.

$$\text{chemical charge} = \frac{\text{mass dry chemical used}}{\text{mass dry material treated}} \times 100\%$$

Liquor to wood ratio

The liquor to wood ratio, liquor:wood, is normally expressed as a ratio; typically, it has a

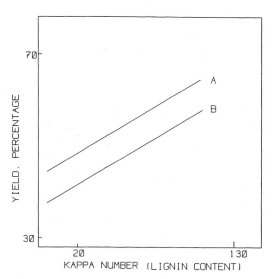

Fig. 3-21. Pulping selectivity curve.

value of 3:1 to 4:1 in full chemical pulping. Rarely it is expressed as a percent. The numerator may or may not include the weight of water coming in with the chips, but either way it must be specified to avoid ambiguity. The liquor to wood ratio is kept as small as possible while maintaining good digester operation, including good liquor circulation for even cooking.

$$\frac{\text{liquor}}{\text{wood}} = \frac{\text{total pulping liquor mass}}{\text{dry wood mass}}$$

Delignification selectivity

Delignification selectivity is an important concept during pulping and bleaching operations where it is desired to remove lignin while retaining as much holocellulose as possible. Delignification selectivity is the ratio of lignin removal to carbohydrate removal during the delignification process. While this ratio is seldom measured directly, it is measured in a relative manner by yield versus kappa plots as shown in Fig. 3-21. Here condition "A" is more selective than condition "B" since for a given kappa number the yield is higher for "A" than for "B".

A high selectivity alone does not mean that pulp "A" is better than "B" since the plot does indicate the condition, such as the viscosity, of the pulp. For example, acid sulfite pulping is more selective than kraft pulping; however, acid sulfite

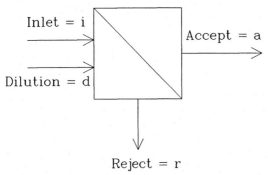

Fig. 3-22. Screening terminology in TIS 0605-04, Screening symbols, terminology, and equations.

pulp is weaker than kraft pulp because the cellulose has a lower degree of polymerization due to acid hydrolysis.

Hot stock refining, defiberating

Chemical pulps and, especially semi-chemical pulps, must be refined after the cooking process to liberate the individual fibers. For chemical pulps, this is relatively easily accomplished in the presence of hot liquor, which is the basis of hot stock refining. Defiberating only separates the fibers for thorough pulp washing; the fibers must be further refined for papermaking.

Rejects, knotter or screener and pulp screener

Rejects are portions of wood, such as knots, that do not sufficiently delignified to fiberate. The hard, poorly cooked lumps of wood are separated

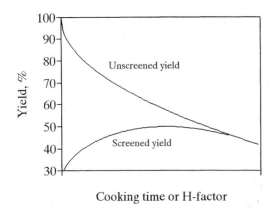

Cooking time or H-factor

Fig. 3-23. Hypothetical relationship of total and screened yields to cooking time or H-factor.

from pulp by coarse screening equipment (3/8 in. holes) prior to pulp washing. Coarse sreenings from sulfite pulping are not reused in bleached grades of pulp. Coarse screening from kraft pulping are repulped. Suggested screen terminology is shown in Fig. 3-22. If pulps are undercooked, the screened yield may actually decrease because much of the (potential) pulp is removed in the screens, as shown in Fig. 3-23.

After washing, pulps are screened to remove shives, dirt, and other contaminants to protect processing equipment and the product's integrity. Shorter fibers screen more easily, thus hardwoods are easier to screen than softwoods. Redwood pulps (long fibers) require about twice the screen capacity as spruce pulps. Similarly, mechanical pulps screen more easily than chemical pulps. Recently, pulp is screened prior to washing for improved washing efficiency. See Section 8.2 for information on pressure screens.

Brown stock washers

Chemical pulps require more washing than mechanical pulps in order to recover process chemicals, but mechanical pulps require more screening to remove shives, knots, and so forth. If chemical pulps are not properly washed, foaming will be a problem, additional make-up chemicals will be needed, more bleaching chemicals will be required, and additional pollutants will result. There are several types of brown stock washers available: 1. *Diffuser washing systems*, which are slow, lead to high liquor dilution, and are therefore being phased out. 2. *Press type washers*. 3. *Countercurrent flow stock washers*. 4. *Rotary vacuum drum washer* with a drop length of pipe to supply a vacuum with two to four stages (Plate 12). 5. *Double wire press washers* (Plates 14-16). 6. *Diffusion washers*.

Rotary vacuum washers

Rotary vacuum drum washers are, by far, the most common type used. These consist of a wire-mesh covered cylinder that rotates in a tub of the pulp slurry with valve arrangements to apply vacuum as is suitable. As the drum contacts the slurry, a vacuum is applied to thicken the stock. The drum rotates past wash showers where the pulp layer is washed with relatively clean water to (more or less) displace black liquor. The vacuum is cut off beyond the wash showers and the pulp mat is dislodged into a pulper. Most of the washing actually occurs at the pulpers between washing stages. Daily loadings for brown stock washers are about 8 t/m^2 (0.8 dry tons/ft^2) for softwood, 6 t/m^2 (0.6 dry tons/ft^2) for hardwood, and 1-2 t/m^2 (0.1-0.2 dry tons/ft^2) for straw pulps. Hardwoods are more difficult to wash than softwoods and have higher chemical losses, partly due to the higher ion-exchange capacity of hardwoods from the carboxylic groups of their hemielluloses.

Typically, about 80% of the contaminants are removed at each stage, so at least three stages are required to thoroughly wash the pulp. Use of additional stages results in lower water consumption, higher recovery of chemicals, and lower pulp dilution. Fig. 3-24 shows a typical configuration of vacuum drum washers for pulp washing. Fig. 3-25 shows an individual vacuum drum washer.

Air is constantly sucked into the pulp sheet and must be separated from the spent cooking liquor to prevent foaming as it is further processed. This causes much foam to form in the seal tanks. Chemical and mechanical defoamers are often used here. The method of applying the wash water influences foam formation.

Diffusion washing

Diffusion washing (Fig. 3-26) was first used by Kamyr for washing chips at the bottom of their digesters. The technique was extended for brown stock washing and washing of pulp between bleaching stages.

Double-wire press

Plate 14 shows a double-wire press suitable for market pulp production, brown stock washing, pulp washing between bleaching stages, or stock thickening. Plate 15 shows a double-wire press designed for brown stock washing. Notice this configuration incorporates the counter-current washing technique.

Brown stock

The washed, screened pulp is called *brown stock*. See Section 4.2 for more information on pulp washing from a chemical recovery point of view where dilution is considered.

Anthraquinone (AQ), soluble anthraquinone (SAQ)

Anthraquinone (whose structure is shown in Fig. 3-27) is a pulping additive used in kraft, soda or alkaline sulfite processes to increase delignification and decrease carbohydrate degradation. It works by going through a cycle which leads to the reduction of lignin and the oxidation of the reducing endgroup of cellulose from an aldehyde to a carboxylic acid as shown in Fig. 3-28. In the latter case the carbohydrates are stabilized against the alkaline peeling reaction, leading to an increase in pulp yield. Because anthraquinone goes through a cyclic process, it is typically used at about 0.1% on wood and results in a 1-3% increase in pulp yield. The use of anthraquinone was first reported in 1977. Modifications of anthraquinone, such as soluble anthraquinone (SAQ, Fig. 3-27), properly called 1,4-dihydro-9,10--dihydroxyanthracene (DDA), have been

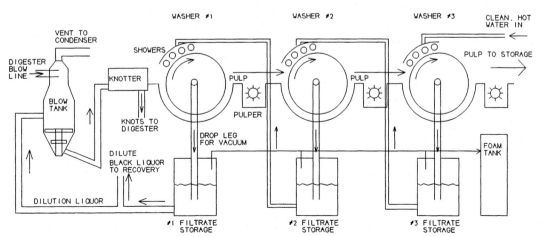

Fig. 3-24. Brown stock washers showing countercurrent flow.

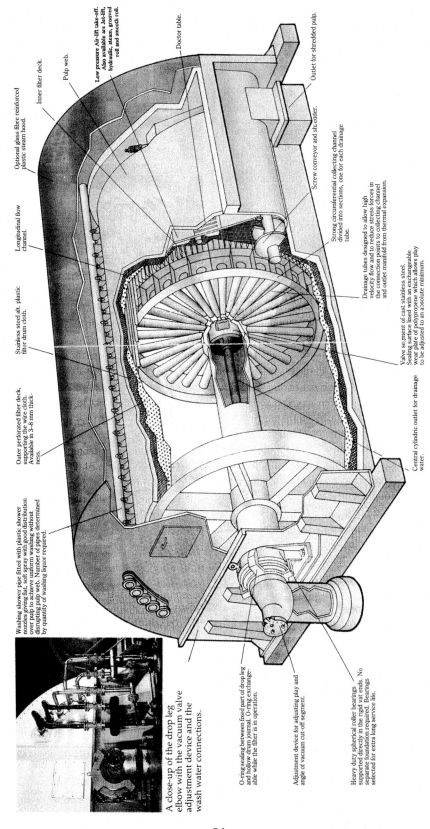

Optional glass fibre reinforced plastic steam hood.

Low pressure Air-lift take-off. Also available are Jet-lift, hydraulic, steam, grooved roll and smooth roll.

Pulp web.

Inner filter deck.

Longitudinal flow channel.

Doctor table.

Outlet for shredded pulp.

Screw conveyor and shredder.

Strong circumferential collecting channel divided into sections, one for each drainage tube.

Stainless steel alt. plastic filter drum cloth.

Drainage tubes designed to allow high velocity flow and to reduce stress forces in the connection points to collecting channel and outlet manifold from thermal expansion.

Valve segment of cast stainless steel. Sealing surface lined with an exchangeable wear plate of polypropene which allows play to be adjusted to an absolute minimum.

Outer perforated filter deck, supporting the wire cloth. Available in 3–8 mm thickness.

Central cylindric outlet for drainage water.

Washing shower pipe fitted with plastic shower nozzles giving flat, soft spray with good distribution over pulp to achieve uniform washing without disrupting pulp web. Number of pipes determined by quantity of washing liquor required.

O-ring sealing between fixed part of drop leg and hollow drum journal. O-ring exchangeable while the filter is in operation.

Adjustment device for adjusting play and angle of vacuum cut-off segment.

Heavy duty spherical roller bearings supported directly in the rigid vat ends. No separate foundation required. Bearings selected for extra long service life.

A close-up of the drop leg elbow with the vacuum valve adjustment device and the wash water connections.

Fig. 3-25. Vacuum drum washer suitable for brown stock washing, bleaching, or pulp thickening. Courtesy of Sunds Defibrator.

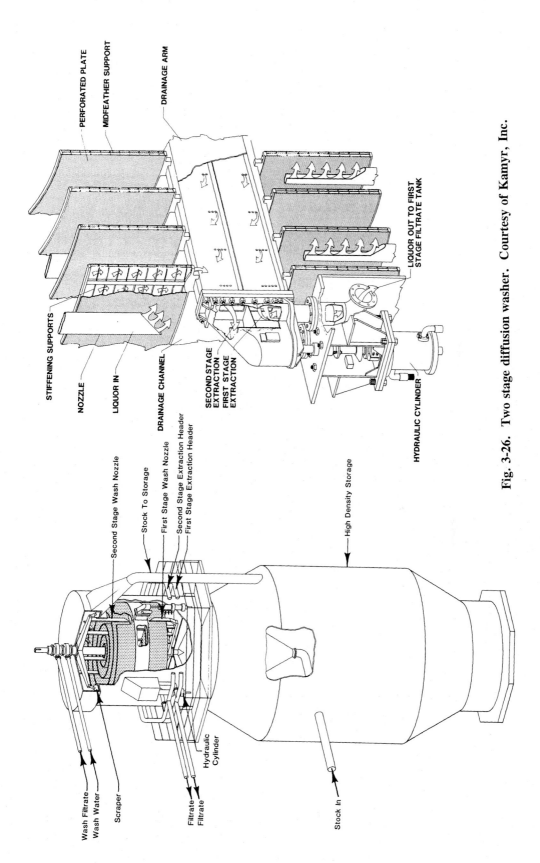

PERFORATED PLATE

MIDFEATHER SUPPORT

DRAINAGE ARM

STIFFENING SUPPORTS

NOZZLE

LIQUOR IN

DRAINAGE CHANNEL

SECOND STAGE EXTRACTION

FIRST STAGE EXTRACTION

LIQUOR OUT TO FIRST STAGE FILTRATE TANK

HYDRAULIC CYLINDER

Second Stage Wash Nozzle

Stock To Storage

First Stage Wash Nozzle

Second Stage Extraction Header

First Stage Extraction Header

High Density Storage

Wash Filtrate

Wash Water

Scraper

Filtrate

Filtrate

Hydraulic Cylinder

Stock In

Fig. 3-26. Two stage diffusion washer. Courtesy of Kamyr, Inc.

85

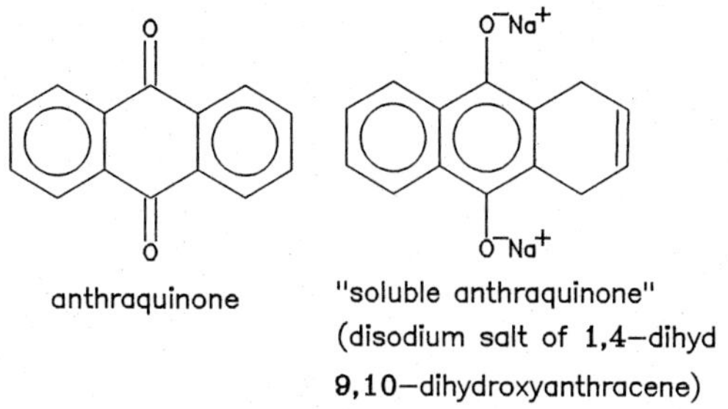

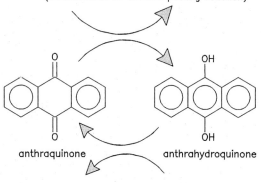

anthraquinone   "soluble anthraquinone"
(disodium salt of 1,4–dihyd
9,10–dihydroxyanthracene)

Fig. 3-27. Structures of AQ and SAQ.

oxidation of reducing end of cellulose
(minimization of alkaline peeling reaction)

anthraquinone   anthrahydroquinone

lignin reduction and cleavage

Fig. 3-28. Cyclic action of anthraquinone.

shown to be even more effective. While not used extensively in North America, anthraquinone and SAQ are widely used in Japan and other countries where fiber supplies are quite limited and expensive.

Prehydrolysis

Prehydrolysis is a dilute acid hydrolysis with 0.5% sulfuric acid for 30 min at 100°C and is used to partially depolymerize the hemicelluloses of wood with little effect on the cellulose. Chips pretreated in this fashion followed by kraft pulping give a pulp with very high cellulose content, but low yield, suitable for dissolving pulp.

3.7 SODA PULPING

Soda pulping, invented in England by Burgess and Watts in 1851, uses sodium hydroxide as the cooking chemical. Finding little enthusiasm in England for this new process, Burgess brought the

method to the U.S. in 1854 and the first mill was started in 1866. Many of the early soda mills converted to the kraft process once it was discovered. The soda process still has limited use for easily pulped materials like straws and some hardwoods, but is not a major process. Anthraquinone may be used as a pulping additive to decrease carbohydrate degradation.

A recent development is the use of oxygen in soda pulping. While oxygen bleaching is not very specific to delignification compared to other bleaching methods, it is fairly specific to delignification relative to other pulping methods.

3.8 KRAFT PULPING

Kraft pulping

In 1879, Dahl, a German chemist, used sodium sulfate as a makeup chemical for soda pulping to regenerate NaOH; actually Na_2S was formed and, unexpectedly, gave much faster delignification and stronger pulps, since shorter cooking times are used resulting in less carbohydrate degradation. This led to the kraft (or sulfate) process, which is now the dominant process. Although related work on the process had been done earlier, Dahl discovered the kraft chemical recovery process, which is perhaps more important than the kraft cooking process. The first kraft mill went into operation in 1890 in Sweden because the German papermaking industry did not accept this new process. The process developed and grew quickly from 1915 to 1930, especially in the southern U.S. where the resinous pine species did not pulp well by the sulfite process with calcium base.

Kraft pulping is a full chemical pulping method using sodium hydroxide and sodium sulfide at pH above 12, at 160-180°C (320-356°F), corresponding to about 800 kPa (120 psi) steam pressure, for 0.5-3 hours to dissolve much of the lignin of wood fibers. It is useful for any wood species, gives a high strength pulp (kraft means strong in German and Swedish), is tolerant to bark, and has an efficient energy and chemical recovery cycle. The disadvantages are the difficulty with which the pulps are bleached compared to sulfite pulps, low yields due to carbohydrate losses, and sulfur in its reduced form provides emissions that are extremely odiferous (they can be detected by the olfactory sense at 10 parts per billion). Another consideration is that a green field (starting in an empty field) mill of 1000 tons per day production including bleaching and paper machines costs about $1 billion. About 75-80% of U.S. virgin pulp is produced by this process. Important variables during kraft cooking are:

1. Wood species (though all species can be pulped) and chip geometry.
2. Ratio of effective alkali to wood weight.
3. Concentration of effective alkali and liquor:wood.
4. Sulfidity.
5. H-factor. (Which, in turn, is a function of cooking time and temperature.)

For example, all other factors being equal in kraft pulping, a higher concentration of cooking chemical shifts the cooking selectivity (Fig. 3-21) from "A" towards "B" (unfavorable), but decreasing the liquor to wood ratio shifts the curve from "B" to "A" (useful, but more evaporation costs).

The importance of uniform wood chips in regards to species, chip thickness, and chip geometry cannot be over stressed. Even with careful screening of wood chips, variability in wood means that some of the pulp will be over-cooked and some of it will be undercooked. Fig. 3-29 and Plate 13 of the same material demonstrates this point. Electron microscopy shows that overthick chips have incomplete fiber liberation (kappa number of 94.7). Fine chips overcook (kappa number of 36.9), leading to lignin condensation and dark, rigid fiber clusters. The pulp from the accept chips has a kappa number of 42.4.

The penetration of liquors at pH above 13 in wood is about the same in all three directions (Fig. 3-30). This is why chips for kraft cooking are separated on the basis of their thickness.

The degree of a kraft cook is often indicated as *soft*, *medium*, or *hard* based on the pulp texture. Soft cooks, for bleachable grades of pulp, have a lignin content of 3.0-5.2% (20-35 kappa number) for softwoods or 1.8-2.4% (12-18 kappa) for hardwoods. Medium softwood cooks for bag and saturating papers have a lignin content of 5.2-7.5% (35-50 kappa). Hard softwood cooks are 9-11% lignin (60-75 kappa) for top linerboards and 12-16.5% lignin (80-110 kappa) for bottom liners.

Fig. 3-29. Over-thick (top), accept (middle), and fine chips (bottom) cooked under identical conditions by kraft pulping.

H-factor

The H-factor is a pulping variable that combines cooking temperature and time into a single variable that indicates the extent of reaction. Although different temperatures might be used, the degree of cook can be accurately estimated by this method provided that other variables such as active alkali, sulfidity, and the liquor to wood ratio remain constant. There is no analogous variable for sulfite or soda pulping.

The rate of delignification approximately doubles for an increase in reaction temperature of 8°C. To obtain a bleachable kraft softwood pulp of about 5% lignin, one typically cooks for about 1.5 hours at 170°C. This corresponds to 0.75 hour at 178°C or 3 hours at 162°C. The integral of the reaction rate with respect to time combines these two parameters into the single parameter, the H factor. By using the H factor, cooks of varying reaction times and temperatures can be compared in a meaningful manner. This important process control variable is discussed in Section 16.6.

White liquor

White liquor is fresh pulping liquor for the kraft process, consisting of the active pulping species NaOH and Na_2S, small amounts of Na_2CO_3 left over from the recovery process, and oxidized sulfur compounds and other chemical impurities from wood, salt water, corrosion, etc.

Black liquor

Black liquor is the waste liquor from the kraft pulping process after pulping is completed. It contains most of the original cooking inorganic elements and the degraded, dissolved wood substance. The latter include acetic acid, formic acid, saccharinic acids, numerous other carboxylic acids (all as the sodium salts), dissolved hemicelluloses (especially xylans), methanol, and hundreds of other components. It is an extremely complex mixture.

About 7 tons of black liquor at 15% solids [about 10% organic chemicals and 5% inorganic chemicals with a total heat content of 13.5-14.5 MJ/kg solid (5500-6500 Btu/lb solid)] are produced per ton of pulp. The black liquor must be concentrated to as high a solids content as possible before being burned in the recovery boiler to maximize the heat recovery. The viscosity rises rapidly with concentration above 50%, with softwood black liquors being more viscous than hardwood black liquors. Black liquor is usually fired at 65-70% solids content. More information on black liquor is in Section 4.3.

Green liquor, chemical recovery

Green liquor is the partially recovered form of kraft liquor. It is obtained after combustion of the black liquor in the recovery boiler. Green liquor is produced by dissolving the smelt from the recovery boiler (Na_2S, Na_2CO_3, and any impurities) in water. Further processing of the green liquor converts it to white liquor. The kraft chemical recovery cycle is fairly complex and detailed. It is considered in detail in Chapter 4.

Pulping chemicals

The active chemicals in the kraft liquor are Na_2S and NaOH. Other chemicals in kraft liquors are important because they are involved in the recovery process; these include Na_2CO_3 and Na_2SO_4. Still other chemicals such as NaCl are important as contaminants that may build up in the system. The amounts of chemicals are reported on an Na_2O basis in North America. Various combinations of these chemicals are given special names because these combinations can predict how the liquor will behave better than considering the amount of chemicals themselves. Pulping chemicals are reported as concentrations in liquor or as a charge on dry wood.

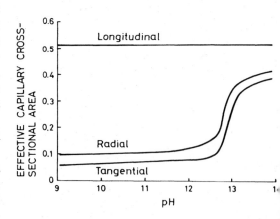

Fig. 3-30. The effect of pH on wood permeability. ©1957 TAPPI. Reprinted from Stone, J.E., *Tappi J.* 40(7):539-541(1957) with permission.

The chemical analysis, method of reporting the cooking chemicals, and calculations of the liquor properties described in the paragraphs below is reported in detail in Chapter 16.

Total chemical or total alkali (TA)

Total alkali is the sum of all sodium salts in the liquors (as Na_2O) that contribute to AA (i.e., NaOH or Na_2S) or are capable of being converted to AA in the kraft cycle. Specifically NaOH, Na_2S, Na_2CO_3, and $Na_2S_xO_y$ (as Na_2O) are included. All chemical amounts may be reported as concentrations of g/L or lb/gal or as a percent relative to oven-dry wood.

$$TA = NaOH + Na_2S + Na_2CO_3 + Na_2S_xO_y$$

Total titratable alkali (TTA)

TTA is the sum of all of the bases in the white liquor that can be titrated with strong acid. Generally this includes NaOH, Na_2S, and Na_2CO_3 (as Na_2O), although small amounts of Na_2SO_3 and other acids might be present. A typical value for TTA is 120 g/L (7.5 lb/ft^3).

$$TTA = NaOH + Na_2S + Na_2CO_3 \quad (as\ Na_2O)$$

Active Alkali (AA)

AA includes the active ingredients in the pulping process, i.e., NaOH + Na_2S (both as Na_2O). A typical value for AA is 100 g/L (6.4 lb/ft^3). Research clearly demonstrates that the active alkali should be kept constant during cooking by adding alkali during the cooking process as it is consumed, but this is difficult to carry out commercially and not practiced.

$$AA = NaOH + Na_2S \quad (as\ Na_2O)$$

Effective alkali (EA)

EA is the ingredients that will actually produce alkali under pulping conditions. The effective alkali is about 12-18% on wood for the production of unbleached kraft and 18-24% on wood for the production of bleached grades, with hardwoods using the lower amounts due to their lower lignin contents. Above 55 g/L EA, cellulose decomposition relative to lignin removal increases dramatically; consequently a liquor:wood ratio above 3:1 should be used.

$$EA = NaOH + ½ Na_2S \quad (as\ Na_2O).$$

Sulfidity

Sulfidity (of white liquor) is the ratio of Na_2S to the active alkali, expressed as percent. Typically a mill runs in the vicinity of 24-28% sulfidity, depending largely on the wood species pulped. Sulfidity increases the rate of delignification which occurs by nucleophilic action of the hydrosulfide anion (HS$^-$). The net effect is cleavage of β-aryl-ether linkages of lignin as well as the methoxy groups, the latter leading to the production of mercaptans. Sulfidity may protect against carbohydrate degradation directly, but the decreased cooking times also result in less carbohydrate degradation. All chemicals are as Na_2O.

$$sulfidity = \frac{Na_2S}{NaOH + Na_2S} \times 100\%$$

If the sulfidity is too low, the lignin content of the pulp may be relatively high and carbohydrate degradation may be severe, leading to low pulp strength. If the sulfidity is too high, emissions of reduced sulfur compounds may increase and corrosion rates in the recovery process may be high.

Causticity

Causticity is the ratio of NaOH to active alkali (as always both chemicals are on an Na_2O basis), expressed as a percentage. Obviously,

$$causticity + sulfidity = 100\%$$

the term sulfidity is used much more commonly than causticity. All chemicals are as Na_2O.

$$causticity = \frac{NaOH}{NaOH + Na_2S} \times 100\%$$

Causticizing efficiency

The causticizing efficiency is the ratio of NaOH to NaOH and Na_2CO_3 (as Na_2O). This is a measure of how efficient causticizing is; it represents the percentage of the Na_2CO_3 from the recovery boiler that is converted back into useful NaOH cooking chemical. A value of 78-80% is typical commercially.

$$causticizing\ eff. = \frac{NaOH}{NaOH + Na_2CO_3} \times 100\%$$

Reduction efficiency

The ratio of Na_2S to Na_2S and Na_2SO_4 in green liquor is known as the reduction efficiency. This is a measure of reduction efficiency of sulfur in the recovery boiler; preferably, this value is high. Typically it is 95% and not routinely measured in the mill. In addition to sodium sulfate, other oxidized forms of sulfur are present such as sodium sulfite and sodium thiosulfate.

$$\text{reduction eff.} = \frac{Na_2S}{Na_2S + Na_2SO_4} \times 100\%$$

Dead load

Inorganic materials, principally NaCl, present in kraft liquor, but cannot be regenerated in the recovery process to AA, contribute to the dead load. Dead load merely circulates through the system thereby decreasing energy recovery and throughput during chemical recovery, increases corrosion, and may interfere with pulping.

Residual Alkali

The effective alkali remaining in a cook at its completion is termed residual alkali. The level of residual alkali determines the final pH of the liquor. At a pH of about 12, some hemicelluloses in the liquor may be deposited on the fibers with a slight increase in yield. If the pH drops much below 12, lignin may deposit on the pulp, increasing the kappa number and bleaching requirements or decreasing the bonding of unbleached pulps. Scaling and lignin precipitation will occur during black liquor concentration if the pH is low.

Polysulfide cooking

One important modification of the kraft process is the use of polysulfide in the cooking liquor, i.e. $NaS(S)_nSNa$ where n is about 4-6, to mitigate the alkaline peeling reaction. The concept of polysulfide kraft cooking dates back to the 1940s. Small amounts of polysulfide apparently increase corrosion, whereas large amounts decrease corrosion by passivation of the digester surface. Few mills use this method because it has several problems. For example, sulfur addition to white liquor generates polysulfide, but subsequent liquor recovery leads to large increases in TRS and SO_2 emissions.

The *MOXY* process is a method whereby some of the sodium sulfide is converted to sodium polysulfide and sodium hydroxide. It is an oxidation with compressed air in a reactor containing a specially treated activated carbon catalyst.

Modifications/developments of kraft pulping

Rapid displacement heating (RDH) for batch digesters and *modified continuous cooking^{TM} (MCC)* for continuous digesters such as the Kamyr (for which it was developed) improve kraft pulping by increasing uniformity, although by slightly different mechanisms. One aspect of the improvements is that alkali is of more constant concentration during cooking. Furthermore, it is important to have adequate sulfide present at the beginning of the cook to improve the selectivity of delignification.

Rapid displacement heating, RDH, is a recent modification of kraft cooking in batch digester that promises high strength pulps and decreased steam consumption. The first installation was at the Nekoosa Packaging mill in Valdosta, Georgia. The process was developed by the Rader Division of Beloit and involves pre-impregnation of steam packed chips with weak black liquors (usually three) of increasing temperatures. The final hot black liquor is then displaced with a mixture of hot white and black liquors at the cooking temperature. Since the liquor is preheated the heating time is minimal and production is increased. The original purpose of this method was to recover the energy of the warm black liquor and decrease digester heating time; however, many other benefits have been obtained.

Cooking is achieved using direct steaming until the desired H factor is achieved. Wash water then displaces the hot, black liquor which passes through a heat exchanger to warm incoming white liquor. Since blowing is at a lower temperature, pulp strength is increased. Good process control is important with RDH.

It is known that the NaOH is depleted to a high extent in black liquor, but large amounts of the sulfide remain. Wood is able to absorb sulfide from black liquor (Tormund and Teder, 1989). This means (hypothetically) the RDH process allows one to cook at high effective sulfidities without the problems of recovering liquor containing high sulfidities. In short, the RDH process allows some direct recovery of sulfide, allowing substantially lower concentrations of sulfide in the prepared cooking liquor. It is also known that the effective concentration of sulfide is low at the initial stage of conventional kraft cooks. Thus, sulfide depletion is less likely to occur during cooking with the RDH process (where extra sulfide is obtained from the black liquor) so that pulping selectivity is improved. Decreased sulfide in black liquor has numerous advantages including decreased foaming, corrosion, and TRS emissions.

Residual alkali in the black liquor is able to cleave acetyl groups from the hemicelluloses and form salts of the phenolic groups of lignin, thereby decreasing the alkali consumption in white liquor that is later introduced. This has a leveling effect on the concentration of alkali during the cook. Better diffusion of cooking chemicals, especially sulfide, before cooking would lead to more selective delignification so that lower kappa numbers may be achieved. Holding black liquor at elevated temperature before concentration also lowers its viscosity so that it can be fired at slightly higher solids content. Another advantage is that the washing action inside the digester from the wash water before blowing the digesters helps keep dilution minimal and improves the washing efficiency. If excess recovery boiler capacity is available, *this technique could be modified by using some green liquor to wash the chips*. This lowers chemical demand during pulping, and the green liquor goes back to the recovery boiler.

Modified continuous cooking, MCC (Fig. 3-14) is a countercurrent process where the concentration of alkali is lower than normal at the beginning of the cook and higher towards the end of the cook. The addition of alkali throughout the cook is an old concept that has been attempted in many ways in the past, but has been difficult to implement. The concentration of dissolved lignin towards the end of the cook is decreased by the countercurrent operation. (One wonders what happens to hemicelluloses that are sometimes redeposited on the fibers towards the end of the cook due to the decreased pH as much of the alkali has been consumed.) Proponents of the process claim higher pulp viscosity, brighter pulps, lower kappa numbers, easier bleaching, etc.

The pulp is said to be suitable for oxygen delignification down to a kappa number of 12. The first commercial demonstration of the process was carried out in Varkaus, Finland.

Another modification of the kraft process is to replace about 20% of the sodium sulfide with sodium sulfite in the *sulfide-sulfite process*. The sulfite presumably is able to effectively cleave some lignin linkages to improve the pulping process, to give higher yields, increased brightness and strength, and increased fiber flexibility.

3.9 SULFITE PULPING

Introduction

The first known patent on work related to sulfite pulping was granted to Julius Roth in 1857 for treatment of wood with sulfurous acid. Benjamin Tilghman is credited with the development of the sulfite pulping process and was granted a U.S. patent in 1867. Numerous difficulties (mainly from sulfuric acid impurities that led to loss of wood strength and darkening of the pulp) prevented commercialization of the sulfite process initially. The first mill using the process was in Sweden in 1874 and used magnesium based on the work of the Swedish chemist C. D. Eckman, although calcium became the dominant metal base in the sulfite pulping industry until 1950.

The sulfite pulping process is a full chemical pulping process, using mixtures of sulfurous acid and/or its alkali salts (Na^+, NH_3^+, Mg^{2+}, K^+ or Ca^{2+}) to solubilize lignin through the formation of sulfonate functionalities and cleavage of lignin bonds. By 1900 it had become the most important pulping process, but was surpassed by kraft pulping in the 1940s. It now accounts for less than 10% of pulp production. Woods with high pitch contents or certain extractives (such as the flavone dihydroquercitin in Douglas-fir) are not easily pulped at the lower pH's. Once the dominant pulping process, now less than 10% of pulp is produced by the sulfite method in this country, partly due to environmental considerations. Some advantages of sulfite pulping are bright, easily bleached pulps, relatively easily refined pulps, pulp that forms a less porous sheet that holds more water than kraft pulps (for use in grease-resistant papers), and pulps with higher yield than kraft.

Disadvantages include a pulp that is weaker than kraft, not all species of wood can be pulped easily, cooking cycles are long, and chemical recovery is fairly complicated or, in the case of calcium, impractical. (In the case of ammonia, the sulfur is recovered, but the ammonia must be replaced as it is burned in the recovery process.) The process involves treating wood chips with the cooking liquor at 120-150°C (250-300°F) from 500-700 kPa (75-100 psig).

Liquor preparation

The liquor is ordinarily made at the mill by burning sulfur to form SO_2 and dissolving this in water to produce sulfurous acid (H_2SO_3). Before sulfur is oxidized to form SO_2, it is handled above 160°C (320°F) since it is then a liquid and easy to convey. During combustion of sulfur at 1000°C (1830°F), the amount of oxygen must be carefully regulated to insure complete oxidation, but not over oxidation that would result in the formation of sulfur trioxide (which forms sulfuric acid when added to water.) The SO_2 gas is cooled fairly rapidly (if cooled too slowly an undesired equilibrium reaction is favored with sulfur trioxide) prior to reaction with water. The sulfurous acid is then treated with alkaline earth element (as the hydroxide or carbonate salt) in an acid-base reaction. The reactions are summarized as follows (where M = Na, K, NH_4, ½Ca or ½Mg and are in solution as their ions):

$$S + O_2 \rightarrow SO_2$$

$$SO_2 + H_2O \rightleftarrows H_2SO_3$$

$$H_2SO_3 + MOH \rightleftarrows MHSO_3 + H_2O$$

$$MHSO_3 + MOH \rightleftarrows M_2SO_3 + H_2O$$

For example:

$$2H_2SO_3 + CaCO_3 \rightleftarrows Ca(HSO_3)_2 + H_2O + CO_2$$

The highly deleterious reactions that produce sulfuric acid that quickly degrades the cellulose and delayed the first commercial use of sulfite pulping are as follows:

$$SO_2 + ½O_2 \rightarrow SO_3$$

$$SO_3 + H_2O \rightarrow H_2SO_4$$

One can easily check for sulfuric acid in cooking liquor using TAPPI UM 600 to see if this is a problem at one's mill. It is accomplished by filtering the liquor, adding formaldehyde, adding $BaCl_2$ to precipitate sulfate as $BaSO_4$, settling overnight, filtering the precipitate, igniting the precipitate, and weighing the precipitate. UM 657 is analysis of sulfur burner gases for SO_2 and SO_3 by the Orsat apparatus.

Sulfite pulping

The sulfite cooking liquor is heated only part way to the desired temperature and held with the wood chips until uniform liquor penetration is achieved, particularly in acid sulfite processes. This is very important, even more important than uniform liquor penetration in the kraft process. Without uniform liquor penetration, the chips will char since there may not be enough base to supply the necessary buffering action. In short, the sulfurous acid is able to penetrate the wood chip very quickly in the vapor form of SO_2, leading to high concentrations of sulfurous acid in the center of the chip. Long impregnation times allow the metal bases to "catch up" and buffer the system. The temperature is then raised quickly to the desired cooking temperature. After the proper cooking time, the digester pressure is reduced from 90 to 40 psig prior to blowing the chips. Sudden decompression during blowing effectively separates the wood fibers. Pressure is reduced before blowing to prevent damage to the fibers. Important variables are wood species, base ion, maximum cooking temperature, cooking time (8-14 hours), acid concentration, and liquor to wood ratio (3.25-4:1) (The magnefite process uses a pH of about 3.5 and cooking time of about 4 ½ hours at 130-135°C. This method has a higher retention of hemicellulose.) In general, the sulfite process produces a medium strength pulp with soft, flexible fibers. The lignin content is low and the pulps are easily bleached using about one-half the amount of bleaching chemicals that kraft pulps require. Yields are from 40-52%. Spruce, balsam fir, and hemlock are the preferred species. Resinous species such as southern pine and Douglas-fir (which contains a compound which inhibits pulping) are not suitable to acid sulfite cooking. Aspen, poplar, birch, beach, maple, and red alder are hardwoods suited to sulfite pulping due to uniformity of structure and low extractive contents. These pulps impart softness and bulk in the final sheet.

Sulfite pulping base metals
    Calcium is the traditional base used in sulfite pulping. Limestone ($CaCO_3$) reacts with H_2SO_3 in pressurized towers to produce the pulping liquor. The liquor is used at a pH of 1-2 (maintained with excess SO_2); at higher pH values, calcium sulfite precipitates. The cooking temperature is 140°C. Pulping is characterized relative to the other bases by intermediate pulping rate, moderate amount of screenings, high scaling tendencies, and no chemical recovery. Developments in sulfite pulping since 1950, including the use of the bases mentioned below, which are soluble in much wider pH ranges and have chemical recovery cycles, have greatly increased the versatility of the sulfite process.
    Magnesium based sulfite pulping is carried out at a pH below 5 and is characterized relative to the other bases by intermediate pulping rate, moderate scaling and screenings, and relatively simple chemical recovery. There are two major processes: acid sulfite at pH below 2 and bisulfite pulping at pH 4.5 (the magnefite process). Due to limited solubility of magnesium sulfite, magnesium based sulfite pulping must be carried out below pH 5. See the *magnefite process* below for more information. The magnesium based sulfite pulping system has a relatively simple recovery process, for example, as shown in Fig. 3-31.
    Sodium based sulfite pulping can be carried out at any pH. NaOH or Na_2CO_3 may be used to form the cooking liquor from H_2SO_3. The process is characterized by a slow pulping rate, low amounts of screenings and low scaling and complex chemical recovery.
    Ammonium based sulfite pulping is very similar to sodium based pulping (since the ammonium ion is very similar to the sodium ion) except for two important differences. The cooking rate is faster than the other three bases mentioned and the ammonium ion is lost in chemical recovery by combustion; however, the use of fresh ammonia (which reacts as ammonium hydroxide) allows the sulfur to be recovered in a relatively simple

process. The upper practical pH limit is about 9; above this pH free ammonia begins to be significant. At pH 9.24 at 25°C half of the ammonium appears as free NH_3.

$$NH_4^+ \rightleftharpoons H^+ + NH_3 \qquad pK_a = 9.24 \text{ at } 25°C$$

    Potassium based sulfite pulping has been studied in the laboratory and gives pulping results similar to those of sodium. It has the advantage that the spent liquor makes a good fertilizer if liquor recovery is not to be practiced as in the case of small mills that may be used for nonwood fibers (Wong and Derdall, 1991).

Cooking liquor
    The cooking liquor is the fresh pulping liquor for the sulfite process, consisting of a mixture of SO_2 together with one of the bases (alkali ions).

Brown (or red) liquor
    Brown or red liquor is the waste liquor from the sulfite pulping process.

Actual forms of sulfite-based cooking chemicals
    The cooking chemicals all start out with SO_2. This is in turn dissolved in water to give H_2SO_3 or alkali solution to give HSO_3^- or SO_3^{2-}.
    Sulfur dioxide is SO_2, a gas formed by burning sulfur; it has an acrid, suffocating odor and limited solubility in water.
    Sulfurous acid is H_2SO_3, the reaction product of SO_2 and water.
    Monosulfite salt or sulfite is the completely "neutralized" form (salt) of H_2SO_3: M_2SO_3, where M = monovalent cation (Na^+, K^+, NH_4^+ or $\frac{1}{2}Ca^{2+}$, $\frac{1}{2}Mg^{2+}$). For example Na_2SO_3 is sodium monosulfite or, simply, sodium sulfite.
    Bisulfite salt is the "half-neutralized" salt of H_2SO_3: $MHSO_3$. Though not used by the industry, the preferred term is *hydrogen sulfite*. For example $NaHSO_3$ is sodium bisulfite or, preferably, sodium hydrogen sulfite.

Cooking chemical terminology
    All sulfite-based cooking chemicals are expressed on a molar-equivalent-to-SO_2 basis, but expressed as a weight percent SO_2. (This is rather confusing and explained in detail in Chapter 16 with examples.)   For example, the formula

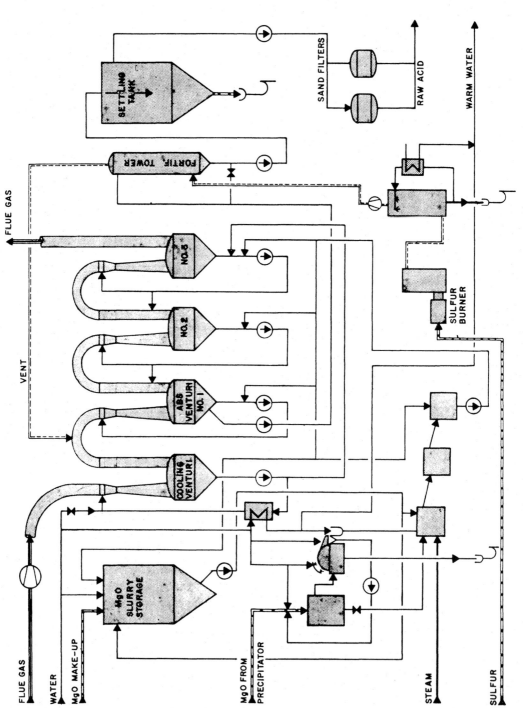

Fig. 3-31. Babcock and Wilcox process for magnesium-based sulfite chemical recovery. Reprinted from EPA-625/7-76-001.

weights for Na_2SO_3 and SO_2 are 126 and 64 g/mol, respectively. Therefore, a solution containing 126 g/L of Na_2SO_3 is equal to 64 g/L on a SO_2 and would be expressed as 6.4% SO_2. The cooking chemicals are expressed as *total, free, or combined SO_2*. The term *true free SO_2* is also used.

Total SO_2 is, for sulfite cooking liquors, the ratio of the total weight of SO_2 to the total weight of the solution containing the SO_2, usually expressed as a percent. Total SO_2 is determined by the Palmrose method, titration with potassium iodate (TAPPI Standard T 604).

$$total\ SO_2 = free\ SO_2 + combined\ SO_2.$$

Free SO_2 is the amount of sulfite chemical, based on SO_2, present in the form of the free acid, H_2SO_3, plus that which can be converted to free acid (i.e., ½ of the HSO_3^-, see reaction below the next definition), as a percent of total liquor weight. It is determined by titration with NaOH.

Combined SO_2 is the difference between the total and free SO_2, expressed as percent of total liquor weight; it is the SO_2 existing in the form of the monosulfite salt, plus that which can be converted to the monosulfite salt (½ of the HSO_3^-). Two moles of bisulfite form one mole of sulfurous acid and one mole of monosulfite by the reaction below; the equilibrium is very, very far to the left.

$$2\ HSO_3^- \rightleftharpoons H_2SO_3 + SO_3^{2-}$$

True free SO_2: In liquors where the free SO_2 exceeds the combined SO_2, the difference is equal to the true free SO_2 and represents the actual amount of sulfurous acid in the pulping liquor.

Square liquor

A square sulfite cooking liquor has equal amounts of *free SO_2* and *combined SO_2*. Since the actual species is all HSO_3^-, there is very little buffering capacity unless a buffer such as sodium carbonate is added.

Acid sulfite process

The acid sulfite process, uses a cooking liquor that is strongly acidic (pH 1.5-2, and, therefore, has a preponderance of free SO_2 in solution), with a pulping temperature of 125-145°C (260-290°F), and a cooking time up to 7 hours. A long

heating time (3 hours) is necessary to prevent diffusion of SO_2 ahead of the base that causes lignin to condense at high temperature in the center of the chip causing a "black cook". The base is usually calcium. Due to acid hydrolysis, the result is a weak pulp, with low hemicellulose content suitable for dissolving pulp, tissue paper, and glassine.

Bisulfite process

A full chemical pulping process with higher liquor pH (3-4), and nearly equal amounts of free and combined SO_2 in the liquor ("square" liquor) is known as the bisulfite process. The temperature is 160-180°C with a cooking time of 0.25-3 hours giving yields of 55-75%. The Arbisco process uses sodium as the base; the magnefite process uses magnesium. Ammonia is also a suitable base, and, more recently, potassium is being used as the base. This pulp is suited for medium grades of paper such as newsprint and writing papers.

Alkaline sulfite

The alkaline sulfite, full chemical pulping process uses a chemical charge containing approximately equal amounts of NaOH and Na_2SO_3 at temperatures of 160-180°C (320-356°F) and 3-5 hours at the maximum temperature. This process produces pulps fairly similar in quality to kraft pulps in terms of yield, brightness, bleachability, and strength.

Magnefite process

The magnefite process is used at pH 4.5 and 160°C (320°F) with Mg^{2+} as the base. The process is often used to produce reinforcing pulps for newsprint. Chemical recovery is necessary due to the high cost of the base. One recovery process is summarized in Fig. 3-31. Spent sulfite liquor is concentrated to 55% solids and burned at 1350°C (2460°F). MgO and SO_2 are recovered. The MgO is slurried in water to give a slurry containing 50-60% $Mg(OH)_2$ which is used to scrub SO_2 from the flue gases using a bubble cap tower or venturi scrubber.

$$MgO + HOH \rightarrow Mg(OH)_2$$

$$Mg(OH)_2 + 2\ SO_2 \rightarrow Mg\ (HSO_3)_2 + H_2O$$

Stora process

The Stora process is a two stage sulfite process with a mild neutral sulfite stage followed by an acid sulfite stage to give a pulp with high yield and low kappa compared to kraft cooking.

Sivola process

The Sivola process is a three stage sulfite process for producing dissolving pulps. The neutral sulfite stage provides sulfonation of lignin for easy removal, the second stage of acid sulfite is then used to lower the molecular weight of cellulose for viscosity control and hemicelluloses for their subsequent removal, and the final stage of basic sulfite removes the hemicelluloses.

Chemical recovery

Chemical recovery in the sulfite process involves 6 steps:

1. Washing of the spent sulfite liquor from the pulp.
2. Concentration of the spent sulfite liquor.
3. Burning of the concentrated liquor.
4. Heat recovery during liquor combustion.
5. Pulping chemical regeneration.
6. By-product recovery (mostly in Ca^{2+} based system).

If ammonia is used as the base, it cannot be recovered in the burning process since it is converted to H_2O and N_2; consequently, fresh ammonia must be used for each batch of pulp. Mg^{2+} and Na^+ bases <u>must</u> be recovered due to their high costs. Magnesium has a well developed, proven recovery system whereby it and sulfur are recovered in their original forms, MgO and SO_2. In this process, the spent sulfite liquor is concentrated to 55% solids, prior to burning at 1370°C (2500°F) in special furnaces. The magnesium is recovered from the gases as MgO which is slurried in water to form $Mg(OH)_2$. The $Mg(OH)_2$ is then used in the scrubbers to trap SO_2, generating $Mg(HSO_3)_2$. The reactions are summarized as follows:

$$MgCO_3 + 2H_2SO_3 \rightarrow Mg(HSO_3)_2 + CO_2 + H_2O$$

$$MgO + HOH \rightarrow Mg(OH)_2$$

$$Mg(OH)_2 + 2SO_2 \rightarrow Mg(HSO_3)_2 + HOH$$

Sulfite byproducts

There are several important byproducts to sulfite cooking. The hexoses, produced by acid hydrolysis of hemicellulose and cellulose, are sometimes fermented to ethanol leaving lignosulfonates (salts of sulfonated lignin) which are water soluble at useful pH's. The lignosulfonates are used in leather tanning, drilling mud dispersants, ore flotation, and resins. Alkaline oxidation of lignosulfonates leads to the production of vanillin, that is, artificial vanilla. It takes only a few mills to meet the world's demand for artificial vanilla.

3.10 OTHER PULPING METHODS

Extended delignification

Chemical pulping and bleaching of chemical pulps arc both delignification reactions. Of course, bleaching reactions are much more specific for lignin removal than pulping, but are much more expensive. Improvements in pulping that allow cooking to lower lignin contents and new processes before conventional bleaching are referred to as extended delignification.

A variety of pretreatment processes (many of which are experimental) applied to pulp lower the amount of bleaching chemicals required, leading to lower levels of chlorinated organic materials.

Organosolv pulping

Organosolv pulping is largely an experimental pulping procedure using organic solvents such as methanol, ethanol, acetic acid, acetone, etc. to remove lignin. It has the advantages of having no sulfur emissions and a simple chemical recovery process which would allow relatively small mills to be constructed.

Klason was the first to try to remove lignin from wood by dissolving it in acidified alcohol solutions in solutions of 5% HCl in ethanol in 1893. Cooking dry spruce chips for 6-10 hours led to dissolution of 28-32% of the wood.

The AlCell (alcohol cellulose) process of Repap enterprises using 50% ethanol and 50% water at 195°C for approximately 1 hour has been demonstrated at 15 t/d. The company plans to build a 300 t/d mill in Newcastle, N.B. to pulp hardwoods. One advantage of the system is a relatively simple recovery system (evaporation of the liquor to recover the alcohol) that allows small mills to be economically feasible. The pulp has tensile and burst strengths and brightness equal to kraft pulp and a higher tear strength by 6-7% (at 400 CSF). For an unknown reason, the pulp appears to be particularly amenable to oxygen delignification without significant strength loss. Also DED bleaching gives a brightness of 90 ISO. Lignin, furfural, and hemicelluloses are obtained as byproducts.

Young (1992) describes a 450 ton per day mill in Germany using the organocell solvent pulping process. The mill uses a single-stage process with 25-30% methanol as the solvent to pulp spruce. The alkali charge is 125 g/L with a liquor to wood ratio of 4.2:1. Anthraquinone, 0.1% on wood, is also used. The cooking temperature is about 160°C (320°F). Bleaching is carried out with an $O_{E/P}PP$ process.

The use of oxygen with methanol, ethanol and propanol has also been investigated as a pulping technique (Deinko et al., 1992).

Biological pulping

Biological pulping is an experimental process whereby chips are pretreated with white rot fungi or lignin degradation enzymes. While researchers (promoting additional research funding) sometimes paint a very favorable picture for this process, the high costs of growth media, biological reactors, and other factors are not mentioned. It appears that this is modestly effective as a treatment before mechanical pulping. When reading publications in this area, see if cellulose viscosities are measured as an indicator of selectivity of lignin attack. Many of the strength improvements and yields observed with these methods do not consider the effect of fungal hyphae.

Other, novel pulping methods

Fengel and Wegener in *Wood* list other, unconventional pulping processes (p. 464) with references. They include nitric acid pulping, organosolv pulping, pulping with a variety of organic solvents, and formaldehyde pulping.

3.11 MARKET PULP

Many mills sell their pulp on the open market. The pulp must be prepared in thick sheets of 50% moisture content if shipping costs are not high (for example, for short distances) or dry lap of 80-86% solids. Plate 14 shows a double wire press suitable for making wet lap, washing pulps, or thickening stock. Plate 16 shows a double wire press used for making wet lap. The machine acts like a slow paper machine, having a notable similarity in purpose. Some mills make wet lap for their own use to keep paper machines running when the pulp mill is shut down.

3.12 ANNOTATED BIBLIOGRAPHY

General mechanical pulping

1. Pearson, A.J., *A Unified Theory of Refining*, No. 6 of Pulp and Pap. Technol. Ser., Joint Textbook Comm., 1990, 128 p., 20 ref.

    This book is a very useful comparison of SGW, RMP, and TMP processes. A small portion of the book (pp 65-85) covers beating and refining of chemical pulps. Mechanical pulping is defined as "pulping brought about by the absorption of energy by the repeated compression and decompression of the fibre." This work mentions much of the early, fundamental work accomplished by the key investigators in this area with few, unfortunately, references.

2. Kayserberg, S.A., High yield pulping--recent trends, *1989 Wood and Pulping Chemistry*, pp 81-84, TAPPI Press, 1989.

    This is an interesting reference on the production of various grades of mechanical pulps (TMP, CTMP, PGW) by region of the world from 1976 to 1988. Included are new developments in mechanical pulping and areas where mechanical pulps are competing against other pulps such as in tissue, fluff, paperboard, and writing paper.

Stone groundwood, SGW

3. White, J.H., Manufacture of mechanical pulp, in *Pulp and Paper Manufacture*, Vol. 1, 2nd ed., MacDonald, R.G., Ed., McGraw-Hill, New York, 1969, pp 148-190. This is a fairly detailed review.

Pressurized groundwood, PGW
4. Burkett, K. and Tapio, M., Super pressurized groundwood (PGW) from southern pine, *Tappi J.* 73(6):117-120(1990).

5. Lucander, M., Pressurized grinding of chemically treated wood, *Tappi J.* 71(1):118-124(1988), no references.

Alkaline peroxide CTMP
6. McCready, M., Millar Western - Meadow Lake making quality APP/BCTMP that's environmentally correct, *PaperAge* 108(1):10-12(1992).

Cold soda process
7. McGovern, J.N. and E.L. Springer, History of FPL cold soda CMP process: 1950-present, *Tappi Proceedings, 1988 Pulping Conference*, pp 641-648.

Kraft pulping
8. Kleppe, P.J., Kraft pulping, *Tappi J.* 53(1):35-47(1970). This is a useful summary article with 115 references on the kraft pulping process.

Modifications of kraft cooking, RDH, MCC
9. Tormund, D. and A. Teder, New findings on sulfide chemistry in kraft pulping, *TAPPI 1989 International Symposium on Wood and Pulping Chemistry Proceedings* pp 247-254.

    This article studied the selective absorption of sulfide by wood. Lignin in wood can absorb large amounts of sulfide from cooking liquors so that the concentration of free HS⁻ is very low. This causes the pulping process to be less selective (like soda cooking), but is the basis of MCC and RDH improvements.

Rapid displacement heating, RDH
10. Vikström, B., M.S. Lindblad, A.-M. Parming, and D. Tormund., Apparent sulfidity and sulfide profiles in the RDH cooking process, *Tappi Proceedings, 1988 Pulping Conference*, pp 669-676.

    This article looks at sulfidity in the RDH process and concludes free hydrosulfide is

four times higher in RDH versus conventional kraft pulping (both at 32% sulfidity), and RDH pulping at 32% sulfidity is equal to 46% sulfidity in conventional kraft.

11. Scheldorf, J.J., L.L Edwards, P. Lidskog, and R.R. Johnson, Using on-line computer simulation for rapid displacement heating control system checkout, *Tappi J.* 74(3)109-112 (1991).

12. Evans, J.W.C., Batch digester heat displacement system reduces steam consumption, *Pulp & Paper* 63(7)132-135(1989). "Installation of rapid displacement heating system also cuts pulping chemical use and improves pulp strength at several mills."

13. Swift, L.K. and J.S. Dayton, Rapid displacement heating in batch digesters, *Pulp Pap. Can.* 89(8)T264-T270(1988).

14. Mera, F.E. and J.L. Chamberlin, Extended delignification, an alternative to conventional kraft pulping, *Tappi J.* 71(1):132-136(1988).

Modified continuous cooking, MCC
15. Johansson, B., J. Mjöberg, P. Sandström, and A. Teder, Modified continuous kraft pulping, *Svensk Papperstidn.* 10:30-35(1984). The first commercial demonstration of the process was carried out in Varkaus, Finland and reported here.

Polysulfide pulping
16. Green, R.P., Polysulfide liquor generation and white liquor oxidation, in Hough, G, Ed., *Chemical Recovery in the Alkaline Pulping Process*, TAPPI Press, Atlanta, 1985, pp 257-268. Details on the MOXY and descriptions of other methods for generating polysulfide in the white liquor are given here.

Sulfite process
17. Kocurek, M.J., O.V. Ingruber, and A. Wong, Ed., *Sulfite Science and Technology*, Joint Textbook Committee, 1986, 352 p.

18. McGregor, G.H., Manufacture of sulfite pulp, in Stephenson, J.N., Ed., *Pulp and Paper Manufacture*, McGraw-Hill, New York, 1950, pp 252-362. This chapter is cited primarily for the extensive information on preparation of SO$_2$ (43 pages of the chapter).

19. Wong, A. and G. Derdall, *Pulp Pap. Can.* 92(7):T184-T187(1991).

Magnesium sulfite chemical recovery

20. Powers, W.E., M.W. Short, J.A. Evensen, and B.A. Dennison, Retrofit options for controlling particulate emissions from a magnesium sulfite recovery furnace, *Tappi J.* 75(5):113-119(1992). Nine North American mills are using the magnesium sulfite process. The emissions of submicron particulates is detailed.

Anthraquinone catalyst

21. Holton, H.H., Soda additive softwood pulping: a major new process, *Pulp Paper Can.*, 78(10):T218-T223(1977). This was the first published work describing the effect of AQ on pulping.

22. Dutta, T. and C. J. Biermann, Kraft pulping of Douglas-fir with 1,4-dihydro-9,10-dihydroxy anthracene (DDA), *Tappi J.* 72(2):175-177(1989).

23. Wandelt, P., The effect of 1,4-dihydro-9,10-dihydroxy anthracene on soda and kraft cooking of pine, *Paperi ja Puu--Papper och Trä*, (11):673-681(1984).

Organosolv pulping

24. Hägglund, E., *Chemistry of Wood*, Academic Press, New York, 1951, pp 237-252. A very good account of the early work on organo-solv lignin is given in this work.

25. Black, N.P., *Tappi J.* 74(4):87(1991). The alkaline sulfite anthraquinone methanol (ASAM) organosolv process claims higher pulp yield and strength than the kraft process.

26. Aziz, S. and K. Sarkanen, Organosolv pulping--a review, *Tappi J.* 72(3):169-175 (1989).

27. Jamison, S. Alcell pulping: world class research right here in Canada, *Pulp Paper Can.* 93(3):16-19(1991). This interesting article summarizes progress in the process.

28. Pyle, E.K. and J.H. Lora, The Alcell™ process: A proven alternative to kraft pulping, *Tappi J.* 74(3):113-118(1991). This article summarizes the process.

29. Goyal, G.C., J.H. Lora, and E.K. Pyle, Autocatalyzed organosolv pulping of hardwoods: effect of pulping conditions on pulp properties and characteristics of soluble and residual lignin, *Tappi J.* 75(2):110-116(1992). Process variables of the title process were studied in this work.

30. Young, J., Commercial organocell process comes online at Kelheim Mill, *Pulp & Paper* 66(9):99-102(1992).

31. Deinko, I.P, O.V. Makarova, and M.Ya. Zarubin, Delignification of wood by oxygen in low-molecular-weight alcoholic media, *Tappi J.* 75(9):136-140(1992).

Novel pulping methods

32. Dahlbom, J., L. Olm, and A. Teder, The characteristics of MSS-AQ pulping--a new pulping method, *Tappi J.* 73(3)257-261(1990).

The mini-sulfide-sulfite-anthraquinone process uses low amounts of sulfide (about 10% of the sulfur cooking chemical) in an alkaline sulfite process. The authors claim an 8% increase in yield over the kraft process with the possibility of achieving a kappa number of 8 if the process is followed by oxygen bleaching with strength properties at least as high as pulp produced by the kraft process.

Pollution aspects

33. U.S. Env. Prot. Agency, Environmental pollution control, Pulp and Paper Industry, Part I, air, EPA-625/7-76-001 (1976).

EXERCISES

General aspects of pulping

1. What are the two principal mechanisms whereby fibers are separated from the woody substrate?

2. What are the four broad categories of pulping processes?

3. Check the appropriate column for the pulp-type exhibiting the <u>higher</u> value for each of the following properties:

	Chemical	Mechanical
a) Yield	—	—
b) Tensile strength of paper	—	—
c) Unbleached brightness	—	—
d) Sheet density	—	—
e) Opacity of paper	—	—

Mechanical pulping

4. What happens if steaming temperatures of 145°C or higher are used during the production of TMP pulp?

5. Chemical pulp has been traditionally used to strengthen newsprint made with stone groundwood pulp. More recently lower amounts of chemical pulp or no chemical pulp has been used with newsprint made from thermomechanical pulp or CTMP. Why is this the case?

Chemi-mechanical pulping

6. Describe the pretreatment processes of wood in the chemi-mechanical pulping methods?

Chemical pulping

7. What are the two major categories of pulping digesters?

8. The yield versus kappa number is the most fundamental relationship for chemical pulping methods. What does this information tell us about the selectivity of a process for lignin removal relative to carbohydrate removal?

9. If an inexpensive pulping process is highly selective for lignin removal, is it necessarily good for making paper?

10. Explain how anthraquinone (a pulping additive) stabilizes cellulose and hemicelluloses during alkaline pulping. (From what does it stabilize the carbohydrates?)

Kraft pulping

11. What are the two active chemicals in the kraft pulping process? Is sulfate an active pulping chemical? Why is the kraft process also called the sulfate process?

12. The H-factor combines what two kraft pulping variables? Does the H-factor alone tell you the degree of cook?

Sulfite pulping

13. Why has calcium-based sulfite pulping, once the dominant pulping process, been replaced by other pulping processes?

14. What are the two forms of SO_2 in sulfite cooking liquors? Give these by their common names and the corresponding chemical forms.

15. Briefly describe how chemical recovery occurs in magnesium based sulfite pulping.

Market Pulp

16. What is market pulp?

4

KRAFT SPENT LIQUOR RECOVERY

4.1 CHEMICAL RECOVERY

Chemical recovery is the process in which the inorganic chemicals used in pulping are recovered and regenerated for reuse. This process results in 1) recovery of the inorganic cooking chemicals, 2) generation of large amounts of heat energy by burning the organic materials derived from the wood, 3) reduction in air and water pollution by converting the waste products into useful (or at least harmless) materials, and 4) regeneration of the inorganic chemicals into pulping chemicals. In summary, the recovery process for kraft pulping is:

1. Concentration of black liquor by evaporation.

2. Combustion of strong black liquor to give the recovered inorganic chemicals in the form of smelt. The smelt, Na_2S and Na_2CO_3, dissolved in water gives green liquor.

3. Preparation of the white cooking liquor from green liquor. This is done by converting the Na_2CO_3 to NaOH using $Ca(OH)_2$, which is recovered as $CaCO_3$.

4. Recovery of byproducts such as tall oil, energy, and turpentine.

5. Regeneration of calcium carbonate, $CaCO_3$, to calcium hydroxide, $Ca(OH)_2$.

Storage at all of the above steps allows the overall operation to continue even though one component requires servicing or is not operating smoothly and also accommodates surges in the system. Storage capacity of 3 to 24 hours is common; longer down times of essential components may result in the shutdown of the entire mill. It is not uncommon, however, for a mill to ship black liquor to a nearby mill for recovery and exchange it for fresh liquor during recovery boiler rebuilding. Fig. 4-1 shows the kraft liquor recovery cycle from the point the pulp and liquor leave the blow tank to the point the liquor is ready for the green liquor clarifiers.

4.2 PULP WASHING

Pulp washers (brown stock washers)
Pulp washers are almost always drum or counter flow washers for separating spent pulping chemicals. Pulp washers use countercurrent flow between stages such that the pulp moves opposite in direction to the flow of wash water as described in Section 3.6. This design allows for the most removal of pulping chemicals (for recovery) and lignin (to reduce bleaching chemical demand or improve papermaking with brown papers) with the least amount of water. The dilution factor is a measure of the amount of water used in washing compared to the amount theoretically required to displace the liquor from the thickened pulp; it is reported as mass of water per mass of dry pulp.

A low dilution factor decreases the energy requirements of the multiple effect evaporators. Using more washers increases removal of pulping chemicals with less water dilution, but increases capital and operating costs. Usually 3 or 4 washers in series are used with about 2 to 3 tons of water being added to the black liquor per ton of pulp to recover over 96% of the pulping chemicals. Washing of some entrained material occurs at the repulpers (shown in Fig. 4-2) used between the washers. The two fundamental controls are drum speed and stock inlet flow rate. Soda loss in the pulp is traditionally measured as lb/ton pulp on a Na_2SO_4 basis. This alone is not a good indicator of washing efficiency since 12-15 kg/t (25-30 lb/ton) soda (as Na_2SO_4) are chemically bound to the pulp, presumably to the carboxylate groups much like an ionic exchange resin.

Resinous species such as the pines tend to foam and require larger washers and filtrate tanks or the use of large amounts of defoamers. Linerboard mills often use excess paper machine white water in the brown stock washers since it is not overly contaminated with filler and additives.

A *drop leg* is used to siphon the water from the washer so that a vacuum pump is not usually required. The bottom end of the drop leg goes to a filtrate storage tank that is designed to prevent

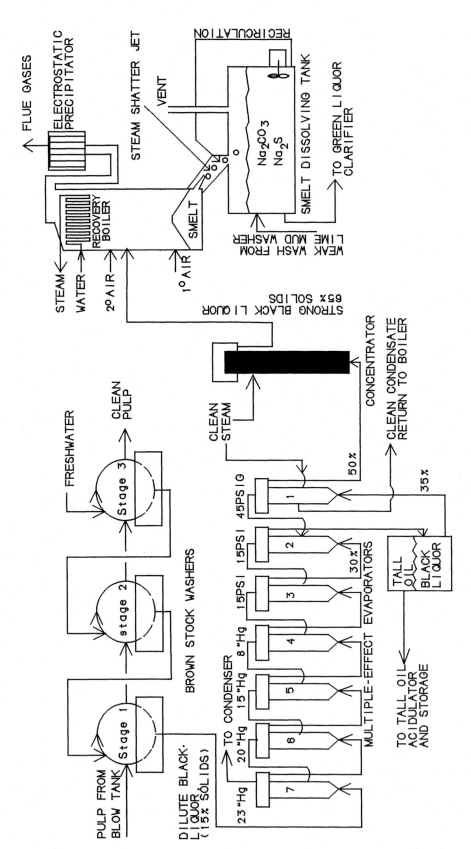

Fig. 4-1. The liquor recovery cycle in a kraft mill before causticizing. (Not to scale and liquor storage points are not shown.)

air from entering the bottom end of the drop leg. The drop leg supplies a vacuum on the order of 24 kPa below atmospheric pressure (7 in. Hg below atmospheric pressure) for the first stage, 30 kPa (9 in. Hg) for the second stage, and 40 kPa (12 in. Hg) for the third stage. The overall drop leg length is about 12-14 m (40-45 ft) with 3 m (10 ft) within the filtrate tank yielding an effective length of 10 m (30-35 ft). The actual vacuum is lowered proportionally to the air bubble content of the pulp slurry. The drop leg is ideally a vertical drop. If a horizontal section is needed, it should be exactly horizontal and placed at least 7 m (22 ft) below the washer. Drop legs that are not entirely horizontal or vertical allow air to separate from the stock and rush upward, thereby accumulating and decreasing the vacuum.

4.3 LIQUOR EVAPORATION

Black liquor behavior during evaporation

It is desirable to concentrate the solids of the black liquor as much as possible to make heat recovery from liquor combustion as efficient as possible, but when the black liquor reaches high solids contents, the viscosity increases drastically (Section 16.4). Combustion of highly concentrated black liquor leads to higher temperatures in the lower part of the furnace, which increases the rate of smelt reduction and decreases sulfur emissions. Black liquor is most often burned at 65-73% solids content in conventional commercial systems. Some systems allow black liquor combustion at 75-80% solids content, but only a few mills use these systems. Sometimes the viscosity of certain black liquors (e.g., from straw pulping) can be decreased by holding the concentrated black liquor at 115°C (240°F) for several hours.

Like any aqueous solution, the boiling point of black liquor increases with increasing solids content. The boiling point rise (relative to water) is about 3°C (6°F) at 33% solids, 8°C (14°F) at 50% solids, 13°C (23°F) at 67% solids, 17°C (30°F) at 75% solids, and 21°C (37°F) at 80% solids. This is the overall solids content; not all of this is actually in solution because, when black liquor is concentrated above 55%, burkeite ($2Na_2SO_4 \cdot Na_2CO_3$) etc. precipitates as scale.

Multiple-effect evaporators, MEE

The multiple-effect evaporators are a series

Fig. 4-2. A repulper after a pulp washer.

of four to eight evaporators with indirect heating for removing water from the dilute black liquor coming from the pulp washers. Traditionally, long tube vertical bodies were used, but recently, falling film evaporators have also been installed. Later effects are operated under vacuum in order to achieve evaporation within the desired temperature range. The black liquor leaves at 50% solids. In North America, the water evaporated from the concentrated black liquor of one effect is used as steam in the previous effect. (In Europe, the operation is often partially co-current to avoid scaling which occurs when the hottest steam contacts the most concentrated liquor). Stated conversely, each effect acts as a surface condenser for the previous effect. Fig. 4-3 shows the arrangement of a set of evaporators.

For example, 100 lb of black liquor containing 15 lb of solids and 85 lb of water is evaporated to 30 lb; with a steam economy of 5.0 it would take 14 lb of steam to remove the 70 lb of water. The steam economy is about 0.8 per effect. An

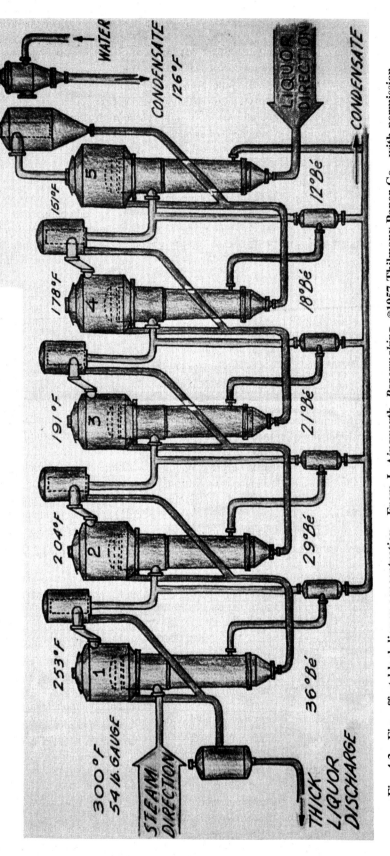

Fig. 4-3. Five-effect black liquor concentration. From J. Ainsworth, *Papermaking*, ©1957 Thilmany Paper Co., with permission.

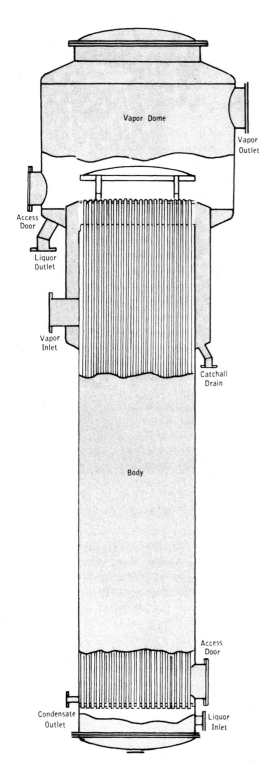

Fig. 4-5. A series of LTV bodies (the multiple-effect evaporators).

overall steam economy of 4-5, depending on the number of effects, is normally good for mill operation. If this stage is a bottleneck, lower steam economy is sacrificed for higher throughput. The initial steam introduced at the first effect comes from the boiler, and its condensate is returned to the boiler. The subsequent condensates are contaminated with volatile black liquor components and are usually sent to the sewer.

$$\text{steam economy} = \frac{\text{water mass evaporated}}{\text{mass of steam used}}$$

Long tube vertical (LTV) bodies

Traditionally, each effect consists of a long tube vertical evaporator (LTV, Fig. 4-4). Several LTV bodies are shown in Fig. 4-5. Black liquor rises up the heat exchanger area until it reaches the vapor dome at the top. In the vapor dome, steam flashes from the black liquor and is pulled by vacuum to the next effect as the steam source.

The shell of the evaporator is usually 12.5 mm (0.5 in.) steel plate. The heat exchanger consists of 2 in. diameter tubes from 14 to 32 feet long. Stainless steel is often used in the first two effects where the liquor is warmest and most concentrated. The first effect is operated at 50

Fig. 4-4. Long tube vertical evaporator. Reprinted from *Making Pulp and Paper*, ©**1967 Crown Zellerbach Corp., with permission.**

psig and 150°C (300°F), while the last effect is about 27 in. Hg vacuum and 46°C (115°F).

Falling film evaporators

Falling film evaporators are used much like conventional evaporators except the mechanism of evaporation in each stage is different. In each stage, steam (or sometimes hot water in the first stage) is used as the heat source and flows between stainless steel plates about 30 mm (1.25 in.) apart. Large banks of these plates are aligned radially outward in each effect. Dimples in the metal plates keep the plates a fixed distance apart and increase the strength of the metal plates. Fig. 4-6 shows a section of the plates.

The black liquor is introduced at the top of the evaporator and flows down the opposite side of the metal plates where the dimples help spread the black liquor into a thin film. The falling film plate design allows for selective condensation of the vapors boiled from the black liquor of previous effects. About 65% of the methanol and BOD is concentrated in 6% of the overall condensate stream in the upper portion of the plates to give *foul condensate segregation*. The remaining condensate is collected from the lower portion of the plates and is suitable for brown stock washing.

Falling film plate evaporators are less subject to fouling than LTV and require boilouts much less often. They also operate at a lower overall vacuum than conventional evaporators with the last stage operating at 26 in. Hg. Black liquor must be recirculated within each stage. For example, the fourth stage of one operation uses 10,000 gal/min recirculation with a liquor feed of 900 gal/min at 14% solids and an outlet of 450 gal/min at 26% solids gal/min. Some of the 4th effect concentrated liquor can be used to concentrate the infeed of the 3rd effect since liquor at low to intermediate concentrations tends to foam, a big problem with large recirculation rates.

Falling film evaporators can be used in conjunction with blow heat recovery at mills using batch digesters since lower temperature gradients are required. The steam discharged during blowing is used to heat large quantities of water. The hot water is then used in liquor evaporation until the next digester blow reheats it. This provides a leveling effect for the intermittent heat generation of blowing with batch digesters.

Fig. 4-6. A small section of a falling film evaporator plate (wood-grain table background).

Direct contact (cascade) evaporator

The direct contact evaporator is a chamber where black liquor of 50% solids content directly contacts the hot flue gases from the recovery furnace. The final black liquor concentration is 65-70% solids. This method is now obsolete because high sulfur emissions result as the hot CO_2-containing flue gases strip sulfide from the black liquor. Also, indirect concentrators allow higher energy recoveries. A few mills that have recovery boilers installed before the early 1960s still use this method. This process, before being replaced by concentrators, required partial *black liquor oxidation*. In this process, the black liquor enters a chamber where it is mixed with air (or oxygen) to convert reduced sulfides, such as $(CH_3)_2S$ and S^{2-}, to oxidized forms of sulfur and thereby minimize sulfur emissions (Section 11.6). The reason this is required is that the CO_2 in the flue gases exists as the acid H_2CO_3 which lowers the pH of black liquor, stripping H_2S from it. This process is not necessary with indirect concentrators, since the pH of the black liquor remains high prior to combustion. Three types of direct contact evaporators have been used: cascade evaporators (CE), cyclone evaporators (B&W), and venturi scrubbers (B&W).

Concentrators, Indirect concentrators

Indirect concentrators (used in so-called *low-odor recovery boilers*) are forced circulation or falling film steam heated evaporators used to concentrate black liquor in the range where burkeite precipitates as scale. Since concentrators are more energy efficient and environmentally sound than direct contact evaporators, they have largely replaced direct contact evaporators since their introduction in the mid 1960s. The final solids content of the black liquor is about 65-70% with a fuel value of 14-16 MJ/kg (6000-7000 Btu/lb) compared to coal, which is 32 MJ/kg (14,000 Btu/lb). Although a very high solids content in black liquor is desired to increase combustion efficiency, the viscosity increases quickly with solids contents above 65-70%, and the black liquor becomes too thick to pump even at elevated temperatures.

Tall oil

Tall oil is a by-product mixture of saponified fatty acids (30-60%), resin acids (40-60%, including mostly abietic and pimaric acids), and unsaponifiables (5-10%) derived from the wood extractives of softwoods. Crude tall oil is isolated from acidified skimming of partially concentrated black liquor. It is collected and refined at special plants. The refined products are sold commercially for soaps, rosin size, etc. Typically 30-50 kg/t (60-100 lbs/ton) on pulp may be recovered from highly resinous species representing about 30-70% recovery. It is recovered from mills pulping resinous species such as the southern pines. The pulp and paper industry recovers about 450,000 tons of crude tall oil annually.

Turpentine

Turpentine is a mixture of volatile extractives (monoterpenes) collected during digester heating.

In batch digesters most of it is collected before the digester temperature reaches 132°C (270°F). It is collected from the digester relief gases and sold for solvents and limited disinfectants used in household pine oil cleaners. Because terpenes are volatile, the recovery can decrease by 50% with outside chip storage of a few weeks. It is often standard practice at mills recovering turpentine to use a portion of green (fresh) wood chips in the digester feed to reduce turpentine loss, while the balance is used from chips rotated in inventory. Fresh wood of loblolly and shortleaf pines yield about 6-12 L/t (1.5 to 3 gal/ton) air dry pulp, while slash and longleaf pine yield 10-18 L/t (2.5 to 4.5 gal/ton). The U.S. pulp industry recovers about 30 million gallons of turpentine annually.

Kraft lignin

Some black liquor is recovered by acidification and used as dispersants, phenol-formaldehyde adhesive extenders, and binders in printing inks. For example, Indulin™ is Westvaco's trade name for kraft lignins of various grades. Dimethylsulfoxide (DMSO, a solvent and controversial healing remedy) can also be recovered from kraft lignin. However, kraft lignin is not isolated and marketed to the same degree that lignosulfonates from sulfite pulping methods have been. Calcium lignosulfonates were a waste problem that were marketed as a "solution in search of a problem."

4.4 RECOVERY BOILER

Recovery boiler or recovery furnace

The development of the recovery boiler by Tomlinson in conjunction with Babcock and Wilcox in the early 1930s led to the predominance of the kraft process. Fig. 4-7 shows a typical recovery boiler design, while Fig. 4-8 compares two types of boilers with widespread use. Fig. 4-9 shows a recovery boiler building at a brown paper mill.

The purpose of the recovery boiler is to recover the inorganic chemicals as smelt (sodium carbonate and sodium sulfide), burn the organic chemicals so they are not discharged from the mill as pollutants, and recover the heat of combustion in the form of steam. The latter is accomplished by large numbers of carbon steel tubes filled with

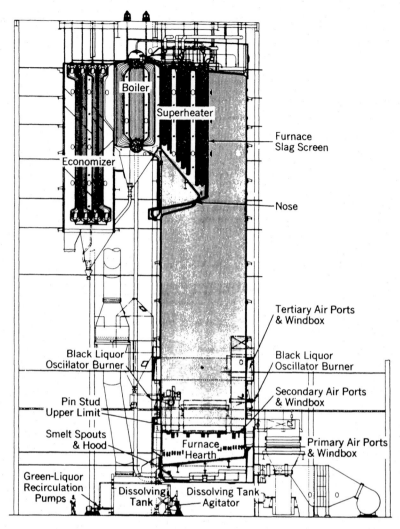

Fig. 4-7. Kraft recovery boiler from Babcock & Wilcox. From Ref. 1, p 10-18.

circulating water or steam to recover heat from the walls of the recovery boiler and the flue gases. Fig. 4-10 shows some banks of tubes to be installed into a recovery boiler. In Finland and Sweden the outer surfaces of these tubes are clad with stainless steel to greatly increase their life. Some recovery boilers in the U.S. are equipped with stainless steel clad tubes as well, but it is not a widespread practice. Combustion Engineering has used a "chromizing" process where chromium is incorporated "in" the surface to produce a stainless-steel-like surface.

The recovery boiler or furnace burns the concentrated black liquor by spraying it into the furnace through side openings (Plate 17). The water evaporates and the organic materials removed from the wood form a char and then burn. There are three zones: The upper section is the *oxidizing zone*, the middle section (where the black liquor is injected) is the *drying zone*, and the bottom section is the *reducing zone* where, in a bottom bed of char, the sulfur compounds are converted to Na_2S. The remaining NaOH and sodium salts of organic acids are converted to Na_2CO_3. These sulfur- and sodium-based inorganic materials leave as molten slag that is directed to the green liquor dissolving tank (Plate 18). Due to the possible adverse reaction of molten smelt with water, all recovery boilers have an *emergency shutdown procedure (ESP)* in the

BABCOCK & WILCOX   COMBUSTION ENGINEERING

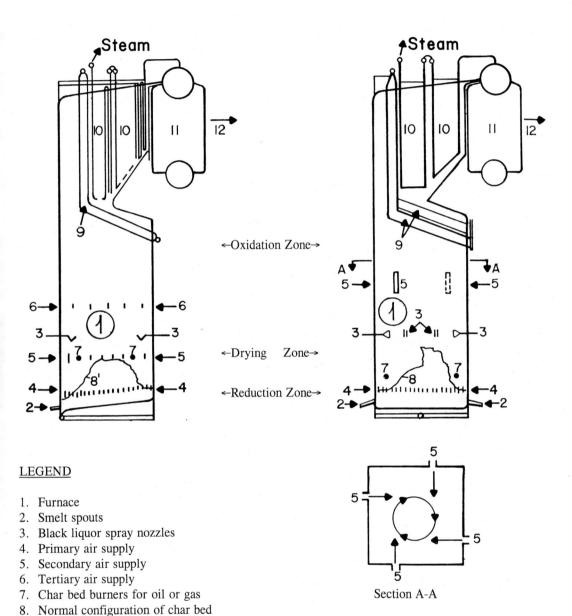

LEGEND

1. Furnace
2. Smelt spouts
3. Black liquor spray nozzles
4. Primary air supply
5. Secondary air supply
6. Tertiary air supply
7. Char bed burners for oil or gas
8. Normal configuration of char bed
8' As 8 but with low primary air supply
9. Screen tubes
10. Superheater
11. Boiler tube bank
12. Exit to economizer section

Section A-A

Fig. 4-8. Principal design and air distribution of two recovery boiler designs (Ref. 1, p 10-15).

Fig. 4-9. Kraft recovery boiler in the background and two lime kilns in the foreground.

Fig. 4-10. A small portion of the heat recovery tubes to install in a recovery boiler rebuild.

event of trouble! The recovery boiler is the largest, single most expensive piece of equipment in a kraft mill costing over $100 million; hence, in many mills the recovery boiler limits the maximum production. The newest recovery boilers may support 2500-3000 tons of pulp production per day. The overall chemical reactions in the recovery boiler in addition to combustion are:

conversion of sodium salts:

$$2NaOH + CO_2 \rightarrow Na_2CO_3 + H_2O$$

reduction of make-up chemical:

$$Na_2SO_4 + 4C \rightleftarrows Na_2S + 4CO$$

The lower zone is deficient in oxygen, so reduction reactions occur. This allows the sulfur in the smelt to occur as Na_2S and not $Na_2S_2O_3$ or Na_2SO_4, which would be unsuitable for fresh liquor. About 40-50% of the air required for combustion is added by forced-draft fans at the primary vents at the bottom of the recovery boiler. The primary air supply is preheated to 150°C (300°F) and burns the organic compounds, leaving smelt, but maintains reduction conditions. The upper zone begins above the region of secondary air supply and has about 10 to 20% excess air above that required for complete combustion of the organic materials. The top zone must be under oxidative conditions to prevent carbon monoxide emissions. The flue gases are drawn away from the recovery boiler at the exit of the electrostatic precipitator; this maintains an air pressure below ambient so that gases will be sucked into the boiler in the vicinity of the black liquor nozzles to improve the safety of the operation. Some example reactions in each zone are:

Oxidation zone:

$$CO + \frac{1}{2}O_2 \rightarrow CO_2$$
$$H_2 + \frac{1}{2}O_2 \rightarrow H_2O$$
$$Na_2S + 2O_2 \rightarrow Na_2SO_4$$
$$H_2S + 1\frac{1}{2}O_2 \rightarrow SO_2 + H_2O$$

Drying zone:

$$Organics \rightarrow C + CO + H_2$$
$$2NaOH + CO_2 \rightarrow Na_2CO_3 + H_2O$$

Reduction zone:

Organics $\rightarrow$ C + CO + H_2

$2C + O_2 \rightarrow 2CO$

$Na_2SO_4 + 4C \rightarrow Na_2S + 4CO$

$C + H_2O \rightarrow CO + H_2$

The low secondary air supply of the B&W boilers is placed about 2 m (6 ft) above the primary air supply. This air acts as secondary air along the walls of the boiler, but as primary air in the char bed and, therefore, controls the height of the bed. Additional secondary air is needed; it is called tertiary air. (By definition all non-primary air is secondary air.) In the CE boilers a tangential air supply is used to produce a rotary movement of the furnace gases (EPA, 1976).

The maximum combustion temperature occurs between the plane of black liquor entry and plane of secondary air entrance. Firing black liquor at 65% solids leads to a maximum combustion temperature of 2000-2400°F while combustion at 70% solids leads to combustion temperatures greater than 2500°F.

Cameras are used to monitor the appearance, size, and position of the char bed at the bottom of the recovery boiler in order to properly control (Section 11.6) liquor combustion.

Heat recovery

The maximum temperature in the recovery boiler is about 1100-1300°C (2000-2400°F) for black liquor burned at 65% solids. The heat of combustion of the organic materials is transferred to tubes filled with water in several areas: in the walls of the recovery boiler, in the *boiler* section, and in the *economizer* section. The economizer section is a final set of tubes (from the point of view of the exhaust gases, but the first tubes the water travels through) used in more recent recovery boilers to warm water for various processes. The *thermal efficiency*, defined below, is the proportion of heat recovered as steam and is about 60%. Most of the heat loss occurs as steam in the flue gases from water in the black liquor.

$$\text{thermal efficiency} = \frac{\text{heat to steam}}{\text{total heat input}}$$

The minimum temperature of the exhaust gases is 130°C (265°F) to prevent condensation of corrosive materials and to insure the exhaust will go upward beyond the smokestack. The combustion gases cool by radiation to about 870°C (1600°F) before entering the convection heating tubes. Temperatures above this, which might result by over-loading the recovery boiler, do not allow complete combustion of the organics, which causes fouling of the screens and superheater tubes by tacky soot particles. The flue gases are cooled to about 450°C (850°F) after the boiler and to 160°C (320°F) after the economizer. (With direct contact evaporation, the flue gases leave the economizer section at 400°C.) About 6000-7000 kg/t (12,000-14,000 lb/ton) steam on pulp are produced by the recovery boiler.

Cogeneration

Cogeneration is the process of producing electricity from steam (or other hot gases) and using the waste heat as steam in chemical processes. In contrast, a stand-alone power producing plant typically converts less than 40% of the heat energy of fuel (coal, natural gas, nuclear etc.) into electricity. The remaining heat is simply lost to the heat sink; the heat sink lowers T_{cold} to increase the efficiency and is usually a large body of water where the effects of thermal pollution must be considered. The *Carnot cycle,* which predicts the maximum possible efficiency for the conversion of heat to work, of a heat engine is:

$$e_{rev} = 1 - T_{cold}/T_{hot} = (T_{hot}-T_{cold})/T_{hot}$$

where T is expressed in an absolute temperature scale such as Kelvin, T_{hot} is the temperature of the steam entering the turbine, and T_{cold} is the temperature of the steam exiting the turbine.

Since pulp mills (both chemical and most mechanical) can use for the steam coming out of the turbine and would produce steam in any case, these mills can essentially convert heat energy into electricity with over 80% efficiency. Surprisingly, many pulp mills do not cogenerate. This is particularly true in the Northwestern U.S. where electric companies and relatively cheap hydroelectricity have discouraged this.

Electrostatic Precipitators (Cottrell)

Electrostatic precipitators (ESP, Fig. 4-11) consist of chambers filled with metal plates, charged with high DC voltage (30,000-80,000 V) through which exhaust gases from the recovery furnace pass. The chambers remove solid materials (particulates) in the gas stream that acquire a charge from the high voltage and collect on the plates, thereby purifying the stream before it is discharged into the atmosphere. Removal of over 99% of the particulates over 0.1 μm can be achieved. The current required is on the order of 500 mA/1000 m^2 (50 mA/1000 ft^2) of plate area. Rapping the plates with a shock wave dislodges the particulates into a collecting tray placed below them. Much of the material collected is sodium sulfate and sodium carbonate, typically 5 to 20 kg (10-40 lb) per ton pulp, which is returned to the recovery boiler. Electrostatic precipitators are now used on many new lime kilns as well.

4.5 COOKING LIQUOR REGENERATION-THE CAUSTICIZING PROCESS

Chemical recovery

The chemical recovery cycle is summarized in Fig. 4-12. Inorganic pulping chemicals are recovered from the furnace as a molten smelt (Na$_2$CO$_3$ and Na$_2$S) that falls to the bottom of the furnace. These are dissolved in water to give green liquor. The combination of molten smelt and large quantities of water in the heat exchanger tubes make recovery boilers potentially explosive, a critical concern at all times. The green liquor is treated with Ca(OH)$_2$ to regenerate the NaOH. The CaCO$_3$ that is formed must go through the lime kiln to generate CaO that is later dissolved in water to regenerate Ca(OH)$_2$.

Green liquor (dissolving) tank

Water (from the dregs washer) fills the green liquor dissolving tank, which is located below the kraft recovery furnace, where the molten slag is added through the smelt spout to form green liquor (mainly Na$_2$CO$_3$ and Na$_2$S). A *steam shatter jet* and recirculated green liquor impinges on the smelt stream to break it into small pieces. If the steam shatter jet fails, major explosions become probable. The green liquor then goes to the green liquor clarifier tank. The density of the green

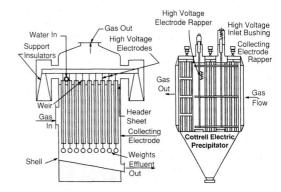

Fig. 4-11. Cottrell electrostatic precipitator. Redrawn from J. Ainsworth, *Papermaking,* ©1957 Thilmany Paper Co., with permission.

liquor at this point is used as a process control variable of green liquor concentration.

Green liquor clarifier

The green liquor clarifier is a settling tank used to remove *dregs* by sedimentation before the green liquor is recausticized as shown in Fig. 4-13. It can also serve as a storage tank for green liquor and should provide at least 12 hours supply of green liquor. Since the mid 1960s single compartment clarifiers have been used in place of the older multi-compartment clarifiers. Overflow rates on the order of 0.6 m/h (2 ft/h) and retention times over 2 hours are used. The dregs settle to the bottom where rakes move them to the outlet. If green liquor clarification is not used or is inadequate, these inert materials build up in the lime, decreasing the *lime availability*. The green liquor clarifier/storage unit should be insulated to limit heat loss of the green liquor coming from the smelt dissolving tank at 85-95°C (185-205°F). Most mills now use polymeric additives to help green liquor clarification.

Green liquor dregs and dregs washer

The green liquor dregs are undissolved materials in the green liquor. The dregs are about 0.1% of the liquor and consist of carbon (50% or more) and foreign materials (mainly insoluble metal carbonates, sulfates, sulfides, hydroxides, and, especially from nonwood fibers, silicates) to give a black bulky material. Incomplete combustion of organic materials in the recovery boiler can lead to inert carbon particles that leave with the

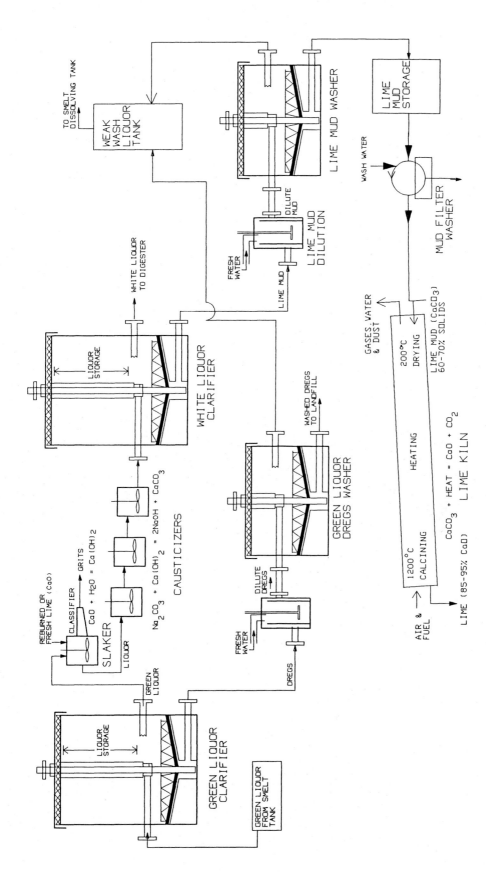

Fig. 4-12. A summary of the causticizing process of a kraft mill.

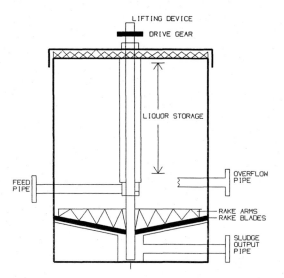

Fig. 4-13. A single compartment gravity sedimentation (settling) clarifier and storage unit suitable for green or white liquor.

smelt and greatly increase the amount of dregs. The metal salts arise from non-process elements (impurities) of the wood and corrosion products of the processing equipment, particularly from the recovery boiler. Improvements in the design of the recovery boiler have decreased the dregs yield to less than 4 kg/t (8 lb/ton) pulp. The dregs are washed in a dregs washer, often a drum filter or sedimentation washer (Fig. 4-14) where about 90-95% of the sodium chemicals are removed, of which there is about 1-4 kg (2-9 lb) (Na₂O basis) per ton of pulp in the dregs before the washers.

Slaker

The slaker is a chemical reactor in which lime is mixed with green liquor (Fig. 4-15). The reaction temperature is 99-105°C (210-220°F). Using a high temperature and lime directly from the kiln gives a lime mud that settles well. The lime, CaO, forms slaked lime, Ca(OH)₂, and much of the causticizing reaction occurs here where the retention time is 10-15 minutes. These two chemical equations are shown below.

slaking reaction:

$$CaO + H_2O \rightarrow Ca(OH)_2$$

causticizing reaction:

$$Ca(OH)_2 + Na_2CO_3 \rightleftarrows 2NaOH + CaCO_3(s)$$

Grits (large, unreactive lime particles and insoluble impurities corresponding to about 0.5-2% of the lime feed) are removed here by the *classifier*, which uses a raking action. The grits are often sent to landfills, although a few mills will send them through a ball mill and use them to neutralize the acid effluent of the bleach plant.

The extent of the causticizing reaction (and therefore the causticizing efficiency in the white liquor) depends on the concentration of the initial Na₂CO₃ and the amount of lime used. With concentrations below 16% of actual chemical, the theoretical conversion is over 90%. At concentrations above this, the theoretical conversion drops off quickly. It is desired to have about 1% excess lime. Much more than this increases the

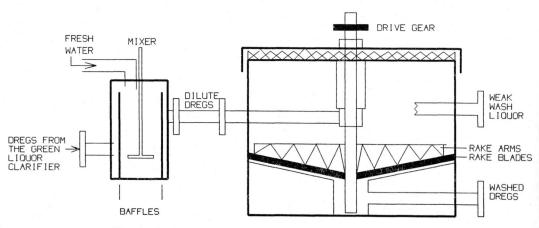

Fig. 4-14. Sedimentation type green liquor dregs washer.

Fig. 4-15. The top (top) and side (bottom) views of a slaker.

turbidity of the white liquor and decreases the filtration rates of the lime mud. Commercially, about 75-85% of the ultimate level of conversion is achieved in the agitator section of the slaker. The causticity efficiency should be 3-4% below the equilibrium value to avoid excess liming.

Causticizers

The causticizers are two to four continuous flow, stirred reactors that are used to complete the causticizing reaction (Fig. 4-16). The contents are stirred with a pitched blade turbine at 70-80 rpm. The liquor/lime slurry flows through them in

Fig. 4-16. The causticizers. The left insert shows a sample drawn for quality analysis. The right insert shows proper sample settling.

series with a total retention time of 1.5-2.5 hours. The internal surfaces must be stainless steel or another corrosion resistant material.

White liquor clarifier

The white liquor clarifiers are settling tanks (gravity sedimentation) used to remove the lime mud ($CaCO_3$) from the white liquor. The clarified liquor should have a turbidity below 100 ppm.

The lime mud leaves with a solids content above 35% in order to minimize entrained soda that is removed in the lime mud washer. Like the green liquor clarifiers, white liquor clarifiers with single compartments with at least 12 hours storage (see Fig. 4-13) are now much more common than the multi-compartment clarifiers.

Poor lime settling may be a result of excess lime to the slaker (more than 1% excess), a low lime availability (below 80-85% that is indicative of a high level of contaminants due to inadequate removal of dregs and/or grits or incomplete slaking due to overloading the lime kiln), a high percentage of low reactivity unburned fresh lime (i.e., purchased lime that has not been through a lime kiln or is aged), or inadequate white liquor clarification or clarifier operation. Lime burned at too low a temperature gives much mud of high viscosity; lime burned at too high a temperature gives a slow causticizing reaction and a slow settling lime mud with entrained alkali. Polymeric additives are often used to help white liquor clarification.

White liquor pressure filters

Recently, some mills have been installing white liquor pressure filters after the white liquor clarifiers to supply additional clarification. Typically the filters have polypropylene filter tubes through which the white liquor flows under a pressure of 140-210 kPa (20-30 psig) to trap further lime mud. A reverse, flushing cycle removes the embedded lime mud that is added to the bulk of the lime mud (from the white liquor clarifiers) prior to the lime mud washers.

Lime mud washer

The lime mud washer removes most of the 15-20% entrained alkali (Na_2O basis) from the lime mud, usually by sedimentation washing. It is typically a settling tank (or two in series) where fresh (make-up) water is used to wash the lime mud. If it were not removed, the Na_2S would cause slagging in the kiln and reduced sulfur compounds would be released as H_2S. About 1% alkali on lime mud remains after washing. The liquor (weak wash or weak white liquor) is then used to dissolve smelt from the recovery boiler. It is important to maintain a proper water balance during lime mud washing to avoid formation of excess weak wash. Excess weak wash must be sent to the sewer with loss of valuable chemical that creates a disposal problem.

Lime mud filter

Thick lime mud from storage is diluted to 25-30% solids (as measured by an X-ray detector) before going to the lime mud filter. The lime mud filter is a rotary drum vacuum filter washer used for final lime washing and thickening to 60-70% solids before the lime enters the kiln. Centrifuges have been installed instead of drum vacuum filters to thicken lime, but they have lower water removal, leading to increased energy costs and lower lime kiln throughput.

Lime kiln

The lime kiln is a chemical reactor in which lime mud (CaCO$_3$) is dried, heated, and converted to CaO, a process known as *calcining* (Plate 19) described by the chemical equation below.

$$\text{calcining reaction: } CaCO_3 \overset{\Delta}{\rightarrow} CaO + CO_2(g)$$

The retention time in the kiln should be on the order of 2-3 hours. CaO is required at about 200-300 kg/t (400-600 lb/ton) for pulp, with the lower amounts for high yield pulps like linerboard and the higher amount for low yield pulps such as bleachable grades. Normally a rotary kiln is used, but some mills use a fluidized bed calciner. Rotary lime kilns (Figs. 4-17, 4-9) are 2.4 to 4 m (8 to 13 ft) in diameter and 30 to 120 m (100 to 400 ft) long and produce 40 to 400 t/day of CaO. However, in mid 1991, the Union Camp mill in

Savannah, Georgia started operation of a lime kiln that is 15 ft in diameter, 440 feet long, and processes 700 tons/day of lime mud, the largest of its kind in the world. The lime kiln is inclined at about 2 to 5° from horizontal to propel the lime along as the kiln rotates at about 1 rpm. The amount of energy required to process the lime is indicated by the specific energy consumption, often reported as Btu/ton, that is defined below.

$$\text{specific energy consumption} = \frac{\text{energy to kiln}}{\text{CaO output}}$$

Fuel oil or natural gas combustion (or occasionally coal or biomass) supplies the fuel to achieve the high temperature (1200°C, 2200°F) required. The combustion gases run counter-current to the lime. Fig. 4-17 shows the steps inside the lime kiln. The rotary kiln requires about 7.8-10 GJ/t CaO (7-9 × 10^6 Btu/ton CaO) while the fluidized bed calciner requires about 8-9 GJ/t CaO. Rotary lime kilns of high capacity (over 300 t/d) and long length (100 m or 330 ft) use the lower amount of energy, while low capacity (less than 100 t/d) and shorter length (less than 60 m or 200 ft) kilns use the higher amount.

The purity of the lime is given by the *lime availability* that is defined as:

$$\text{lime availability} = \frac{\text{CaO}}{\text{lime}} \quad \text{(as mass ratio)}$$

Low lime availability results if the kiln temperature is too low or if the entering lime mud is overly wet, since the calcining reaction does not occur to the extent that is should.

Some mills with short lime kilns, or where the lime kiln is the bottleneck and higher through-

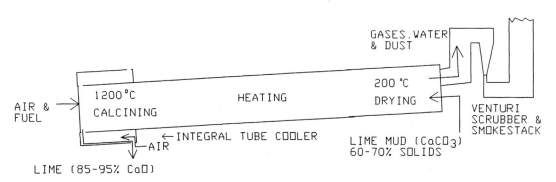

Fig. 4-17. Function and operation of the lime kiln.

put is required, use *flash driers* to dry the lime mud to less than 1% moisture content before it enters the kiln. In flash drying, the exit gases of the kiln (much hotter than 200°C since dry mud is being fed into it) are used to dry the lime mud. The lime mud is combined with the gases in a mixer. The flue gases and water vapor are then separated from the lime mud in a cyclone.

Heat is recovered from the flue gases and hot burned lime of the lime kiln. A development of the early 1980s is the discharging of the hot, burned lime product through about 8-10 *integral tube coolers* forming a ring around the lime kiln on the outer side of the discharge end. Secondary air flows through these tubes in the opposite direction as the lime so that it is preheated, thereby increasing the combustion temperature or lime throughput. However, the lime must remain above 320°C (600°F) so it will slake well.

Sulfur oxides are not emitted to a large extent because the CaO is a good scavenger for the highly acidic forms of these compounds and $CaSO_3$ and $CaSO_4$ are formed. Emissions of nitrogen oxides are high due to the excess air and high combustion temperatures. See Section 11.6 for more information. Pure oxygen gas has been used as a partial substitute for air in order to decrease H_2S generation from the lime mud and increase lime mud throughput rates.

Ring formation within the lime kiln is part of the territory. If it becomes a severe problem, expensive shutdowns may result. Ring formation can result (Tran and Barham, 1991) when non-condensable gases are introduced to the lime kiln and the ring occurs from the formation of $CaSO_4$. The use of a TRS scrubber column for the NCG prior to their introduction to the lime kiln can drastically decrease ring formation.

The exhaust gases must be treated in order to remove particulate matter. Usually a venturi scrubber (Fig. 4-18) is used. The flue gases flow through a constriction on the order of 1 ft^2 with the introduction of a water spray to trap the particulates. In the mid 1970s, a few mills started using electrostatic precipitators (ESP, Fig. 4-11) to treat the lime kiln exhaust gases. ESP have many advantages over the venturi scrubbers, including cleaner exhaust gases, easier control of the system, and low exhaust fan power requirements since the exhaust is not forced through a narrow constriction. Now most new systems use electrostatic precipitators to capture the particulates.

Salt cake (Na_2SO_4)

Salt cake is the make-up chemical for the kraft process that is used to replace chemicals lost (mostly through the pulp) in pulping. About 50-100 lb/ton pulp is normally dissolved in the concentrated black liquor just before spraying into the recovery furnace. The name "sulfate" process was derived from the use of this salt as a make-up chemical; however, sodium sulfate is *not* involved in the actual pulping process. The make-up chemical can be other chemicals as well.

Cross recovery

Cross recovery is the collection of waste NSSC liquor and burning it in a nearby kraft mill. Kraft mills recover the sodium and sulfur as Na_2S and Na_2CO_3 after the recovery furnace. The kraft mill pays the NSSC mill for Na_2SO_4, which allows the NSSC mill to purchase fresh cooking chemicals (sulfur and either NaOH or Na_2CO_3).

4.6 ANNOTATED BIBLIOGRAPHY

1. U.S. Environmental Protection Agency, Environmental pollution control, Pulp and Paper Industry, Part I, air, EPA-625/7-76-001 (1976).

Brown stock washing

2. Crotogino, R.H., N.A. Poirier, and D.T. Trinh, The principles of pulp washing, *Tappi J.* 70(6):95-103(1987). This is an excellent overview of pulp washing from the theoretical and equipment perspectives.

3. Perkins, J.K., Ed., *1982 Brown Stock Washing, Short Course Notes*, TAPPI Press, Atlanta, GA, 1982, 49 p.

    This is a very useful introduction to brown stock washing. Descriptions of Chapters 1-9 follow. Perkins, J.K and K.M. Jackson, The washing of brown stock, pp 1-6 (chap. 1), gives a historical overview of the basic washing methods; Perkins, J.K., Rotary filter washing equipment, pp 7-9 (chap. 2) has 10

Fig. 4-18. The top view shows the lime kiln duct carrying the flue gases to the venturi scrubber (bottom), and from the scrubber to the chimney.

figures with descriptions of the title equipment; Hough, G.W., Auxiliary components and design for vacuum filter washing systems, pp 11-16 (chap. 3) addresses stock flow, drop legs, seal chambers, and seal tanks; McSweeney, J.M., Control and instrumentation, pp 17-20 (chap. 4); Smyth, Jr., J.A., The mathematics and variables of counter-current brown stock washing, pp 21-26; Guillory, A.L., Sorbed soda and its determination, pp 27-31 (chap. 6); Guillory, A.L., Brown stock washer efficiency and physical testing, pp 33-37 (chap. 7); Perkins, J.K., Analyzing and troubleshooting a washing system, pp 39-45 (chap. 8); and Perkins, J.K., Newly introduced methods of brown stock washing, pp 47-48 (chap. 9).

4. Phillips, J.R., and J. Nelson, Diffusion washing system performance (plus a simplified method for estimating saltcake loss from any equipment combination, *Pulp Paper Can.* 78(6):T123-T127(73-76)(1977). This article includes a brief review of the equipment.

5. Nordén, H.V., V.J. Pohjola, and R. Seppänen, Statistical analysis of pulp washing on an industrial rotary drum, *Pulp Paper Mag. Can.* 74(10):T229-T337(83-91)(1973). Regression analysis showed the dependence of the underflow leaving the drum upon various factors.

Black liquor properties
6. Löwendahl, L., G. Petersson, and O. Samuelson, Formation of carboxylic acids by degradation of carbohydrates during kraft cooking of pine, *Tappi J.* 59(9):118-120(1976). The amounts of 26 organic acids in black liquor are given.

7. Venkatesh, V and X.N. Nguyen, 3. Evaporation and concentration of black liquor, in Hough, G, Ed., *Chemical Recovery in the Alkaline Pulping Process*, TAPPI Press, Atlanta, 1985, pp 15-85. A general presentation of black liquor properties is given, including specific gravity as a function of solids content, boiling point rise, specific heat, viscosity, chemical composition, and thermal conductivity.

8. Harvin, R.L. and W.F. Brown, *Tappi J.*, 36(6):270-274(1953). The specific heat of black liquor is fairly linear from 1 cal/g at 0% solids to 0.70 at 60% solids.

Black liquor evaporation
9. Gudmundson, C., H. Alsholm, and B. Hedström, Heat transfer in industrial black liquor evaporator plants, *Svensk Papperstidning*, Part I, 75(19):773-783(1972); Part II, 75(22):901-908(1972). The overall heat transfer coefficient in climbing film evaporators is dependent on the heat flux, boiling point of the liquid, and the liquid feed rate, temperature, viscosity, and surface tension. Foaming flow produces remarkable heat transfer, especially at low heat flux, compared to non-foaming flow. The error between calculated and measured data was less than 15% for 95% of the data.

Falling film evaporators
10. Shalansky, G., D. Burton, and B. Lefebvre, Northwood pulp & timber's experience using the falling film plate type evaporator for high solids concentration of kraft black liquor, *Pulp Paper Can.* 93(1):51-56(1992). This is a **very** useful article on the subject with some nice figures to demonstrate the concepts.

11. Burris, B. and J.F. Howe, Operational experiences of the first eight-effect tubular falling-film evaporator train, *Tappi J.*, 70(8):87-91,94(1987). This claims a steam economy of 8.4 when evaporating black liquor from 13 to 50% solids for a system that handles 1 million pounds per hour.

Recovery boiler
12. Taylor, S.S., Good firing practices improve boiler safety, *Am. Papermaker*, 52(4, mislabeled as 3 in the front):39-42(1989).

13. Koncel, J.A., Preventive maintenance can end smelt-water explosions, *American Papermaker*, 54:(8):26-27(1991). The cause

of 125 BLRB explosions are itemized along with a six-step emergency shutdown procedure. Useful guidelines for preventing smelt-water explosions are found here. "Industry statistics show that smelt-water explosions take place in one of every 72 operating BLRBs each year."

14. Bauer, D.G., and W.B.A. Sharp, The inspection of recovery boilers to detect factors that cause critical leaks, *Tappi J.* 74(9):92-100(1991). This is a technical article along these lines that should be in the library of anyone responsible for recovery boiler safety.

Casticizing

15. Mehra, N.K, C.F. Cornell, and G.W. Hough, Chap. 5. Cooking liquor preparation, in Hough, G, Ed., *Chemical Recovery in the Alkaline Pulping Process*, TAPPI Press, Atlanta, 1985, pp 191-256. A good discussion of the theoretical causticizing efficiency as well as material and energy balances for liquor and lime cycles is presented here.

16. Daily, C.M. and J.M. Genco, Thermodynamic model of the kraft causticizing reaction, *J. Pulp Paper Sci.* 18(1):J1-J10(1992). The thermodynamics of the causticizing reaction has been modeled based on K.S. Pitzer's method to predict the ratio of the activity coefficients of hydroxyl and carbonate anions in strong electrolyte solutions.

17. Lewko, L.A. and B. Blackwell, Lime mud recycling improves the performance of kraft recausticizing, *Tappi J.* 74(10):123-127(1991). This claims to improve the causticizing process by recycling some of the lime mud to the slaker. This might be important at a mill that is limited by the lime kiln or lime mud washing. It also will decrease the dead load slightly in the recovery boiler and lead to a small increase in boiler capacity. In Table II the "common overflow" units should be mg/L not g/L. Reports after this paper tell of a mill whose free lime mud content in the white liquor clarifier

overflow decreased from 1.35% to 0.26%, while the slurry solids concentration of the underflow increased from 32% to 44% using a 100% recycle rate [*Tappi J.* 75(3):20(1992)].

Cogeneration

18. Price, K.R. and W.A. Anderson, New cogeneration plant provides steam for Oxnard papermaking facility, *Tappi J.* 74(7):52-55(1991). Here cogeneration using the GE LM2500, which is a modification of the GE CF6-6 aircraft engine used in DC-10 jet aircraft, is described.

Lime kilns

19. *Amer. Papermaker*, 55(1):34-35(1992). This article describes the use of ESP in a new lime kiln.

20. Tran, H.N. and D. Barham, An overview of ring formation in lime kilns, *Tappi J.* 74(1):131-136(1991).

Miscellaneous considerations

21. Sell, N. and J.C. Norman, Reductive burning of high-yield pulp by the addition of pulverized coal, *Tappi J.* 75(10):152-156. Pulverized coal was added to pulp liquor (from a mill that makes 78% yield sodium bisulfite hardwood pulp) to help recover the sulfur as sulfide (77%) and sulfite (20%) instead of sulfate, which is the typical recovery product for sulfur at these mills.

Although not discussed in this paper, it would seem that high sulfur-content coals might actually be burned in a recovery boiler with sulfur recovery in the smelt.

EXERCISES

General

1. What are the three main objectives in the chemical recovery cycle?

2. Write the chemical equations as indicated.

   a) Reaction of NaOH in the recovery boiler:

b) The formation of white liquor from green liquor:

c) The regeneration of calcium so it can be used in reaction b):

d) Concentration of black liquor from 15% to 50% solids:

3. Where is turpentine recovered during kraft pulping? Where is tall oil collected?

4. What is cross recovery?

Pulp washing

5. Indicate the effect of a high dilution factor in pulp washing on the following:

		Increase	Decrease
a)	sodium loss	—	—
b)	evaporation costs	—	—
c)	bleaching cost	—	—

6. The steam economy of the multiple effect evaporators is 4.24; suppose the dilution factor in the brown stock washers goes from 3 tons water per ton pulp, to 4 tons water per ton pulp. How much more steam is required to process a ton of pulp?

Liquor concentration

7. What are the two types of bodies used in multiple-effect evaporators?

8. Black liquor is first concentrated to about 45-50% solids using the multiple-effect evaporators. Final concentration is achieved by direct contact evaporators or by concentrators. Since the late 1960s, direct contact evaporators have not been installed in pulp and paper mills. Why is this the case?

9. Why is the concentration of black liquor usually limited to about 65-75% solids prior to combustion in the recovery boiler? Why would a higher solids content be desirable?

Recovery boiler

10. Describe the location of the oxidation and reduction zones in the recovery boiler. Give some examples of chemical reactions that occur in each of these zones.

11. What is the purpose of the electrostatic precipitator after the recovery boiler?

12. What would happen to the sodium sulfide if it was oxidized in the bottom of the recovery boiler? Is this beneficial?

Lime cycle

13. What are the three zones in the lime kiln in terms of what is happening to the lime?

14. What are the consequences of not removing the dregs from the green liquor?

Causticizing

15. Of what is the casticizing efficiency of white liquor a measure?

Miscellaneous

16. It has been a long practice at many mills to add some black liquor to the white liquor and chips before pulping. What are some possible advantages of adding black liquor to the digester as part of the digester charge?

5
PULP BLEACHING

5.1 INTRODUCTION

Bleaching

Bleaching is the treatment of wood (and other lignocellulosic) pulps with chemical agents to increase their brightness. The percentage of bleached pulp to total pulp production in the U.S. has risen from 32% in 1950 to 40% in 1990.

Bleaching of chemical pulps involves a much different strategy than bleaching mechanical pulps. Bleaching of chemical pulps is achieved by lignin removal. Lignin removal in chemical pulps leads to greater fiber-fiber bonding strength in paper, but the strong chemical used in bleaching chemical pulps decreases the length of cellulose molecules, resulting in weaker fibers.

Bleaching mechanical pulps is achieved by chemically altering the portions of the lignin molecule that absorb light (i.e., have color). Obviously, lignin removal in mechanical pulps is counterproductive; bleaching mechanical pulps is referred to as *lignin-preserving*. Sometimes bleaching mechanical pulps is called "brightening" to distinguish it from bleaching of chemical pulps.

Brightness

Brightness is a term used to describe the whiteness of pulp or paper, on a scale from 0% (absolute black) to 100% (relative to a MgO standard, which has an absolute brightness of about 96%) by the reflectance of blue light (457 nm) from the paper. See Section 7.8 for more details on these tests. The approximate brightness levels of some pulps are as follows:

Unbleached kraft	20%
Unbleached sulfite	35%
Newsprint	60%
Groundwood	65%
White tablet paper	75%
High grade bond	85%
Dissolving pulp	90%

Color reversion

Color reversion is the yellowing of pulps on exposure to air, light, heat, certain metallic ions, and fungi due to modification of residual lignin forming chromophores. Mechanical pulps are particularly susceptible to color reversion, though chemical pulps may experience this when exposed to high temperatures.

Consistency

Bleaching stages are carried out at consistencies from 3-20%. Higher consistencies of 10-20% are used with chemicals such as oxygen, peroxide, and hypochlorite, which react with the lignin slowly. By using high consistencies, higher concentrations of the bleaching agent are realized for a given chemical loading, which increases the reaction rate.

5.2 BLEACHING MECHANICAL PULPS

Mechanical pulps are bleached with chemicals designed to alter many of the chromophores. Chromophores are most often conjugated double bond systems arising in the lignin of pulps. Other chromophores such as sap-stain induced by microorganisms, dirt, metal ions such as ferric that complex with lignin, and water impurities may impart color to paper. Bleaching of mechanical pulps involves masking the lignin that is present, instead of removing the lignin as is the case for bleaching chemical pulps. To emphasize this distinction, bleaching mechanical pulps is often referred to as *brightening*.

Brightening mechanical pulps is accomplished with reducing agents, such as dithionite, or oxidants, such as hydrogen peroxide, often in a single stage process. Fairly limited brightness improvements are realized (6-12% typically) with a maximum brightness of 60-70% in a single stage, or up to 75% in a two-stage process. The pulp brightness also dependents on the wood species from which the pulp was derived. Brightened mechanical pulps are subject to *color reversion*. Since the lignin is largely decolored, but not removed, there is only a small loss of yield. If two stages are used, the oxidative stage is used before the reductive stage or else the oxidant will undo what the reductive compound accomplished.

Dithionite, hydrosulfite bleaching

Hydrosulfite (the common name in the industry, but dithionite is the preferred name) bleaching is carried out with 0.5-1.0% dithionite on wood. Previously, zinc dithionite was used because it is very stable. However, zinc is toxic to fish, therefore, the sodium form has replaced the zinc form. Zinc dithionite was prepared in the pulp mill from zinc and sulfur dioxide as follows: $Zn + 2SO_2 \rightarrow ZnS_2O_4$. Bleaching is carried out at pH 5-6 with chelating agents such as ethylenediaminetetraacetic acid (EDTA) or sodium tripolyphosphate (0.1-0.2% on pulp) to prevent metal ions such as iron(III) from coloring the pulp. Bleaching is often carried out in the refiners. The reaction time is on the order of 10-30 minutes. The brightness gain is only 5-8%.

Dithionite reacts with oxygen, so bleaching with it is carried out at 4% consistency; consistency below this is unnecessarily dilute, so reaction with dissolved oxygen consumes dithionite. Consistency above 4% leads to entrained air that consumes dithionite. [The solubility of oxygen in the atmosphere is only a few milliliters of gas per liter of water at 25°C (77°F), and decreases with increasing temperature; thus, entrained oxygen is more significant than dissolved oxygen at high consistencies for dithionite degradation.] Using temperatures as high as 70°C (158°F) reduces the oxygen solubility in water. Fig. 11-1 shows the solubility of oxygen (from air) in water as a function of temperature.

Sometimes dithionite is added between the refiner plates where temperatures above 100°C (212°F) cause steam to displace air. Dithionite ion reduces lignin and is itself oxidized to sulfite ion. If hydrogen peroxide and dithionite are used in a two-stage process, the dithionite must be the second stage or hydrogen peroxide will reoxidize those moieties reduced by the dithionite. The reaction of dithionite is shown below.

$$S_2O_4^{2-} + 2H_2O \rightarrow 2HSO_3^- + 2H^+ + 2\ e^-$$

Peroxide bleaching

Some metal ions, such as Fe^{3+}, Mn^{2+}, and Cu^{2+}, catalytically decompose hydrogen peroxide,

so peroxide bleaching is carried out with agents that deactivate these metal ions.

$$H_2O_2 \xrightarrow{Fe^{3+}} H_2O + 1/2\ O_2$$

Chelating agents, such as EDTA, have the added gain of preventing pulp discoloration by binding with ferric ion that would otherwise form a colored complex with the phenolic lignin structure.

Sodium silicate (5% on wood) is also used (usually after the addition of magnesium ion). The mechanism for inactivating the ions by sodium silicate is not clear; it may precipitate the ions, but, strictly speaking, it is not a chelating agent. Buffering action is required to keep the pH high even as organic acids are produced as a result of some carbohydrate degradation. Sodium silicate acts as a buffering agent.

Bleaching conditions are 0.5-3% peroxide and 0.05% magnesium ion (to mitigate carbohydrate degradation by oxygen under alkaline conditions) on pulp, temperature of 40-60°C (104-140°F) (about 20°C lower than with chemical pulps since lignin removal is not the goal), pH of 10.5-11, consistency of 10-20%, 1-3 hour retention time, with a brightness gain of 6-20%.

Hydrogen peroxide with sodium hydroxide and/or sodium peroxide (NaOOH) is used to produce the high pH that is necessary to produce the active perhydroxyl ion, HOO^-. The formation of the perhydroxyl ion is as follows:

$$H_2O_2 \rightleftharpoons HOO^- + H^+ \qquad pK_a = 11.65\ @\ 25°C$$

Some carbohydrate degradation occurs and is responsible for about half of the peroxide consumed. Pine and fir are difficult to brighten. Color reduction occurs by altering chromophoric groups such as orthoquinones. The pulp is sometimes subsequently treated with SO_2 to neutralize OH^- and reduce any residual peroxide.

5.3 MEASUREMENT OF LIGNIN CONTENT

General considerations

The measurement of lignin in chemical pulps is a vital tool to monitor the degree of cook (extent of delignification during pulping) or to measure residual lignin before bleaching and between various stages of bleaching to monitor the process, although pulp brightness between bleaching stages is often used to control the bleaching operation. Since most of the lignin remains in mechanical pulps and bleached mechanical pulps, it is not measured in these pulps.

Lignin is easily measured indirectly by measuring the amount of an oxidant (such as chlorine or potassium permanganate) consumed by lignin in a sample of pulp of known mass. Fig. 5-1 is a comparison of several cooking degree numbers. Methods based on consumption of potassium permanganate (kappa and K numbers) are the most common ones used. The idea (Fig. 5-2) is to treat the pulp with an excess of permanganate ion (or chlorine for Roe or C number) for a specified period of time to oxidize the lignin. After the specified reaction time, the residual permanganate (or chlorine) is determined by indirect titration; i.e., the excess permanganate converts iodide to iodine, and the iodine is titrated with thiosulfate (Section 14.5).

Other methods of lignin measurement are included here since they appear in some of the older literature or have laboratory uses.

Kappa number

The kappa test is an indirect method for determining lignin by the consumption of permanganate ion by lignin. The kappa number is the number of milliliters of 0.1 $KMnO_4$ consumed by one gram of pulp in 0.5 N sulfuric acid after a ten minute reaction time at $25\,^{\circ}C$ ($77\,^{\circ}F$) under conditions such that one-half of the permanganate remains unreacted. The 50% residual permanganate is titrated to determine the exact consumption. Experimentally, 30-70% excess is common with factors to convert this to 50%. TAPPI Standard T236 is based on this procedure. The kappa number test can be used on bleached pulps, unbleached pulps, and high yield chemical pulps by use of a single scale with bleached pulps always giving low numbers and unbleached pulps giving high numbers, unlike the K number. A variety of lignin contents are accommodated by varying the amount of pulp used in the test, but keeping the amount of $KMnO_4$ constant.

It is useful to know how these tests correspond to actual lignin content. This is achieved by measuring the lignin content of pulps gravimetrically and relating this to the kappa number. The

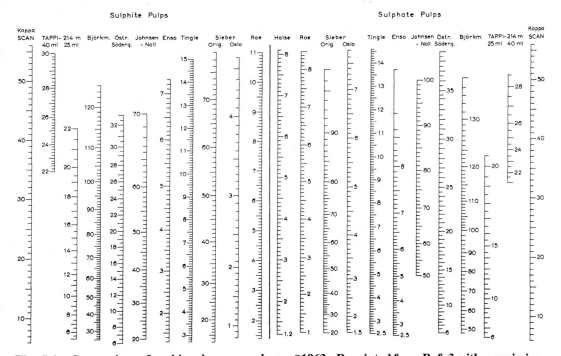

Fig. 5-1. Comparison of cooking degree numbers. ©1963. Reprinted from Ref. 3 with permission.

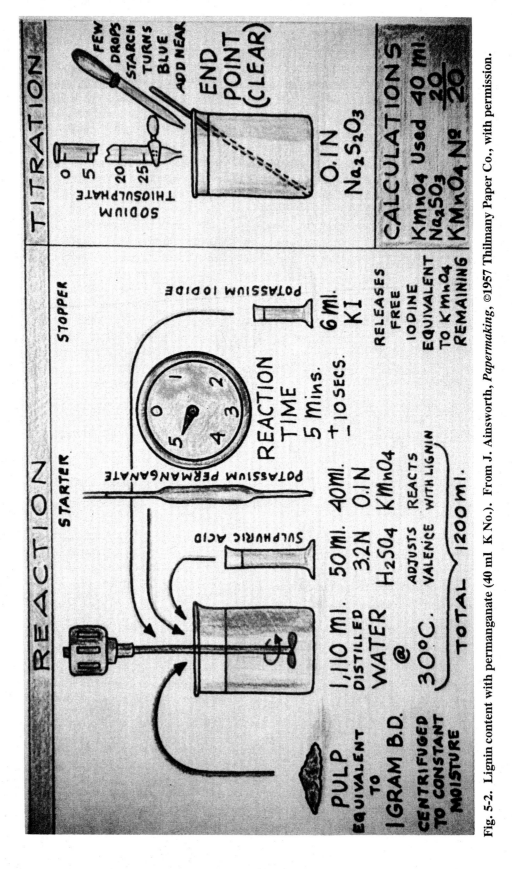

Fig. 5-2. Lignin content with permanganate (40 ml K No.). From J. Ainsworth, *Papermaking*, ©1957 Thilmany Paper Co., with permission.

Klason lignin procedure is described below. Klason lignin is considered to be essentially the same as the actual lignin content. Ålander, Palenius, and Kyrklund (1963) give the following relationship for sulfite and kraft chemical hardwood pulps:

Klason lignin, % = 0.15 × kappa number

This relationship is approximately correct for softwoods as well. Chiang et al (1987) found the coefficient of 0.159 for Douglas-fir and 0.168 for western hemlock. The authors related kappa number to the sum of Klason lignin and acid soluble lignin (which are defined below).

Permanganate number, K number

The permanganate (or K) number, is really four different tests. A constant amount of pulp is used with either 25 ml (for bleached pulp), 40 ml (Fig. 5-2), 75 ml, or 100 ml (for high yield pulps) of permanganate. Results of the 100 ml K number test are not easily compared to the results of the 75 ml (or any other) K number test, so there is no continuum or results for all types of pulps as with the kappa test. Guillory (1982) gives the relationship in the equation:

log (kappa no.) = 0.837 + 0.0323 (40 ml K no.)

Therefore, a 40 ml K number of 10 corresponds to a kappa no. of 14.5, 20 (40 ml K number) corresponds to 30.4 kappa number, and 30 (40 ml K number) corresponds to 64.1 kappa number.

Roe number

The Roe number is a measure of lignin content by the number of grams of gaseous Cl_2 consumed by 100 grams dry pulp at 25°C (77°F) in 15 minutes. TAPPI Standard 202 (now withdrawn) was one method. Ålander, Palenius, and Kyrklund (1963) give the following relationship for hardwood pulps:

Roe number = 0.158 × kappa - 0.2   (kraft)

Roe number = 0.199 × kappa + 0.1   (sulfite)

Chlorine number, C, hypo number

The chlorine number is a test method similar to that of Roe, except the ClO_2 is generated *in situ*

by acidification of sodium hypochlorite. TAPPI Standard T 253 uses this method to determine a hypo number. The following empirical equation relates the chlorine number to the Roe number.

Chlorine number = 0.90 × Roe number

Klason lignin, acid insoluble lignin

Klason lignin is the residue obtained after total acid hydrolysis of the carbohydrate portion of wood. It is a gravimetric method for determining lignin directly in woody materials, for example, by TAPPI Standard T 222. This method is not used for routine quality control in the mill, but has uses in the laboratory. Wood meal or pulp is treated with 72% sulfuric acid at 20°C (68°F) for 2.0 hours followed by dilution to 3% sulfuric acid and refluxing for 4 hours. The lignin is filtered in a tared crucible, washed, dried, and weighed. The isolated lignin in this manner is degraded considerably; nevertheless, it corresponds (by weight) closely to the original amount of lignin in the sample. Some of the lignin, especially in sulfite or hardwood pulps, remains soluble (called acid soluble lignin) and can be estimated spectrophotometrically in the UV region.

In the wood chemistry literature many modifications of the hydrolysis conditions exist, but the difference in the amount of lignin isolated is probably not appreciable. For example, the primary hydrolysis may be carried out for 1.5 h at 25°C (77°F) or 1 h at 30°C (86°F); the secondary hydrolysis is often carried out at 4 or 6% H_2SO_4 for 3 to 4 hours, but TAPPI Standards seldom offer such flexibility.

5.4 BLEACHING CHEMICAL PULPS

Cellulose viscosity

The cellulose degree of polymerization in low yield pulps and bleached pulps is very important since these processes lower the degree of polymerization of cellulose to the point where the paper strength properties are adversely affected. Cellulose viscosity of mechanical pulps and high yield pulps are not measured since it is usually quite high and not a factor in the strength properties of papers derived from these pulps. Cellulose viscosity is measured by dissolving the pulp in

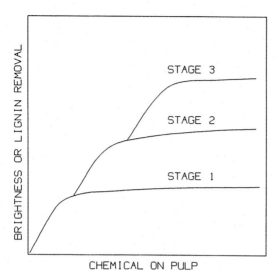

Fig. 5-3. Hypothetical increased brightness vs. chemical usage for three stages of bleaching.

cupriethylenediamine solution and measuring the viscosity of the solution (TAPPI Standards T 254 and T 230). See Section 6.3 for more detail.

Full chemical pulp bleaching

Chemical pulp bleaching is accomplished with various compounds containing chlorine or oxygen and alkali extractions in several stages. Fig. 5-3 indicates why several stages with different compounds are used. The use of three to seven stages increases the efficiency of bleaching by reducing the amount of chemical required. This is due to the complex nature of lignin; each bleaching chemical is going to react differently with lignin. Since lignin is a complex molecule with different types of linkages, the use of different chemicals will break various types of bonds. For example, a large increase in brightness is achieved by using relatively small amounts of ClO_2 in a later stage that could only be achieved using massive amounts of additional Cl_2 in stage 1; use of large amounts of chlorine in stage 1 would also cause much carbohydrate degradation. Plate 20 shows handsheets made from pulp at various stages of the bleaching process. Plate 21 shows several washers of a four stage bleach plant.

Since chemical pulps are dark to begin with, bleaching increases brightness up to 70% with a maximum brightness of about 92%. Unlike bleached mechanical pulps, the high brightness is stable since the lignin is removed. Bleached chemical pulps are insensitive to color reversion, but high temperatures may induce some color reversion. Lignin removal is accompanied by significant losses of pulp yield and strength of the individual fibers. However, the strength of fiber-fiber bonding increases after bleaching.

Bleaching of chemical pulps involves the use of chemicals which are more specific to lignin removal than to carbohydrate degradation compared with the chemicals used in pulping. If this were not true, one could simply continue the pulping process in order to remove more lignin. On the other hand, bleaching is much more expensive than pulping for a given amount of lignin removal. Within the range of useable bleaching chemicals, some are very specific to lignin removal while others are much less specific and cause appreciable carbohydrate degradation and diminished yield. For example, oxygen and chlorine are relatively inexpensive, but not particularly selective for lignin removal. These chemicals are used in the early stages of bleaching to remove most of the lignin. Residual lignin is removed in later stages with expensive, but highly selective bleaching agents like chlorine dioxide, hypochlorite, and hydrogen peroxide.

Table 5-1 is a summary of conditions used in various bleaching stages. Bleaching letters are explained below. Typical sequences are CEH to a brightness of 84-86% or CEHD or CEHHD to a brightness of 92%. Fig. 5-4 shows the layout of a CEHDP sequence. Each stage consists of a pump to mix the chemical with the pulp, a retention tower to provide time for the bleaching chemical to react with the pulp (Plate 22), and a washer (Fig. 5-5) to remove the bleaching chemicals and solubilized pulp components. Diffusion washers may also be used (but are much less common) and are similar to those discussed in Chapter 3 in regards to pulp washing.

C stage, Chlorine (Greek chloros, greenish yellow)

Normally, chlorine is the first bleaching stage, where unbleached pulp is treated with elemental chlorine, Cl_2, that is either gaseous or in solution, at a pH of 0.5-1.5. Bleaching is carried out with a pulp consistency of 3-4%, ambient

Table 5-1. Summary of conditions used in various bleaching stages of chemical pulps.

General conditions ↓	C stage	E_1 stage	H stage	D stage	P stage	O stage
Chemical addition (on pulp)	3-8%	2-3%	2% (as Cl_2)	0.4-0.8%	1-2% Na_2O_2; Mg^{2+}; silicate	2-3% 0.4-.8 MPa 60-120 psi Mg^{2+}
Pulp consistency	3-4%	10-18%	4-18%	10-12%	10%	20-30% 10-12%
pH	0.5-1.5	11-12	8-10	3.5-6	8-10	10-12
Temp. °C	20-30°	50-95°	35-45°	60-80°	60-70°	90-110°
Time, hours	0.3-1.5	0.75-1.5	1-5	3-5	2-4	0.3-1.0

temperature since the reaction is quick, and a retention time of 0.3-1.0 hour. Pressurized, upflow reactors are used since the solubility of chlorine in water is low (4 g/L at STP). Chlorine application is 6-8% on softwood kraft pulps or 3-4% on sulfite or hardwood kraft pulps.

Elemental chlorine was the first agent used to chemically bleach cellulose fibers (cotton, not wood) and has been used since shortly after its discovery in 1774 by Scheele. Its large-scale use

commercially for pulp bleaching had to wait until stainless steel was available in the 1930s.

Chlorine is manufactured concomitantly with sodium hydroxide by electrolysis of sodium chloride; since these two chemicals are produced together, one often speaks of the chlor-alkali industry. One method of producing these chemicals is shown in Fig. 5-6. The production of chlorine is summarized as follows:

$$2NaCl + 2H_2O + elect. \rightarrow Cl_2 + 2NaOH + H_2$$

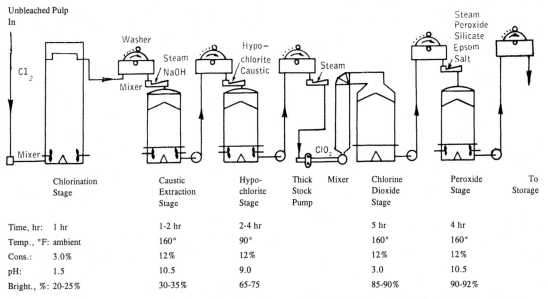

	Chlorination Stage	Caustic Extraction Stage	Hypo-chlorite Stage	Thick Stock Pump	Mixer	Chlorine Dioxide Stage	Peroxide Stage	To Storage
Time, hr:	1 hr	1-2 hr	2-4 hr			5 hr	4 hr
Temp., °F:	ambient	160°	90°			160°	160°
Cons.:	3.0%	12%	12%			12%	12%
pH:	1.5	10.5	9.0			3.0	10.5
Bright., %:	20-25%	30-35%	65-75			85-90%	90-92%

Fig. 5-4. CEHDP sequence used to bleach pulp to 90-92% brightness. Reprinted from *Making Pulp and Paper*, ©1967 Crown Zellerbach Corp., with permission.

Fig. 5-5. An open, bleach washer of the vacuum drum type. Modern washers are enclosed.

Chlorine is not overly specific to lignin, and much carbohydrate degradation occurs through its use. The chlorine reacts with lignin by *substitution* of hydrogen atoms for chlorine atoms (particularly on the aromatic ring), *oxidation* of lignin moieties to carboxylic acid groups, and, to a small extent, *addition* of chlorine across carbon-carbon double bonds (Fig. 5-7). The substitution reactions (Fig. 5-8) are probably the most important in the ultimate lignin removal.

It has been known for a long time that chlorine is first rapidly depleted by the pulp and then is depleted much more slowly. This indicates several types of reactions are occurring. Fig. 5-9 shows how chlorine reacts with pulp as a function of time. The substitution and addition reactions are much faster than the oxidation reactions and do not leave Cl⁻ in solution, whereas oxidation reactions do. This is the basis for determining the type of reaction occurring.

Oxidation includes reactions with both lignin and carbohydrates. Oxidation of the carbohydrates leads to a decreased cellulose viscosity and decreased pulp strength. The species HClO (Fig. 17.1 shows its formation versus pH) especially contributes to oxidative degradation of the carbohydrates; therefore, chlorination is carried out above pH 0.5 (to avoid acid hydrolysis of cellulose) and below pH 1.5 (to avoid carbohydrate degradation by oxidation). Also sodium hypochlorite bleaching is carried out with excess NaOH to avoid carbohydrate oxidation by HClO. According to Giertz (1951) the actual species causing oxidation may be a complex of hypochlorous acid and hypochlorite ion as the pH of maximum degradation is 6.5 to 7.5 and not 4-5, where hypochlorous is at a maximum (Fig. 17-1).

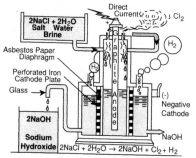

Fig. 5-6. A method of NaOH and Cl₂ manufacture. Redrawn from J. Ainsworth, *Papermaking*, ©1957 Thilmany Paper Co., with permission.

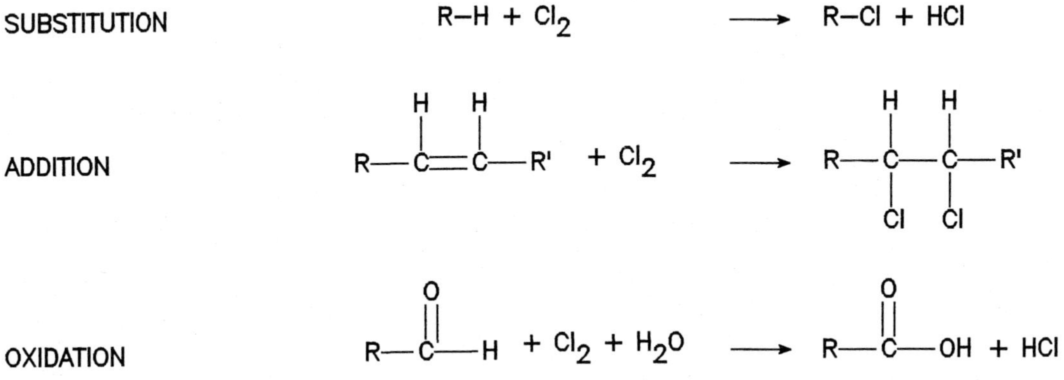

SUBSTITUTION   R–H + Cl$_2$ ⟶   R–Cl + HCl

ADDITION
$$R–\overset{\displaystyle H}{\underset{}{C}}=\overset{\displaystyle H}{\underset{}{C}}–R' + Cl_2 \longrightarrow R–\overset{\displaystyle H}{\underset{\displaystyle Cl}{C}}–\overset{\displaystyle H}{\underset{\displaystyle Cl}{C}}–R'$$

OXIDATION
$$R–\overset{\displaystyle O}{\overset{\|}{C}}–H + Cl_2 + H_2O \longrightarrow R–\overset{\displaystyle O}{\overset{\|}{C}}–OH + HCl$$

Fig. 5-7. Examples of Cl$_2$ reactions with lignin. Oxidation reactions occur with both lignin and carbohydrates. In the case of the latter, significant decreases in pulp strength result.

In the past, the amount of chlorine added to pulp was controlled by measuring the residual chlorine after the chlorination stage. In fact it is better to avoid adding excess chlorine, particularly at elevated chlorination temperatures, so that the substitution reactions occur to completion (rapidly), but oxidation reactions are controlled by not allowing residual chlorine to be present. As Fig. 5-9 shows, the oxidation reactions do not increase the amount of lignin removed in the alkali extraction of sulfite pulp; however, this is less true for kraft pulps because of their highly condensed nature (O'Neil, et al, 1962). For the same reason kraft pulps require longer chlorination periods than sulfite pulps, 60-90 minutes versus 15-45 minutes, respectively. The practice of substituting about 10% of Cl$_2$ with ClO$_2$ (C$_D$) has been practiced for a long time since it results in a stronger pulp by avoiding over-chlorination.

Lignin is not removed to a large degree in this stage, and the pulp actually gets darker (with a characteristic orange color). The pulp is diluted to 1% consistency and washed to remove acid which would otherwise consume alkali in the next stage. The lignin removal and brightness increase actually occur in the alkali extraction stage that invariably follows the chlorination stage.

Chlorination produces chlorinated organic materials including a very small amount of dioxin, which has led to the use of other chemicals to replace a part or all of the chlorine use in bleaching (e.g., O and CD bleaching stages). Dioxin and bleaching is discussed in Chapter 11. Many mills have already replaced up to 50% of the Cl$_2$ with ClO$_2$. Chlorine-free bleaching sequences are used commercially for sulfite pulps at a few mills.

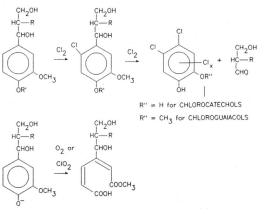

Fig. 5-8. Example reactions of bleaching agents with lignin.

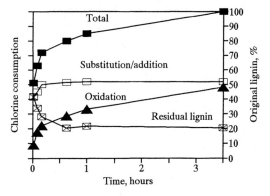

Fig. 5-9. Reaction of excess Cl$_2$ and sulfite pulp with time. The top 3 curves use the left scale. The bottom curve (right scale) is residual lignin after NaOH extraction. After Giertz (1951).

Papers made from these pulps are promoted as "environmentally friendly" or words to that effect. It is difficult to make bright grades of kraft pulps without the use of chlorine, but it is possible if a system is designed specifically for this purpose. The future will bring many changes in this area.

CD stage

The CD or C_D stage is a modification of C stage bleaching, where some of the chlorine is replaced with ClO_2. ClO_2 acts as a scavenger of chlorine radicals in this stage. Substitution of 10% of the chlorine with chlorine dioxide is used to prevent over chlorination. Substitution of 50% or more of chlorine with chlorine dioxide at many mills is becoming common to reduce production of dioxins and other chlorinated organic chemicals. One problem is that since Cl_2 and NaOH are both derived from electrolysis of NaCl, a slackening in the Cl_2 market will mean much higher NaOH prices. Other industries have taken advantage of excess Cl_2 supplies, such as the polymer industry in the manufacture of polyvinyl chloride and other chlorinated polymers, which has mitigated this effect.

E stage

The E stage is *extraction* of degraded lignin compounds, which would otherwise increase the chemical usage in subsequent bleaching stages, with caustic (NaOH) solution. It follows the C stage and sometimes other stages of bleaching. When it follows the C stage (E_1), it is used at 2-3% on pulp, often with a downflow tower due to the high consistency of pulp (10-18%), with a temperature of 50-95°C (120-200°F), and a reaction time of 0.75 to 1.5 hours. High temperatures and alkali loading up to 5% are used to remove hemicelluloses for dissolving pulps or absorbent pulps. In later E stages, alkali is used at less than 1% on pulp. The alkali displaces chlorine and makes the lignin soluble by reactions such as:

$$Lignin-Cl + NaOH \rightarrow Lignin-OH + NaCl$$

The lignin in the E_1 effluent gives a dark color that is ultimately responsible for much of the color of the final mill effluent. Recently oxygen gas has been incorporated into this stage (0.5% on pulp) at many mills and the term E_O applies.

H stage

The H stage consists of bleaching with hypochlorite solution, usually as the sodium salt NaClO; this is the same chemical found in household liquid bleach. This stage is carried out at 4-18% consistency, 35-45°C (95-113°F), 1-5 hours, and pH 10. The process is often controlled by measuring the oxidation-reduction potential. It is important to maintain the pH above 8 because below this pH hypochlorite is in equilibrium with significant amounts of hypochlorous acid (Fig. 17-1), which is a powerful oxidant of carbohydrates, with a E° of 1.63 V. Since the pH is high, lignin is continuously extracted as it is depolymerized. Hypochlorite reacts principally by oxidation. About 1% on wood (based on chlorine) is used. This chemical is more selective than elemental chlorine, but less selective than chlorine dioxide; consequently, the use of hypochlorite has decreased since the advent of chlorine dioxide. Cellulose is oxidized along its chain, forming aldehyde groups and making random cleavage more likely. Calcium hypochlorite was discovered in 1798 by dissolving Cl_2 in an aqueous slurry of calcium hydroxide and was the principal bleaching agent of the industry for a century. Sodium hypochlorite, which is now used since it leads to less scaling, is made from chlorine as follows:

$$Cl_2 + 2NaOH \rightarrow NaOCl + NaCl + H_2O$$

D stage

The D stage involves bleaching with chlorine dioxide. Chlorine dioxide (first studied for pulp bleaching in 1921 by Schmidt and used for commercial pulp bleaching in the mid 1940s) is relatively expensive, but highly selective for lignin. This makes it very useful for the latter bleaching stages where lignin is present in very low concentrations. It is explosive at concentrations above 10 kPa (1.5 psi, or 0.1 atm); hence, it cannot be transported and must be manufactured on site. Its solubility is 6 g/L at 25°C with a partial pressure of 70 mm Hg. It is used at consistencies of 10-12%, 60-80°C (140-176°F), for 3-5 hours at a pH of 3.5-6. It is used at 0.4-0.8% on pulp. Downflow towers are used to decrease the risk of gas accumulation. The D stage is useful for reducing shive contents.

$ClO_2 \rightarrow \frac{1}{2}Cl_2 + O_2$ (undesirable breakdown above 100 torr pressure)

$2ClO_2 + H_2O \rightarrow$ (slow, undesired de-
$HClO_3 + HClO_2$ composition in soln.)

Chlorine dioxide may react in two steps. In the first step ClO_2^- is formed; this then reacts under acidic conditions to form Cl^-. Thus, $ClO_2 \rightarrow ClO_2^- \rightarrow Cl^-$. It reacts by oxidation; one reaction is shown in Fig. 5-8.

P stage

Bleaching with hydrogen peroxide, H_2O_2, is not common for chemical pulps. (But this is changing somewhat as mills look for chlorine-free systems.) It is usually used for brightening mechanical pulps, but when it is used to bleach chemical pulps it appears as the last stage of a sequence such as C-E-H-P or C-E-H-D-P. It is used at 10% consistency, 60-70°C (140-160°F), pH of 8-10, for 2-4 hours. It is an expensive bleaching agent, but may be used more frequently as the use of elemental chlorine decreases. Peroxide oxidizes carbonyl groups of carbohydrates (produced by oxidants such as hypochlorite) to carboxylic acid groups. Its use with chemical pulps is fairly similar to that with mechanical pulps, except for a higher temperature, as described in Section 5.2.

O stage, oxygen pulping and bleaching

Oxygen bleaching or pulping is the delignification of pulp using oxygen under pressure (550-700 kPa or 80-100 psi) and NaOH (3-4% on pulp). Oxygen bleaching has been used commercially since the late 1960s. This is an odorless, relatively pollution-free process used prior to chlorination at high consistencies (20-30%) or medium consistencies (10-15%). Delignification is carried out at 90-130°C (195-266°F) for 20-60 minutes. The key to the use of O_2 delignification was the discovery that small amounts of magnesium ion (0.05-0.1% on pulp) must be present to protect the carbohydrates from extensive degradation. This is the most inexpensive bleaching chemical to use, but also the least specific for lignin removal. A considerable decrease in cellulose viscosity accompanies this process.

Bleaching may be thought of as extended delignification, that is, an extension of the pulping process. Many mills are considering an oxygen delignification step after pulping but before the traditional chlorination first step of bleaching. Some call it oxygen delignification; others call it oxygen bleaching. Many mills have added oxygen to the first alkali extract step, designated as the E_o step, because it is not a major change to the process and saves bleaching chemicals in subsequent stages. Oxygen is even less specific at lignin removal than chlorination so it is used to decrease the softwood pulp kappa number from 30-35 after pulping (20-24 for hardwoods) to 14-18 after the oxygen bleaching stage. Bleaching to kappa numbers below this leads to unacceptable losses of the cellulose viscosity.

The effluent of oxygen delignification can be used in the brown stock washers or otherwise ultimately sent to the recovery boiler because there is no chloride ion present that would lead to high dead loads and corrosion in the recovery boiler. Reducing the amount of lignin removed by the chlorination step by half reduces by half the color and BOD from the bleach mill water discharge after secondary treatment. Environmental considerations are the main reason for using oxygen bleaching. The bleach plant produces mush of the color of the final mill discharge.

There are two main methods of oxygen bleaching: medium and high consistency. The high consistency process is at 30% consistency and 90-110°C (195-230°F) in a pressurized reactor. In the medium consistency process pulp at 10-15% consistency (the normal consistency as it comes off the drum washers) is used. While high consistency oxygen bleaching is more common, there are some difficulties with it. The gas in the reactor contains high concentrations of oxygen and volatile organic chemicals such as methanol, ethanol, and acetone that are potentially explosive; it is difficult to evenly distribute NaOH and O_2; and the pulp is saturated with oxygen which tends to cause foaming on the pulp washers.

Studies indicate that the temperature of oxygen bleaching is important to the selectivity of the process with better selectivity at 100°C (212°F) than 130°(266°F). Partial pressure of oxygen above 200 kPa (30 psig) leads to little increase in the rate of delignification.

Ozone bleaching

Young (1992) describes a mill in Austria that uses ozone bleaching. The mill is able to use ozone since magnesium sulfite pulp is produced (as opposed to kraft pulp that is inherently difficult to bleach), hardwood is pulped, which has less lignin than softwood, and the mill makes dissolving pulp where a low pulp viscosity is desired. Ozone bleaching promises to become increasingly important and is discussed in Section 34.10.

Other stages

Nitrogen oxides, and other chemicals are being considered for bleaching stages, but do not enjoy commercial significance as yet.

Biobleaching

Biobleaching is a term used to describe each of two completely different methods. One method is the use of decay fungi or enzymes isolated from decay fungi to selectively degrade lignin. It is akin to biopulping, and, like biopulping, is still only in the laboratory stages.

Biobleaching is also used to describe the application of hemicellulases to pulp to help the bleaching process. One example is the use of xylanase enzymes. These enzymes hydrolyze the xylans into short chains. Their mode-of-action is explained as the breaking of lignin carbohydrate linkages so that the lignin is removed from the pulp. The method has been largely promoted by manufacturers of enzymes. It is difficult to know if this method will ever have widespread use.

Effluent reuse in the bleach plant

Up to 20,000 gallons of water per ton pulp is required to bleach pulp in a five-stage sequence if there is no reuse of effluents. The use of counter-current washing of pulps in the bleaching stages has reduced this value to less than half. The reuse of chlorination filtrate in the chlorination plant has reduced the figure further. Some mills use the E1 effluent in the brown stock washers, but relatively high amounts of chlorine will cause corrosion and dead load in the recovery cycle. Use of the effluent from first stage oxygen bleaching in the brown stock washers, since it does not contain large amounts of NaCl, also conserves water.

5.5 ANNOTATED BIBLIOGRAPHY

General aspects
1. O'Neil, F.W., K. Sarkanen, and J. Schuber, Bleaching, in *Pulp and Paper Science and Technology*, Vol. 1, Libby, C.E., Ed. McGraw-Hill, New York, 1962, pp 346-374. A good overview of bleaching chemistry.

2. Singh, R.P., Ed., *The Bleaching of Pulp*, Tappi Press, Atlanta, 1979, 694 pp. While many recent developments in oxygen delignification, decreased use of elemental chlorine, dioxin formation, etc. are not covered, this text is an excellent reference for the topic. It is well illustrated and also covers equipment.

Kappa number, permanganate number, K number
3. Ålander, P., I. Palenius and B. Kyrklund, The relationship between different cooking degree numbers, *Paperi ja Puu* (8):403-406(1963).

4. Chiang, V.L., H.J. Cho, R.J. Puumala, R.E. Eckert, and W.S. Fuller, *Tappi J.* 70(2):101-104(1987).

5. Guillory, A.L., *1982 Brown Stock Washing*, Tappi Press, Atlanta, Georgia, p 29.

Chlorination, organochlorine compounds
6. Giertz, H.W., Delignification with bleaching agents, in E. Hägglund, Ed., *Chemistry of Wood*, Academic Press, New York, 1951, pp 506-530.

7. Pryke, D.C., Mill trials of substantial substitution of chlorine dioxide for chlorine: Part II, *Pulp Paper Can.* 90(6):T203-T207(1989).

8. Reeve, D.W., Organochlorine in paper products, *Tappi J.* 75(2):63-69(1992). The chlorinated organic compounds in bleached papers do not seem to be a health hazard due to the low concentration of mobile species and low fraction that might be transferred to the user.

9. Hise, R.G., R.C. Streisel, and A. M. Bills, The effect of brown stock washing, split addition of chlorine, and pH control in the C stage on formation of AOX and chlorophenols during bleaching, *Tappi J.* 75(2):57-62(1992). (Also in *1991 TAPPI Pulping Conference Proceedings*.) Split addition of chlorine and other modifications of pulping and bleaching can decrease the production of dioxins and other chlorinated organic compounds. This is a useful work with 20 references.

Oxygen bleaching and delignification
10. Coetzee, B., Continuous sapoxal bleaching--operating, technical experience, *Pulp Paper Mag. Can.* 75(6):T223-T228 (1974).

The first commercial mill to use oxygen delignification was the Enstra mill of SAPPI in South Africa where, notably, water is in short supply. The successful operation of this process at SAPPI led to the widespread use of oxygen delignification and is reported in detail here.

There are well over 100 articles on the subject in the primary literature of the last 20 years. Some recent general articles on oxygen bleaching follow.

11. Shackford, L.D. and J.M. Oswald, Flexible brown stock washing eases implementing oxygen delignification, *Pulp & Paper* 61(5):136-141(1987); no references.

12. Kleppe, P.J. and S. Storebraten, Delignifying high-yield pulps with oxygen and alkali, *Tappi J.* 68(7):68-73(1985). This publication reports the results of four years production experience at a mill using medium consistency oxygen bleaching.

13. Enz, S.M. and F.A. Emmerling, North America's first fully integrated, medium-consistency oxygen delignification stage, *Tappi J.* 70(6):105-112(1987). This is a very good summary of operating conditions and results for hardwood pulp.

14. Althouse, E.B., J. H. Bostwick, and D.K. Jain, Using hydrogen peroxide and oxygen to replace sodium hypochlorite in chemical pulp bleaching, *Tappi J.* 70(6):113-117(1987). This is a review article with ten references that concludes, in general, that using less hypochlorite at equal brightness gives higher pulp viscosity for kraft and sulfite pulps.

15. Tench, L. and S. Harper, Oxygen-bleaching practices and benefits: an overview, *Tappi J.* 70(11):55-61(1987). This is a worldwide survey of mills using the title method.

16. Idner, K., Oxygen bleaching of kraft pulp: high consistency vs. medium consistency, *Tappi J.* 71(2):47-50(1988). This article concludes that a slightly higher pulp viscosity is achieved with medium consistency than with high consistency oxygen bleaching.

17. Hornsey, D., A.S. Perkins, J. Ayton, and M. Muguet, A mill survey of oxygen utilization in extraction and delignification processes, *Tappi J.* 74(6):65-72(1991). This is a practical discussion on oxygen delignification from mill experiences, especially for medium consistency mills, the growth of which has been spurred by the development of mixers for dispersing oxygen into pulp.

The chemistry of oxygen delignification
18. Hsu, C.L. and J.S. Hsieh, Advantages of oxygen vs. air in delignifying pulp, *Tappi J.* 69(4):125-128(1986).

19. van Lierop, B., N. Liebergott, and G. J. Kubes, Pressure in an oxidative extraction stage of a bleaching sequence, *Tappi J.* 69(12):75-78(1986).

20. Gellerstedt, G. and E.-L. Lindfors, Hydrophilic groups in lignin after oxygen bleaching, *Tappi J.* 70(6):119-122(1987).

21. Hsu, C.L and J.S. Hsieh, Oxygen bleaching kinetics at ultra-low consistencies, *Tappi J.* 70:(12):107-112(1987). This is a laboratory study conducted at 0.4% consistency.

Chlorine dioxide bleaching

22. Schmidt, E., *Ber.* 54:1860(1921); ibid., 56:25(1923).

23. Wilson, R., J. Swaney, D.C. Pryke, C.E. Luthe, and B.I. O'Connor, Mill experience with chlorine dioxide delignification, *Pulp & Paper Can.* 93(10):T275-T283(1992). This article is a good starting point with 42 references.

Miscellaneous topics

24. Young, J., Lenzing mill bringing second ozone bleaching line onstream, *Pulp & Paper* 66(9):93-95(1992).

25. Senior, D.J. and J. Hamilton. Biobleaching with xylanases brings biotechnology to reality, *Pulp & Paper* 66(9): 111-114(1992).

EXERCISES

1. What is the fundamental difference in approach to bleaching chemical pulps versus mechanical pulps?

2. In what types of pulps is color reversion a problem and why?

3. How is the lignin content of pulp measured as a quality control measure for the pulping process?

4. To approximately what percentage of lignin in pulp does a kappa number of 30 correspond for a softwood pulp?

5. What is cellulose viscosity, and what is its significance?

6. Why is chlorine bleaching always followed by alkali extraction?

7. Would chlorine be used as a bleaching agent in stage 5 of a bleaching sequence? Why or why not?

8. Why are pH control and chlorine addition of critical concern during chlorine bleaching?

9. At what consistencies, temperatures, chemical usage etc. is oxygen delignification carried out for the medium consistency and high consistency methods? What are the advantages of oxygen delignification? What are the disadvantages?

10. What are the advantages and disadvantages of bleaching in several stages?

11. Why is buffering action important during bleaching with sodium hypochlorite?

12. Why is the effluent from E_1 after chlorination not normally used in the brown stock washers?

13. Discuss how water can be conserved in the bleach plant.

14. Discuss how oxygen bleaching can conserve water, decrease color from the mill effluent, and save on chemical costs.

6

REFINING AND PULP CHARACTERIZATION

6.1 INTRODUCTION TO REFINING

Introduction

Pulp *refining* is a mechanical treatment of pulp fibers to *develop* their optimum papermaking properties. The optimum paper properties, of course, depend on the product being made. Furthermore, there is always a tradeoff between various properties. Refining of fibers is very important before making paper from them. Refining increases the strength of fiber to fiber bonds by increasing the surface area of the fibers and making the fibers more pliable to conform around each other, which increases the bonding surface area and leads to a denser sheet. During refining, however, individual fibers are weakened and made shorter due to cutting action. With very long fibers of a few species this cutting action improves the formation of the sheet on the paper machine, but in most cases, it is an undesired effect of refining (even if formation is improved somewhat); consequently, refining is generally a tradeoff between improving fiber to fiber bonding and decreasing the strength of individual fibers.

Most strength properties of paper increase with pulp refining, since they rely on fiber to fiber bonding. The tear strength, which depends highly on the strength of the individual fibers, actually decreases with refining. After a certain point the limiting factor of strength is not fiber to fiber bonding, but the strength of the individual fibers. Refining beyond this point begins to decrease other strength properties besides tear. Refining of pulp increases their flexibility and leads to denser paper. This means bulk, opacity, and porosity values decrease with refining.

Fig. 6-1 shows four paper types made from increasingly refined fibers. Two of these samples are glassine papers which are typically refined to a much higher degree than most printing and packaging papers. It is apparent from these figures that fiber to fiber bonding increases with refining. Also, the volume of space between the fibers decreases; hence, refining increases the density of the sheet.

Fiber Brushing

Refining at high consistency with a relatively large distance between the refiner plates increases fiber-fiber interactions that are termed fiber brushing. This tends to roughen the fiber surface, with minimal fiber cutting for improved fiber-fiber bonding.

Fiber Cutting

Operating refiners at low consistencies with a minimal distance between the refining surfaces increases fiber-bar contact, resulting in fiber cutting. This is desired with certain woods (especially redwood with 7 mm fibers and cotton with long fibers) and other materials containing long fibers to increase the quality of formation on the paper machine. In most cases, however, it is desirable to minimize fiber cutting to maintain high paper strength.

Drainage

Drainage is the ease of removing water from pulp fibers, either by gravity or mechanical means. CSF is a measure of drainage and a useful means for determining the level of refining.

Fibrillation

Fibrillation is the production of rough surfaces on fibers by mechanical action; refiners break the outer layer of fibers, i. e., the primary cell wall, causing the fibrils from the secondary cell wall to protrude from the fiber surfaces.

Average fiber length

The average fiber length is a statistical average length of fibers in pulp. Fiber length is measured microscopically (number average), by classification with screens (weight average) or by optical scanners (number average). Two such averages are expressed here, akin to molecular weight averages of polymers. L_i is the length of the fraction of fibers with a length of i, N_i is the number of fibers of length i, and W_i is the weight of the fraction of fibers with length i. The weight average fiber length is equal to or larger than the number average fiber length.

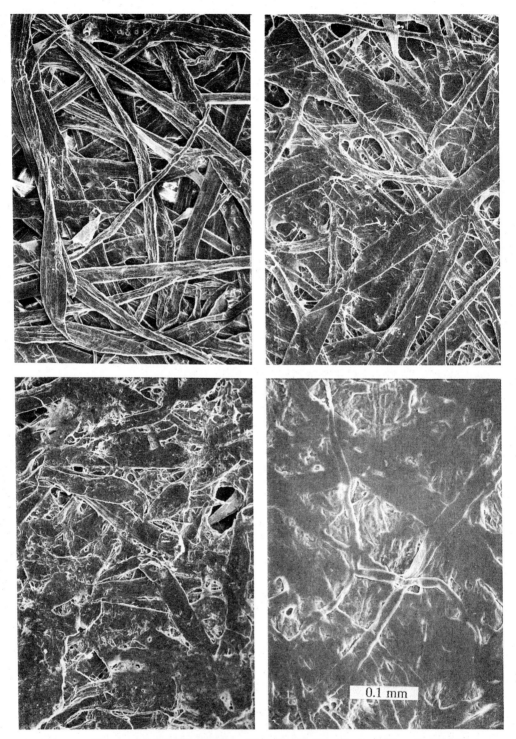

Fig. 6-1. Paper of increasingly refined pulps. Bleached kraft softwood fibers (top left); same pulp with refining (top right); machine glazed florist tissue (bottom left); and glassine weighing paper.

$$\text{number average length} = \frac{\sum N_i \times L_i}{\sum N_i}$$

$$\text{weight average length} = \frac{\sum W_i \times L_i}{\sum W_i}$$

Theory

One action of refining is the "pumping" of water into the cell wall making it much more flexible. This is sometimes compared to the softening of spaghetti upon cooking, which is relevant in that in both cases internal hydrogen bonds are broken with the addition of water molecules. While this would occur whenever fibers are thoroughly wetted, to the extent that refining disrupts the crystallinity of fibers, more water can be added to cellulose fibers.

A second action of refining is *fibrillation*, that is, exposure of cellulose fibrils to increase the surface area of fibers, thereby improving fiber-fiber bonding in the final sheet. For example, the surface area of kraft, softwood fibers is on the order of 1 m^2/g at 750 CSF, but it increases to about 5 m^2/g at 350 CSF.

A third action in refining is delamination of the cell wall, such as between the primary and secondary layers, which increases the fiber flexibility. The analogy of greased leaf springs, which are much more flexible than the equivalent amount of solid metal, has been used to explain this effect.

Refining decreases the pulp freeness, the rate at which water will drain through the pulp. Refined pulp, therefore, has a low freeness. The extent of refining is monitored by measuring the pulp freeness and the strength properties of handsheets made from the refined pulp.

Pulp used to be treated in beaters, such as the Hollander beater that is explained below, but now refiners are used. The terms *beating* and *refining* are often used interchangeably, but refining is applicable to most modern equipment.

Refining variables

Pulps low in lignin (low yield pulps) and high in hemicelluloses are relatively easy to refine in terms of development of strength properties. High temperature and high pH during refining decrease the refining energy requirements. Refining at high pH offers the advantages of an increase in the rate of hydration due to fiber swelling (Scallan, 1983), an increase in the ultimate strength of the paper, and an increase in sheet bulk. It also produces a lower stock freeness, reduces equipment corrosion, and produces a softer sheet. Refining at low pH hardens the fibers, leads to a denser sheet that is hard and snappy, and runs better on the paper machine. Refiner plate speed, plate clearance, and type of tackle are other important variables.

Major changes in refining have resulted in the development of equipment that allows refining to occur at high consistency. High consistency refining leads to more fiber fibrillation and less fiber cutting. Refining at high consistency, however, may lead to fiber curling. Fig. 6-2 shows two pulps, one refined at low consistency (3%) in a Jordan refiner and one refined at high consistency in a double disk refiner. The fibers refined at high consistency have much more fibrillation and less fiber cutting.

Canadian Standard freeness, CSF

Refining is easily monitored by the drainage rate of water through the pulp. A high drainage rate also means a high freeness. Obviously freeness is of utmost importance in the operation of a paper machine. A low freeness means that the paper machine will have to operate relatively slowly, a condition that is usually intolerable.

The CSF test measures the drainage of 1 liter of 0.3% consistency pulp slurry through a calibrated screen. The test is shown in Fig. 6-3. The device is shown in Fig. 6-4. The 1992 TAPPI Standard T 227 revision describes a change in design that was made in 1967, where the angle and shape of the side orifice was changed. The CSF test was developed for use with groundwood pulps and was not intended for use with chemical pulps; nevertheless, it is the standard test for monitoring refining in North American mills. It tends to be most influenced by the fines content of the pulp and to a smaller extent by the degree of fibrillation and other fiber properties. The water is diverted around a spreader cone where, if it drains quickly, some of it overflows to a side orifice and escapes collection from the bottom

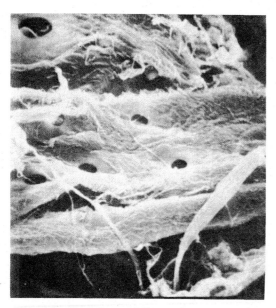
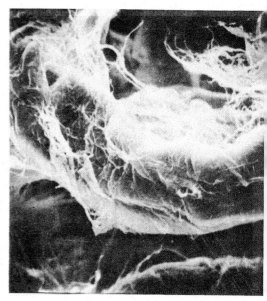

Fig. 6-2. Refining at 3% consistency (left) shows more fiber cutting with less fibrillation than at 30% consistency. From *Making Pulp and Paper*, ©1967 Crown Zellerbach, with permission.

orifice. The water from the side orifice is collected and measured in a graduate cylinder and reported in ml. Unrefined softwood might be 700 ml CSF, whereas mechanical pulp is about 100-200 ml CSF (Fig. 6-5).

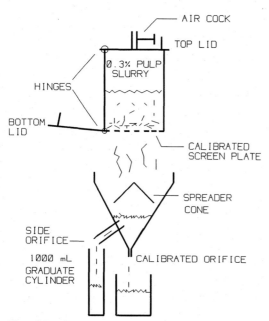

Fig. 6-3. Schematic diagram of the Canadian Standard freeness test in progress.

Fig. 6-4. The Canadian Standard freeness device.

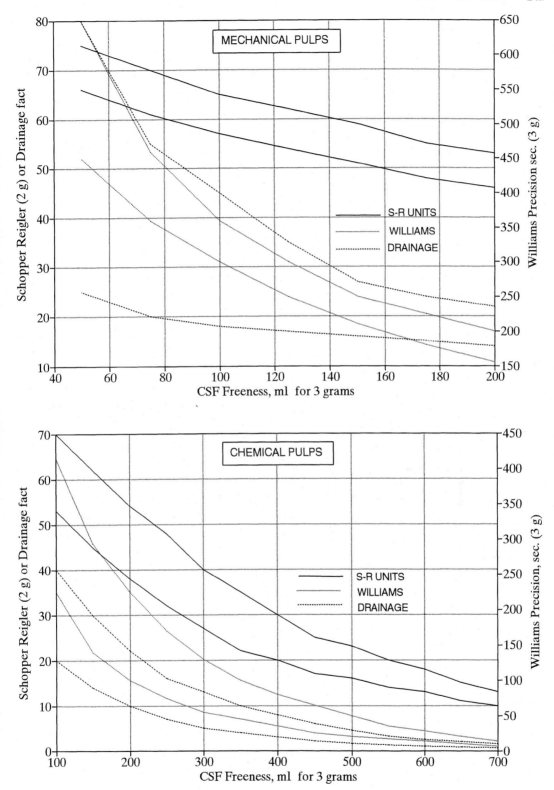

Fig. 6-5. Comparison of freeness scales for mechanical and chemical pulps. Data from TIS 0809.

There are other freeness tests that are used around the world. Perhaps the most common one is the *Schopper Reigler* test, which is similar in concept to the CSF test. TAPPI TIS 0809-01 gives inter-conversions of CSF, *Schopper Reigler units*, *Williams Precision*, and *Drainage Factor* for various types of pulp. Fig. 6-5 has two graphical presentations from the first table of TIS 0809. Since the conversions are not exact, the two lines of each set represent the bounds of the conversion. See Section 17.5 for CSF correction equations for temperature and consistency.

6.2 REFINING

Beater

A beater is an early device (the *Hollander Beater* was invented in the 1700s) used to treat pulp to improve the papermaking properties. Beating is a batch process where the pulp slurry circulates through an oval tank around a midsection and passes between a revolving roll with bars and a bedplate with bars. The pulp is at about 6% consistency and emphasizes fiber brushing. The use of these had been phased out at most mills by the late 1970s because they are slow and expensive

to operate. Fig. 6-6 shows a Hollander beater that was in use until the mid 1970s. Some mills retain these for use in mixing stock for small paper machines.

Refiners

Refiners are machines that mechanically macerate and/or cut pulp fibers before they are made into paper. There are two principal types: *disk* and *conical refiners*. Disk refiners have superseded the conical refiners for many purposes, as they offer many advantages.

The operation of a *conical refiner* (Fig. 6-7) is similar to the operation of a disk refiners, except for the geometry of the refiners. In conical refiners, the refining surfaces are on a tapered plug. These surfaces consist of a *rotor* that turns against the housing and the *stator*, both of which contain metal bars mounted perpendicularly to rotation. The *Jordan refiner*, patented by Joseph Jordan in 1858, is one type of conical refiner with a 12° angle on the rotor (with respect to the longitudinal axis); it is suited to low consistency refining (2% consistency) with much fiber cutting. The *Claflin* refiner uses a rotor with a 45° angle with respect to the longitudinal axis that revolves

Fig. 6-6. The Hollander beater for mill beating of pulp.

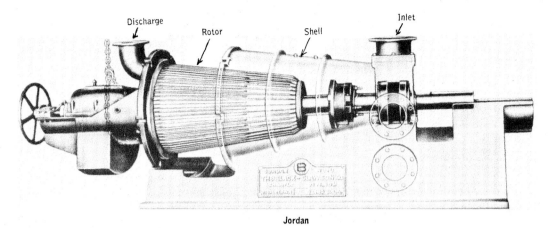

Jordan

Claflin

Fig. 6-7. The Jordan and Claflin conical refiners. Reprinted from *Making Pulp and Paper,* ©1967 **Crown Zellerbach Corp., with permission.**

inside the mating shell. The inlet is at the small end of the taper. The Claflin refiner is intermediate in operation to the Jordan disk refiners. These two refiners are shown in Fig. 6-7.

Disk refiners became available for papermaking in the 1930s, after the conical refiners. Disk refiners had enjoyed widespread use in food processing. For example, they are used to process peanuts into peanut butter and corn into cornstarch and flour. The pulp slurry makes one pass between rotating plates equipped with teeth or bars. There are three common configurations: one fixed plate with one rotating plate (Fig. 6-8), two rotating plates that turn in opposite directions, or a set of two pairs of plates formed by a double sided rotating disk between two stationary plates (Plate 23.)

Disk refiners are able to operate at high consistency, which favors fiber fibrillation with minimal fiber cutting. They have lower no-load energy requirements (an indication of energy that does not contribute to refining), are more compact, and are easier to maintain. Disk refiners are also used for production of mechanical pulp from wood chips. *Tackle* (the plates) is easily replaced; a wide variety of tackle metals (Table 6-1) and designs (Fig. 6-9) exist for pulping and refining.

It is interesting to consider the historical development of beating and refining in terms of the angle of the bars with respect to the axis of

Fig. 6-8. Single-disk refiner (notice the plate segments). The ribbon feeder (for pulping or high consistency refining) is observed through the right disk. Courtesy of Andritz Sprout-Bauer.

rotation. The Hollander beater has an angle of $0°$, the early conical refiners have $12°$ angles, the Claflin refiner has a $45°$ angle, and disk refiners have $90°$ angles. Generally, the higher the angle, the higher the consistency at which refining can occur, leading to lower fiber cutting.

Table 6-1. Typical industry refiner plate metallurgies. Courtesy of Andritz Sprout-Bauer.

Metallurgy	Hardness (Rc)	Corrosion Resistance	Abrasion Resistance	Impact Resistance	Elonga-tion	Fluidity	Cost
Ni–Hard White Iron	55–62	> Carbon and < SS	Good	Extremely brittle	None	Good	1 X
X–C (Hi–C) White Iron	50–55	Lower than SS	Good	Brittle	None	Fair	1.5 X
MCK & K–Alloys White Iron	50–55	> Ni–Hard < Hi–C	Good	> Ni–Hard & X–C	None	Fair	1.5 X
440–C High–Carbon Stainless Steel	55–60	Better than white iron	Less than white iron	Tougher than white iron	1%–2%	Poor	3 X
SA1 High–Carbon Stainless Steel	50–55	Same as 440–C	Same as 440–C	Same as 440–C	1%–2%	Poor	3 X
17–4 PH Stainless Steel (SS)	32–40	Excellent	Less than others	Best resis–tance of all	10%–15%	Very poor	4 X

Low Consistency Refiner Plates for Pulp & Paper Stock Preparation

(Twin Flo Refiners)

High Consistency Refiner Plates for Pulp, Paper, Board & Industrial

Mechanical Pulping

Wet Processing

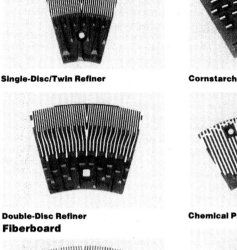

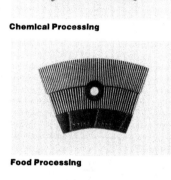

Cutting — Coarse Bar

Single-Disc/Twin Refiner

Cornstarch

Fiber Development — Medium Bar

Double-Disc Refiner

Fiberboard

Chemical Processing

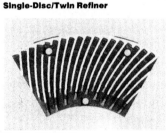

Maximum Development — Fine Bar

Single-Disc/Twin Refiner

Food Processing

Double-Disc Refiner

Fig. 6-9. Representative refiner plate designs. Courtesy of Andritz Sprout-Bauer.

Disk refiner plates

Calderon, Sharpe, and Rodarmel (1987) provide an informative summary of low consistency refining for fiber approach or hot stock refining. The consistency should be 3.5-5%; consistencies below 2.5% causes undue wear and short plate life. Fibrillation (separation of the S-1 cell wall layer) occurs at the bar edge, so more, narrow bars means higher fibrillation. Smaller

volume in the groves promotes refining action, but decreases the volume that flows through the refiner. Increasing the bar angle increases refining, but also increases the refining power required. *Dams* are used to inhibit channeling of the pulp, but are not required at low consistencies with the proper selection of the plate pattern. Fig. 6-10 shows the effect of many refining variables on the refining process.

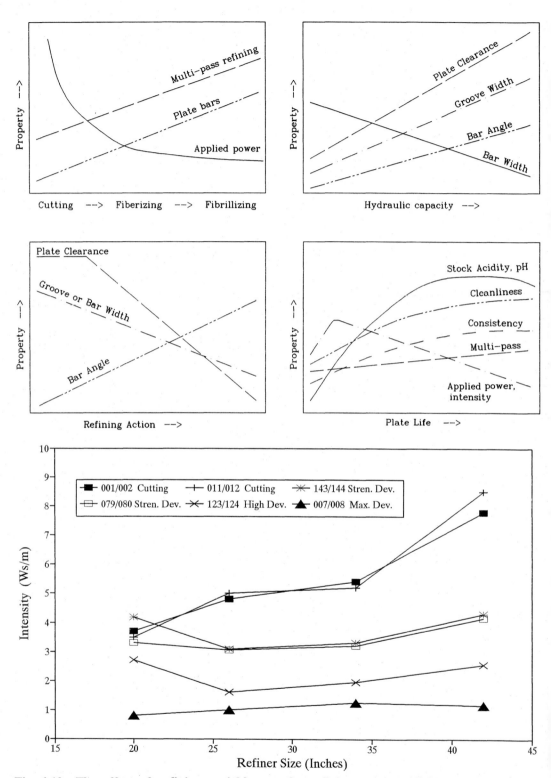

Fig. 6-10. The effect of refining variables on the refining process (Calderon, Sharpe, and Rodarmel, 1987). Courtesy of Andritz Sprout-Bauer.

Fig. 6-11. Valley beater used for laboratory refining of pulp.

Laboratory refining

The *valley beater* (Fig. 6-11) is a laboratory version of the *beater* used to evaluate pulps on a small scale (TAPPI Standard T 200). It requires frequent calibration and adjustments to maintain standardization. For this reason it is facing competition from the *PFI mill* (Fig. 6-12), a laboratory refiner using bars on the edge of a rotating disk against a smooth bed (TAPPI Standard T 248). The amount of refining on a PFI mill may be reported in revolutions or PFI counts. One PFI count is 10 revolutions. Neither method gives results which are directly comparable to commercial scale refining, though relative results can be obtained. Other standard laboratory refiners are (or have been) also used such as the *Kollergang* (TAPPI UM 258), *Lampén*, and *Jokro*. For non-standardized refining, double disk refiners as small as 12 in. in diameter are available.

Fig. 6-13 shows a refining curve from PFI mill refining of an unbleached commercial pulp. Fig. 6-14 shows a beating curve of a commercial pulp using the Valley beater. Both figures are the

Fig. 6-12. PFI mill used for laboratory refining of pulp. The insert shows the refining area.

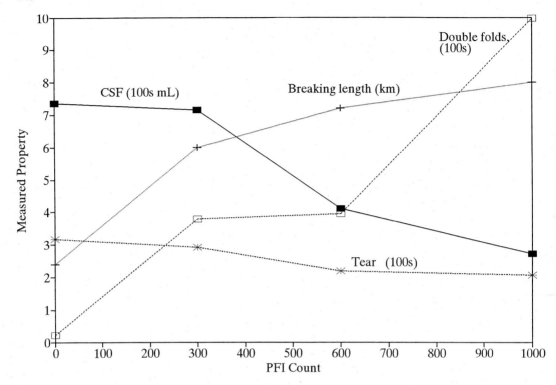

Fig. 6-13.   PFI mill refining of commercial, unbleached Douglas-fir.   (TAPPI standard handsheets.)

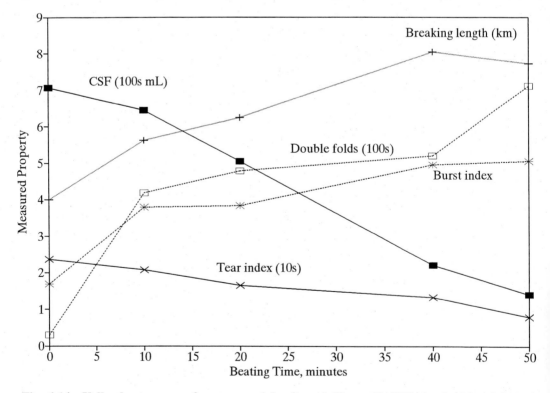

Fig. 6-14.   Valley beater curve for commercial softwood fiber.   (TAPPI standard handsheets.)

test results of laboratory handsheets. These figures represent the strength qualities of paper made from pulp refined at various levels, although it is preferable to have more levels of refining to give smoother curves. Also, one usually plots strength versus CSF, which gives smoother curves.

The tradeoff between pulp properties is demonstrated by looking at the burst and tear strengths with refining. Increased refining decreases the tear strength but increases the burst strength. In brown paper bags both the burst and tear strengths are important, although increasing refining to raise the burst strength leads to lower tear strength; hence, the tear strength of brown paper grocery sacks tends to be low (as most people have experienced at one time or another), but they can hold heavy objects without bursting.

Refining power

Refining power is a measure of the power input to the motors of the refiner based on amount of pulp processed. It is an indirect measure of the energy expended in cutting and fibrillating the pulp fibers, although only a small percentage of the power actually is consumed by these processes. Obviously, these values are important in design and economic calculations and so forth. Refining power is commonly expressed in units of kilowatt-hours per ton or horsepower-days per ton of pulp processed. Smook, *Handbook for Pulp & Paper Technologists*, gives typical refining energy requirements for different paper grades on page 189. For example, tissue and toweling (lightly refined paper using low yield, bleached pulp) require about 100-120 kWh/t (5-6 hp-day/ton) pulp, fine papers require about 230 kWh/t (12 Hp-day/ton) pulp, while greaseproof glassine (a very highly refined, very dense sheet) uses 400-500 kWh/t (20-25 Hp-day/ton) pulp. (There are 17.904 kWh per HP-day and 1.1 ton per t.)

6.3 PULP CHARACTERIZATION

Pulp characterization is very important to determine the effects pulping, bleaching, refining, etc. on the properties of the pulp and, therefore, on the final paper properties. Some of these methods such as Canadian Standard freeness, cellulose viscosity, and lignin content have been

Fig. 6-15. Standard pulp disintegrator. The insert shows the mixing blade.

discussed in more appropriate sections. Additional specialized tests that are not included here are available in the TAPPI Standards or similar resources. In order to disperse pulp fibers into solution a standard disintegrator is often used as shown in Fig. 6-15, which is used in a wide variety of TAPPI Standard methods.

Moisture content

The moisture content of pulps is determined by drying a weighed portion of pulp in an oven at $105 \pm 3°C$ ($221 \pm 5°F$) to constant weight (TAPPI Standard T 210). The sample is weighed in the oven or cooled in a desiccator or other closed container before weighing. The sample is reheated for at least three hours until two successive weights show a variation of 0.1% or less.

Physical properties

To determine what effects pulp modifications will have on the final paper product, laboratory handsheets are made. Fig. 6-16 shows some of the steps involved in making laboratory handsheets on a square mold. Round handsheets 15.9 cm (6.25 in.) in diameter are made according to TAPPI Standard T 205 and tested by methods in TAPPI Standard T 220. This handsheet former is called a *British sheet mold* (Fig. 6-17). Mechanical pulps such as PGW, RMP, TMP, and CTMP have significant fiber curling. These pulps are prepared according to TAPPI Standard T 262, circulating 2% consistency stock at 90-95°C (194-203°F), to fully develop their strength properties by removing the curl. Other handsheet formers are also used.

Fig. 6-16. Preparation of laboratory handsheets for pulp characterization. The pulp slurry is added; the pulp is mixed; the sheet is pressed after draining; and the sheet is dried.

diameter holes) and the sample is 500 ml of 0.5% consistency stock.

Fig. 6-17. British sheet mold for preparing laboratory handsheets.

Fiber length and fines content

The fiber length of pulp is traditionally measured by projection, which is a very tedious procedure (TAPPI Standard T 232). Fiber lengths can also be determined by fiber classification using a series of at least four screens of increasingly smaller openings (TAPPI Standard T 233); two common instruments are the Clark type (Fig. 6-18) or Bauer-McNett type (Fig. 6-19). More recently it is measured automatically in dilute solutions using optical methods. One common instrument for optical analysis is the Kajaani (Fig. 6-20).

TAPPI Standard T 261 is a means of measuring the fines content of pulps with a single screen classifier, the so-called *Britt jar test* (Fig. 6-21). Usually the screen is 200 mesh (76 μm

Pulp viscosity

The pulp viscosity is a measure of the average chain length (*degree of polymerization, DP*) of cellulose. It is determined after dissolving the pulp in a suitable solvent such as cupriethylenediamine solution (Plate 24). Higher viscosity indicates a higher average cellulose DP that, in turn, usually indicates stronger pulp and paper. Decreases in viscosity result from chemical pulping and bleaching operations and to a certain extent are unavoidable. Loss of pulp viscosity must be minimized by proper attention to important process parameters. Cellulose viscosity has little merit in mechanical pulps since the cellulose chains are not appreciably degraded by pulping or bleaching processes that form these. Fig. 6-22 is a comparison of pulp viscosities determined from a variety of methods and cellulose DP.

Bleached chemical pulp and paper deterioration

Brightness reversion of bleached chemical pulps is determined by TAPPI Standard T 260 by measuring the brightness before and after exposing the pulp to 100% humidity at 100°C (212°F) for one hour. TAPPI Standard T 430, the copper number of bleached pulp, paper, and paperboard, measures the amount of $CuSO_4$ reduced by the fiber source. This is an indication of oxycellulose, hydrocellulose, lignin, and sugars that reduce the copper sulfate. These materials may be markers for deterioration and may be an indicator for paper permanence. Papers containing ZnS, $CaSO_3$, melamine resins, and other materials that reduce $CuSO_4$ may be used, provided the total reducing power of the additives is known and no more than about 75% of the total reducing power of the sheet.

Miscellaneous tests

Dirt (foreign matter in a sheet that has a marked contrasting color to the rest of the sheet) in pulp is numerically quantified by TAPPI Standard T 213 (Section 34.18). Foreign particulate matter (particles and unbleached fiber bundles which are embedded in wood pulp as viewed by transmitted light) is measured by TAPPI Standard

Fig. 6-18. Clark pulp fiber classifier.

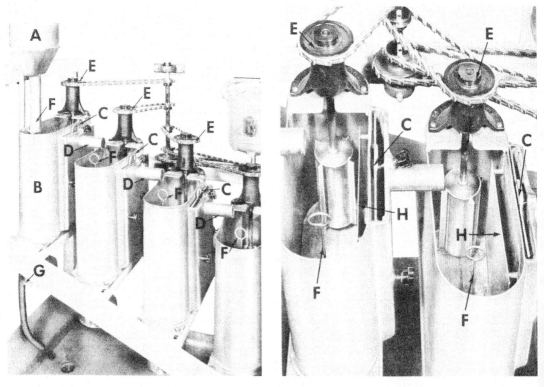

A, Constant level funnel; B, first tank; C, screen; D, outlet; F, stopper; G, drainage cup; H, midfeather.

Fig. 6-19. Bauer-McNett pulp fiber classifier. Courtesy of Andritz Sprout-Bauer.

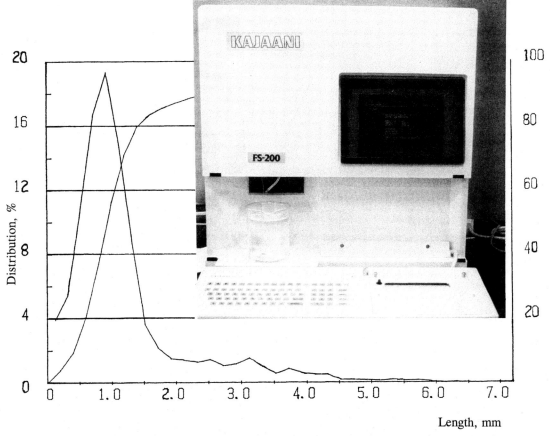

Fig. 6-20. Kajaani optical fiber classifier with analysis of 19,060 fibers (60% hardwood, 40% softwood by weight). Averages are number, 0.55 mm; length, 1.16 mm; and, mass, 1.88 mm.

T 246. The specific surface area of lightly beaten pulps can be measured by the method of Clark using silver deposition (TAPPI Standard T 226). Fig. 6-23 shows the relative surface area of kraft and sulfite pulps as a function of freeness.

6.4 PULP PROPERTIES VERSUS PERFORMANCE

The morphology of paper fibers is important to the properties of the final sheet. Dadswell and Watson (1961, 1962) reviewed the results of 35 references in their discussion of the influence of the morphology of wood pulp fibers on paper properties. They have many interesting points. For example, when determining the effect of fiber length on paper properties, one should not attempt to fractionate pulp. With hardwoods one might get vessels in one fraction that differ by more than

just fiber length from the other fibers; there may be differences in the chemical composition of various fractions too. An often used method they cite (Brown, 1932) is to cut paper into narrow strips, repulp the cut paper, and form handsheets. The main difference in such new hand-sheets is the average fiber length as determined by the width of the cut (or uncut) strips.

Another point they make is that wood fibers with thin cell walls, which are from woods of low specific gravity, tend to make dense well-formed paper (with well bonded fibers) having high burst, fold, and breaking length, compared to fibers with thick cell walls from dense woods. The thickness of fibers is also dependent on the ratio of latewood to earlywood, the ratio of compression to normal wood, and other factors. Softwoods with large amounts of latewood, on the other hand, give bulky, porous sheets, with poorly bonded fibers,

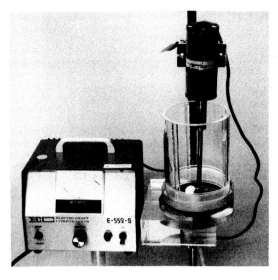

Fig. 6-21. Britt jar test.

having high tear resistance. The authors cite the work of Runkel who says that if the thickness of the cell wall is less than the radius of the lumen of fibers in wood, the wood makes a good paper. The relative thickness of cell walls is reflected in the *fiber coarseness* (mg/100 m of fiber). These workers minimized the importance of the fiber length/diameter ratio that many workers feel is

more important than the cell wall thickness; however, all of these fiber properties are interrelated, and no single variable means much alone.

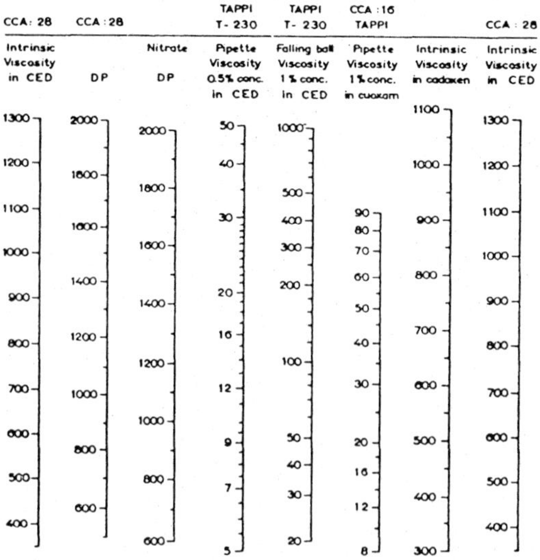

Many studies (for example, Kellogg and Gonzalez, 1976) show that mature boles, i.e., outer growth rings that are 10-50 rings from the center, have fibers with the longest length, the largest radial and tangential diameters, and the smallest fibril angles. Once again, these facts indicate that short rotation, plantation grown wood may have decreased strength properties.

6.5 ANNOTATED BIBLIOGRAPHY

General aspects of refining (Also see references 14, 15 for the effect of pH on refining)

1. Clark, J. d'A., *Pulp Technology and Treatment for Paper*, Miller Freeman Pub. Co., San Francisco, 1978, 751 p. (The 2nd ed. 1985, 878 p., is updated for some of the introductory, basic material, but is otherwise very similar).

Numerous chapters in this book relate to refining including Chapter 8, Fibrillation and fiber bonding, pp 160-180, not a look at the refining process, but a fundamental look at fibrillation of fibers induced by various treatments and how this promotes fiber bonding; Chapter 7, Bonding of cellulose surfac-

Fig. 6-22. Conversion nomograph for pulp viscosities. ©1963 . Reprinted from Siholta et al, (1963) with permission.

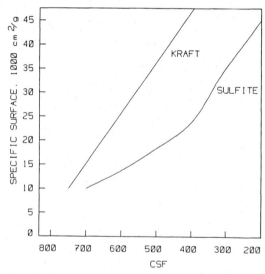

Fig. 6-23. Specific surface area versus CSF for laboratory pulp. After Robertson and Mason (1949).

es, pp 145-159 and early refiners, pp 205-215 are of related interest; Chapter 12, Nature and effects of beating, pp 257-280; Chapter 13, Mill beating and refining, pp 281-316; Chapter 14, Laboratory beating, pp 317-362; and Chapter 25, Control of beating and refining, pp 560-568.

Many chapters from the first edition relate to pulp and fiber quality and testing. Chapter 9, Properties of pulps, pp 181-200; Chapter 11, Moisture content of bales, pp 230-256; Chapter 15, Test sheet making, pp 363-379; Chapter 16, Hand sheet testing, pp 380-400; Chapter 17, Fiber length, pp 402-437; Chapter 18, fiber coarseness, pp 438-449; Chapter 19, Intrinsic fiber strength, pp 450-464; Chapter 20, Cohesiveness, pp 465-487; Chapter 21, Wet fiber compactability, pp 488-503; Chapter 22, Drainage of water from pulp, pp 504-532, Chapter 23, Surface measurements, pp 533-540; Chapter 24, Wet-web strength, pp 541-559; Chapter 27, Characterization and control of mechanical pulps, pp 579-602; Chapter 28, Chemical and microscopical analysis, pp 603-613; Chapter 29, Optical characteristics, dirt, and shives, pp 614-638; and Chapter 32, Formulas for pulp properties, pp 680-698.

Clark describes the properties of fibers (discussed in chapters 17-21) for papermaking in terms of five of their principal properties: the length-average fiber length, the coarseness (mg per 100 m of fiber, measured by Tappi Standard T 234, with coarse fibers giving lower strength due to poorer bonding), wet compactability (measured by the apparent sheet density, for example, TAPPI Standards T 205 and T 220), the intrinsic fiber strength, and the fiber cohesiveness.

2. Calderon, P., P.E. Sharpe, and J.L. Rodarmel, Low consistency refiner plate design and selection, *Spectrum* (Sprout-Bauer), Summer, 1987, 7 p.

3. Root, E.M., Stock preparation, in *Pulp and Paper Science and Technology*, Vol. 2, Libby, C.E., Ed. McGraw-Hill, New York, 1962, pp 1-39. A lot of useful information on early refining methods such as beaters and the Jordan refiner. This also has a good overview on the physical properties of fillers.

4. Spencer, H.S., N.G.M. Tuck, and R.W. Gordon, Beating and refining, in *Pulp and Paper Manufacture, Vol. 3*, MacDonald, R. G., Ed., McGraw-Hill, New York, 1969, pp 131-185. Scanning electron micrographs, fundamental aspects of refining, equipment, and refiner plates are included in this article.

Pulp surface area
5. Casey, J.P., Ed., *Pulp and Paper Chemistry and Chemical Technology*, 2nd ed. (1960), volume 2, p 621 has a discussion on the surface area of pulps. The more sophisticated types of studies mentioned use silver deposition or gas adsorption to measure the total available surface area of pulps. These studies give reasonably close values to each other (within a factor of two) for surface area. Fig. 6-23 is an approximation of the results of Robertson and Mason (1949) cited in this work.

6. Robertson, A.A. and S.G. Mason, Specific surface area of cellulose fibres by the liquid penetration method, *Pulp Paper Mag. Can.* 50(13):103-110(1949).

Fiber morphology and performance
7. Dadswell, H.E. and A.J. Watson, Influence of the morphology of woodpulp fibres on paper properties, Transactions of the "Formation and Structure of Paper" Symposium, 25-29 Sept., 1961, pp 537-572 (1962). (See also *Appita* 14(5):168-176(1961) and ibid. 17(6):146-156(1964).)

8. R.B. Brown, *Paper Trade J.* 95(13):27-29(1932).

9.   Kellogg, R.M., and J.S. Gonzalez, Relationship between anatomical and sheet properties in western hemlock kraft pulps. Part I. Anatomical relationships, *Transactions Tech. Soc.* 2(3):69-72(1976).

10. Smith, W.E. and Von L. Byrd, Fiber bonding and tensile stress-strain properties of earlywood and latewood handsheets, USDA, For. Res. Ser. Res. Pap. FPL 193(1972), 9 p.

11. Horn, R.A., Morphology of wood pulp fiber from softwoods and influence on paper strength, USDA, For. Res. Ser. Res. Pap. FPL 242(1974), 11 p. This study concluded that the fibril angle of fibers was the most important factor in stretch properties of paper made from unrefined, softwood pulp and an appreciable factor in stretch properties of paper made from refined, softwood pulp.

12. Dinwoodie, J.M., The relationship between fiber morphology and paper properties: A review of the literature, *Tappi J.* 48(8):440-447(1965). This includes 115 references.

13. Megraw, R.A., *Wood Quality Factors in Loblolly Pine*, Tappi Press, 1985, 88 p. This book is subtitled "The influence of tree age, position in tree, and cultural practice on wood specific gravity, fiber length, and fibril angle".

Fiber chemical properties and performance
14. Scallan, A.M., The effect of acidic groups on the swelling of pulps: a review, *Tappi J.* 66(11):73-75(1983). This has 31 references.

Acidic groups of pulp, i.e., carboxylic acids from hemicelluloses and possibly sulfonate groups in sulfite-pretreated mechanical pulps or unbleached or semi-bleached sulfite chemical pulps, contribute to fiber swelling in water. Fiber swelling contributes to ease of refining and development of paper strength. At pH below 7 the carboxylate groups are protonated and no longer dissociated, decreasing fiber swelling. At higher pH, the metal counter ion plays a big role in fiber swelling with $M^+ > M^{2+} > M^{3+}$. Also, it is recognized that swelling increases with the series $H^+ < Ca^{2+} < Mg^{2+} < NH_4^+ < Na^+$.

Pulps bleached with hydrogen peroxide are often treated magnesium. It may be useful to wash these pulps with some dilute alkali to improve their papermaking properties. Alum would have a detrimental effect on pulp swelling and should be kept out of the process until after refining.

15. Lindström, T. and G. Carlsson, The effect of chemical environment on fiber swelling, *Svensk Papperstidn* 85(3):R14-20(1982).

Swelling of pulps under a variety of pH and metal ion conditions was determined by the water retention values (WRV). Bleached softwood sulfate pulp was fairly insensitive to changes in WRV under these conditions since it had only 2.5 meq/100 g (acidic groups on pulp). An unbleached softwood sulfate pulp (47% yield with 38 kappa number and 7.4 meq/100 g acidic groups on pulp) had larger changes in WRV going from a low of 200 to a high of 260. The maximum WRV occurred at pH 8-9.5 with NaCl concentration below about 0.01 M. 0.05 M NaCl decreased the WRV to 220 at pH 9, and a NaCl concentration of 0.5 M decreased it to 200, the base value. The tensile index was directly proportional to WRV. These results indicate that one should keep the ionic strength of pulp low during pulping and pulp storage before and during papermaking.

Pulp viscosity
16. Siholta, H., B. Kyrklund, L. Laamanen, and I. Palenius, Comparison and conversion of viscosity and DP-values determined by different methods. This is the most extensive and about the only publication on the subject. *Paperi ja Puu* (4a):225-232(1963).

EXERCISES

Refining

1. Plot representative tensile strength, tear strength, CSF, and specific surface area of paper made from pulp as a function of refining energy (time) applied to the pulp.

2. Describe how refining helps fiber to fiber bonding in the final sheet. Give three mechanisms for its action.

3. Describe the effect of pulp consistency on the refining process in terms of bar-fiber contacts. Use Fig. 6-24 to help you.

4. What is the major drawback of any laboratory refining process?

5. Give two methods whereby the level and effectiveness of refining is measured.

6. One is visiting a mill using redwood fiber and sees a Jordan refiner. Is this unusual considering very few mills use Jordan refiners anymore?

7. Using the discussion following references 14 and 15, describe why refining and papermaking at elevated pH (say pH 8 versus pH 4) often leads to a significant increase in the strength of the paper. When is this especially true?

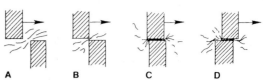

As the upper bar passes the lower, low consistency will favor either trapping single fibers and cutting them, as in B, or stretching and breaking them, as in Views C and D.

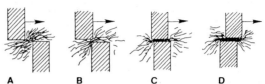

At higher consistencies, the fibers will tend to form a blanket or cushion on the bars, leading to crushing rather than a cutting action on the fibers.

Fig. 6-24. Effects of consistency on refining. **©1991 James E. Kline. Reprinted from** *Paper and Paperboard* **with permission.**

Pulp characterization

8. How are fiber size distributions measured?

9. What is the significance of fiber size distributions?

10. What is the purpose of making laboratory handsheets?

7
PAPER AND ITS PROPERTIES

7.1 INTRODUCTION

Paper

Paper is a pliable material used for writing, packaging, and a variety of specialized purposes. Paper consists of a web of pulp fibers (normally from wood or other vegetable fibers), usually formed from an aqueous slurry on a wire or screen, and held together by hydrogen bonding. Paper may also contain a variety of additives and fillers (Section 8.4).

The first paper is credited to a Chinese man named Ts'ai Lun in the early part of the second century A.D. and was made from a mixture of mulberry bark and hemp. The art of papermaking was a closely guarded secret and only very slowly moved westward. The first American paper mill was built in 1690 in the northeastern U.S. The process was very slow, because it was performed entirely by hand. In the late 18th century, a Frenchman named Louis Robert invented a papermaking machine that used a continuous, moving wire screen belt. He sold the invention rights to the Fourdrinier brothers, who made improvements in the machine's design. It came into widespread use from 1830-1850.

Fiber

A fiber is a tubular or cylindrical element, obtained from plant matter, containing cellulose as the principal constituent. Fibers are the basic components of paper. The properties of pulp fibers determine many of the characteristics of the final paper and have been described (Chapter 6).

7.2 GENERAL GRADES OF PAPER

Paper materials are classified as paper (newsprint, stationary, tissue, bags, towels, napkins, etc.) or paperboard (linerboard, corrugating media, tubes, drums, milk cartons, recycled board used in shoe and cereal boxes, roofing felt, fiberboard, etc.) The industry typically divides paper into broad categories based on the types of fibers used in paper and the weight of the paper. Table 7-1 shows the production quantities of some grades of paper and their approximate cost.

Tissues

Tissue papers are a class of lightweight papers of 15-60 g/m^2. They are made primarily of bleached chemical softwood pulps and may contain bleached softwood sawdust or hardwood pulps to impart smoothness. Sanitary tissues are the most important type by production amount. Increasingly, tissue paper is made with some or all deinked secondary fiber. Some mechanical pulp may be used as well.

Uncoated groundwood

This grade is made from mechanical pulps, although small amounts of chemical pulps are used to increase the strength. TMP pulps, being stronger than groundwood, have decreased the chemical pulp requirements of these grades. Newsprint accounts for about 80% of this grade, but it also includes directory, computer, catalog, and similar grades. The term *uncoated groundwood* is a misnomer, since other mechanical pulping processes have largely replaced groundwood. (Fig. 9-62 contains figures of uncoated groundwood, coated groundwood, and supercalendered, coated groundwood papers.)

Coated groundwood

Coated groundwood by definition includes at least 10% mechanical pulps, although 50% is more typical. This grade is used in magazines, catalogs, some No. 4 and all No. 5 enamel grades, letterpress, offset, and LWC. Coated groundwood was used in almost 70% of the 1.6 million tons of paper used in U.S. magazines produced in 1990.

Uncoated wood-free paper (printing or writing)

Uncoated wood-free paper is predominantly made from kraft or sulfite softwood pulps and may contain limited amounts of mechanical pulp or

Table 7-1. Capacities (thousands of tons) and price of paper by grade.[1,2]

Paper type	U.S. 1988	U.S. 1976	1988 US$/MT	Global, 1989
Newsprint	6116	4015	$540	35 000
Coated groundwood	4348	2298	$1100	
Coated free-sheet	3437	1903	$880-1600	
Uncoated free-sheet	12 062	7465	$650-860	
Bristols	1170	1165	$1000	
Industr. packaging	5404	6199	unbleach. $550	
Tissue	5667	4500		12 500
TOTAL	40 205	29 219		
Paperboard type				
Unbleach. kraft liner	20 269	14 568	$420	
Solid bleached	4732	3998	$730	
Semi-chemical	5789	4758	$360	
TOTAL	40 326	32 082		88 000

[1]The principal sources are the *1990 North American Factbook and Pulp and Paper Week*, Nov 21, 1988. Paper costs are approximate. Price ranges indicate a wide range of possible products and grades.

[2]The global production of printing and writing papers in 1988 was 60,358 tons with percentages follow: North America, 37.3%; EEC, 21.5%; Asia, 17.8%; Scandinavia, 10.6%; the rest of Europe, 7.3%; Latin America, 4.1%; Australia, 0.52%, and Africa, 0.85%. The per capita consumption of all paper and board in the same year in pounds was: North America, 669; Scandinavia, 537; EEC, 321; Australia, 279; rest of Europe, 96; Latin America, 55; Asia, 39; and Africa, 12. *PPI, July 1989.*

recycled fiber. (Wood-free implies less than 10% mechanical pulp.)   It is used for envelopes, photocopy, bond, and tablet papers.

Coated wood-free paper

The base sheet of coated, wood-free paper is made from kraft or sulfite softwood pulps. Coating is applied on one or both sides. Coated paper is supercalendered to produce smooth, glossy surfaces for good printing. This paper is used for high grade (No. 1-3 and some 4) enamel papers for magazines, books, and printing.

Kraft wrapping or bag

Kraft bag or wrapping papers are made from bleached or unbleached kraft softwood pulp of southern pine or Douglas-fir. They are made in various weights from 50 to 134 g/m^2.

Cast-coated paper, machine glaze, MG kraft

Cast-coated paper is a very high gloss paper made by allowing the coating on paper to dry on a large, chrome plated dryer with a polished surface. These papers are used for wrapping, carbonizing, wax base, and specialty papers.

Specialty papers

Specialty papers are made for specific uses. They are produced in small volumes, but have the potential for high profit margins. These papers are used in packaging, printing, manufacturing, electronics, etc. They include capacitor, cigarette, bible, glassine, and greaseproof papers.

Kraft paperboards

Paperboards are any of the heavyweight papers, generally above 134 g/m², used in packaging. *Bleached paperboards*, about half of which are coated, are made from bleached kraft pulp and are used in folded milk cartons, index cards, cups, and plates. *Unbleached paperboards* are used in linerboard and corrugating medium for the production of corrugated boxes. Recycled fiber is being used in larger amounts in the production of unbleached paperboards.

Chipboard, recycled paperboard

Chipboard is a thick paper of low density and quality used for making solid fiber boxes and papers that require low strength. It has multi-ply construction and is often made on cylinder machines. The furnish is often recycled newspapers and other inexpensive pulps, which is known as recycled paperboard. Grades include gypsum liner, core (tube) stock, and clay coated folding boxboard. These grades have a characteristic grayish cast because the furnish is not deinked.

Market pulp

Some mills produce pulp but do not have paper machines. Their pulp is sold on the open market as wet lap (about 50% solids) or dry lap, which, is carefully dried to no more than about 80-85% solids to avoid irreversible loss of hydrogen bonding sites that occurs when pulp is over-dried. Other mills produce excess pulp that is also sold on the market. Newsprint makers that use bleached kraft pulp usually purchase it on the open market. Dissolving pulp is also produced as market pulp since a handful of mills provide pulp to the many users of dissolving pulp. Fig. 7-1 and Plate 16 show the production of market pulp.

7.3 SPECIFIC TYPES OF PAPER

TAPPI TIS 0404-36 lists the general grades of paper that are used to make specific types of papers with designated end uses. Over 400 types of paper are included in this reference, which is an indicator of the diversity of the pulp and paper industry. Specific descriptions of a few high production, commodity types of paper are listed here. Electron micrographs of some of these papers may be found in Fig. 3-2.

Newsprint

Newsprint is paper used for printing newspapers and other low cost, short lived pub-

Fig. 7-1. Wet lap production for market pulp or for use within the same mill at times when the pulp mill may shut down but it is important to keep the paper machines running.

lications. Newsprint contains a high percentage of mechanical pulp and a small percentage of full chemical pulp to increase the strength. Newsprint must have moderate strength, good printability, and low cost. Today, the basis weight of newsprint is 48.8 g/m² (30 lb/3000 ft²), down from 52.2 g/m² (32 lb/3000 ft²) that was typical before 1974. This decrease in basis weight has been made possible by improved manufacturing techniques that produce a stronger, more opaque paper. Over 70% of newsprint in North America is made on twin-wire machines at speeds up to 3500 ft/min. The caliper of newsprint is nominally 0.085 mm (3.3 mil or 0.0033 in.)

Bond paper

Bond papers are a broad category of high quality printing or writing papers. They are made from bleached chemical pulps and cotton fibers and may be watermarked. *Rag bonds* are made with a minimum of 25 percent rag fiber from cotton. They are high quality, long lasting, strong, expensive papers. Nominal weights of bond are 48, 60, 75, or 90 g/m² (corresponding to 13, 16, 20, or 24 pounds per 17 in. by 22 in. by 500 sheet ream.)

Fine papers

Fine papers are intended for writing, typing, and printing purposes. They may be white or colored, are made from bleached kraft or sulfite softwood pulps, and may contain hardwood pulp for smoothness and opacity.

Tissue

Sanitary tissues are made from bleached sulfite or kraft pulps and are soft, bulky, absorbent, and moderately strong. *Wrapping tissues* are used for wrapping clothes, flowers, etc. and are made of bleached kraft or sulfite softwood pulps that impart high strength. *Tracing tissue* is a dense, transparent grade of paper that is used for tracing.

Glassine and greaseproof paper

Glassine papers are made from highly refined chemical pulps, resulting in a very dense, translucent sheet. Often a special grade of un-bleached sulfite pulp with a high hemicellulose content is used to make a nonporous sheet. These papers are used in laboratory weighing papers, greaseproof papers, and food handling papers like the fluted paper cups for chocolates. The nominal caliper of glassine is 0.03 mm (1.2 mil or 0.0012 in.)

Linerboard

Linerboard is an unbleached kraft softwood sheet of southern pine or Douglas-fir (when made in the U.S.), made in various weights. Frequently, linerboard is a two ply (duplex) (or even triplex) sheet, made on a flat wire paper machine with two headboxes. It is used in making corrugated boxboard; the two outer visible layers of corrugated boxes are linerboard. The compression strength and burst strength are important. Some common structural linerboards have basis weights of 205, 229, and 337 g/m² (42, 47, and 69 lb/1000 ft²), with calipers of 12, 13, and 19 thousandths of an inch (0.001 in.) and burst strength of 80, 90, and 130 psi (*1990 Pulp and Paper Factbook*). Other basis weights are 127, 161, 186, and 439 g/m² (26, 33, 38, and 90 lb/1000 ft²) with 9-24 point caliper (0.23-0.64 mm). Linerboard is assembled to corrugating medium with the machine direction perpendicular to the flutes of corrugating medium since the flutes contribute strength and stiffness in the weaker cross machine direction.

Corrugating medium

Corrugating medium is made from unbleached, semi-chemical pulp, especially NSSC hardwood pulp, and recycled fiber from corrugated boxes. It is formed into a fluted (wavy) structure and sandwiched between two plies of linerboard to form the corrugated structure for boxes. Occasionally the green liquor process is used as are softwoods to make the pulp furnish. This paper must be stiff and inexpensive. This board normally has a basis weight of 127 g/m² (26 lb/1000 ft²) and a caliper of 0.23 mm or 0.009 in., so-called 9 point board. There are four standard flute shapes commonly used in the U.S. Letting F denote the number of flutes (thickness does not include the linerboards), they are:

Flute A with 118 F/m (36 F/ft) with a flute thickness of 4.8 mm (3/16 in.)

Flute B with 168 F/m (51 F/ft) with a flute thickness of 2.4 mm (3/32 in.)

Flute C with 130 F/m (40 F/ft) with a flute thickness of 3.6 mm (9/64 in.)

Flute E with 316 F/m (96 F/ft) with a flute thickness of 1.2 mm (3/64 in.)

Corrugating medium must have relatively high stiffness and resistance to crush. Also, corrugating medium should be made from adequately refined pulp, since shives will cause paper breaks during its conversion into boxes. Plasticity of corrugating medium also contributes to good performance during conversion without paper breaks and is enhanced by drying it under only moderate restraint.

No carbon required (NCR)

Carbonless paper was developed by the 3M company in the 1950s and is manufactured by several companies today. This paper uses encapsulated chemicals on the backside of one sheet that react with a chemical on the front of the adjacent sheet to give color. The capsules and second chemical may be on the same sheet so that any paper can be used as the original.

The chemicals are encapsulated using emulsion polymerization techniques. When the capsules containing the chemicals are crushed, the chemicals react to form a colored substance and leave a mark. Fig. 7-2 shows a scanning electron micrograph of carbonless paper where pressure from a pen has crushed the chemical containing capsules.

Miscellaneous paper types

Construction boards are thick boards used in building construction (insulation board, etc.). *Molded pulp products* include egg cartons, flower pots, and spacers for packaging. These products are molded primarily from recycled newspapers, although other papers can be used. *Forms papers* are fine papers normally in roll form for

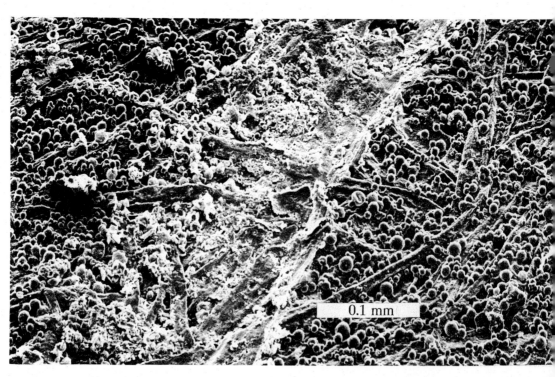

0.1 mm

Fig. 7-2. SEM of NCR carbonless paper showing destruction of the encapsulated pigments.

conversion into continuous, multiple-copy, business forms. These are sometimes called *register bond*. *Construction paper* is a heavy, colored paper that consists primarily of mechanical pulps and that is used for drawing and art work. *Bristols* are heavy, stiff papers of high quality made from bleached kraft pulp, and they are used for postcards, greeting cards, invitations, and announcements.

7.4 BASIC PAPER PROPERTIES

Paper may be characterized by moisture content, physical characteristics, strength properties, optical properties, and other criteria depending on its end use. It is **very important** to obtain a sufficient number of representative samples by using a sampling procedure such as that recommended in TAPPI Standard T 400.

Chapter 15 discusses the detailed calculations for paper tests, the interconversion of metric and English units, and other aspects of paper test calculations. Table 15-2 shows some typical values for paper tested by OSU pulp and paper students in a laboratory exercise. These values are included as "typical" values of a variety of papers and are used for examples of calculations.

When measuring strength properties MD refers to the test force applied in the *machine direction* and CD is force applied in the *cross machine direction*. Many paper properties depend on the orientation of the paper since the MD and CD properties are not identical.

Effect of moisture content

The moisture content of paper has an important effect on paper quality; therefore, paper properties must be measured under standard conditions of temperature and relative humidity [generally 23°C (73°F) and 50% relative humidity] since these affect the equilibrium moisture content (EMC). As stated above, the EMC affects strength and other properties of paper. According to TAPPI T 402, paper should be placed in a hot, dry room (20-40°C or 68-105°F, at 10-35% relative humidity) before placement in the standard room so that the moisture content of paper approaches its EMC by adsorbing water from the atmosphere. Due to the

Table 7-2. EMC values of various papers at 50% RH and 72°F.

Paper type	EMC, %, at 50% RH (based on dry weight of paper)	
	From 35% RH	From 80% RH
Filter, 100% rag	6.6	7.2
Bond, 100% rag	6.6	7.5
Bond, 100% softwood	7.4	8.4
Wrapping, kraft	7.8	9.3
Newsprint	9.0	10.9

hysteresis effect, a slightly higher EMC would result if paper approached the EMC for a given relative humidity and temperature, by giving off water. (Laboratory prepared handsheets are usually not treated in this way as they are air dried without ever going into a hot, dry room.)

Table 7-2 gives some EMC values at 50% relative humidity for several paper samples approaching from below 50% and above 50% relative humidity to demonstrate the hysteresis effect. Fig. 7-3 shows EMC with increasing and decreasing relative humidities. In these figures the softwood bond is paper made from acid sulfite pulp; thus, much of the hemicellulose has been removed by acid hydrolysis, making the paper similar to rag paper. The bond paper of today, made from bleached kraft pulp, probably behaves somewhere between kraft wrapping and sulfite-derived bond paper. There is no such thing as an average paper when it comes to moisture content and the changes in properties as a function of moisture content. Throughout this subsection four or five papers of widely differing qualities are used to demonstrate the differences one expects to encounter among all of the possible papers.

It is important to know the rate of moisture gain or loss of paper to achieve equilibrium in order to understand how long to condition samples. For example, how long should paper be

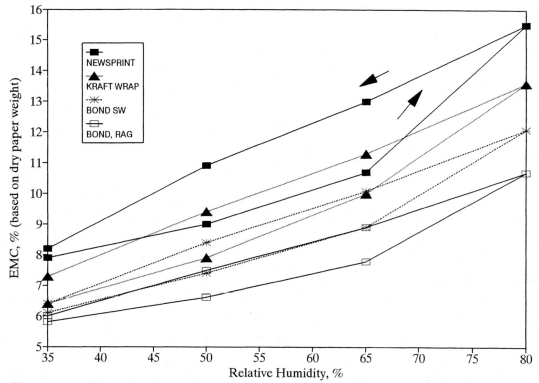

Fig. 7-3. The EMC of four types of paper with increasing or decreasing RH. The lower line for a given paper is increasing RH. The data are from Jarrell (1927).

conditioned when you perform routine quality control at a paper mill? Clearly, if the paper is not conditioned to a known moisture content, then the results may be misleading. Fig. 7-4 shows the moisture content of paper preconditioned at 35% RH as a function of time the paper is at 65% RH and 21°C (70°F). Under these conditions paper (the average of several types) has gained 36% of the weight in 0.5 h, 52% in 1 h, 86% in 4 h, and 92% in 6 h. When paper is conditioned at 35% RH from 65% RH, on the average 62% of the water is lost after 2 h, 72% is lost after 4 h, 86% is lost after 24 h, and 94% is lost after 48 h.

The most recent extensive publication giving paper properties as a function of moisture content is from the staff of the Institute of Paper Chemistry (IPC) from 1937. Data from this study for changes in structural properties of four types of paper as a function of RH at 21°C (70°F) are presented in Table 7-3. (The original data were presented with 65% RH as the baseline, the standard for testing paper at that time.) Data on

sizing and linear expansion are also available in this reference. While some of the data in this table are inconsistent, the trends are valid and are a good approximation for certain grades of paper if one has no other information available. The results of the study by Carson (1944), which investigated folding endurance as a function of relative humidity, are presented in Table 7-4. As indicated in these studies, strength properties may vary by as much as 40% with relative humidity changes that are within the range of normal conditions. Indeed, fold values may change by a factor of five.

Considering the importance of moisture content on paper properties, one would assume this would be well studied so producers would be able to help their customers to know how their materials will behave under various conditions.

Manufactured paper immediately off the reel that is not conditioned before testing may have a moisture content that is several percent below paper that has been conditioned in a standard

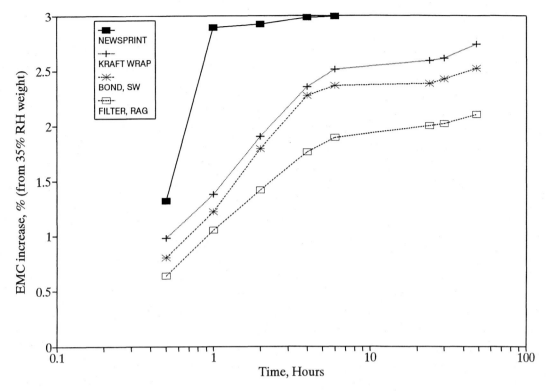

Fig. 7-4. Moisture gain of paper conditioned at 35% RH vs. time at 65% RH at 21°C (70°F). From Jarrell (1927). The x-axis is a logarithmic scale.

room for an hour. Since paper properties may differ significantly over such a moisture content range, one might expect considerable problems in routine mill measurement and quality control of paper properties. The importance of controlling the relative humidity during paper and textile processing has been known for some time. Willis Carrier installed several of his early air conditioning units in textile and printing shops at the turn of this century for this very reason. These were air *conditioning* units and not air *cooling* units. The three papers cited in this section are excellent, thorough works and should be consulted for more information on this topic.

7.5 PHYSICAL PROPERTIES OF PAPER

Ream

A ream is a specified area of paper or paperboard, most frequently expressed in terms of a specified number of sheets (usually 500) of a certain dimension. For example, a 24 × 36 × 500 ream means the combined area of 500 sheets, each 24 in. by 36 in., for a total of 3000 ft^2. Other ream sizes are expressed in area, such as 1000 ft^2 for paperboard. The Tappi Standard ream is 1 square meter of paper or board. For other standard paper ream sizes see Table 12-8.

Basis weight, grammage

The weight of paper per ream, normally expressed on an air-dry basis (about 6-10% moisture, with 9% the standard) is known as the basis weight. For example, a printing paper is reported at 20 pounds per (24 in. × 36 in. × 500 sheet) ream; newsprint is 49 gsm (g/m^2 is the Tappi Standard unit); and one grade of lightweight linerboard is 26 pounds per 1000 ft^2. The basis weight is measured either directly during paper manufacturing with the β-ray gauge or off-machine by weighing a precisely cut piece of paper with a balance (TAPPI Standard T 410, which includes a table of ream sizes with conversion factors.) Electronic balances (in preference to mechanical balances) should always be used to increase speed, accuracy, and precision.

Table 7-3. Paper properties as a function of RH at 70°F (IPC, 1937). Paper samples were allowed to reach equilibrium at each relative humidity.

Base property values at 50% RH increasing from 30% RH at 70°F								
			Tensile, lbs		Stretch, %		Tearing g/16 sheets	
Sample	Basis weight g/m^2	Burst, psi	MD	CD	MD	CD	MD	CD
Bond, rag	92.5	97.3	35.7	20.7	5.8	9.8	97	100
Bond, sulfite	74.8	35.7	21.4	9.8	2.5	4.6	43	41
glassine, sulfite	37.7	16.2	10.4	4.3	1.8	3.2	15	13
newsprint	54.7	8.3	6.0	1.7	1.6	2.1	29	23

Relative humidity	Bond, 100% rag	Bond, sulfite	Glassine, sulfite	Newsprint
	% Change in bursting strength from 50% RH (preconditioned at 30% RH)			
40%, increasing	-0.9	+1.5	+1.1	+3.6
65%, increasing	+1.9	-4.7	-2.0	+3.1
75%, increasing	-10.3	-10.1	-3.9	-15.3
75%, decreasing	-12.4	-11.2	-11.1	-29.4
65%, decreasing	-8.1	-3.4	-1.4	-27.1
50%, decreasing	-4.6	-1.6	+0.7	-40.7
40%, decreasing	-10.1	-1.1	+3.7	-18.6

Table 7-3 Cont'd. Paper properties as a function of RH at 70°F (IPC, 1937).

Relative humidity	Bond, 100% rag	Bond, sulfite	Glassine, sulfite	Newsprint
	%Change in tensile, MD from 50% RH (conditioned from 30% RH)			
40%, increasing	+4.5	-2.6	+7.1	-0.6
65%, increasing	-8.1	-3.8	-6.5	-7.4
75%, increasing	-10.3	-22.1	-14.4	+4.1
75%, decreasing	-17.1	-20.2	-21.1	-16.8
65%, decreasing	-11.1	-9.5	-11.0	-7.8
50%, decreasing	-1.1	-6.0	-4.7	+1.0
40%, decreasing	-1.4	-4.1	-3.4	+8.1
	%Change in tensile, CD from 50% RH (conditioned from 30% RH)			
40%, increasing	+2.5	+4.9	+0.9	-0.7
65%, increasing	-3.4	-8.3	-10.7	-5.7
75%, increasing	-13.4	-22.1	-18.1	-15.6
75%, decreasing	-15.5	-23.5	-21.0	-19.3
65%, decreasing	-7.3	-18.0	+5.5	-11.6
50%, decreasing	-3.4	-8.0	-5.8	-3.7
40%, decreasing	-0.7	-2.2	-2.8	-3.7
	%Change in stretch, MD from 50% RH (conditioned from 30% RH)			
40%, increasing	+3.4	-10.5	-8.8	-0.7
65%, increasing	+8.6	+10.4	-2.9	-5.7
75%, increasing	+22.1	+14.6	+23.6	-15.6
75%, decreasing	+24.1	+24.9	+17.7	-19.3
65%, decreasing	+18.9	+29.0	+17.7	-11.6
50%, decreasing	+7.7	+16.7	+8.8	-3.7
40%, decreasing	+8.6	+6.2	+11.8	-3.7

Table 7-3 Cont'd. Paper properties as a function of RH at 70°F (IPC, 1937).

Relative humidity	Bond, 100% rag	Bond, sulfite	Glassine, sulfite	Newsprint
	% Change in stretch, in CD from 50% RH (conditioned from 30% RH)			
40%, increasing	+1.6	-3.3	-15.3	-9.5
65%, increasing	+20.5	+30.4	+4.0	+7.2
75%, increasing	+29.2	+76.0	+72.8	+23.9
75%, decreasing	+37.3	+41.2	+50.4	+23.9
65%, decreasing	+35.3	+41.2	+56.8	+21.4
50%, decreasing	+8.2	+8.6	+18.4	-7.1
40%, decreasing	+6.7	+5.3	+10.4	-11.9
	%Change in tear, MD from 50% RH (conditioned from 30% RH)			
40%, increasing	-2.4	+39.3	-6.5	-6.8
65%, increasing	+16.4	+79.5	+11.7	+13.1
75%, increasing	+20.5	+78.1	+29.3	+12.8
75%, decreasing	+28.8	+81.0	+55.3	+22.6
65%, decreasing	+20.5	+78.1	+20.2	+18.6
50%, decreasing	+17.2	+45.8	+3.9	+9.8
40%, decreasing	+10.2	+76.8	+9.8	+7.8
	%Change in tear, CD from 50% RH (conditioned from 30% RH)			
40%, increasing	-6.5	-2.9	-7.9	-5.6
65%, increasing	+15.5	+9.9	+11.9	+14.8
75%, increasing	+16.7	+15.8	+12.6	+24.0
75%, decreasing	+33.0	+24.6	+41.2	+31.3
65%, decreasing	+17.6	+12.9	+19.0	+20.1
50%, decreasing	+15.0	+6.9	+12.6	+11.7
40%, decreasing	+11.4	+1.0	+12.6	+11.4

Table 7-4. Fold endurance of paper from Carson (1944).

Relative Humidity	Cotton writing		Index bristol, 50:50 cotton:bleached chemical		Softwood kraft, unbleached		25:75 softwood: unbleached manila and jute	
	MD	CD	MD	CD	MD	CD	MD	CD
15%	59[1]	42	253	51	4	9	260	163
25%	84	66	612	111	20	25	498	280
35%	115	160	870	235	58	140	914	504
45%	219	171	1070	253	132	227	2126	835
55%	276	229	1262	247	152	358	2486	1156
65%	468	366	1139	270	582	1252	5007	1427
75%	697	444	1257	827	933	1378	10^4+	1596

[1] *Schopper units*, no. of double folds required to reduce the breaking strength of 15 mm strips to 1 kg$_f$.

Fig. 7-5 shows a mechanical and an electronic balance for basis weight measurements.

Caliper

The nominal thickness of paper is known as its caliper. Caliper is measured in *mils* for paper or *points* for paperboard, both of which are 0.001 in. The metric unit is mm. The caliper of paper is measured by using a micrometer with circular contact surfaces of 16 mm (0.63 in.) diameter as described in TAPPI Standard T 411. Fig. 7-6 shows two examples of micrometers for measuring the caliper of paper.

Density, bulk

The density of a sheet of paper is its mass per unit volume, which is also called the *apparent density*. Normally density is expressed as g/cm³. The density of papers is typically 0.5-0.8 g/cm³. For comparison, cell wall material has a density of 1.5 g/cm³, which means most paper has a high volume of air space. The density is an indication of the relative amount of air in the paper which, in turn, affects optical and strength properties of paper. Density is calculated from the basis weight and caliper of paper as shown below. *Bulk* is the reciprocal of density and is expressed as cm³/g.

Fig. 7-5. Mechanical (left) and electronic basis weight scales (right).

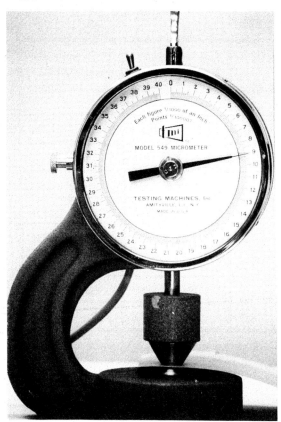

Fig. 7-6. Mechanical (left) and electronic calipers (right) for paper.

$$\text{density} = \frac{\text{basis weight}}{\text{caliper}} = \frac{\text{g/cm}^2}{\text{cm thickness}} \quad (7-1)$$

$$= 0.001 \frac{\text{mm} \cdot \text{m}^2}{\text{cm}^3} \times \frac{\text{g/m}^2 \text{ basis weight}}{\text{mm thickness}}$$

Smoothness

Smoothness is a measure of the surface contour of paper. Smoothness is measured by the *Bekk* method, TAPPI Standard T 479, by measuring the time to draw 10 mL of air radially over the surface of paper clamped in an anvil.

Fig. 7-7 shows the apparatus. The insert shows the anvil in the reflection of the glass. Air is introduced through four small holes inside of the inner ring. Air escaping through the first ring is trapped within the second ring and diverted to one of the flow meters. The less smooth the paper surface, the faster the air will escape through the inner ring.

Gloss finish

Paper with a gloss finish has a very smooth surface (see Fig. 9-62) giving a high degree of specular light reflectance. TAPPI Standard T 480 describes specular gloss at 75°. For lacquered, highly varnished, or waxed papers the gloss at 20° should be measured (TAPPI Standard T 480.)

Transport properties (air resistance and water vapor transmission)

The air resistance of paper is measured by the *Gurley densitometer* (TAPPI Standard T 460) as shown in Fig. 7-8. It depends on the level of pulp refining, the paper's absorbency, specific gravity, and other variables. TAPPI Standard T 536 describes a high pressure method for papers with high levels of air resistance. The *Sheffield* test of air porosity is best suited for smooth, reasonably incompressible papers (TAPPI Standard T 547 which also includes a conversion table for estimating Gurley-Hill values from Sheffield

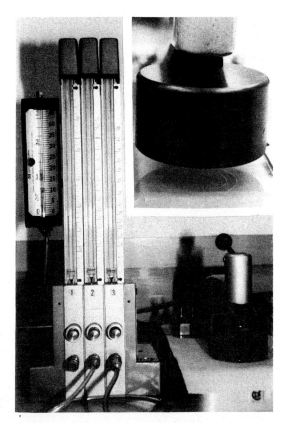

Fig. 7-7. Paper smoothness by the Bekk method. The insert shows the paper test area.

units). In this test, the specimen is exposed to air pressure of 10.3 kPa (1.5 psi).

The rate of water vapor transport through paper is measured with 90% RH on one side of the paper and desiccant on the other side (TAPPI Standards T 464 at 37.8°C or 100°F and T523.) It is significant in packaging papers that must retain moisture of their contents.

Wire side and felt side (air side)

The wire side is the side of the paper formed against the moving wire of the paper machine. Paper formed on twin wire machines has two wire sides. The felt side is the opposite side of the paper from the wire side (for paper from single wire machines). TAPPI Standard T 455 helps with the identification of the wire side using direct observation, carbon smudge, wetting, wetting and charring, tearing, and marking with a soft pencil. It may be difficult to ascertain the wire side of coated and certain specialty grades.

Trim

Trim is the width of the paper on the machine. The web is usually trimmed twice, once at the couch roll by water squirts and once at the rewinder with circular knives.

Formation

Formation is a term used to describe the uniformity of the sheet structure and orientation of the fibers. *Poor* or *wild* formation means irregular distribution of fibers in the plane of the sheet, resulting in many thick and thin spots. *Good* formation indicates uniform fiber and filler distribution in the sheet.

Watermark

A watermark is a pattern that is deliberately put on some bond papers. It becomes visible by looking through the sheet as it is held up to light. It is imparted by a raised pattern on a dandy roll near the end of the fourdrinier wire.

Deckle

The two edges of the paper web on the forming wire are called the deckle. *Deckle edge* means untrimmed paper with its original rough textured edges on the side.

Finish and cockle

Finish refers to the appearance or feel of the paper surface. Cockles are irregularities in the sheet conformation leading to a rough, pock-marked surface.

Curl

Papers tend to curl, especially when each side of the paper is at a different moisture content. A paper that has different properties or compositions on the top and bottom sides (two-sided papers) also tend to curl, acting much like a bimetallic strip does in a thermostat coil. The edges usually tend to contract in the cross machine direction. The inclination of paper to curl is measured by TAPPI Standard T 466 which involves treating one side of paper with water and measuring the curl as a function of time.

Sizing (or lack thereof)

Most printing papers and other papers used in contact with liquids must be able to resist water

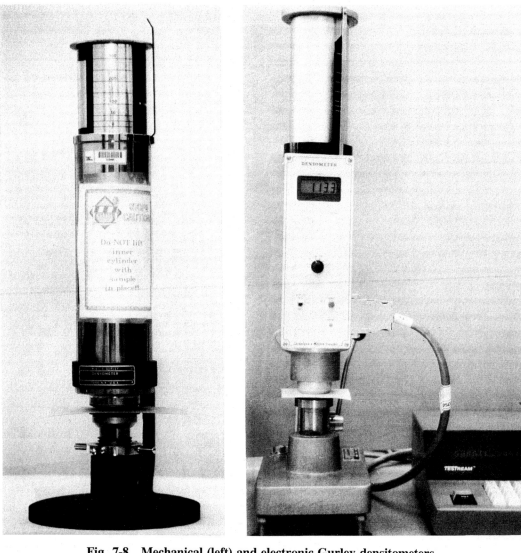

Fig. 7-8. Mechanical (left) and electronic Gurley densitometers.

penetration. (See sizing agents in Section 8.4.) Sizing, however, tends to reduce paper strength and brightness, and increases the cost if ASA and AKD are used. The more commonly used sizing tests are reported here, but there are many more as well. The surface wetability of paper is determined by measuring the contact angle of a drop of water on the paper. It is used to measure sizing and printability of glassines, printing papers, and film laminates (TAPPI Standard T 458.) *Holdout* is related to sizing and is the ability of paper to resist penetration by liquids.

Non-bibulous or *hard-sized* papers have a high degree of resistance to water penetration. They are characterized by the *Cobb* size test (TAPPI Standard T 441) or the *Hercules Size Test* (TAPPI Standard T 530). In the Cobb size test (Plate 25), a 1 cm thick pool of water is placed over 100 cm^2 of weighed paper restrained by a ring collar (11.28 cm in diameter.) After a period of time (usually 2 minutes), the water is drained, and the sheet is treated with blotting paper and then weighed. The absorbed water is reported in mg/m^2. Low Cobb size values indicate good sizing. This method does not distinguish between hard sized and very hard sized paper.

With the Hercules size test (HST, Plate 26), 10 ml of green dye (with 1% formic acid, 10% formic acid or other solution depending on how well sized the paper) is added to a reservoir contained by a metal collar with paper on the bottom. A timer is started and the reflectance of the bottom surface of the paper is monitored. Once the reflectance decreases to a pre-determined point, such as 80%, the timer stops and the time in seconds is recorded. High HST values indicate good sizing. Its reproducibility and sensitivity makes it very useful for determining subtle differences in hard sized papers that the Cobb test would not be able to distinguish.

Slack-sized (moderately sized) papers can be characterized by the so-called *sugar-boat* or *dry indicator* test (Fig. 7-9, TAPPI Standard T 433). A sample of paper is formed into a "boat" (with the inside of the bag outward) that is filled with powdered sugar containing a dye. The boat is placed on the surface of water and the time is measured for water to travel through the paper until the sugar absorbs some water, which causes the dye to appear abruptly. Although there is variability from operator to operator, this is a U.S. government specification for some grades of paper. Slack sized papers may also be characterized by the water immersion test (TAPPI Standard T 491) that involves 10 minute immersion into water, blotting, and weighing.

Water absorbency of *bibulous* papers (unsized papers such as blotting, tissue, or towel) is described in TAPPI Standard T 432. The absorptive quality of blotting paper can be quantified by ink absorbency according to TAPPI Standard T 431.

Printability, Bristow test

Often the traditional sizing tests have been used to indicate how well paper will respond to ink. However, ink jet printers print on the surface of paper and so conventional sizing techniques, which measure liquid penetration, are not suitable indicators of print quality for these papers. The *Bristow test* gets around this limitation. The test measures the track length produced when a liquid is spread over paper at different speeds corresponding to 0-2 seconds distribution time to give a roughness index and absorption coefficient.

Fig. 7-9. The sugar-boat test. The sample on the left has not passed the point of water absorption, but the sample on the right has absorbed water.

Fig. 7-10. The Bristow printability test. The insert shows the "headbox" where ink is applied.

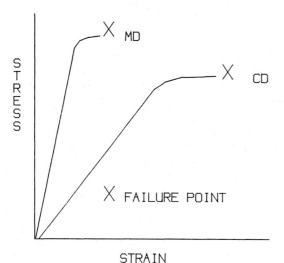

Fig. 7-11. The relative tensile behavior of paper in the machine (MD) and cross machine directions (CD). Stress is the applied force, strain is the sample extension (increase in length). Stiffness = stress/strain.

The test device is shown in Fig. 7-10. The absorption of oils occurs proportionally to the square root of time, but aqueous liquids have an initial wetting time before absorption begins. The test is carried out using paper attached to the 25 mm (1 in.) wide circumference of a wheel 1 m (3.3 ft) in diameter; a stationary "headbox" is filled with a known amount of ink that is drawn over the sheet under a light pressure.

7.6 MECHANICAL PROPERTIES OF PAPER

Machine- and cross-direction of paper

The machine direction of paper is parallel to the direction of travel during the manufacture of the paper; cross direction is perpendicular to machine direction. Paper properties such as tensile, fold, and compression vary significantly between these directions. This is often attributed to fiber orientation, but often of even more importance is the fact that in the machine direction paper is dried under much higher restraint than in the cross machine direction. If fiber alignment were the principle reasons for these differences in strength properties, one would expect to see a large difference in MD and CD tear strength

values, since tear strength in highly dependent upon the properties of the individual fibers (i.e. fiber length and fiber strength. The first part of Table 7-3 indicates the tear strength is about the same in the MD and CD directions, although the tensile strength varies considerably in these two directions.

Fig. 7-11 shows a representation of the difference in tensile strength of paper in the machine and cross machine directions (the direction the testing force is applied). Paper tends to be stiffer and stronger in the machine direction than the cross machine direction. TAPPI Standard T 409 lists methods for determining the machine direction of paper and paperboard.

Paper strength versus loading rate

The strength of wood and paper (and any material) depends upon the rate of loading (how quickly the force increases). A relatively large load can be supported for a short period of time. The relationship between load and loading rate is typically logarithmic. For example, the edgewise compressive strength of corrugated fiberboard was found to increase about 7.5% for each tenfold increase in loading rate. Also, the strength of an A-flute was 40.2 lb/in. at 0.01 in./min loading rate, 45.1 lb/in. at 0.1 in./min, and 48.1 lb/in. at 1.0 in./min (Moody and Konig, 1966). For this reason, standard test methods for measuring paper strength always specify loading rates.

Tensile strength

Tensile strength is measured on paper strips 20 cm (7.9 in.) long by 15-25 mm (0.6-1.0 in.) wide using a constant rate of elongation according to TAPPI Standard T 494. The tensile test is shown in Fig. 7-12. The ultimate force is reported in lb/in., kg/m, or N/m. The tensile index (obtained by dividing the tensile strength by the basis weight) and breaking length are alternate means of reporting tensile strength where the basis weight is normalized. The tensile strength of other materials is reported in units of force/area not force/width as is done for paper.

As with all other strength properties dependent on the paper direction, the tensile strength should be measured separately in the machine and cross machine directions of paper.

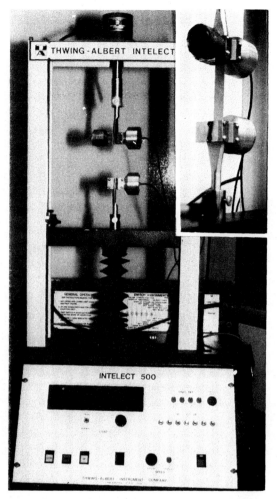

Fig. 7-12. The tensile test. The insert shows the paper sample during a test prior to fracture.

The tensile strength can also be measured on pendulum type machines (Standard T 404), which may give slightly different results. The tensile strength of wet strength papers is measured by Standard T 456. A zero-span tensile test can be used to measure the strength of fibers (as opposed to interfiber bonding) within the sheet.

Breaking length

The tensile strength of paper (or any other material) can be reported as a hypothetical length of paper that just supports its own weight when

supported at one end. The breaking length is discussed in detail in Section 15.3 with examples for many materials. It compares the tensile strength without regard to paper thickness, density, or width. The breaking length of most papers varies from 2.5 to 12 km.

Burst strength

The burst strength, which is also called the *Mullen* or *pop strength*, measures the amount of hydrostatic pressure (which increases at a specified rate) required to rupture a piece of paper. The *Mullen tester* is shown in Fig. 7-13; inserts show the test area. The paper is ruptured by a small, circular (30.5 mm, 1.20 in. diameter), rubber diaphragm pushed against the paper by glycerin. The maximum pressure before rupture is measured in psi or kPa. The burst test is highly correlated to the tensile strength. TAPPI Standard T 403 describes the test for paper, T 807 for paperboard and linerboard, and T 810 for corrugated and solid fiberboard. The burst may also be reported in units of force divided by basis weight to give the burst index, burst factor, or burst ratio.

$$\text{index} = \frac{\text{kPa}}{\text{g/m}^2}$$

$$\text{factor} = \frac{\text{g/cm}^2}{\text{g/m}^2}$$

$$\text{ratio} = \frac{\text{psi}}{\text{lb/ream}}$$

U.S. Federal regulation Rule 41, item 222 requires the burst strength and basis weight of corrugating medium and linerboard to be reported on most corrugated boxes used in the U.S. A recent modification effective since January, 1991 allows edge crush strength to be used instead (Kroeschell, 1991). Many people think that the linerboard STFI compression test is a much better indication of the final box strength than the burst test in most applications.

Fig. 7-13. The mechanical (top) and electronic burst testers. The top insert shows a sample breaking during the test. The bottom insert shows the sample test area.

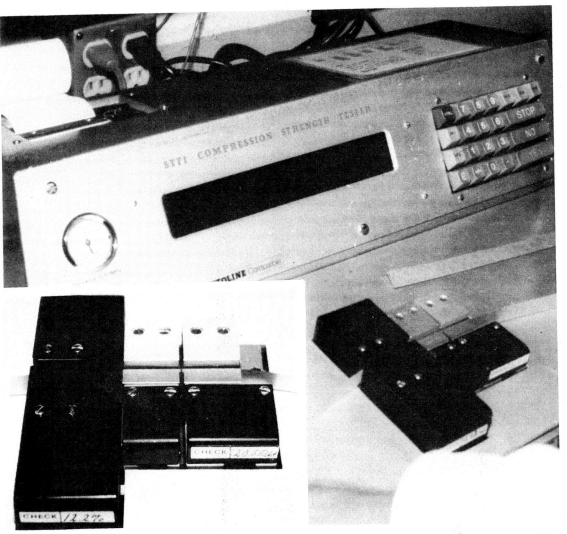

Fig. 7-14. The STFI compression strength tester. The insert shows the sample test area with moisture meter. The moisture content correction is calculated automatically.

Edgewise compression strength, STFI

The short span compression (edgewise compression) strength of paper gives values similar to those of ring crush tests (described below). The results can be correlated with tensile and burst strengths of paper. It is very important in predicting how a corrugated box will perform with a particular linerboard. This test is becoming the standard for container board papers. The instrument is shown in Fig. 7-14 and may contain a moisture meter (TAPPI Standard T 826.)

Strength determination by sound velocity

A recent development (Fig 7-15) is the use of devices that measure the sound velocity in paper. Ultrasound (frequencies above that which humans can here) is used. Physics tells us that the speed of propagation (*c*) of a transverse wave is a function of the mass per unit length (μ) and the tension or stiffness (*S*) of the material for a perfectly flexible string. The relationship is:

$$c = (S/\mu)^{1/2} \qquad \text{(for a transverse wave)}$$

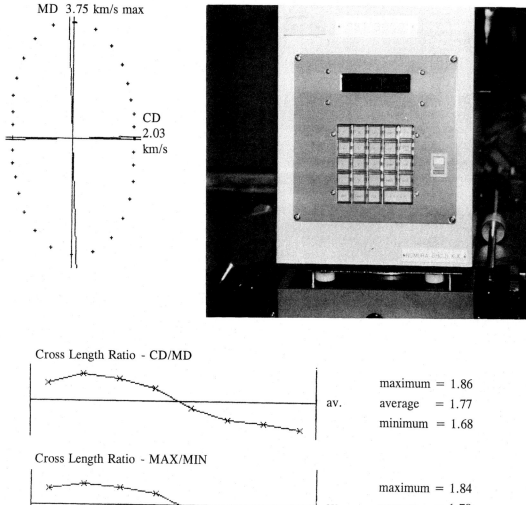

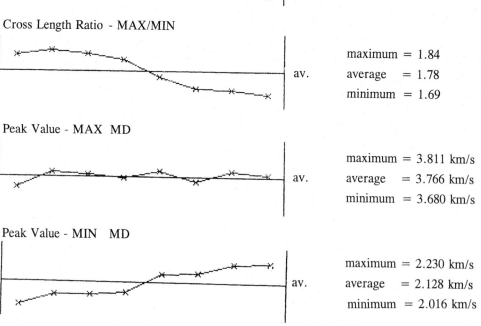

Fig. 7-15. Device to automatically measure the speed of sound in paper as a function of angle at a point or position across the reel with printouts at one position (top) and across the reel (bottom).

The speed of sound in paper is incorrectly attributed only to fiber alignment. In fact, the stiffness is also a function of drying under tension and the properties of the individual fibers themselves, which also depends upon whether they were dried under tension or not. Fig. 7-15 shows a device (with some sample printouts) used to measure sound velocity in paper. The device automatically scans across a sample of paper previously cut from the width of the reel. Time will determine the usefulness of this method.

Ring crush, concora, other container board tests

A variety of ring and other crush tests are used on paperboards. The test apparatus for measuring the strength of samples is shown in Fig. 7-16, using concora flat crush (Tappi Standard T 808.) Most standard paper tests such as tensile, burst, and tear do not relate well to how corrugating medium will perform in actual boxes, so medium is fluted in the laboratory and tested to determine its corrugated board properties.

Concora is an acronym from the *CON*tainer *COR*poration of America. There is a concora fluting machine for forming corrugated samples in the laboratory. There are concora ring and crush tests that are performed on corrugating medium that is fluted in the laboratory.

The ring crush correlates well with the STFI edgewise compressive strength. The procedure is given in TAPPI Standard T 818 without rigid support and T 822 with rigid support.

Stiffness

The *Taber* stiffness (Fig. 7-17) measures the bending moment of a vertically clamped 38 mm (1.5 in.) wide paper specimen at 15° from its center line (TAPPI Standard T 489.) The test has been commonly used with other materials besides paper products. Taber stiffness is important for papers used in wrapping, structural uses, and printing. TAPPI Standard T 453 describes the *Gurley* tester for measuring stiffness.

Of historical interest is the flexing resistance of paper and lightweight paperboard, which is measured by the *Clark stiffness* test (TAPPI Standard T 451). It is a measure of how paper resists bending when handled and is particularly applicable to newsprint.

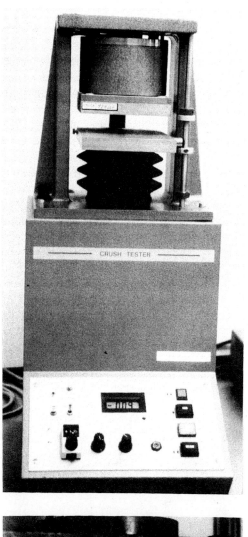

Fig. 7-16. The crush tester (top) and a flat crush test in progress (insert).

Fold endurance

The fold endurance (TAPPI Standard T 511) is a measure of the number of double folds a piece of paper 15 mm wide will endure before its tensile strength falls below a standard value, usually 1 kg$_f$. It measures the strength and flexibility of paper. Because only a very small area, perhaps 15 mm^2 is measured, there is a large variation between measurements from the same piece of paper. The MIT folding endurance is shown in Fig. 7-18. The *Schopper* method is described in Standard T 423. Since heat is developed during the test, and since the folding endurance depends heavily on the paper moisture content (Table 7-4), a fan (supplied in newer machines) should be used to direct air at the testing surface during the test.

Abrasion tester

The abrasion test is conducted by rubbing the surface of paper with a rotating wheel. The weight loss with time is measured. The abrasion test apparatus is shown in Fig. 7-19.

Tear

The internal tear resistance measures the energy required to propagate an initial tear through several sheets of paper for a fixed distance. The value is reported in g-cm/sheet. The difference in potential energy of a pendulum before and after the test is measured by an angle along the pendulum. Fig. 7-20 shows the *Elmendorf* device (TAPPI Standard T 414). Edge tear is measured by the *Finch* method, Standard T 470; highly directional papers are measured as described in Standard T 496. The edge tearing resistance of paper by the edge-tear stirrup method is described in Standard T 470.

Internal bond strength of paperboard, z-direction tensile, Scott bond test

Plate 27 shows the Scott bond test for

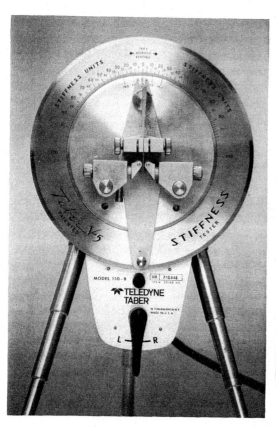

Fig. 7-17. The mechanical (left) and electronic Taber stiffness testers.

Fig. 7-19. The abrasion tester.

Fig. 7-18. The MIT fold tester. The insert is a side view of the test sample. Several samples are clamped at the top and tested one at a time.

measuring the tensile strength in the z direction (TAPPI Standard T 541.), the direction of the paper thickness. The stress is applied by pulling the top and bottom surfaces of paper in opposite directions. This is accomplished by applying double sided, pressure-sensitive adhesive tape to both sides of a 1 in. by 1 in. square test specimen. The specimen is pressed between two platens to bond the tape. The platens are then pulled apart while the energy required to cause the specimen to fail is monitored.

Coefficient of static friction

The coefficient of static friction is important for many packaging papers. Sufficient friction in these papers is important to keep boxes from sliding off of each other during shipping. It is measured by the inclined plane method (Fig. 7-21)

and reported as the tangent of the angle required for like surfaces placed crosswise to allow sliding to commence. TAPPI Standards include T 503 (sack papers), T 548 (writing and printing papers), T 542 (packaging papers), and T 815 (corrugated and solid fiberboard.) The horizontal plane method (for example, TAPPI Standard T 816) gives identical results. The friction of paper can be increased during production by surface treatment at the calender stack.

Surface strength, wax picking

The surface strength of paper is measured by the wax pick test (TAPPI Standard T 459) shown in Plate 28. Calibrated waxes of various adhesive powers are used to quantify the surface strength of paper. The waxes are melted and pushed against the paper while they cool. The wax stick is then removed from the paper. Waxes with high adhesive powers will pull the surface fibers away from the paper.

Other tests

There are numerous other tests of paper covered in the Tappi Standards and other

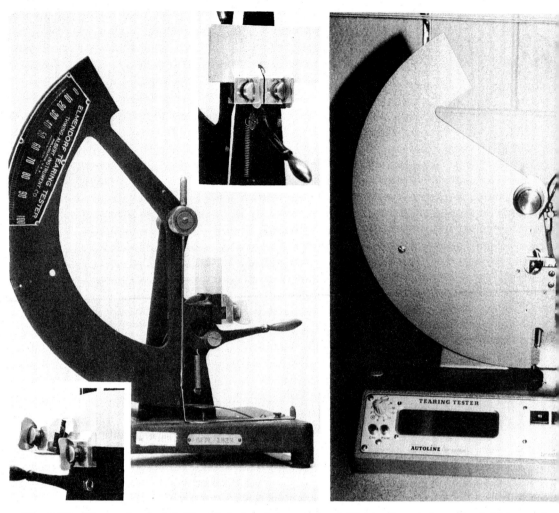

Fig. 7-20. The mechanical (left) and electronic tear testers. The top insert shows the initial sample cut. The bottom insert shows the sample tearing during the test.

standards. Only some of the more common tests have been discussed in this chapter.

7.7 CHEMICAL ANALYSIS OF PAPER

Many of the chemical analysis tests that are applied to paper can be applied to pulp. Section 6.3 and others regarding chemical analysis of pulps should be consulted for additional information.

Ash content, silica

The amount of mineral residue left after the complete combustion of organic material

represents the ash content. It is a measure of filler content in papers containing clay, calcium carbonate, or titanium dioxide fillers and soda content in brown papers. Several modifications of this method are used. The ash content (TAPPI Standard T 413) is determined by weighing the residue after igniting the paper at $900 \pm 25°C$. Standard T 244 describes the ash content after an acid extraction to remove the metal salts, leaving mostly silica and silicates.

Content of selected elements

Many TAPPI Standards are available to measure elemental content of paper. T 266

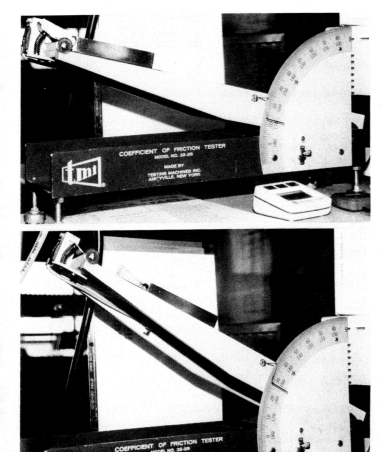

Fig. 7-21. The coefficient of friction (inclined plane test). The bottom picture shows the weight starting to move down the paper.

SEM with elemental analysis

Scanning electron microscopy (SEM) with energy dispersive x-ray analysis (EDAX) is a very useful tool for qualitative and semi-quantitative analysis of elements with atomic numbers of 13 or higher (aluminum and heavier) on the surface of paper. Electrons have very little penetrating power, so only the surface is characterized. It is very useful for looking at element distributions on the surface or through the thickness of paper if cross-sections are made.

When elements are bombarded with electrons, they give off X-rays at frequencies (measured in energy units of electron volts) characteristic to the element. With proper calibration, the number of X-rays given off show the relative amount of that material.

Fig. 7-22 shows an analysis of a 3M Post-it™ note from 1987 (before most mills went to alkaline papermaking). The base sheet contains aluminum (probably from rosin sizing and as a general retention aid) silicon, and small amounts of magnesium (probably this paper has pulp bleached with hydrogen peroxide where magnesium and silicates were used to protect the fiber.) The coated sheet contains these elements in addition to zinc and chloride (possibly from a zinc oxide filler.) The adhesive on top of the coating does not alter the spectrum except to attenuate the signal, since the adhesive contains only light elements (C, H, and O) and absorbs some of the electrons before they can reach other elements beneath them.

This method has numerous important widespread applications to scale residues, corrosion products, and sludge analysis. The method is simple, fast, and reliable.

describes the determination of sodium, calcium, copper, iron, and manganese content of pulp and paper by atomic absorption spectroscopy. Acid soluble iron (which excludes iron that is an integral part of fillers or silicates that probably would not interfere with dyes, etc.) is measured according to TAPPI Standard T 434. Sodium is of interest from the point of view of chemical recovery, while iron and copper may cause discoloring and interfere with dyes. Silica and silicates are determined by the wet ash method (T 245). Cadmium and zinc fillers are used in highly opaque pigments and measured according to T 438. Organic nitrogen is determined by the Kjeldahl method (T 418).

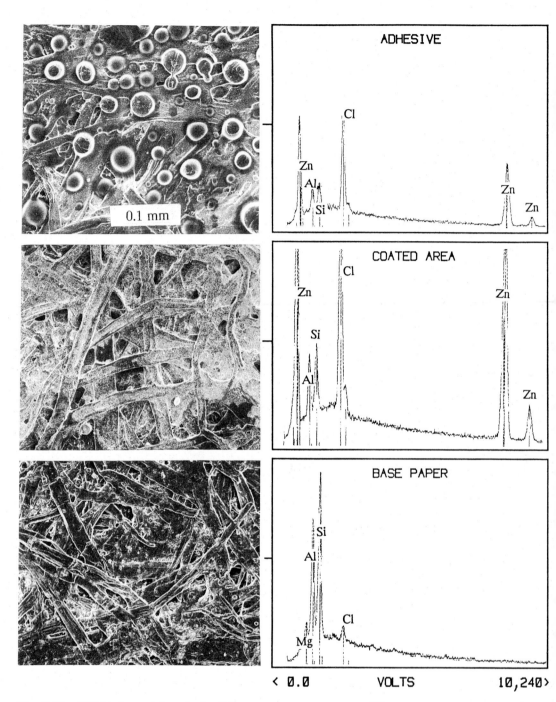

Fig. 7-22. SEM / elemental analysis of three areas on a 3M Post-it™. Courtesy of R. L. Krahmer.

Starch

The use of various types of starch is becoming increasingly important in paper in order to increase its strength and to increase the retention of fines, fillers, and other materials. In the latter case, cationic starches are commonly used. Starch may be measured qualitatively by an iodine solution or quantitatively by HCl extraction followed by colorimetric determination with iodine (TAPPI Standard T 419). Starch consumption in corrugated board is measured by enzymatic hydrolysis followed by either gravimetric determination of the dissolved solids (TAPPI Standard T 531) or colorimetric assay of the liberated glucose (TAPPI Standard T 532).

Other chemical analyses of paper

There are numerous other tests of paper. For example, TAPPI Standard Tests are used to measure rosin (T 408 by gravimetric analysis of an ethanol extract), dirt (T 437), melamine resin (T 493), and other components of paper. The pH is measured at the surface (T 529), in hot water extracts (T 435), or in cold water extracts (T 509). Fiber analysis of paper and paperboard is described in T 401. Qualitative analysis of mineral fillers and coatings by optical microscopy and chemical analysis is described in TAPPI Standard T 421, which includes several photomicrographs.

7.8 BASIC OPTICAL TESTS OF PAPER

Introduction

The optical properties of paper can be extremely complex. Some of the rudimentary concepts are presented here, but the reader should be aware that this field is complex. Dirt is considered in Section 34.18.

Color

Color is the measure of the hue or chroma of light reflected from the surface of paper. It cannot be easily put in numbers, and it is fre-quently expressed descriptively, as "red" or "blue". Color of paper can be expressed by a series of three numbers in the International scale (CIE) system, x, y, and z. Details of this system are given in Chapter 24. The spectral reflectance, transmittance and color of paper are measured as a function of wavelength (poly-chromatic) in T 442. Related tests are T 524 and T 527. The perception of color is dependent upon the light source used to view the object.

Brightness

Brightness is a measure of the "whiteness" of paper. These methods are not applicable to colored papers that are characterized with the tests

described above. Precisely, brightness is the percentage of diffuse reflected light from a thick pad of paper to visible light at a wavelength of 457 nm. This brightness is designated as R_∞. 457 nm is in the blue range of light, where yellowish paper would absorb this light instead of reflecting it. Therefore, brightness is a good indicator of the yellowness of paper or the degree of bleaching. If a black body is used behind a single sheet of paper while the diffuse reflectance is measured, a lower value is obtained (unless the paper is opaque), which is designated R_0. If the diffuse reflectance of a single sheet of paper is measured backed by a surface with reflectance of 0.89 (or 89%), the result is designated as $R_{0.89}$.

Two main types of instruments are used. In the *G.E. brightness* (TAPPI Standard T 452), shown in Fig. 7-23, the light is illuminated on the paper at a 45° angle and the reflected light at 0° is measured. (If an ultraviolet source is used, the effect of fluorescent dyes, the optical brighteners, may also be measured.) This technique uses a reference cell to compensate for slight changes in the light output of the source. A simple orifice directs some of the light from the source to a reference cell, but this is less desirable than having an identical light path for the reference beam as in the Elrepho test.

In the *Elrepho test* (TAPPI Standard T 525), the light source is diffuse and the reflected light is measured at 90° from the surface of the paper (Fig. 7-24, Plate 29). The diffuse light source is a sphere coated with titanium dioxide from which light from two light bulbs reflects. A reference photocell measures the light on the diffuse white surface near the sample. Because the reference photocell detects changes in the light level, large changes in the light source have almost no effect on the reading of paper brightness; it is an extremely robust method compared to the G.E. brightness. (I have taken out one of the light bulbs with a test in progress and observed less than a 0.2% difference in brightness.) A variety of filters can be used with this instrument, but the 457 nm filter is the most commonly used filter.

Opacity

Opacity is the ability of paper to hide or mask a color or object in back of the sheet. A

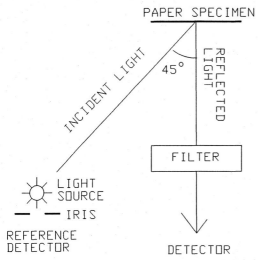

Fig. 7-23. Brightness measurement by directional reflectance (G.E. brightness).

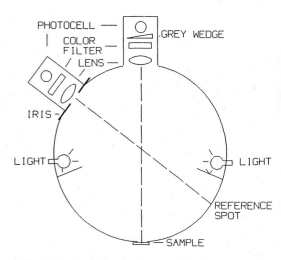

Fig. 7-24. Brightness by diffuse reflectance using the Elrepho brightness meter.

high opacity in printed paper allows one to read the front side of the page without being distracted by print images on the back side. Printing and TAPPI opacity are defined below; the higher the opacity, the better the hiding power, with opaque paper having an opacity of 100%.

$$\text{printing opacity} = \frac{R_0}{R_\infty} \times 100\%$$

$$\text{TAPPI opacity} = \frac{R_0}{R_{0.89}} \times 100\%$$

TAPPI Standard T 425 describes the measurement of opacity $R_{0.89}$, which is sometimes called printer's opacity. TAPPI Standard T 519 is used for diffuse opacity (paper backing).

Opacity is a very important property in printing papers. It is important to consider what properties of paper contribute to its opacity. Opacity is the result of light scattering in many directions when the path of light is bent. This occurs as light goes from material of one index of refraction to a material with a different index of refraction. The classic example is putting a wooden stick into clear water at an angle; the stick appears to bend at the water-air interface. Rough, non-flat surfaces help scatter the light in different directions.

Cellophane is almost identical in chemical structure to paper, since it is fairly pure cellulose. Cellophane, unlike paper, is transparent since it is solid and has two uniform, parallel surfaces. The air spaces in paper provide interfaces for light to scatter. Thus, bulky sheets have much higher opacity than dense sheets. Unfortunately, bulky sheets are less strong than dense sheets since fiber to fiber bonding is poorer in bulky sheets. Since refining increases fiber to fiber bonding and leads to a denser sheet, increased refining lowers opacity. The highly refined glassine papers are very dense and translucent (partially transparent in that they let a large amount of light through, but images are not as clear as they are in transparent materials).

Pulp fines have very high surface areas and irregular shapes that contribute to high opacity. Stone groundwood pulps are of especially high opacity. Fillers (Table 8-1) tend to have high indices of refraction and provide interfaces for light scattering as well. The very high index of refraction of titanium dioxide allows thin papers to have high opacity as in the case of bible papers.

Gloss

Gloss is a measure of the sheen or polish of paper. It is measured by illuminating the paper at a very low angle and reading the reflectance at a similar low angle.

Fig. 7-25. Beloit sheet splitter.

7.9 SHEET SPLITTING OF PAPER

Introduction

Sheet splitting refers to the separation of a sheet of paper in the *z*-axis. In this delamination process, the top of the sheet is separated from the bottom of the sheet. This has always been difficult to do without altering the sheet to the point that many of its aspects could not be studied. Past methods were time consuming, complex, and/or only applicable to small areas. It has been used as an important tool for analysis of fillers across the thickness of the sheet, although SEM might be an easier, less time consuming, and more direct tool for this.

The first method of sheet splitting involved the use of razor blades. One other method of doing this is to use a precision grinder to remove the portion of paper that is not of interest. A vacuum table can hold the sheet down so that it is not altered. Still another method is to use double - sided adhesive tape on both sides of the paper. The exposed area of tape is attached to some surface and the paper is pulled apart. Toluene can then be used to dissolve the tape and leave the paper (to some degree) intact. This is often used, with repeated applications, to measure the filler location across the thickness of the paper.

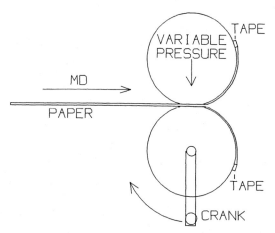

Fig. 7-26. Minter's method of sheet splitting.

Beloit sheet splitter

In 1963, the Liberty Engineering Company invented what is now the Beloit sheet splitter. This method (TAPPI UM 576, Fig. 7-25) works by soaking the paper in water and then putting it through a nip made by two metal rolls that are chilled below the freezing point of water. Each side of the paper is frozen to one of the rolls, so it is effectively split. The paper is doctored from the chilled roll. Unfortunately, the paper is altered to a large degree by this method, but it is useful for mineral contents by ashing.

Minter's method

Minter developed a method (Fig. 7-26) where two pieces of double-sided adhesive tape are applied to the leading edge (the edge first off the paper machine). The paper is then put in the nip formed between two rubber press rolls. Each side of the paper is anchored to the roll at the leading edge so that the split is started. Turning the roll propagates the split indefinitely. The width of rolls determines the CD length that is split, and the circumference of the roll determines the MD length that can be split. The nip pressure is not high, but should be adjustable to give the best split. Two large pieces of unaltered paper are obtained. Strength, sizing, ash contents, and most any other test can be applied to the individual "sides" of the paper. This should prove to be an invaluable tool for determining the two-sidedness of paper and formation characteristics on the paper machine. The one disadvantage of the method is that only a single split may be made.

7.10 ANNOTATED BIBLIOGRAPHY

Paper properties--physical and mechanical
1. Jarrell, T.D., Effect of atmospheric humidity on the moisture content of paper, *Paper Trade J.* 85(3):47-51(1927).

2. Institute of Paper Chemistry, Instrumentation studies. XII, Effect of relative humidity on physical properties with respect to the hysteresis effect in changes from one humidity to another, *Paper Trade J.* 104(15):45-48(1937).

3. Carson, F.T., Effect of humidity on physical properties of paper, U.S. Dept. Commerce, Circular National Bureau of Standards C445 (1944).

4. Moody, R.C. and J.W. Konig, Jr., Effect of loading rate on the edgewise compressive strength of corrugated fiberboard, U.S. Forest Service Research Note FPL-0121 (April, 1966), 11 p.

5. Page, D.H., A theory for the tensile strength of paper, *Tappi J.* 52(4):674-681(1969). The tensile strength of paper is considered in terms of fiber strength and fiber bonding. Paper made from pulps of high freeness have a tensile strength limited by fiber to fiber bonding. Paper made from highly refined pulps have a tensile strength limited by the strength of the individual fibers.

6. Sachs, I.B. and T.A. Kuster, Edgewise compression failure mechanism of linerboard observed in dynamic mode, *Tappi* 63(10):69-73(1980). This is a fascinating visual picture of paper compression failure. The authors state that failure occurs by delamination at the S_1/S_2 cell wall interface. Buckling of the fibers follows failure. There are 13 electron micrograph photographs.

Item 222/Rule 41
7. Kroeschell, W.O., New carrier regulations for corrugated shipping containers are now in effect, *Tappi J.* 74(7):63-65(1991).

8. Three consecutive articles in *Tappi J.* are of interest in regards to Item 222/Rule 41 and follow:

   Batelka, J.J., Compliance statistics for the edge-crush specifications of Item 222/Rule 41, *Tappi J.* 75(1):75-78 (1992).

9. Kroeschell, W.O., The edge crush test, *Tappi J.* 75(1):79-82(1992).

10. McNown, W.J., Short-span compressive strength testing: procedures and tools for improved tester accuracy, *Tappi J.* 75(1)83-86(1992).

Bristow test
11. Bristow, J.A., Liquid absorption into paper during short time intervals, *Svensk Papperstid.* 70(19):623-629(1967).

12. Lyne, M.B. and J.S. Aspler, Wetting and the sorption of water by paper under dynamic conditions, *Tappi J.* 65(12):98-101(1982). This article presents some modifications to the Bristow test.

13. Bares, S.J. and K.D. Rennels, Paper compatibility with next generation ink-jet printers, *Tappi J.* 73(1):123-125(1990).

General chemical analysis
14. B.L. Browning, *Analysis of Paper*, 2nd ed., Marcel Dekker, Inc., New York, New York, 1977, 366 pp. This volume covers many spectrometric, colorimetric, and other methods for the analysis of paper constituents including fillers, additives, coatings, polymers, etc.

SEM with elemental analysis
15. Gibbon, D.L., G.C. Simon, and R.C. Cornelius, New electron and light optical techniques for examining papermaking, *Tappi J.* 72(10):87-91(1989).

16. Ormerod, D.L., G.C. Simon, and D.L. Gibbon, Paper mill additives and contaminant distribution mapping using energy dispersive X-ray analysis, *Tappi J.*

71(4):211-214(1988). Both of these papers show color diagrams that indicate the importance and usefulness of this method. Neither paper has references. This method has widespread applicability to corrosion and scale analysis.

Optical properties
17. TAPPI Technical Information Sheet TIS 0804-02 1986 gives an extensive glossary of optical measurements terminology.

Sheet splitting
18. Minter, S., Delamination as a sheet splitting method, *TAPPI 1992 Papermaking Conference*, pp 81-84.

General interest
19. Mark, R.E., Ed., *Handbook of Physical and Mechanical Testing of Paper and Paperboard*, Marcel Dekker, Inc., New York, 1983, 2 volumes, 640 p., 508 p., respectively. This is useful for background knowledge of a variety of paper and fiber tests of mechanical, optical, wetting and penetration, electrical, thermal, structural, and bonding properties.

20. Dahl, H. H. Holik, and E. Weisshuhn, The influence of headbox flow conditions on paper properties and their constancy, *Tappi J.* 71(2):93-98(1988). (As shown in Fig. 7-15) the maximum strength in paper is not always exactly in the machine direction, due to headbox factors explained by a simple fluid model in this work.

EXERCISES

1. Give four specific types of paper and their properties and uses.

2. Why is it important to put paper in a hot, dry room before putting it in the TAPPI Standard room?

3. What are the problems associated with measuring paper strength properties of paper immediately off the reel during quality control at a paper mill?

4. Which property of paper can change by a factor over ten depending on the EMC of paper?

5. Define these terms: formation, gloss, caliper, density, bulk, and sizing.

6. Describe how sizing is measured by the Cobb and Hercules size tests.

7. Give two possible reasons why tensile strength and other paper properties vary depending upon the direction of paper?

8. The basis weight variation in 12 in. by 12. square sheets might be on the order of 0.2% while the fold endurance of these sheets might vary by 40%. Give two contributing reasons to explain this phenomenon.

9. Give two general methods whereby the amount or type of metal atoms/ions can be determined.

10. What is paper brightness? (Be precise in your answer.)

11. Describe two ways of measuring brightness.

12. What is paper opacity? Describe how the following factors influence opacity:

    a) bulk

    b) presence of fines

    c) presence of fillers

    d) level of refining

    e) thickness of paper

13. Use equation 7-1 (page 170) to determine the density of corrugating medium that has a basis weight of 26 lb/1000 ft² and caliper of 0.009 in. If the cell wall material has a specific gravity of 1.50, what percent air is this sheet?

8

STOCK PREPARATION AND ADDITIVES FOR PAPERMAKING

8.1 INTRODUCTION

This chapter discusses preparation of the stock (Fig. 8-1) prior to the headbox of the paper machine. Included is a discussion of the numerous additives and fillers used to make paper. Wet end chemistry, the chemistry of the dilute aqueous solution of fibers, fillers, and additives, is also discussed. Knowledge of wet end chemistry helps predict retention of additives, drainage during

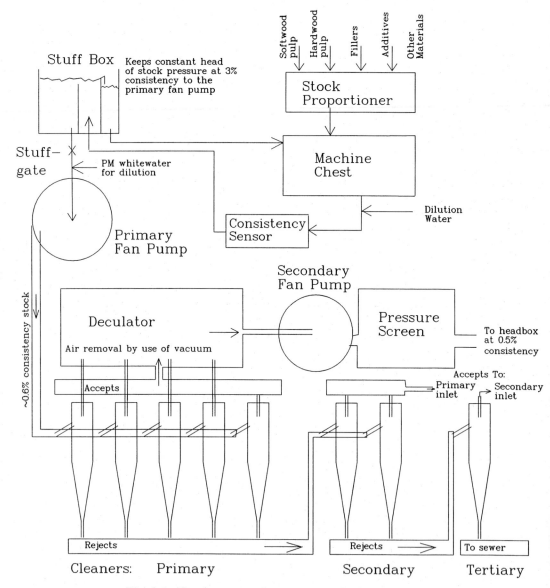

Fig. 8-1. Stock preparation system to the headbox feed.

papermaking and the level of sizing of the final paper. The operation of the size press and coating processes are discussed in Chapter 9.

8.2 FIBER PREPARATION AND AP-PROACH

Fig. 8-1 shows a typical stock preparation system for paper machines. The system includes stock proportioning and stock cleaning. Some auxiliary equipment, such as pumps between the primary and secondary and secondary and tertiary clarifiers, is not shown.

Consistency

Consistency is a measure of the solids content; it is the dry weight of fibers and other solids divided by the total wet weight of stock weight and expressed as a percentage. The consistency of stock at the headbox may be as low as 0.3%.

$$\text{consistency} = \frac{\text{weight of dry material}}{\text{weight of suspension}} \times 100\%$$

Hydrapulper

Hydrapulpers are large mixing vessels used to disintegrate purchased pulp (either wet lap at 50% solids or dry sheets at 80-85% solids), broke (paper that must be reprocessed within the mill), and recycled paper into a relatively dilute slurry which can be processed within the mill. Pulps and papers with high lignin contents and low moisture contents require relatively high energy to be repulped.

Machine chest

The machine chest is an agitated storage test that holds stock prior to being sent to the paper-making process. If the pulper or other equipment must be shut down temporarily, the paper machine can continue running by drawing pulp from the machine chest. Being forced to shut down the paper machine is something of a disaster since it may take several tons of production before high quality paper is again formed, not to mention lost production time.

Stuff box

The stuff box is a small box that pulp enters prior to the fan pump for use in the paper ma-

chine. Pulp enters at a regulated 3-4% consistency through a stock or stuff valve and maintains a constant head to precisely control the flow rate to the fan pump.

Fan pump

Stock and recirculated white water are mixed together at the primary fan pump, a very large pump. For example, a paper machine producing 200 tons per day of paper requiring the equivalent of a 65 foot head of water uses a 10,000 gal/min pump powered by a 200 hp motor.

Vortex cleaners (also *centrifugal, hydrocyclone, centri* or *cyclone cleaners*)

Introduced around 1950, vortex cleaners are plastic or stainless steel cylinders 0.5-1.5 m (20-60 in.) long tapered from 10-20 cm (4-8 in.) at the top to less than 1 cm (0.4 in.) at the bottom. Stock enters the side near the top tangentially to the cylinder causing a vortex to form (Fig. 8-2). Clean stock is removed from the top and stock containing dense contaminants is removed from the bottom. Centrifugal action causes the dense materials to lose their momentum on the inside walls of the cleaner. This allows the dense material to settle much more quickly than fibers.

The vortex cleaners used in the paper machine approach are operated at low consistency (0.5-1.5%) and remove contaminants of high specific gravity. Rejects are removed continuously with high percentages (10-30%) of usable fiber.

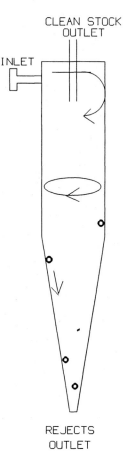

Fig. 8-2. Operation of a centrifugal cleaner.

Fig. 8-3. Two methods of arranging vortex cleaners. The left setup is prior to a fine paper machine and the right is at a secondary fiber plant that employs deinking.

To recover the fiber lost with the rejects, the rejects from the first bank of cleaners (*primary cleaners*) are treated in a bank of *secondary cleaners*; the secondary rejects are treated in the *tertiary cleaners*; and sometimes the tertiary rejects are treated in a set of *quaternary cleaners*. Fig. 8-1 shows a system where the secondary accepts are returned to the primary cleaners and the tertiary accepts go to the secondary inlet. In practice, many diverse configurations are possible. See Section 17.6 for more detail on calculating optimal configurations.

Flow in the vortex cleaners is controlled by pressure drops. For example, by decreasing the pressure in the rejects outlet, thereby increasing the pressure drop from the input to the reject output, a higher flow to the reject output will result. This may lead to better removal of the rejects, but with additional fiber loss as well.

Vortex cleaners can be grouped together in several fashions. Fig. 8-3 shows two common configurations for arranging vortex cleaners.

Medium (1-3%) or high consistency (2-5%) vortex cleaners are used as junk removers in secondary fiber plants (Section 10.3). In secondary fiber plants there are lightweight contaminants such as plastics that must be removed. This is effected by using the same type of cleaners except that the rejects are removed at the top while the cleaned fibers are removed at the bottom. These are called *reverse cleaners* or *through-flow* and are discussed in more detail in Section 10.3.

Deculator

The deculator is a device for subjecting the stock going to a paper machine with a partial vacuum. This causes small air bubbles that are entrained in the stock to expand, separate from the stock, and come to the surface of the stock. Dissolved gases will also be removed to some extent. This reduces the formation of foam and pits in the papermaking process. De-aeration increases the drainage rate and decreases fiber flotation and flocculation.

Screens, pressure screens

Pulp is screened through small holes or slots to remove shives, dirt, and other large particles. The reject material is usually refined and screened, and the final rejects discarded to the sewer system. *Vibratory flat screens* and *rotary screens* are open, gravity flow systems, which may have high frequency vibration to help the screening action. Usually they have round holes and are often used in knotter screens at the pulp mill or course screens in the paper machine. They are now obsolete because they are expensive to operate and maintain, although they are quite useful in the laboratory. *Centrifugal screens* are enclosed,

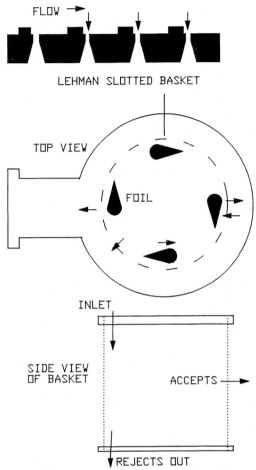

FLOW →

LEHMAN SLOTTED BASKET

TOP VIEW

FOIL

INLET

SIDE VIEW
OF BASKET

ACCEPTS →

REJECTS OUT

Fig. 8-4. Radial outward flow pressure screen. The arrows show the direction of pulp flow.

non-pressurized screens with a horizontal cylinder containing round holes. Centrifugal rotation of the stock by a rotor with large paddles facilitates the screening action by keeping the screen clean; these are used as fine screens in the pulp mill.

Pressure screens, introduced in the 1950s for headbox approach systems, are enclosed, pressurized screens of high capacity. Fig. 8-4 demonstrates the principle of operation as do Plates 30 to 32. Improvements in designs, such as the use of slots instead of holes in the early 1970s, allow them to be used in any fine screening application throughout the pulp and paper mill. One important improvement was the reduction in the screen slot width from 0.50 mm (0.020 in.) prior to 1970, which is still used for corrugating medium,

to as low as 0.20-0.30 mm (0.008-0.012 in.), or less, now used for fine papers. The holes or slots are tapered through the thickness of the screen so that the wider portion is on the exit side (the accepts) to limit the trapping of dirt.

The operation of pressure screens is similar to that of centrifugal screens where the screen is an enclosed, horizontal cylinder. Fiber flow may be outward, inward, or both if two concentric screens are used. Screening action and screen cleaning is most often carried out by non-contacting, rotating hydrofoil surfaces as shown in Fig. 8-4, a major advance in pressure screens. The hydrofoil surface causes the debris, which constantly plugs the holes of the screen, to be lifted away by the back-flushing action of the liquid flow around the hydrofoil surfaces. The frequency and magnitude of the pulse generated by the hydrofoil surface are two of the primary design criteria. The efficiency of pressure screens is a function of the basket (hole size and hole shape), while the fiber reject rate is a function of the rotor (configuration and speed). Pressure screens require about 20-40 kWh/t (0.5-2 hp-day/ton) of pulp for slotted screens and less for screens with holes, and they use an inlet consistency of 2-5%. Black Clawson uses a roughened screen surface to induce microturbulence, which is called a Lehman slotted plate (Fig 8-4), while other manufactures use other profiles to do the same thing. In the future, screen baskets will be made in sections that are bolted together so that individual sections can be replaced without replacing the entire screen.

8.3 RAW MATERIALS

Stock, furnish, broke

Stock is the mixture (slurry) of pulp, fillers, other papermaking materials, and water. *Furnish* is the combination of all of the materials used to make paper. *Broke* consists of paper trimmings and waste, paper not up to specification, and other paper that is often reused in the paper mill. A mill producing more than 10-15% broke is not operating efficiently. (Broke is generally not considered recycled paper.)

Additives

Additives are materials used to improve the finished paper itself or aid in the process of

Table 8-1. Summary of the properties of paper fillers and coating pigments.

Name	Index of Re- fract.	Bright. 457 nm	Spec. Grav.	Particle Size, μm^1	Remarks
Clay, filler	1.56	82	2.58	0.5-10	abrasive, very cheap
Clay, No. 2 coating	1.56	86-92	2.58	80-82%	most common type used
Clay, No. 1 coating	1.56	86-92	2.58	92%	better gloss and opacity
PCC	1.65	97-98	2.65	0.1-2.5	cheap, bright pH > 7
Ground $CaCO_3$	1.65	94-97	2.65	0.1-0.4	cheap, bright pH > 7
Anatase TiO_2	2.55	97-98	3.9	0.15-0.30	excellent opacity, bright
Rutile TiO_2	2.70	97-98	4.2	0.15-0.30	excellent opacity, bright
Talc	1.57	96	2.7	0.5-5	Used in Europe
Cell wall material	1.53		1.50	10×1000	

[1]A percentage indicates the percentage of particles that pass through a 2 μm mesh.

papermaking. *Functional additives* such as dyes, internal sizing agents, adhesives to increase wet or dry strength, and fillers are used to improve or impart certain qualities to the paper product and must be retained on the sheet to be effective. *Control additives* such as biocides, drainage aids, retention aids, pitch control agents, and defoamers are added to improve the papermaking process, but do not directly affect the product and are not necessarily retained on the product. Of coarse, many additives have several effects at the same time; for example, alum is required for rosin sizing under acid conditions but also serves as a drainage and retention aid. Both types of additives are added to the stock prior to papermaking.

Metering and pumping of additives

Liquid additives are expensive chemicals that must be used at exact levels; thus, they are usually metered into the system from positive displacement metering pumps. Like most positive displacement pumps, be sure to use appropriate filters (such as a 10 mesh or finer screen) in front of the pump inlet to keep the pump from fouling by clogging the check valves in an open (and useless) position. Use a pump inlet line of at least 12 mm (½ in.)

for additives with viscosity above 100 cps. Use a flow measurement device of some sort to insure flow of each additive. Since it is often more effective to pump above ambient pressure, a pressure gauge can be used with a pressurized output to at least give a qualitative flow rate, which is better than no indication of flow. Surprisingly, many vendors of additives supervise the addition of their additives even to the extent that they set the flow rates at mills. While one might rationalize that they know their product best, they do not know how their product will affect other (unknown) wet end additives.

8.4 FUNCTIONAL ADDITIVES

Fillers

Fillers (Table 8-1) are *pigments* that are added to stock for opacity and brightness improvements of printing papers. The ideal properties of pigments for printing papers are high brightness, high index of refraction (to help scatter light and increase opacity), small and uniform particle size for smooth paper, low water solubility, inertness, low cost, low abrasion, low specific gravity, and high retention levels. About 50% of the filler is retained in the sheet. Fillers are often ground or precipitated calcium carbonate (with paper machines operating at pH of 7 or higher), titanium dioxide, or clay. Filler is used at 10-30% to replace expensive fiber; at the higher levels of addition, the paper becomes limp since fillers interfere with fiber-fiber bonding and do not impart strength themselves. Fillers are not used in linerboard or other papers where strength is the principal desired property. Table 8-1 summarizes the properties of paper fillers and coating pigments.

Clay (or *kaolin*) is an inexpensive filler, mined from natural deposits, used in magazine and book paper. It is not as bright (80-92%) as calcium carbonate or titanium dioxide, has an index of refraction of 1.55, a specific gravity around 2.58, and is abrasive. Typically 40-90% of clay particles are less than 2 μm. More clay is used than any other filler in paper traditionally, accounting for 90% of all fillers and coating pigments. About 10 million tons are used world-wide and 4 million tons are used in North America annually. Clay consists of hydrated SiO_2 and Al_2O_3.

Calcium carbonate (*chalk* or *limestone*) is becoming an extremely important filler to the industry. The mineral form is called calcite and occurs in limestones, chalks, marbles, and other forms. Since it reacts with HCl to give $CaCl_2$ and CO_2, it must be used in alkaline papermaking systems at pH 7.0 or higher. The brightness is about 92-95%, with an index of refraction of 1.65, and a specific gravity of 2.7 to 2.85.

Median particle sizes for the various forms of calcium carbonate are as low as 0.5 μm to as high as 3 μm or larger depending on how fine it is ground. Finer particle sizes give higher gloss and may improve brightness. Calcium carbonate is inexpensive (about 25-50% of the cost of bleached pulp) and has low abrasion. It is made by grinding one of the naturally occurring mineral forms or by precipitation from solution to form $CaCO_3\downarrow$, *precipitated calcium carbonate, PCC*. PCC has a high scattering coefficient (compared to ground calcium carbonate) that increases the opacity of paper. Calcium carbonate compares well at $200-$500/ton to bleached wood pulp at $600-$800/ton and titanium dioxide which is as high as $2000/ton.

PCC is not a new product, but is new to the North American pulp and paper industry. Fig. 8-5 shows some examples of PCCs with different properties. Pfizer chemical is the major supplier and has installed 11 of the 17 satellite mills near paper mills as of April 1991. This number has probably increased at the time you read this. PCC is produced by mixing CaO with water to form $Ca(OH)_2$ in solution. Addition of CO_2, for example, from flue gases of the lime kiln, causes formation of the insoluble $CaCO_3$. PCC offers advantages over chalk because of its finer particle size and uniform size distribution, higher chemical purity, and control over the zeta potential (ionic charge) of the surface. The shape of the particle can be controlled so as to give plate-shapes that are ideal for papermaking. It is said that the pH of PCC tends to be somewhat higher than ground calcium carbonates.

Titanium dioxide is an expensive filler which costs slightly more than bleached pulp and is very bright (98%). It has a high index of refraction (2.56-2.70) and a particle size (typically only 0.20 to 0.25 μm) that both contribute to high opacity. As anatase it has a specific gravity of 3.90 and as rutile, 4.2. It is used in bible, airmail, bond, and other printing papers, which are lightweight, expensive, and require high opacity. It is also used in most white paints as the pigment.

Talc (hydrated magnesium silicate, refractive index of 1.57, brightness of 90-95%, specific gravity of 2.7, particles 0.5-5 μm) is used in the U.S. mostly as a pitch control agent; it is used as a filler in Europe. *Diatomaceous earth* is normally used for pitch control. Some other fillers are used, including zinc oxide and calcium sulfate.

Dyes and brighteners

Dyes are water soluble colors added to stock to impart color to the final product. Dyes are absorbed on the fiber surfaces imparting their colors to the paper fibers. They have structures involving large conjugated double bond systems with azo, metallic azo, anthraquinone, triarylmethane, quinoline, and similar structures.

Basic dyes are cationic organic dyes (containing amine groups) that are used with inorganic anions to fix them to the surface of fibers. They have strong affinity for lignin but not for bleached

Fig. 8-5. PCC types: rhombohedric (UL), high surface area (UR) for high opacity and brightness, large spherical for matte coated papers (LL), and scalenohedral (LR). Courtesy of Pfizer.

pulps and have poor fastness to light. They are used in newsprint, in "yellow pages" of phone books, and similar products. *Acid dyes* are anionic organic dyes (containing sulfonate groups) that are used with organic cations such as alum to fix them to the surface of fibers. (The term *mordant* is used by the dyeing industry for the role of alum, although this terminology has not caught on in the pulp and paper industry.) They have good fastness to light. *Direct dyes* are similar to acid dyes, but have higher molecular weights and direct affinity for cellulose (in part, due to their lower solubility in water and the fact that they tend to

form colloids in the wet end). They have excellent fastness to light and are used extensively for coloring tissue paper, blotting paper, and fine papers. Acid or direct dyes are generally not compatible with basic dyes, although occasionally small amounts of basic dyes are used to improve the effect of acid or direct dyes.

Fluorescent brightening agents (sometimes called *optical brighteners*) are used to brighten paper to, as the laundry detergent commercials used to say, "brighter than bright". They are colorless (technically not dyes) and convert invisible UV light to lower energy visible light, espe-

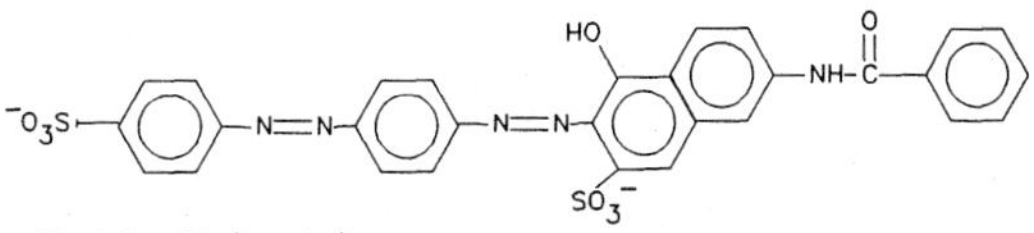

Direct Red 81 (Azo dye)

Azo dyes: R-NH=NH=R'   Stilbene dyes: R-CH=CH-R'

Fig. 8-6. Representative dyes used in paper.

cially blue light, which masks the inherent yellowness of paper. This is the same effect used in black light posters. The class includes stilbenes, azoles, coumarins, pyrazenes, or naphthalimides. Some representative dyes are shown in Fig. 8-6. Dyes must be made colorless in broke. This is accomplished by the use of oxidative chemicals (hydrogen peroxide or chlorine) or reducing chemicals (sodium hydrosulfite or formamidine sulfinic acid, FAS).

Blue dye is often added to pulp to offset the tendency for pulp to be yellow. The blue dye does not make the pulp brighter, it only makes the yellow color look gray, and gray has the perception of being brighter than yellow.

Internal size and alkaline papermaking advantages

Internal sizing develops resistance to penetration of aqueous liquids throughout the sheet. Internal sizing is accomplished by adding materials to the stock before the headbox to retard water penetration into the final paper. Water penetration is retarded by the nonpolar portions of the size molecule. A reactive portion of the size molecule anchors it to the surface of the fiber.

Surface sizing works by a different mechanism and occurs at the size press where an application of starch (or other material) fills the capillaries of paper, making water penetration much more difficult. Starch is not hydrophobic as are internal sizing agents.

Rosin sizing with alum, used since the early 1800s, is only effective below pH of 7 or so. Alkaline sizing was developed in the 1940s and 1950s. It has recently enjoyed widespread use in the U.S. in printing papers filled with calcium carbonate that must be used at pH of 7 or higher, because in acid conditions it decomposes to carbon dioxide gas that causes pitting of the sheet.

Paper may be *hard-sized* (high resistance to liquid penetration, such as many printing and packaging papers), *slack-sized* (low resistance to

liquid penetration, such as newsprint), or *no-sized* paper (toweling, blotting, and sanitary papers). There are two common methods used in internal sizing: rosin sizing, which must be carried out at pH 4-6 and alkaline sizing, which is carried out at pH 8 or higher with ASA or AKD. There are several important advantages to alkaline papermaking. The papers have longevity since no residual acid is present to degrade carbohydrates; calcium carbonate (an inexpensive, bright filler) can be used; paper is stronger and less brittle; there is less corrosion on the paper machine; and there are fewer problems using recycled fiber containing calcium carbonate filler. Thus, while less than 20% of printing papers were manufactured under alkaline conditions just a few years ago, the percentage is expected to be over 80% by 1995.

With most internal sizing methods (rosin or alkaline), hardwood pulps are much easier to size than softwood pulps; sulfate pulps are easier to size than sulfite pulps, which are both easier to size than thermomechanical pulps, which are easier to size than groundwood pulps; semibleached pulps are easier to size than bleached pulps; and pulps with moderate levels of α-cellulose are easier to size than pulps that are almost pure α-cellulose. The explanation centers around the number of carboxylate groups as anchor points in sizing. It is known that rosin sizing decreases the strength of refined pulp for papermaking.

Internal sizing with rosin

The most common internal size traditionally is rosin at 2-9 kg/t (4-20 lb/ton) rosin solids on pulp, precipitated with alum, $Al_2(SO_4)_3$, a process developed by Illig in 1807 when it was called engine sizing because the chemicals were added to the stock at the beating engine.

Abietic acid and homologues are now fortified by addition of maleic anhydride to give a tricarboxylic acid via the Diels-Alder reaction shown in Fig. 8-7. This process was invented by Wilson and Duston (1943). Rosin may be used in the salt form as a solution at pH 10-11 (*soap size*), as a solid salt form with 20-30% water (*paste rosin size*) or in the free acid form as an emulsion (*emulsion size*), which is effective at slightly higher headbox pH than the others. There are several TAPPI standards for rosin characterization.

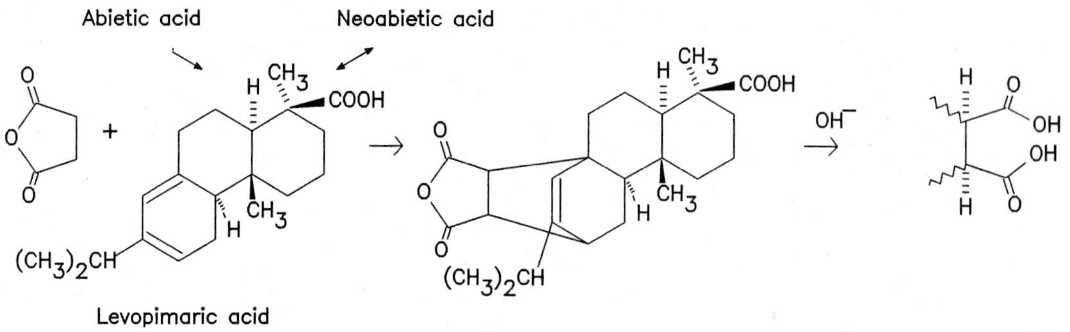

Fig. 8-7. The Diels-Alder reaction and hydrolysis yielding a tricarboxylic acid (fortified rosin size).

Normally the rosin is added before the alum. If alum is added before rosin, the process is called *reverse sizing*. Reverse sizing might be useful, for example, in water containing high amounts of calcium which might precipitate the rosin before it could react with alum. Reverse sizing is also recommended in alkaline systems.

Internal sizing with ASA or AKD

Paper manufactured under alkaline conditions uses synthetic sizing agents (Fig. 8-8) like alkylketene dimer (AKD, one trade name is *Aquapel*) or alkyl succinic anhydride (ASA), which were developed in the 1930s and became available in the 1950s, at the rate of about 0.5-1.5 kg/t (1-3 lb/ton). AKD is prepared by dimerization of the acyl chlorides of fatty acids. ASA is made by reacting mixtures of C-16 to C-20 olefins with maleic anhydride.

Unlike fortified rosins, the anhydride functionality must not be hydrolyzed with water or else the size will be ineffective and cause pitch problems. Like other anhydrides and esters, these agents are hydrolyzed more quickly with increasing alkalinity and temperature, so emulsions of ASA should be stored for short periods of time under slightly acidic conditions (pH 3.5-4 for ASA) at relatively low temperatures. Hydrolysis

is also acid catalyzed, but to a much smaller extent. AKD is much more stable and can be stored as a dilute emulsion for several weeks to three months (although occasional inspection by FTIR analysis may be useful to verify this). Hydrolysis of ASA and AKD produces carboxylic acids that decrease the pH.

These agents are not water soluble and must be used as emulsions. Cationic starch is used to stabilize these emulsions. According to Markillie (1989), the cationic starch solution used to emulsify ASA must be cooled below 38°C (100°F) and have a pH of 4-4.5 to minimize ASA hydrolysis. If the starch viscosity is excessively high, poor emulsification may result leading to poor ASA retention, but generally low starch viscosity leads to poor sizing with ASA. A low starch solution pH for emulsification of AKD can interfere with the process so the pH should be around 5-7. The starch intrinsic viscosity (4% starch solution at 150°F) should be low, around 1.0 to 1.1, when emulsifying AKD to give maximum sizing which is the opposite of ASA. Quaternary ammonium starches work better than tertiary amino starches for alkaline sizing. Generally, 3:1 cationic starch (about 0.30% N) to ASA or AKD is used for emulsification with an additional 8 lb/ton cationic starch for papermaking.

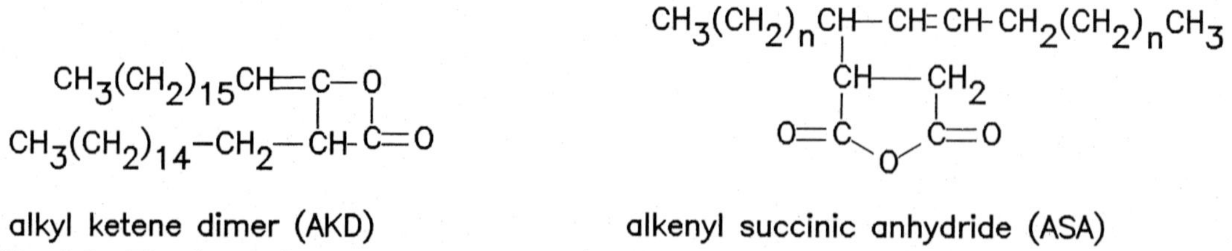

alkyl ketene dimer (AKD)   alkenyl succinic anhydride (ASA)

Fig. 8-8. The chemical structures of AKD and ASA, the so-called "synthetic" sizing agents.

The anhydride functionality reacts with the hydroxyl groups of cellulose, which is catalyzed by relatively high pH and temperature to give sizing by the formation of an ester linkage, although many argue that these agents are actually held to the fiber by weak bonds. Certainly only a small portion of the size molecules (<30%) form the covalent ester linkage; on the order of 0.01-0.03% size on pulp must be retained for sizing. Cationic polymers such as cationic starch (DS ≈ 0.03 cationic amines) help retention and sizing with these materials. Some synthetic sizes incorporate cationic groups within the size. ASA is used with 4 to 5 kg/t (8 to 10 lb/ton) alum and has rapid on-machine curing. There is literature indicating that alum has a detrimental effect on AKD sizing, and AKD sizing fully develops off the machine over a period of 1-2 weeks.

Sizing with any of the these synthetic sizes is not well understood, and there are numerous problems in the mills using these agents. Papers tend to be slippery (lower coefficient of friction) especially with AKD sized paper using excess AKD to develop sufficient sizing for size press holdout, forming fabrics may have a decreased life using calcium carbonate fillers at high levels, there is increased picking at the press section, and there may be poorer sizing at the size press with alkaline sizes. Because water drains more quickly from calcium carbonate than clay, calcium carbonate filled papers may have poorer formation, and some of the foils may be removed from the wet end to help formation for these grades.

Internal sizing with other chemicals

Other chemicals are available for internal sizing. Stearic acid is used in specialty papers such as photographic papers requiring high whiteness where rosin might otherwise oxidize and darken the paper. Stearic acid is used in a manner similar to that of rosin size. Fluorochemicals, $CF_3(CF_2)_nR$, where R may be an anionic, cationic, or other functional group are used in some greaseproof and solvent-resistant papers; it is added in the wet end or the size press. Teflon (one trade name is Gortex™) is a fluorochemical.

Starch

Starch is the third most important paper furnish based on the weight used, with pulp and clay the first two. It is used as a retention aid, dry strength agent, surface sizing agent, coating binder, and adhesive in corrugated board and other converting operations. There are two types of starch: amylose and amylopectin. Starches from various sources contain various amounts of each. For example, untreated corn starch is about 25% amylose and 75% amylopectin, potato starch is about 20% amylose and 80% amylopectin, and waxy maize is almost 100% amylopectin. Amylose is a linear polymer of glucose with alpha 1,4 linkages making it a $(1{\rightarrow}4)$-α-D-glucan with a relatively low degree of polymerization (DP) on the order of 100-1000 (molecular weights of 15,000-150,000). Amylopectin differs from amylose in that it has a very high DP, up to several million, and about 4% of the linkages are $(1{\rightarrow}6)$-α-D-glucose links to give a highly branched polymer.

Various treatments of starch with enzymes, acid hydrolysis, hot water treatment, etc. are used to alter its physical form and lower its molecular weight for a variety of commercial applications. One reason it is important to lower the molecular weight of starch is so that is can be used in solutions with high solids content without having a very high viscosity. In the paper mill, starch is typically heated in water at 95-110°C (205-230°F) for 30 minutes with gentle stirring to form *pasted* starch that is diluted and stored around 60°C (140°F). Cooking longer than this is usually not detrimental, but insufficient cooking results in incompletely gelatinized starch that does not give its maximum benefit.

Starch is reacted with materials such as $RCH_2OCH_2CH(OH)CH_2\text{-}N(CH_3)_3^+$ at high pH to introduce cationic groups. Cationic starch contains tertiary amines or quaternary ammonium salts on 3-5 per 100 anhydroglucose units of starch, i.e., a degree of substitution (DS) of 0.03 to 0.05. The quaternary ammonium salt seems to work better at high pH than the tertiary amine as it is necessarily cationic. A useful conversion for DS from percent nitrogen for nitrogen contents below about 5 percent nitrogen (0.5% N = DS 0.06) is:

$$DS = 0.117 \times \text{Percent Nitrogen}$$

Cationic starches are used to help form stable emulsions from ASA and AKD sizing agents. Some cationic starches also carry anionic groups, notably phosphate, that form coordinate bonds easily with alum and other materials. Potato starch, with about 0.1% phosphorous, has a slight negative charge due to phosphate groups.

Dry strength additives

Polyacrylamides are used to increase the dry strength of papers by hydrogen bonding. Polyacrylamide of molecular weight of 100,000 to 500,000 is used at about 0.5% on pulp and also acts as a retention aid. It is sometimes used to allow increased use of lower quality pulps such as secondary and hardwood fiber. Cationic (containing some amine functional groups) polyacrylamide is used at 0.2-0.5% on pulp, is suitable for use at pH 4-8.5, and reduces the alum requirement, while anionic (containing some carboxylate functional groups) polyacrylamide is used at a pH of 4.5-4.8 (with alum) to increase strength properties.

Starch is also used to improve strength, drainage, retention, and sizing of paper. Cationic, anionic, and ampholytic (cationic and anionic) starches are available. Guar gum, carboxymethyl cellulose, methyl cellulose, etc. can be used, but tend to be more expensive for a given effect.

Wet strength resins

Wet strength agents are thermosetting resins that are added to stock to impart wet strength to the paper. Curing of the resin leads to the formation of covalent fiber-fiber bonds. The tensile strength of wet paper is from 0-5% of the dry tensile strength with no wet strength resin to as much as 15-50% of the dry tensile strength of the paper with wet strength resins. Wet strength paper is tested by measuring the resistance to rupture of the paper saturated with water. Papers made with wet strength adhesives are used for produce boxes, paper towels, maps, outdoor posters, food wraps, photographic papers, filter papers, tea bags, and disposable garments and bed sheets (such as for hospital use). Three wet strength resins are used commonly, especially the polyacrylamide resins (Fig. 8-9).

Urea-formaldehyde (UF) is the traditional wet strength resin. It can be used with permanent papers and gives paper repulpable under acidic conditions. UF has largely been replaced by *melamine-formaldehyde (MF)*, which uses melamine in place of urea, largely due to concerns about the health aspects of formaldehyde emission. *Glyoxal-polyacrylamide copolymers*, which can be made cationic, form labile bonds and are not useful for permanent papers such as paperboards. *Epoxidized polyamine-polyamide* resins are also used in permanent papers and result in papers that are not easily repulped. This material is formed in two steps. In the first step a dicarboxylic acid

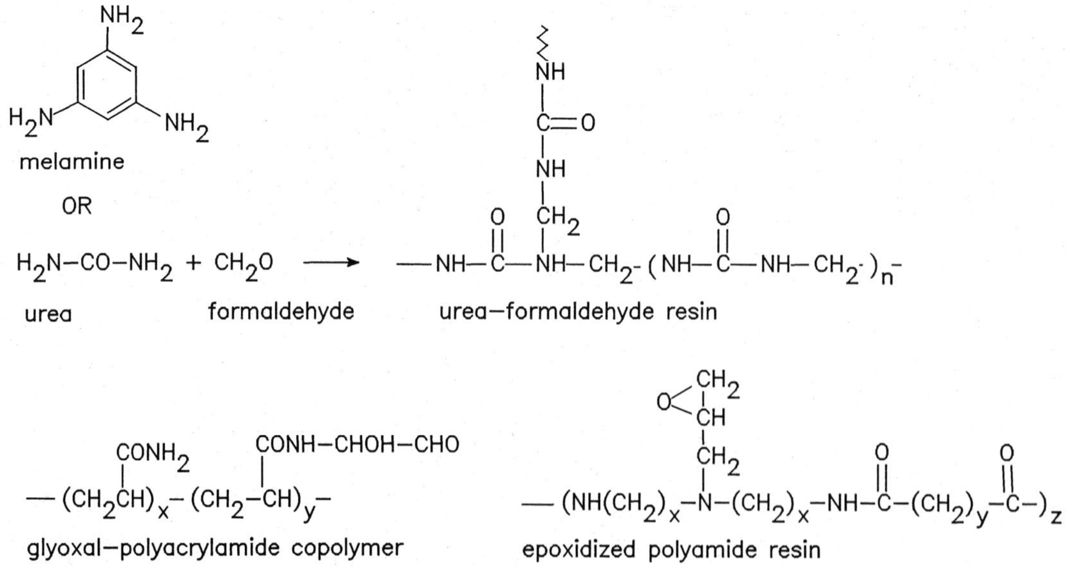

Fig. 8-9. Common wet strength resins.

Table 8-2. The relative surface area of fibers and fiber fragments.

Fiber or fine dimensions (box-shape)	Gross outer surface area, m^2/g	Relative surface area
4000 μm $\times$ 40 μm $\times$ 40 μm	0.100	1.00
1000 μm $\times$ 10 μm $\times$ 10 μm	0.402	4.02
40 μm $\times$ 40 μm $\times$ 40 μm	0.150	1.49
10 μm $\times$ 10 μm $\times$ 10 μm	0.600	5.97
100 μm $\times$ 2 μm $\times$ 2 μm	2.020	20.10

such as adipic (1,6-hexanedioic) acid is reacted with triethylamine to give a polyamine-amide backbone. In the second step, available amines are then reacted with epichlorohydrin to produce the epoxidized polyamide.

Glyoxal based resins are borrowed from the textile field and are of the form R-CO-CHO; at pH above 9, the inactive form predominates by the Cannizzaro reaction to give R-CH(OH)-COO⁻ which does not crosslink with starch or the fibers. Glyoxal resins require curing time on the warm paper reel for up to 30 minutes.

Specialty chemicals

Some other additives include flame retardants, anti-tarnish chemicals (for example those used in tissue papers to wrap silver, etc.).

8.5 CONTROL ADDITIVES

Retention aids

Retention is a measure of how much material remains on the paper machine wire and is incorporated into the final sheet. Two types of retention are considered in the industry: overall retention and first-pass retention. These are defined below in terms of filler, although the retention of fiber fines, sizing agents, and other materials is also important.

$$\text{overall retention} = \frac{\text{filler in sheet}}{\text{filler added to furnish}} \times 100\,\%$$

$$\text{first-pass retention} = \frac{\text{filler in sheet}}{\text{filler in headbox}} \times 100\,\%$$

A high first-pass retention is important for many aspects of wet end chemistry and sheet quality. For example, a high first pass retention of the sizing agents ASA and AKD will limit their hydrolysis. Hydrolysis of these expensive chemicals means they do not contribute to sizing but can cause pitch problems. Overall retention of fillers is important to their economical use. The overall retention really is a measure of first-pass retention and the amount of paper machine excess white water.

Retention occurs by two mechanisms. Particles larger than 200 μm (or the size of the largest openings of the paper machine wire), which are fibers and fiber fines, are primarily retained by *filtration*, particles smaller than 10 μm, which includes all additives, are primarily retained by adsorption onto the fibers via formation of various secondary chemical bonds. The retention aids form bonds with both the fiber surfaces and all of the additives to be retained. The retained additives are preferentially retained onto fiber fines due to their inordinately high surface area as shown in Table 8-2. This should be considered as an advantage because, without a high surface area on the fibers, there may not be enough fiber surface area to achieve adequate retention. Particles of intermediate size (which is generally only the fines) are retained by a combination of both mechanisms. Retention aids function on the small and mid-size materials.

Retention aids are often polymers that are added to improve the retention of fines (small fiber fragments), fillers, internal sizing agents, etc. They may rely on high charge density for the polyamines such as polyethylenimine (PEI) or poly(diallyldimethylammonium chloride) (DADMAC). If the charge density is low, then higher molecular weight is required in the case of cationic or anionic polyacrylamide or even cationic starch, although the latter is usually used in dual polymer systems. Polyacrylamide of high molecular weight, 500,000 to several million, is a common retention aid. Polyethyleneoxide (PEO), starches, gums, alum and aluminum polymers, are also used. Retention is measured on a *first-pass* basis, which is the amount of filler retained in the sheet compared to the amount of filler in the headbox, or on an *overall retention* basis, which is the filler retained in the sheet compared to the amount of filler added in the white water. The difference is that in the second case filler in recirculated white water has several more chances for retention.

Wood fibers have a negative surface so other fibers and negatively charged particles repel the fibers. On the other hand particles or polymers with a positive surface are attracted to the negative charge of the fiber surface (like charges repel each other, unlike charges attract each other) and are large enough to attract several fibers, fines, or fillers and cause flocculation, resulting in an increase in retention and drainage. The *zeta* potential is a measure of the electrical potential between the fiber-liquid boundary and the bulk liquid. The zeta potential has been used as an indication of retention, however its success has been limited because ionic interactions are only a small factor in retention.

Materials like alum are good retention aids but are shear sensitive. The shear sensitivity is probably related to the size of the material to be retained, with large particles such as fillers more shear sensitive than small molecules such as rosin. Large polymers are less shear sensitive as retention aids since they have more anchor points; consequently, many retention aids are used, especially in light of modern, high speed machines with high turbulence that cause high sheer.

Dual component systems for retention include cationic and anionic agents. One important system is PEI with anionic polyacrylamide. The folklore goes that low molecular weight highly (highly is a relative term, remember ultra high yield pulps have a yield of about 75%) cationic polymers

make positive "patches" on the anionic surfaces of fibers, fillers, etc. The high molecular weight anionic polymers then attach to several of these patches, holding them all together. One other system of retention is the microparticle system using cationic starch or cation PAM with colloidal silica.

Drainage aid

Drainage aids are materials that increase the drainage rate of water from the pulp slurry on the wire. Almost any retention aid is apt to improve the drainage rate as fines and fillers are removed from the whitewater, which decreases the solids content of the whitewater; consequently, the effects of drainage and retention may be indistinguishable, and the two are usually considered together. The drainage aid also influences the moisture content of the web going to the press section. A moisture increase of 1% can reduce the web strength by over 10%, leading to press picking and breaks. Drainage aids are also used on broke or waste deckers to increase capacity.

Formation aid

Formation aids are additives used to promote dispersion of fibers. This improves formation and may allow higher headbox consistencies. There is very little information on formation aids in the literature. According to Wasser (1978) the best formation aids are linear, water soluble polyelectrolytes of ultra high molecular weight. He found the most effective agents to be anionic polyacrylamides. Traditionally locust bean gum, de-acetylated karaya gum, and guar gum have been shown to be effective dispersants for fibers. Other work has demonstrated that polyacrylamide (MW > 5 million) with moderate anionic character (15-25% hydrolysis) used at 3-5 lb/ton gives the best results, although drainage rates were decreased. Furthermore, like any high molecular weight polymer, these materials are subject to shear degradation, and should be added after the fan pump, centicleaners, etc., although adequate mixing is required. These materials should be handled in solutions below 1% to reduce viscosity and should not be mixed or added through tanks and pipes containing metal which might complex with the anionic moieties of the polymer.

Defoamers, antifoamers

Defoamers are used to control foaming. Foam exists when air or some other insoluble gas is mixed into water containing *surfactants*, that is, surface active agents. Soaps and detergents are good examples of surfactants; they have a hydrophobic tail and a hydrophilic portion which allows them to form surfaces of much lower surface tension (i.e., of high stability) than water alone. These materials also form micelles, small globules of hundreds of molecules with the nonpolar tails intertwined in the middle of the globule and the polar heads exposed to the solution on the outer surface. The center region can dissolve nonpolar materials like grease and oil and suspend it in water, which is how soap removes grease. Other surfactants found in various portions of the pulping and papermaking process include rosin and synthetic sizing agents (particularly if the latter are hydrolyzed), free fatty acids (from saponified triglycerides and other fatty acids), lignosulfonates, and similar materials. Traditionally defoamers were aliphatic chemicals such as kerosene, fuel oil, or vegetable oils used up to 0.5%; these materials, being water insoluble, would stay at the surface and act at the surface of water. Their activity is fairly limited, but they are still often used as carriers.

Strictly speaking, defoamers are used to destabilize (i.e., break apart) existing foams, whereas antifoamers are used to prevent the formation of foam. In practice, foam control agents work by both mechanisms, although to various degrees, and so the terms are often used interchangeably. Agents which act predominantly as antifoamers are added relatively early in the system before foaming occurs, whereas agents which act predominantly as defoamers are added late in the system near the origin of the foam.

Foam control agents work by a several mechanisms. 1) Newer agents that work by spreading on the surface of water include alkylpolyethers and silicone oils such as polydimethylsiloxanes. The silicone products do not enjoy widespread use because of their high cost and insolubility which make them difficult to disperse into systems, but new formulations appear frequently.

2) Defoaming surfactants include oligomers (with a degree of polymerization of 3-8 or so) of ethylene oxide (EO) or polypropylene oxide (PO) which are attached to an alcohol, an amine, or an organic acid. Three moles of EO attached to a C_{14}-C_{22} alcohol gives a good defoamer, with a structure of $CH_3(CH_2)_{12\text{-}20}CH_2OCH_2CH_2O\text{-}CH_2CH_2OCH_2CH_2OH$, but if 12 moles of EO were used instead the material would exhibit detergent behavior and actually contribute to foaming. Ethoxylated nonylphenols with 9 or more moles of EO give severe foaming even at levels of 10 ppm. (Reaction of EO onto materials is often called ethoxylation.) One can see that numerous formulations are possible (and sold). These materials are described by their *HLB* number (hydrophilic-lipophilic balance) which is a scale of 1-20 where 1 is 100% lipophilic (such as a pure C_{22} alcohol) and 20 is 100% hydrophilic (100% EO).

3) Hydrophobic particles such as hydrocarbon or polyethylene waxes, fatty alcohols, fatty acids, fatty esters, hydrophobic silica, or ethylenebisstearamide (EBS) suspended in a vehicle of oil, hydrocarbon, or water emulsion also control foam. Hydrophobic silica and EBS can be used at elevated temperatures where the others of this series would otherwise melt. More information on defoamers is available in Keegan (1991), where some of this information was gleaned.

Defoamers should be stored between 18-32°C (50-90°F) to avoid product degradation. Should defoamers need to be heated do not use direct steam since the high temperature will damage the defoamer. Instead, use insulated tanks with steam or hot water coils on the inside or external electric heat tracing. Never let overly hot or cold surfaces contact the defoamer.

Deposit control agents

Chemicals are sometimes used to control organic or inorganic deposits. Inorganic deposits can sometimes be controlled with sequestering agents for the metal ions to keep them in solution by binding them to polar molecules. (These are very similar to those used in bleaching mechanical pulps where metal ions can decompose hydrogen peroxide or cause discoloring of the pulp by reacting with phenolic compounds.) Sequestering agents include *chelants* and *threshold inhibitors*. Chelants such as EDTA, NTA, and DTPA react stoichiometrically (on an equal mole basis) with ions. Since the molecular weight of EDTA is 292 and calcium sulfate is 136, fairly large amounts of EDTA must be used. On the other hand, polyphosphates, polyacrylates, and phosphonates work by the *threshold effect* where less than stoichiometric amounts offer protection against scaling. Organic deposits, i.e., stickies or pitch, are treated by emulsification with nonionic surfactants or dispersants, attached to fibers with cationic fixatives such as polyamines or ammonium zirconium carbonate, or made not sticky by adsorbents such as talc or bentonite clays, which are magnesium and aluminum silicates, respectively. Sequestering agents for excess aluminum and addition of synthetic fibers such as polypropylene have also been explored for stickies control. Biological deposits (slime) are controlled by biocides.

Biocides

Microorganisms, particularly bacteria and fungi, will grow around the paper machine and produce slime consisting of proteins and polysaccharides. This slime may break off in pieces and lead to pitting of paper, actual holes in lighterweight papers, and even breaks in the web which lead to very expensive downtime. A number of bacteria and fungi will grow in various raw materials used to make paper. To grow, microorganisms need the proper pH (as a rule, pH 4-6.5 is ideal for fungi and pH 6-8 is ideal for bacteria), the proper temperature (40°C or 105°F is ideal, but little slime is produced above 60°C or 140°F), water, food, and oxygen. Some *anaerobic* microorganisms grow in conditions where oxygen is absent and produce methane, hydrogen sulfide, or hydrogen, which have caused fatal explosions at mills; anaerobic bacteria produce a characteristic foul odor, and the sulfate using bacteria, living under sulfate scales, can cause considerable corrosion. Recycled fiber and large amounts of starch may aggravate the problem of microorganism growth, as does poor housekeeping. Too, alkaline papermaking may require more expensive or higher levels of biocides to achieve control.

Good housekeeping and biocides can be used to control slime. Typically they are added from 0.05-1 kg/t (0.1 to 2 lb/ton) on paper. Oxidizing biocides (which are usually inorganic chemicals) include chlorine or chlorine dioxide. As shown in Fig. 17-1, in papermaking systems below pH 6 chlorine exists as the highly effective HOCl species; in alkaline systems approaching pH 9 the active species is the less active species OCl$^-$. Chlorine dioxide is added at 500-2000 ppm at various application points, although it quickly disappears from the system. There are several commonly used organic biocides whose structures are shown in Fig. 8-10. *Quaternary ammonium salts* contain alkyl, aryl, or heterocyclic substituents of C_8-C_{25}; they are most effective under alkaline conditions, although they lose their effect in systems with lots of contaminants, especially nonpolar oils and related materials. *Methylene bis-thiocyanate* is particularly effective against the sulfate-reducing bacteria of the *Desulfovibrio* genus but decomposes in whitewater systems if the pH is much above 8. *Brominated propionamides* are potent, broad spectrum biocides. *Carbamates*

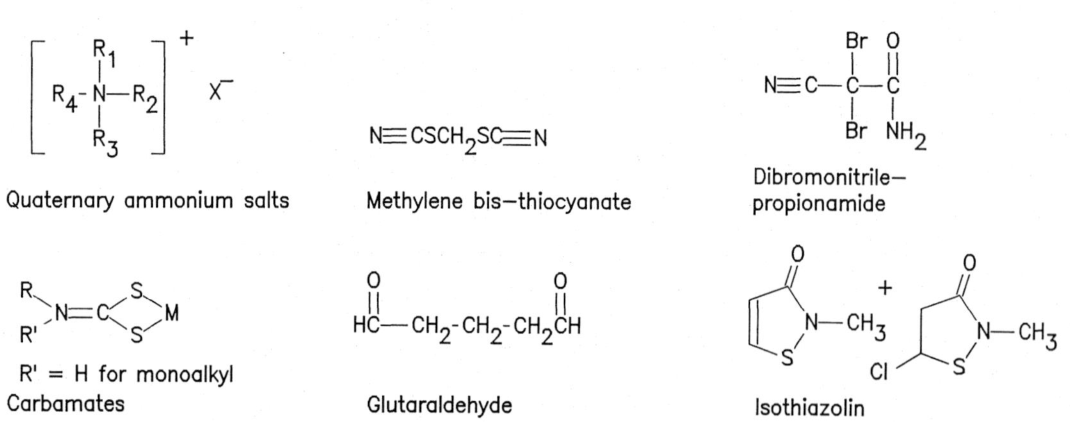

Fig. 8-10. Structures of some commonly used organic biocides.

are particularly effective above pH 7 and occur as dialkyl or monoalkyl, with longer alkyl chain members having lower toxicity than those containing short chain alkyl groups. *Glutaraldehyde* is an important crosslinking agent, as it has two reactive aldehydes that readily react with amines, and may react with proteins of the microorganisms. *Isothiazolin* is a mixture of two compounds; it has acute and broad activity. Because of the complex chemistry of all of these materials, one should be very careful before saying a certain microorganism is actually resistant to a particular biocide. Sometimes two or more biocides are used together to develop a synergistic effect. For more information on biocides is available in Purvis and Tomlin (1991).

Biocides are, by their very nature, toxic chemicals; as when handling any toxic chemical be sure to follow the manufacturer's directions, including suggestions for protective equipment and clothing, and read the *Materials Safety Data Sheet*. Occasional *boilouts* of some paper machines using hot water at 60°C (140°F) and pH 12 with a dispersant can reduce the amount of biocides required to control slime.

Other additives

Other additives include pH control agents and corrosion inhibitors. Sulfuric acid is often used to lower the pH of stock for the paper machine. Carbon dioxide (a byproduct of fertilizer manufacturing and refineries) is being used at some locations to replace sulfuric acid.

8.6 WET END CHEMISTRY

Flocculation

There is relatively little information on fiber flocculation in the literature. Formation is clearly dependent on fiber flocculation, which makes it very important. One fundamental study with a good evaluation of the literature is the work of Jokinen and Ebeling (1985). They used a pilot plant machine to evaluate numerous factors on fiber flocculation. Mechanical properties of fibers were much more important than chemical aspects of the slurry. For example, temperature from 18-35°C had no effect, nor did pH from 3.9 to 10.7. Fiber length had an important effect although the

magnitude depends on the pulp consistency; under one set of typical commercial conditions, flocculation would change by 1.2 times the relative change in fiber length. A flow velocity change of 100% would cause a 14% effect on flocculation in the range of 0.15 to 0.6 m/s. Fiber flexibility had little effect on flocculation, although there was more of an effect at high consistencies and flocs formed from stiff fibers were more difficult to break apart. The effects of beating on pulp are similar to the effects of the component changes of beating--namely the changes in fiber length and flexibility. A variety of deflocculation aids were tested, and anionic polyacrylamides (PAM) of high molecular weight had the most pronounced effect, with 0.5% causing up to 65% decrease in flocculation. Chemicals could not be used alone to increase flocculation above 5%. With fillers, however, retention aids can greatly increase flocculation.

Alum chemistry, coordination chemistry

The behavior of alum in aqueous solutions is critical to many aspects of wet end chemistry. Alum is used in rosin and ASA sizing, as a retention aid, and as a drainage aid. Alum is the central chemical in wet end chemistry and has received much attention. There are numerous papers describing the behavior of alum during papermaking that generally contradict each other. Many of these studies are meaningless or simply wrong. Most aspects of wet end chemistry are attributed to ionic charge interactions; however, coordination chemistry is probably more important, at least in the final product where free water is absent. The classic *Advanced Inorganic Chemistry* text by Cotton and Wilkinson (4th edition, 1980) shows the absurdity of many published works describing the chemistry of alum in the pulp and paper "literature".

This role of alum coordination chemistry in retention has been clear for decades to textile chemists that describe the mordant action of alum in textile dyeing--the same thing that occurs with rosin sizing. In the textile field a mordant is a substance that fixes a dye to a material. A mordant works by combining with the dye to form an insoluble complex that fixes the dye to the material. Mordants are metal atoms that attach to the dye at the oxygen and nitrogen atoms through formation of coordinate bonds. Actually alum is a common mordant used in textile dyeing. In pulp and paper literature the alum rosin complex is depicted in terms of ionic interactions (which must form exceedingly hydrophilic linkages). In fact coordinate structures are formed that are nonpolar and hydrophobic in nature, although ionic interactions are important for retention.

In a series of experiments Subrahmanyam and Biermann (1992) showed unequivocally that coordination chemistry is central to how alum achieves its function in rosin sizing. Rosin sizing with numerous highly coordinating transition and lanthanide elements was efficacious at a variety of acid and alkaline pH's depending on the formation constant with hydroxide of each individual element. Formation constants with rosinate, sulfate, chloride, and other ligands were also relevant to the degree of sizing.

Aluminum ions in solution form polymers by sharing hydroxide groups. Two adjacent aluminum ions share two hydroxide groups to form the chain; these linkages are termed *hydroxo* linkages and are relatively nonpolar linkages, unlike the polar linkages often shown that would be destroyed by water. The chemistry of aluminum is developed in more detail in Section 14.7.

The results of the first study allowed Biermann (1992) to take rosin sizing to a pH of 10 using polyamines with very high charge densities. This indicates that alkaline rosin sizing is possible, something that used to be debated.

Zeta potential

The zeta potential is a colloidal effect having to do with charge distributions on the surface of suspended particles. The zeta potential is the charge density on the surface of colloids and suspended particles. It varies from about -50 mV to +50 mV. Retention is often at a maximum when the charge density is near zero. This is probably where the solubility is lowest. At one time, it was thought to be the panacea for wet end control. While not meeting this expectation, it does have some use for wet end control. Wet end reactions such as retention are looked at primarily from the point of view of charge distributions. As discussed in several sections of this book, however, this is only part of the picture.

8.7 ANNOTATED BIBLIOGRAPHY

Auxiliary equipment

1. Ingraham, H.G. and E.E. Forslind, Auxiliary apparatus and operations preliminary to paper machines, in *Pulp and Paper Manufacture*, Volume 3, MacDonald, R.G., Ed., McGraw-Hill, New York, 1969, pp 185-243. This is a good description of equipment and operation of vacuum systems, stock chests, white water flow with material balance, savealls, centrifugal separators, deculators, and screens.

2. TAPPI TIS 0410-01 gives fan pump calculations.

3. Gooding, R.W. and Craig, D.F., The effect of slot spacing on pulp screen capacity, *Tappi J.* 75(2):71-75(1992). Blinding, accumulation of fibers near the screen plate apertures, occurs when more than 7% of the fibers are longer than the slot spacing. When 30% of the fibers are of this length severe blinding may occur, decreasing the capacity of the screens appreciably.

Fillers

4. Schwalbe, H.C., Fillers and loading, in *Pulp and Paper Science and Technology*, Volume 2, Libby, C.E., Ed. McGraw-Hill, New York, 1962, pp 60-89. This is a good overview on the physical properties of fillers.

5. TAPPI TIS 0106-05 (1988) gives trade names, % solids, brightness, and median particle sizes on a variety of North American calcium carbonates from various suppliers. TAPPI TIS 0106-06 (1988) does this for kaolin clays, and TAPPI TIS 0106-07 (1989) does this for titanium dioxide.

Internal sizing--general aspects.

6. Reynolds, W.F., Ed., *The Sizing of Paper*, 2nd ed., TAPPI, Atlanta, 1989. 156 p. This is a good overview of internal sizing with rosin, AKD, ASA, stearic acid, and fluorochemicals. Surface sizing and testing of paper and board for sizing are also covered. The volume is fairly general.

7. Crouse, B.W. and D.G. Wimer, Alkaline papermaking: an overview, *Tappi J.* 74(7):152-159(1991). This has a good introductory discussion on sizing agents used in the alkaline region including ASA, AKD, and rosin, with 53 references.

Rosin sizing
8. Davidson, R.W., Retention of rosin sizes in papermaking systems, *J. Pulp Paper Sci.* 14(6):J151-J159(1988). This is a fundamental study on rosin retention during sizing.

9. Wilson, W.S. and H.E. Duston, *Paper Trade J.* 117(21):223-228(1943).

Alkaline sizing agents
10. Markillie, M.A., Effects of starch properties on sizing in alkaline papermaking systems, *TAPPI Notes, 1989 Advanced Topics in Wet End Chemistry Short Course*, pp 7-13. This is a very useful article to help form a good emulsion without hydrolysis of the sizing agents.

AKD size
11. Davidson, R.W. and A.S. Hirwe, Retention of alkyl ketene dimer size in papermaking systems, *TAPPI Notes, 1985 Alkaline Papermaking*, pp 7-16.

This reference, which uses AKD labeled with the radioisotope carbon-14, describes AKD retention under alkaline papermaking conditions, using the Dynamic Drainage Tester with 15% kaolin clay or calcium carbonate filler or no filler. The study found that cationic potato starch (0.35% N) or an unspecified cationic resin (5.2 meq/g) (both used at 0.1% on pulp) was suitable for AKD (0.1% addition) retention. Use of post-addition of an unspecified high molecular weight anionic polymer (at 0.01-0.02%) greatly improves retention of AKD and fines. The level of sizing (HST) in handsheets correlates well to AKD retention, which, in turn, correlates well to fines retention. The presence of mineral fillers, especially the clay, greatly reduces sizing, which for calcium carbonate can be partially circumvented by higher levels of AKD addition.

ASA size
12. Wasser, R.B.. The reactivity of alkenyl succinic anhydride, *TAPPI Notes, 1985 Alkaline Papermaking*, pp 17-20. This is a useful paper on hydrolysis of ASA and development of sizing in paper.

13. E. Strazdins, Role of alum in alkaline papermaking, *TAPPI Notes, 1985 Alkaline Papermaking*, pp 37-42. Alum is a useful additive to develop a high level of sizing with ASA. Some useful data on the use of alum in ASA sizing is presented here.

Formation aids
14. Wasser, R.B., Formation aids in paper, An evaluation of chemical additives for dispersing long-fibered pulps, *Tappi J.* 61(11):115-118(1978).

Defoamers
15. Keegan, K.R., Defoamer theory and chemistry, *Notes, 1991 Chemical Processing Aids Short Course*, Tappi Press, pp 27-36.

Biocides
16. Purvis, M.R and J.L. Tomlin, Microbiological growth and control in the papermaking process, *1991 Chemical Processing Aids*, TAPPI press, pp 69-77.

Flocculation
17. Jokinen, O. and K. Ebeling, Flocculation tendency of papermaking fibres, *Paperi ja Puu - Papper och Trä* (5):317-325(1985). This is highly recommended reading and is a fundamental study with a good evaluation of the literature.

18. Allen, L.H. and I.M. Yaraskavitch, Effects of retention and drainage aids on paper machine drainage: a review, *Tappi J.* 74(7):79-84(1991) with 42 references. Retention and drainage aids can increase or decrease the rate of dewatering on paper machines. The effect of a candidate additive on your furnish

should be determined by performing laboratory tests prior to a mill trial.

Alum behavior
19. Subrahmanyam, S. and C.J. Biermann, Generalized rosin soap sizing with coordinating elements, *Tappi J.* 75(3):223-228(1992).

Polyamines in wet end chemistry
20. Biermann, C.J., The use of polyamine mordants in rosin sizing from pH 3 to 10, *Tappi J.* 75(5):166-171(1992).

EXERCISES

Stock approach
1. Name two methods of cleaning pulp prior to the paper machine.

Additives
2. What are the differences between functional and control additives?

Fillers
3. Give several advantages for using PCC in papers.

4. Which filler is used in very thin papers of high opacity? Why is this filler used?

5 What are the major advantages of alkaline papermaking?

6. What filler enjoys the most use by weight?

7. What filler would you use in linerboard?

Sizing
8. Penetration (sizing) of liquids is determined by what properties of the liquid and base sheet?

9. What sizing agent would one use for tissue and towel grades?

10. What is the mechanism whereby rosin and alkaline sizing agents size paper? How do the alkaline sizing agents differ from rosin size?

11. Can rosin size be easily used in papers filled with PCC? What is the contradiction between these two?

12. What is the mechanism of sizing with starch on the size press? Is starch hydrophobic?

Miscellaneous
13. What material is used as an adhesive in corrugated boxes? Hint: this material differs from the principal constituent of paper by one chemical linkage. What is the mechanism of adhesion?

14. What is the difference between wet strength additives and dry strength additives.

15. Why is it important to limit growth of microorganisms? What problems do they cause?

9
PAPER MANUFACTURE

9.1 INTRODUCTION

Paper

Paper consists of a web of pulp fibers (normally from wood), usually formed from an aqueous slurry on a wire or screen, and held together by hydrogen bonding. The same basic steps are involved for either hand- or machine-made paper and were shown in Figs. 1-5 for handmade paper and 6-16 for laboratory handsheets. The steps are as follows:

1. Forming - applying the pulp slurry to a screen.

2. Draining - allowing water to drain by means of a force such as gravity or a pressure difference developed by a water column.

3. Pressing - further dewatering by squeezing water from the sheet.

4. Drying - air drying or drying of the sheet over a hot surface.

The first American paper mill was built in the late 17th century in the Northeast. The process was very slow, being performed entirely by hand. The early development of the fourdrinier paper machine allowed the production of paper to increase tremendously and has been chronicled (page 1). By 1850 there were 191 machines and Byran Donkin was honored at the Great Exhibition for his contribution. The width of machines was stabilized at 160 in. until an American engineer, William Hulse Millspaugh, developed the centrifugally cast couch with a cycloidal vacuum pump in 1911. From this development the width, length and speed of paper machines increased rapidly.

Maximum paper machine speeds by grade

It is interesting to compare paper machine speeds and increases in machine speed over the last 10 years by paper grade and basis weight (Ely, 1981; Mardon et al, 1991) as shown in Table 9-1. With increasing basis weight, a slower machine speed is necessary. However, the production weight tends to increase with increasing basis weight. Glassine paper has a low production speed for its basis weight because the fibers are highly refined.

9.2 THE PAPER MACHINE

Fiber slurry preparation

Fibers must be properly slurried and mixed with additives. The slurry must be treated to remove contaminants and entrained air. Consistency regulation is also important. These subjects were discussed in the previous chapter.

Overview of the paper machine

The paper machine is a device for continuously forming, dewatering, pressing, and drying a web of paper fibers. Until recently, the most common type of wet end was the *fourdrinier*, where a dilute suspension of fibers (typically 0.3-0.6% consistency) is applied to an endless wire screen or plastic fabric. Water is removed by gravity, or the ΔP developed by table rolls, foils or suction equipment, and the drilled couch. The web at this point is 18-23% consistency. More water is squeezed out in the press section to a consistency of 35-55%. Finally the sheet is dried with steam heating in the dryer section.

Machines that use two wires to form and drain water from the dilute, pulp slurry are called *twin wire formers*. These have become popular since the late 1960s for printing and lightweight papers. During the 1970s, many multi-ply formers were developed for heavyweight board. These formers use up to seven wires consecutively and are modifications of another type of paper machine, used for heavy-weight board, the *cylinder* machine, which has been used since the early 1800s. Modern paper machines can be quite expensive, costing up to several hundred million dollars each.

Table 9-1. Maximum paper machine speeds and production by paper grade (Ely, 1981; Mardon, Vyse, and Ely, 1991).

Paper grade	Basis weight		Speed, fpm		Production[1], ton/in./d	
	lb/3000 ft[2] [2]	g/m[2]	1991	1981	1991	1981
Towel, dry–creped	8	13	6560	5750	1.05	0.91
Towel, wet–creped	25	41	3000	3000	1.49	1.49
Glassine	25	41		1215		0.60
Towel, wet–creped	27	43	2800	2800	1.52	1.52
Coating rawstock	24	39	4000	3100	1.96	1.47
Newsprint	30	49	4503	3700	2.70	2.22
Printing, writing	54.5	89	3650	2175	3.97	2.37
Corrugated medium	78	127	2505	2180	3.90	3.40
Kraft linerboard	78	126	2550	2300	3.98	3.59
Kraft linerboard	126	204	2290	2100	5.77	5.29
Kraft linerboard	207	337	1700	1630	7.29	6.75
Kraft linerboard	270	438	1260		6.80	
Liquid packing	470	765	330		3.10	

[1]Production is a direct function of paper machine speed and basis weight assuming no down time.

[2]Many of these grades normally use basis weights for 1000 ft[2], so divide the table value by three. It appears there were errors from this difference in some of the basis weights from the 1991 paper.

Pulp must be applied to the screen at low consistencies to give good formation, that is, an even distribution of fibers so the paper has uniform thickness. Softwood pulp slurries at 3% consistency do not even flow well. Therefore, the entire purpose of the paper machine is to remove all of this water that one is forced to use to give paper that is uniform. For every 1 lb of fiber 200 lb of water are used. For this one lb of fiber about 195 lb of water (98%) are removed at the wet end with the web leaving at a consistency of 20%. Another 2 to 3 lb of water are removed in the press section (1 to 1.5%) with the web leaving the press section at 35-50% consistency. The remaining 1 to 1.5 lb (0.5 to 0.75%) of water are removed by the dryer section. Although the dryer section removes the least water, the high energy requirements make this the most expensive section to operate. The white water removed in the wet end of the paper machine is reused to dilute pulp slurry for the headbox. The fiber and fillers of excess white water are removed by *savealls* such described in Section 10.3. Pulp *sweetener stock* is sometimes added to the excess white water to help recover the fines and fillers by filtration.

Vacuum systems use about 30% of the power of the paper machine with almost 75% of the vacuum power going to the press section.

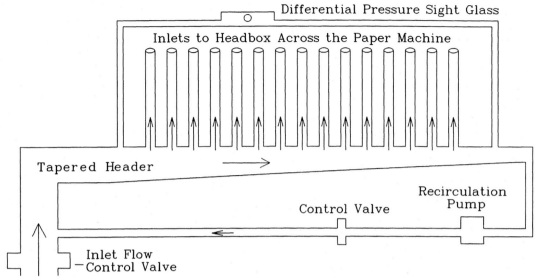

Differential Pressure Sight Glass

Inlets to Headbox Across the Paper Machine

Tapered Header

Recirculation Pump

Control Valve

Inlet Flow
Control Valve

Fig. 9-1. Tapered inlet headbox approach now used on virtually all paper machines.

The dewatering process

The key to good papermaking with long fabric life, good retention, and minimized sheet two-sidedness is control of the process (Hansen, 1991). It is important to delay sheet sealing when forming since this will lead to extra drag. Sheet sealing occurs at around 0.8-1.4% consistency unless precautions are taken.

9.3 THE HEADBOX

The headbox approach

The inlet to the headbox must insure an even fiber slurry consistency and pressure across the width of the headbox to insure cross direction uniformity of paper. A variety of techniques in the past has been used to insure this, including complex distribution piping or various types of manifolds. The invention of the tapered inlet by J. Mardon in the 1950s greatly simplified this goal. All modern headboxes use a tapered inlet pipe (Fig. 9-1) to accomplish this purpose.

An even pressure across the headbox is maintained by controlling the rate of recirculation. For example, by increasing the recirculation, i.e., by reducing the pressure at the narrow end of the tapered inlet, the pressure and, therefore, flow rate at the headbox over the narrow part of the tapered inlet is decreased.

Headbox

The headbox is a pressurized device that delivers a *uniform* pulp slurry on the wire, through the *slice*, at approximately the same velocity as that of the wire. Original headboxes were open, unpressurized, and used a hydrostatic head for the necessary pressure

Almost all headboxes on paper machines operating below speeds of 2500 ft/min have two to five perforated *rectifier rolls (holey rolls)* inside the headbox that create microturbulence to keep the fibers in suspension, giving even formation. Fig. 9-2 shows this type of headbox with the standard three-roll design. New headboxes use the three-roll design, never the five-roll design that was once common. The rolls are 0.2-0.8 meters (8-32 in.) in diameter with holes 1.5-2.5 cm (0.6-1.0 in.) in diameter that occupy 40-52% of the roll area and rotate at 5-20 rpm. Fig. 9-3 shows a rectifier roll.

A *secondary headbox* may be used part of the way down the table to give a top coat of high quality fiber relative to the rest of the sheet. This is done, for example, to produce a white printing surface on linerboard or to put secondary fiber in the middle layer of linerboard, where contaminants such as polymers and wax are hidden. Fig. 9-4 shows a fourdrinier paper machine with a secondary headbox.

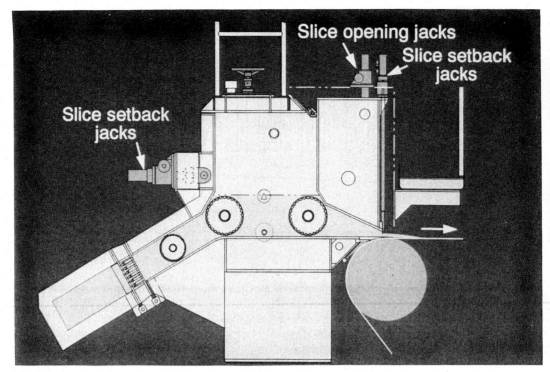

Fig. 9-2. A closed, pressurized headbox with air cushion for relatively slow paper machines. This is a standard three-roll rectifier roll headbox. Courtesy of Beloit Corp.

Originally, the pressure to accelerate the stock to the speed of the paper machine was supplied by a hydrostatic head of liquid. But as Fig. 9-5 shows, high speed paper machines require too much pressure for this to be practical, so paper machines now used closed, pressurized headboxes to supply the necessary pressure.

Paper machines operating above 2500 ft/min require a special, high pressure headbox known as a *hydraulic headbox*. These headboxes do not use rectifier rolls because the high turbulence that is generated with these rolls causes formation problems beyond the headbox.

Slice

The slice is a rectangular slit in the headbox where the pulp slurry is applied to the wire. It consists of a lower, fixed apron and an upper, adjustable lip controlling the slice height. The slice height controls how much stock is applied to

the wire (and therefore the basis weight); the slice height is variable across the width of the paper machine to insure uniformity of the paper across the width of the paper machine. Paper machines now use *velocity formation* where the lower apron

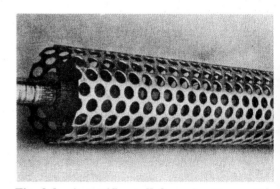

Fig. 9-3. A rectifier roll for a headbox. Reprinted from *Making Pulp and Paper*, ©1967 Crown Zellerbach Corp., with permission.

Fig. 9-4. A two headbox fourdrinier machine for linerboard.

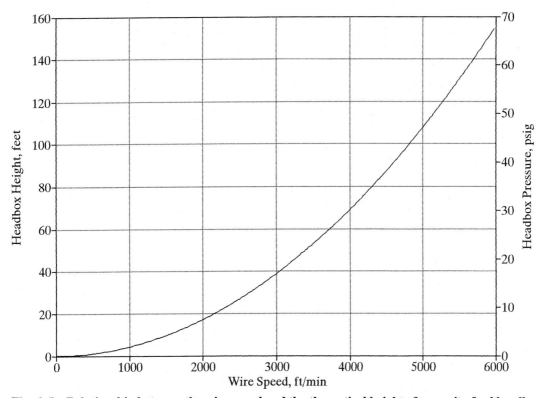

Fig. 9-5. Relationship between the wire speed and the theoretical height of a gravity feed headbox (left) or pressure of a pressurized headbox (right). Frictional losses are not included. $H = V^2/(2g)$

protrudes beyond the headbox to direct the slice outward. With *pressure formation*, the upper lip protrudes beyond the apron and the jet is directed toward the wire, which may cause poor formation, two sidedness, poor retention, and wire mark. As a result of the development of water removal equipment, it is no longer necessary, or desirable, to use pressure forming except for tissue grades.

The equation, $F = h \sin \beta$, for the forming force, F, shows that the force will increase with the speed (since h, the head, increases with speed) unless the approach angle, β, can be reduced (Fig. 9-6. If the forming pressure reaches 34 kPa (5 psi or 10 in. Hg) then the sheet is "welded" to the wire, and cannot be removed. When $\beta = 8°$, this will occur at a speed of 23 m/s (4500 ft/min).

The energy of the jet is used to displace the air accompanying the return fabric into the pans. As speeds increase, the energy of this air film increases, so the angle cannot be reduced. Therefore, the forming force increases with advancing speed. To circumvent this contradiction that occurs at high speeds, the breast roll can be lowered to decrease β by allowing a flat delivery as depicted in Fig. 9-7.

9.4 THE FOURDRINIER WET END

Fourdrinier

The fourdrinier, or flat wire machine, is a paper machine with a horizontal, moving, fine mesh, woven wire cloth or plastic fabric upon which the pulp slurry is deposited, forming the web (Fig. 9-8). The *front side* or *tending side* is the side from which the paper machine is controlled, whereas the other side is the *backside* or *drive side*. The wire forms a continuous belt that picks up fiber at the breast roll from the headbox, runs over the table rolls, foils, suction boxes, and then over a couch roll, where the web of fibers leaves the fourdrinier table. The wire, however, continues around the couch roll, under the machine, to the breast roll where more fiber is received. A wire might make several hundred thousand trips in its life. The position and tension of the wire are controlled by special rolls. Fig. 9-9 shows the guide roll with wire showers and Fig. 9-10 shows the tensioning roll from a linerboard machine.

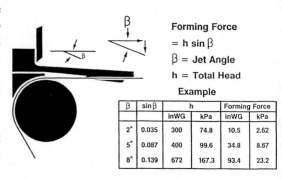

Forming Force
$$= h \sin \beta$$
β = Jet Angle
h = Total Head
Example

β	$\sin\beta$	h		Forming Force	
		inWG	kPa	inWG	kPa
2°	0.035	300	74.8	10.5	2.62
5°	0.087	400	99.6	34.8	8.67
8°	0.139	672	167.3	93.4	23.2

Fig. 9-6. The forming force.   ©1991 TAPPI. Courtesy of V. E. Hansen.

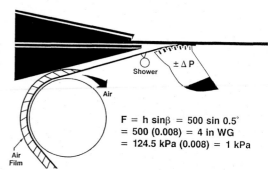

Shower

$\pm \Delta P$

Air

Air Film

$$F = h \sin\beta = 500 \sin 0.5°$$
$$= 500 (0.008) = 4 \text{ in WG}$$
$$= 124.5 \text{ kPa} (0.008) = 1 \text{ kPa}$$

Fig. 9-7. Reducing pressure formation by lowering the breast roll. ©1991 TAPPI. Courtesy of V. E. Hansen.

Clothing

Paper machine clothing consists of forming fabrics (wires), press felts, and dryer felts.

Wire, forming fabric

The wire, or now more precisely called the forming fabric, is a continuous loop or belt of finely woven screen made from wire or plastic; the mesh size varies from 40 to 100 mesh (openings per inch). A coarse wire allows faster drainage but gives a coarser paper; as in most aspects of pulp and paper, there is always a tradeoff in goals. Before 1960, wires were made from metals such as bronze, but they are now almost invariably made from plastic such as polyester which lasts much longer and is corrosion resistant, although it stretches more and cannot handle highly abrasive furnishes. The forming media has three functions:

1. to transport the fiber.

2. to permit draining the sheet.

3. to transmit power.

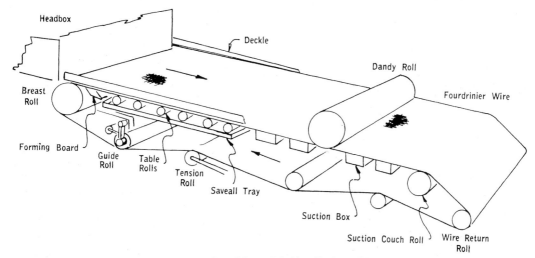

Fig. 9-8. A fourdrinier wet end. Reprinted from *Making Pulp and Paper*, ©**1967 Crown Zellerbach Corp., with permission.**

The power input is usually through the couch and wire turning roll.

The geometry, weaves, mesh, wear patterns, void volume, and numerous other aspects of forming fabric can be quite complex. The three references listed at the end of the chapter for this area are a good introduction to the topic.

Wire showers

Wire showers (Fig. 9-9) are high pressure showers on the underside of the wire used to remove fillers and other material which may plug the wire. Usually, they move back and forth slowly to clean all parts of the wire and prevent wire wearing in one spot.

Fig. 9-9. Sensor paddle (left, on the machine backside) and guide roll (right, on the machine frontside) for linerboard machine. Wire showers are visible to the top left of the guide roll.

Fig. 9-10. The tensioning roll for linerboard.

Web

The continuous mat of fibers that is in the process of forming or which has already formed the final paper is known as the web.

Breast roll

The breast roll is located under the headbox and serves to return the fabric to the forming area to receive the stock once again. It must be rigid enough to resist deflection. Fig. 9-11 shows the breast roll from a linerboard machine. During the era of pressure forming, it served as a water removal device acting like a table roll. Pressure forming is now not advised except in tissue forming.

Forming board

The forming board (Fig. 9-12) consists of a large leading blade to reach in close to the slice followed by several smaller blades, usually with gaps between them. The blades must be located at intervals such that the harmonic (pulse frequency) developed agrees with the harmonic of the rest of the drainage equipment. Several designs of surfaces give configurations that serve to retard drainage as well as to provide agitation that prevents flocculation. Retarding early drainage facilitates uniform water removal and improves

Fig. 9-11. Breast roll from a linerboard machine also showing the forming board and headbox slice adjustments.

formation. The blade material can be plastic, usually polyethylene, or a ceramic (such as aluminum oxide, zirconium oxide, silicon nitride, or a combination of these) to minimize wear. The blades are usually attached to the heavy structure by Tee bars to simplify maintenance. Forming boards often cause problems and will likely not be used on newer designs machines.

Deckle board

Deckle boards are used to prevent the stock from flowing off the two sides of the forming fabric when a thick layer of stock is delivered to the fabric by a large slice opening. Often the two edges of the fabric are curled upward by means of plastic edge curlers. Stationary deckle boards cause a deckle wave due to the friction between the moving stock and the stationary board. Curling the edge of the fabric helps to alleviate this problem. Often water jets are used in place of boards or curlers to provide a hydraulic "wall" to contain the stock for small slice openings.

Trim squirts, edge squirts

The width of the sheet delivered to the press section is controlled by trim squirts. The narrow bands of stock left on the fabric are knocked off by edge squirts. On newer designs of fabric there may have to be two trim squirts in line on each side to get a clean "cut". Economics dictate that a minimum be wasted in trimming, about 5-7 cm (2-3 in.), although this material is returned to the paper machine.

Table roll

Table rolls (Fig. 9-13, top) are freely revolving rolls under the fourdrinier wire that support the weight of the wire and wet web. Water is removed from the bottom of the web by a partial vacuum [45 kPa below ambient pressure or 14 in. of mercury at machine speeds of 10 m/sec or (2000 ft/min) over a distance of about 1 cm] at the meniscus that forms at the nip where the roll leaves the wire. Burkhard and Wrist (1956) showed the maximum suction is about $\frac{1}{2}\rho U^2$, where ρ is the density of water and U is the machine speed. Table rolls are not useful at speeds much higher than this, because at higher speeds, water on the incoming side of the table

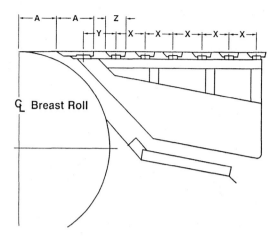

Fig. 9-12. Forming board with variable open cover. ©1991 TAPPI. Courtesy of V. E. Hansen.

roll goes back into the sheet; this pumping action causes stock jump and tends to remove fines and fillers leaving the paper's wire side deficient of these items. Also, since the white water may be fairly warm, it may actually boil at the reduced pressure at the nip of the table roll. Thus, a two sided sheet forms, one with different properties on the two sides. Foils have, therefore, replaced table rolls on all paper machines.

Foils

A foil (Fig. 9-13, bottom photo), introduced by Wrist and Burkhard (1956), is a stationary blade 5-10 cm (2-4 in.) wide with a divergent surface so that an angle forms between the stationary fabric and the foil surface. The angle is usually between 0.5 to 3°. With stock on the fabric, and the fabric in motion, the suction that develops causes the fabric to draw down towards the foil surface. As the fabric leaves the suction area, it must rise to the height of the next foil blade. In order to develop fine scale turbulence, the *frequency* should be 40 to 90 foils per second with a maximum of 120 per second. (This does not mean there are that many foils.) The movement of the fabric in the vertical plane must not be so rapid as to cause the stock to separate from the fabric and be "thrown" down the machine, because, due to air resistance, it will land on a different area of the fabric from which it left. Foils have two functions apart from supporting the fabric. These are:

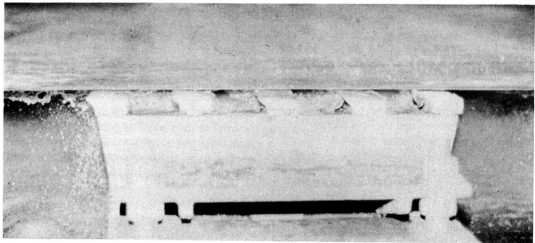

Fig. 9-13. Comparison of table rolls (top) and foils on fourdrinier machines.

1. to provide hydraulic shear (i.e., activity, as shown in Fig. 9-14).

2. to give uniform, controlled water removal.

In his classic work on paper formation from dilute fiber slurries, J. D. Parker (1972) maintained that the diameter of the "spouts" should equal the fiber length to prevent flocculation.

Fig. 9-15 shows a foil removing much water shortly after the headbox. Foils are effective at relatively high speeds and form a smaller vacuum force, but over a longer distance than table rolls (about 10-13 kPa below ambient pressure or 3-4 in. of mercury over a distance of 5-10 cm or 2-4 in.). The leading edge of a foil doctors water pulled down from the previous foil.

A foil table must be designed carefully to avoid the common problem of *sheet sealing*. This condition arises from rapid drainage from 0.8 to 1.4% (overall) consistency (Hansen, 1991). A high consistency (between 5 and 10%) layer of fiber is "laid" on the fabric, but the stock above is still at headbox consistency. Further controlled drainage is not possible because of the short duration of foil pulses. A high dragload and a poor moisture profile results when drainage is attempted by hivacs (suction boxes). Thus, the table should be divided so that foils activate the sheet and carefully advance the consistency through 1.4% to about 1.6 or 1.8%. Then drainage is taken over by the lovacs and hivacs.

High speed designs must control the activity so that stock jump does not occur. Activity must

Fig. 9-14.   Activity on the forming fabric with proper operation of the foils. Courtesy of V.E. Hansen.

be fine scale to insure the best formation. In other words, the headbox must deliver a well-formed sheet, and the foil table must develop shear so the formation is not destroyed. Uniform foil spacing is necessary to maintain the frequency pulses until 1.6 to 1.8% consistency is reached.

Lovacs

A lovac (Hansen, 1991) develops suction using water-filled drop legs (barometric legs like those used in brown stock washers) to provide siphoning action (Figs. 9-16 to 9-18). These are sometimes called *wet boxes*. The pressure gradient) is proportional to the height of the water column. Four distinct types of covers are used, all of which are slotted. To cause drainage only, a series of flat blades (7 to 13 is common) is used. To impart activity as well as drainage, three distinct cover shapes are in use: one uses foil blades called a Vacufoil™; a second uses a shaped blade, called an Unfoil™; and a third uses blades at different elevation, called an ISOFLO™ (Fig. 9-17). The Vacufoil allows imposing a ΔP greater than the water leg permits as a result of

the suction from the foil. The Unfoil allows activity to be generated using a ΔP generated from the water leg. The disadvantage of these two is that the white water consistency is greater than

Fig. 9-15.   Foil action on a fourdrinier machine.

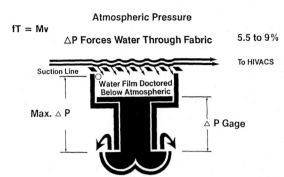

fT = Mv

Atmospheric Pressure

ΔP Forces Water Through Fabric   5.5 to 9%

To HIVACS

Suction Line

Water Film Doctored
Below Atmospheric

Max. Δ P

Δ P Gage

Fig. 9-16. Water removal by lovac. ©**1991 TAPPI. Courtesy of V.E. Hansen.**

Fig. 9-18. A lovac on a paper machine.

when the flat blades are used. The ISOFLO uses flat blades employing ceramic blades for fabric support. The control blades limit the vertical downward travel of the fabric, thereby control the level of activity. Sheet sealing can be prevented by both Unfoil and the ISOFLO. The unfoil uses the return water to the bottom side of the sheet (similar to the action of table rolls), and the ISOFLO uses the pulse from the sudden stopping of the downward travel of the fabric.

Controls are essential for the successful operation of lovacs. These entail proper size of piping and using an exhauster to generate the suction. Automatic valves are necessary to hold the ΔP values selected (Fig. 9-19).

On free draining grades, it is usual to have a series of lovacs, often four, operating at increasing ΔP as the sheet advances towards the couch.

Fig. 9-17. An ISOFLO™ lovac with flat blades at different levels and ceramic supports. ©**1991 TAPPI. Courtesy of V. E. Hansen.**

Fig. 9-19. Automatic valves to control the ΔP on lovacs. Courtesy of V. E. Hansen.

Both the wetline (5.5% consistency) and dryline (10% consistency) can be developed economically on lovacs (Fig. 9-20).   Until the dryline is reached, water is removed in film form as with foils.   After the dryline, air passes through the sheet to transport the water in droplet form.

In slow draining stocks, such as mechanical pulps, a wetline cannot be developed economically on lovacs. For some chemi-mechanical pulps with freeness above 230 CSF, a wetline can develop on a lovac. When the CSF is below 190, the maximum consistency obtained is about 4% by lovacs. The lovac blades take on the shape of hivac blades with doctoring edges (Fig. 9-21). This doctoring edge is necessary because water is removed in films until at least 9% consistency is reached. Actually the unit is a special design of lovac. If the usual blade spacing for lovacs is used at these higher ΔP values on these grades, the fabric "sags" into the slots, and more power is required to pull the fabric out of the slots.

Very good automatic control for ΔP is necessary to minimize power consumption and maxi-mize fabric life. The area of the waterlegs for draining the box is not less than 20% of the open area of the cover. Lovacs operate between 4 and 50 in. water gravity (WG) (1 to 12 kPa) generally. However, for high speed machines where the lovacs are used for activities (ISOFLO) the ΔP may be as low as 1 in. WG (0.25 kPa).

Suction boxes, flat boxes, hivacs

In the progression along the forming table the stock is subjected to increasing ΔP to drain the sheet.   After the lovacs, which rarely operate above 12 kPa (50 in. WG), the stock encounters the suction boxes or hivacs. The term "flat box" is no longer appropriate since curved covers are used regularly on two-wire machines.

Drilled ceramic became popular in the development of suitable cover material for these boxes. As machine speeds increased, however, these solid ceramic covers were not strong enough for the increasing ΔP. In addition, drilled holes presented too much surface area inside the hole, creating excessive friction for the water to pass through.

Fig. 9-20. Generation of wetline and dryline on a lovac. Courtesy of V.E. Hansen.

A slotted technology developed wherein ceramic pieces are bonded to stainless steel bars. Each of these pieces are 15 and 20 mm wide in the machine direction, with the former being more efficient. Slot sizes vary from 25 mm (1 in.) to 19 mm (3/4 in.) to 16 mm (5/8 in.) for maximum ΔP of 17 kPa (5 in. Hg), 36 kPa (10.5 in. Hg), and 61 kPa (18 in. Hg.), respectively. Eight is the optimum number of slots, but as many as ten are used for a particular ΔP (Fig. 9-22). The bottom of the fabric must be kept wet to minimize wear and power requirements; consequently, if more slots are used at a given ΔP, fabric wear and power requirements increase. To work within these limits ΔP must increase exponentially. The result is fewer hivacs are used (Fig. 9-23). Again, the need for accurate, automatic control is obvious.

Adequate water removal at the hivacs, before the drilled couch roll, is essential on high speed machines since the couch has less time to remove

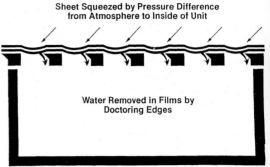

Fig. 9-21. Water removal between the wet- and dryline for slow draining stocks. ©1991 TAPPI. Courtesy of V.E. Hansen.

water; therefore, the hivacs must present a dryer sheet to the couch in order to avoid pinholes.

The design of piping is critical to ensure that water droplets can be transported from the sheet to the box to the separator. After the separator, when water flows down to the seal tank and the air goes to the header and to the vacuum pump, the air velocities can increase (Fig. 9-24). Pipe diameters must be large enough to control the air velocities to a maximum of 17.8 m/s (3500 ft/min) from the hivac to the separator, 0.45 m/s (750 ft/min) through the separator, and 25.4 m/s (5000 ft/min) through the risers and headers.

To dry different grades of paper, different flow densities are required. Kraft bags are very "open", and require the highest at 290 cm³/s/cm² (4 ft³/min/in²). Most other grades require 110-145 cm³/s/cm² (1.5 to 2 ft³/min/in²). The objective is to dry the sheet as much as possible with a suitable power consumption (drag-load).

Free draining grades can be "dried" to 10% on lovacs at low dragload. Three or four hivacs can be used to increase the consistency to 19% before the couch. Slow draining grades can be raised to 4% on lovacs, but then require two special chambers to arrive at 10% (dryline). Four more graduated chambers with proper airflow density will normally raise the sheet consistency to 19% before the couch. Some grades may require six chambers.

Because of the development of multi-layer fabrics that carry large volumes of air (Table 9-2), multi-chambered hivacs are useful to minimize the increase in airflow. As the fabric and sheet enter

Fig. 9-22. Multi-chambered (two, of eight bars each) hivac unit. Courtesy of V. E. Hansen.

the slotted area, the air must be removed from the fabric before moisture can be removed from the sheet. As the fabric leaves the hivac, air reenters the fabric. The use of a multi-chambered unit minimizes the number of times the fabric is evacuated of air (Fig. 9-25). For example, a triple layer fabric on a machine 5.1 m (200 in.) wide at

12.2 m/s (2400 ft/min) at 38 kPa (10 in. Hg) requires 40.1 L/s (85 ft³/min) more airflow than a single layer for each hivac. On a machine 7.62 m (300 in.) wide running at 18.3 m/s (3600 ft/min) can waste 0.755 m³/s (1600 ft³/min) pumping out the fabric for three chambers. Thus, it is important to minimize hivac chambers by combining

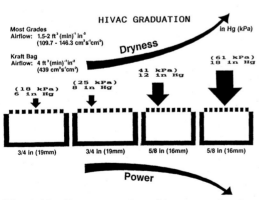

Fig. 9-23. Representative hivac graduation. ©1991 TAPPI. Courtesy of V. E. Hansen.

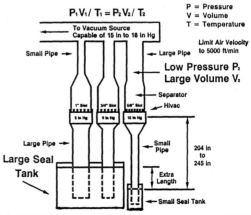

Fig. 9-24. Piping for a hivac system. ©1991 TAPPI. Courtesy of V. E. Hansen.

Table 9-2. Wasted vacuum requirements (ft³/min) for various fabrics. Courtesy of V.E. Hansen.

	Single wire	Double	Double X	Triple	Kraft D
Void volume, %	58.6	56.1	57.0	63.0	67.0
Coefficient V	7.5	9.9	10.3	19.3	23.2
W vol. @ 10" Hg 1C	54	71.3	74.2	139	167
6C	324	428	445	834	1002
300 in @ 3600	729	962	1001	1876	2255

X is an extra strand;   D is a double layer;   ©1991 TAPPI.
V is standard atmosphere cfm/(100 in. width)/(100 fpm)/box
1C is 1 chamber @ 2400 fpm/(200 in. wide) @ 10 in.Hg;   6C is 6 chambers (units) @ 10 in.Hg

them into multi-chambered units that have to be evacuated only once.

In summary, free draining grades require three or four hivacs, depending on speed. Mechanical pulps may require as many as eight, which is conveniently carried out with four two-chambered units. Linerboard is a special case that requires careful design to control the dragload. Before the secondary delivery, one hivac chamber is necessary if the primary sheet requires 15% dryness because whitetop is being manufactured. The secondary layer is accepted on a chamber at a low ΔP. Then four chambers are required where the ΔP is slowly increased at a linear rate to raise the sheet to the dryline at about 9% consistency. After the dryline is achieved, four of five hivacs are used in the usual fashion. [Suction boxes with showers are also used for cleaning felts in the press section (Fig. 9-26).]

A note about ceramic materials

Slow paper machines still use polyethylene surfaces to resist wear on moving parts. Fast paper machines are using more ceramic materials for surfaces that contact their clothing. Nevertheless, there are no established ceramic standards for the paper industry and few for the ceramic industry, and those few are applied differently by each supplier of ceramics, so technical specifications supplied by the manufacturers of ceramics must be supplemented by specific tests relevant to paper machine operation to be useful to those fabricating ceramics into devices. **This area must be addressed by the pulp and paper industry.**

Dandy roll

The dandy roll (Fig. 9-27) is a hollow, light, wire covered roll that rides on top of the fourdrinier wire just ahead of the suction boxes. This roll has four purposes:

1. to impart a *water mark* to the sheet. Water marks are visible in high quality bond papers when held to the light and can be used to show a company's logo, for instance.
2. to improve the top surface for printing.
3. to improve formation by mechanical shear.
4. to increase the drainage capacity of the flat wire.

The dandy must be axially mounted, like any roll, or end ring mounted so that it will rotate without vibration. During the 1980s, dandy diameters have increased, and the mounting mechanism

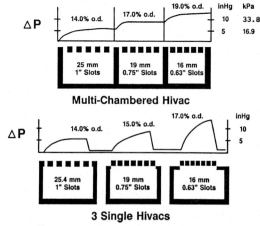

Fig. 9-25. Multi- vs. single-chamber hivac efficiency. ©1991 TAPPI. Courtesy of V.E. Hansen.

Fig. 9-26. Profile of vacuum box and showers for cleaning a press felt.

Fig. 9-27. The dandy roll of a machine operating in 1976.

has improved. A minimum diameter for purposes (3) and (4) is 1.09 m (42 in.) but 1.2 to 1.5 m (4 to 5 ft) will impart more shear. The roll must be driven by a timing belt and monitored for speed to impart shear. It is most effective when mounted between lovacs, with the lovac separation equal to the diameter of the dandy. If the objective is to improve formation, the consistency should be 2.5 to 3%, controlled by a gamma gauge. For drainage improvement, about 1.3 to 1.4% consistency will cause "dandy throw". This stock is collected and returned to the stock tank. This "throw" sometimes succeeds in evening out the two sides of the sheet.

Couch roll

The couch roll (Fig. 9-28) is the guide or turning roll for the fourdrinier wire, where the paper web leaves the wire and the wire returns to the breast roll. It has two purposes: to transmit power to the fabric and to increase the dryness of the sheet. In an open draw machine the fabric wraps the couch at least 180°. In a machine that uses a pickup felt, the fabric wraps the couch for about 40° and then passes on to the wire turning roll. The fabric wraps this roll for about 135°. These two rolls are responsible for driving the fabric, with the wire turning roll providing 55-80% of the drive load. The couch roll is really a drilled shell with countersunk holes 5.5 mm (7/16 in.) in diameter for 28-65% of its surface. Inside the shell is a close fitting "box" containing seal strips on the ingoing and outgoing edges. This box is connected to a vacuum pump capable of generating as high as 85 kPa (25 in. Hg) at sea level, depending on the grade of paper (Fig. 9-29). For linerboard the box often has two chambers. The first chamber may offer 34-41 kPa (10-12 in. Hg) starting at top dead center, and the second chamber offers up to 85 kPa (25 in. Hg). The ratio of the size of the first to second chamber is usually 2:1. Airflow density in the couch varies from 440-1460 cm^3/s/cm^2 (6 to 20 ft^3/min/in^2).

The limitation of the couch's ability to dry the sheet is its ability to "throw off" the water it draws into the shell in the suction zone. As each

Fig. 9-28. A couch roll (bottom left) and lump breaker roll of a linerboard machine. Note the water being thrown off the couch roll (through the drilled holes) after the felt separates from it.

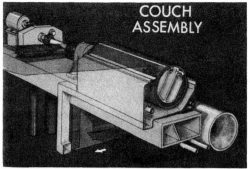

COUCH ASSEMBLY

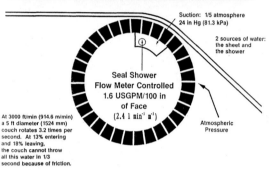

Suction: 1/5 atmosphere
24 in Hg (81.3 kPa)

2 sources of water:
the sheet and
the shower

Seal Shower
Flow Meter Controlled
1.6 USGPM/100 in
of Face
$(2.4 \, 1 \, min^{-1} \, m^{-1})$

Atmospheric
Pressure

At 3000 ft/min (914.6 m/min) a 5 ft diameter (1524 mm) couch rotates 3.2 times per second. At 13% entering and 18% leaving, the couch cannot throw all this water in 1/3 second because of friction.

Wv = Wasted Volume

D.A. = Drilled Area = 50%
W = Drilled Width = 286 in (7.26 m)
t = Shell Thickness = 2.5 in (6.35 cm)
U = Machine Speed = 3000 ft/min (914.6 m/min)
Δ P = 24 in Hg (81.3 kPa)
$Wv = D.A. (W) U (t) \dfrac{\Delta P}{P}$

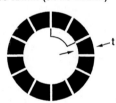

t

Example

$Wv = \dfrac{50}{100}\left(\dfrac{286}{12}\right)\left(\dfrac{2.5}{12}\right)(3000)\left(\dfrac{24}{30}\right) = 5958 \text{ ft}^3/\text{min}$

$Wv = \dfrac{50}{1000}\left(\dfrac{7.26}{60}\right)\left(\dfrac{6.35}{100}\right)(914.6)\left(\dfrac{81.3}{101.6}\right) = 2.81 \text{ m}^3 \text{ s}^{-1}$

Fig. 9-29. The couch assembly (top); courtesy of Beloit. Operation of the drilled couch roll and wasted volume of the couch. ©1991 TAPPI. Courtesy of V. E. Hansen.

roll of holes enters the suction zone the water and air must be reduced to make room for the water to be removed from the sheet and the water from the seal shower. This water is pulled into the shell until the row of holes passes the seal strip on the exit edge of the box. Now the pressure is equal-

POWER FOR DE-WATERING TO A PARTICULAR DRYNESS
CALCULATED AS FABRIC TENSION INCREASE (ΔT)

SI Units

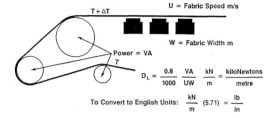

$T + \Delta T$ U = Fabric Speed m/s

W = Fabric Width m

Power = VA

T

$D_L = \dfrac{0.8}{1000} \cdot \dfrac{VA}{UW} = \dfrac{kN}{m} = \dfrac{\text{kiloNewtons}}{\text{metre}}$

To Convert to English Units: $\dfrac{kN}{m}$ (5.71) $= \dfrac{lb}{in}$

Fig. 9-30. Dragload. ©1991 TAPPI. Courtesy of V. E. Hansen.

ized at atmospheric pressure at each end of these holes. During the remainder of the revolution back to top dead center, the water must be thrown out by centrifugal force. As the diameter the holes has been reduced in an attempt to increase the open area, the friction in the holes has been increased. (Consider the flow through a pipe is proportional to r^4.) Consequently, as machines run faster, the couch runs wetter and does less drying of the sheet.

This repeated pumping out of the couch leads to its mechanical inefficiency (Fig. 9-29). As speeds increase, it is necessary to improve the hivac operation to make up for the loss of couch efficiency. This makes it necessary to monitor dragloads, the power required to keep the paper machine moving per unit of velocity.

Dragload

The term dragload resulted from the necessity of fabric manufacturers to monitor the power used to drive the fabric. Dragload is reported in units of lb/in. or N/m (Fig. 9-30). It is a measure of the increase in tension (ΔT) of the fabric as a result of the suction forces pulling the fabric against the foil surfaces, the lovac surfaces, and the hivac surfaces. It is a calculated number, although it can be measured with an instrument. Dragload is equivalent to drive power divided by machine speed and can be calculated from the product of the volts and amps supplied to the drive motor divided by the machine speed. An on line readout of dragload is recommended as useful information for the paper machine operator.

Fig. 9-31 shows typical equipment and the

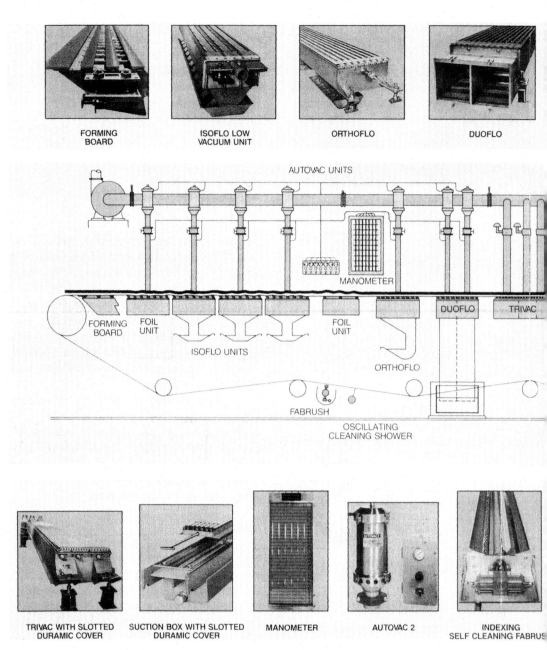

Fig. 9-31. Equipment and controls for dewatering. Courtesy of JWI Group, Johnson Foils.

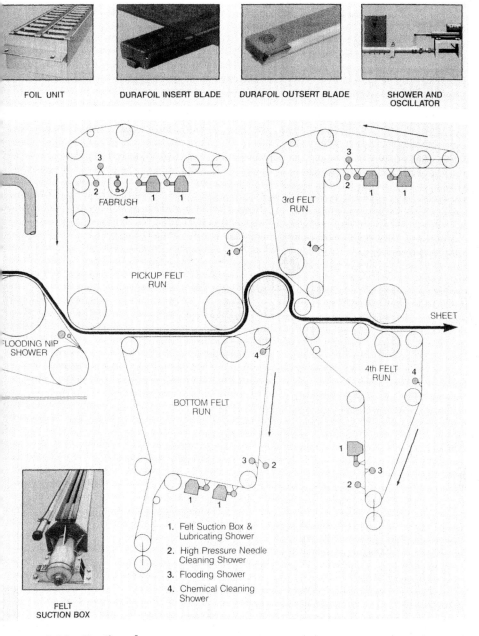

FOIL UNIT   DURAFOIL INSERT BLADE   DURAFOIL OUTSERT BLADE   SHOWER AND
                                                                                  OSCILLATOR

3
2
FABRUSH   1   1

3rd FELT
RUN

3
2
1   1

4   4

PICKUP FELT
RUN

SHEET

FLOODING NIP
SHOWER

4   4

4th FELT
RUN   4

BOTTOM FELT
RUN

3   2

3
1
2   3

1   1

1. Felt Suction Box &
   Lubricating Shower
2. High Pressure Needle
   Cleaning Shower
3. Flooding Shower
4. Chemical Cleaning
   Shower

FELT
SUCTION BOX

Fig. 9-31. Continued.

layout of that equipment used in dewatering at the wet end. This figure summarizes many of the concepts discussed in this section.

Lump breaker roll

The lump breaker roll (Fig. 9-28) is a solid roll with a suitable soft cover that is mounted over the couch roll to assist in drying the sheet by pressure. If the couch has two chambers, the lump breaker roll is mounted over the separating wall between them. As the sheet passes under the lump breaker, it is squeezed by the lump breaker. Actually, this arrangement is a press. The soft cover causes a relatively wide nip to be formed. As the sheet is squeezed, water flows in reverse to enter the holes of the first chamber. Since some of the holes of the second chamber are covered by the nip formed by the soft cover of the lump breaker, the air velocity in the second chamber is increased. This combination of actions will remove an additional 5% of the water.

Lump breaker rolls are always present on linerboard machines and machines using mechanical pulps. As the speeds increase, the lump breaker efficiency will decrease unless the couch can be improved as a result of hivac improvement. Usually a misting shower, which must be absorbed by the couch, keeps the cover wet so that fibers will not adhere to it.

Pickup (transfer) felt

The pickup felt is a traveling felt (blanket) designed to pick the wet paper web off the wire and transfer it to press section. This means that the web is supported at all times by either the wire or felt. In the past, paper machines that operated at slow speeds with heavy paper grades used to have an *open draw*, where the web was actually unsupported during this transfer. A supported transfer is known as a *closed draw*. Some examples of felt transfers are given in Section 9.7.

9.5 TWIN WIRE FORMERS

Twin wire formers

Twin wire formers are machines that use a jet of stock imparted on two converging wires to accelerate water removal and maintain better web uniformity. These are particularly useful for high speed machines, where the fourdrinier wet end would tend to give a two-sided sheet, since both sides are wire sides and the sheet is formed symmetrically on the two sides. D. Webster is credited with the invention of the twin wire former. The first twin wire machine for board, the inverformer, was commercially available in 1958 and for paper in 1965.

Some examples (Fig. 9-32) are the Black Clawson Verti-Former®, the Papriforma, the Beloit Bel Baie II, the Inverformer, and the Duoformer. Fig. 9-33 shows a Bel Baie II making computer printing paper. Fig. 9-34 shows the Verti-former. In the early verti-formers, the stock flowed downward from the headbox mounted above and between the two wires traveling vertically downward. In the Bel Baie former, the stock is sprayed as a high speed jet in an upward direction from the headbox that is mounted below and between the two wires that move upward.

Many flat wire machines use a second wire part way down the table to help with dewatering and formation. The Top Flyte™ "C" former is shown in Fig. 9-35 and Plate 33. Note that this type of machine has two flat wire parts: one leading to the nip and one after the former. These

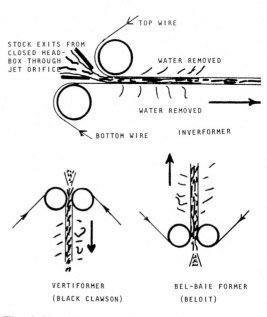

Fig. 9-32. Diagrams of several types of twin-wire formers. Courtesy of W. Bublitz.

Fig. 9-33. Beloit BelBaie II twin wire former (forming section) for making writing paper. The insert is a close-up of the slice where the dilute pulp slurry is applied upward between the wires.

two areas must obey flat wire operation. In the flat section after the headbox, the table consists of a forming board, foils, and one or two lovacs. If this section is very short, the danger of sheet sealing is present. If the sheet enters the nip in a sealed condition, the formation is unsatisfactory and the retention is low. The flat wire must be managed carefully if foils are used as they tend to seal the sheet between 0.8 and 1.4% consistency. Fig. 9-36 is an audit of a news retrofit with a bladed shoe. Note that the sheet is 7.25% consistency upon entering the hivacs. The foil unit

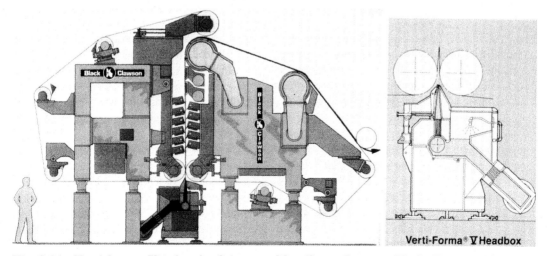

Fig. 9-34. Verti-former V twin wire former and headbox. Courtesy Black Clawson-Kennedy.

Fig. 9-35. Top Flyte™ "C" Former. Courtesy of Black Clawson-Kennedy.

shown is for the fabric support only. With four hivacs (two drilled and two slotted) the sheet enters the couch at 11.8% and leaves at 14.0% consistency. This leaves much work for the first press to perform. Here, the first press vibrated. Fig. 9-37 shows this machine after some modification. The main change occurred at the hivac area: the two drilled covers were changed to two slotted; the piping was enlarged; automatic valves were added; the airflow density was increased from 60 to 130 cm^3/s/cm^2 (0.8 to 1.8 ft^3/min/in^2; and the maximum ΔP raised to 60 kPa (18 in. Hg). The remarkable increase in dryness to and from the couch eliminated press section vibration

and resulted in 15% savings in steam to the dryers. The foils were changed to lovacs that pulsate the fabric. As a result, more water was added and formation improved. Note that the consistency to the nip was reduced below the 1.4% danger level for sealing, and so extra shear resulted in the two wire portion.

Twin wire formers can exert high pressures between the two sides of the web, whereas single wire machines are limited to a pressure differential of one atmosphere. Shear forces occur that are not possible on fourdrinier machines. These shear forces greatly help the formation of the sheet and improve the overall quality of paper.

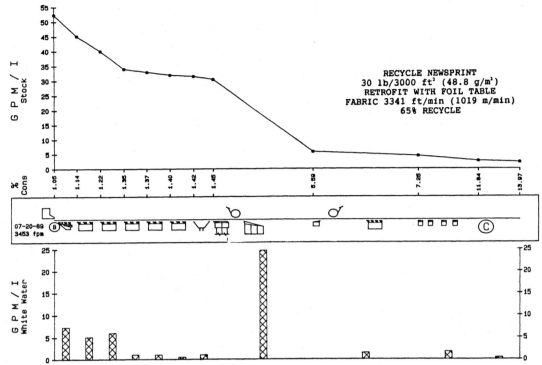

Fig. 9-36. Retrofit of a newsprint machine using 65% recycled fiber. The fabric speed is 17.0 m/s (3341 ft/min) and the basis weight is 48.4 g/m² (30 lb/3000 ft²). Courtesy of V. E. Hansen.

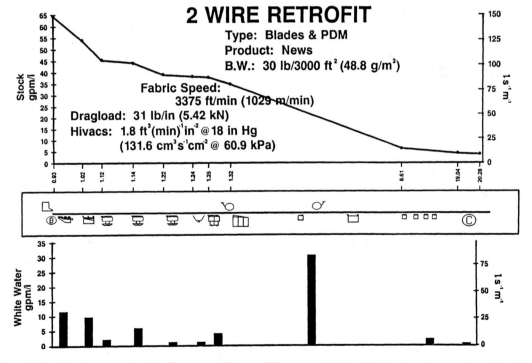

Fig. 9-37. Blade-type retrofit of news machine with a remarkable 19% before the couch. Suitable risers and header insure air velocity < 25.4 m/s (5000 fpm). ©1991 TAPPI. Courtesy V.E. Hansen.

9.6 THE CYLINDER MACHINE

Cylinder

The cylinder paper machine (Fig. 9-38) with a cylindrical forming wet end was invented in 1809. The web is formed on a rotating cylindrical screen 36 to 60 in. in diameter and immediately picked off. Typically, 5 to 10 of these operate in series to make a multi-layered sheet. This method is usually used to make heavyweight board from secondary fiber, which is not de-inked, for folding boxboard, chipboard and gypsum board. A high quality fiber surface may be added for printing upon. Heavy weight paper and centrifugal force limit speeds to less than 7.5 m/s (1500 ft/min).

The cylinder is a large hollow roll covered with a wire screen (similar to a fourdrinier wire) which revolves in a bath or tub of stock; as the wire comes out of the vat, a mat of fibers collects on the wire. The water passes through the wire and is carried away by a vacuum from the inside of the rotating cylinder. A horizontal felt is pressed against the top side of the cylinder with a press roll, which further dewaters the mat, to pick up the web. Several other webs are added as the felt picks up other webs from additional cylinders in tandem. Normally there are no more than six or seven cylinders in a single machine.

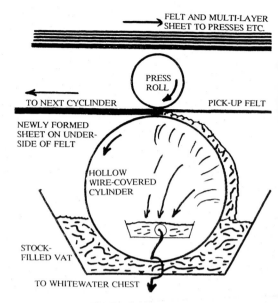

Fig. 9-38. Diagram of a cylinder machine. Courtesy of W. Bublitz.

Many new designs of cylinder machines are available (Anon., 1978). Modern versions of cylinder machines, such as the *roto former*, control the stock flow to the cylinder with channels and baffles to give improved control over formation of the web. The Ultraformer (Fig. 9-39) is one of the few machines that applies the stock to

Fig. 9-39. The Ultraformer. Courtesy of Kobayashi Engineering Works, Ltd.

the cylinder and the plies are then picked up by the topside of the felt.

9.7 THE PRESS SECTION

Press, press section

A press is a pair of squeeze or wringer rolls designed to remove water mechanically and smooth and compress the sheet (Fig. 9-40). A suction press is shown in Fig. 9-41. A variety of press configurations are possible depending on the desired goals. Fig. 9-42 shows two configurations. Some of the methods include:

1. The use of a pickup felt while the web is still in contact with the forming material so that there is no unsupported web during the transfer. An open draw is delayed until several press nips have been passed. An unsupported web is called an *open draw*.

2. Double felts (one felt on each side of the web) in the first press because of the large amount of water being removed. This avoids sheet crushing (described below) and allows more water to be removed.

3. Shoe presses or other special presses to promote good fiber bonding and decreased web moisture contents.

Fig. 9-41. Suction press of Fig. 9-43.

4. A design that gives the ability to have high machine speeds with operation stability and low vibration.

5. Allowing both sides of the web to be pressed against a solid roll (at different times).

6. Designs that minimize space and cost.

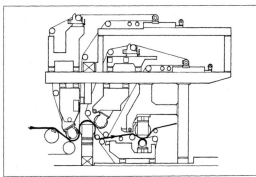

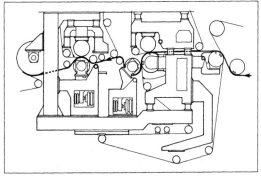

Fig. 9-42. Two press designs. Courtesy Black Clawson-Kennedy. Guarding omitted for clarity.

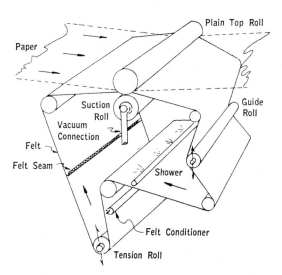

Fig. 9-40. Diagram of a press. Reprinted from *Making Pulp and Paper*, ©1967 Crown Zellerbach Corp., with permission.

Fig. 9-43. The press section of the machine shown in Fig. 9-33.

Often three presses (Fig. 9-43) are used in a press section. Water from the sheet transfers to the press felt which is then cleaned with water jets and dewatered by vacuum boxes (Fig. 9-26).

In a *plain press*, two smooth rolls are used and the capacity to remove water depends on the void volume of the press felt(s) to receive the water (Fig. 9-44). The first press is often a plain press and is often double-felted to receive more water. A *grooved roll press* uses helical grooves cut into the bottom press roll that is in contact with the press felt (Fig. 9-45). These grooves provide additional volume and a path for water transfer from the sheet through the press nip. A grooved roll press is much less expensive to install and operate than a suction press.

In a *suction press*, one of the press rolls is equipped with a suction box (Fig. 9-46 and Plate 34) to remove water and hold it into the felt against centrifugal force until the sheet has separated from the felt. The material of the shell is important. A variety of suction shell materials is available to tolerate the high cyclic pressures.

Pressing can cause some compaction of the sheet, but the level of refining is the principal factor in final sheet density. Press rolls are *crowned* to give uniform pressure across the press; that is, the middle of the roll might be on the order of 0.05 in. thicker so that pressure will be even across the width of the paper machine.

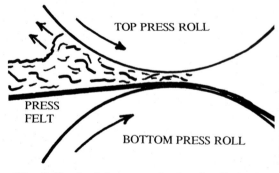

Fig. 9-44. A plain press nip showing limitation of water removal. Courtesy of W. Bublitz.

Pressures of 400-8500 N/cm (50-1000 pounds per linear inch, PLI) are used depending on the press location and the paper grade; fine papers use 3000-5000 N/cm (350-600 PLI), kraft bags use 5000-7000 N/cm (600-800 PLI), and linerboard uses up to 13,500 N/cm (1600 PLI) in the final press. If too much pressure is used, particularly at the first press where the consistency is low, *crushing* marks appear (on the dried sheet) where water escapes too quickly, creasing the web; consequently, later press sections are normally operated at higher pressures.

Several techniques can be applied to increase water removal. Often the web is heated prior to the press section to reduce the viscosity of water and increase pressing effectiveness. [The viscosity of water in centipoise is 1.79 at 0°C (32°F), 1.00 at 20°C (68°F), 0.65 at 40°C (104°F), 0.47 at 60°C (140°F), 0.35 at 80°C (176°F), and 0.28 at 100°C (212°F).] Double felting or extended nip presses are used on heavy grades of paper. Well designed felts that run without distortion and are effectively cleaned improve dewatering. It is important to minimize rewetting of the sheet just beyond the press nip by quickly separating the web and felt from each other. The more water removed in the press section, the less water that must be removed in the dryers, meaning faster paper machine speeds and lower energy costs.

Fibers may stick to the top press roll or even to the felt. Sticking, or *picking*, may be caused by weak internal bonding of the web and by tacky contaminants (called *stickies*) arising from wood pitch, rosin or synthetic sizes, paper machine additives, polymers from secondary fiber, etc. Various additives such as clay or talc can be used to counteract the tackiness of the contaminant.

Shoe presses, Beloit extended nip press

Shoe presses use a long nip that is formed against a stationary shoe. The extended nip press of Beloit, the first and best known of the shoe presses (Fig. 9-47), is a special press used in the last position of the press section on paper machines producing heavy paper grades requiring high strength and stiffness, such as linerboard or

Fig. 9-45. A grooved press roll press from a linerboard machine. The paper and felt are quickly separated after the nip. Notice the water being thrown off the bottom press roll.

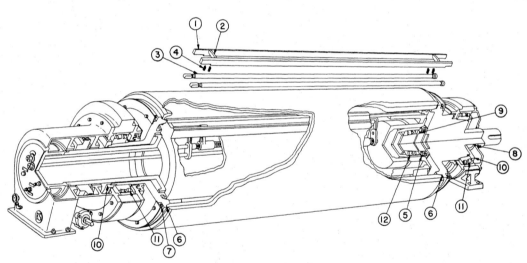

1) side seals, 2) end deckle seals, 3) pneumatic tubing for strips, 4)coil springs for deckles, 5) neoprene seal ring, 6) head shims, 7) studs and nuts, 8) drive end housing, 9) pilot bearing housing seals, 10) front and rear bearing retaining plates, 11) two main bearings, 12) one suction box pilot bearing.

Fig. 9-46. A suction roll (top) and diagram of a suction roll of the type in Plate 34 (bottom). Courtesy of Black Clawson-Kennedy.

corrugating medium. A plastic, impermeable belt and a special shoe increase the nip to about 25 cm (10 in.) wide, allowing better compaction of the web and increased water removal. The first extended nip press was installed at the Weyer-haeuser mill in Springfield, Oregon in the late 1970s to make linerboard. The increase in web consolidation gives papers with high stiffness, which is especially important in linerboard. Dryer requirements are reduced as the consistency from

Fig. 9-47. Beloit extended nip press. The bottom is a close-up of the shoe where the plastic belt, felt, and sheet are visible. Note the sheet is quickly separated from the felt at the nip exit.

extended nip presses may be above 50%, allowing faster machine speeds and decreased energy requirement per unit production. The pressure applied to paper is approximately 4000 kPa (600 psi) or 40,000 N/cm (5000 PLI). Some machines now use two extended nip presses in a row.

The extended nip press uses a plastic, impermeable belt next to the shoe. Oil lubricant is added between the belt and the shoe to decrease friction. The web is carried by a felt that separates it from the belt and picks up the water discharged from the web during pressing.

Other suppliers provide shoe presses as well. Fig. 9-48 is a diagram comparing various shoe and conventional presses in terms of nip pressures and nip lengths.

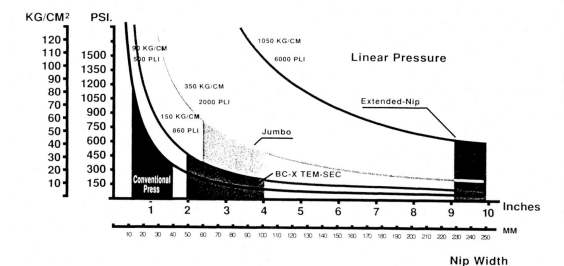

Fig. 9-48. Comparison of various types of press nips. Courtesy Black Clawson-Kennedy.

Press felt

The press felt normally runs underneath the paper web going through a press section. It is designed to absorb water squeezed out of the sheet and to support the sheet through the presses. As with forming fabrics, a coarse felt allows for more water removal but gives a coarser sheet. The felt is separated as quickly as possible after the press nip in order to avoid *rewetting* of the sheet from the felt.

Reverse press

A reverse press is mounted in such a way that the paper web goes through in a direction that is opposite to its original travel direction so that both sides on the web ultimately contact a smooth roll, thereby providing smoothing action on the opposite side of the paper web. This can also be accomplished by using two smooth press rolls, one on the bottom of the web for one press and one on the top for another press.

9.8 THE DRYER SECTION

General aspects

Removing water from the web is accomplished by adding heat to the web and circulating the air. Heat is applied by the pressurized, steam filled circular steel [up to 1000 kPa (150 psi)] or cast iron [up to 500 kPa (75 psi), older designs]

dryer drums (or cans). Circulation of air over the web is very important since the air quickly becomes saturated with water vapor that prevents further removal of water from the web's surface. Most dryer sections are completely enclosed, with side doors that open for dryer access, to allow efficient air circulation. Fig. 9-49 shows a dryer section where the doors are open in order to thread the paper through the section after a web break. Plates 35 and 36 show close-ups of several dryer cans with the area requiring pocket ventilation.

The rate of water removal depends on many factors: 1) the temperature and amount of steam entering the dryer can. 2) adequate removal of the steam condensate and air from the interior of the drying can. 3) the amount of sheet--dryer surface contact, contact time, and contact pressure. 4) Cleanliness of the drum's exterior and interior surfaces (mainly in regards to the amount of scale on the surfaces). 5) Type of felt and condition of felt. 6) Circulation of hot, dry air.

Dryers

Dryers are hollow, revolving, steam filled drums (about 2 m or 6 ft in diameter) designed to heat the web by direct contact and remove water by evaporation. Fig. 9-50 is a diagram of a dryer drum with a stationary (does not rotate) siphon. About 1.2-1.5 kg of steam are used to remove 1

Fig. 9-49. A dryer section on a machine making white printing paper. The doors are open to allow the paper to be threaded after a web break.

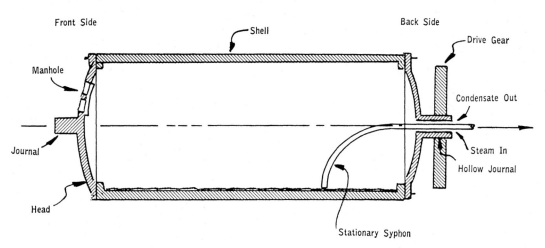

Fig. 9-50. Diagram of a dryer can. Reprinted from *Making Pulp and Paper*, ©**1967 Crown Zellerbach Corp., with permission.**

kg of water from the web in the dryer section. The web is dried on one side than the other side, alternating between drums, making numerous felt transfers necessary. The initial drums cannot be too hot, otherwise large amounts of liberated steam will cause the paper to cockle.

The paper often stretches a few percent in the machine direction while contracting in the cross machine direction through the course of drying, causing the machine direction to be stiffer and to have a higher tensile strength than the cross machine direction. Some of this is due to some preferential fiber alignment in the machine direction, but mostly it is due to differential shrinkage in the dryer section. Sophisticated control of individual dryer motors controls the tension throughout the dryer section and reduces web breaks.

Pocket ventilation

The ventilation system is designed to keep hot, dry air over the free paper surface, especially in the pockets between dryer drums where water quickly evaporates. Pocket ventilation systems blow hot, dry air into the pockets through air ducts and nozzles at high velocity. Hot, dry air may also be used to dry the dryer felts to improve web drying. Exhaust fans over the dryer hoods draw the moisture-laden air out.

Original felts were made of wool and were impermeable to air and made ventilation very difficult. With the advent of synthetic felts of much higher permeability in the 1960s, ventilation has improved significantly.

The last top dryer can be run with cold water to condense water vapor on its exterior surface to add some water to the sheet prior to calendering.

Carrier ropes carry the paper tail through the dryer section when this section is threaded with paper during startup or after paper breaks. These ropes run in grooves on each side of the dryer section.

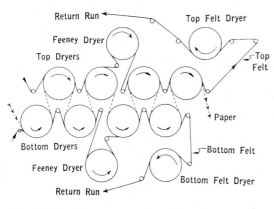

Fig. 9-51. An arrangement of dryer felts. Reprinted from *Making Pulp and Paper*, ©1967 Crown Zellerbach Corp., with permission.

Dryer felts

Dryer felts hold the paper web against the drums to improve heat transfer between the drum and web and absorb a portion of the water evaporated from the web. They help keep the sheet flat to minimize cockle formation on the web and improve drying. Even tension on the felt across the machine is important. Ideally, the felt is non-absorbent, strong, flexible, and (as always) inexpensive. Monofilament fabrics tend to be less absorbent; tend to avoid pick-up of fillers, fines, and pitch; and tend to deliver a more uniform cross machine tension than multifilament fabrics. Fig. 9-51 is a diagram of dryer felt runs.

Most dryer felts are constructed of polyester monofilament. However, in hot, moist areas of paper machines, polyester is not suitable as it is subject to hydrolysis of the ester. This is particularly true on machines making heavy grades of paper, such as linerboard, where temperatures tend to be high. In these areas multifilament fabrics containing nylon, fiberglass, acrylic polymers, and other materials resistant to hydrolysis are used. Research is necessary in the development of monofilament fabrics of reasonable cost that can be used in adverse conditions.

Steam control systems

The first step in controlling the heating of dryer drums is to remove the steam condensate from the drum. Fig. 9-52 (top, left) shows the possible positions of the condensate. The condensate tends to form pools or puddles at low machine speeds and rims at speeds above about 6 m/s (1200 ft/min). Larger amounts of condensate increase the speed required for rimming. At intermediate speeds, the condensate will splash or cascade. Therefore, older, slow machines used *stationary siphons* that continuously collected condensate from the bottom of the dryer drum. Modern, faster machines use rotating siphons that continuously removes rimming water. Rotating siphons are mechanically more reliable because they do not impact against stationary pools of water. Too much condensate, especially when rimming, decreases the heat transfer.

A pressure gradient across the inlet and outlet of the dryer, known as the *differential pressure* or DP, is used to control the rate of condensate

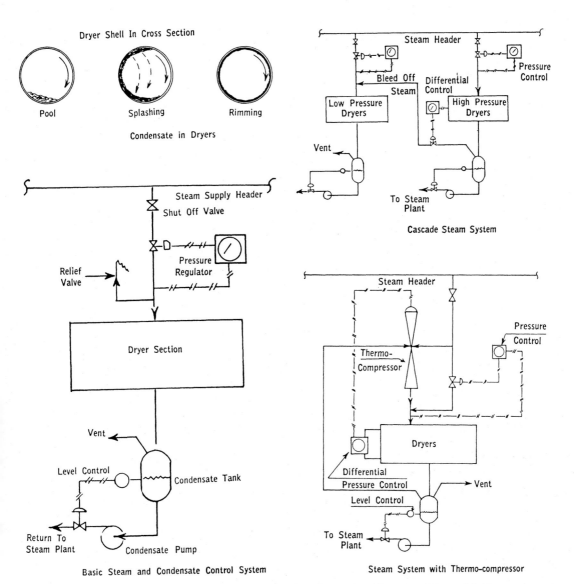

Fig. 9-52. Dryer condensate behavior and steam control systems. Reprinted from *Making pulp and Paper*, ©**1967 Crown Zellerbach Corp., with permission.**

removal. A differential pressure of 3-5 psi is used at machine speeds of 1000-1500 ft/min and is well over 25 psi at speeds above 4000 ft/min. Rotating siphons generally operate with a higher differential pressure than stationary siphons due to the centrifugal force the condensate must overcome on its way out. TAPPI TIS 0404-31 gives recommended differential pressures for given dryer speeds and condensing rates for rotating and stationary siphons. The differential pressure must be high enough so that some steam *blows through* without condensing; otherwise condensate will collect in the system and the dryer fills with water (becomes flooded), which decreases its efficiency and increases the drive load.

After the drying rate has been selected by the operator, it is important to maintain it at the desired level using a steam control system. Fig. 9-52 shows three control systems used to maintain the proper amount of steam in the dryer section.

The simplest design used on older machines is the *basic steam and condensate control system*, which supplies steam at controlled pressure and removes condensate for reuse in the steam plant.

The *cascade steam system* is slightly more complex and is useful for a system using high and low pressure dryers. The bleed-off from the high pressure dryers is reused through the low pressure dryers. The *thermo-compressor* takes low pressure bleed-off steam and puts it into the high pressure inlet header to recover its heat value. The action of the thermo-compressor is based on the venturi effect where the pressure of a flow of gases is decreased by moving it at high speed (i.e., through a constriction).

Dryer hood

The dryer hood is an enclosure around the dryer section and is used to improve drying efficiency by removing the moist air near the surface of the web. If this air is not removed, it quickly becomes saturated with water, preventing further water removal from the web.

Yankee dryer

The Yankee dryer is a large dryer drum (3.5-4.5 m or 12-15 ft in diameter) for drying tissue papers that are not strong enough to endure numerous felt transfers (Fig. 9-53). The Yankee dryer is normally the only dryer used to dry tissue. A Yankee dryer with a smooth chrome surface is used in some printing papers (prior to other dryers) to impart a machine glaze on the surface of these papers. An *air cap* is a hood mounted close to the surface of the Yankee dryer that forces heated air on the web, improving the drying efficiency. The *creping blade (doctor blade)* is a thin metal blade that scrapes the dry tissue off the Yankee dryer. It also compresses or *crepes* the tissue for increased flexibility in the machine direction. Fig. 9-54 shows an example of creped paper. Because of its large size, the yankee dryer must be heated with care when cold.

Impulse drying

The use of a long press nip with one roll at 175-400°C (350-750°F) to remove water from the web is known as impulse drying. The objective is to vaporize some of the water quickly so that the

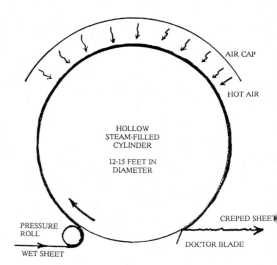

Fig. 9-53. A Yankee dryer with air cap and doctor blade. Courtesy of W. Bublitz.

generated steam front forces much of the remaining water out as a liquid. This eliminates the energy required to vaporize much of the water and reduces the dryer section size. Impulse drying is still an experimental technique. It was developed at the Institute of Paper Science and Technology. The method leads to sheet delamination, although a recent publication claims to have solved this problem.

Size press

The size press is located between dryer sections and consists of a pair of squeeze rolls mounted horizontally, vertically, or at a 45° angle (Fig. 9-55). It is used to apply surface size (usually a starch solution) to papers. In the horizontal size press, the bottom roll rotates in a pan filled with size solution, and the same solution is delivered to the top surface of the web by a shower pipe. The excess solution is squeezed off the sheet. The base sheet is then redried, using, of course, much energy. Fig. 9-56 is a size press using a flooded nip. Fig. 9-57 is a gate roll (transfer roll) size press.

Defoamers are typically used in the size press at 10-100 ppm. Defoamers that contain glycols, modified glycols, or polyalkyloxylated phenols must be avoided in the size press because small amounts can greatly decrease sizing. One should

Fig. 9-54. Creped giblet paper. The insert is magnified 4X.

also consider the possibility of defoamer carry-over through pulp from the deinking plant.

Surface sizing

Surface size uses polymeric materials, commonly starch, but sometimes starch reacted with ethylene oxide (hydroxyethyl starch with a relatively low degree of substitution), carboxymethyl cellulose, polyvinyl alcohol, methyl cellulose, alginates, or wax emulsions using alum that are applied as a solution to the surface of the web at the size press, followed by drying.

Surface sizing improves the water resistance of paper and bonds surface fibers to it to improve printing properties. Except for wax, surface sizes are generally not hydrophobic like internal sizes; instead, they act by sealing the sheet, that is, filling the capillaries and pores with sizing formulation. This reduces the absorption of water by capillary action. (Fig. 17-5 demonstrates capillary action.)

However, in the last several years, nonpolar synthetic polymers have been used. The three major types include polystyrene maleic anhydride copolymers (SMA, which may be partially hydrolyzed to convert about 10% of the maleic anhydride groups to maleic acid) that are used with *tackifiers*, polystyrene acrylic acid copolymers, and polyurethanes. (Tackifiers are usually coordinating cations, such as alum, that bind the carboxylate groups so the polymers are not "sticky".) These polymers are effective in providing ink holdout. Ammonium zirconium carbonate is another sizing agent that has been used for many years in certain applications.

Typically about 30-50% wet pick-up is used or 300-500 kg/t (600-1000 lb/ton). Papers without internal size may pick up higher levels. Sizing efficacy may by measured by the Gurley porosity or the sizing tests.

9.9 POST DRYING OPERATIONS

On-line paper testing

There are a number of on-line (Fig. 9-58) paper tests that can be accomplished and the

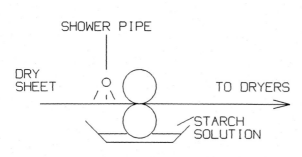

Vertical Size Press

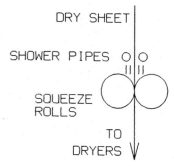

Horizontal Size Press

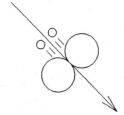

Inclined Size Press

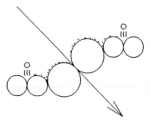

Gate Roll Size Press

Fig. 9-55. Size press configurations.

Fig. 9-56. Flooded nip size press.

Fig. 9-57. Gate roll size press. The top insert shows the roll coated with sizing solution. The bottom insert shows the size press nip.

Fig. 9-58. Basis weight and moisture content measurement with a scanning detector.

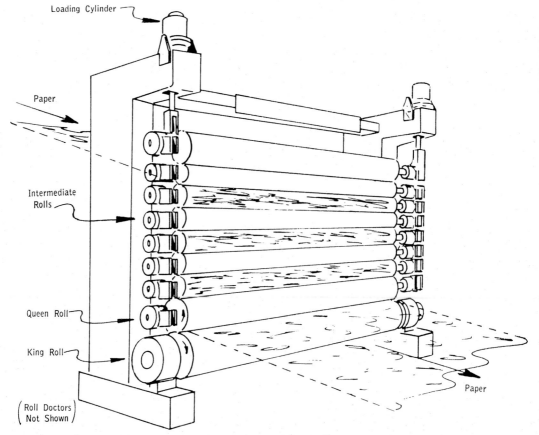

Loading Cylinder

Paper

Intermediate
Rolls

Queen Roll

King Roll

$\left(\begin{array}{c}\text{Roll Doctors}\\\text{Not Shown}\end{array}\right)$

Paper

Fig. 9-59. Paper machine calender stack. Reprinted from *Making Pulp and Paper*, ©1967 Crown Zellerbach Corp., with permission.

number is increasing. The most common ones used are basis weight measured by a β (beta) ray detector (it measures the absorption by paper of electrons emitted by a radioisotope) and moisture content by infrared (IR absorption is taken at about 4 different wavelengths to determine moisture content with a correction factor for IR absorbed by the paper fibers). These devices are combined into a unit which scans the paper across the paper machine just before the calender stack. Other devices measure brightness, strength properties (by ultrasonic properties of the paper), thickness, color, opacity, gloss, ash content (by low energy X-ray analysis), etc.

Calender, calender stack

The calender stack is a series of solid rolls, usually steel or cast iron, mounted horizontally and stacked vertically (Fig. 9-59). During machine calendering, the dry paper passes between the rolls under pressure, thereby improving the surface smoothness (for example, from imperfections caused by felt marks, cockle, lumps, fibrils, etc.) and gloss and making the caliper more uniform, if not decreasing the caliper and porosity. These improvements make the paper better suited for printing and decrease problems during printing. (If uniform caliper is the only goal, one nip is often sufficient to achieve this.) Pressures of 850-85,000 N/cm (100-1000 PLI) are used. There is some loss of opacity, brightness, and stiffness on calender finished papers and paperboards.

The bottom roll is called the *king roll*. It is usually larger in diameter than the other rolls and *crowned* (it has a larger diameter in the center) to permit even pressure when loaded on the ends. A

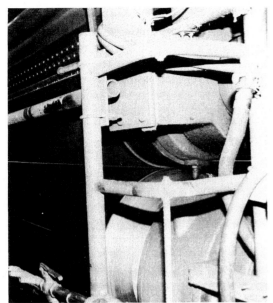

Fig. 9-60. A single-nip calender for caliper control of linerboard.

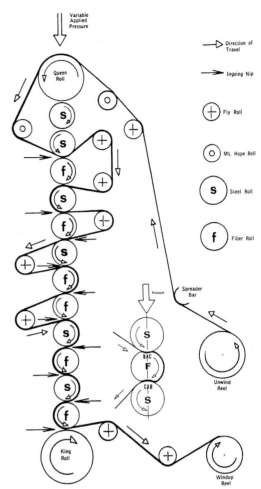

Fig. 9-61. Supercalender. Reprinted from *Making Pulp and Paper,* **©1967 Crown Zellerbach Corp., with permission.**

variable crown can be achieved using a pressurized, oil-filled roll where the oil pressure controls the degree of crown. The roll above the king roll is the queen roll. The trend is towards the use of calender stacks with fewer rolls. Fig. 9-60 shows a calender with a single nip used for caliper control of linerboard.

Cooling air is supplied through a series of controllable jets across one of the rolls for caliper control in the cross machine direction. This arrangement is known as the *actuator*. The air cools the roll and causes a local decrease in roll diameter and an increase in sheet thickness. Recent work on a newsprint machine (Journeaux et al., 1992) shows that the actuators might be most effective on the queen roll or one of the rolls immediately above the queen roll, especially if they are unheated, thin walled rolls. Using a roll with a single nip, such as the king roll or top roll, was not very effective. Since the caliper control is in the early nips, the actuator is more effective on the roll above the queen roll compared with rolls farther away from the queen roll.

Because supercalendering (described below) has high operating costs, many paper machines are using hot calendering with soft rolls. These rolls differ from those of supercalenders and are not filled rolls. Generally, two nips are used to calender each side of the paper or, if necessary, four nips are used, which may be through two calenders with three rolls to make two nips. Soft calendering of papers made from mechanical pulps has been discussed by Tuomisto (1992). This method can be done on-line.

Supercalender

A supercalender is similar to the calender but uses alternate hard and soft, heated rolls (Fig. 9-61). While considered here for the purpose of comparing it to the calender section, it is usually used off-machine such as after paper coating. For example (Fig. 9-62), it is used to impart high gloss

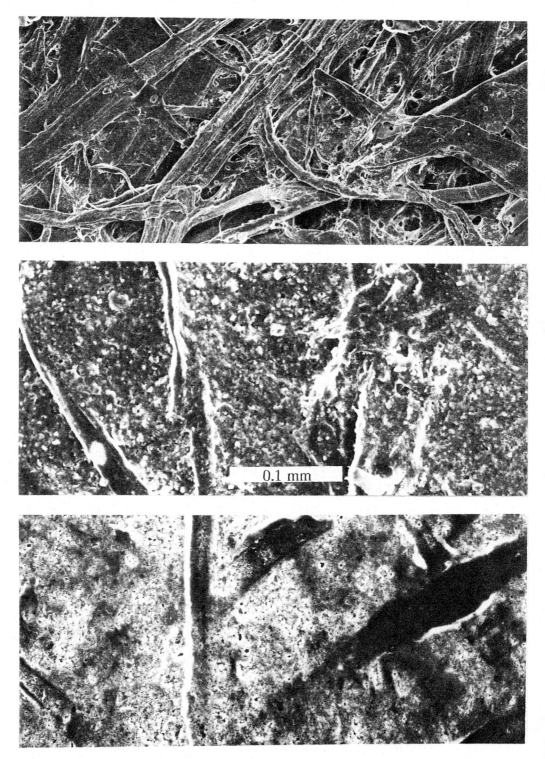

Fig. 9-62.   Uncoated groundwood paper (top), coated groundwood paper (middle), and supercalendered, coated groundwood paper (bottom).

to clay coated papers and to produce glassine paper from grease-proof paper. Pressures of 17,000-34,000 N/cm (2000-4000 PLI) and temperatures of 40-70°C (105-160°F) are used. Supercalendering may actually increase fiber bonding with an increase in strength, but the compact sheet has decreased opacity. It is usually done off-machine because of the relatively low speeds required.

Reel

The reel is the last unit on the paper machine that collects the paper (Fig. 9-63). The paper is wound on a spool that rotates against a drum. A new spool is started on a secondary arm before the filled reel is discharged, for changing reels "on the fly". Plate 37 shows reel handling.

9.10 PAPER MACHINE BROKE SYSTEM

During startup, threading, and the (hopefully rare) event of paper machine breaks, broke (unusable paper) will be produced at various parts of the paper machine. Fig. 9-64 shows a representative broke handling system. Broke at the wet end is easily repulped and metered back into the system, such as by use of the Agi-Pulper™. Broke in the dryer sections requires more severe pulping. Coated papers, of course, may have to be handled separately.

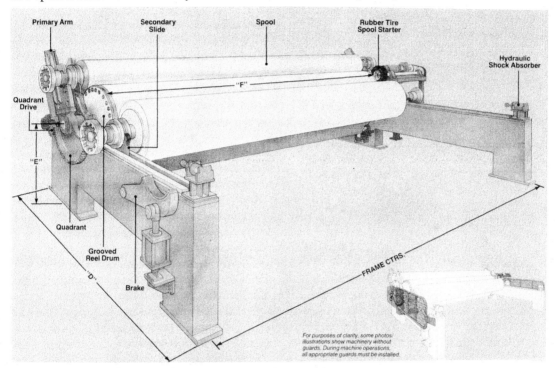

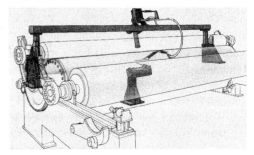

Fig. 9-63. Autoflyte reel. The gooseneck and cutter knife (lower right) are used for positive automatic sheet transfer. Courtesy of Black Clawson-Kennedy.

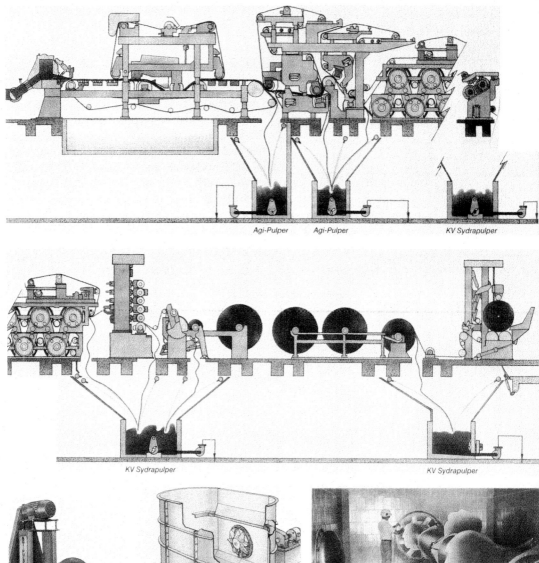

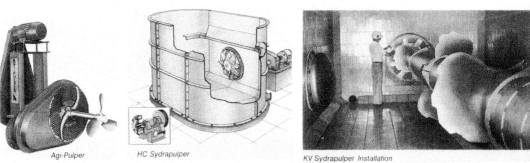

Fig. 9-64. A paper machine broke system in two segments. Courtesy of Black Clawson-Kennedy.

9.11 CONVERSION

Converting

Conversion is a general term for operations on paper following its manufacture. Converting operations include trimming, rewinding, sheeting, coating, printing, saturation, and box making.

Winder

The winder is a device that receives the large reels off the paper machine and slits the paper into smaller widths while rewinding it onto smaller rolls for shipping to customers (Fig. 9-65) or for further conversion within the paper mill.

Fig. 9-65. Banding a roll for shipping after rewinding from the reel.

Slitter knives

The slitters cut the paper into narrower rolls during rewinding (Fig. 9-66). Sharp circular knives for trimming reels into rolls are called slitter knives.

Sheeter

A sheeter is a machine holding multiple rolls of paper that cuts webs into large sheets and stacks the sheets on a pallet.

Guillotine

The guillotine is a device for cutting stacks of large sheets from the sheeter into commercial size sheets (for example, 8½ in. × 11 in. sheets).

Saturation

Saturation involves treatment of paper whereby the saturating material is distributed completely through the sheet, in contrast to a coating, which lays on the surface. The process is used to increase the strength and durability of the paper. Thermosetting resins (such as phenol-formaldehyde or melamine-formaldehyde) are used to make hard, stiff papers for use in hardboard overlay, countertops, and laminates with high

Fig. 9-66. Slitting during rewinding. The insert is a close-up of a knife slitting linerboard.

electrical resistance.   Thermoplastic resins are used to make tough, flexible papers used in masking tape, book covers, artificial leather, etc.

Lamination

Lamination is the combination of two or more films or sheets of paper, normally held together with adhesives, to make a composite.

Corrugated boxboard production

Box production is almost always done in operations (called box plants) away from the paper mill since paper mills tend to be located in rural areas with large amounts of wood, whereas boxes are used in metropolitan areas with high populations.   Furthermore, corrugated boxes are necessarily bulky and do not transport efficiently.

About 450,000 tons per year of cornstarch is used as corrugated box adhesive alone in the U.S. Cornstarch is treated with metaborate (formed by the action of caustic on borax) to obtain the adhesive for corrugated boxboard.   It is important that the viscosity of the starch-metaborate solution not be allowed to increase too much during starch cooking.

9.12   COATING

Coating

Coating is the treatment or application of pigments, polymers, or other materials to one or both surfaces of paper.   Coating can be applied *on-machine* or *off-machine*.   On-machine coating has the obvious advantages associated with decreased handling of the paper.   Off-machine has the advantages of providing greater flexibility in the system including the speed of coating.

About 20% of paper is coated.   Coating improves brightness, gloss, smoothness, caliper, and uniformity of the base paper.   Coating also improves many aspects of the printing process. Improvements in printing are improved gloss of the ink film (ink holdout), freedom from mottling (uniform absorption of ink), improved resolution of printed images, improved contrast due to ink holdout, and decreased show-through since the ink remains on the surface.

Fig. 9-67 shows scanning electron micrographs from a base sheet at low magnification and

four types of coating papers, applied to the same base sheet, at very high magnification.   Coatings that have high gloss must be smooth relative to the wavelength of light (about 0.5 μm).   The center picture is at low magnification and is from the base sheet. Its surface is quite rough compared to the coated samples. The upper left micrograph is from a *matte coated* paper (similar to the paper used in this book) that has little gloss.   The upper right micrograph is from a *dull coated* paper that has some gloss.   The lower left micrograph is from a *gloss coated* paper that typical of high quality magazines. The lower right micrograph is from a *cast coated* paper (machine glazed with a Yankee dryer) that would be used in the highest quality magazines and annual reports of large companies. As the surface becomes more uniform (on a μm scale), the gloss increases.

Coaters

Coatings are applied at the size press or with a roll coater (Fig. 9-68), air knife, or, the most common method, blade coater (Fig. 9-69).   Coating speeds are 3000-4000 ft/min for lightweight papers and 1200-1300 for board materials.

Coating with transfer *roll coaters* has the advantages of giving an even coat weight, even when the surface of the base sheet is somewhat rough.   It is used at speeds of 600 m/min (2000 ft/min), with high solid content coatings with high viscosity.   These coaters have low maintenance. The disadvantages are the large space requirements, film splitting, high installation costs, and the requirements of numerous adjustments.   Coating weights are limited to about 10 g/m^2, although two layers are used sometimes.   Reverse roll coaters have a lower speed rate, about 50% of other roll coaters.

Coating with an *air knife* has the advantages of giving an even coat weight, being able to use high coat weights, not leaving scratches, and being versatile.   These coaters have several disadvantages.   They can only be used with coatings containing low solid contents and having low viscosities, they must be used at relatively low speeds around 500 m/min (1500 ft/min), and they may cause streaking and patterning.

Blade coaters offer the advantages of low cost, the ability to coat two sides at once, and

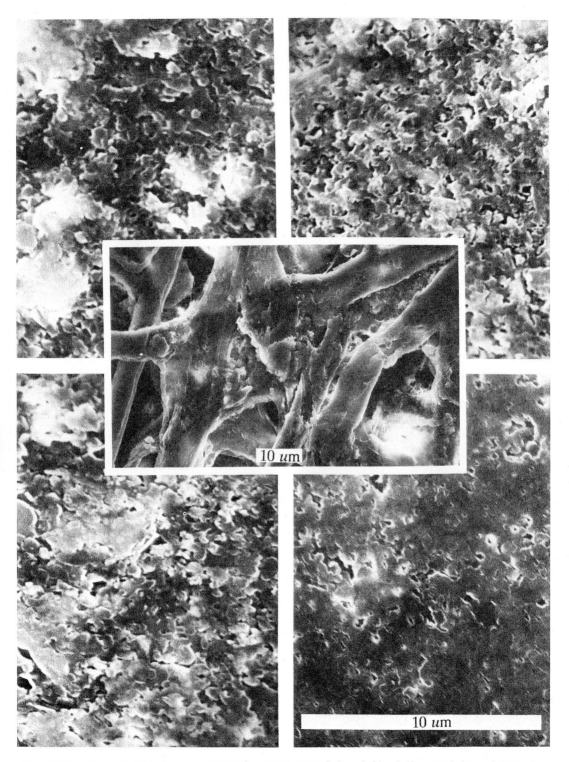

Fig. 9-67.   Uncoated base paper (center), matte coated (top left), dull coated (top right), gloss coated (bottom left), and cast coated (bottom right) papers.

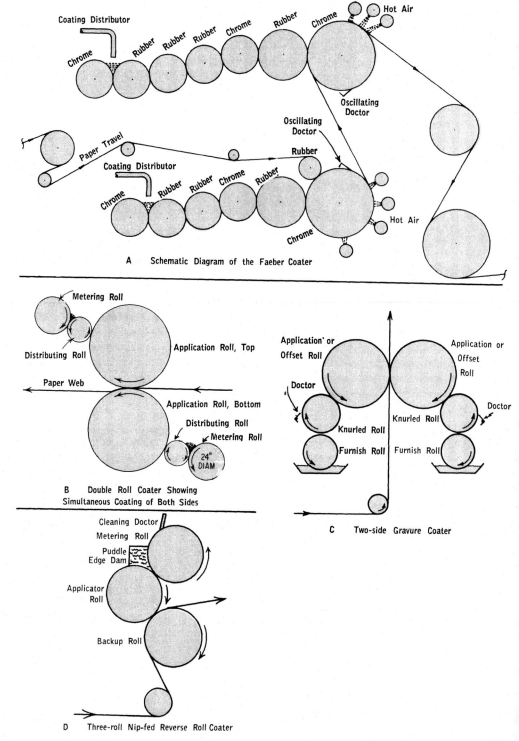

Fig. 9-68. Four variations of the roll coater. Reprinted from *Making Pulp and Paper*, ©1967 Crown Zellerbach Corp., with permission.

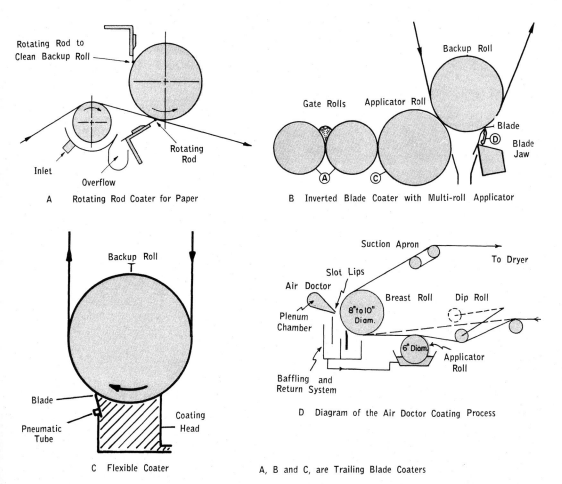

Fig. 9-69. Four variations of blade coaters. Reprinted from *Making Pulp and Paper*, ©1967 Crown Zellerbach Corp., with permission.

adaptability to on-machine coating. The disadvantages include an *orange peel* surface on the product, low speeds, the use of coatings with low solids and viscosity, and low coating levels.

Coating components

The coating usually consists of three categories of compounds. The pigment, the binder, and additives. Pigments are materials such as clay, calcium carbonate, titanium dioxide (the previous three materials are described in Section 8.4 as fillers), silicas, and other materials. The binders are classified as *natural binders* or *synthetic binders*. The natural binders include starch, carboxylated soy protein, and casein, a white milk protein containing phosphorus, which was the mainstay

before 1955. Starch may be treated in one of a variety of methods to give a binder. These modifications include oxidation to give carboxylic acid groups, addition of amines to make it cationic, partial hydroxyethylation, or cyanoethylation. Synthetic binders include styrene-butadiene rubber (SBR) latex, polyvinylacetate latex, or vinylacrylic latex. (Latexes are defined below.) SBR is usually carboxylated and must be used with a tackifier to make it less sticky. Crosslinking agents such as melamine-formaldehyde, glyoxal-based resins, and, especially for the synthetic binders, ammonium zirconium acetate.

Latex

A latex is a water emulsion of a synthetic

polymer (rubber or plastic). Latexes are used in paints, and many coatings are really much like paints. An emulsion is a suspension of mutually insoluble materials. One of the materials is dispersed as small droplets throughout the other. Milk is an emulsion of fat and in water.

Drying of coatings

The coating is dried by electrical or gas infrared dryers (with heat transfer by radiation), convection dryers, or conduction dryers (with heat transfer by hot air impingement). Infra-red dryers are usually gas fired, although electric heaters are used occasionally. Hot air impingement uses hoods containing air moving at high velocities.

Printing coating

A printing coating consists primarily of a pigment (clay, TiO_2, $CaCO_3$, etc.) and a binder (starch, protein, synthetic resin, etc.) and is designed to impart a good printing surface to the paper. Printing coatings are used for *slick* magazines because of the high gloss and high quality printing reproduction imparted to the paper.

Decorative coating

A decorative coating is applied mainly for the sake of appearance or to act as a base for printing. Some examples are decorative wrapping papers and coated magazine and book stock.

Functional coatings and other coatings

Functional coatings are designed as barriers to liquids and vapors such as for frozen food containers. Often, pure polymers such as polyethylene are used to make laminates, milk cartons, etc. Other coatings are used to generate images such as in photographic papers, carbon paper, or carbonless papers (NCR or 3M).

9.13 ANNOTATED BIBLIOGRAPHY

Introduction

1. Ely, D.A., How to rate your paper machine and mill performance, competitiveness, *Pulp & Paper* 55(9):89-93(1981).

2. Mardon, J., R. Vyse, and D. Ely, Paper machine efficiency: The most important operating parameter; how to get it and how to keep it, *Pulp Paper Can.* 92(12):87-100 (1991). This reference includes a separate three page commentary by G.P. Chinn.

Dewatering

3. Hansen, V.E., Water management for the wet end, *1991 Wet End Operations Short Course*, TAPPI Press, pp 131-192. This is an excellent, highly detailed resource with numerous photographs of paper machines. This article must be in the library of anyone responsible for dewatering at the wet end of flat wire machines. The TAPPI copyright figures on dewatering elements presented in this chapter was from this article through the courtesy of the author.

4. Burkhard, G. and P.E. Wrist, Investigation of high speed paper machine drainage phenomena, *Pulp Paper Mag. Can.* 57(4):100-118(1956). This is the landmark paper that introduced foils.

5. Parker, J.D., The sheet forming process, TAPPI STAP No. 9, 104 p. (1972). This monograph with 140 references describes wet-process sheet formation from a dilute fiber suspension as a function of drainage, oriented shear, and turbulence.

Headbox design

6. Waller, M.H., Recent developments in headboxes, *Tappi J.* 70(1):33-42. (A similar paper is *ibid.,* Twin-wire headboxes, *TAPPI Notes 1989 Twin-Wire Seminar* pp 17-28.) This excellent article is a 10-year survey of headbox development with pictorial examples from several suppliers.

7. Newcombe, D., Demanding customers have meant dramatic changes in headboxes, *PIMA* 69(10):27-34(1987). No references. The historical development of the headbox and present limitations of the headbox are chronicled here.

8. TAPPI TIS 0410-02, -03, and -04 (1988). These are 4, 4, and 3 pages, respectively.

These information sheets mathematically describe the flow from headboxes with vertical, 45° angle, and nozzle with inclined upper lip slices, respectively.

9. TIS 0410-06, Headbox piping system--general design guide. 6 p.

Forming fabrics
10. Kobayashi, T., Latest trend of forming fabric for PM, *Japan Pulp Paper* 27(4):73-87(1990). This article, with numerous electron micrographs, discusses new forming fabrics including 2 and 2.5 layer fabrics.

11. These three references appear consecutively in *TAPPI Seminar Notes, 1989 Wet End Operations*. Fliss, T., Forming fabric design fundamentals, pp 207-227.

12. Pitt, R. and E.M. Cordon, Wear of forming fabrics, *TAPPI Seminar Notes, 1989 Wet End Operations* pp 229-236.

13. Givin, W.R., Understanding fabrics through computer technology, *TAPPI Seminar Notes, 1989 Wet End Operations* pp 237-248.

Fourdrinier paper machine
14. Kennedy, W.H. and P.E. Wrist, The fourdrinier paper machine, in *Pulp and Paper Science and Technology*, Volume 2, Libby, C.E., Ed. McGraw-Hill, New York, 1962, pp 163-208.

This chapter covers fundamental aspects of the wet end including flow of fiber slurries in circular pipes, stock distribution to the headbox, headbox operation, slices, slice velocities as a function of geometry, fiber orientation, suction profiles of table rolls and drainage foils, sheet formation, sheet two sidedness, and a flow balance exercise.

15. Strauss, R.W., Papermaking machines; the fourdrinier, in *Pulp and Paper Manufacture*, Volume 3, MacDonald, R.G., Ed., McGraw-Hill, New York, 1969, pp 245-297. A similar type of article to the preceding one, with many identical figures, but including more history of the paper machine and the twin wire formers.

16. TAPPI TIS 0410-07 (1988), Evaluating the drainage performance of a fourdrinier. 4 p.

17. TIS 0410-08 (1968), Control of stock jump on a fourdrinier wire. 3 p. This is a dated discussion with table rolls.

Twin wire machines
18. Thorp, B.A., Fundamental and commercial overview of twin wire and multiple wire forming, *TAPPI Notes 1989 Twin-Wire Seminar*, pp 1-15 (the identical article, on glossy coated paper, is found in *Tappi Notes 1989 Wet End Operations Seminar* pp 191-206). This is an extremely useful overview of twin wire machines. Several other articles from the twin wire seminar also describe twin wire machines, but with a lot of repetition. The entire notes are 147 pages.

19. Baker, D.E., and K.N. Riemer, Twin-wire gap forming: leading edge technology for newsprint in the 90s, *Pulp Paper Can.* 93(6):T170-T175(1992). Many diagrams and a recent list of startups are included.

20. Sinkey, J.D. and D. Wahren, Quality comparisons of twin-wire and fourdrinier papers, *Tappi J.* 68(9):110-114(1986). This paper, with 60 references, discusses the advantages of paper made on twin wire machines, compared to paper made on fourdrinier machines, including improved formation, smoothness, basis weight, and moisture profiles; reduced two-sidedness and linting; and somewhat higher porosity.

Multi-ply formers, cylinder machine
21. Nutter, J.C., Papermaking machines; cylinder, in *Pulp and Paper Manufacture*, Volume 3, MacDonald, R.G., Ed., McGraw-Hill, New York, 1969, pp 297-365. This is a detailed chapter on the equipment and operation of the cylinder wet end.

22. Anon., *Pulp & Paper* 52(14):89-104 (Dec., 1978). "Within the last 13 years, machine producers have introduced 20 new types of multi-ply formers." This is perhaps the best starting point on the subject of twin wire formers and multi-ply board forming.

Pressing, press felts, dewatering

23. Wahlström, B., A long term study of water removal and moisture distribution on a newsprint machine press section, Part I, *Pulp Pap. Mag. Can.* 61(8):T379-T401(1960); *ibid.*, Part II, 61(9):T418-T451(1960). The author won the 1992 TAPPI Gunnar Nicholson Gold Medal largely for this classic work on pressing.

24. Busker, L.H., The effects of wet pressing on paper quality, *TAPPI Seminar Notes, 1985 Practical Aspects of Pressing and Drying*, pp 83-95. Wet pressing has an effect on the final sheet density and fiber bonding, so numerous sheet properties are affected by it. Also, movement of fines, sheet crushing, consolidation, and felt marking are discussed. The response to wet pressing is very dependent on the furnish.

25. Lewyta, J. and P. Geoghegan, Recent developments in paper machine clothing: wet felts, *Tappi J.* 70(4):57-62(1987). A historical look at wet felts is given. Many photographs show the construction of modern wet felts.

26. The next four papers are more technical aspects of press felts.

    Ballard, J., Press felt characterization, *Tappi J.* 70(7):57-61(1987).

27. Beck, D.A., Re-examining wet pressing fundamentals: a look inside the nip using dynamic measurement, *Tappi J.* 70(4):129-133(1987).

28. Jaavidaan, Y., W.H Ceckler, and E.V. Thompson, Re-wetting in the expansion side of press nips, *Tappi J.* 71(3):151-155(1988).

29. Fekete, E.Z., The importance of felt dewatering, *Tappi J.* 69(12):147-148(1986).

30. Reese, R.A., Linerboard survey shows that drying rates drop as press loading rises, *Am. Papermaker* 54(4):42-43(1991). The results of a survey indicate that as press loading increases the rate of water removal in the dryer section decreases.

31. TIS 0404-01 (1983), Determination of water removal by wet presses. There are about 10 other TIS references on pressing, press felts, wet felt void volumes, etc. that are of related interest.

Suction roll shells

32. TIS 0402-10 (1985), Guide for evaluation of paper machine suction roll shells. This is an extensive reference of 15 pages on the title topic.

33. TIS 0402-11 (1987), Liquid dye penetrant testing of new suction roll shells. 4 p.

Drying

34. Coveney, D.B. and G.A. Robb, in *Pulp and Paper Manufacture, Vol. 3*, MacDonald, R.G., Ed., McGraw-Hill, New York, 1969, pp 405-551. This is an extensive treatise on drying paper, with engineering and economic calculations.

35. Chalmers, G.J., Today's dryer section: an overview, *PIMA* 70(6):45-51(June, 1988). This is a good, introductory, nontechnical article on drying.

36. Siegler, D.M., Reviewing dryer section drive methods, operating considerations, and costs, *PaperAge* 107(2):12, 13, 30(1991). This is a useful article on *drive methods* for the dryer section.

37. TAPPI TIS 0402-16, Guidelines for the inspection and nondestructive examination of paper machine dryers (9 p.) and about 20 other TIS references.

Yankee dryers

38. Sloan, J.H., Yankee dryer coatings, *Tappi J.* 74(8):123-126(1991) with no references. This article discusses coatings and how they relate to drying and creping.

39. Corby, W.G., Jr., *Guidelines for the Safe Operation of Yankee Dryers*, Tappi Press, 1986, 33 p. The introduction describes yankee dryers as the class of grey cast iron drying cylinders 10-20 ft in diameter and up to 300 in. wide that weigh up to 380,000 lb and operate with up to 140 psi steam pressure at speeds up to and beyond 6000 ft/min. The introduction indicates that they may contain an energy equal to 100 lb of TNT, so safety is very important. Since 1950, the article indicates, there has been only one disastrous failure in the U.S.

Moisture content profile control

40. Roth, R.C., Electromagnetic radiation for CD moisture profile control, *Tappi J.* 69(1):57-61(1986).

41. Higham, J.D., D.A. Stockford, and J. Egierski, Locating infra-red drying equipment for moisture profiling, *Appita* 41(5):398-400(1988).

Impulse drying

42. Crouse, J.W., Y.D. Woo, and C.H. Sprague, Delamination--a stumbling block to implementing impulse drying technology for linerboard, *Tappi J.* 72(10):211-215(1989).

43. Sprague, C.H., Impulse drying and press drying: a critical comparison, *Tappi J.* 70(4):79-84(1987).

44. Bach, E.L., Why is press drying/impulse drying delayed?--A critical review, *Tappi J.* 74(3)135-147(1991), with 55 refs. A discussion of this article by J. D. Lindsay with rebuttal by the author is in *Tappi J.* 74(9):238-241(1991).

45. Orloff, D.I., Impulse drying of linerboard: control of delamination, *J. Pulp Paper Sci.* 18(1):J23-J32(1992). The author purports to have controlled delamination, one reason this process is not yet commercial.

On-line measurement of paper properties

46. Mercer, P.G., On-line instrumentation for wet-end control, *Appita* 41(4):308-312(Jul 1988). This is a useful article on measuring grammage, moisture, mineral content, caliper, optical properties, and fiber orientation.

Calendering, reeling, and rewinding

47. MacDonald, R.G., Ed., *Pulp and Paper Manufacture*, Volume 3, McGraw-Hill, New York, 1969, pp 553-641. Three chapters spanning these pages cover these areas.

48. Schiller, F.E., *Manual of Supercalender Operations*, Miller Freeman, San Francisco, 1976, 136 p. This is a operator guide.

49. Journeaux, I., R. Crotogino, and M. Douglas, The effect of actuator position on performance of a CD calender control system, *Pulp Pap. Can.* 93(1):T17-T21(1992).

50. Tuomisto, M.V., Upgrading groundwood printing papers: What can soft calendering do?, *Pulp Pap. Can.* 93(2):T37-T40(1992).

Coating and other converting

See references 4 and 5 in Chapter 8.

51. Satas, D., Ed., *Web Processing and Converting Technology and Equipment*, Van Nostrand Reinhold Co., New York, 1984, 537 p. There are 24 chapters including 9 chapters on coaters and coating. There are also chapters on some hard-to-find topics including calendering, extrusion, saturators, laminating, surface treatment, drying, IR and electron beam curing, film orientation, winding, slitting, splicing, web handling, sheeting, die cutting, embossing, and static electricity.

52. Gunning, J.R., Coating, in *Pulp and Paper Manufacture*, Volume 2, MacDonald, R.G., Ed., McGraw-Hill, New York, 1969, pp 457-532. This is a good introduction to the historical equipment development, methods of coating, properties of coated paper, pigments used in coating, lists of dyes, starch, additives, coating formulations, and finishing.

53. TAPPI TIS 0102-01 (1985), Literature references on the measurement of water retention of paper coatings. 6 p. Over 40 abstracts on are given on the title topic. The function of water retention is "to improve the ability to hold the vehicle (water) within coatings after application and to decrease the migration to the surface or into the base sheet."

54. TAPPI TIS 0110-01 (1987), Factors to be considered in selecting screening media for coating processes. 2 p. TIS 0110-02 (1986), Viscosity of coating colors at high shear rates. 6 p.

EXERCISES

Wet end

1. Why is an "S"-shaped curve used as the path of the wires in some twin wire machines?

2. Why is fiber sometimes added to excess white water prior to *save-alls*, i.e., what is the purpose of sweetener stock?

3. Is turbulence desired in the papermaking process? Discuss carefully!

Pressing

4. What are two advantages of the extended nip press?

5. Why are the web and felt separated as quickly as possible after a press nip?

6. What is roll crowning and why is it used?

7. Do press felts usually "wear out" per se?

8. Give at least three sources of contaminants that may contribute to "stickies" formation in the press felts and paper machine wire.

9. Often, the paper web is heated before the press section for a particular purpose. What do you think is the purpose, and what is the theoretical basis for this action?

Drying

10. Why are Yankee dryers used to make tissue papers?

11. Give two reasons why Yankee dryers are used with various grades of paper.

12. How do the temperatures of the early and later dryer sections compare. Give some actual reasonable values for them, and explain why their is a difference.

13. What is the major advantage of a coarse dryer felt and what is the major advantage of a fine dryer felt?

14. What is the typical steam economy for the dryer section? How does this compare with the steam economy of the multiple effect evaporators?

Coating

15. Besides additives, what are the two main categories of coating constituents?

Paper properties

16. What operations contribute to the final caliper of paper (for a given basis weight) and to what extent is each important (relatively)?

17. Where does the ash content tend to be higher on filled sheets, on the wire side or on the felt side of paper made on single-wire fourdrinier machines? Why?

18. What is one possible disadvantage of having a two-sided sheet? Hint: think of how a thermostat coil of metal ribbon works to indicate and control temperature or how a Bourdon tube works to indicate pressure.

19. How do strength and bulk relate to each other. Stated another way, how do they interact and why do they interact that way? (Think of refining curves and paper strength.) Why are both desired in printing papers? How does the use of adhesives affect the picture?

20. What is the difference between bonds formed with wet-strength adhesives and those formed between fibers without added adhesive or using starch as an adhesive?

21. How is a corrugated box like an I-beam?

Sizing

22. What is the mechanism whereby starch acts as a sizing agent?

10

FIBER FROM RECYCLED PAPER

10.1 INTRODUCTION

Obtaining fiber from recycled paper is a matter of separating impurities from the usable fiber. This involves extensive systems for removing foreign materials including skimmers to remove floating items, removing heavy items at the bottom of a repulper, and removing stringy items such as rope and wet strength papers. Coarse screening is used for further cleaning before using fine screens. Vortex cleaners, through-flow cleaners, and, for printing and tissue grades, deinking systems are used. The steps used in recovering fiber can be accomplished in a wide variety of different orders.

Recycled or secondary fiber

The source of recycled fiber is recycled paper or paperboard arising outside of the mill. It is distinguished from *broke*, which is off-specification paper produced at the mill and reused within the mill. There may be similar considerations, however, when using broke in a mill.

It is very important to have "pure" sources of paper from which to recover high quality fiber. Newspapers should not be mixed with magazines, brown paper, or boxes. Office papers should not be contaminated with newsprint or brown papers. For example, in 1988 mixed waste paper was worth about $5-25/ton, newspaper was worth $25-45/ton, corrugated containers were worth $36-60/ton, while clean, hard, white paper clippings from an envelope factory were worth $390-470/ton. In the U.S. almost 30% of the paper consumed in 1989 was recycled, while Japan had a recovery rate of 50%. Other developed nations tend to have recovery rates between 30-50%.

The product with the largest recovery by amount and percentage in the U.S. is old corrugated containers (OCC). Over 50% of OCC is recovered in the U.S. One reason is that large amounts of OCC are generated at specific sites such as grocery stores and other retail outlets. Newspapers and other post-consumer wastes are much more expensive to collect and tend to be highly contaminated with unusable papers and trash. Still, 33% of newsprint is recovered, but only one-third of this ends up in new newsprint. Many U.S. states, however, have enacted (or are working on) legislation that demands large amounts of recycled fiber (10-50%) in newsprint.

The Garden State Paper Co. in New Jersey was the first U.S. company to use deinked newsprint in newsprint. They started with a 90 ton per day facility using a washing process in the 1960s.

Use of recycled paper

About 80% of all waste paper comes from three sources: corrugated boxes, newspapers, and office papers. Less than 20% of waste paper is deinked to be used in newsprint, tissue, or other bright grades. Most waste paper is used in paperboards, chipboard, and roofing materials where color is not important. The percentage of deinked newsprint and other grades is expected to increase considerably over the next few years.

Shrinkage

Shrinkage represents the loss of material from the original feedstock to the recovered product. A plant that recovers 90 lb of dry fiber for every 100 lb of dry paper has a shrinkage of 10%. A plant recovering OCC may have a shrinkage of about 8%. A plant recovering newsprint for use in newsprint (news to news) may have a shrinkage of 15-20% for flotation deinking and up to 30% using combined flotation and washing deinking. A plant recovering office paper and producing pulp of high brightness (above 85%) may have shrinkage of 35%. Shrinkage represents the loss of fillers, fines, plastics, dirt, ink, and useful fiber.

10.2 RECYCLED FIBER PREPARATION

The nature of the contaminants and their removal

One of the biggest obstacles to using recycled or secondary fiber is effective contaminant removal. Recycling of paper is a process of removing

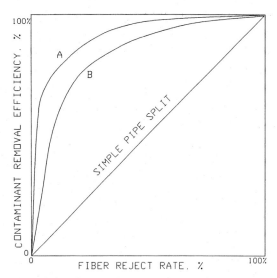

Fig. 10-1. Contaminant removal efficiency and fiber reject rate.

contaminants from the useful fiber. It is useful to compare various steps used for contaminant removal by diagrams such as that in Fig. 10-1. The efficiency of a process is the percentage of rejects removed. The reject rate is the percentage of useful fiber that is discharged with the rejects. Consider a simple pipe split with equal flows to rejects and accepts; 50% of the contaminants are removed (50% efficiency), but with a fiber reject rate of 50%, nothing has been accomplished. In Fig. 10-1, two hypothetical systems are presented, "A" and "B". It should be apparent that system "A" is superior because it can be run at the same efficiency for a lower fiber reject rate or can be run at a higher efficiency with the same fiber reject rate compared to "B". For any given system, this diagram shows the trade-offs between improved efficiency and fiber reject rates that one will encounter with cleaning, screening, deinking, and other processes.

Contaminants can be classified according to their source. *Stickies* are a diverse group of materials that are tacky to the touch. They may enter the process through the polymeric contaminants of secondary fiber including hot melt adhesives (typically consisting of ethylene vinyl acetate-wax-hydrogenated resin acid combinations), wax and polyethylene from coated boxes, contact adhesives (polybutylene, natural rubber, etc.),

pressure sensitive adhesives [styrene-butadiene rubber (SBR), carboxylated polybutadiene, and vinyl acrylates], polystyrene, and so forth. Some of these structures are shown in Fig. 10-2. Stickies may also arise on the paper machine from wood extractives that polymerize during pulping and, especially, bleaching, additives used within the mill such as fatty acids and other defoamers, rosin and synthetic sizing agents, and polymers used within the mill such as natural rubber splicing tapes. They tend to be soluble in nonpolar solvents such as diethyl ether, toluene or methylene chloride, have a tacky feel to them, and are very difficult to remove from the process.

Fillers include calcium carbonate, which can interfere with rosin/alum sizing, clays, and titanium dioxide. *Films and laminates* include polyethylene, aluminum foil, etc. *High density materials* include materials such as glass, grit, sand, and metallic objects. *Other materials* include small amounts of unbleachable dyes, wet strength resins, and stringy materials such as carpets.

Ink consists of pigments, such as carbon black or titanium dioxide, to supply color and opacity and a vehicle to carry the pigment and bind it to the paper. The vehicle consists of solvent and resin to bind the pigment to the paper. Traditionally the vehicle was vegetable oil. This is significant since vegetable oil is easily saponifiable with alkali, allowing the ink to be dispersed during deinking. Saponification converts the triglyceride plant oil to glycerol and the salts of the three constituent fatty acids (see Fig. 2-31).

The type of resin depends on the type of ink and how it attaches to the paper. Inks may "set" on the paper by one of several mechanisms according to Horacek (1978). 1) *Absorption* of the hydrocarbon vehicle into the paper substrate; this type of ink is used in newsprint and tends to smudge. 2) *Evaporation* of the ink vehicle; this is used in magazine and catalog grades using letterpress or offset printing with rosin esters and metal binders. 3) *Oxidation* of a drying oil with multifunctional carboxylic acids and alcohols left on the paper surface after the vehicle is absorbed into the paper. 4) *UV or electron beam radiation curing* of monomers or prepolymers in the vehicle into polymers such as acrylics. *Infrared hardening*, *precipitation* of binders, *gelation*, and *cooling*

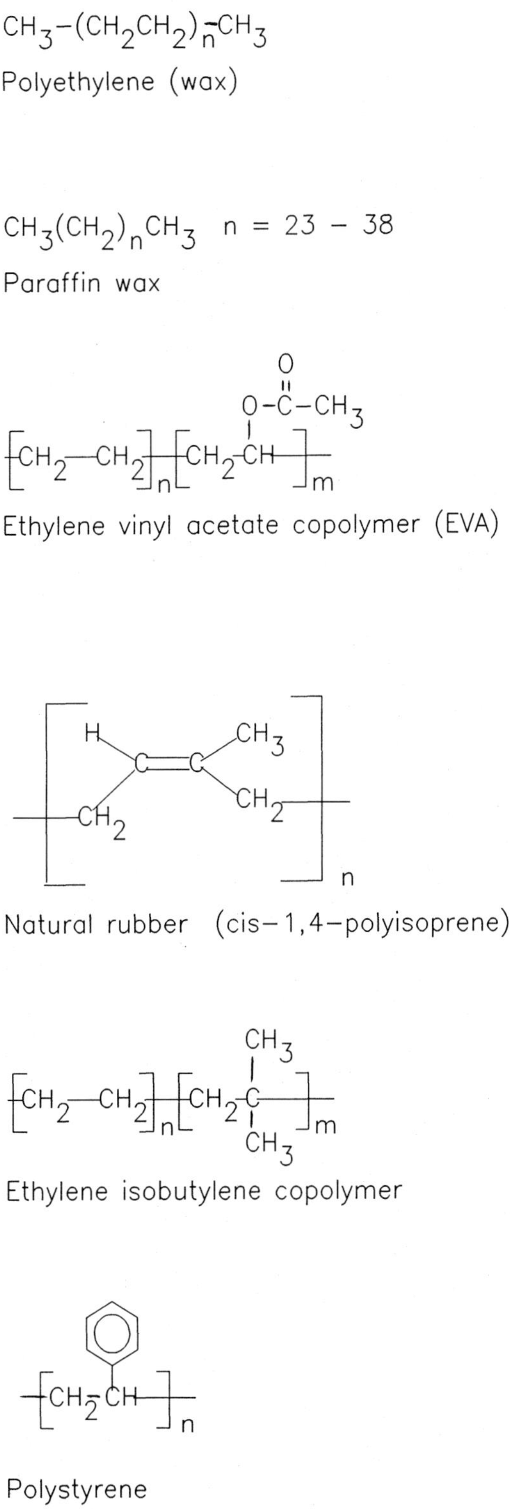

$$CH_3-(CH_2CH_2)_n^-CH_3$$

Polyethylene (wax)

$$CH_3(CH_2)_nCH_3 \quad n = 23 - 38$$

Paraffin wax

Ethylene vinyl acetate copolymer (EVA)

Natural rubber (cis−1,4−polyisoprene)

Ethylene isobutylene copolymer

Polystyrene

Fig. 10-2. Several waste paper contaminants.

of hot thermoplastic inks used in electrostatic printing (photocopy and laser printers) are other mechanisms used less often, but are increasing.

The first two types of ink are readily dispersible by emulsification. Oxidation cured inks are often saponified by alkali, but radiation cured inks do not easily break into pieces smaller than 50-100 μm. The vehicles of inks that dry consist of mineral oil, waxes, alkyd resins, and other hydrocarbon resins; inks which cure by infrared or ultraviolet radiation are made with epoxy, urethane, vinyl, or polyol acrylates; xerographic inks contain SBR, polyesters, and acrylates. Latex binders are SBR, carboxylated butadiene, polyvinyl acetate, etc. As ink, and the paper to which it is attached, ages, it becomes more difficult to remove. This should be kept in mind when deciding how to handle wastepaper inventory. Also, the size of the ink particles is very important; typically, newsprint ink has very small particles, on the order of 1-10 μm, ledger has large particles, on the order of 50-1000 μm, and mixed news/magazine grades have intermediate particle sizes on the order of 1-50 μm.

Aside from being classified by their source, contaminants are often classified by their method of removal. Some examples are large, dense materials; small, dense materials; "ultradense" materials; fibrous, light materials; small, light flakes; and so on until you are sick. Such classification schemes identify categories by density, size, composition, shape, etc. Since these classifications are arbitrary and the contaminants change as the fiber is processed, they are not mentioned here. Keep in mind, however, what types of contaminants each process removes, to what degree each is removed, and how does the process alter the contaminants so that they may or may not be removed in subsequent steps. For example, too much agitation will break stickies apart so that they may not be removed by screening; however, remaining stickies are sometimes deliberately dispersed so that they will have limited impact during papermaking.

One way to limit the amount of contaminants in the secondary fiber is to pay close attention to the waste paper coming to the mill. The Paper Stock Institute of America circular PS-83 specifies that newspapers for recycling should be almost as clean as when they were delivered to the consumer. Newspapers that are wrapped in plastic (even strappings) or kraft paper or that have more than 2% rejects should be rejected. Ideally, little useful fiber should be removed with the contaminants when processing waste paper.

10.3 RECYCLED FIBER RECOVERY

Introduction

Recycling fiber is the process of separating useful fiber from the contaminants of waste paper. A series of processes can be used to accomplish this task. There are also many possible ways of arranging these steps. Each mill has its own needs, its own hypotheses about what is the best method, and its own operating experiences. In this section we will consider many of the processes available and how they work.

Crow and Secor (1987) list 10 steps for deinking, including pulping, pre-washing heat and chemical loop, screening (coarse and fine), through-flow and reverse cleaning, washing, flotation, dispersion, bleaching, and water recirculation and makeup, with brief descriptions of each of these processes. While other methods are available, these represent most of the tools available for fiber recovery from waste paper. Not all steps are used in all recycled fiber plants.

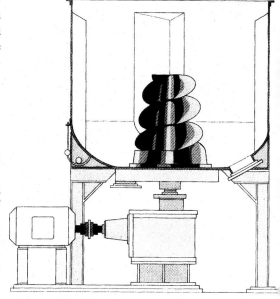

Fig. 10-3. HI-CON™ high consistency pulper. Courtesy of Black Clawson, Shartle Division.

Continuous pulper

Recycled fiber recovery begins at the pulper, Figs. 10-3 and 10-4, which is nothing more than a large blender to disperse pulp into an aqueous slurry. Pulping may be done at high or low con-

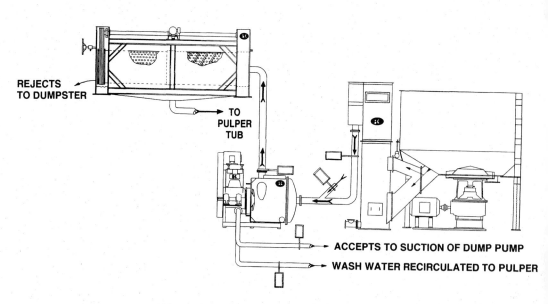

REJECTS TO DUMPSTER

TO PULPER TUB

ACCEPTS TO SUCTION OF DUMP PUMP

WASH WATER RECIRCULATED TO PULPER

Fig. 10-4. Hydrapurge™ detrashing system. Courtesy Black Clawson, Shartle Division.

sistencies using different rotors. Fig. 10-3 shows a high consistency pulper and Fig. 10-4 shows a low consistency pulper with trash removal.

During pulping, some of the gross contaminants are removed. A *ragger* is a chain which is pulled out slowly over a long period of time (perhaps a meter every few hours). The ragger catches baling wire, wires, plastic sheeting, stringy materials, tapes, wet strength paper, and other long materials that wrap around it as the stock rotates in the pulper. The accumulated material forms a *rag rope* that is over a foot in diameter upon removal. Fig. 10-5 shows a diagram of a ragger with cutter.

Relatively large heavy contaminants such as nuts, bolts, rocks, pipe fittings, and other material do not pass through the holes of the pulper and eventually are forced outward to the *junk chute* at the lower peripheral of the pulper. This material is removed by one of several possible means. A

junk tower is a long tube with an overhead grapple that is used to manually remove debris that accumulates at the bottom of the tower. A *continuous junk remover* is an enclosed bucket conveyer used to automatically remove heavy and floating debris on a continuous basis (Fig. 10-6). It has often been used when processing old corrugated containers (OCC). Because of the high maintenance due to chain breakage and inability to keep the pulper clean, bucket and chain trash-removing configurations are no longer used in new equipment. The *junk trap* consists of a vertical tube with valves on the top and bottom to isolate it from the pulper. The top valve is normally open to allow junk to enter, and the bottom valve is closed to retain the contents. The junk trap is emptied by closing the top valve and opening the bottom valve to let the debris out. The bottom valve is then closed, the top valve opened, and the processes repeated as necessary.

On the order of 18 kWh/t (1 hp-day/ton) is required to disperse the fibers, and about the same amount is used for cleaning and screening. The pulp from the pulper then usually goes to pressure screens and high density centrifugal cleaners.

Sometimes high temperatures and agitation in the pulper might break up contaminants into very small particles and must be avoided. These conditions would be avoided if fine screening is the only subsequent method of contaminant removal. These conditions might also be avoided if ink flotation is being used. In other cases, as with ink washing, these conditions might be desirable.

Most deinking plants operate their pulpers in batch mode. Deinking conditions in the pulper are typically 6-15% consistency, 40-70°C, pH 9-11 (generally below 10 for groundwood pulps to limit yellowing), for one hour. The trend is towards higher consistency pulping, which has the advantages of enhanced ink separation, reduced pulping time, and lower specific energy consumption.

Pulping at high consistency requires dilution of the pulp to about 5% consistency before it will pass through the extraction plate at the bottom of the pulper (Plate 38) or other coarse screening device [6-16 mm (1/4 to 5/8 in.) hole size]. A tub extension on the pulper is used so dilution water can be added, or the pulper contents are dumped into a stock chest. Typically white ledger paper is

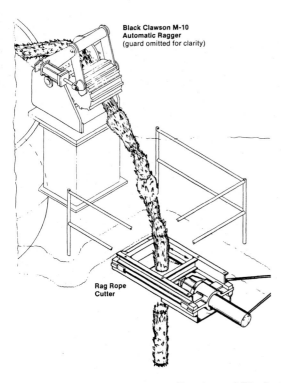

Black Clawson M-10
Automatic Ragger
(guard omitted for clarity)

Rag Rope
Cutter

Fig. 10-5. Rag rope cutter. Courtesy of Black Clawson, Shartle Division.

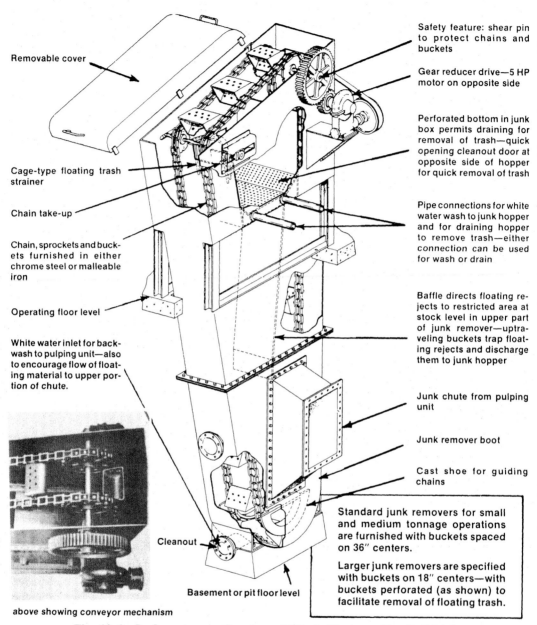

Removable cover

Cage-type floating trash strainer

Chain take-up

Chain, sprockets and buckets furnished in either chrome steel or malleable iron

Operating floor level

White water inlet for backwash to pulping unit—also to encourage flow of floating material to upper portion of chute.

Safety feature: shear pin to protect chains and buckets

Gear reducer drive—5 HP motor on opposite side

Perforated bottom in junk box permits draining for removal of trash—quick opening cleanout door at opposite side of hopper for quick removal of trash

Pipe connections for white water wash to junk hopper and for draining hopper to remove trash—either connection can be used for wash or drain

Baffle directs floating rejects to restricted area at stock level in upper part of junk remover—uptraveling buckets trap floating rejects and discharge them to junk hopper

Junk chute from pulping unit

Junk remover boot

Cast shoe for guiding chains

Standard junk removers for small and medium tonnage operations are furnished with buckets spaced on 36" centers.

Larger junk removers are specified with buckets on 18" centers—with buckets perforated (as shown) to facilitate removal of floating trash.

Cleanout

Basement or pit floor level

above showing conveyor mechanism

Fig. 10-6. Junk remover. Courtesy of Black Clawson, Shartle Division.

pulped at 15-20% consistency, newsprint at 10-12% consistency, and OCC at 4-5% consistency.

Screening

The stock is then sent to a screening system. Because primary screen rejects may be as high as

50%, the primary screen rejects are often sent to a secondary screen to recover usable fiber. Pressure screens are coming into widespread use for this purpose. The basket has a life of 6-12 months for deinking grades and 2-4 months for nondeinking grades, which tend to have more abrasive particles in them. Wear occurs at the rejects end where the consistency is higher and water does not lubricate as well. The trend is towards modular assembly of pressure screens so that the reject end section can be replaced while reusing most of the basket. Fiberprep uses an inward-flow design and claims that centrifugal force keeps the heavy debris away from the screen and thereby reduces screen basket wear and increases throughput. More information on pressure screens may be found in Section 8.2.

Centrifugal, vortex, or cyclone cleaners

The basic principles of low density vortex cleaners for cleaning pulp prior to the paper machine were described in Section 8.2. In forward cleaners a tangential input flow causes a vortex to form inside the cleaner; heavy particles move to the outside of the vortex and eventually drop to the bottom of the cleaner with the rejects while the pulp accepts move to the center of the

Fig. 10-7. Two high density cleaners.

vortex where they are removed at lower pressure. By this means, small heavy debris such as sand, glass, and metal fragments are removed.

Medium (1-3%) and high (2-5%) density centrifugal cleaners are used at 2-5% consistency and low pressure drops, 70-210 kPa (10-25 psi), to remove heavy contaminants of particle size 50-1000 μm (0.002-0.040 in.) that escape with the pulp through the pulper coarse screen and other screens. Fig 10-7 shows two high density cleaners used in a 500 ton/day secondary fiber mill. Fig. 10-8 shows a diagram of a high density cleaner and its operation. Rejects consist of small pieces of tramp metal, glass, and stones. Reject rates are below 1%. Water (10-40 gal/min) may be used in the throat of the cleaner to decrease the consistency in this region and help the heavy particles settle. These cleaners are typically about 2.5 to 6 m (8-20 ft) tall (although low profile cleaners are available) with a diameter of 0.20-0.65 m (8-26 in.) at the wide portion. The reject trap consists of a pair of valves that are manually or automatically operated. The top valve is left open to collect debris while the bottom valve remains closed. When the trap gets full, the top valve is closed and the bottom valve is opened to remove debris. The bottom valve is then closed, and the top valve is then opened. A single high density cleaner can process several hundred tons of pulp per day.

If very fine grit remain, a low density cleaning system is used that operates below 1% consistency (high pressure, 200-275 kPa or 30-40 psi) as described in Section 8.2.

Black Clawson's trademark for high and medium density cleaners is the Ruffclone™. The high density cleaners use 5-10 kWh/t (0.3-0.5 hp/d/t) pulp. Their fine paper forward cleaners are Ultra-clones that use about 30-80 kWh/t (1.7-4.5 hp/d/t).

Reverse cleaners, through or parallel flow cleaners

Lightweight contaminants were first removed by *reverse-cleaners*, where normal forward cleaners were used and merely operated in reverse. These cleaners required very large pressure drops on the order of 500 kPa (70 psi) and used large amounts of energy. Recent developments allow both the rejects and the accepts to be removed at

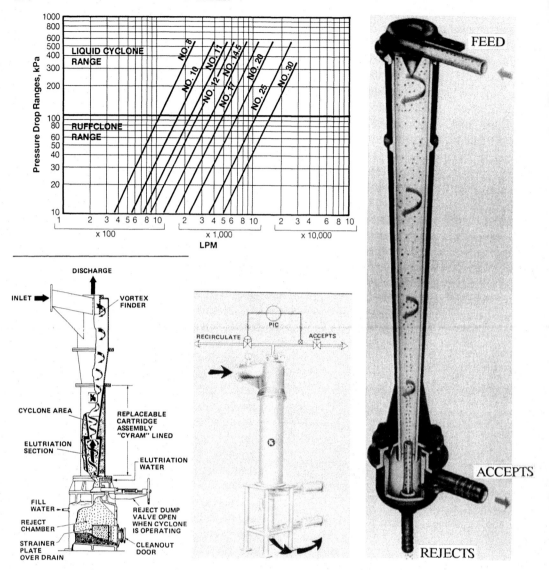

Fig. 10-8. Ruffclone™ and Liquid Cyclone™ high density cleaners with capacity chart and recirculation loop for constant discharge pressure. Courtesy Black Clawson, Shartle Division.

Fig. 10-9. A through-flow cleaner designed to remove low density contaminants. Courtesy of Beloit.

the bottom. Such cleaners, while called reverse cleaners, are better termed *through* or *parallel flow* cleaners. The reject rates have been reduced from 20-30% for the original reverse cleaners to 2-3% for modern designs. This means that numerous stages are not required. Fig. 10-9 shows the operation of a through flow cleaner.

Beloit (LeBlanc and McCool, 1988) has developed a number of centrifugal cleaners that are as examples of recent improvements. Beloit developed these cleaners using image analysis to track contaminants in the isolated fiber, transparent cleaners with high speed video cameras to observe the separation process, and analysis of the internal flow patterns. The Uniflow™ cleaner, which appeared in 1982, is a *through flow cleaner* that removes lighter-than-water contaminants such as plastics and other polymers, stickies, and pitch. The feed of the cleaner is at the top, and both the rejects and accepts flow through the cleaner and out the bottom, hence the term uniflow. An internal core with a small annulus around the air core collects the rejects. This circumvents the upward moving, inward spiral flow that interfered with lightweight reject separation and led to high reject rates in earlier reverse flow cleaners. These changes were claimed to reduce the rejects from 20% to 1-2% with a two-fold improvement in lightweight contaminant removal.

Black Clawson's through flow cleaners are the 3 in. X-clone™ cleaners, which were first operational at the Stone Container mill in Florence, South Carolina in May of 1984, and have been described by Bliss (1986). These cleaners use 1% consistency pulp with a 60-100 kPa (10-15 psi) pressure drop. The power requirements are 6-10 kWh/t (0.3-0.5 hp/d/t). The reject rate is normally 1-4% at a consistency of 0.1 to 0.4%. High stock temperatures improve the efficiency. Two banks of primary cleaners in series may be used, and the rejects from the primary bank are sometimes treated in secondary cleaners. The Beloit Posiflow™ centrifugal cleaner, introduced in 1986, is a *forward cleaner* that removes heavier-than-water contaminants. It is claimed that in some systems the conventional three-stage cleaning systems are replaced with a single stage.

Gyroclean™

An alternative to reverse cleaners is the Gyroclean of Fiberprep shown in Fig. 10-10. The unit was developed by Lamort and Centre Technique du Papeterie (CTP) in France. The manufacturer claims the unit can operate at up to 2% consistency. Fiber loss is said to be minimal, and no secondary units are required. Units are available to process 70-520 m^3/hr (300-2300 gal/min). The acceleration experienced by the stock is said to be up to 700 times normal gravity.

Deinking chemistry

Less than 20% of secondary fiber is deinked in the U.S. In addition to high wood costs, recent legislation will change this picture as many states will require recycled fiber in new newsprint as a means of reducing the large quantities of material sent to landfills, of which about 40% is paper. The ink is about 0.5-2% of the mass of the waste paper to be deinked.

The overall deinking process from the point of view of the fiber can be broken down into four steps. 1) Repulping with concomitant ink removal from the fibers, 2) cleaning and screening operations to remove the bulk of the ink from the stock, 3) separation of residual ink contaminants from the fiber stock, and 4) bleaching, if necessary. Other steps are necessary for the process such as recovering the wastes from the effluents to allow water reuse.

A variety of chemicals are used in ink removal including the sodium salts of hydroxide (for fiber swelling, saponification of ester-containing resins, and ink dispersion), carbonate (as a buffering agent), silicate (peroxide stabilizer via metal ion sequestering, wetting agent, pH buffer, and ink dispersant), polyphosphate (0.2-1% on pulp as a metal ion sequestering agent and ink dispersant), peroxide or hydrosulfite, fatty acid soaps, nonionic surfactants, and other materials. The nonionic surfactants are usually alkyl phenol or linear alcohol ethylene oxides as described in Section 8.5. Petroleum ether, a mixture of C_6-C_{10} alkanes, is sometimes used in small amounts to soften the ink vehicle. If any of these materials carry over into the papermaking system, there may be difficulties with foaming, scale formation, wet end chemistry, or surface sizing at the size press.

There are two commonly used methods of ink removal: *ink washing* and *ink flotation* or *froth flotation*. Ink can also be removed by *solvent extraction* in a process that resembles dry-cleaning, but it is too expensive to have widespread use.

Ink washing

Ink washing involves ink removal by washing it from the fiber using sodium hydroxide, sodium silicate, and hydrogen peroxide with a suitable dispersant in the pulper. Often the dispersant is stearic acid and micelle formation occurs in the classic mechanism by which soap is able to make grease and oils water "soluble". A water-ink emulsion system is formed with particle sizes averaging below 1 μm. The emulsion is washed from the pulp, and the ink is removed from the washwater by flocculation so the wash water may be reused. Hard water should not be used during ink washing since this will precipitate the soap and the complex will not be water soluble. The clarified water is then reused to wash more pulp. This general method has been used since the nineteenth century. It is suitable for traditional inks with vegetable oil vehicles that are readily saponifiable and dispersible.

Fig. 10-11 shows the theoretical ink removal for a washing system. This graph indicates the percentage of water removed in a single stage of washing based on the water going in and the water going out. For example, consider 100 lb of a pulp slurry with an inlet consistency of 3% and an

outlet consistency of 12%. There are 3 lb of fiber and 97 lb of water going in. The 3 lb of fiber corresponds to 25 lb of slurry at 12% outlet consistency or 22 lb of water going out. This means that 75 lb of the original 97 lb were removed to give a washing efficiency of 77.3%.

In fact, this is only true for materials that are water soluble and have no affinity for the fibers. Ink particles of finite size will be trapped by fiber networks that form as the slurry is concentrated on a screen. Larger ink particles are more easily retained by this mechanism.

Fig. 10-10. Gyroclean™ system for removal of lightweight contaminants. Three units in service (top) and the operating principle (bottom) are shown. Courtesy of Fiberprep.

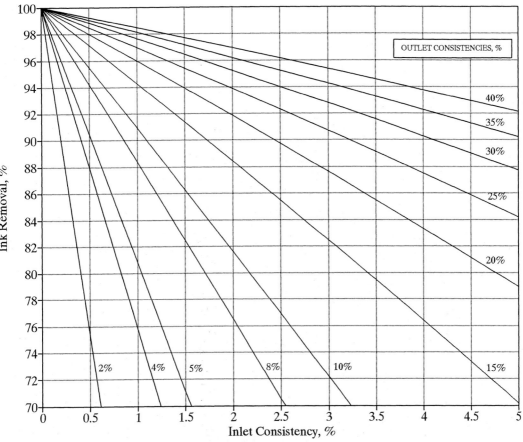

Fig. 10-11.   Theoretical washing efficiency based on inlet and outlet consistencies.

Washing equipment may take on a variety of forms such as rotary vacuum washers and double belt washers discussed in brown stock washing. Some manufacturers have specific products for this task, such as Black Clawson's DNT belt washer (Gilkey and McCarthy, 1988). This washer uses a endless synthetic wire with headbox, breast roll, and couch roll to dewater and thicken pulp up to 15% consistency.

Ink flotation

Ink flotation borrows from technology developed by the metallurgy industry for mining, which started being used to deink paper in the 1960s. *Perry's Chemical Engineer's Handbook, 6th edition* is a good reference for the basic principles of this process. Flotation has been used extensively in Japan and Europe, but only recently came into widespread use in the U.S. Flotation is a process that separates materials based on the property of wetability. Under appropriate conditions, non-polar (hydrophobic) materials are able to adhere to air bubbles and rise to the surface.

The process is carried out in ink flotation cells (several cells may be used in series) using sodium hydroxide, sodium silicate, and hydrogen peroxide with a *collector system* consisting of a surfactant. With ink flotation, large ink particle sizes are desired (at least 5 μm, but 10-50 μm is ideal) so the ink can agglomerate and be skimmed from the slurry. Many new inks and inks on office papers incorporate crosslinked resins and flake off the fibers in relatively large particles, making flotation the preferred method. Since the ink is already flocculated, it is fairly easily separated from the process water.

Ink flotation can be divided into three phases: collision of ink particles with air bubbles, attachment of ink particles to air bubbles, and separation of the ink particle-air bubble complex from the pulp. The reject rate of any system depends on how the ink-foam complex is removed. The fate of the collector chemical seems to be of great importance in some paper grades. The amount of deinking chemical in the pulp accepts is called the *carryover*. Ideally very little collector must leave with the stock accepts. Flotation deinking is done at lower consistency than washing and is done at about 1% consistency.

One type of flotation cell is shown in Fig. 10-12 and Plate 39. In this system, water and air are pumped into various points of the cell. The surface is skimmed to remove the ink. In 1988, Beloit introduced pressurized ink flotation (Fig. 10-13) with the claim that it removes more ink over a wider particle size range. Details of this method are found in Carroll and McCool (1990).

The collector system is based on fatty acids such as stearic acid with calcium or nonpolar surfactants. Occasionally cationic or ampholytic compounds have been used. In the first system, fatty acid salts (0.5-5% on pulp with a known

Fig. 10-12. Verticel™ deinking flotation cell. Air and water are introduced under pressure. The inserts show the skimmer and water inlet holes (an empty cell on the left). Courtesy of Fiberprep.

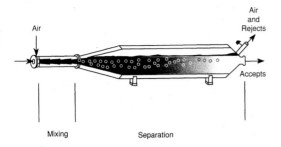

Fig. 10-13. Pressure Deinking Module™. Redrawn from a photograph supplied courtesy of Beloit.

degree of unsaturation, that is, number of double bonds as determined by the iodine number) and calcium ions are used. Calcium (like taking a shower in hard water) causes the fatty acids to precipitate, thereby supplying a hydrocarbon nucleus upon which the nonpolar ink particles can collect. Calcium is used at about 200 ppm as $CaCO_3$. The nonionic collector systems use an alkyl (C-9 is common) phenol or relatively large primary alcohols linked to oligomers (6-16 monomers) of ethylene oxide. The presence of inorganic fillers such as clay or talc assists in the ink flotation method and thus they are usually used. Without the use of inorganic fillers, additional, larger bubbles may be required, lower rise distance should be used, and larger flotation cell capacity may be needed. However, most of the filler will not be removed by flotation.

Flotation methods are also used to recover the water from waste water streams. This is described below.

Hybrid deinking processes

New deinking mills may use combined ink flotation and washing, especially for recovery of bleached chemical pulps. The bulk of the ink is removed by flotation, and residual ink is removed by washing. In this system, the dispersants are added after the pulper but before washing, so that residual ink that is not removed by ink flotation is emulsified and removed during the washing process. This process is the best of both worlds and is particularly useful for paper wastes that contain many different types of ink formulations. As ink formulations become more difficult to disperse by

saponification and emulsification, such as radiation cured resins, ink flotation is necessary, but removal of residual ink during washing is best accomplished by emulsification. Mills in Europe and Japan even use flotation-wash-flotation systems. About 10-20% of mills using a dual system use washing followed by ink flotation. This method has the disadvantage that many large ink particles that might have been removed by flotation would be dispersed prior to flotation, but has the advantage of more favorable treatment of various washing solutions in some mills.

Post-treatment processes are developing as well for even further ink removal. One method is to use disk dispersion above 20% consistency with about 80-140 kWh/t (4-6 hp-day/ton) to reduce particle sizes from several hundred microns to less than 50 microns so they are not visible as individual specks to the eye. This is followed by additional washing or flotation. Another approach is acidification of the pulp slurry followed by additional treatment of cleaning, very fine screening, etc. The state of the art for deinking newsprint is 92-95% or higher ink removal.

Solvent extraction of ink

At the risk of confusing the issue, there is a third method for ink removal, but it is generally uneconomical. The method is solvent extraction by use of nonpolar solvents (Hiraoka, 1990). The process resembles dry cleaning of clothing and was used, although not fundamentally for deinking, by Riverside Paper Co. in the U.S. and Tagonoura Sangyo in Fuji City, Japan for recovery of waxed paper, milk carton grades, etc. where both the paper and polyethylene resin are reusable. Riverside recovered 99% of its trichloroethylene solvent, and Tagonoura recovers 90% of its hexane solvent.

Water recycling and sludge recovery

The water containing the ink particles that have been removed from the fiber must be cleaned for reuse in the mill. One method of doing this is to use a flotation method similar to that used in flotation deinking.

Krofta Engineering Corp. is one manufacturer of a device that does this (Fig. 10-14). The raw liquid is pressurized with air under 60-80 psig

1) RAW WATER INLET
2) CLARIFIED WATER OUTLET
3) FLOATED SLUDGE OUTLET
4) CLARIFIED WATER RECYCLE OUTLET
5) PRESSURIZED WATER INLET
6) ROTARY JOINT
7) RUBBER PIPE CONNECTION
8) PRESSURIZED WATER PIPING
9) PRESSURIZED WATER DISTRIBUTION HEADER
10) RAW WATER DISTRIBUTION HEADER
11) DISTRIBUTION HEADER OUTLET PIPES
12) FLOW CONTROL CHANNELS
13) TURBULENCE REDUCTION BAFFLES
14) ADJUSTABLE HEIGHT BAFFLE ATTACHMENT
15) FLOW CONTROL CHANNEL OUTER WALL
16) ROTATING CARRIAGE GEARMOTOR DRIVE
17) CARRIAGE DRIVE WHEEL
18) WHEEL SUPPORT RIM
20) TANK WALL
21) TANK FLOOR SUPPORT STRUCTURE
22) ROTATING CLARIFIED WATER CONTAINMENT WALL
23) SLUDGE WELL
24) LEVEL CONTROL OVERFLOW WEIR
25) ROTATING CARRIAGE STRUCTURE
26) REVOLVING SPIRAL SCOOP
27) SPIRAL SCOOP GEARMOTOR DRIVE
28) CLARIFIED WATER EXTRACTION PIPES
29) ELECTRICAL SLIP RING
30) TANK WINDOW
31) SEDIMENT REMOVAL SUMP
32) FINAL DRAIN
33) SEDIMENT PURGE OUTLET
34) LEVEL CONTROL ADJUSTMENT

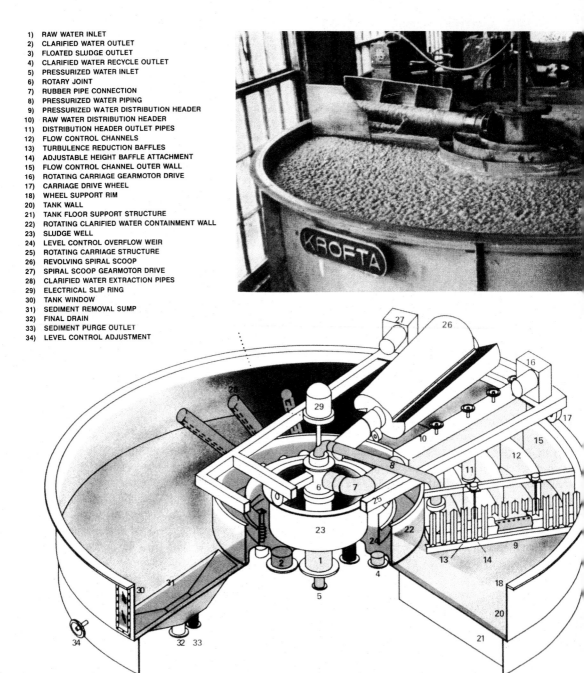

Fig. 10-14. Dissolved air flotation deinking effluent clarification. Courtesy of Krofta Eng. Corp.

Fig. 10-15. A sludge press in the foreground. The mill's primary and secondary effluent treatment is in the background.

pressure. When the raw liquid is introduced into the tank, it goes to ambient pressure so that small air bubbles are formed that attach to the suspended solids to be removed. The lightweight sludge then floats to the surface. Since the air was dissolved in the water, the process is called *dissolved air flotation.* The retention time is typically 3-4 minutes.

Polymer systems are used to help the separation. The sludge is skimmed from the top by the rotating skimmer and collected in the middle for removal. The treated water escapes underneath. The sludge collector rotates around the cleaner in this continuous process. The sludge that is obtained from the water to be recycled must be pressed to remove much of the water so that the sludge can be sent to a landfill or disposed of otherwise. Fig. 10-15 shows a sludge press.

Some problems with recycling paper are mentioned by Hoekstra (1991). For example, using 210 t/d old magazine (OMG) paper to produce 150 t/d secondary fiber generates 300 t of

waste at 21% solids content. With high tipping fees at the landfill, this is quite a problem. Burning of the sludge gives 6.5 t ash if 22% of the solids are ash. Analysis of sludge from five deinking plants gave the following level of heavy metals (parts per million): cadmium, <0.2; chromium, 16-118; copper, 31-400; lead, 3-210; manganese, 31-880; nickel, 1-25; and zinc, 36-1200. These were generally below those of municipal sludge. Also, the printing industry is accepting new ink formulations without heavy metals. BOD load is about 15-30 kg/t (30-60 lb/ton) pulp and TSS is about 10-50 kg/t (20-100 lb/ton) for deinked grades with washing methods producing more than flotation methods. Effluent volume is about 5 m^3/t.

Evaluation of the deinking process

The process of deinking may be evaluated by brightness or ink speck counts of the pulp or handsheets from the pulp. Solvent extraction of deinked pulp gives a quantitative method of ink content provided interfering compounds are not present; however, appropriate selection of spectroscopic methods of analysis (IR, UV, VIS, NMR, etc.) of the extract could easily give accurate ink determinations.

Brightness alone cannot be used to quantify ink since brightness depends on a variety of factors including the distribution of the ink on the fibers. Small particle sizes decrease brightness much more than large particle sizes of the same amount of ink. This is due to the increased surface area to volume ratio of smaller objects. This also accounts for the decrease in brightness often observed to occur with mechanical disintegration. Still, the overall pulp brightness gives a good indication of deinking effectiveness, but does little to indicate the source of trouble should the brightness go down if the system is not operating well. Furthermore, one should not allow any sort of washing of the pulp during formation of the brightness pad. Finally, brightness does not indicate the presence of ink specks visible to the unaided eye, i.e., larger than 50-70 μm (about 0.003 in.) in diameter, which may alone be unacceptable.

Image analysis techniques offer a good method to observe the deinking process. Modern image analysis systems controlled by computers offer a reproducible method for characterizing deinked furnish without undue operator time and bias which can be quite significant in an otherwise quite time-consuming and subjective procedure. Not only can ink speck counts be made, but ink speck size distributions can be determined so that information on the degree of emulsion or flocculation in the deinking process can be gathered over a period of time at a given mill. Fig. 10-16 shows an automated system used in a mill for quality control.

Image analysis systems are available that require very little theoretical knowledge on their operation. Earlier models are not as automatic and require attention to the light source which must not vary in intensity and spectral composition. Image analysis has also been used in several studies (Hacker, 1991) to measure lightweight contaminants. Here it is necessary to dye the paper with a water-based black ink (such as Parker Super Quink Ink) to enhance contrast by leaving the lightweight (nonpolar) contaminants amber in color while the paper turns black.

Slurry concentration

After deinking and cleaning, the dilute pulp slurry must be concentrated for further processing and storage. Fig. 10-17 shows a large disk filter used to concentrate a pulp slurry from 0.6 to 10% consistency at a mill that recovers 500 tons/day of secondary fiber.

Bleaching of secondary fiber

The philosophy of bleaching deinked pulp is very similar to that of bleaching virgin pulps (Chapter 5). Mechanical pulps (or so called wood-containing pulp) are bleached with peroxide (about 1% on pulp, with 4% sodium silicate, 50°C) or dithionite (hydrosulfite, about 1% on wood, 50-60°C at pH 5-6 to mitigate air oxidation of dithionite). Wood free pulps are usually bleached with a single stage hypochlorite treatment (1% on wood as chlorine), although an initial chlorination before hypochlorite has also been used. Sometimes these bleaching agents are added to the pulper to help with ink removal, but more efficient chemical usage is realized if the pulp is bleached after cleaning and screening.

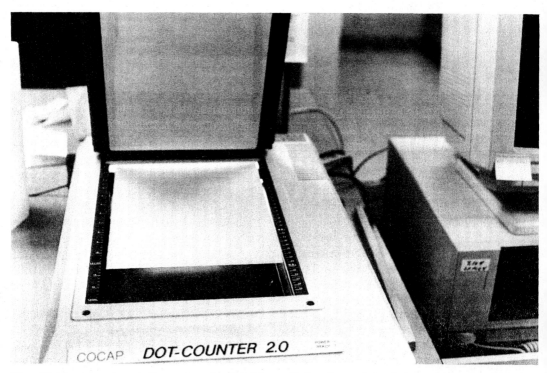

Fig. 10-16. An automated counter at a deinking mill for determining ink spot size distributions.

Fig. 10-17. A large disk filter for concentrating the pulp slurry at a deinking mill (500 t/d). The inserts show the filter disks when cleaned, with fiber, and with fiber removal.

Example of a process

In order to summarize this chapter, a recycling plant is presented in Fig. 10-18. This is a news to news plant using 70% news and 30% magazine for the feedstock. Numerous plant layouts are possible; this represents one choice.

10.4 ANNOTATED BIBLIOGRAPHY

General deinking
1. Woodward, T.W., Deinking chemistry, *Notes, 1991 Chemical Processing Aids Short Course*, Tappi Press, pp 85-105. This is an excellent, highly recommended reference on deinking chemistry with 85 literature citations that includes information on inks, printing, deinking chemistry, and bleaching.

2. Crow, D.R. and R.F. Secor, The ten steps of deinking, *Tappi J.* 70(7):101-106(1987).

3. Shrinath, A., J.T. Szewczak, and L.J. Bowen, A review of ink-removal techniques in

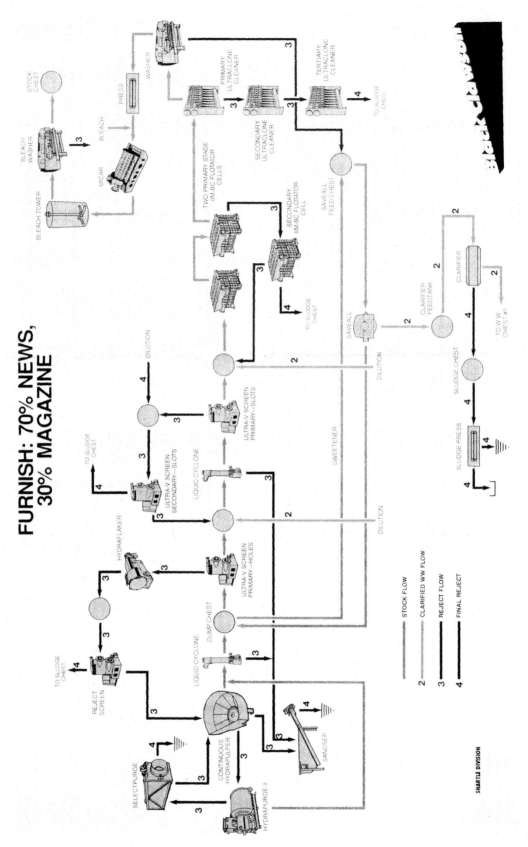

FURNISH: 70% NEWS, 30% MAGAZINE

Fig. 10-18. Layout of a newsprint to newsprint plant. Courtesy of Black Clawson, Shartle Division.

current deinking technology, *Tappi J.* 74(7):85-93(1991). The information in this reference is better presented in the articles it cites or those cited in this chapter. The article's strong point is the coverage of printing processes and ink removal strategies for specific types of ink.

Contaminants

4. Horacek, R., Deinking, Beloit (Jones Div.), Dalton, Mass., 1978. This reference is considered out of date by Beloit and is no longer available from them.

Centrifugal cleaners

5. LeBlanc, P.E. and M.A. McCool, Recycling and Separation Technology, *1988 TAPPI Pulping Conf. Proceedings*, pp 661-667.

6. Bliss, T., Through-flow cleaners offer good efficiency with low pressure drop, *Pulp & Paper* 59(3):61-65(1985).

Ink washing and flotation

7. Gilkey, M.W. and C.E. McCarthy, A new device for high efficiency washing of deink furnishes, *Proceedings, 1988 Pulping Conference*, TAPPI Press, pp 649-654.

8. Carroll, W.P. and M.A. McCool, Pressurized deinking module, *Proceedings, 1990 TAPPI Pulping Conference*, pp 145-152.

Surfactants

9. Mah, T., Deinking of waste newspaper, *Tappi J.* 66(10):81-83(1983), no references. Mill experience is given regarding the deinking of waste newspaper. Paper of 48-52% brightness was obtained by washing. The two problems were foaming and deinking waste news; both problems are related to the surfactant. The results of several alkyl and alkyl phenol oligo-ethyleneoxides are presented.

10. Okada, E., Deinking agent and its research development, *Japan Pulp Paper* 27(4):45-49(1990). Deinking surfactants for flotation deinking were studied. Also the fatty acid method was compared to nonionic surfactants, with the fatty acid method generally working better.

11. Balos, B, J. Patterson, K. Hornfeck, and M. Liphard, Flotation deinking chemistry, *PaperAge* 106(Recycl. Annual):15-17(1991). This study compares classes of surfactants in flotation deinking based on a laboratory setup. Here again the fatty acid soaps compared well to compounds of 6 other classes. Water hardness of at least 180 ppm $CaCO_3$ was recommended for complete flotation.

Solvent extraction of ink

12. Hiraoka, M. Latest waste paper treatment process, *Japan Pulp Paper* 27(4):33-42(1990).

Sludge, environmental considerations

13. Hoekstra, P.L., Paper recycling creates its own set of environmental problems, *Amer. Papermak.* 54(4):30,33(1991). The problems of sludge disposal are considered.

Metals in ink

14. Donvito, T.N., T.S. Turan, and J.R. Wilson, Heavy metal analysis of inks: a survey, *Tappi J.* 75(4):163-170(1992). Heavy metals in inks are an environmental concern, especially in deinking sludge. This paper investigates methods of determining heavy metals in inks. Many states will require the sum of Pb, Cd, Hg, and Cr to be below 100 ppm by 1994.

Decolorizing methods

15. Cheek, M.C., A practical review of paper decolorizing methods - present and future, in *Proceedings, 1991 Papermaker's Conference*, TAPPI Press, pp 71-78. This is a useful, general reference on paper decolorizing chemicals and processing conditions (i.e., for decoloring broke for reuse within the mill).

Ink removal test methods

16. McKinney, R.W.J., Evaluation of deinking performance, a review of test methods, *Tappi*

J. 71(1):129-131(1988). A brief evaluation is made of the following test methods: brightness measurement, ink removal-ink speck analysis, yield, and solvent extraction.

Image analysis

17. M.P. Hacker, 1991 *Pulping Conference Proceedings*, pp 455-468. This study measured lightweight contaminants in paper with a water-based black ink (such as Parker Super Quink Ink) to enhance contrast by leaving the lightweight (nonpolar) contaminants amber in color while the paper turns black.

18. McCool, M.A. and L. Silveri, Removal of specks and nondispersed ink from a deinking furnish, *Tappi J.* 70(11):75-79(1987).

19. Zabala, J.M. and M.A. McCool, Deinking at Papelera Peninsular and the philosophy of deinking system design, *Tappi J.* 71(8):62-68(1988)

20. Klungness, J.H, L.E. Fernandez, and P.L. Plantings, Image analysis for measuring adhesive contaminants in pulp, *Tappi J.* 71(1):89-93(1989).

21. Lowe, G., B.H. Licht, and G. Leighton, Deinking plant optimization using image analysis, *Tappi J.* 74(1):125-129(1991).

22. Dallard, D.H and C.M. Brown, *Computer Vision*, Prentice-Hall, Englewood Cliffs, New Jersey, 1982, 523 p. This is an often cited general text on image analysis.

23. Gonzalez, R.C. and P. Wintz, *Digital Image Processing*, 2nd ed., Addison-Wesley, Reading, Mass., 1987, 503 p. This is an often cited general text on image analysis.

Flotation clarification of effluents

24. Mahony, L.H., The role of flotation clarification in de-inking and secondary fiber operations, *PaperAge* 107(Recycling Update):46-47(1991). This paper describes the Krofta Engineering Corp. dissolved air flotation system.

Screening stickies

25. Heise, O., Screening foreign materials and stickies, *Tappi J.* 75(2):78-81(1992). This article is a review of the current technology. Charts show the screening efficiency versus reject rate, slot velocity, rotor velocity, and screen plate surface.

EXERCISES

1. What types of contaminants are encountered in recycled paper?

2. What are some of the tools used to remove contaminants from paper?

3. What is the relationship between contaminant removal and fiber reject rate?

4. What are the two mechanisms of deinking in widespread use?

5. Deinked newsprint generally would not be bleached with sodium hypochlorite. Why not?

6. How do through-flow cleaners operate? What type of contaminants do they remove?

11
ENVIRONMENTAL IMPACT

11.1 INTRODUCTION

Pollution is a subjective term; therefore, pollution "is in the eye of the beholder". First what does one mean by a *pollutant*. Generally, one might say a pollutant is a material which has an adverse effect on the environment. What is an adverse effect? For example, much color in the effluent of a pulp and paper mill may not cause "harm" to the environment in terms of toxicity per se, but it is considered aesthetically unpleasing. Does this make it a pollutant? It doesn't matter; it is still undesirable.

Toxicity is also relative. At one time, a toxic material was considered one that in small amounts, perhaps less than a gram, was enough to kill half of the people who might be exposed to it. The LD_{50} is the required dose, often expressed as milligrams toxin per kilogram body weight, that is lethal to 50% of the population exposed to such a dose. Usually the population is made up of rats when the LD_{50} is determined in the laboratory. For example, sodium cyanide has an LD_{50} of 15 mg/kg (orally in rats), which means a 100 kg (220 lb) person would need about 1.5 g to cause death, presuming that a human reacts proportionally to a rat. Sodium chloride, or table salt, has an LD_{50} of 3.75 g/kg (orally in rats), which means our 100 kg person would need to ingest 375 g of salt (over three-fourths of a pound) to kill him. Table 11-1 shows some compounds with their LD_{50} doses.

Some people (read mercenaries who are hired to have a particular view) say "the dose makes the poison" and would argue sodium chloride (a nutrient) and sodium cyanide are both toxic! While literally true, this is not grounds for discharging toxins into the environment.

More recently, government organizations are considering exposure to a material over a lifetime that causes more than a single cancer in one million people as an unacceptable level of toxicity. For example, let us say an individual consumes six ounces of fish caught from a lake every day for 20 years. If it is estimated that doing so would cause

Table 11-1. LD_{50} doses in mice or rats of some materials[1].

Chemical	LD_{50} mg/kg	LD_{50} g/100 kg
Na cyanide, oral	15	1.5
NaCl, oral	3750	375
TCDD, dioxin, oral	0.02	
arsenic acid, i.v.	6	0.6
2,4,5 T herbicide	390	39
phenol, human		1-25
nicotine, i.v.	0.3	0.03
caffeine, oral	130	13

[1]Data from the *Merck Index*, 10th Edition.

two or more people per million people to get cancer from this water (or more precisely, the contaminants in that water), it is an unacceptable risk. It is absurd that decisions are made on the basis of such calculations. Dioxin discharges in water effluents at the parts per quadrillion or even parts per quintillion basis have become a major concern to government agencies, and therefore the industry, in the late 1980s. There are numerous difficulties with such regulations.

First, long term toxicity is often determined in animal studies at levels many orders of magnitude higher than actual exposure would be. If one is looking at cancer rates of one in a million, over 100 million rats would have to be used (unless toxicity is extrapolated from higher doses). Clearly this is not possible. More likely the researcher is using 100-1000 rats and an exposure over 1000 times the expected ambient level.

Second, do the results on rats at high levels correspond to anything meaningful in humans at low levels. The fact is that these estimates must be considered accurate within only a few orders of magnitude. Anyone who tells you they can predict with certainty an exact level of a chemical which will cause one cancer in a million people exposed to this chemical over a lifetime at low levels is not being reasonable. Science is simply not that precise by any means. For example, if two people each smoke 40 cigarettes per day for 40 years, about one million cigarettes are smoked. We might expect one of these people to develop cancer. In other words about one million cigarettes might induce cancer in a person. Does that mean a person who smokes a single cigarette in his life has a one in a million chance of getting cancer from it? We will never know until 25 million sets of identical twins are willing to subject themselves to a life-long experiment where every detail of their lives are controlled. However, it is known that the risk of cancer in people who quit smoking decreases significantly and after many years approaches the risk level of people who have never smoked, compared to their colleagues who continue to smoke. This indicates that a certain levels of toxins are required to induce cancer.

In the case of a compound like dioxin which was found to be highly toxic in laboratory studies, only later were studies done on humans who have been exposed to it through industrial exposure over many years, accidents, etc. Even these studies are often done on small sample sizes by people overly eager to advance their careers by being able to report something "controversial". By the time responsible, long term studies of reasonably large populations are finished, the damage has long been done, such as in the overly high estimated toxicity of dioxin.

Third, some chemicals behave much differently in rats than in humans. The chemistry of cancer is much more complex than the chemistry of cyanide ion, which binds to hemoglobin and prevents oxygen exchange by the blood where one can expect the same result in a rat as in a human as in any other animal with hemoglobin. But that is certainly not true with long term carcinogens, especially at very low rates of cancer mortality.

Fourth, the fact is that people in the medical community tend to be aggressive; that is a factor in the selection factor. Many studies get published with statistical analysis, but statistically insignificant results. Let me explain. If one correlates 20 different would-be "dependent" variables to one

independent variable and finds there is a 95% chance that one of these dependent variables correlates to the independent variable, does this mean anything? The answer is probably not since there is a 5% chance that the correlation is just due to chance. Since you started with 20 variables the chances are that one of these 20 (which is 5%) is bound to show a correlation just by chance at the 95% confidence level. Stated another way, one out of 20 researchers can expect to observe a correlation at the 95% confidence level that is merely due to chance for each correlation tested. And these are the studies that use statistical analysis! Mark Twain is quoted as saying "there are lies, damn lies, and statistics!"

In the news media, which should not be construed as scientifically (or otherwise) meaningful, one week coffee (caffeine) increases risk of heart attach and cancer, and one week it does not increase your risk. Along these lines, an article in the *New England Journal of Medicine* (Fingerhut, 1991) indicates that earlier studies indicating the toxicity of dioxin overstated the problem due to small sample sizes and other reasons.

The EPA level of permissible dioxin emission is 13 parts per quintillion in 1991; yet it is undetectable below 10 parts per quadrillion (10,000 parts per quintillion). Do not shrug off these incredibly small numbers without thinking about them. One part in a quadrillion is only this much: 0.000000000000001! An olympic-sized pool might have on the order of 250,000 gallons of water or 1 billion grams. The level of arsenic in drinking water is on the order of 0.01 parts per million corresponding to 10 grams in the pool. The measurable total amount of dioxin in the pool corresponds to 0.00001 g; the proposed regulation would be 0.000000013 g, 0.000013 mg, or 0.013 μg in the entire pool. (It would take more than 10 lifetimes for a person to drink the water in this pool.) The area of low-level, long-term toxicity is all very new ground. It is difficult to know what rules, regulations, and technologies will come out of all of this. Certainly, however, it must be handled in a more rational manner in the future.

11.2 WATER POLLUTION

It is useful to trace some of the important U.S. Federal legislation in the area of water. The first federal law was the *Water Pollution Control Act of 1948*, which dealt with conventional forms of water pollution (BOD, TSS, and pH), such as municipal wastes. The *Water Pollution Control Act Amendments* of 1956 and 1965 funded municipalities to build sewage treatment plants. The *Federal Water Pollution Control Act of 1972* increased grants to municipalities for sewage treatment plants, established the *Environmental Protection Agency (EPA)* as the organization to approve discharge permits, and set deadlines for pollution control. The *Clean Water Act of 1977* clarified conventional versus toxic water pollutants and postponed some of the deadlines of the 1972 legislation. The *Water Quality Act of 1987* (or the *Clean Water Act*) continued federal grants to municipalities, established a program for toxic pollutants, postponed some deadlines, and set new requirements for non-point sources such as storm water runoff for industrial and urban sites.

Undesirable characteristics of pulp and paper mill effluents include:

1. The presence of a food source for microorganisms which would deplete oxygen in rivers if not at least partially removed (BOD).
2. Color discharges which detract from the appearance of a river, lake, etc.
3. Suspended solids which make water appear murky.
4. Other materials such as fillers, acids, bases, and sludges, which may be sent to the sewer during mill upsets.

The severity of the problem depends on the type of mill. Mills producing bleached kraft pulp discharge over 30,000 gallons per ton of pulp. The digester and evaporator condensates of kraft mills have BOD producing compounds that cause water pollution and odorous compounds that cause air pollution. Mechanical pulp mills discharge about 5,000 to 7,000 gallons of water per ton of pulp. In TMP mills the principal pollutants are derived from wood extractives including resin acids. Resin acids may be as high as 50-200 mg/L in untreated effluent, but may be decreased to 0.5 mg/L with a high level of secondary treatment. In the U.S., mill effluents are typically diluted at least 1:100 into the receiving body of water.

Because of the importance of oxygen in water in the environment, the solubility of oxygen in water over air (0.2098 atm O_2/atm pressure) with and without the effect of water vapor displacement is shown in Fig. 11-1; for example, the vapor pressure of oxygen at atmospheric pressure in the atmosphere over water at 100°C (212°F) is 0 since the atmosphere is all steam.

11.3 WATER QUALITY TESTS

Biological oxygen demand, BOD

BOD is the amount of oxygen necessary for bacteria to consume the organic material in water. Organic material that is discharged into natural waters causes a rapid increase in the growth of microorganisms that deplete the oxygen required for other aquatic life. BOD is determined in the laboratory by measuring the oxygen depletion of a diluted aqueous sample incubated with microorganisms. Oxygen is measured with a probe much like when determining a pH. BOD_5 is the BOD determined after 5 days incubation.

Regulatory agencies in North America decrease the allowable levels of BOD discharge in summer months for most mills. There are two important reasons for this. First, at this time of year, the water is warmer than other times of the year and the oxygen solubility is decreased. Second, the amount of water in most rivers and lakes is much lower in summer than other times of the year.

Chemical oxygen demand, COD

Chemical oxygen demand is a measure of the components that can be oxidized in a chemical reaction. The amount of sodium permanganate or sodium dichromate as a chemical oxidant indicates the oxygen demand. This procedure can be done quickly and gives an estimate of the BOD, but, because it does not involve a biological system, it is only loosely correlated to BOD_5.

Total suspended solids, TSS

Total suspended solids are all of the materials in water that are not soluble. They contribute to the turbidity (cloudiness or murkiness) of water according to their amount and size distribution.

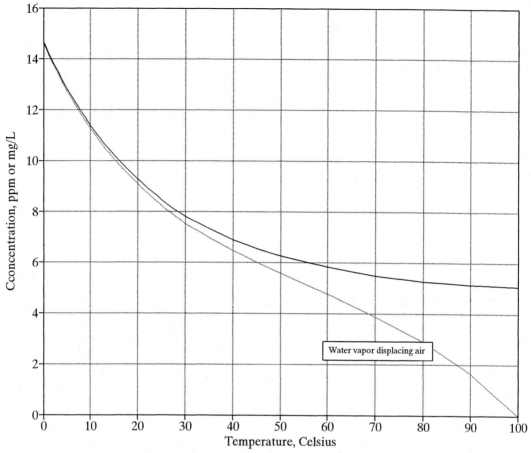

Fig. 11-1. The solubility of O$_2$ in water. The upper curve is a constant partial pressure of 0.2098 atm. The lower curve has partial pressure at 0.2098 atm minus oxygen displaced by water vapor.

Total organically bound chlorine, TOCl

TOCl is a measure of chlorine in organic material (chlorinated organic compounds), which are considered to be toxic pollutants, although the toxicity varies tremendously among the millions of possible compounds. TOCl can be decreased by improving delignification during pulping, proper washing of the pulp before bleaching, using substitutes for chlorine, avoiding over-chlorination (by efficient mixing of chlorine added to pulp or using several smaller additions of chlorine to avoid localized high chlorine concentrations), or effectively removing TOCl from effluents. For example, TOCl decreases linearly with chlorine dioxide substitution of chlorine in bleaching. Dioxin production decreases if chlorine is added in small amounts during bleaching; this effect is probably a result of decreasing over-chlorination.

Color units

Color in water is a measure of the absorbed light, by definition, at 465 nm wavelength. The color of waste water is measured using color units; 500 color units corresponds to 0.14 absorbance (70% transmittance, Section 14.6 explains this in more detail) at 465 nm. Allowable color discharge (of nontoxic materials) should take into account local geography. In some areas many of the rivers tend to be muddy to begin with, so color discharge should not be as much of a concern as in other areas where water is very clear.

Dioxins

Dioxins deserve special mention only because of the notoriety they have received in the media. This is an example of the media creating the news in terms dioxin and the pulp and paper industry.

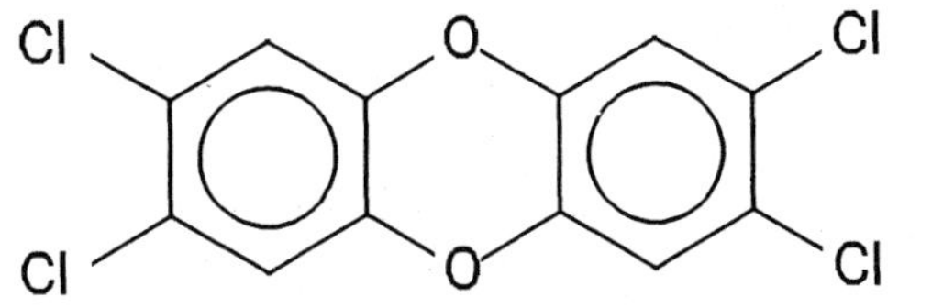

Fig. 11-2. The structure of the dioxin 2,3,7,8-TCDD.

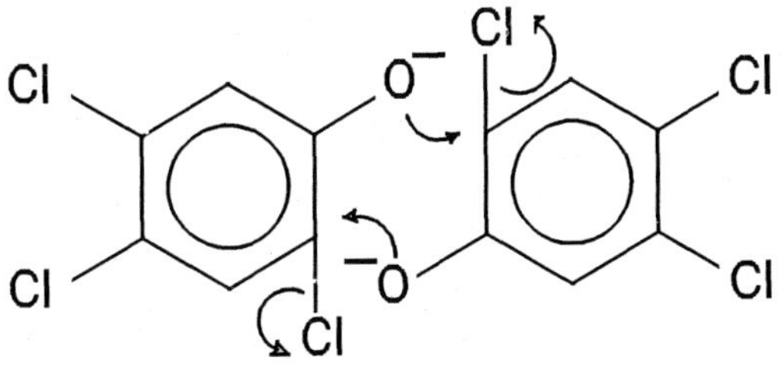

Fig. 11-3. Formation of 2,3,7,8-TCDD from two molecules of trichlorophenol.

Dioxins are chlorinated derivatives of dibenzo-*p*-dioxin. Indeed, there are 75 different dioxins with one to eight chlorine molecules differing in toxicity and other properties. By far, the most toxic dioxin is 2,3,7,8-tetrachlorodibenzo-*p*-dioxin (2,3,7,8-TCDD); it is this compound that is called dioxin, although the term is misleading as dioxins are a family of related compounds. Fig. 11-2 shows the structure of 2,3,7,8-TCDD.

Dioxins occur naturally and as by-products of man's activities. They are not made deliberately. The production of trichlorophenol and 2,4,5 T involves side reactions leading to undesired contaminants. For example, trichlorophenol is the precursor for the production of the herbicide 2,4,5 T (2,4,5-trichlorophenoxyacetic acid). (This herbicide was used in the defoliant *Agent Orange* by the U.S. during the Vietnam war and was implicated as the cause of symptoms described by veterans.) Trichlorophenol is reacted under alkaline conditions with monochloroacetic acid to make this herbicide. Unfortunately this allows the side reaction (Fig. 11-3) to occur, making significant amounts of 2,3,7,8-TCDD as an impurity (on the order of one to several parts per million). This has resulted in a decrease in the use of 2,4,5-T and 2,4,5-trichlorophenol in the U.S. Incineration of municipal wastes is one of the largest human-made sources of dioxins, but this tends to

be octachlorodioxin, which is millions of times less toxic than 2,3,7,8-TCDD. Forest fires are a large source of natural dioxins.

In the pulp and paper industry most dioxins are formed by the action of chlorine during bleaching on lignin. Processes not using elemental chlorine do not produce measurable levels of dioxin. Certain impurities associated with additives such as defoamers have also been implicated as dioxin precursors. When identified, these materials are removed from the process. Many mills have substituted part of the chlorine with chlorine dioxide to reduce dioxin formation. Cooking to lower kappa numbers, using oxygen delignification, and other techniques have further decreased the amount of dioxins produced at bleached pulp mills. There is much evidence that the precursors to formation of dioxin and chlorinated dibenzofurans in chlorine bleaching are the unchlorinated compounds dibenzo-*p*-dioxin and dibenzofuran, which means these two compounds may be present in the lignin before chlorine bleaching or formed during bleaching.

Presently dioxin can be detected in water at the level of 13 parts per quadrillion (13 parts in 10^{15}). This is no small feat and each sample costs about $1000 to process. The ability to measure such small amounts of materials is a tribute to the highly selective and sensitive instrumentation of modern physics. Yet many U.S. states have imposed regulations requiring diluted mill effluents to be below 13 parts per quintillion (13 parts in 10^{18}) of dioxin.

11.4 AQUEOUS EFFLUENT TREATMENTS

Pretreatment

Water is pretreated by screening with coarse filters, grit removal, and pH adjustment. The adjustment of pH is made by metering effluents from various parts of the mill together. The acid line from the chlorination stage raises the pH of other lines that tend to be alkaline. Washed grits from kraft liquor recovery may be ground and used to raise the pH.

Primary treatment, primary clarifiers

In primary treatment, solids are removed from water by allowing them to settle from the water. Primary clarifiers came into widespread use in the pulp and paper industry in the late 1950s in the U.S. A relatively coarse screen may be used to remove large materials that might otherwise clog pumps and other equipment. The most commonly used device is the mechanical clarifier. This is a circular, concrete tank with a rotating sludge scraper mechanism which rakes the sludge to the center sump of the tapered bottom. The clarified waste water is removed at the outer edge of the tank. Floating debris may be collected by a surface skimmer. The sludge is collected and dewatered. It is then disposed of in a landfill. Occasionally the sludge is incinerated or spread on agricultural fields. Some mills use settling ponds for primary clarification. The ponds must be dredged periodically (perhaps once each year).

A second method of solids removal employs air flotation where fine air bubbles attach themselves to suspended solids and cause them to float. The floating layer is then skimmed. The method is very effective; however, the method is also expensive. This method is used to reuse effluents from the processing of recycled fiber with deinking operations and was discussed in Chapter 10. In most cases effluents then undergo a secondary treatment process.

Secondary treatment, biological treatment

Secondary treatment came into widespread use by the U.S. pulp and paper industry during the 1960s. Secondary treatment involves reaction of the effluent (after primary treatment) with oxygen and microorganisms (bacteria and fungi) to remove oxygen-consuming materials. The effluent is prepared by pH balancing and addition of nutrients. In principle, microorganisms use the food source before the effluent is discharged; otherwise the rapid growth of microorganisms in the body of water in which the effluent is discharged leads to rapid oxygen depletion and the death of other aquatic life. The BOD is typically reduced by 80-90% during secondary treatment. The "toxicity" of the effluent is also decreased considerably, although this is much more subjective and difficult to define let alone measure. Secondary treatment is usually accomplished by oxidation basins, aerated stabilization basins, or the activated sludge process. The words *basin, pond*, and *lagoon* are used to describe a body of water which is usually three to five feet deep. This is a pond constructed so sunlight, microorganisms, and oxygen interact to break down the organic materials.

The *oxidation basin* is merely a treatment pond that relies on natural aeration from the atmosphere to supply oxygen. This is suited to mills with low production, processes producing relatively low amounts of BOD, and large areas for treatment ponds. Long retention times are required.

Aerated stabilization basins, ASB, are the most common method used (Plate 40). These are similar to oxidation basins except that oxygen transfer into the water is improved by use of aerators. This greatly improves the efficiency of the system. Retention times for treatment are typically 6-14 days. Periodic dredging (on the order of once in several years) is required to remove the sludge that tends to build.

For maximum growth of the organisms and BOD reduction, the nitrogen to phosphorus ratio should be about 5:1 in the nutrients added prior to treatment (assuming neither of these are present already), and the ammonia level should be at least 0.5 ppm in the final effluent discharge. In some mills energy reductions have led to decreased effluent temperatures that have decreased microorganism growth, especially in winter. Increasing the retention time of the effluent may help this condition. It is always important to make sure the ASB is not "short circuited" so that some parts of the basin are not being fully used and the retention time is not what one would calculate based on the size of the basin. High levels of dissolved oxygen and nutrients in some parts of the basin are indicators that this may be the case. The use of baffles or several basins in series are useful to prevent this condition. Proper operation of the ASB will mean fewer expensive dredgings will be required.

In the *activated sludge process* (Fig. 11-4), decomposing bacteria are purposely added along with agitation to dissolve oxygen. The microorganisms are then removed in a secondary clarifier. Some of this supplies the source of additional bacteria. The remaining sludge represents a considerable disposal problem and is either landfilled or sometimes dried and burned. This method is relatively expensive and generally used where space is limited or where BOD$_5$ removal above 90% is required. This method works best with a constant feed source of fairly highly concentrated BOD. Retention times of 6-8 hours are typical for secondary fiber effluent.

Rotating biological filters or *rotary disks* are occasionally used for areas with space limitations. The partially submerged disks provide surfaces for the microorganisms to prosper with a high surface area for oxygen exchange. These systems are very efficient but require high capital investment and operating costs. They look a bit like large disk filters that are used to thicken or recover stock.

Tertiary treatments

Tertiary treatments are advanced treatments following secondary treatment. They are often considered for color removal, since primary and secondary treatments are not particularly effective in this area. Secondary treatments may remove on the order of 50% of the color. Much research has been done on decoloring mill effluents, but very little is done commercially. In bleached kraft mills, most of the color comes from the first alkali extraction and the chlorination stages. A few mills decolorize these streams specifically by ultrafiltration or sedimentation with polymers. Most mills do not practice decolorization. Oxygen delignification offers the potential for reducing color emission by allowing the first bleaching stream to be used in the brown stock washers so it ultimately goes to the recovery boiler (if a mill is not recovery limited).

Ultrafiltration uses a semipermeable membrane and reverse osmosis. The effluent is forced through the membrane under high pressure while large molecules remain behind. Flocculation with alum removes much material but creates a large sludge-disposal problem. Massive lime addition is similar to flocculation with alum. Polymers can be used for flocculation but are very expensive to use. Carbon adsorption can be used for decolorization, but it is also very expensive.

Fig. 11-4. An activated sludge process with retention time of 48 hours. One of the sludge presses is shown to the right and the clarifier is in the middle.

Fate of paper machine waste water

Water from the paper machine must be recycled to recover fiber, to reduce treatment of fresh water, and to reduce the pollution load. Much of the fiber is recovered in *disk filter save alls* such as that shown in Fig. 10-17. This also cleans the water for reuse. Thus it may take 30,000 gallons of water to make a ton of paper, but most of this water is recycled water. The consumption of water may be as low as 1000 gallons per ton of paper.

11.5 AIR POLLUTION

Materials contributing to air pollution include particulate matter from the boilers, lime kiln, and smelt dissolving tank; gaseous combustion pollutants such as carbon monoxide, nitrogen oxides, and volatile organics from power boilers, recovery furnaces, and the lime kiln; odor consisting of reduced sulfur compounds in the kraft process arising from the digesters, evaporators, recovery boiler, smelt dissolving tank, and lime kiln; sulfur dioxide from the recovery furnace and boilers using fuel containing sulfur; and volatile organic chemicals from miscellaneous sources. In the sulfite process the primary pollutants are organics, oxides of sulfur, and particulates.

11.6 AIR QUALITY TESTS AND CONTROL

Total reduced sulfur, TRS

Reduced sulfur compounds include hydrogen sulfide (H_2S), the mercaptans, including methanethiol (CH_3SH, alias methyl mercaptan or methyl sulfhydrate), methyl sulfide (CH_3SCH_3, alias dimethylsulfide), dimethyl disulfide (CH_3SSCH_3), and other compounds. They are formed during kraft pulping (Fig. 11-5) by reaction of sulfides with methoxy groups of lignin via nucleophilic substitution reactions.

At low concentrations, TRS is more of a nuisance than a serious health hazard. Hydrogen sulfide and organic sulfides are detectable at a concentration of a few parts per billion by the olfactory sense, so air pollution is always a concern. Reduced sulfides originate from the black liquor of the kraft process. Sources of odor pollution from the kraft process in decreasing

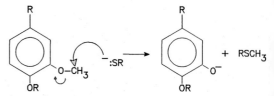

Fig. 11-5. The formation of mercaptans and organic sulfides.

order of importance are 1) black liquor combustion, 2) black liquor evaporation, and 3) the digestion process. The EPA limits emission for NSPS lime kilns to below 8 ppm in new equipment. High emission levels in these new systems have been traced by NCASI to 1) high soluble sulfides in particulate control scrubber solutions, 2) low fresh water addition rates at the precoat filter, and 3) changes in combustion conditions due to firing of noncondensible gases in the lime kiln. About 0.1-0.4 kg of TRS is emitted per dry ton of pulp (0.2-0.8 lb/ton) at 5 ppm in the recovery boiler flue gases. The total TRS production has gone from 3 kg/t (6.5 lb/ton) pulp in 1960 to less than 0.5 kg/t (1 lb/ton) in the early 1990s.

Oxidation of most of the sulfides before liquor concentration, *weak liquor oxidation*, was first practiced in 1956 in the Soviet Union. Introduction of oxygen at 7.5 g/L black liquor at 70°C (158°F) with a retention time of about one minute converts most of the sulfides to nonvolatile thiosulfates. The concentrations of the mercaptans and methyl sulfides are decreased as well. Partial oxidation is required with direct contact evaporators but not with concentrators. As the solids content increases, the vapor pressure of the sulfides increases. The pH of the black liquor is another parameter which determines how much sulfide will be liberated in the atmosphere. A pH of 12 or higher decreases hydrogen sulfide and methyl mercaptan emission by keeping these compounds in their salt forms, decreases lignin precipitation as a cause of evaporator plugging, and decreases scaling. This pH is maintained in indirect contact evaporation, but not in direct contact evaporators. In the case of direct contact evaporation, residual NaOH of the black liquor reacts with CO_2 to produce Na_2CO_3; this lowers the pH by several units. Black liquor oxidation does cause heating of the black liquor and increas-

Table 11-2. The effect of black liquor oxidation on its chemical composition (EPA, 1976).

Location	Concentration, (g/L)				Account, %
	Na_2S	$Na_2S_2O_3$	Na_2SO_4	Total	
Reactor inlet	4.56	0.80	0.53	5.89	100.0
Reactor outlet	0.45	4.35	0.70	5.50	93.7
Storage outlet	0.19	4.61	0.63	5.43	92.8

es the temperature up to 10°C. A representative change in liquor composition is shown in Table 11-2. The ratio of sulfide converted to thiosulfate depends on the oxidation method, the purity of the oxygen, and the amount of oxygen used.

Final odor treatment of gases with alkaline scrubbing liquids can remove compounds like H_2S, $HSCH_3$, SO_2 (which forms H_2SO_3), and sulfur trioxide (which forms H_2SO_4) that have ionizable protons. For example $HSCH_3 + NaOH$ forms $NaSCH_3$, which is a nonvolatile salt. Other organic sulfur compounds such as $(CH_3S)_2$ or $(CH_3)_2S$ cannot be removed by alkali scrubbing as they do not have ionizable protons and do not form nonvolatile salts.

Nuisance odors of hydrogen sulfide result from small amounts of H_2S in aqueous effluents that sometimes can be overcome by the addition of iron ions to precipitate the H_2S as FeS, which has a very low water solubility.

Sulfur dioxide, SO₂

Sulfur dioxide and, to a much smaller extent, sulfur trioxide are formed by oxidation of sulfur compounds in the recovery boiler and the burning of sulfur-containing fuels in power boilers. When properly operated, modern recovery boilers should have emissions below 100 parts per million. Sulfur emissions from power boilers are controlled by using fuels of low sulfur content. Some sulfur dioxide is emitted by the lime kiln and smelt dissolving tank.

Nitrogen oxides (NOₓ)

The majority of the oxides of sulfur and nitrogen originate under the oxidative conditions of the lime kiln and kraft recovery furnace. Oxides of nitrogen can form in any combustion process, but particularly during high temperature combus-

tion. The principal oxides of nitrogen formed during combustion processes are nitric oxide (NO) and nitrogen dioxide (NO_2), with the latter accounting for 5-10% of the total NO_x by volume. These form by the reactions $N_2 + O_2 \rightleftarrows 2NO$ and $2NO + O_2 \rightleftarrows 2NO_2$. Table 11-3 shows the production of nitric oxide NO as a function of combustion temperature. Typical NO_x emissions by volume are 32 ppm for the kraft recovery boiler and 200 ppm for the rotary lime kiln. The recovery boiler produces about 0.5 kg (1.2 lb) of NO_x per ton black liquor solids. The lime kiln emissions are probably relatively low for the high combustion temperature because the gases are slowly cooled before exiting the kiln allowing the equilibrium to shift back to N_2. Power boilers emit from 100-450 ppm NO_x by volume depending on the temperature of combustion and percentage of excess air.

Table 11-3. Flame temperature vs. nitric oxide equilibrium concentration and reaction time.*

Gas temperature of combustion↓		NO [1] Conc., (ppm, vol.)	Reaction time, (s)
°C	°F		
2430	4400	19,000	0.004
1930	3500	14,000	0.090
1760	3200	4,000	0.700
1430	2600	500	21.0
1090	2000	10	162.0

[1]Reactant gas of 77% N_2, 15% O_2, and 8% inert gases. Ref. 1 with data from Ref. 3.

Noncondensible Gas, NCG

Noncondensible gases consist of TRS and smaller amounts of other volatile organic chemicals such as methanol, terpenes, phenols, and hydrocarbons liberated from the wood. Liberated organic chemicals may directly be pollutants or may form pollutants as they undergo photochemical reactions (formation of smog).

Because pulp mills must run at full capacity, and because the limiting factor in batch digesting is the condensing or heat recovery unit, often pollutants emitted during surges of gases formed during blowing are not fully collected in the condensers. For example, during blowing of a $200 \ m^3$ digester from 78 psig to atmospheric pressure, 20 tons of pulp and 20 tons of steam $(34,000 \ m^3)$ are produced in about 20 minutes, a huge amount. There is also loss of recoverable heat if the condensing system is not adequate to meet these needs. Factors which may compromise the necessary condensing rate include large digesters, short blow periods, high blow pressures, dirty or fouled condensers, and high water temperatures. Mills with pollution problems caused by inadequate condensation of blow gases can limit the problem by the usual methods for increasing condensation efficiency, including increasing pump capacity for the condensers, increasing the heat exchanger surface area, adding a secondary condenser, and so forth. Because the time scale is longer for continuous pulping processes, condensation is usually not a critical problem. Most of the TRS and much of the BOD can be removed from the condensates by steam stripping of the condensates or other treatments.

In some cases noncondensible gases are sent to the primary air inlet of the lime kiln [with at least 50:1 dilution and an inlet velocity of over 9 m/s (30 ft/s) to prevent flameout] where they are oxidized to sulfite and form the calcium salt. This is not possible in a highly closed recovery system.

Particulate matter

Particulate matter originates primarily in the kraft recovery furnace, the smelt dissolving tank, and the lime kiln. The first of these is the largest potential source, on the order of 1 kg/t (2 lb/ton) pulp. Sodium sulfate is the major constituent of particulates, with smaller amounts of sodium car-bonate, sodium chloride, and other salts. Particulates vary in size from $0.1 \ \mu m$ to 1 mm for uncontrolled emissions to 0.1 to $10 \ \mu m$ for highly controlled emissions. Two major types of particulate control devices are the electrostatic precipitators (ESP, Section 4.4) and venturi-type evaporator-scrubbers (Fig. 4-18).

Operation of the kraft recovery boiler

Proper design and operation of the kraft recovery boiler is necessary to limit air pollution (Van Heiningen, 1992). The upward flow of gases should be uniform, while combustion air mixes completely with combustion gases. Computational fluid dynamics (CFD) helps achieve this goal.

Continuous monitoring of emissions should be done for SO_2 (below 150 ppm), NO_x (below 80 ppm), CO (50-200 ppm), and TRS (below 5 ppm). Stack opacity should be below 20%.

Firing liquor above 75% solids leads to high emissions of particulates and high levels of NO_x caused by high combustion temperatures. Combustion temperatures can be lowered by using less preheating of the combustion air intake.

Attempting to limit NO_x emissions from the recovery boiler by limiting the excess oxygen leads to a large increase in carbon monoxide and TRS emissions, so CO should be monitored also. NO_x emissions can be decreased, sometimes, by increasing tertiary air and decreasing air.

SO_2 emission is controlled by generating sodium fume by reducing secondary air to develop a large bed.

Gas analysis for pollution control and correlations between gas constituents

In order to control pollution, one has to measure the composition of various components of stack gases. The basis of measuring certain gases is given here. The following gases can be measured continuously, on-line to measure at the source of pollution (*point sampling*) as follows: oxygen is measured by paramagnetic torque, carbon dioxide and carbon monoxide are measured by infrared absorption, total hydrocarbons are measured by flame ionization (although different hydrocarbons, alcohols, etc. have different detector responses for a given amount of material), nitrogen oxides are measured by chemiluminescence, sulfur dioxide is measured by flame photometry, and sulfur dioxide and total reduced sulfur are measured by UV absorbance. Sulfur dioxide can be measured separately from the TRS if a citrate scrubber is used to remove TRS.

More sophisticated tests can be used to quantify individual components (i.e., of hydrocarbons and TRS) without interference or to measure low levels of pollutants downwind of the mill or from diffuse sources of pollution (*ambient sampling*), such as TRS emissions from aeration basins. For example, hydrocarbons and reduced sulfur compounds can be trapped from a portion of effluent and measured by gas chromatography.

There is some correlation among the individual constituents of gaseous effluents. For example, an elevated level of CO indicates a somewhat reducing environment and is expected to be accompanied by high levels of TRS from recovery boiler gases. Conversely, one might see a high level of TRS, CO, and hydrocarbons with only 4% excess oxygen, but a relatively low level of these materials with 5% excess oxygen, although this might mean a high level of SO_2 emissions.

11.7 SOLID WASTE DISPOSAL

Sludges that are produced at paper mills must be disposed. These include the sludge from the primary clarifiers (usually a high percentage of wood fiber and calcium deposits from the kraft recovery process), sludge removed from secondary treatment of effluent, and sludge from clarification of deinking mill effluent. Sludges can be thickened by use of screw presses (Fig. 11-6). Chemical additives (alum or polymers) are also used.

Usually these sludges are sent to landfills. About 20% are spread on agricultural fields, and some is burned. In 1991, after being sued by "pro-environmental" organizations, the EPA proposed a new set of rules that would affect mills that use chlorine bleaching and use land application of sludge. To summarize the rules, the concentration of dioxin and furan in sludges would have to be measured to insure that the soil concentrations would not exceed 10 ppt. Sludge application within 10 m (33 ft) of bodies of water will not be allowed. There are numerous other rules that apply to land application of sludges.

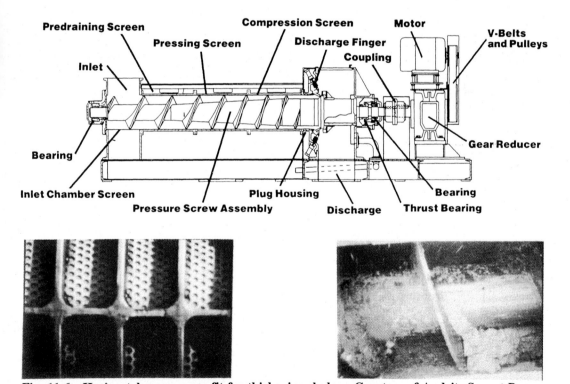

Fig. 11-6. Horizontal screw press fit for thickening sludge. Courtesy of Andritz Sprout-Bauer.

11.8 U.S. REGULATIONS

U.S. government regulations for effluent requirements can be fairly complex. The EPA sets maximum guidelines for the states. The states may impose stricter regulations and regulate additional items, but they cannot be less stringent than EPA regulations. It can be quite difficult to obtain all of the information one needs to determine compliance, learn how to apply for permits, find state regulations, etc. Fortunately there are some resources available that collect this information into sets of large volumes with loose-leaf bindings that allow sections to be replaced as they are updated. Each section is dated for the time of its printing. Unfortunately, the annual subscription rates for these resources is fairly high, making them fairly hard to find in libraries, but this resource provides important information.

The information from this section was obtained from the *BNA Reporter* (*Environmental Reporter*, Bureau of National Affairs, Inc., 1231 25th Street, Washington, D.C.). There are volumes on Federal Regulations, State Water Laws, and State Air Laws for many industries.

EPA regulations for pulp and paper effluents fall into 26 categories by type of mill, such as unbleached kraft and semi-chemical. BOD_5, TSS, and pH are covered by BPT (best practicable control technology) regulations, while pentachlorophenol and trichlorophenol are covered by BAT (best available technology economically achievable). The phenols occur from biocides that may be used in the various processes and will not be considered further. New mills must meet NSPS (new source performance standards) regulations. Generally speaking 1 day averages may not exceed about twice the 30 day average. The pH levels are reported as allowable ranges; no discharges outside these ranges is ever allowed. Table 11-4 gives some example effluent regulations for several types of mills from the January 30, 1987 listing of the *BNA Reporter*.

While the air quality standards (as of June 23, 1989) are not broken down into as many subsections as the effluent standards, they are slightly more complicated and detailed and have

Table 11-4. Thirty day average of EPA allowable effluent discharges for several types of mills. Values are kg per 1000 kg air dry pulp or pH range.

Type of mill	BOD_5	TSS	pH
BPT Unbleached kraft	2.8	6.0	6-9
NSPS linerboard	1.8	3.0	5-9
BPT Semi-chemical	4.4	5.5	6-9
Dissolving kraft	12.3	20.0	5-9

many exceptions. Units are reported in g/m^3 or grains/ft^3; note that 1 g/m^3 = 0.43700 grains/ft^3.

Performance standards for kraft pulp mills include particulate matter and TRS. Particulate matter must be below 0.10 g/dscm (0.044 gr/dscf) corrected to 8% oxygen with less than 35% opacity for the recovery boiler, below 0.1 g/kg black liquor solids for the smelt dissolving tank, and 0.15 g/dscm for gaseous fossil fuel (0.30 g/dscm for liquid fossil fuel) fired lime kilns corrected to 10% oxygen.

Total reduced sulfur (TRS) for most sources must be below 5 ppm by volume corrected to 10% oxygen. However, if the overall TRS emissions are below 5 g/MT (0.01 lb/ton) air dry pulp then point sources are allowed to be higher than 5 ppm. Cross recovery furnaces are to be below 25 ppm by volume at 8% oxygen. The smelt dissolving tank limit is 16 g/MT (0.033 lb/ton) black liquor solids, the lime kiln limit is 8 ppm by volume at 10% oxygen, and the recovery boiler limit is 5 ppm by volume at 8% oxygen. The air quality standards section also includes information on monitoring emissions and operations. To obtain the corrected pollutant concentration at the corrected oxygen concentration from the measured oxygen concentration, use the following formula:

$$C_{corrected} = C_{measured} \times (21 - X/21 - Y)$$

where X is 8% or 10%

   Y is the 12 hour average volume O_2 concentration

11.9 ANNOTATED BIBLIOGRAPHY

General

1. U.S. Environmental Protection Agency, *Environmental Pollution Control Pulp and Paper Industry, Part I, Air*. EPA-625/7-76-001 (1976). This is a useful reference for amounts and sources of various air pollutants in the pulp and paper industry, especially from kraft pulping. Control measures such as partial liquor oxidation before direct contact evaporation are described with emissions levels given for operations with and without such systems.

Weak black liquor oxidation

2. Galeano, S.F. and C.D. Amsden, Oxidation of weak black liquor with molecular oxygen, *Tappi J.* 53(11):2142-2146(1970).

Nitrogen oxide formation

3. Ermenc, E.D., Controlling nitric oxide emissions, *Chem. Eng.* 77(12):103-05(Jun 1, 1970).

Kraft recovery boiler operation

4. Van Heiningen, A.R.P., Chemical recovery conference highlights CFD modelling and black liquor gasification, *Pulp Paper Can.* 93(8):9-16(1992). The highlights of the Seattle, Washington conference of 67 technical presentations.

Electrostatic precipitators

5. White, H., *Industrial Electrostatic Precipitation*, Addison-Wesley, 1963. The author describes the industrial operation and theory of electrostatic precipitators.

Bleaching

6. Neilson, A.H., et al., The environmental fate of chlorophenolic constituents of bleachery effluents, *Tappi J.* 73(3):239-247(1990), 43 references.

Dioxin

7. Harris, F.E., Dioxins--an overview, *Tappi J.* 73(4):267-269(1990). This is a nontechnical overview about dioxins, their sources, the history of their measurement, and why we can never reach "zero concentration" of dioxins.

8. Fingerhut, M.A., Cancer mortality in workers exposed to 2,3,7,8-tetrachlorodibenzo-p-dioxin, *N. Engl. J. Med.* 324:212-218(1991). The results of a retrospective cohort study of 5172 workers at 12 plants in the U.S. that produced chemicals contaminated with TCDD does not confirm the high relative risks reported for many cancers in previous studies.

9. Bailer, III, J.C., How dangerous is dioxin?, *N. Engl. J. Med.* 324:260-262(1991). This is an editorial on the toxicity of dioxin with 9 references.

10. Luthe, C.E., P.E. Wrist, and R.M. Berry, An evaluation of the effectiveness of dioxins control strategies on organochlorine effluent discharge from the Canadian bleached chemical pulp industry, *Pulp & Paper Can.* 93(9):T239-T248(1992). This paper is an excellent introduction to dioxins in chemical pulp mills. It is a "must have in your library" article due to its very direct, practical approach.

Anaerobic treatment of mill wastewaters

11. Lee, Jr., J.W., D.L. Peterson, and A.R. Stickney, Anaerobic treatment of pulp and paper mill wastewaters, *Tappi Proceedings, 1989 Environmental Conference*, pp 473-496. An extensive article on this subject was given by CH2M Hill personnel. About 37 paper mills around the world were either using or were installing the process as of the article's printing.

EXERCISES

Water pollution

1. What makes water look cloudy?

2. What are four undesirable aspects of pulp and paper mill aqueous effluents?

3. Which of the following is the most difficult to remove from the water effluent of a bleached kraft pulp mill?

   a) fibers   b) BOD   c) color

4. A closed can of a carbonated beverage with contents at 25°C is much more apt to bubble out the top than one with contents at 5°C. Why is this so, and what does this have to do with water pollution?

5. Why are permissible levels of BOD emission often lower during hot months of the year?

6. How is color removed from mill effluent?

7. Why does the determination of BOD levels take several days?

8. If a mill discharges dioxin at 10 parts per quadrillion in 20 million gallons of effluent every day, what is the mass of dioxin discharged in one year?

Air pollution

9. Which of the following are removed by alkali scrubbing of gases and why? H_2S, CH_3SH, $(CH_3)_2S$, and CH_3SSCH_3? What effect does mild oxidation of black liquor prior to concentration have on this problem?

10. Where is methyl sulfide produced during chemical pulping? Is it a desirable product?

11. Where do nitrogen oxides originate?

12. What are ambient and point sampling?

13. Describe the tradeoffs involved in controlling air pollutants from the kraft recovery boiler.

12

METRIC AND ENGLISH UNITS AND UNIT ANALYSIS

12.1 METRIC UNITS

The metric system (called SI from the French for système internationale) is the common system used in the scientific literature to describe measurements. All measurements can be derived from the seven base units and two supplementary units shown in Table 12-1. Since the U.S. pulp and paper industry (unfortunately!) uses English measurements extensively, some of the base SI units are given in terms of their approximate English system equivalents. Because only seven fundamental units are used in the metric system, it is extremely easy to use in problem-solving. Even when going from English to English units, it is sometimes helpful to use metric units as intermediates.

The multipliers that are used as prefixes before the units are given in Table 12-2. Here the English equivalents of the multipliers greater than one are given.

Complex measurements are combinations of the base units. For example, area is derived from length and is actually length^2 such as m^2; volume is also derived from length and is length^3 such as m^3; density is derived from mass and volume with units such as kg/m^3 or g/cm^3. Table 12-3 gives the list of derived SI units which have special names.

Table 12-4 gives some values of the universal constants that are used in various aspects of engineering and science. Table 12-5 gives some additional metric based units with special names that have traditional use.

In the following section, the numerous English units will be presented with conversion factors to metric units. Also, the relationships between many of the English units are presented.

12.2 ENGLISH AND METRIC UNITS

Tables 12-6 and 12-7 are designed to define most commonly encountered English units in pulp and paper without having unnecessary conversion factors for all of the possible derived units. For example, if one knows the conversion factor for cm to in. then cm^2 can be converted to in.^{2} and cm^3 to in.^{3} Similarly, gal/min can easily be converted to ml/s if one has a conversion factor for gallons to milliliters (or liters) and knows there are 60 seconds in a minute. The technique for this is shown in Section 12.3. Table 12-8 gives units for paper analysis. Table 12-9 gives DIN format sizes of paper.

Tables in this chapter include:

12-1.   Base units of the metric system.

12-2.   Prefix units with corresponding multipliers for SI units.

12-3.   Derived SI units with special symbols and names.

12-4.   Table of universal constants.

12-5.   Other metric units with special names.

12-6.   Fundamental and derived English units with special names and their conversion factors.

12-7.   Common compound units with conversion factors.

12-8.   Paper analysis conversion factors.

12-9.   DIN format sizes of paper.

Table 12-1 Base units of the metric system.

Base Quantity	SI Unit	Symbol	Equivalent English units
length	meter	m	39.3701 inches
mass	kilogram	kg	2.2046 pounds ($g = 1.00$)
time	second	s	second
temperature	kelvin	K	$K = (5/9)\cdot(^\circ F\text{-}32) + 273.15$
amount by number	mole	mol	$6.0220\cdot10^{23}$
electric current	ampere	A	$6.2422\cdot10^{18}$ electrons/s
luminous intensity	candela	cd	

Supplementary Units

plane angle	radian	rad	$360^\circ/2\pi = 57.296^\circ$ of a circle
solid angle	steradian	sr	$360^\circ/2\pi = 57.296^\circ$ of a sphere

Table 12-2. Prefix units with corresponding multipliers for SI units.

Name	Symbol	Multiplier	English Equivalent	Name	Symbol	Multiplier
deca	da	10		deci	d	0.1
hecto	h	100		centi	c	0.01
kilo	k	1000	thousand	milli	m	0.001
mega	M	10^6	million	micro	μ	10^{-6}
giga	G	10^9	billion[1]	nano	n	10^{-9}
tera	T	10^{12}	trillion[2]	pico	p	10^{-12}
peta	P	10^{15}	quadrillion	femto	f	10^{-15}
exa	E	10^{18}	quintillion	atto	a	10^{-18}

[1]British define a billion as a million millions or 10^{12}

[2]British define a trillion as a million billions or 10^{18}

Table 12-3. Derived SI units with special symbols and names.

Quantity	Name	Symbol	Base Units	Other Units
SI units with special symbols				
frequency	hertz	Hz	$(\text{cycle})\cdot s^{-1}$	
force	newton	N	$m\cdot kg\cdot s^{-2}$	
pressure; stress	pascal	Pa	$m^{-1}\cdot kg\cdot s^{-2}$	N/m^2
energy; work; heat	joule	J	$m^2\cdot kg\cdot s^{-2}$	$N\cdot m$
power, radiant flux	watt	W	$m^2\cdot kg\cdot s^{-3}$	J/s
electric charge	coulomb	C	$A\cdot s$	
electric potential	volt	V	$m^2\cdot kg\cdot s^{-3}\cdot A^{-1}$	W/A
electric capacitance	farad	F	$m^{-2}\cdot kg^{-1}\cdot s^4\cdot A^2$	C/V
electric resistance	ohm	Ω	$m^2\cdot kg\cdot s^{-3}\cdot A^{-2}$	V/A
electric conductance	siemens	S	$m^{-2}\cdot kg^{-1}\cdot s^3\cdot A^2$	A/V
magnetic flux	weber	Wb	$m^2\cdot kg\cdot s^{-2}\cdot A^{-1}$	$V\cdot s$
magnetic flux density	telsa	T	$kg\cdot s^{-2}\cdot A^{-1}$	Wb/m^2
inductance	henry	H	$m^2\cdot kg\cdot s^{-2}\cdot A^{-2}$	Wb/A
luminous flux	lumen	lm	$cd\cdot sr$	
luminous flux density	lux	lx	$m^{-2}\cdot cd\cdot sr$	lm/m^2
radionuclide activity	becquer	Bq	s^{-1}	
absorbed dose of ionizing radiation	gray	Gy	$m^2\cdot s^{-2}$	J/kg
Other derived SI units				
area	square meter	m^2		
volume	cubic meter	m^3		
speed, velocity	meter per second	m/s		
acceleration	meter per second squared	m/s^2		
wavenumber	per meter	m^{-1}		
concentration	molarity	mol/m^3		mol/L
density of mass	kilogram per cubic meter	kg/m^3		g/L
moment of force	newton meter	$N\cdot m$		
surface tension	newton per meter	N/m		
specific heat capacity	joule per kilogram kelvin	$J/(kg\cdot K)$		

Table 12-4. Table of universal constants.

Constant	Symbol	Value
Avogadro constant	N_o	$6.0220 \cdot 10^{23}$ mol^{-1}
Gas constant	R	8.314 J·mol^{-1}·K^{-1}
		0.08206 L·atm K^{-1} mol^{-1}
		62.37 L·Torr·K^{-1}·mol^{-1}
Faraday constant	F	96485 C·mol^{-1}
Planck constant	h	$6.6262 \cdot 10^{-34}$ J·s
Speed of light (vacuum)	c	$2.99792 \cdot 10^{8}$ m·s^{-1}
Acceleration, gravity	g	9.80665 m·s^{-2} (by definition)
Pi	π	3.141592654
e	e	2.718281828

Table 12-5. Other metric units with special names (not SI).

Quantity	Name	Symbol	Base Units	Other Units
CGS System of derived units (using base units of cm, g and s)				
force	dyne	dyn	cm·g·s^{-2}	10^{-5} N
energy	erg	erg	cm^2·g·s^{-2}	10^{-7} J
viscosity	poise	P	cm^{-1}·g·s^{-1}	
kinematic viscosity	stokes	St	cm·s^{-2}	
acceleration, gravity	gal	Gal	cm·s^{-2}	
Miscellaneous				
distance (obsolete)	micron	μ	μm	
distance (obsolete)	millimicron	mμ	nm	
distance (obsolete)	angstrom	A or Å	10^{-10} m	0.1 nm
area	hectare	ha	10,000 m^2	0.01 km^2
volume	liter[1]	l, L	0.001 m^3	1000 cm^3
mass, metric ton	tonne	t or MT	10^6 g	1000 kg
radionuclide decay	Curie	Ci	$3.7 \cdot 10^{10}$ Bq	
absorbed dose	rad	rd	0.01 Gy	
exposure	roentgen	R	2.58×10^{-4} C/kg	

[1]One liter of water weighs exactly 1 kg at 25°C; 1 ml is then 1 g.

Table 12-6. Fundamental and derived English units with special names and their conversion factors to other units. The reciprocal is listed for reverse conversions. (Less than five significant figures indicates exact to at least six significant figures.)

Unit	Abbrev.	Multiplier	Other Unit	Reciprocal
Dimensionless values				
dozen	doz	12		0.083333
score		20		0.05
gross		144		0.0069444
thousand	M	1000		0.001
Units of length				
mil	mil	0.001	inch	1000
caliber		0.01	inch	100
inch	in., "	2.54	cm	0.39370
foot	ft, '	30.48	cm	0.032808
		0.3048	m	3.2808
		12	inch	0.083333
yard	yd	0.91440	meter	1.0936
		3	foot	0.33333
		36	inch	0.027778
fathom	fath	6	foot	0.16667
rod	rod	16.5	foot	0.060606
chain	ch			
(Gunter's)		66	foot	0.015152
(Ramsden's)		100	foot	0.01
furlong	fur.	660	foot	0.0015152
mile	mile, mi			
(statute)		1.6093	km	0.62137
		5280	foot	0.00018939
(nautical)		1.852	km	0.53996
		6076.1	foot	0.00016458
league				
(statute or nautical)		3	mile	0.33333

Table 12-6. Fundamental and derived English units with special names and their conversion factors. Cont'd.

Unit	Abbrev.	Multiplier	Other Unit	Reciprocal
Force				
kip		4448.2	N	0.00022481
poundals		0.13826	N	7.2327
Units of area				
acre	acre	0.0015625	mi^2	640
		0.40469	hectare	2.4711
Units of volume				
teaspoon, U.S.	tsp	4.93	ml	0.20284
tablespoon	Tbs	14.79	ml	0.067613
ounce	oz			
U.S. fluid oz		0.029574	L	33.814
pint[1]		0.5	qt	2
quart	qt			
(U.S. liquid)		0.94635	L	1.0567
(U.S. dry)		1.1012	L	0.98794
(British)		1.1365	L	0.87988
gallon[1]	gal	4	qt	0.25
peck[1]	pk	8	qt	0.125
bushel[1]	bu	32	qt	0.03125
barrel	bbl	42 (petrol.)	gallon (U.S. liquid)	0.023810
Wood, volumetric				
board foot	fbm	0.083333	ft^3	12
cord, stacked wood	cd	128	ft^3	0.0078125
unit, usu. chips		2400	lb, dry	0.00041667
cunit, solid wood		100	ft^3	0.01
stere		1	m^3	1

[1] The British pint, gallon, peck, and bushel are made up of 0.5, 4, 8, and 32 British quarts, respectively, while the U.S. pint, gallon, peck, and bushel have the same number of U.S. dry quarts. One U.S. liquid gallon of water has a weight of 8.349 lb. There are 7.4805 gallons per ft^3, and 1 ft^3 of water weights 62.45 lb.

Table 12-6. Fundamental and derived English units with special names and their conversion factors. Cont'd.

Unit	Abbrev.	Multiplier	Other Unit	Reciprocal
Units of mass[2]				
grain		64.799	mg	0.015432
carat, metric	c	0.2	g	5
ounce	oz			
avoirdupois		28.350	g	0.035274
troy (for gold)		31.103	g	0.032151
pound	lb, #			
avoirdupois		453.59	g	0.0022046
ton (short ton)		2000	pound	0.0005
ton (long ton)		2240	pound	0.00044643
metric ton, tonne	t, MT	2204.6	pound	0.00045359
		1000	kg	0.001
Units of pressure				
atmosphere	atm	$1.0133 \cdot 10^5$	pascal	$9.8692 \cdot 10^{-6}$
		760	mm of Hg	0.0013157
		14.696	psi	0.068046
		33.899	feet water	0.029450
		29.921	in. Hg	0.033421
		1.0332	technical atm	0.96787
psi (lb/sq in)	psi	6894.8	pascal	0.0001450
Units of time				
minute	min	60	s	0.016667
hour	hr, h	60	min	0.016667
		3600	s	0.00027778
day		24	hour	0.041667
week	wk	7	day	0.14286
		168	hour	0.59524
year (solar)	yr	365.25	day	0.0027379
		$3.1558 \cdot 10^7$	s	$3.1688 \cdot 10^{-8}$

[2] Most English units of "mass" such as the pound (except the slug which really is mass) are really measures of weight at the earth's surface. The mass of an object is constant; the weight depends on the acceleration of gravity. The English units are commonly set equal to SI mass units, although this is not strictly correct.

Table 12-6. Fundamental and derived English units with special names and their conversion factors to other units. Cont'd.

Unit	Abbrev.	Multiplier	Other Unit	Reciprocal
Units of velocity				
knot	kn	1.852	km/h	0.53996
Units of energy				
calorie, Internat.	cal	4.1868	J	0.23885
(thermochemical)		4.184	J	0.23901
British thermal unit	Btu	1055.1	J	0.00094782
(International)		252.00	cal	0.0039683
		0.00029307	kWh	3412.1
kilowatt-hour	kWh	3.6	MJ	0.27778
Units of power				
horsepower	hp	746	J/s	0.0013405
		550	ft-lb/s	0.0018182
Angular values				
Circumference		2π	rad	0.15915
Degree	deg °	0.017453	rad	57.296
Minute	min '	0.016667	deg	60
Second	sec "	0.016667	min	60
Temperature intervals[1]				
Fahrenheit	°F	0.55556	°C or K	1.8
		1	Rankine	1
Celsius	°C	1	K	1

[1]Temperature scale conversions are as follows:

   °F = 9/5 °C + 32°

   °C = K - 273.15°

   Rankine = 9/5 K = °F + 491.67

Table 12-7   Common compound units with conversion factors.[1]

Compound Unit	Abbrev.	Multiplier	Other Unit	Reciprocal
Units of area				
inch^2	in.^{2}	6.4516	cm^2	0.15500
foot^2	ft^2	0.0929	m^2	10.764
mile^2	mi^2	2.5898	km^2	0.38613
Units of volume				
inch^3	in.^{3}	16.387	cm^3	0.061024
foot^3	ft^3	0.028317	m^3	35.314
Units of density or concentration				
lb/ft^3		0.016018	g/cm^3	62.428
lb/U.S. liq. gallon		0.11983	g/cm^3	8.3454
		119.83	g/l	0.0083454
Energy				
horsepower·hour	hp-hr	0.746	kWh	1.3405
horsepower·day	hp-day	17.904	kWh	0.055853
kilowatt-hour	kWh	3.6	MJ	0.27778
Specific energy				
Btu per pound	Btu/lb	2.3260	kJ/kg	0.42991
Pressure				
lb/sq in	psi	6894.8	pascal	0.00014504
Velocity, speed				
miles per hour	mph	0.44704	m/s	2.2369

[1]All of these conversion factors could be derived from information in the previous tables; they are presented here as a reference tool.

Table 12-8. Paper analysis conversion factors.

Compound Unit	Type	Multiplier	Other Unit	Reciprocal
Basis weight in pounds per ream of 500 sheets				
17 in. by 22 in.	printing	3.7597	g/m^2	0.26598
19 in. by 24 in.	blotting	3.0836	g/m^2	0.32429
20 in. by 26 in.	cover	2.7041	g/m^2	0.36981
22 in. by 28 in.	cardboard	2.2827	g/m^2	0.43808
22 in. by 34 in.		1.8799	g/m^2	0.53196
22.5" by 28.5"	bristol	2.1928	g/m^2	0.45604
24 in. by 36 in.	news	1.6275	g/m^2	0.61445
25 in. by 38 in.	book	1.4801	g/m^2	0.67561
25 in. by 40 in.	old standard	1.4061	g/m^2	0.71117
Basis weight by sheet area				
$lb/1000 \ ft^2$	paperboard	4.8824	g/m^2	0.20482
$lb/3000 \ ft^2$	ISO	1.6275	g/m^2	0.61445
Selected paper tests (to four significant figures)				
lb/in^2	burst	6.895	kPa	0.1450
$(g_f/cm^2)/(g/m^2)$	burst index	0.09807	$kPa \cdot m^2 \cdot g^{-1}$	10.20
in	caliper	25.40	mm	0.03937
lb	concora crush	4.448	N	0.2248
$lb/in.^2$	flat crush	6.895	kPa	0.1450
$ft\text{-}lb/in.^2$	internal bond	2.102	$kJ \cdot m^{-2}$	0.4758
lb	ring crush	4.448	N	0.2248
mg_f	Stiffness, Gurley	0.009807	mN	102.0
Taber units	Stiffness, Taber	2.03	mN	0.493
$100g_f \cdot m^2/g$	tear index	0.09807	$mN \cdot m^2 \cdot g^{-1}$	10.20
g_f	tear strength	9.807	mN	0.1020
$kg_f/(15 \ mm)$	tensile strength	6.538	$kN \cdot m^{-1}$	0.1530
lb/in.	tensile strength	0.1751	$kN \cdot m^{-1}$	5.710
$g_f \cdot cm$	toughness index	0.09807	mJ	10.20

Japan, China, and many European countries use the DIN standards for paper size. DIN is an acronym for Deutsche Industrie Normen, which might be thought of as the German Industry Standard. The standard sized sheet for a particular DIN series is indicated by a capital letter such as A, B, C, or D. The A series is usually used for printing and writing papers. Envelopes, file folders, and so on use the B and C series. The standard sheet size of a series is indicated by 0 such as A0. Subsequent designations within a series are indicated by the number of folds needed to make the sheet. For example, A1 is defined as a sheet of A0 folded in half.

The base sheet of paper has a length to width ratio of $(2)^{1/2} : 1$, or approximately 1.414:1. Since smaller sheets are defined by folding the larger sheet in half successively, the length to width of any sheet of paper will maintain the 1.414:1 ratio. The most common series is the A series where the base sheet is exactly 1 m^2. Since the area and the ratio of the length to width are both stipulated, the base sheet is "forced" to have the dimensions of 0.841 m by 1.189 m. The A series in millimeters and inches is given in Table 12-8. If one knows the weight of a single sheet of A4, the basis weight in g/m^2 can be determined by multiplying by 4^2 (or 16).

Table 12-9. DIN format sizes of paper.

Designation	Metric Dimensions, Millimeters		English Dimensions, Inches	
	Length	Width	Length	Width
A series				
A0	1189	841	46.82	33.11
A1	841	595	33.11	23.41
A2	595	420	23.41	16.55
A3	420	297	16.55	11.70
A4[1]	297	210	11.70	8.28
A5	210	149	8.28	5.85
A6	149	105	5.85	4.14
A7	105	74	4.14	2.93
A8	74	53	2.93	2.07
B series				
B0	1414	1000	55.68	39.37
C series				
C0	1297	917	51.06	36.10

[1]This is the standard letter size that corresponds to the English system 8.5 in. by 11 in. sheet.

12.3 UNIT ANALYSIS

Unit analysis (sometimes called dimensional analysis) is a powerful way of solving problems by keeping track of the units (mass, length, area or length^2, density or mass/length^3, time, etc.) of numbers in order to solve problems. By using this method, one can solve a wide variety of problems by memorizing very few things. All solutions to problems should show units throughout the solution in order to prevent "silly" oversights and to be considered correct; a number without the proper units is meaningless. In many of the examples throughout this book, conversion factors are listed for convenience. In other examples it will be necessary to find conversion factors from the tables presented in this chapter. One important tip for solving word problems is that "of" usually implies multiplication while "per" usually implies division.

One straightforward application of unit analysis lies in multiplying a starting figure by conversion factors which are equal to unity (one) to convert the original number to new units. These conversion factors merely substitute one set of units for another set of units without making any changes in the quantity considered.

EXAMPLE 1. You are a Canadian buying a car in the U.S. and the salesman tells you it gets 30 miles per gallon, but you are not really sure this is good gas milage since your point of reference is kilometers per liter. Convert 30 mi/gal (mpg) to km/L.

SOLUTION. One realizes that both miles and km are units of distance and comparable, and that gallons and liters are units of volume and comparable. We can look up conversion factors and find out that 1 mile = 1.609 km and 1 gallon = 3.637 liters. Thus, we can take 30 mpg and multiply by appropriate conversion factors that cancel out the units we have and leave us with the units desired:

$$30 \, \frac{\text{mile}}{\text{gallon}} \times \frac{1.609 \text{ km}}{1 \text{ mile}} \times \frac{1 \text{ gallon}}{3.785 \text{ } \ell} = 12.75 \, \frac{\text{km}}{\ell}$$

One can use any conversion factor (with the correct units) that is equal to unity. For example, we could just as easily use 1 km = 0.6215 mile and 1 L = 0.2642 gal.

$$30 \, \frac{\text{mile}}{\text{gallon}} \times \frac{1 \text{ km}}{0.6215 \text{ mile}} \times \frac{0.2642 \text{ gallon}}{1 \text{ } \ell} = 12.75 \, \frac{\text{km}}{\ell}$$

Of course, there are an infinite number of equally valid ways to achieve the same result. In the above case one might remember the following:

1 mile = 5280 feet	1 foot = 12 in.	1 inch = 2.54 cm
1 cm = 0.00001 km	1 gal = 3.785 L	

Thus, within rounding error, the same result is achieved as follows:

$$30 \, \frac{\text{mile}}{\text{gallon}} \times \frac{5280 \text{ ft}}{1 \text{ mile}} \times \frac{12 \text{ in.}}{1 \text{ ft}} \times \frac{2.54 \text{ cm}}{1 \text{ in.}} \times \frac{10^{-5} \text{ km}}{1 \text{ cm}} \times \frac{1 \text{ gal}}{3.785 \text{ } \ell} = 12.76 \, \frac{\text{km}}{\ell}$$

To find a general conversion factor to convert mpg to km/L start with 1 mpg as follows:

$$1 \, \frac{\text{mile}}{\text{gallon}} \times \frac{1.609 \text{ km}}{1 \text{ mile}} \times \frac{1 \text{ gallon}}{3.785 \text{ } \ell} = 0.4251 \, \frac{\text{km}}{\ell}$$

This is often abbreviated by saying "multiply mpg by 0.4251 to give km/liter"; it is preferable, but alas not common, to indicate "multiply mpg by 0.4251 gal km mile^{-1} liter^{-1}" to give km/L. The difference is that 0.4251 is not equal to 1.000 (unity); however, the term 0.4251 gal km mile^{-1} liter^{-1} is equal to unity.

EXAMPLE 2.   The basis weight of one type of linerboard is 79 lb/1000 ft^2. Convert this to g/m^2.

SOLUTION.   This involves two conversion factors to get appropriate units. Notice how the conversion factor for area is derived by squaring the linear conversion factor. Instead of squaring the numerator and denominator separately, the entire fraction could be squared; the result, of course, is the same since $(a/b)^2 = a^2/b^2$.

$$79 \, \frac{\text{lb}}{\text{ft}^2} \times \frac{453.6 \text{ g}}{1 \text{ lb}} \times \frac{(1 \text{ ft})^2}{(0.3048 \text{ m})^2} = 38.6 \frac{\text{g}}{\text{m}^2}$$

EXERCISE.   In Table 12–8 the conversion factor for basis weight in terms of pounds per 17 in. by 22 in. printing ream (500 sheets) to g/m^2 of individual paper sheets is given as 3.7597; show how this factor is derived.

Unit analysis is very important when using equations. The units of the answer must be appro– priate for those used in the equation; in short, they must match.

There are often difficulties in unit analysis in pulp and paper because units may be mixed be– tween English and metric, or even unmatched within either system.

EXAMPLE 3.   Velocity (v) is equal to distance (s) traveled divided by the time (t) it took to travel that distance ($v = s/t = s \times t^{-1}$). Suppose we know that 60 miles per hour is equal to 26.82 meters per second. Does a paper machine that travels 140 feet in two seconds travel faster or slower than this?

SOLUTION. $s = v/t = 140 \text{ ft}/(2 \text{ s}) = 70 \text{ ft/s}$

The units of the answer are not directly comparable, but we can compare the results after the use of conversion factors.

$$70 \, \frac{\text{ft}}{\text{s}} \times \frac{1 \text{ m}}{3.281 \text{ ft}} = 21.3 \frac{\text{m}}{\text{s}} \quad \text{or} \quad 70 \, \frac{\text{ft}}{\text{s}} \times \frac{3600 \text{ s}}{1 \text{ hr}} \times \frac{1 \text{ mi}}{5280 \text{ ft}} = 47.7 \frac{\text{mi}}{\text{hr}}$$

In either case, whether we convert ft/s to m/s or mi/hr, one obtains values which are comparable to (and less than) the original speeds given.

EXAMPLE 4.   Use Table 2–2 (page 32) to determine an approximate conversion factor of cords to units.

SOLUTION.   Since a direct relationship is not available, two relationships are used in series.

$$1 \text{ cord} \times \frac{3.62 \text{ m}^3 \text{ stacked wood}}{1 \text{ cord}} \times \frac{1 \text{ unit}}{3.28 \text{ m}^3 \text{ stacked wood}} = 1.10 \text{ units}$$

EXAMPLE 5.   The relationship for converting mass into energy is the famous equation of Albert Einstein, $E = mc^2$. While this is applicable to any change of energy, it is usually applied to nuclear reactions where very large amounts of energy are released. In normal chemical reactions the changes in mass are minuscule. For the sake of argument, assume one could completely convert coal into energy via a nuclear reaction. How many horsepower–days of energy would be produced from 1 gram of coal?

c = speed of light in a vacuum = 3×10^8 m/s.

hp = 746 J/s; thus 1.00 hp $\times$ s = 746 J

SOLUTION.   (Note that if one were to try to solve this problem in English units one might be tempted to use the units of pounds; however, pounds are a unit of weight, not mass, and this would lead to incorrect results. This is just one of many advantages to the metric system.) Since energy is in units of joules, which in turn is units of m, kg, and s, it is convenient to solve as follows:

$$E = 0.001 \text{ kg} \times (3 \cdot 10^8 \frac{\text{m}}{\text{s}})^2 = 9 \cdot 10^{13} \frac{\text{kg} \cdot \text{m}^2}{\text{s}^2} = 9 \cdot 10^{13} \text{ J}$$

$$9 \cdot 10^{13} \text{ J} \times \frac{1 \text{ hp} \cdot \text{s}}{746 \text{ J}} \times \frac{1 \text{ hr}}{3600 \text{ s}} \times \frac{1 \text{ day}}{24 \text{ hr}} = 1,396,000 \text{ hp–day}$$

The amount of energy generated is enough to keep a 4000 hp motor operating for almost one year. This calculation indicates the tremendous amounts of energy released in nuclear reactions. Although coal is not converted to pure energy on the face of the earth, nuclear reactions can occur. For example, one mole (235.0439 grams) of uranium–235 decomposes (fission) to form 234.8286 grams of products while 0.2153 grams of mass is converted to energy. Analogously, 140.4634 grams of hydrogen can be fused (fusion) to form 139.4635 grams of helium while 1.000 gram of mass is converted to energy. The larger amount of energy given off and the high availability of hydrogen and deuterium makes fusion a very good potential source of energy, although the conditions required for fusion reactions are extremely difficult to obtain under controllable conditions. For this reason fusion has not been used as an energy source to generate electricity. Energy of nuclear reactions is given off as heat. About 30% of the heat energy is converted to electricity.

EXAMPLE 6.   How much work in horsepower is accomplished by pumping 3000 gallons (22,500 lb) of water per minute up 70 vertical feet (equal to a pressure of 30 psi)?

SOLUTION.   The horsepower is a rate of energy. One horsepower is the work accomplished by lifting 550 pounds one foot high (perpendicular to gravity) per second. This is equal to 33,000 foot–pounds per minute. The work required to lift 3000 gallons of water every minute a height of 70 feet is easily converted to foot pounds of work. This is about the amount of work the primary fan pump must accomplish for a paper machine that produces 80 tons/day with a speed of 4000 ft/min.

$$22400 \text{ lb} \times 70 \text{ ft} \times \frac{1 \text{ hp}}{33000 \text{ ft} \cdot \text{lb}} = 47.4 \text{ hp}$$

EXERCISES

1.   Calculate the number of gallons of paint required to paint the walls of a room 6 m by 8 m by 3.25 m high with a dry paint thickness of 0.1 mm. Use these conversion factors:

   1 gal paint = 0.30 gallons solids when dry

   1 gal = 3.785 liters

   1 L   = 0.001 m^3

   1 m   = 100 cm

   One approach is to calculate the wall area to be painted. Convert this to the volume of dry paint (surface area times thickness) in a sepa–rate calculation, then convert to wet paint gallons. The details remain.

2.   Refining energy is often reported in units of horsepower–days/ton; however one buys electricity by kilowatt hours.   Convert hp–days/ton to kWh/ton.

3.   Derive a conversion factor to convert Btu/hr to watts.

4.   Derive a conversion factor to convert pounds/(24 in. by 36 in. ream) to g/m^2 of a single sheet of paper. Remember one ream has 500 sheets, 1 lb is 453.59 g, and 1 cm is 2.54 in.

5.   Derive a conversion factor for kg·s·m^{-2} to A/mol.

6.   A certain type of paper is 78 lb/3000 ft^2 with a thickness of 0.009 in. What is its density?

7.   How many seconds would a one ampere flow of current take to move one mole (6.0220 × 10^{23}) of electrons?

8.   Keeping in mind that one watt is one J/s, if electricity is $0.10 per kilowatt hour, how much is the energy generated by one gram of mass (Example 5) worth?

9.   If a mill makes 1000 tons/day of paper and a certain additive worth $1.15/lb is used at 0.03% on paper, how much will this mill spend on this additive in one year?

13
INTRODUCTORY CHEMISTRY REVIEW

This book presumes a knowledge of introductory chemistry. Basic concepts especially related to the pulp and paper industry are briefly reviewed here. One should consult a good introductory chemistry book if more detail is required.

13.1 THE ELEMENTS

Unless one works in the field of subatomic particle physics, all matter may be considered to be made up of the 103 naturally occurring elements. About one dozen of these are important to pulp and paper products. Another dozen are important to the metallurgy of equipment, an area covered in this volume in regards to corrosion.

Each element is made up of *protons*, *electrons*, and *neutrons* arranged to form atoms. Protons and neutrons together are called *nucleons* because they are the constituents of the *nucleus*, the center of the atom. Protons and neutrons have about the same mass and make up most of the mass of an atom. Electrons have 1/1836 the mass of protons and a negative charge assigned at -1 unit, since the electron is the basis of all charges. Electrons orbit the nucleus in *shells*. The outermost shell of electrons are the *valence electrons*, which are the electrons primarily involved in forming chemical bonds. Protons are positively charged (opposite in charge of the electron) with exactly the same magnitude of charge as the electron, that is +1. Like charges repel each other, and opposite charges attract each other; therefore, electrons repel each other and electrons are attracted to protons.

Each atom of a pure element has equal numbers of electrons and protons and, therefore, is electrically neutral. The particular type of element an atom forms (and its chemical properties) depends on the number of protons in the nucleus; this is called the *atomic number*. The periodic table of the elements (on the inside cover) shows the elements in increasing atomic numbers. There are many reasons why the table has this arrangement. For example, vertical columns of elements have similar properties.

Atoms tend to have approximately equal numbers of neutrons and protons. The *relative masses* of individual atoms have approximately whole number relationships to each other because protons and neutrons constitute most of the mass and have almost identical masses. For this reason it is useful to define an *atomic mass unit (amu)*. An amu is precisely 1/12 the mass of the carbon-12 isotope. (Carbon-12 is made up of 6 neutrons, 6 protons, and 6 electrons.) If one looks up the atomic weight of carbon, it is actually 12.01 amu as shown in the periodic table of the elements. The reason is that carbon in nature exists as 98.9% of carbon-12 and about 1.1% of carbon-13. Carbon-13 atoms have an extra neutron, which has a slight effect on the physical properties but little effect on the chemical properties of carbon. Carbon-12 and carbon-13 are said to be *isotopes* of each other; an alternate way of writing isotopes is by adding a superscript to the left of the chemical symbol as in ^{12}C or ^{13}C. Isotopes have equal atomic numbers but differ in atomic weights and the number of neutrons in the nucleus.

While much more information is available in chemistry textbooks on the arrangement of the orbiting electrons in atoms and molecules, the important features of orbitals will be briefly summarized here. The arrangement of electrons determines the reactivity of the elements with each other. Electrons can occupy *orbitals* in pairs. Numbers before orbitals designate the shell number; shells of electrons may be thought of as layers of electrons around the nucleus. In any shell the single *s* orbital can accommodate 2 electrons; the three *p* orbitals, 6 electrons; the five *d* orbitals, 10 electrons; and the seven *f* orbitals, 14 electrons. (The first shell only has a single *s* orbital, the second shell has one *s* and three *p* orbitals, the third shell has *s*, *p*, and *d* orbitals, and the remainder of the shells have *s*, *p*, *d*, and *f* orbitals.) The orbitals are filled (with some exceptions involving shifting of electrons within the outermost *d*, *f*, and *s* orbitals) in the order that describes the arrangement of the elements in the periodic table of the elements (bottom of Table 13-1). Table 13-1 shows allowable orbitals in each of the first four shells to demonstrate the concept.

This means that the first row of the periodic table of the elements (H and He) will have two electrons in the outer shell when it is full, while all of the other elements have eight electron "positions" available in the outer, bonding shell *valence shell*. (In coordinate chemistry and certain compounds, the inner *d* orbital electrons are also available for forming bonds, but we will overlook this for now.) Outer shells can be filled by sharing electrons with other elements or taking electrons from elements that give them up easily; the *octet rule* states that atoms tend to share electrons to fill their octets (or duplet in the case of hydrogen). Hydrogen requires only one additional electron to fill this shell, while helium already has its shell filled. Hydrogen will share one electron when forming bonds with other elements, while helium is very unreactive and will not form chemical bonds. For elements above helium, stable compounds are usually formed when these shells are completely filled or completely empty.

The second row of the periodic table of the elements shows this relationship. Lithium easily gives up an electron and, when combined with other elements, exists as a *cation* [a positively charged ion (an ion is a species with a charge)] of +1. (It is easy to remember cations are positively charged because "I like cats" is a positive thought.) Fluorine has seven valence electrons and needs only one to fill its valence shell. Fluorine will take an electron from lithium or other metals to complete its octet and then exists as an *anion* (a negatively charged ion) of -1 charge, such as in the compound LiF. Neon (like any inert gas) already has a complete octet and is unreactive, generally.

The third row elements of sodium to argon behave in a similar fashion as the second row elements of lithium to neon. Sodium reacts with

Table 13-1. Orbitals occupied by the first four shells and number of electrons per orbital.

Shell	Subshell	Orbital	Number of Electrons		Corresponding Elements as Filled
			Orbital	Shell	
1	0	1s	2	2	H, He
2	0	2s	2	10	Li, Be
	1	2p	6		B to Ne
3	0	3s	2	18	Na, Mg
	1	3p	6		Al to Ar
	2	3d	10		Sc to Zn
4	0	4s	2	32	K, Ca
	1	4p	6		Ga to Kr
	2	4d	10		Y to Cd
	3	4f	14		Ce to Lu

Orbital:	1s	2s	2p	3s	3p	4s	3d	4p	5s	4d	5p	6s	4f	5d	6p	7s	5f
No. electrons:	2	2	6	2	6	2	10	6	2	10	6	2	14	10	8	2	14

chlorine to give NaCl. There is a complication that $3d$ orbitals can be filled with bonding electrons so that an element like sulfur commonly has two *valence* (*sharing* or *combining*) electrons but can also have four or six valence electrons; indeed in the most stable compounds of sulfur, the sulfur atoms share six electrons as in the case of H_2SO_4. This, of course, also happens in the higher rows. Fig. 13-1 shows the valence electrons for some atomic, ionic, and molecular structures; the next section helps one predict the types of bonds formed as shown in this figure. Depiction of electrons in this fashion is credited to G.N. Lewis.

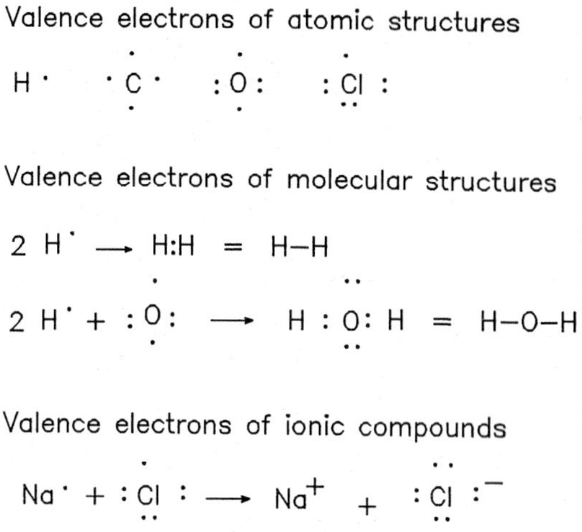

13.2 IONIC AND COVALENT BONDS

The elements of the lower left of the periodic table donate electrons the most easily and are the *electropositive elements*. The metallic elements (element to the lower right and including the elements of the diagonal formed by Al, Ge, Sb, and Po) donate electrons easily. When electrons are donated, the atom remains as a cation such as Na^+, Ca^{2+}, Al^{3+}, and so on. The elements in the upper right, not including the inert gases, attract electrons with high affinity and are the *electronegative elements*, including (in order of decreasing electronegativity) F, O, N, Cl, Br, Se, S, and I. The periodic table of the elements gives values for electronegativity of the elements (the values are listed in the lower right of each element's box). When electrons are accepted, the atom becomes an anion, such as Cl^-, I^-, S^{2-}, and so on. Chemical bonds cover a spectrum of equally sharing electrons to forming ion pairs or ionic bonds. Bonds between the electronegative elements and the electropositive ions are highly ionic in nature; such compounds when dissolved in water will conduct an electric current and are described as strong electrolytes. Some examples of these are NaCl, $CaBr_2$, and Na-OH. Bonds between elements of intermediate electronegativity share the electrons, forming *covalent bonds*, and do not disassociate into individual ions and will not conduct electricity in water (*nonelectrolytes*) or are weak conductors (*weak electrolytes*). Some examples are CH_4, CH_3OH, H_2S, and CO_2. Generally, if the difference in electronegativity of two elements sharing electrons is 1.7 or greater, the bond formed

Valence electrons of atomic structures

$$H \cdot \quad \cdot \overset{\displaystyle \cdot}{C} \cdot \quad : \overset{\displaystyle \cdot}{\underset{\displaystyle \cdot}{O}} : \quad : \overset{}{\underset{\displaystyle \cdot \cdot}{Cl}} :$$

Valence electrons of molecular structures

$$2\ H \cdot \longrightarrow H:H \ = \ H-H$$

$$2\ H \cdot + : \overset{\displaystyle \cdot \cdot}{O} : \longrightarrow H : \overset{\displaystyle \cdot \cdot}{O} : H \ = \ H-O-H$$

Valence electrons of ionic compounds

$$Na \cdot + : \overset{\displaystyle \cdot}{\underset{\displaystyle \cdot \cdot}{Cl}} : \longrightarrow Na^+ \ + \ : \overset{\displaystyle \cdot \cdot}{\underset{\displaystyle \cdot \cdot}{Cl}} :^-$$

Fig. 13-1. Lewis structures showing the valence electrons of various structures.

between the two elements behaves in a highly ionic manner. If the difference is less than 1.0, the bond behaves in a covalent manner. The actual amount of ionization in a chemical containing intermediate bonds depends on many things, such as the presence and type of solvent.

There are several particularly important polyatomic ions encountered. For example, the ammonium cation NH_4^+ behaves very similarly to the Na^+ ion, except the ammonium ion decomposes at pH above 9 and is subject to combustion. Many polyatomic anions with several atoms of oxygen have *delocalized* (that is, spread out over two or more atoms) negative charges, which increase the stability of the oxygen anion such as in the cases of RCO_2^-, CO_3^{2-}, NO_3^-, NO_2^-, SO_4^{2-}, and SO_3^{2-}. Resonance structures are Lewis structures which differ only in the position of a double bond and other electrons. The actual structure is a *resonance hybrid*, which is an average of all of the resonance structures. Fig. 13-2 shows how two resonance structures contribute to the carboxylate anion. Experiments indicate that the two C-O bonds are identical and intermediate in properties between C-O and C=O bonds.

$$R-C\!\!\overset{\displaystyle O^-}{\underset{\displaystyle O}{<}} \quad \longleftrightarrow \quad R-C\!\!\overset{\displaystyle O}{\underset{\displaystyle O^-}{<}}$$

Fig. 13-2. Resonance structures for the carboxylate anion.

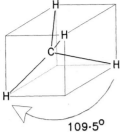

Fig. 13-3. Tetrahedron formed by CH₄.

It is useful to consider the strength of chemical bonds. The strength of chemical bonds is measured as the enthalpy (energy) required to break the bond between two elements. It is reported in units of energy per mole. For example, one mole of Cl_2 gas (70.9 g) can be separated (hypothetically) into two moles of chlorine radicals Cl· by the addition of 242 kJ of energy as follows: Cl_2 + energy → 2 Cl· . Other types of chemical bonds occur that are weaker than these primary bonds. These will be discussed in other sections. Table 13-2 gives some average values for bond energies. This table shows that the shorter the bond length, the more stable it is (that is, the more energy it takes to break it).

Electrostatic repulsion of electrons means the valence electron pairs are more stable when farther apart from each other. This causes the bonds and lone electron pairs of an octet to, approximately, form the corners of a tetrahedron. Fig. 13-3 shows the *tetrahedral* shape of the four carbon-hydrogen bonds of methane.

The tetrahedron form is characteristic of elements that complete an octet with just eight electrons. In the case of the oxygen atom of water, two pairs of lone electrons occupy two corners of the octet. Since lone electron pairs have high repulsion, the H-O-H bond angle of water is only 104.5°. While the carbon backbone of materials like hexane is often drawn as C-C-C-C-C-C, it is more closely approximated as:

$$C \diagdown^C \diagdown_C \diagdown^C \diagdown_C \diagdown^C$$

13.3 HYDROGEN BONDING

Hydrogen bonding is one example of a secondary bond and is one of the stronger secondary bonds. Secondary bonds are much weaker than

Table 13-2. Approximate bond energies for interatomic bonds.

Bond type	Bond energy		Bond length, nm
	kJ/mol	kcal/mol	
Cl-Cl	242.	57.8	0.2
H-H	436.	104.	0.074
H-Cl	431.	103.	0.135
C-H	420	100	0.11
C-C	350	85	0.154
C=C	620	150	0.13
C=C benzene	518	124	0.139
C-O	360	85	0.14
C=O	740	175	0.12
C-N	305	75	0.15
O-H	463	110	0.10
O-O	220	55	0.15
N-H	430	105	0.10

the primary bonds, being on the order of 1-10 kcal/mol compared to 50-200 kcal/mol for the primary covalent chemical bonds. Hydrogen bonds are on the order of 4-5 kcal/mol, although hydrogen bonding of carboxylic acids to each other is about 13 kcal/mol. Many hydrogen bonds, however, can add up to a strong structure; most papers are held together without adhesives, relying only on hydrogen bonding to hold the fibers together. *Hydrogen bonding occurs when hydrogen that is bonded to one of the four highly electronegative elements (F, O, N, or Cl) comes near a second highly electronegative element.* Partial charges develop between the hydrogen and electronegative atom since the bond is fairly ionic. The partial positive charge on the side of the hydrogen atom away from the shared electrons is attracted to the partial negative charge of the unshared electron pairs of the second electronegative element, as shown in the following diagram where δ designates a partial charge:

$$\delta^+ \quad \delta^-$$
$$\text{R-O:H} \quad \text{:OR}$$
$$\text{H}$$

For water, the difference in electronegativities is 1.4 units, corresponding to a bond that is about 40% ionic. While the length of the covalent bonds of water is 0.1 nm, the length of a hydrogen bond in water is almost 0.2 nm.

EXAMPLE 1. Generally the boiling points of substances increase with increasing formula weights for a related series. The boiling points of the hydrides of the elements of row 6 of the periodic table are as follows (note the temperatures are given in Kelvin):

Element	Hydride	b.p., K
Oxygen	H_2O	?????
Sulfur	H_2S	212.5
Selenium	H_2Se	231.6
Tellurium	H_2Te	271.0

From this, predict the boiling point of water, H_2O. What factor should be considered to explain its *unusual* boiling point?

SOLUTION. From the data, one might predict a boiling point around 200 K for water. In fact the boiling point is 373 K, due to relatively strong hydrogen bonding. This means that a water molecule has a high affinity for the liquid, thereby decreasing its vapor pressure and increasing the boiling point.

EXAMPLE 2. One mole of acetic acid at 273 K occupying 22.4 liters should have a pressure of 1.00 atmosphere according to ideal gas theory. In fact, there is a huge deviation and the pressure is about 0.60 atmospheres. Why? One mole of water under the same conditions has a pressure of about 0.90 atmosphere. Why? What does this say about the magnitude of hydrogen bonding in hydroxyl groups relative to that of carboxylic acids? Could this help explain why hemicelluloses increase the strength of papers?

SOLUTION. The strong hydrogen bonding of carboxylate groups (about 13 kcal/mol) means that most of the acetic acid under these conditions will exist as dimers. The relatively weak hydrogen bonding of water (about 5 kcal/mol) means that a significant, but smaller portion than acetic acid, of water will exist as dimers. The carboxylic acid groups of hemicellulose could form strong hydrogen bonds (with other carboxylate groups or with hydroxyl groups). Carboxylation of fibers or addition of carboxymethyl cellulose to fibers is known to increase the strength of paper made from these fibers.

PROBLEM: Ethanol (CH_3CH_2OH) has a boiling point of 79°C; its isomer, dimethyl ether (CH_3OCH_3) has a boiling point of -25°C. Explain this phenomenon with the use of a diagram showing partial charges.

13.4 THE MOLE AND MASS PERCENTAGE

Atoms almost always react with each other in whole number ratios to form chemical compounds. For instance carbon reacts with oxygen gas to give carbon dioxide during combustion of charcoal.

$$C + O_2 \rightarrow CO_2$$

It is very useful, therefore, to have a measure of atoms and molecules that reflects the number of particles. (Of course, we usually deal with weights when actually handling chemicals.) Since the gram is the basis of the metric system, as well as a convenient amount of material to handle in the laboratory, it is used as the basis of the amu, which is the scale used to identify the mass of individual atoms and molecules. By using the weight of an amu, we can calculate the number of amu per g; this number, $6.0220 \cdot 10^{23}$, is called the Avogadro constant and is denoted as N_A.

$$1 \text{ amu} = 1.6606 \cdot 10^{-24} \text{ g};$$

$$\text{therefore,} \quad 1 \text{ g} = 6.0220 \cdot 10^{23} \text{ amu}$$

This number of objects is known as the *mole*. The concept of the mole is *extremely important* because chemicals react in whole number relationships when compared on a molar basis. Since carbon is 12.01 amu, one mole of carbon is 12.01 g of carbon. This is often called the *gram atomic weight* since it is the atomic weight of a substance in grams. Similarly, one mole of carbon dioxide (CO_2) is $12.01 + (2 \times 16.00) = 44.01$ grams. In the case of molecules with covalent bonds we speak of *gram molecular weights*, and in the case of compounds with ionic bonds we speak of *gram formula weights*. The gram molecular weight of CO_2 is 44.01 grams, while the gram formula weight of NaCl is 58.44 (from $22.99 + 35.45$). It is not so important to remember atomic versus molecular versus formula weights as it is to understand the concept of the mole, because one can speak of 1 mole of carbon, carbon dioxide, and sodium chloride.

The mass percentage of a substance is a convenient method of expressing its composition. Each component (such as "A") of the whole may be expressed as a mass percentage as follows:

(13-1)

$$\text{mass \% A} = \frac{\text{mass of A in sample}}{\text{total mass of sample}} \times 100\%$$

EXAMPLE 3. Calculate the formula weight of $Al_2(SO_4)_3$ and the mass percentages of the individual elements.

SOLUTION: This compound consists of two aluminum cations and three sulfate anions. The formula weight of this material is calculated as follows:

```
 2  Al is 2 × 27 =  54
 3  S  is 3 × 32 =  96
12  O  is 12 × 16 = 192
                    342 = formula weight
```

$$\%Al = 54/342 \times 100\% = 15.8\%$$

$$\%S \ = 96/342 \times 100\% = 28.1\%$$

$$\%O \ = 192/342 \times 100\% = 56.1\%$$

PROBLEM: Calculate the molecular weight for glucose and the mass percentage of carbon in glucose. The formula for glucose is $C_6H_{12}O_6$. Answers: 180 g/mol and 40.0% carbon.

EXAMPLE 4. Given the (quantitative) chemical equation below, which is the conversion of the kraft makeup chemical sodium sulfate (Na_2SO_4) into the active cooking species sodium sulfide (Na_2S) in the recovery boiler, how many grams of Na_2S will be produced from 100 grams of Na_2SO_4?

$$2C + Na_2SO_4 \rightarrow Na_2S + 2CO_2$$

SOLUTION: The formula weight of sodium sulfate is 142 g/mole and that of sodium sulfide is 78 g/mole.

$$100 \text{ g } Na_2SO_4 \times \frac{1 \text{ mol } Na_2SO_4}{142 \text{ g } Na_2SO_4} \times \frac{1 \text{ mol } Na_2S}{1 \text{ mol } Na_2SO_4}$$

$$\times \frac{78 \text{ g } Na_2S}{1 \text{ mol } Na_2S} = 54.9 \text{ g } Na_2S$$

PROBLEM: In this example, how many grams of carbon were reacted? Answer: 16.9 g C.

13.5 EQUIVALENCY, MOLARITY AND NORMALITY

Molarity (*M*) is a measure of concentration in terms of the mole and is moles of a substance per liter (mol/L) of solution. One *M* solution of NaCl would be 58.44 g/L of NaCl. An equivalent is one mole of reactive species, whether it be an H^+ of acid-base reactions, an electron of reduction-oxidation reactions, or some other unit. In acid-base reactions the basis of the equivalent is one mole of hydrogen ions, H^+. For a *monoprotic acid*, which has one ionizable proton, such as HCl, one equivalent is the same as one mole. For a *diprotic acid*, which has two ionizable protons, such as $H_2SO_4 \rightleftarrows 2 H^+ + SO_4^{2-}$, the equivalent weight is one-half the molecular weight. One mole of H_2SO_4 is 98 g; one equivalent of H_2SO_4 is 49 g. In electron-transfer (oxidation/reduction) reactions the electron is the basis of the mole. Potassium permanganate ($KMnO_4$) transfers five electrons in most reactions where it is used. The molecular weight of $KMnO_4$ is 158.0, and the equivalent weight is 31.6 g/eq. *Normality* is analogous to molarity and is concentration expressed as equivalents per liter. For example, 1 *N*

H_2SO_4 is 49 g/L of H_2SO_4 in an acid-base reaction. In titrations, the volume is measured in ml. Note that 1 eq/L = 1 meq/ml and 1 mol/L = 1 mmol/ml; these are useful relationships for titrations.

EXAMPLE 5. If 17 grams of sodium chloride is dissolved in water to make 600 ml of solution, what is the molarity of this solution?

SOLUTION: The formula weight of NaCl is 58.44. Start with 17 grams NaCl per 0.6 L and convert this to mol/L:

$$\frac{17 \text{ g NaCl}}{0.6 \text{ } \ell} \times \frac{1 \text{ mole NaCl}}{58.44 \text{ g NaCl}}$$

= 0.485 mol/ℓ NaCl = 0.485 M NaCl

PROBLEM: 2.0 g HCl is dissolved in water to make 1 L of solution. What is the normality of this solution? Same question with sulfuric acid? Answers: 0.0549 N HCl and 0.0408 N H_2SO_4.

13.6 ACIDS, BASES, AND THE pH SCALE

For the purposes of pulp and paper science it is useful to use the *Brønsted-Lowry theory* of acids and bases. According to this theory, an acid is a compound that is a proton (H^+) donor and a base is a compound that is a proton acceptor. There must be a proton acceptor to accept a donated proton since protons are not stable in an uncombined form. Thus, an acidic solution has a relatively high concentration of protons (H^+ ions) in hydrated form, since water is the proton acceptor. The hydrated proton is known as the *hydronium ion*, H_3O^+. H^+ and H_3O^+ are often used interchangeably, although H^+ does not exist in water.

Alkaline or basic solutions have very low concentrations of protons which, in water, necessarily means a high concentration of OH^- ion (to be shown in Eq. 13-4). The empirical formula of acids are often written with leading protons. For example, CH_3CH_2OH, C_2H_4, and C_3H_8 are not acids, whereas H_2S, H_2SO_3, and $HC_2H_3O_2$ are

acids. There are many exceptions to this rule. For example, $HC_2H_3O_2$ is acetic acid and is often written as CH_3COOH to show that it is a carboxylic acid.

Another rule of thumb is that the elements H, Al, Ga, Sn, and Pb form oxides that are *amphoteric*, that is, compounds that are both acidic and basic. The first two of these elements are particularly relevant to pulp and paper. Elements (not including the noble gases) to the upper left of these on the periodic table of the elements form acidic oxides (for example, SO_2 forms the acid H_2SO_3 in water), while those elements to the lower right of these form basic oxides [for example, CaO forms the base $Ca(OH)_2$ in water].

Examples of proton donors are HCl, which becomes Cl^-; HNO_3, which becomes NO_3^-; and RCOOH, which becomes $RCOO^-$. Examples of proton acceptors other than water are OH^- from bases such as NaOH, which becomes water since $H^+ + OH^- \rightleftarrows H_2O$; carboxylate anion $RCOO^-$, which becomes RCOOH; and NH_3, which becomes NH_4^+. A proton donor and the corresponding base formed after the donation of a proton are called *conjugate pairs*. The compound left after donating a proton is called a *conjugate base* since, in principle, it should be able to accept a proton. A compound like HCl is highly ionic and completely dissociates in water to form H^+ (as H_3O^+) and Cl^- ions; consequently, it is a very good hydrogen donor (*strong acid*). On the other hand, the conjugate base, Cl^-, would not be expected to have a high affinity for protons. Therefore, *the conjugate base of a strong acid is a weak base*. In the same manner *the conjugate acid of a strong base is a weak acid*. For this reason Na^+ is a very weak acid because its conjugate base (NaOH) is a very strong base. The group IA ions of Li to Fr are very poor bases and act as *spectator ions* in acid-base reactions, that is, they are ions not involved in a reaction.

In the case of acetic acid, CH_3COOH, there is partial ionization to form the acetate and proton ions in solution. This is termed a *weak acid*. Table 13-3 lists all of the commonly encountered strong acids, which are completely ionized in water, and some weak acids with their ionization constants. The *acid ionization constant*, K_a, is the equilibrium constant for the degree of ionization.

Table 13-3. Strong and weak electrolytes in water at 25°C (77°F).

Substance	K_1	pK_1	K_2	pK_2
Strong acids	1000^+	Below -3		
HCl	10^6	-6		
H_2SO_4	1000	-3	0.012	1.92
HNO_3	21	-1.32		
$HClO_4$				
Weak acids of general interest				
Abietic acid	$2.40 \cdot 10^{-8}$	7.62		
$HC_2H_3O_2$ (acetic acid)	$1.75 \cdot 10^{-5}$	4.76		
NH_4^+	$5.75 \cdot 10^{-10}$	9.24		
C_6H_5OH (phenol)	$1.1 \cdot 10^{-10}$	9.96		
H_3PO_4 (phosphoric acid)	0.011	1.96	$7.5 \cdot 10^{-8}$	7.12
$K_3 = 4.8 \cdot 10^{-13}$ $pK_3 = 12.32$				
CF_3COOH (trifluoroacetic acid)	1.698	0.23		
Weak acids of significance to pulping				
H_2CO_3	$4.3 \cdot 10^{-7}$	6.37	$4.8 \cdot 10^{-11}$	10.32
H_2SO_3	$1.3 \cdot 10^{-2}$	1.89	$2 \cdot 10^{-7}$	6.7
(at 140°C)	$8.0 \cdot 10^{-4}$	3.10		
H_2S	$9.1 \cdot 10^{-8}$	7.04	$1 \cdot 10^{-17}$ [1]	17^1
H_2S_4 polysulfide	$1.6 \cdot 10^{-4}$	3.8	$5.0 \cdot 10^{-7}$	6.3
Weak acids of significance to bleaching				
HOCl	$1.1 \cdot 10^{-8}$ [1]	7.96^1		
$HClO_2$	0.011	1.95		
H_2O_2	$2.2 \cdot 10^{-12}$	11.65		
Strong bases				
LiOH				
NaOH				
KOH				
Strong electrolytes				
All salts of group IA and IIA elements				

[1] Sizeable variation was found for these values. Older sources report pK_a of 12 for Na_2S, another around 15, but the most recent work indicates it is at least 17. CRC's *Handbook of Chemistry and Physics* reports a value of 4.53 for HOCl, although Cotton and Wilkinson give 7.50.

In the case of polyprotic acids, they are listed as K_1, K_2, K_3, ... There may be considerable variation in the reported values of some acid dissociation constants! Since the concentration of water is not included in the definition of the equilibrium constant (since it is approximately constant in dilute solutions), using A^- as a general term for the conjugate base (such as Ac^- for CH_3COO^-, from acetic acid, HAc) gives the following relationship:

$$(13\text{-}2)$$
$$HA + H_2O \rightleftharpoons H_3O^+ + A^-$$

$$K_a = \frac{[H_3O^+][A^-]}{[HA]}$$

Analogously, the *base ionization constant, K_b,* is the equilibrium constant for the production of OH^-, such as from NH_3 (abbreviated as B) to give NH_4^+ (abbreviated as B^+) as shown in the following equation:

$$(13\text{-}3)$$
$$B + H_2O \rightleftharpoons BH^+ + OH^-$$

$$K_b = \frac{[OH^-][BH^+]}{[B]}$$

A base can begin as a negative ion such as HSO_3^-, in which case BH^+ is H_2SO_3. These ionization constants are examples of the law of mass action, which is explained in more detail in the following section. Strictly speaking, the *effective concentration* of ionic species is lower than the actual concentration. The ratio of effective concentration to the actual concentration is known as the *activity coefficient* of the ion. Generally, as the ionic strength of a solution increases, the activity coefficient of a given species decreases from unity. Most pH calculations assume activity coefficients of unity. The *activity* is the *effective concentration* of an ionic species. All equilibrium constants are correctly stated only in terms of activity. Also, pH meters and indicators actually measure hydrogen ion activity. Fortunately, in dilute solutions the activity coefficients are approximately equal to unity, and one can use concentration directly instead of activities to simplify the calculations. Even concentrated solutions are analyzed (such as by titration) under relatively dilute conditions, al-

though in pulp and paper mills the pH of concentrated solutions is sometimes measured.

Water is both an acid and a base since it can accept a proton to form H_3O^+ or donate a proton to become OH^-. The ionization constant for water, K_w, is 1×10^{-14} at $25°C$; pK_w is 14.00. Substituting water as the acid into Eq. (13-2), and remembering that water as the solvent is not included in the definition of an equilibrium constant, one obtains:

$$[H_3O^+][OH^-] = 10^{-14} = K_w \qquad (25°C) \quad (13\text{-}4)$$

Lange's Handbook gives the pK_w for water as 14.94 at $0°$, 13.26 at $50°$, 12.26 at $100°$, and 11.91 at $130°C$, while *CRC's Handbook* gives 11.64 at $150°C$ and 11.29 at $200°C$.

This means that for pure water at $25°C$, the $[H_3O^+] = [OH^-] = 10^{-7}$. The concentration of H_3O^+ in water can vary from above 10^1 to less than 10^{-15} molar. Because the acidity can vary by over 16 orders of magnitude, it becomes convenient to use a logarithmic scale to describe how acidic or basic a solution is! The pH scale was devised for that purpose. The pH is *the negative value of the base-10 logarithm of the concentration of hydrogen ions*. The pH of pure water is 7, the pH of $0.1 M$ ($10^{-1} M$) HCl is 1, and the pH of $10^{-1} M$ NaOH is 13. By definition the pH scale goes from 0 to 14, but it is sometimes useful to go beyond a pH of 0 or 14. For example, one can (secretly) imagine $10 M$ HCl as having a pH of -1.

In general, the p function is quite useful as will be seen in the next chapter. The p function is the negative logarithm of a value. For example:

$$p(x) = -\log(x)$$

We will use the $p(OH)$, the $p(K_a)$, and the $p(K_b)$ functions. In water $pK_a + pK_b$ (of the conjugate base) = 14; for example, the pK_a of ammonium ion (4.76) plus the pK_b of ammonia (9.24) equal 14. Table 13-4 shows the relationship between pH, pOH, $[H^+]$, and $[OH^-]$ in dilute, aqueous solutions.

By taking the logarithm of both sides of Eq. (13-4) and using the definition of the p function, we obtain the relationship pH + pOH = 14 for water at $25°C$.

Table 13-4A. From strong, concentrated acid decreasing to neutral pH.

pH	-1	0	1	2	3	4	5	6	7
pOH	15	14	13	12	11	10	9	8	7
$[H^+]$	10	1	0.1	10^{-2}	10^{-3}	10^{-4}	10^{-5}	10^{-6}	10^{-7}
$[OH^-]$	10^{-15}	10^{-14}	10^{-13}	10^{-12}	10^{-11}	10^{-10}	10^{-9}	10^{-8}	10^{-8}

Table 13-4B. From neutral pH increasing to strong, concentrated base.

pH	7	8	9	10	11	12	13	14	15
pOH	7	6	5	4	3	2	1	0	-1
$[H^+]$	10^{-7}	10^{-8}	10^{-9}	10^{-10}	10^{-11}	10^{-12}	10^{-13}	10^{-14}	10^{-15}
$[OH^-]$	10^{-7}	10^{-6}	10^{-5}	10^{-4}	10^{-3}	10^{-2}	0.1	1	10

EXAMPLE 6. What is the pH of 0.025 M NaOH?

SOLUTION: NaOH is a strong electrolyte, so $[OH^-] = 0.025$ M. From Eq. 13-4 it follows that $[H^+] = 4.16 \cdot 10^{-13}$, so the pH = -log $[H^+] = 12.38$. This is a reasonable answer if one looks at the pH versus $[OH^-]$ table above.

PROBLEM: What is the pH of 0.33 M HCl? Answer: pH = 0.48

The pH of a "pure" weak acid can be approximated by Eq. 13-5 as long as less than 10% to 15% of the acid is ionized. Example 7 shows why this works.

For a solution of weak acid alone

$$[H^+] = \sqrt{K_a(C_{HA} - [H^+])} \approx \sqrt{K_a C_{HA}} \quad (13\text{-}5)$$

An interesting rearrangement of the approximation is:

$$pH \approx \frac{pK_a - \log C_A}{2}$$

for a 1 M solution of weak acid: $pH \approx \dfrac{pK_a}{2}$

Eq. 13-5 shows that the pH of a weak acid is dependent upon its concentration. The $[OH^-]$ may be determined for a "pure" conjugate base of a weak acid with K_a and is shown in Eq. 13-6. Remember that $K_a = K_w/K_b$ where b is the conjugate base of the acid a.

For a solution of weak base alone $\quad(13\text{-}6)$

$$[OH^-] = \sqrt{\frac{K_w}{K_a}(C_A - [OH^-])} \approx \sqrt{\frac{K_w}{K_a} C_A}$$

EXAMPLE 7. The pH of a weak acid by itself. Calculate the pH of 0.01 M acetic acid.

SOLUTION: If x equals the concentration of acetic acid that ionizes, then $x = [H^+] = [Ac^-]$; $[HAc] = (0.01$ $M - x)$. Using Eq. 13-5 leads to the following equation.

$$x^2/(0.01 - x) = 1.75 \times 10^{-5}$$

By rearrangement and use of the quadratic equation (see Example 9), this could be easily solved precisely. Since this is a weak acid, however, the amount of ionization will be small and $[HAc] \approx 0.01$ M; therefore,

$$x^2 = 1.75 \times 10^{-5} \times 0.01;$$

$$x = [H^+] = 4.18 \times 10^{-4}; \quad pH = 3.38$$

Notice that the assumption that x will have a negligible effect on [HAc] is approximately true; otherwise x would have to be recalculated. Use of the quadratic equation would have given a pH of 3.39, which is not appreciably different.

EXAMPLE 8. What is the pH of 0.1 M sodium acetate?

SOLUTION: Using the approximation in Eq. 13-6 gives:

$$[OH^-] = ((10^{-14}/1.75 \times 10^{-5})(0.1))^{0.5}$$

$$[OH^-] = 7.56 \times 10^{-6}; \quad pH = 8.89$$

Since the [OH$^-$] is small relative to sodium acetate, the approximation is warranted.

EXAMPLE 9. K_3 for phosphoric acid, H_3PO_4, is 4.8 $\times$ 10^{-13}. What is the pH of 0.1 M trisodium phosphate (TSP)?

SOLUTION. Use of Eq. 13-5 with the approximation that [H]$^+$ is small gives:

$$[OH^-] = ((10^{-14}/4.8 \times 10^{-13})(0.1))^{0.5}$$

[OH$^-$] = 0.046; Note that the approximation is not valid since this acid is nearly 50% ionized!

Since the approximation cannot be used, the quadratic equation is used to get a more exact answer. A *quadratic equation* is in the form of $ax^2 + bx + c = 0$ where a, b and c are coefficients and x is the variable. The general solution to the quadratic equation is:

$$x = \frac{-b \pm \sqrt{b^2 - 4ac}}{2a}$$

Therefore, let x = [TSP] that reacts with H$^+$, so that x = [OH$^-$] = [C$_A$]. At equilibrium, [TSP] = (0.1 M - x). Using Eq. 13-5 leads to the following equation.

$$x^2 = (0.1 - x) K_w/K_b = (0.0208)(0.1 - x)$$

$$x^2 + 0.0208\,x - 0.00208 = 0$$

By use of the quadratic equation this is solved as x = 0.0364 (the negative solution or root, -0.057, is discarded since concentrations must be positive). This corresponds to a pH of 12.56, which is very alkaline.

For polyprotic acids the determination of pH of the intermediate salt, such as NaHSO$_3$, is more complex since the bisulfite ion can act as a base or acid. In this case, the [H$^+$] for a solution of the salt alone can be shown to be equal to $(K_1K_2)^{0.5}$ as long as $K_1 >> K_2$. (K_1 should be 10^4 or 10,000 times as large as K_2.) By taking the log of both sides this is rewritten as pH = (pK_1 + pK_2)/2. Notice that the concentration of the salt does not affect the pH; however, this is not a buffer solution as it represents an equivalence point of a titration. See the section on pulp liquor analysis for more details on this point. Therefore, for a solution of a conjugate base which itself is an acid and $K_1 >> K_2$:

$$[H^+] = \sqrt{K_1 K_2}; \quad pH = \frac{pK_1 + pK_2}{2} \quad (13\text{--}7)$$

EXAMPLE 10.   Determine the pH of 0.1 M KHSO$_3$?

SOLUTION: Using Eq. 13-7 gives: pH = (pK_1 + pK_2)/2 = (1.89 + 5.3)/2 = 3.59.

EXAMPLE 11.   Determine the pH of 0.10 M HK$_2$PO$_3$?

SOLUTION: Use of Eq. 13-7, but with K_2 and K_3, gives a pH of 9.72.

13.7 THE LAW OF MASS ACTION

All chemical reactions are reversible (although many to an infinitesimally small degree), and an equilibrium condition is eventually established where the rate of the forward reaction is equal to the rate of the reverse reaction. A generalized form of a chemical reaction is given as $aA + bB \rightleftarrows cC + dD$, where the lowercase letters in italics are stoichiometric coefficients and the capital letters are reacting species or products.

Strictly speaking, the *effective concentration* of ionic species is lower than the actual concentration. The ratio of effective concentration to the actual concentration is known as the *activity coefficient* of the ion. Generally, as the total ionic strength of a solution increases, the activity coefficient of a given species decreases from unity. The *activity* is the *effective concentration* of ionic species. All equilibrium constants are correctly stated only in terms of activity. Fortunately, in dilute solutions the activity coefficients are approximately equal to unity, and one can use concentration directly instead of activities to simplify the calculations. Even concentrated solutions are analyzed (such as by titration) under relatively dilute conditions. Assuming activity coefficients of unity, the molar concentrations, [A], [B], etc. are related at equilibrium as follows:

$$K_c = \frac{[C]^c\,[D]^d}{[A]^a\,[B]^b} \qquad (13\text{-}8)$$

K_c is called the *equilibrium constant* and varies with temperature. Solubility products and acid ionization constants are specific examples of the law of mass action. This law applies to gaseous or solution reactions. A key concept is that the law of mass action does not predict the reaction rate, only the final equilibrium concentration. For example, a gaseous mixture of hydrogen and oxygen at 25°C is predicted to be almost completely reacted to form water; however, at 25°C the reaction will take thousands of years to reach equilibrium. *Catalysts* are used to increase reaction rates, but do not affect equilibrium constants. Sometimes it is difficult to know whether there is no change observed in the concentration of reactants because equilibrium is achieved or because the reactions are very slow.

With gases, it is often easier to express concentrations as partial pressures (the pressure exerted by an individual species). When this is done, the term K_p is used and often the units are atmospheres. K_p is usually highly dependent on the temperature. For example, the equilibrium constant for the oxidation of sulfur dioxide to sulfur trioxide is $4.0 \cdot 10^{24}$ atm^{-1} at 298 K, $2.5 \cdot 10^{10}$ atm^{-1} at 500 K, and 3.4 atm^{-1} at 1000 K. (Of course, K_p can be determined from K_c using the ideal gas law.) When the reaction species are expressed as their pressures, P, Eq. 13-8 becomes:

$$K_p = \frac{(P_C)^c\,(P_D)^d}{(P_A)^a\,(P_B)^b} \qquad (13\text{-}9)$$

EXAMPLE 12.   Nitrogen gas reacts with hydrogen gas at elevated temperature in the presence of a catalyst to produce ammonia in the Haber synthesis. At 500°C and equilibrium the concentration of ammonia is 0.43 M, the concentration of hydrogen gas is 1.09 M, and the concentration of nitrogen gas is 1.3 M. Calculate the equilibrium constant for the formation of ammonia.

SOLUTION: The reaction is $N_2 + 3H_2 \rightleftarrows 2NH_3$. Using Eq. 13-8 with $a = 1$, $b = 3$ and $c = 2$, and substituting in the concentrations in molarity, it is found that $K_c = 0.11\ M$.

PROBLEM: During the production of SO_2 for use in sulfite pulping, if excess oxygen is present, SO_3 may be formed and interferes with the pulping process. At 1000 K, $K_p = 3.4$ atm^{-1} for the reaction $2SO_2 + O_2 \rightleftarrows 2SO_3$. If the partial pressures of SO_2 and O_2 are 0.25 atm and 0.001 atm, respectively, what is the partial pressure of SO_3 at equilibrium? Answer: 0.0146 atm.

13.8 SOLUBILITY PRODUCTS

Many salts have limited solubility in water. This leads to scale formation on processing equipment. Conditions are chosen to minimize scale formation by understanding the solubility properties of these compounds. For a salt of the form A_aB_b, the following relationship can be written: $A_aB_b(s) \rightleftarrows aA + bB$. The effective concentration of a solid in dilute aqueous solutions is a constant, provided there is excess solid. The solubility constant can be derived from the law of mass action where the concentration (strictly speaking activities) of the reactant (the solid, which is of constant activity) is incorporated into the equilibrium constant and defined as follows:

$$K_{sp} = [A]^a [B]^b \qquad (13\text{-}10)$$

If one of the ions is already in solution, the solubility of the original solid will be much lower. This is known as the *common ion effect*. For example, the solubility of AgCl is much lower in 0.1 M HCl than it is in pure water. Example 13 demonstrates this concept.

Table 13-5 gives some values of solubility product constants. One should realize that different sources give different values that may vary by as much as a factor of ten, especially for extremely small values. *Lange's Handbook of Chemistry* lists many values and the *CRC Handbook of Chemistry and Physics* gives calculated values of K_{sp}. K_{sp} values can be calculated from the relationship of $\ln K_{sp} = \Delta G/RT$, where ΔG is the Gibbs free energy of the dissolution reaction. This is the same relationship for any equilibrium constant, of course.

EXAMPLE 13. $BaSO_4$ from wood causes much scaling on processing equipment in the southern U.S. K_{sp} for barium sulfate is 1.1×10^{-10} (mol/L)^2 at 25°C. What is the concentration of Ba^{2+} and SO_4^{2-} in distilled water containing excess $BaSO_4(s)$?

SOLUTION: $BaSO_4(s) \rightleftarrows Ba^{2+} + SO_4^{2-}$; so

$[Ba^{2+}] = [SO_4^{2-}] = x;\ x^2 = 1.1 \times 10^{-10};$

$x = 1.05 \times 10^{-5}\ M.$

Table 13-5. Solubility product constants at 25°C.

Substance	Formula	K_{sp}
Aluminum hydroxide	$Al(OH)_3$	1×10^{-33}
Barium carbonate	$BaCO_3$	5×10^{-9}
Barium sulfate	$BaSO_4$	1.1×10^{-10}
Calcium carbonate	$CaCO_3$	4×10^{-9}
Calcium hydroxide	$Ca(OH)_2$	5.5×10^{-6}
Calcium sulfate	$CaSO_4$	1×10^{-5}
Calcium sulfite	$CaSO_3$	7×10^{-8}
Iron(II) hydroxide	$Fe(OH)_2$	1×10^{-14}
Iron(II) sulfide	FeS	6.3×10^{-18}
Iron(III) hydroxide	$Fe(OH)_3$	3×10^{-39}
Magnesium carbonate	$MgCO_3$	1×10^{-5}
Magnesium hydroxide	$Mg(OH)_2$	1.5×10^{-11}
Magnesium sulfite	$MgSO_3$	3.2×10^{-3}

PROBLEM: The concentration of sulfate from sulfuric acid in whitewater is 0.0001 M. What is the equilibrium concentration of Ba^{2+} ion if excess $BaSO_4$ is present?

Answer: $1.1 \times 10^{-6}\ M$. This shows that the presence of sulfate ion in solution reduces the solubility of the barium sulfate, which is predicted by the common-ion effect.

EXAMPLE 14. Calcium hydroxide is important in the kraft process as it is used to regenerate sodium hydroxide from sodium carbonate. The K_{sp} for $Ca(OH)_2$ is 5.5×10^{-6} (mol/L)^3 at 25°C. What is the solubility of calcium hydroxide in water?

SOLUTION: $Ca(OH)_2 \rightleftarrows Ca^{2+} + 2 OH^-$; so

$[Ca^{2+}][OH^-]^2 = 5.5 \times 10^{-6} M^3$

$[Ca^{2+}] = x; \quad [OH^-] = 2x$

$x \cdot (2x)^2 = 5.5 \times 10^{-6} M^3$

$x = 0.0111 M; \quad [OH^-] = 0.0222$

the pH is 12.34 and the original concentration of OH^- (10^{-7}) does not interfere with its solubilization by the common ion effect.

Salts containing anions that are basic (for example, OH^-, CO_3^{2-}, or S^{2-}) have more complicated chemistry. For example K_{sp} for calcium carbonate is 4×10^{-9} (mol/L)^2 at 25°C. However, the carbonate anion is a weak base and at pH 10.32 half of the CO_3^{2-} is actually in the form of HCO_3^-. Since calcium bicarbonate is fairly soluble in water, below pH 7 calcium carbonate is fairly soluble and exists as calcium bicarbonate.

Salts may contain cations that are acidic because they complex with OH^- groups (leaving H^+ behind). The solubilities of such salts may be strongly dependent on pH. Aluminum will be considered in detail because of its importance in wet end chemistry. Aluminum hydroxide has very limited solubility, as shown in Fig. 13-4, since:

$$Al(OH)_3 \text{ (s)} \rightleftarrows Al^{3+} + 3\ OH^-; \quad K_{sp} = 1 \times 10^{-33}.$$

Since the solubility of the aluminum ion is of concern in wet end chemistry, it will be determined as follows: $[Al^{3+}] = K_{sp} \times [OH^-]^3$ Since $K_w = 10^{-14} = [H^+][OH^-]$, the molar solubility of aluminum ion can be written as a function of pH as follows: $[Al^{3+}] = K_{sp} \times ([H^+]/K_w)^3$

Additionally, one can consider the solubility of other aluminum species. At higher pH's the formation of a higher complex occurs to an appre-

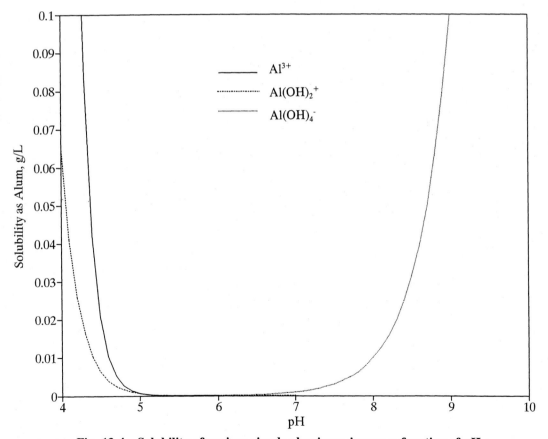

Fig. 13-4. Solubility of various simple aluminum ions as a function of pH.

Table 13-6. Assigning oxidation numbers (ON) to atoms of molecules or ions.

No.	Rule
1.	The total ON of all of the atoms in the molecule or ion is equal to the total charge of the molecule or atom.
2.	Atoms in their elemental form have an ON = 0
3.	Group I elements in combined form have ON = +1 Group II elements in combined form have ON = +2 Group III elements (except boron) have ON = +1 or +3 depending on the charge of the metal ion. Group IV elements (except carbon and silicon) have ON = +4 or +2 depending on the charge of the metal ion.
4.	Hydrogen ON = +1 in combination with nonmetals and -1 in combination with metals.
5.	Fluorine in combined form has an ON = -1
6.	Oxygen has ON = -1 in peroxides (O_2^{2-}, for example, HOOII) ON = -1/2 in superoxides (O_2^-) ON = -1/3 in ozonides (O_3^-) ON = -2 unless combined with F

ciable extent. $Al(OH)_3$ (s) + OH⁻ ⇌ $Al(OH)_4^-$ (aq) with K_c = 40 at 25°C. The concentration of aluminum is solved as $Al(OH)_4^-$(aq) = K_c × (K_w/[H⁺]). At intermediate pH the aluminum ion complexes according to $Al^{3+} + H_2O \rightarrow Al(OH)^{2+} + H^+$ with K_a = 9.78 × 10⁻⁶; $[Al(OH)^{2+}]$ = $K_a[Al^{3+}]/[H^+]$. Other complexes (Section 14.7) exist in the pH range of 4-6, but their concentrations are too complicated to calculate. Fig. 13-4 shows the concentration of Al^{3+}, $Al(OH)^{2+}$, and $Al(OH)_4^-$ as a function of pH in terms of g/L hydrated aluminum sulfate (alum, with a formula weight of 666 grams per 2 moles of Al^{3+}.)

13.9 OXIDATION-REDUCTION REACTIONS

Oxidation-reduction (redox) reactions are similar to acid-base reactions except the unit of currency is not the proton but the electron. They are important in combustion, redox titrations such as, determination of lignin contents using potassium permanganate, and many other processes. Like protons, electrons cannot exist in solution by themselves so both electron donors and electron acceptors are required for a reaction to occur. Redox reactions, however, are not confined to solvents such as water, but can occur in any form of matter. *Oxidation* is defined as the loss of electrons by an element during a chemical reaction, and *reduction* is defined as the gain of electrons. (This is easily remembered as LEO growled GERRR; loss-electrons-oxidation, gain-electrons-reduction.) The *oxidizing agent*, or *oxidant*, is the substance that causes oxidation, while the *reducing agent* is the substance that causes reduction.   Oxidation was originally a reaction between a compound and oxygen gas, where oxygen itself is reduced but causes oxidation. Oxygen is the most electronegative element (except for fluorine which is not nearly as ubiquitous as oxygen). *Disproportionation* is a redox reaction where a compound is both oxidized and reduced, such as NO_2 in the production of nitric acid:

$$3NO_2 + H_2O \rightarrow 2HNO_3 + NO$$

Electrons are assigned to atoms of molecules or ions according to the rules in Table 13-6.

Oxidation numbers (ON) are assigned to elements with these rules. The lower number rule always takes precedence over a later numbered rule. Rule 6 is summarized as follows: except for hydrogen peroxide where oxygen has a charge of -1, oxygen almost always has a charge of -2. While these rules are somewhat arbitrary, it allows reactions involving transfer of electrons to be balanced. Let us consider the reaction of sodium metal with water used to make sodium hydroxide. $2Na + 2H_2O \rightarrow 2Na^+ + 2OH^- + H_2$. Clearly sodium has gone from neutral to a 1+ charge by losing an electron; in this reaction sodium was the reducing agent (for hydrogen) and was oxidized. Some of the hydrogen of water went from 1+ to neutral by accepting an electron; hydrogen was the oxidizing agent, and it was reduced in the reaction.

EXAMPLE 15. Calculate the charge on sulfur in each of the following compounds: S_8, H_2S, SO_2, H_2SO_3, and SO_3.

SOLUTION: In each case hydrogen has an ON = +1 and oxygen = -2; therefore the charges on sulfur are 0, -2, +4, +4, +6. The chemistry of sulfur is relatively complex due to all of the possible oxidation numbers in a variety of compounds.

PROBLEM: What are the oxidation numbers of carbon in each of the following compounds: C, CO, CO_2, and H_2CO_3?

Answer: 0, +2, +4, +4.

One can determine whether or not a redox reaction has occurred by determining whether the oxidation number of any element has changed in the course of a reaction. For example, the reaction of $H^+ + OH^- \rightarrow H_2O$ is not a redox reaction, but $C + O_2 \rightarrow CO_2$ is a redox reaction.

13.10 ELECTROCHEMISTRY

Many materials are capable of both oxidation and reduction reactions. The type of reaction depends on what the other reactant is. For example, sulfur may be reduced with hydrogen to form H_2S or oxidized with oxygen to form SO_2. Oxygen,

being more electronegative than sulfur, causes oxidation, while hydrogen is more electropositive than sulfur and causes reduction. Therefore, redox reactions are conveniently split into two *half reactions* with one half reaction corresponding to oxidation and the other half reaction corresponding to reduction. Zinc metal dissolves in hydrochloric acid with the *overall reaction* as follows:

$$Zn(s) + 2\ HCl \rightarrow ZnCl_2 + H_2(g) \quad \text{(overall)}$$

$$Zn(s) \rightarrow Zn^{2+} + 2e^- \quad \text{(oxidation)}$$

$$2H^+ + 2e^- \rightarrow H_2 \quad \text{(reduction)}$$

It is useful to prepare a table of half reactions with the energy released or absorbed in the reaction to predict whether or not overall reactions will occur based on thermodynamics. This technique also simplifies the balancing of complex chemical equations. *Lange's Handbook of Chemistry* is one good source for half reaction potentials. Some half reactions are listed in Table 13-7 with compounds on the left in decreasing order of oxidation power. This means the higher a compound is on the left hand portion of the table, the more likely it is to induce oxidation of another compound or ion. Similarly, compounds or ions on the lower right are powerful reducing agents, the compounds higher up on the right hand side of the table are lower in reduction potential. Remember these are thermodynamic values; they say nothing about the reaction kinetics.

EXAMPLE 16. Balance the following equation, which represents the Palmrose method of titrating sulfite liquors:

$$H_2SO_3 + KIO_3 \rightarrow H_2SO_4 + KI$$

SOLUTION: First write the two half reactions. Each mole of sulfur in sulfurous acid looses two electrons as it goes from +4 to +6 valence. Each mole of iodine gains six electrons as it goes from +5 to -1. So,

$$KIO_3 + 6e^- + 6H^+ \rightarrow 3H_2O + KI$$

$$3H_2SO_3 + 3H_2O \rightarrow 3H_2SO_4 + 6H^+ + 6e^-$$

$$3H_2SO_3 + KIO_3 \rightarrow 3H_2SO_4 + KI$$

Table 13-7. Reduction half reaction potentials at 25°C.

Oxidizing agent	Reducing agent	$E°$, V at 25°
↓ *Strongly oxidizing*		
$HClO + H^+$	$+\ e^- \rightarrow\ 1/2\ Cl_2 + H_2O$	$+1.63$
$MnO_4^- + 8H^+$	$+\ 5e^- \rightarrow\ Mn^{2+} + 2H_2O$	$+1.51$
$HClO + H^+$	$+\ 2e^- \rightarrow\ Cl^- + H_2O$	$+1.49$
Cl_2	$+\ 2e^- \rightarrow\ 2\ Cl^-$	$+1.36$
$ClO_2 + H^+$	$+\ e^- \rightarrow\ HClO_2$	$+1.27$
$IO_3^- + 6H^+$	$+\ 5e^- \rightarrow\ ½\ I_2 + 3H_2O$	$+1.19$
$IO_3^- + 6H^+$	$+\ 6e^- \rightarrow\ I^- + 3H_2O$	sum of two reactions
ClO_2	$+\ e^- \rightarrow\ HClO_2^-$	$+0.95$
$ClO^- + H_2O$	$+\ 2e^- \rightarrow\ Cl^- + 2\ OH^-$	$+0.89$
H_2O_2	$+\ 2e^- \rightarrow\ 2\ OH^-$	$+0.88$
Fe^{3+}	$+\ e^- \rightarrow\ Fe^{2+}$	$+0.77$
$ClO_2^- + 4H^+$	$+\ 4e^- \rightarrow\ Cl^- + 2H_2O$	$+0.75$
I_2 (aq)	$+\ 2e^- \rightarrow\ 2\ I^-$	$+0.620$
I_3^-	$+\ 2e^- \rightarrow\ 3\ I^-$	$+0.534$
$O_2 + 2H_2O$	$+\ 4e^- \rightarrow\ 4\ OH^-$	$+0.41$ (1 N NaOH)
Cu^{2+}	$+\ 2e^- \rightarrow\ Cu$	$+0.345$
$SO_4^{2-} + 4H^+$	$+\ 2e^- \rightarrow\ SO_2 + 2H_2O$ (or H_2SO_3)	$+0.17$
$S_4O_6^{2-}$	$+\ 2e^- \rightarrow\ 2S_2O_3^{2-}$ (thiosulfate, hypo)	$+0.08$
$2\ H^+$	$+\ 2e^- \rightarrow\ H_2$	0.00 (by definition)
Fe^{3+}	$+\ 3e^- \rightarrow\ Fe$	-0.04
Ni^{2+}	$+\ 2e^- \rightarrow\ Ni$	-0.250
$2S$	$+\ 2e^- \rightarrow\ S_2^{2-}$	-0.43
Fe^{2+}	$+\ 2e^- \rightarrow\ Fe$	-0.440
Cr^{3+}	$+\ 3e^- \rightarrow\ Cr$	-0.71
Zn^{2+}	$+\ 2e^- \rightarrow\ Zn$	-0.762
$2H_2O$	$+\ 2e^- \rightarrow\ H_2 + 2\ OH^-$	-0.83
Mn^{2+}	$+\ 2e^- \rightarrow\ Mn$	-1.05
$2SO_3^{2-} + 2H_2O$	$+\ 2e^- \rightarrow\ S_2O_4^{2-} + 4OH^-$ (hydrosulfite)	-1.12
Al^{3+}	$+\ 3e^- \rightarrow\ Al$	-1.66
Mg^{2+}	$+\ 2e^- \rightarrow\ Mg$	-2.36
Na^+	$+\ e^- \rightarrow\ Na$	-2.712
Ca^{2+}	$+\ 2e^- \rightarrow\ Ca$	-2.87
K^+	$+\ e^- \rightarrow\ K$	-2.922
	↑ *Strongly reducing*	

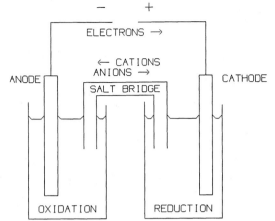

ELECTRONS →

← CATIONS
ANIONS →

ANODE   CATHODE

SALT BRIDGE

OXIDATION   REDUCTION

Fig. 13-5. Diagram of the Daniell cell, a representative electrochemical cell.

Two half reactions can form a redox *couple* which produces a voltage. By convention, the oxidation cell is written first. The classic electrochemical cell is the Daniell cell invented in 1836 to provide power for the telegraph industry. It is a zinc-copper cell. A representative cell is constructed as shown in Fig. 13-5.

Cell diagrams are written with the anode $\|$ cathode half reactions. The chemical equation and cell diagram of the Daniell cell are:

$$Zn + Cu^{2+} \rightleftarrows Cu + Zn^{2+}$$

$$Zn\,|\,Zn^{2+}(aq)\,\|\,Cu^{2+}(aq)\,|\,Cu$$

If a solid metal is not an electrode for a particular half reaction, platinum should be specified to provide the necessary electrical contact. For example, the reaction of zinc metal with hydrochloric acid is specified as:

$$Zn\,|\,Zn^{2+}(aq)\,\|\,H^{+}(aq)\,|\,H_2(g)\,|\,Pt$$

$E°$ is the *standard cell potential*, or voltage, generated when the activity of all of the species is equal to unity and the pressure is 1 atm, i.e., the concentrations of all (non gas) species are essentially 1 M. The *cathode* is the electrode where reduction occurs, and the *anode* is the electrode where oxidation occurs. The anode is the source of electrons and is the negative terminal. $E°$(anode) is $-E°$ of the reduction reaction.

$$E°(cell) = E°(anode) + E°(cathode) \quad (13\text{-}11)$$

EXAMPLE 17. What is $E°$ of the voltaic cell $Zn\,|\,Zn^{2+}(1\ M)\,\|\,Fe^{3+}(1\ M)\,|\,Fe$?

SOLUTION:

$$\begin{array}{ll} 3(Zn \rightarrow Zn^{2+} + 2e^-) & +0.76 \\ 2(Fe^{3+} + 3e^- \rightarrow Fe) & \underline{-\ 0.04} \\ & +0.72 \end{array}$$

PROBLEM: What voltage is generated by a lithium cell operating under standard conditions with the reaction:

$$2Li + Cl_2 \rightarrow 2Li^+ + 2Cl^-$$

Answer: 4.4 V.

The equilibrium constant for any reaction which can be expressed as two half reactions of known potential can be derived by substituting $\Delta G = -nFE°$ into the equation $\Delta G = -RT\ \ln K$ to give:

$$\ln K = nF/RT \times E° \quad (13\text{-}12)$$

where F is the number of coulombs in a mole of electrons or 96,485 C/mol and n is the number of electrons transferred in the reaction. Since 1 J = 1 C $\times$ 1 V, the term $nFE°$ is the energy of the reaction in joules. The term F/RT appears frequently in electrochemistry and is equal to 0.02569 V at 25°C. Eq. 13-12 is usually written as a common logarithm and takes on the form:

$$\log K = nE°/0.0592\ V \quad 25°C \quad (13\text{-}13)$$

The *Nernst equation* relates the actual voltage of the cell to the standard voltage of the cell and the concentration of reactants and products. The reaction quotient, Q, is the actual concentration of products raised to the stoichiometric coefficients divided by the concentration of reactants raised to the stoichiometric coefficients in the usual manner.

$$(13\text{-}13a)$$

$$E = E^o_{(cell)} - \frac{0.0592\ V}{n} \times \log Q$$

EXAMPLE 18.   Give the equilibrium constant for the reaction: $2Ag^+ + Zn \rightarrow Zn^{2+} + 2Ag$

SOLUTION: $E° = 1.56$ V;
   $\log K = 2(1.56)/0.0592 = 52.7$;
   $K = 5 \cdot 10^{52}$

EXAMPLE 19.   What is the potential of a Zn-Cu^{2+} cell with $[Cu^{2+}] = 2$ M and $[Zn^{2+}] = 0.2$ M?

SOLUTION: $E = 1.10 - 0.0592/2 \times \log 0.2/2$
   $= 1.13$ V.

13.11   PRACTICAL ASPECTS OF ELECTROCHEMISTRY

Corrosion

Corrosion of metals is the result of electrochemical reactions. For example, in the zinc-copper cell, the zinc metal is lost or corroded. Since metal becomes soluble (corroded) as it goes from a neutral to cationic state, corrosion occurs at the anode where electrons are given up by metal. These electrons must be used by a cathodic reaction to complete the cell. An obvious corrosion reaction occurs when zinc or iron is put into hydrochloric acid. Numerous bubbles form as the metal displaces hydrogen from water. Parts of the metal act as the anode where electrons are given up and the metal goes into solution as ions, while other parts of the metal act as the cathode where electrons are accepted to convert hydrogen ions into hydrogen gas. As the anode and cathode areas move around the surface, the entire piece of metal is ultimately dissolved.

In water near neutral pH, hydroxide ion is produced at the cathode (by loss of hydrogen ions) which precipitates ferrous ion in solution to give ferrous hydroxide. The ferrous hydroxide is further oxidized to ferric hydroxide, which is the rust observed on the surface of iron. The behavior of the rust (whether it forms on the surface or away from the surface) depends on the pH and oxygen content of the water and gives rise to several types of corrosion mentioned below.

When two metals of different composition (dissimilar metals) are electrically connected and placed in a solution with an electrolyte, a voltage is generated. The resultant corrosion is called *galvanic action*. This can occur as potentials exist across various parts of the same piece of metal due to differences in surface compositions (like variations in oxygen concentrations caused by dirt that excludes oxygen, leading to pitting), solution composition at the surface, stress in the metal, etc. When two dissimilar metals are in contact, the more strongly reducing metal (Table 13-7) will corrode. A large difference in potential may lead to severe corrosion. For example, zinc will corrode if attached to iron. When the surface area of the metal that corrodes is small relative to the other metal, corrosion is particularly quick. For example, iron rivets holding copper sheets together will corrode very quickly. Also, pitting represents a rapid corrosion of a small area made anionic by oxygen starvation. (These are thermodynamic considerations, they do not say anything about the rate of corrosion.)

Often an easily corroded metal is deliberately attached to a large piece of metal to protect the large piece of metal from corroding. The metal that is corroded is called a *sacrifice anode* and keeps the metal it protects cathodic, thereby not allowing it to corrode. This technique is known as *cathodic protection*. Galvanized (zinc coated) iron, magnesium blocks inside hot water tanks, and zinc anodes attached to the submerged metal parts of marine vessels are three examples of sacrifice anodes. It is used in water heaters, marine vessels and engines, pipelines, and bridges. The sacrificial anode (usually magnesium or zinc) corrodes in preference to the material that it protects. The sacrificial anode is periodically replaced to maintain protection.

A related method of protection is to apply a small, negative direct current voltage to the device being protected if it is electrically insulated from its environment. It must be insulated from its environment or else it would take too much power (amperage) to maintain the voltage. The applied voltage makes the metal surface cathodic. This technique has been used to protect underground metal pipes (if they are electrically insulated from ground).

A more strongly reducing metal can displace another metal from solution. A copper film forms over most of a piece of iron that is dipped in copper sulfate. If a surface of a less strongly reducing metal is desired, the reaction can be forced by the application of an external voltage in the process of *electroplating*. Electroplating occurs when reduction of a metal from solution is forced by an external voltage, much like the reaction at the cathode in a Daniell cell.

Some materials, like aluminum, would seem to be unstable based on their position in the electrochemical series. However, aluminum and chromium both form oxides on their surfaces that are very hard, insoluble, nonporous, self-healing, and inert. The formation of such films renders metals passive. These surfaces protect the underlying metal from corrosion. For example, 12-18% chromium in stainless steels protects the accompanying iron from corrosion. Actually, stainless steel (active) is vulnerable to corrosion; however, the coating (passive) that forms on stainless steel is inert. *Crevice corrosion* results when oxygen is kept from the surface of an active material. Oxygen gains electrons during its reaction so high areas of oxygen react at the cathode and low areas of oxygen provide an area of decreased potential for the anode reactions. This occurs under deposits, barnacles, or plastic washers used to insulate fastenings on stainless steel plates used in sea water. Table 13-8 shows metals presented in increasing resistance to galvanic corrosion in sea water. In other solutions the order may change.

Corrosion can be decreased in a number of ways. Painting prevents air, water, and salts from reaching the surface of the metal. *Galvanization* is the process of coating the entire surface with an unbroken coat of zinc by immersion in molten zinc or by *electroplating*. The zinc is protected by passivation with a zinc oxide coating. The iron is protected since zinc is oxidized in preference to it.

Electrolysis

Some other practical aspects of electrochemistry include production of NaOH and Cl_2 and production of ClO_2.

Ion-selective electrodes

Ion-selective electrodes are cells that obey the

Table 13-8. Galvanic series of metals and alloys in sea water with increasing resistance.

Metal or alloy	Nominal composition
Magnesium	Mg
Zinc	Zn
Aluminum alloys	
Aluminum	Al
Cadmium	Cd
Mild Steel	Low carbon
Alloy Steel	Various below 5%
Cast Iron	3-4% C, 0.5-3.5% Si
Type 410 SS, Active	11.5-13.5% Cr
Type 430 SS, Active	12-16% Cr
Type 304 SS, Active	18% Cr, 8% Ni
Type 316 SS, Active	18% Cr, 8% Ni, 3% Mo
Nickel Cast Iron	0.5-10% Ni, 3-4% C, Cr
Yellow Brass	65% Cu, 35% Zn
Copper	Cu
Nickel	Ni
Type 304 SS, passive	18% Cr, 8% Ni
Type 316 SS, passive	18% Cr, 8% Ni, 3% Mo
Monel	67% Ni, 28% Cu
Silver	Ag
Hastelloy, Alloy C	Ni-Cr-Mo
Titanium	Ti
Platinum	Pt

rules of electrochemistry. The Nernst equation shows that the electrochemical potential of a cell depends on the chemical concentration of the reactive species. This principle is used in pH electrodes and other ion-specific electrodes for sodium, sulfide, carbonate, fluoride, etc. Ion-selective electrodes, other than for pH, have not been used extensively by the pulp and paper industry, but they offer the potential for rapid laboratory analysis or on-line sensors for process control.

13.12 PROPERTIES OF GASES

Ideal gas law

The *kinetic theory* of gases is a set of three assumptions about the behavior of ideal gases. While the assumptions are not completely true, they are approximately true at low pressures, generally below one atmosphere pressure. (Much of the non-ideal behavior of gases is a result of the nonzero volume of molecules, which increases the pressure compared with ideal behavior, and the formation of dimers, trimers, and clusters, which decrease the actual pressure compared with ideal behavior.) From these assumptions, much about gas behavior can be predicted. The assumptions are:

1. A gas consists of a collection of molecules or atoms in continuous random motion.

2. The molecules or atoms are infinitely small particles which move in straight lines until they collide.

3. Gas molecules do not affect each other except during collisions.

The ideal gas law relates the temperature, pressure, volume, and number of moles of an ideal gas. T is always expressed in the absolute temperature scale of Kelvin, while volume is usually expressed in liters. When pressure is in atmospheres, $R = 0.08206$ L·atm·K^{-1}·mol^{-1}; when pressure is in torr, $R = 62.37$ L·torr·K^{-1}·mol^{-1}; when pressure is in Pascals, $R = 8.314$ J·K^{-1}·mol^{-1}. R is called the *gas constant*. The ideal gas law is usually written as Eq. 13-14. *Standard temperature and pressure, STP*, conditions are defined as 273.15 K and 760 torr (or 1 atm); under these conditions one mole of an ideal gas occupies 22.41 L.

$$PV = nRT \qquad (13\text{-}14)$$

The compressibility factor, Z, is defined in Eq. 13-15, and is equal to 1 for an ideal gas. Z is a measure of the deviation from ideal behavior and is usually less than one except for highly volatile gases at low pressures.

$$Z = PV/nRT \qquad (13\text{-}15)$$

EXAMPLE 20. 1.5 g of argon occupies 4.03 L at 298 K. What is the pressure exerted by this gas inside the vessel, in atm, if argon behaves as an ideal gas?

SOLUTION: First the number of moles of argon is calculated to be 0.0375 based on its atomic weight of 40. Using Eq. 13-14 with $R = 0.08206$ L·atm·K^{-1}·mol^{-1}, the pressure is determined to be 0.228 atm.

PROBLEM: What is the volume of 0.154 mol of an ideal gas at 745 torr and 30.0°C? Answer: 3.91 L.

Often one desires to compare a single system under two different states of pressure, volume, and temperature. In this case the two states will be designated as 1 and 2 with subscripts. Rearrangement of Eq. 13-14, with nR a constant, leads to Eq. 13-16.

$$P_1 V_1/T_1 = P_2 V_2/T_2 \qquad (13\text{-}16)$$

EXAMPLE 21. 5.0 liters of a gas are collected at a temperature of 300 K and a pressure of 720 torr. What would the volume be at STP conditions?

SOLUTION:

$$V_2 = V_1 \times \frac{P_1}{P_2} \times \frac{T_2}{T_1} = 5\,\ell \times \frac{720\ \text{torr}}{760\ \text{torr}} \times \frac{273\,\text{K}}{300\,\text{K}} = 4.31\,\ell$$

The root-mean-square velocity of the molecules v is related to the molar mass M by the relationship $v = (3RT/M)^{1/2}$. Since diffusion is proportional to velocity, it is also inversely proportional to the square root of molecular weight (Graham's law). Since kinetic energy is ½ mv^2, it follows that the internal energy (not to be confused with work done by an expanding gas which is an external energy) of a gas is 3/2·RT. The latter equation can be used to calculate the heat capacity of an ideal gas.

Real gas equations

There are many empirical equations that approximate the behavior of real gases at elevated pressures where deviations from ideal become appreciable. The most famous is the *van der Waals equation* developed by the nineteenth century Dutch chemist. The constant *a* represents the attractive forces which decrease the actual pressure from ideal, and the constant *b* represents the effect of repulsions or the finite volume of the molecules which increase the actual pressure from ideal. The equation is not useful for the liquid phase.

$$(P + \frac{an^2}{V^2}) \times (V-nb) = nRT$$

The equation is not useful at very high pressures; for example, while the van der Waals equation predicts a compressibility factor, Z, of 0.375 at the critical point of a gas, in fact, gases usually have a compressibility factor on the order of 0.25 to 0.30 at their critical point. The *CRC Handbook of Chemistry and Physics* gives values of *a* and *b* for many gases (Table 13-9). The constants may also be obtained from the critical data of gases, usually using P_c and T_c as follows:

$$a \approx 27R_2T_c^2/64P_c; \quad b \approx RT_c/8P_c$$

One of the most accurate equations of state is the *Redlich-Kwong equation*, which is useful over wide ranges of temperature and pressure. The constants *a* and *b* are different than those of the van der Waals equation.

$$(P + \frac{an^2}{V(V+b/n)\,T^{1/2}}) \times (V-nb) = nRT$$

13.13 ANNOTATED BIBLIOGRAPHY

Corrosion

1. *Perry's Chemical Engineers' Handbook* has

Table 13-9. Van der Waals' constants of several gases.

Gas	a L²·atm·mol⁻²	b, L·mol⁻¹
CO_2	3.59	0.0427
Cl_2	6.49	0.0562
N_2	1.39	0.0391
O_2	1.36	0.0318
$(CH_3)_2CH_2$	8.66	0.0845
SO_2	6.71	0.0564
H_2O	5.46	0.0305
CH_4	2.25	0.0428

a very useful, detailed chapter on the subject (section 23 of the 6th ed.).

2. *Corrosion in Action*, International Nickel Co., 1961. This was used as a source of information about corrosion for this chapter.

EXERCISES

The elements

1. Draw the Lewis structures for the following compounds showing all valence electrons.

    a) Na_2S b) CO_2
    c) CCl_4 d) CH_4

Ionic and covalent bonds

2. Which of the following statements is true of a solution containing an electrolyte?

    a) An electrolyte solution contains ions.

    b) Water is a good solvent to use to produce an electrolyte solution.

    c) An electrolyte solution will conduct electricity.

    d) All of the above.

3. Which of the following is the most ionic compound?

    a) BF_3 b) CCl_4
    c) Na_2S d) SiO_2

Hydrogen bonding

4. In which of the following compounds is hydrogen bonding possible?

    a) RNH_2 b) RCHO   c) RCOOH
    d) H_2S e) CO_2 f) HF
    g) EtOH h) RCH_2Cl i) H_2SO_3
    j) RCH_2OH

5. Draw the partial charges involved with hydrogen bonding of HCl in water.

6. Which of the following three compounds is expected to have the highest boiling point? Why?

    a) $CH_3CH_2OCH_2CH_3$
    b) $CH_3CH_2CH_2OCH_3$
    c) $CH_3CH_2CH_2CH_2OH$

The mole and mass percentage

7. Calculate the number of moles in each of the following:

    a) 98.3 g $C_6H_{12}O_6$
    b) 17.3 g $Na_2SO_4 \cdot 10H_2O$
    c) 210 g Na_2S

8. Calculate the mass percentage of sulfur in the following chemicals:

    a) Na_2S b) Na_2SO_4 c) H_2SO_3
    d) SO_2 e) $NaHSO_3$

9. 50 g of a mixture of NaCl and sucrose ($C_{12}H_{22}O_{11}$) was burned leaving 30.5 grams of NaCl ash. How much CO_2 was produced?

Equivalency, molarity, and normality

10. 150 g of water contains 20 g of glucose, $C_6H_{12}O_6$, with MW = 180 g/mol. What is the molality of the solution.

11. How many grams of NaOH are required to make 700 ml of 1.25 M NaOH solution?

What volume of 0.50 M H_2SO_4 is required to neutralize this solution?

Acids, bases, and the pH scale

12. What is the pH of the following solutions?

    a) 0.047 M HCl
    b) 0.0013 N NaOH
    c) 0.01 M H_2SO_4

13. What is the pH of a solution containing 0.5 M acetic acid and 0.2 M sodium acetate?

14. What is the pH of 0.1 M $NaHCO_3$?

Oxidation-reduction reactions

15. What is the oxidation number of chromium in $K_2Cr_2O_7$?

16. In the reaction $2Na + 2 H_2O \rightarrow 2NaOH + H_2\uparrow$, what species is the reducing agent?

Electrochemistry and corrosion

17. The voltage required to produce electrolysis of NaCl is about 4 volts. How much electrical energy is required to produce one pound of NaOH?

18. Describe how pitting leads to high localized corrosion.

19. Why is it important to periodically replace the magnesium strip in hot water heaters equipped with these?

20. What is galvanized metal?

Properties of gases

21. A pure liquid is analyzed as 85.6% C and 14.4% H. Under STP conditions the density of its vapor is 3.75 g/L. What is the molecular formula of the substance? Show a possible structure.

14

ANALYTICAL AND COORDINATE CHEMISTRY

This chapter presents concepts of analytical chemistry. Acid-base chemistry is important for pulping and bleaching operations. Redox reactions are important to understand bleaching, corrosion, and ion-specific electrodes. Coordinate chemistry is central to alum and wet end chemistry and involves many principles of analytical chemistry. For more detail on these concepts one should consult an introductory text on quantitative analysis or analytical chemistry.

14.1 STRONG ACID--STRONG BASE TITRATIONS

Neutralization is the formation of water from a *strong* acid and a *strong* base to form a neutral compound, i.e., a compound that is neither an acid nor a base, such as NaCl. Consider the reaction of HCl + NaOH → H_2O + NaCl. Since HCl, NaOH, and NaCl are all completely ionized in solution, this reaction can be written as: H^+ + Cl^- + Na^+ + OH^- → Na^+ + Cl^- + HOH. Na^+ and Cl^- are *spectator ions*, that is, they are not involved in the reaction, so the overall reaction is written simply as H^+ + OH^- → HOH; this is the essence of neutralization reactions. The enthalpy of neutralization of a strong acid and strong base is about -57.2 kJ/mol. The term "neutralization" is sometimes used for the reaction of any acid and base, such as the reaction of acetic acid and sodium hydroxide to form sodium acetate and water, but this is not strictly correct use of the word, since sodium acetate is a weak base.

During the start of a titration of a strong base with a strong acid (Fig. 14-1), all of the acid will be consumed by the base since there is more base than acid. The concentration of the base will be reduced slightly and the pH will decrease slightly. As the base becomes depleted near the equivalence point, its concentration decreases by several orders of magnitude and the pH decreases rapidly. At the equivalence point, all of the base is reacted with all of the acid, and the pH is equal to 7. As more acid is added, the pH decreases less and less rapidly. Fig. 14-1 shows the pH during the titration of a strong base with a strong acid. For a strong acid being titrated with a strong base, the figure can be read with the pH scale inverted.

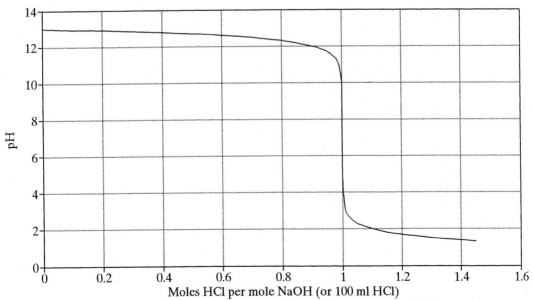

Fig. 14-1. Titration of 100 ml of 0.1 *M* NaOH with 0.1 *M* HCl.

The endpoint is detected with a pH meter or suitable indicator (usually phenolphthalein) as discussed in the next section.

Since normality is equivalents per liter, if the normality of a solution is multiplied by the volume of solution, the number of equivalents is determined for that solution: normality × volume = equivalents. Also, at the endpoint of a titration, by definition, the number of equivalents in the sample being analyzed (the analyte) is equal to the number of equivalents in the titrant (the *standard solution* used in the buret). This leads to the very useful titration equation of

$$N_1V_1 = N_2V_2 \qquad (14\text{-}1)$$

The normality of titrant multiplied by the volume of titrant equals the normality of analyte multiplied by the volume of analyte. *During titrations the volume is measured in ml; therefore it is useful to remember that* $1\ N = 1$ eq/L = 1 meq/ml.

EXAMPLE 1.   5.00 ml of HCl solution of unknown strength required 15.35 ml of 0.500 N NaOH to achieve the endpoint (neutralization). Calculate the following:

1. Milliequivalents of NaOH consumed.
2. Milliequivalents of HCl consumed, or neutralized by the NaOH.
3. Equivalents of HCl consumed.
4. The equivalent weight of HCl.
5. The milliequivalent weight of HCl.
6. The weight of HCl neutralized.
7. The normality of the original HCl solution.
8. The strength of the original HCl solution in g/L.
9. The strength of the original HCl solution in lb/gal.
10. The weight of the NaOH consumed.

SOLUTION:
1. 15.35 ml NaOH × 0.5 meq/ml = 7.67 meq NaOH.
2. By definition of equivalency, eq NaOH = eq HCl = 7.67 meq.
3. 7.67 meq/(1000 meq/eq) = 0.00767 eq HCl.

4. The equivalent weight of HCl is equal to the formula weight; therefore, the equivalent weight = 1 + 35.5 = 36.5.
5. (36.5 g/eq HCl)/(1000 meq/eq) = 0.0365 g/meq HCl.
6. 0.00767 eq HCl × 36.5 g/eq = 0.280 g HCl.
7. Since $N_1V_1 = N_2V_2$, with rearrangement one obtains the following:

$$N_1 = \frac{N_2 \times V_2}{V_1} = \frac{7.67\ \text{meq}}{5.0\ \text{m}\ell} = 1.534\ N\ \text{HCl}$$

8. (0.280 g HCl)/(0.005 L) = 56 g/L HCl
9. Dimensional analysis of the concentration in g/L gives:

$$56\,\frac{\text{g}}{\ell} \times \frac{1\ \text{lb}}{454\,\text{g}} \times \frac{1\ \ell}{0.264\,\text{gal}} = 0.467\,\frac{\text{lb}}{\text{gal}}\ \text{HCl}$$

10. 7.67 meq NaOH × 40 mg/meq × 1 g/(1000 mg) = 0.306 g NaOH

EXAMPLE 2.   How many ml of concentrated sulfuric acid (density = 1.86 g/ml) should one add to make 1.00 L of 0.05 N H_2SO_4 solution? Note the notation of normality!

SOLUTION:

$$0.05\,\text{eq}/\ell \times 1\ell \times \frac{1\ \text{mol}\ H_2SO_4}{2\ \text{eq}\ H_2SO_4} \times 98\,\text{g/mol}\ H_2SO_4$$

$$\times \frac{1\ \text{m}\ell\ H_2SO_4}{1.86\ \text{g}\ H_2SO_4} = 1.32\ \text{m}\ell\ \text{conc.}\ H_2SO_4$$

14.2 pH PROPERTIES OF WEAK ACID-CONJUGATE BASE PAIRS

The pH of a weak acid can be expressed in terms of the ionization constant and the concentrations of the weak acid, HA, and its conjugate base, A⁻. The conjugate base is sometimes referred to as the salt of the weak acid. Rearrangement of Eq. (13-2) gives the following equation that is applicable when $pK_a/[C] > 0.1$; i.e., greater than 10% ionization of A⁻:

$$[H^+] = K_a [A]/[A^-]$$

Taking the logarithm of each side gives:

$$pH = pK_a - \log([A]/[A^-])$$

Using the properties of logarithms, this equation becomes:

$$pH = pK_a + \log([A^-]/[A]) \qquad (14\text{-}2)$$

Eq. 14-2 is called the *Henderson-Hasselbalch equation*. This equation should be used with some care if the ratio of the salt to acid is very high or very low. A 10^{-3} M solution of acetic acid without sodium acetate might be expected to have a very low pH as the ratio approaches 0 (and the log approaches $-\infty$). However, as shown in Example 13-7, the acid does ionize and appreciable amounts of the conjugate base are present. Thus, Eq. 14-2 cannot be used near equivalence points (endpoints) of titrations. Eq. 14-2 can be written in the form of a weak base as follows:

$$pOH = pK_b + \log([BH^+]/[B])$$

$$= pK_b + \log([salt]/[base]) \qquad (14\text{-}3)$$

The Henderson-Hasselbalch equation is very useful for describing the pH behaviors of mixtures of weak acids and their conjugate bases in solution. If equal molar concentrations of an acid and its conjugate base are in solution, the pH will be equal to the pK_a of that acid, since $-\log(n/n) = 0$ and $pH = pK_a$. This is a useful fact to remember because it gives a rough estimate of pH of a solution containing an acid and its conjugate base. Furthermore, even if this solution is diluted, the pH will not change. Also this solution can absorb small amounts of acid or base without appreciably changing the ratio of acid and salt form. Therefore, such a solution would be a good *buffer*, that is, a solution which resists changes in pH by dilution or by addition of acid or base. Therefore, any solution of a weak acid and its conjugate base is a buffer. The amount of buffering capacity is proportional to the concentration of these species. The maximum buffering action is achieved for a given concentration of weak acid when the ratio of

the concentration of salt to acid is unity (i.e., both the acid and its conjugate base are of equal concentration). Later, it will be observed that this is the flattest part of a titration curve for a weak acid or base.

EXAMPLE 3. What is the pH of a solution of 0.1 M H_2SO_3 and 0.2 M $NaHSO_3$?

SOLUTION: Table 11-3 gives a pK_a of 1.89 for sulfurous acid at 25°C. Therefore:

$$pH = 1.89 + \log (0.2/0.1) = 2.19$$

EXAMPLE 4. A solution that is 1 M acetic acid and 1 M sodium acetate has a pH of 4.76, which is the pK_a for acetic acid. What will the pH be after 0.5 mole of HCl is added to 1 liter of this solution?

SOLUTION: The strong acid reacts completely with the weak base to give 0.5 mol/L HAc and leaving 0.5 mol/L Ac⁻. The total [HAc] = 1.5 M. Therefore, pH = 4.76 + $\log(0.5/1.5) = 4.28$, a small change.

PROBLEM: What would be the pH of 1 L of water with 0.5 moles HCl? Answer: The pH would be 0.30, which shows the effect of a buffer solution when compared to this example.

14.3 pH INDICATORS

The pH of solutions can be monitored using *indicators*. Indicators are themselves weak acids or bases whose ionized forms differ in color from the un-ionized forms. The color of these species should be intense so only small amounts of indicator are used; after all, the indicators are also reacting with the titrant. Usually these compounds involve conjugated aromatic systems. A commonly used indicator is phenolphthalein. Fig. 14-2 shows the un-ionized, colorless form and the ionized, deep red-purple form. The pK_a of phenolphthalein is about 9. Using Eq. 14-2, this

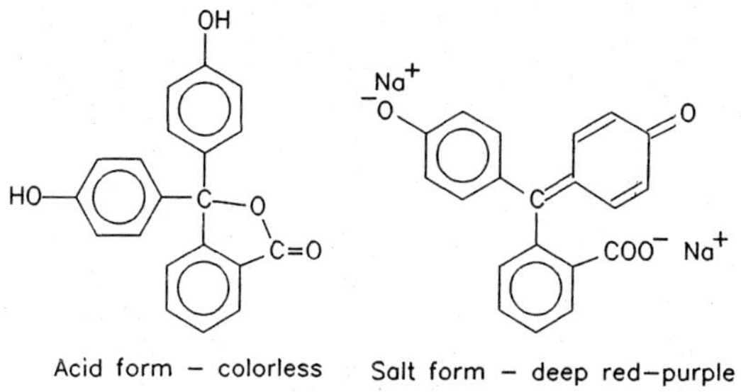

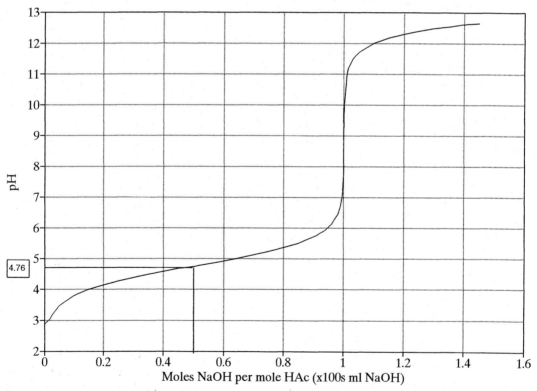

Fig. 14-2. Phenolphthalein indicator forms.

Acid form – colorless   Salt form – deep red–purple

means that at pH 9 about half the indicator will be in the ionized form, while at pH 8 only 1/10 the indicator will be ionized. This is the range in which one observes the most color change. Most pH indicators change color over a pH range of about 2, going from 90% of one color to 90% of the other color. If one of the forms is colorless, the color change is over one pH unit.

Phenolphthalein is a suitable indicator for strong acid-strong base reactions. It reacts with CO_2, so the CO_2 must be boiled off near the endpoint to give a sharp endpoint for titrations involving carbonate. This will be discussed in greater detail in Chapter 16 on pulping liquor analysis. Phenolphthalein is suitable for titration of borax if glycerol is present to avoid color fading. The color change goes from colorless up to pH 8.5 to deep red-purple above pH 9. Phenolphthalein is usually used as a few drops of a 1% ethanol-water solution, since it is only slightly soluble in water in its un-ionized form.

Methyl orange is another common pH indicator and goes from red below pH 3.1 to yellow above pH 4.4; it is used as a 0.1% aqueous solution. There are about 60 additional pH indicators available.

14.4 TITRATION OF A WEAK ACID WITH A STRONG BASE OR WEAK BASE WITH A STRONG ACID

The tools are now available to describe the titration curve of a weak acid by a strong base or, more likely in kraft pulping liquor analysis, the titration of a weak base by a strong acid. Consider the titration of 100 ml of 0.1 N acetic acid by 0.1 N NaOH (Fig. 14-3). With no addition of

Fig. 14-3. Titration of 100 ml of 0.1 M acetic acid with 0.1 M NaOH.

base Eq. 13-5 shows $[H^+] = (K_a \cdot C_{HA})^{0.5}$ or 1.32 $\times 10^{-3}$, a pH of 2.88; thus, the first part the titration curve is concentration dependent. From 2 ml to 98 ml of NaOH the pH is determined from the Henderson-Hasselbalch equation, and the pH is not concentration dependent. At the endpoint the pH is determined from Eq. 13-6, which is solved for this case in Example 13-8 and is pH of 8.89. Beyond the endpoint, excess alkali suppresses abstraction of protons by sodium acetate so the hydroxide concentration is equal to the excess equivalents of hydroxide divided by the volume of solution. At 101 ml, $[OH^-] = (101-100 \text{ ml})(0.1 \text{ mmol/ml})/(200 \text{ ml}) = 5 \times 10^{-4}$ mmol/ml (or mol/L); this corresponds to a pH of 10.70. Since the endpoint occurs at pH 8.9, phenolphthalein is an ideal pH indicator to use to detect this endpoint. Be sure to notice that when the acetic acid is half titrated, the pH = 4.76 = pK_a and, as this is the maximum buffer region, the titration curve is flattest here. This curve is similar to those of other carboxylic acids, such as fortified rosin and carboxylate salts in black liquor.

In a similar fashion, ammonia can be titrated (Fig. 14-4). The format of the *x*-axis is going to be modified here, and the modified format will be used throughout the remainder of the book, unless stated otherwise. Instead of the ratio of acid to base, we will assume all of the acid reacts with all of the base. The *x*-axis will be reported in terms of percent of acid and percent of base form. One can use this to see where endpoints occur, but more generally, one can quickly determine the pH of a solution of mixed composition.

Mixtures of acids or bases can be titrated sequentially if they differ by a factor of about 10^4 in their K_a or K_b values. Sometimes a species may be removed from solution to prevent interference. Other times a species may react with an added compound to change its acid or base characteristics. Examples of such titrations are given in the section on analysis of pulping liquors.

14.5 REDUCTION-OXIDATION TITRATIONS

Reduction-oxidation (redox) titrations are performed in a manner similar to those of acid-

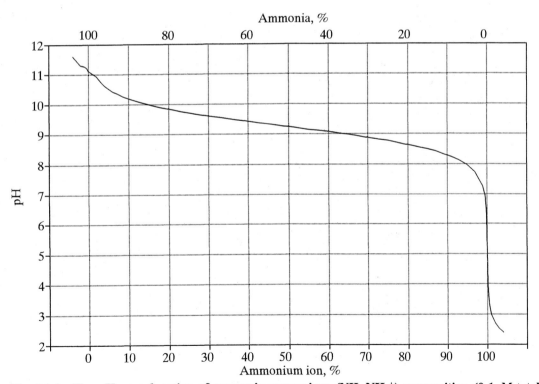

Fig. 14-4. The pH as a function of ammonia-ammonium (NH_3-NH_4^+) composition (0.1 *M* total concentration) at 25°C.

base titrations. The titration is monitored by the potential (voltage) with respect to a standard electrode or by the use of indicators that are oxidized or reduced after the analyte. In redox titrations, equivalents are based on moles of electrons transferred in the reaction. A compound with a certain equivalent weight for an acid-base reaction may have a different equivalent weight in a redox reaction; it may even have an equivalent weight that varies depending on the particular redox reaction.

In order for an analyte to be titrated with a titrant there should be at least 0.2 to 0.3 V difference between the standard two electrode potentials. This does not say the reaction will occur quickly, only that thermodynamically the reaction is predicted to occur.

When the two materials have reacted, i.e., they are in equilibrium, both of their actual half-reaction potentials are equal. The general form of the titration reaction with species (1) and (2) with stoichiometric reaction coefficients of a, b, c, and d is:

$$aOx_1 + bRed_2 \rightleftarrows cRed_1 + dOx_2$$

Activity coefficients are ignored here since the concentrations are low and the change in potential at the endpoint will be sharp.

Consider the titration of I_2, which usually occurs as I_3^-, with $Na_2S_2O_3$ (sodium thiosulfate) that is used in lignin determinations with permanganate or determination of bleaching chemicals with iodine. The reaction becomes:

$$I_3^- + 2S_2O_3^{2-} \rightleftarrows 3I^- + S_4O_6^{2-}$$

At any time during the titration the E of the iodine/iodide cell is equal to the E of the thiosulfate/tetrathionate cell. The Nernst equation for the half reaction of the compound that is reduced (iodine) is :

$$E = E^o - \frac{0.05915}{n} \times \log\frac{[Red]^c}{[Ox]^a} \qquad \text{at } 25°C$$

The iodine is generated from excess iodide that is oxidized by permanganate or bleaching chemicals in many pulp and paper laboratory

determinations. Therefore, let us consider the titration of 100 ml of 0.1 M I_3^- in the presence of 0.2 M iodide with 0.1 M thiosulfate (Fig. 14-5). $E°$ for the iodine/iodide cell is 0.534 V. The $E°$ for the thiosulfate cell is 0.08 so the $E°$ (cell) is 0.454 V, enough of a difference to give a sharp endpoint. As soon as the first 0.5 ml or so is added, the [I⁻] will be known (if none had been present initially) and the cell potential can be calculated.

EXAMPLE 5. What is E of the iodine/iodide cell just described after 0.5 ml of thiosulfate is added in the reaction above?

SOLUTION: For each I_3^- reacted 3 I^- form. Ignoring the small dilution:

$E = E° - 0.05915/2 \times \log(0.2015)^3/0.0995$

$= 0.534 + 0.032 = 0.566$ V.

EXAMPLE 6. Knowing that the E_{cell} of the thiosulfate cell is 0.566 V, what is the concentration of thiosulfate?

SOLUTION: Assume all of the thiosulfate reacted except for a minuscule amount in equilibrium.

$0.566 = 0.08 - 0.05915/2 \times \log(x/0.0005)$

$3.69 \times 10^{-17} = (x/0.0005)$
$x = 1.83 \times 10^{-20}$ M (correct assumption)

PROBLEM: Redo this example with 2 ml of titrant added. The answers are 0.565 V and 2×10^{-20} M

EXAMPLE 7. Shortly before the endpoint, Fig. 14-5 shows the potential as 0.40 V. What is the concentration of thiosulfate if tetrathionate is 0.1 M?

$0.40 = 0.08 - 0.05915/2 \times \log(x/0.1000)$

$1.51 \times 10^{-11} = (x/0.1000)$
$x = 1.51 \times 10^{-10}$ M

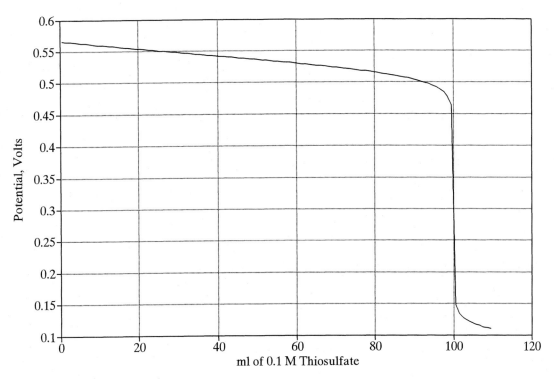

Fig. 14-5. Titration of 100 ml of 0.1 M I_3^- in the presence of 0.2 M I^- with 0.1 M $S_2O_3^{2-}$.

During the titration up to the equivalence point the potential is calculated from the cell reaction of the analyte. Beyond the endpoint the potential is calculated from the cell reaction of the titrant.

Iodometry

The technique just demonstrated of using the conversion of iodide to iodine by moderately strong oxidizing agents followed by the titration of the iodine is termed iodometry. This indirect titration is a common technique in pulp and paper measurements as well. One might ask why iodide is not used directly as the titrant to measure these oxidizing agents. The reason is that starch cannot be used for the endpoint determination and the reduction of iodide is fairly slow unless excess iodide is present.

The level of a wide variety of oxidants such as hydrogen peroxide, ozone, oxygen, Cl_2, permanganate, and HClO can be determined by this method.

Indicators

The potential of a solution during a redox reaction can be followed with electrodes. The potential is usually measured across a platinum electrode and an electrode that supplies a standard potential. Chemical indicators can be used and are compounds that are oxidized or reduced and change colors between the two forms. The endpoint of iodine titrations is conveniently followed with *starch indicator*, since small amounts of iodine form a dark blue complex with it. (This is also the basis of a qualitative test for the presence of starch in paper.) Titrants or analytes that are highly colored can be used as *self indicators*.

Miscellaneous aspects

This section has only been a cursory introduction to redox titrations. Many standard potentials vary with pH, the presence of complexing species, and other factors. These considerations may cause deviations from expected results in experiments.

14.6 COLORIMETRIC ANALYSIS

Colorimetric analysis is important in the area of effluent color of mill discharges and some determinations of concentrations of chemical species. Glucose liberated by enzymatic hydrolysis of starch is measured colorimetrically in the assay of starch in corrugated board (TAPPI Standard T 532). In liquid solutions, the amount of monochromatic light absorbed by a solution is described by *Beer's law*. Consider Fig. 14-6 where incident light P_0 travels through a sample of liquid of path length b and concentration c. *Transmittance, T*, is defined as the fraction of light transmitted through the sample and is equal to P/P_0. The transmittance of light decreases exponentially. For example, if 90% of the light is transmitted through cell length *b*, then 90% of 90% (or 81%) will be transmitted through length 2*b*. Concentration behaves similarly.

A constant is introduced called *the molar absorptivity*, ϵ, and the term *absorbance, A*, is defined to give Eq. 14-4 (below) where *c* is usually expressed in moles per liter and *b* is

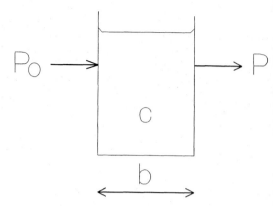

Fig. 14-6. Absorbance of light in a solution.

usually 1 cm, the pathlength commonly used in spectrophotometers. Fig. 14-7 shows the relationship between absorbance and transmittance. Sometimes the *absorptivity, a*, is expressed in g/L if the molecular weight is unknown; in this case c is expressed in g/L. The *percent transmittance, %T*, is equal to 100% $\times$ *T*, and *percent absorbance* = 100% - %T.

Optical density is an obsolete term of colorimetric analysis and should not be used.

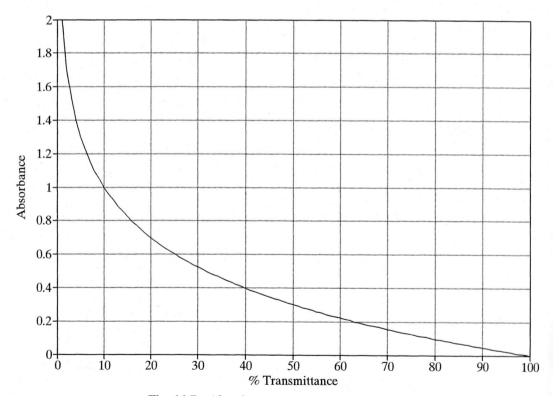

Fig. 14-7. Absorbance versus % transmittance.

Absorbance, and not transmittance nor percent absorbance, is proportional to the concentration and path length. Absorbance much above 2.5 cannot be measured accurately since this corresponds to only 0.31% transmittance.

$$A = - \log T = \log 1/T$$

$$= \log P_0/P = \epsilon bc$$

$$\log 1/T = \log P_0/P = abc \qquad (14\text{-}4)$$

EXAMPLE 8. A sample in a 1.0 cm cell absorbs 10% of the light at a certain wavelength. If the absorptivity of the material is 3.0 L(g·cm) at this wavelength, what is the concentration of the material?

SOLUTION: The transmittance is 0.90,
-log 0.90 = 3.0 L(g·cm) × 1.0 cm × c;
c = 0.0153 g/L.

EXAMPLE 9. 10% of the light of a certain wavelength of a pulp mill effluent is transmitted through a distance of 1 feet. What will the percent transmittance be at the same wavelength through the same distance be if the effluent is diluted 1:100?

SOLUTION: $A_1 = abc_1 = 1$; $A_2 = abc_2 = abc_1/100$; $A_2 = 0.01$; $\%T_2 = 97.7\%$

PROBLEM: For this same problem what would the apparent transmittance be for a river 2.5 feet deep? Remember the effective distance will be 5 feet since the light will reflect off the bottom and back through the water before reaching the eye of the observer. Answer: 89.1%.

Beers law almost always applies. However, apparent deviations may occur. For example, the color of pulp mill effluents are pH dependent. If, in the problem above, the pH were different before and after dilution, then Beer's law may not appear to hold. This is why the pH's of mill effluents are adjusted to that at which they will ultimately be discharged before the absorbance (using color units, Section 11.3) is measured.

14.7 COORDINATE CHEMISTRY

Introduction

Coordinate chemistry involves the formation of complexes when two atoms share an electron pair. One species makes the donation and the second species accepts the electron pair. Coordinate covalent bonds are formed when one atom donates both electrons of the electron pair; with other covalent bonds each atom donates one electron to the pair. The distinction is arbitrary, but has historical precedence. More information and detail on this important subject may be found in advanced inorganic chemistry textbooks.

Coordinate chemistry explains the behavior of alum (which is central to many aspects of wet end chemistry) in aqueous solutions. Aluminum ions will be used to demonstrate some aspects of coordinate chemistry. Also, many size press formulations contain zirconium compounds or synthetic polymers containing carboxylated polymers that are "set" with metal ions. Applications of coordinate chemistry will undoubtedly receive much wider recognition in the future by the pulp and paper industry.

The first explanation of the actual nature of complexes is credited to the classic 1906 work of Werner, for which Werner received the Nobel Prize in 1913. Werner showed that neutral compounds were bound directly to metal atoms in many complexes. Thus, $CuCl_2 \cdot 6NH_3$ is correctly written as $Cu(NH_3)_6Cl_2$. In general, coordination complexes are formed upon the reaction *of Lewis acids*, compounds which accept electron pairs, with *Lewis bases*, compounds which donate electron pairs, to form the *complex*. Acid-base reactions involving proton transfers are examples of coordinate chemistry reactions. The reaction of H^+ and OH^- to form the product H_2O proceeds as $H^+ + :OH^- \rightarrow H:OH$, where the electron pair is shared in the product. Lewis greatly expanded the class of acid-base reactions over that of the Brønsted-Lowry theory by making electron pairs of central importance rather than the proton.

Many other reactions besides acid-base reactions are included as well. One classic example is the reaction of the two gases BF_3 and $:NH_3$ to form the complex $H_3N:BF_3$, which is a white solid:

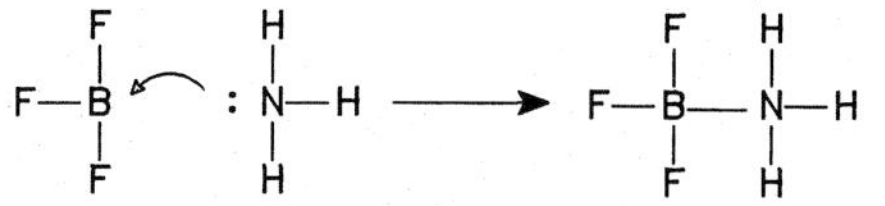

Ligands

The term *ligand* denotes the species donating the electron pair(s). Ligands are Lewis bases. In the previous chemical equation the ligand was $:NH_3$. Ligands that have only one electron pair involved with complex formation are known as *unidentate*. Ligands that can donate two electron pairs are known as *bidentate*. If a ligand has two or more pairs of electrons that combine with the same ion, the ligand is a *chelating* agent. Chelating agents are used in pulp and paper processing to bind metal ions that might otherwise interfere with the process. EDTA is often used to bind iron ions that would otherwise color mechanical pulps or lead to catalytic hydrogen peroxide decomposition. Its structure, shown below, has six possible binding electron pairs, one on each of the nitrogen atoms and one on each of the four carboxylic acids.

"EDTA"

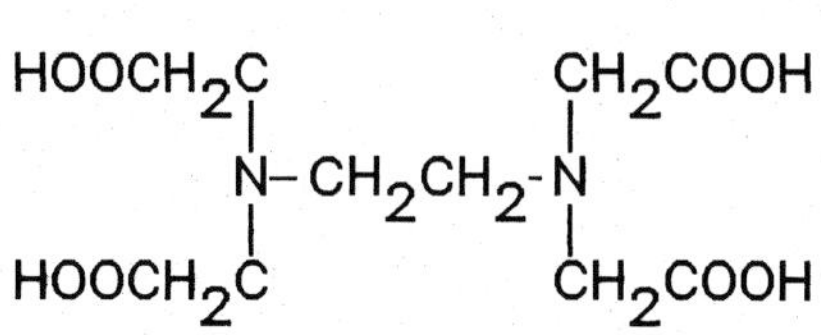

Coordinate chemistry of aluminum

The Al^{3+} cation forms coordination complexes with many ions or molecules in aqueous solutions. The aluminum cation has a *coordination number* of 6 in aqueous solutions. This means aluminum will react with six electron pairs of ligands. In the absence of other ligands, Al^{3+} forms a hydrated complex by reacting with six water molecules to form $[Al(H_2O)_6]^{3+}$ with the octahedral shape shown as a reactant in Fig. 14-8. Because of steric (spatial) hinderance, only one pair of electrons from one atom is donated to a particular atom. Thus, while the oxygen atom of water has two free electron pairs, only one pair acts with a particular aluminum cation.

Substitution reactions occur when one ligand is replaced by another. Metallic complexes of water may undergo hydrolysis, a type of substitution reaction that is equivalent to acid dissociation. For aluminum, the first two acid dissociations are as follows:

$$[Al(H_2O)_6]^{3+} \rightarrow [Al(H_2O)_5(OH)]^{2+} + H^+$$

$$K_1 = 10^{-5.01}; \quad pK_1 = 5.01 \quad (14\text{-}5)$$

$$[Al(H_2O)_5(OH)]^{2+} \rightarrow [Al(H_2O)_4(OH)_2]^+ + H^+$$

The value of the second acid dissociation has little practical significance because of formation of polyaluminum complexes (to be discussed). When the pH of a solution is equal to the pK_a of most acids, then both the acid and the conjugate base are present in equal amounts. This is a consequence of the *Henderson-Hasselbalch equation*. With aluminum and other highly coordinating species the chemistry is more complex. The pK_a for the first ionization of aluminum is 5.01. Rather than the predicted 0.5 equivalents, however, it takes about 2.5 equivalents of OH^- per equivalent of aluminum to increase the pH to 5.01 in water. The reason is that competing reactions involving higher substitutions and polyaluminum complexes occur. (Since the monosubstituted aluminum ion product undergoes additional reactions with the hydroxide anion, the equilibrium in Eq. 14-1 is forced further to the right. The pH is lower than anticipated (based on its pK_a) for a given amount of hydroxide ion added to alum.

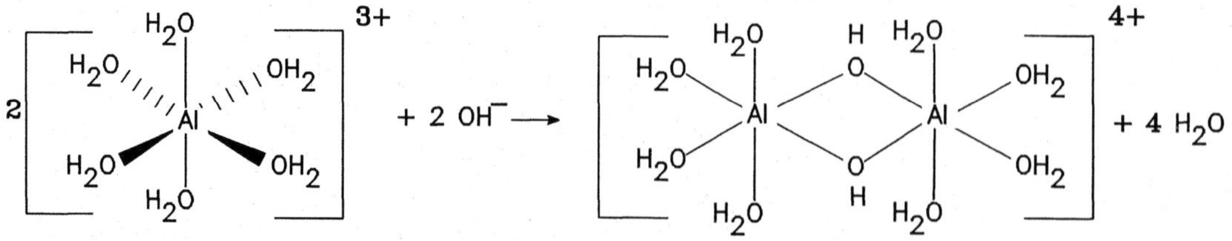

Fig. 14-8. Formation of aluminum ion dimer by OH⁻ bridging.

The oxygen atom of the OH⁻ ion, which has three unshared electron pairs, shares only one pair of electrons with a particular aluminum cation. However, the oxygen atom of the OH⁻ anion is known to form bonds with two different coordinating cations and, thereby, forms a *hydroxo bridge*; double bridges involving two hydroxide groups, as shown in Fig. 14-8 with aluminum as an example, are the most common. These bridges are crucial to rosin sizing with alum as they increase the size of the aluminum-rosin complex and decrease water solubility. Water solubility is decreased because the charge on each aluminum ion is decreased from +3 to +1 and because the bonds that are formed are hydrophobic. At pH of 5 as much as 90% of the aluminum is in the form of polymers. At high degrees of hydrolysis the main species may be $[Al_{13}O_4(OH)_{24}(H_2O)_{12}]^{7+}$ (Cotton and Wilkinson, 1980d); therefore, the average degree of polymerization varies with pH.

The loss of the hydrogen ion from the hydroxide of the hydroxo bridges leads to the formation of *oxo compounds*, the double bridged species of which are almost always symmetrical as shown to the right where M indicates a metal cation such as aluminum. These linkages are extremely insoluble in water. For example, the K_{sp} of $Al(OH)_3$ is on the order of 10^{-33}. An aluminum compound with a single oxo bond occurs at pH > 13 in dilute aqueous solutions, has the formula $[(HO)_3AlOAl(OH)_3]^{2-}$, and occurs in the crystalline, potassium salt of this compound (Cotton and Wilkinson, 1980e).

The role of aluminum in rosin sizing

While much attention has been paid to the reactions in dilute aqueous solutions of rosin sizing, the equilibrium mixture of coordination species in the press and dryer sections is much different, and probably much more important in the overall sizing process. Coordinate chemistry in dilute aqueous solutions is important principally as a means of rosin and aluminum retention in the web. Many coordination complexes will form in the press and dryer sections where the concentrations of the species involved are relatively high as is the temperature.

Other ligands that are of interest in wet end chemistry include carboxylate groups of cellulose fibers and rosin, alcohol groups of cellulose fibers, and sulfate ions to name a few. Polycarboxylic acids can coordinate with more than one metal ion to give nonpolar compounds; indeed, aluminum naphthenic acids were used as a gelling agent in "napalm". Sulfate anion is capable of being unidentate, bidentate, chelating, or bridging.

During rosin sizing, alum forms bonds with water, rosinate anions, hydroxyl groups, sulfate anions, and other ligands present to form a relatively large complex of complicated structure called the *aluminum-rosinate complex*, which is depicted in a representative fashion in Fig. 14-9. The aluminum-rosin complex must have limited water solubility so that it will adhere to fibers with relatively weak bonds. Limited water solubility (the formation of a colloidal state) is achieved by increasing the size of the complex, forming hydrophobic linkages, and reducing the charge by coordinating the aluminum cation with anions. Alum at pH below 4 has very little sizing ability

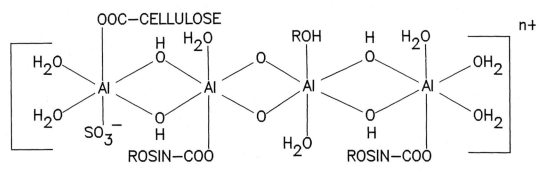

Fig. 14-9. A representative structure of the aluminum rosin complex.

because the Al^{3+} ions have not complexed with OH^- species. Therefore, it has high solubility. While rosinate certainly coordinates with aluminum at pH of 4, the complex is not retained on the fiber as it is at higher pH's.

The cellulose fiber probably forms bonds with the aluminum-rosinate complex as the sheet dries. Water removal in the press section and, more importantly, in the dryer section at elevated temperatures, allows new, stronger bonds to form that are not possible in dilute aqueous solutions. This is probably where oxo linkages form. Oxo linkages during rosin sizing were first hypothesized in 1935 (Thomas), but there was no experimental evidence until 1992 (Subrahmanyam and Biermann) to support this hypothesis.

Elevated temperature is not necessarily required with some coordinating cations, but may speed up coordination reactions with other cations that may react more slowly. For example, work in my laboratory shows heating at 120°C (250°F) for approximately two minutes is necessary to fully develop surface sizing with zirconyl acetate.

Carboxylate groups on the surface of fibers, being anionic, would form strong bonds with the aluminum-rosinate complex. Aside from carboxylate groups, another functional group available as a ligand on the surface of the fiber is the hydroxyl group of cellulose and hemicelluloses. Hydroxyl groups are available for hydrogen bonding or coordination, but, of course, these are not as effective under papermaking conditions since it is known that pulps with higher concentrations of carboxylate groups are more effectively sized than pulps with lower concentrations or without carboxylate groups. Experience has shown that hardwoods are easier to size than softwoods, and dissolving pulps are not easily sized due to the concentration of carboxylate groups. But, is this only a matter of retention?

Ions that replace the role of aluminum

In theory, any coordinating, trivalent cation can give rosin sizing. Subrahmanyam and Biermann (1992) showed that gallium, cerium, and lanthanum gave good sizing. Also, the chemistry of chromium is very similar to that of aluminum. However, none of these materials have commercial viability in this area.

Ferric ion behaves very similarly to aluminum ion. In fact, it has an even higher affinity for OH^- so that rosin sizing with it occurs at pH values below that of aluminum. The ferric ion was used to replace alum for rosin sizing in the U.S. during World War II when alum was in relatively short supply. Biermann (1992) discovered that highly protonated polyamines are extremely effective mordants for rosin sizing from pH 3 to 10 because they have properties very similar to the polyaluminum complexes. This study describes in detail the expected behavior of polyamines in wet end chemistry using rosin sizing as an example.

Some mills can use rosin sizing with alum at very low pH's, even as low as 3.8. Work in my laboratory has shown that dicarboxylic acids can partially replace the role of OH^- so that succinic acid increases sizing at pH 3.5 in the laboratory. Other work showed that small amounts of fluoride (0.1 times the molar concentration of aluminum), which is known to complex highly with aluminum, has adverse effects on rosin sizing. This may be important in areas where fluoride has relatively high concentrations.

14.8 MISCELLANEOUS CONSIDERATIONS

Standard T 610 contains information on the preparation of indicators and standard solutions for a variety of experiments. It also contains correction factors for volumetric glassware used from 15 to 25°C. For every 1°C below the standard temperature, the glassware volume will be 0.02% too high since water has a higher volumetric coefficient of thermal expansion than glass.

14.9 ANNOTATED BIBLIOGRAPHY

Coordinate chemistry
1.   Cotton, F.A. and G. Wilkinson, Advanced Inorganic Chemistry, Interscience Publishers, U.S.A., 2nd ed., 1962. a) p 715, b) p 338, and c) p 715; 4th ed., 1980 d) pp 152-153, e) pp 333-334, f) p 992.

2.   Thomas, A.W., Solutions of basic salts of aluminum, *Tech. Assoc. Papers* 18:242-245(1935).

3. Subrahmanyam, S. and C.J. Biermann, Generalized rosin soap sizing with coordinating elements, *Tappi J.* 75(3):223-228(1992).

4. Biermann, C.J. Rosin sizing with polyamine mordants from pH 3 to 10, *Tappi J.* 75(5):166-171(1992).

EXERCISES

Strong acid/strong base titrations

1. In order to determine the concentration of an unknown sodium hydroxide solution, a 25.00 ml aliquot of the unknown solution was diluted to 1 L in a volumetric flask. A 20.00 ml aliquot of the diluted solution was pipetted into 50 ml of water and titrated with 21.37 ml of 0.1027 *N* HCl solution to a phenolphthalein endpoint. What was the concentration of the original NaOH solution in g/L?

2. In question 1, would it be important if the 20.00 ml diluted solution were pipetted into 60 ml of water instead of 50 ml? Could one use a graduated cylinder to measure the 20.00 ml aliquot of diluted sample to save time?

pH of weak acid-conjugate base pairs

3. What is the pH of a solution containing 0.5 *M* acetic acid and 0.2 *M* sodium acetate?

Reduction-oxidation titrations

4. Which of the following are examples of redox titrations.

    a) Analysis of sulfite liquors with periodate.

    b) Titration of sulfurous acid sodium hydroxide.

    c) Reaction of residual chlorine with iodide followed by titration with thiosulfate.

    d) The measurement of pH in black liquor.

    e) The measurement of pH with color indicators such as in a test strip.

Colorimetric analysis

5. Is it easy to disguise color in effluents with dilution?

6. Why is pH adjustment important when measuring the color of bleach plant effluents?

Coordination chemistry

7. Show how aluminum ions may form a chain in solution.

8. Indicate how the conversion of hydroxo linkages to oxo linkages of aluminum in paper might account for the long term disintegration of paper by acid hydrolysis.

15

CALCULATIONS OF WOOD, PAPER, AND OTHER MATERIALS

15.1 WOOD MOISTURE CONTENT AND DENSITY

Moisture content

There are several definitions regarding wood and properties of wood that are useful to understand. One of the most fundamental of these is the moisture content of wood. The moisture content of wood is a measure of the water content relative to either the total wet weight of material (the *green* weight of wood) or to the weight of oven-dried wood material (the *oven-dry* basis). The oven-dry weight is determined by drying the wood to constant weight at 103-105°C (217-221°F). The pulp and paper industry almost invariably reports the moisture content of wood and other materials in terms of the total weight of material. Note that the maximum moisture content in this case is 100% for the case of pure water. The other forest products industries almost invariably report moisture contents in terms of the weight of oven-dry wood. Here moisture contents over 100% are possible and commonly encountered. The two definitions follow (keep in mind the relationship: weight of water in wood = wet weight of wood - oven-dry weight of wood):

$$MC_{GR} = \frac{\text{weight of water in wood}}{\text{wet weight of wood}} \times 100\% \quad (15\text{-}1)$$

$$MC_{OD} = \frac{\text{weight of water in wood}}{\text{ovendry weight of wood}} \times 100\% \quad (15\text{-}2)$$

The moisture content of wood (green basis) is typically 50%, but varies from 30 to 60%. This corresponds to 0.43 to 1.5 kg water per kg dry wood, (43-150% MC_{OD}). The two moisture contents are interchangeable as shown in Eqs. 15-3 and 15-4 and Fig. 15-1.

$$MC_{GR} = \frac{MC_{OD}}{100\% + MC_{OD}} \times 100\% \quad (15\text{-}3)$$

$$MC_{OD} = \frac{MC_{GR}}{100\% - MC_{GR}} \times 100\% \quad (15\text{-}4)$$

EXAMPLE 1. Convert 20% MC_{OD} to MC_{GR}.

SOLUTION. One could use Fig. 15-1 and come up with about 17% MC_{GR}. Or else, one could use Eq. 15-3 and solve it for 16.7% MC_{GR}. If one uses the two definitions in Eqs. 15-1 and 15-2, it is not necessary to remember Eqs. 15-3 and 15-4. For example, 20% MC_{OD} means 20 parts water to 100 parts dry wood; this, in turn, means 20 parts water for 120 parts wet wood or 16.7% MC_{GR}.

Fiber saturation point, FSP

Below MC_{GR} of about 25%, free water disappears and the remaining water is chemically adsorbed to the wood, by hydrogen bonding of water with the hydroxyl groups of cellulose and hemicelluloses. The moisture content corresponding to the disappearance of *free water* is called the *fiber saturation point, FSP*. As the moisture content of wood decreases from 25% towards 0%, the energy required to remove an aliquot of water increases from 540 cal/g (the heat of vaporization of water) to about 700 cal/g (125%) midway and approaches 1100 cal/g (200%) near 0% MC.

Consequently, *air dry* wood is not really dry since wood is a hygroscopic material (as are pulp and paper). This means that wood absorbs or gives off moisture with the atmosphere until an *equilibrium moisture content (EMC)* is achieved. Fig. 15-2 shows the EMC value of wood as a function of ambient temperature and relative humidity. There is a *hysteresis* effect with the moisture content of wood, pulp, paper, and any lignocellulosic material. The actual EMC in a given environment will depend upon whether the material is losing or gaining water to achieve the EMC. Paper below 6% MC put into an environment of 72°F and 50% relative humidity might achieve 7% EMC, whereas paper above 9% MC might achieve 8% EMC in this environment. A difference in the physical properties of these papers result. TAPPI T 402 specifies paper should be put in a warm, dry room before conditioning at standard testing conditions.

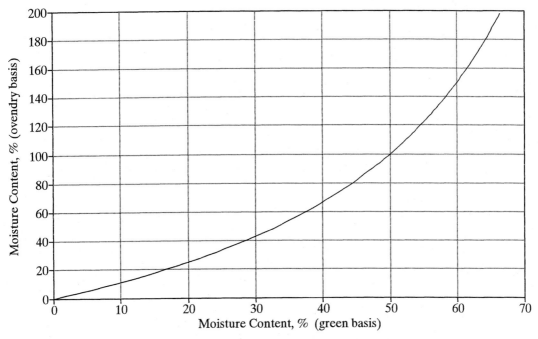

Fig. 15-1. Comparison of green and oven-dry basis moisture contents.

Wood shrinkage with decreasing moisture content

As the moisture content of wood decreases from the FSP to 0% moisture content, the wood shrinks. *Shrinkage* is defined as the change in a dimension of wood (from the FSP to a specified moisture content) from the swollen state relative to the swollen dimension. Wood tends to shrink from the FSP to the oven-dry moisture content by the following rule-of-thumb amounts.

Longitudinal	0.5%
Radial	4%
Tangential	6%
Volumetric	9%

In fact, the volumetric shrinkage varies from 7-15% depending on species, growth rate, and so forth, and is loosely related to the specific gravity of wood as shown in Fig. 15-3, which gives shrinkage values for several species as examples.

Wood specific gravity

The *specific gravity* (*sp gr*) of wood is the oven-dry weight of wood divided by the weight of displaced volume of water. This produces a unitless number. The displaced volume of water can be measured by calculating the volume of the wood if it is of even shape such as a rectangular block. If the wood is of uneven shape, it may be coated with a thin layer of wax and the water it displaces when immersed is measured. While the oven-dry weight of wood is always used, it is possible to use a volume corresponding to any moisture content; however, the green volume is that most commonly used and this gives the special term *basic specific gravity*. The basic specific gravity of woods is commonly between 0.35 and 0.60, but can vary from 0.2 to 0.7. The basic specific gravity of cell wall material is approximately 1.50.

The density of a material is defined as the mass per unit volume, mass/volume. When the units of pounds or ounces are used one actually obtains a weight density. For wood, it is customary to take the total mass (or weight) divided by the volume both at the same moisture content. Since pulp and paper mills seldom process wood below the fiber saturation point, it is preferable to use the basic specific gravity when calculating amounts of wood material.

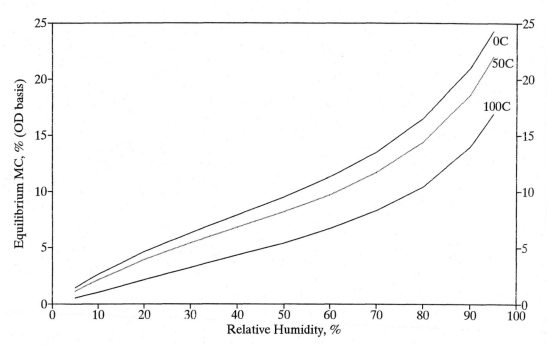

Fig. 15-2. Equilibrium moisture content of wood versus relative humidity.

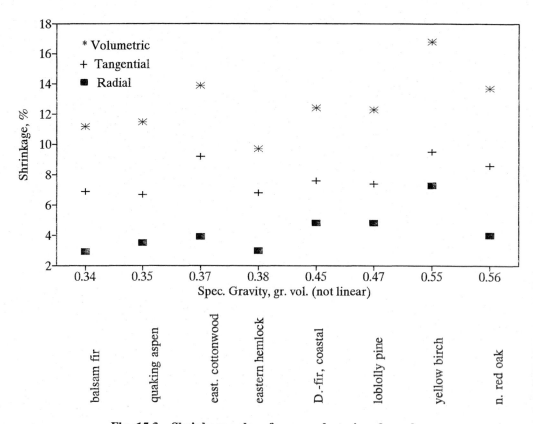

Fig. 15-3. Shrinkage values for several species of wood.

EXAMPLE 2.   A sample of wet Douglas-fir wood 2 cm thick, 5 cm wide and 10 cm long weighs 90.21 g. It is then dried at 105°C to constant weight (to insure complete water removal) and reweighed after cooling in a desiccator. The oven-dry weight is 44.37 g. What are the MC_{GR}, MC_{OD}, and basic specific gravity of the wood sample?

SOLUTION:

$$MC_{GR} = \frac{90.21 \text{ g wet weight} - 44.37 \text{ g oven-dry weight}}{90.21 \text{ g wet weight}} \times 100\% = 50.8\%$$

$$MC_{OD} = \frac{90.21 \text{ g wet weight} - 44.37 \text{ g oven-dry weight}}{44.37 \text{ g oven-dry weight}} \times 100\% = 103.3\%$$

$$\text{sp gr} = \frac{\dfrac{44.37 \text{ g dry wood material}}{2 \text{cm} \times 5 \text{ cm} \times 10 \text{ cm}}}{1 \text{ g/cm}^3 \text{ (density of water)}} = 0.444$$

PROBLEM: A sample of wood with green dimensions of 3 cm $\times$ 5 cm $\times$ 10 cm weighs 105 g green (wet) and 62 g when oven-dry. Calculate:

1. Specific gravity
2. MC_{OD}
3. MC_{GR}
4. Oven-dry weight in kg/m^3

Answers: 0.413; 69.4%; 41.0%; 413 kg/m^3

15.2 PULPWOOD MEASUREMENT

It is the responsibility of the woodyard operations to obtain a steady supply of suitable wood. Wood sources are:

1. Roundwood (logs at least 4 in. diameter)
2. Chips
3. Slabs
4. Edgings
5. Sawdust and planar shavings

Measurements of wood can be based on weight (with moisture content correction to determine a reliable estimate of oven-dry weight), by solid wood volume (for example, the cunit), by gross, stacked volume (for example, the cord), or as volume of chips (for example, the unit). Traditionally, a measure of the gross roundwood volume was used since barking and chipping were done at the mill.

The standard measure was the cord, a pile of wood with 4 ft long logs stacked 8 ft across and 4 ft high, or 128 cubic feet of stacked wood. The amount of solid wood in a cord must be determined by sampling the cord and measuring the solid wood, decayed wood, bark, and air contents.

The cunit is 100 ft^3 of solid wood material of pulp logs. Purchased wood chips are measured in *units* (200 ft^2 of chips) or *bone dry units* (*BDU*, 2400 lb oven dry wood). The unit and BDU are terms first used on the U.S. West Coast where mills first started using residual chips. One unit of Douglas-fir chips is approximately the same as one BDU. [It takes about 6 kWh/t (0.33 hp·day/ton) dry wood basis to chip wood for pulping.] A reasonable approximation is that 1 ft^3 of wood will make about 2.6 ft^3 of chips. Thus, if one knows the basic specific gravity of the wood, one will have a reasonable figure for the density of the wood chips. Sections 2.4 and 2.5 describe units of wood measurement. Some conversion factors are listed in Table 2-2.

EXAMPLE 3.   A mill produces 700 metric tons of wet wood chips at 50% moisture content (green basis) per day. How many railroad cars are needed to transport them?

Conversion factors: 1 kg wet chips = 0.5 kg dry chips
1 rail car holds 18 units
1 unit = 200 ft^3
10 pounds dry chips = 1 ft^3
1 kg = 2.2 pounds
1 metric ton = 1000 kg
Others of your choosing.

SOLUTION:

$$700{,}000 \text{ kg wet chips} \times \frac{0.5 \text{ kg dry chips}}{1 \text{ kg wet chips}} \times \frac{2.2 \text{ lb}}{1 \text{ kg}} \times \frac{1 \text{ ft}^3}{10 \text{ lb dry chips}}$$

$$\times \frac{1 \text{ unit}}{200 \text{ ft}^3} \times \frac{1 \text{ rail car}}{18 \text{ units}} = 21.4 \text{ rail cars}$$

PROBLEM: A sawmill produces 87.3 BDU (one bone dry unit is 2400 pounds of oven-dry wood equivalent) of chips per day with 50% moisture content green basis. (The solid wood specific gravity is 0.44.) The chip bulk density is 10 pounds dry wood per cubic foot. These rail cars have a rated capacity of 30 units. (One unit is 200 cubic feet.) How many rail cars are needed per day. (Give the final result to 3 significant digits.) Answer: 3.49 rail cars.

15.3 TENSILE STRENGTH AND BREAK-ING LENGTH OF MATERIALS

It is useful to have a short discussion on *force*, *weight*, and *mass*. *Mass* is a measure of the amount of material present in an object, that is, the quantity of matter. It is not influenced by gravitational fields, and an object's mass is constant whether on the earth's surface or in outer space. *Force* is that which changes the motion (that is, the momentum) of an object. (*Pressure* or *stress* is *force per unit area*.) A force may act on a body without changes in motion if an equal force in the opposite direction is also acting. For example, we feel the force of gravity against the ground, but the ground exerts an equal force on us in the opposite direction of gravity to keep us stationary. If one walks off a cliff the force of gravity continues to act, but the force of the ground in the opposite direction is no longer there

and one begins to accelerate at the rate of 9.81 m/s^2 (32 ft/s^2). [This means after one second the velocity is 32 ft/s, after two seconds 64 ft/s, after three seconds 96 ft/s, etc. Since air is a viscous fluid, it exerts force on a moving object, and when a person in free fall reaches a speed of about 180 ft/s (120 miles per hour), the air resistance force is equal to the force of gravity, and one stops accelerating. This is known as *terminal velocity*. The terminal velocity of an object depends on its mass, shape, and surface area; the terminal velocity of a feather of low mass and high surface area might be substantially less than 1 ft/s.]

Weight is the measure of the force a gravitational field exerts on an object. The relationship is: weight = $m \cdot g$; that is, weight equals mass times acceleration of gravity. The units of mass are kg (or g) and slug and the units of weight are newton and pound for the metric and English systems, respectively. In practice, weight always refers to the force exerted by gravity at the earth's surface.

Let's examine how these terms are used (and often misused) in speaking. An 80 kg person at the surface of the earth will be 80 kg on the surface of the moon (whose gravitational pull is 1/6 of that on the earth's surface). Since 1 kg = 2.2 lb (at the earth's surface) this person weighs 176 pounds on the earth. But this person weighs 176/6 or 29.3 pounds on the surface of the moon. Pounds are a unit of weight and kilograms are a unit of mass and they are not comparable. People get around this by defining the pound$_{force}$ (lb$_f$) and the kilogram$_{force}$ (kg$_f$). Thus, an object that weighs 2.2 lb is the same as the force exerted by gravity at the earth's surface on a 1 kg object, or 1 kg$_f$. One kg$_f$ = mg = 1 kg × 9.81 m/s^2 = 9.81 N. Unfortunately, when the kg is used as a force the subscript is usually omitted. It is vastly preferable to use the units of N rather than kg$_f$; one reason is that one must be very careful using dimensional analysis with kg$_f$.

When materials are subjected to tensile forces, they behave in the manner shown in Fig. 15-4, that is, they undergo changes in dimensions called deformation or strain. The exact shape of the curve depends on whether the material is brittle or ductile. Of course, any type of force causes deformations; however, we will consider only tensile forces here. *Strain* is defined as the deformation per unit (original) length. For example, a 10 cm long rod that stretches 0.1 cm under a tensile load would have a deformation of 0.1 cm/10 cm = 0.01 = 1%; notice that strain is unitless. *Stress* is the force per unit area. *Strength* is a measure of the stress that causes failure. The energy absorbed in the process (*toughness*) is the integral of force times distance and, therefore, is the area under the curve. Brittle materials such as glass can withstand high forces but fail with low strain; therefore, brittle materials are *strong*, but not *tough*.

The *breaking length*, L, is a measure of tensile strength by calculating the length of a piece of material such that it breaks under its own weight. The breaking length evens out the differences in density and geometry so that comparisons are made on a "pound for pound" basis. This is particularly useful in the paper industry since

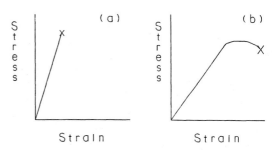

Fig. 15-4. Stress strain diagrams of (a) a brittle, nonductile material such as glass or cast iron and (b) a ductile material such as aluminum or paper.

tensile strengths are usually reported in units of force per width of paper. In order to calculate the breaking length we need to know the specific gravity of a material and the tensile strength in units of force/area.

In the metric system the tensile strength is often given in units of N/cm^2 or MPa = 10^6 Pa = 10^6 N/m^2 = 100 N/cm^2. Let us work out the conversion of N/cm^2 tensile strength into breaking length. First calculate the force exerted by a column of the material one kilometer high per square centimeter, which is stress/km/cm^2, which we will designate as x.

$$x = \text{sp gr} \times \frac{1\,\text{g}}{1\,\text{cm}^3} \times \frac{1\,\text{kg}}{1000\,\text{g}} \times \frac{10^5\,\text{cm}}{1\,\text{km}} \times \frac{9.81\,\text{N}}{1\,\text{kg}}$$

$$x = \text{sp gr} \times 981\ \text{N} \cdot \text{cm}^{-2} \cdot \text{km}^{-1} \qquad (15\text{-}5)$$

The breaking length in kilometers, L$_{km}$, is determined by dividing the tensile strength by the stress per kilometer of material (with both stresses in identical units). This is done in the following equation.

$$L_{km} = \frac{\text{tensile strength in N} \cdot \text{cm}^{-2}}{\text{sp gr} \times 981\ \text{N} \cdot \text{cm}^{-2} \cdot \text{km}^{-1}} \qquad (15\text{-}6)$$

If the tensile strength is reported in Pa, then making the substitution of 9.81 MPa/km for 981 N·cm^{-2}·km^{-1} in the previous equation gives the following equation for breaking length:

$$L_{km} = \frac{\text{tensile strength in MPa}}{\text{sp gr} \times 9.81 \text{ MPa} \cdot \text{km}^{-1}} \quad (15\text{-}7)$$

The tensile strength of wood and many other materials is often reported in units of lb/in.^{2} (psi). The following equation can be used to determine the breaking length in kilometers.

$$L_{km} = \frac{\text{tensile strength in psi}}{\text{sp gr} \times 1422 \text{ psi} \cdot \text{km}^{-1}} \quad (15\text{-}8)$$

By using the conversion factors given in the three previous equations, we can compare the breaking length of a variety of materials. Table 15-1 is a comparison of many materials to give an overall picture of the strength of available materials. Specific gravity and tensile strength values of small, knot-free, wood samples at 12% MC_{OD} were obtained from the *Wood Handbook*. See Section 2.7 for the maximum theoretical tensile strength of cellulose fibers, which is about 100 km. Table 15-1 shows that the tensile strength of pulp fibers is quite high in comparison with many materials.

EXAMPLE 4.   The tensile strength of black willow is about 15,800 psi parallel to the grain. It has a specific gravity of 0.41. What is its breaking length?

SOLUTION. Using Eq. 15-8 the solution is:

$$L = \frac{15,800 \text{ psi}}{0.41 \times 1422 \text{ psi} \cdot \text{km}^{-1}} = 27.1 \text{ km}$$

15.4 PAPER PROPERTIES

Table 15-2 shows some properties of paper samples tested by OSU students in a 1976 laboratory exercise. These results can be used for comparison between grades of paper. They also supply data for example exercises. Table 12-8 gives some conversion factors of paper properties from English to metric units. Definitions of some paper properties are given in Chapter 7.

EXAMPLE 5.   The breaking length of the filter paper listed in Table 15-2 in the machine direction is given as 2.60 km. What was the original force in pounds to break a 15 mm wide specimen?

SOLUTION.   Using Eq. 15.8, and a specific gravity of 0.47 for this paper, one obtains a tensile strength of 1737 psi. The width of the paper is 15 mm or 0.591 inches. The caliper of the paper is determined by the relationship: density = mass/volume. For 1 m^2 of paper, the mass is the basis weight (in g/m^2) and the volume (in cm^3) = (100 cm)^2 $\times$ caliper (cm). Therefore,

$$0.47 \text{ g/cm}^3 = 84 \text{ g/}(10,000 \text{ cm}^2 \times \text{caliper})$$

$$
\begin{aligned}
\text{caliper} \quad &= 0.0178 \text{ cm} \\
&= 0.178 \text{ mm} \\
&= 0.00704 \text{ in.}
\end{aligned}
$$

So, 1737 psi $\times$ .591 in. $\times$ 0.00704 in.
   = 7.22 lb.

Table 15-1. The breaking lengths of various materials.

Material	Tensile Strength, psi	Sp gr	Breaking, length, km
Wood, parallel to wood grain			
Black willow	15,800	0.41	27.1
Yellow poplar	22,400	0.46	34.2
Sweetgum	17,300	0.52	23.4
Overcup oak	14,700	0.63	16.4
Interior Douglas-fir	18,900	0.48	27.7
Eastern white pine	11,300	0.35	22.7
Englemann spruce	13,000	0.34	26.9
Western larch	19,400	0.55	24.8
Wood, perpendicular to grain			
Yellow poplar	540	0.42	0.9
Sweetgum	760	0.52	1.0
Interior Douglas-fir	390	0.48	0.6
Englemann spruce	350	0.35	0.7
Paper, etc.			
Newsprint			2-5
Bleached softwood			8-10
Cellophane	21,000 maximum	1.50	9.85
Other materials			
Aluminum	13,000	2.7	3.4
Aluminum alloy, aircraft	75,000	2.7+	19.5
Al_2O_3, ceramic	50,000	3.8	8.4
Acrylics, plexiglas	10,000 maximum	1.2	5.9
Epoxy	10,000 typical	1.2	5.9
E-glass	500,000	2.4	160
Graphite fiber	100,000	1.9	37.0
Nylon, monofilament	50,000	1.15	30.6
Phenol-formaldehyde	6,000	1.27	3.3
Steel (0.15% C)	50,000	7.85	4.5
Wood fiber	210,000 maximum	1.50	100

Table 15-2. Results of OSU student testing of laboratory samples from 1976.

Paper Sample	Basis weight g/m²	Density g/cm³	Tensile Strength brk. length, km MD	CD	Tear Strength mN·m²·g⁻¹ MD	CD	MIT Fold Double folds MD	CD	Burst Index kPa·m²·g⁻¹	Gurley Porosity s/(100 cm²)	Cobb Size g/m²	Elrepho Brightness	Printing Opacity
Onionskin, yellow	47	0.63	4.56	1.90	4.4	6.6	21	4	1.4	44	10	32	95
Cascade duplicator	76	0.79	4.60	2.80	6.4	6.9	28	12	1.5	25	50	81	92
Filter paper	84	0.47	2.60	1.70	7.8	7.5	7	4	0.9	9	-	76	81
Vellum tracing	70	0.96	11.40	5.40	6.9	7.3	740	740	5.0	∞	20	82	57
Pamphlet cover	103	0.58	6.50	3.30	8.8	9.8	58	34	2.6	12	34	36	100
Lightweight letterhead	36	0.65	7.03	4.73	20.3	22.0	454	529	3.0	1150	54	84	66
Kraft wrapping postal	63	0.55	8.21	4.50	10.5	14.0	322	228	3.5	6	27	20	99
Straw corrugating medium	126	0.51	5.20	3.30	8.0	8.2	4	8	1.7	60	38	25	99
Black liteproof	103	0.47	5.05	2.43	5.6	3.1	5	3	0.8	22	-	5	99
Yellow duplicating	75	0.84	8.45	3.46	6.3	7.5	158	59	2.9	36	17	28	91
Recorder paper	46	0.82	8.49	5.32	5.3	5.2	514	437	4.1	315	17	70	75
Letterhead	76	0.68	6.97	4.03	3.1	3.5	150	99	3.2	52	17	88	87
Manila envelope	105	0.69	6.20	4.00	12.1	13.0	450	211	3.3	15	29	37	92
Summit bond	75	0.67	5.10	3.10	8.4	6.1	29	25	1.3	16	28	78	91
Newsprint	58	0.68	4.00	3.20	6.3	7.4	21	5	1.2	74	76	54	95
Kraft bag	65	0.44	6.90	3.00	13.8	17.4	350	100	3.0	11	30	19	99
Note pad	58	0.72	7.20	2.90	8.2	10.5	25	9	2.1	8	36	78	82
Action paper	45	0.44	5.90	2.20	6.7	8.4	41	36	1.4	44	35	35	77
Strathmore report cover	181	0.73	5.90	3.00	10.8	12.2	129	77	2.4	235	45	9	100
Straw bond	71	0.76	8.10	2.90	10.0	13.6	146	67	2.9	368	35	82	83
Index card	180	0.80	5.50	3.20	6.9	6.0	14	15	2.3	77	21	77	99
Green notepad	72	0.66	5.00	3.20	6.4	6.1	34	27	1.9	19	25	51	97
Gibraltar onionskin	32	0.48	8.00	3.20	6.9	9.1	730	120	3.0	185	10	78	60
Simpson vellum	104	0.57	5.10	1.20	9.7	12.7	4	4	1.0	17	23	71	99
Fore duplicator	72	0.85	7.10	3.70	5.7	6.5	145	103	2.5	27	21	47	96

EXAMPLE 6. The burst index is given in Table 15-2 for kraft bag as 3.0 kPa·m²·g⁻¹. What is the burst strength in psi?

SOLUTION.   The burst index is obtained by dividing the burst strength by the basis weight. To solve for the burst strength it is necessary to take the burst index and multiply by the basis weight. The burst strength obtained by this method is in kPa. It is necessary to convert kPa to psi to obtain the desired units. The conversion factor is available in Table 12-8 under flat crush. Although a flat crush was not performed here, the conversion factor from kPa to psi is always the same. Therefore,

$$3 \text{ kPa·m}^2 \cdot \text{g}^{-1} \times 65 \text{ g·m}^{-2} = 195 \text{ kPa}; \quad 195 \text{ kPa} \times \frac{1 \text{ psi}}{6.895 \text{ kPa}} = 28 \text{ psi}$$

This exercise shows that to convert tensile or burst indexes to tensile or burst strength the basis weight must be known. It is possible to convert a tensile or burst index back and forth from the English to the metric system without knowing the basis weight.

EXERCISES

Wood

1. A sawmill is *nominally* rated at 150,000 board feet (a board foot is 1/12 of a ft³) of 2 in. × 4 in. lumber per day. The *actual* dimensions of a *two by four* are 1.5 in. × 3.5 in. (The actual board feet production is only 65.6% of 150,000; why?) Douglas-fir wood is used with a specific gravity of 0.44 and a moisture content of 120% on an oven-dry basis. One uses log volume tables to learn that typically (at this mill) for every 100 lb of solid wood coming in, 50 lb come out as lumber, 33.3 lb come out in the form of chips, and 16.7 lb come out in the form of sawdust. Chips have a bulk density of 10 lb/ft³ (oven-dry wood material), and sawdust 8 lb/ft³. Rail cars have a rated capacity of 18 units (3600 ft³). Chips are worth $160 per BDU, and sawdust $60 per BDU. Calculate the following:

   Lumber: Volume_____ft³; Wet weight_____tons; Dry weight_____tons.

   Chips: O.D. weight _____tons; _____BDU; Green weight_____tons.
                       Number of rail cars needed_____; Chips per car_____tons.

   Sawdust: O.D. weight _____tons; _____BDU.

   Revenue to mill: $_____for chips; $_____ for sawdust; $_____total.

2. Given: a piece of wood with a specific gravity of 0.42 and a moisture content (MC_{GR}) of 45%. Calculate the following on a green volume basis:

	kg/m³	lb/ft³
Oven-dry weight		
Green (wet) weight		
Weight of water contained		

3. Douglas-fir wood with a basic specific gravity of 0.45 is converted into chips that have a bulk density of 10 lb/ft^3 (oven-dry weight of chips). Assume the moisture content on a green basis is 45%. Calculate:

Volume of chips from 1 ft^3 of solid wood: _____ ft^3
Volume of chips from 1 m^3 of solid wood: _____ m^3
Oven-dry weight of wood from 1 ft^3 of solid wood: _____ lb
Green weight of wood from 1 ft^3 of solid wood: _____ lb
Oven-dry weight of wood from 1 m^3 of solid wood: _____ kg
Green weight of wood from 1 m^3 of solid wood: _____ kg

Breaking length
4. Show that 9.81 MPa/km is equal to 981 N·cm^{-2}·km^{-1}.

5. A plastic material has a tensile strength of 3000 psi and a density of 0.85. What is its breaking length? Is it intrinsically stronger than paper?

6. Calculate the breaking force (lb/15 mm) of the newsprint listed in Table 15-2.

Miscellaneous paper tests
7. Calculate the caliper of the kraft bag listed in Table 15-2.

16
PULPING CALCULATIONS

16.1 GENERAL CHEMICAL PULPING DEFINITIONS

Introduction

Brief definitions are given for the various pulp liquor terms in this chapter. The significance and other information about these terms is found in Chapter 3. Here the terms are presented as mathematical variables to be manipulated algebraically.

Chemical concentration

Chemical concentration is a measure of the concentration of the pulping chemical in the liquor. For example, in sulfite pulping the liquor may be 6% SO_2, indicating 6 grams of sulfite chemical (SO_2 basis) per 100 ml of liquor. If the liquor:wood ratio is 4:1, the percent chemical on wood is 24% as SO_2. The following is an important relationship, not the definition of chemical concentration.

$$\text{chemical concentration in liquor} = \frac{\text{percent chemical on wood}}{\text{liquor:wood ratio}} \quad (16\text{-}1)$$

Chemical charge (to a process), percent chemical (on wood or pulp)

The chemical charge is a measure of the weight of chemical used to process (i.e., pulp or bleach) a material. For example, kraft pulping is carried out with about 25% total alkali on wood. This would indicate 500 pounds of alkali for 2000 pounds of dry wood. Chemicals in sulfite pulping are expressed on an SO_2 basis. When bleaching mechanical pulp, one might use "0.5% sodium peroxide on pulp". This means that 10 pounds of sodium peroxide are used per ton of dry pulp.

$$\text{chemical charge, \%} = \frac{\text{dry weight of chemical used}}{\text{dry weight of material treated}} \times 100\% \quad (16\text{-}2)$$

Liquor to wood ratio

The liquor to wood ratio is normally expressed as a ratio, not as a percent. It is typically 3:1 or 4:1 in full chemical pulping. The numerator may or may not include the weight of water coming in with the chips, and it must be specified to avoid ambiguity.

$$\frac{\text{liquor}}{\text{wood}} = \frac{\text{total weight of pulping liquor}}{\text{dry weight of wood}} \quad (16\text{-}3)$$

16.2 KRAFT LIQUOR--CHEMICAL CALCULATIONS

Total chemical or total alkali (TA)

The total alkali is the sum of all of the sodium salts in the liquors (as Na_2O) that contribute to AA or are capable of being converted to AA in the kraft cycle, specifically NaOH, Na_2S, Na_2CO_3, and $Na_2S_xO_y$ (as Na_2O). All chemical amounts may be reported as a concentrations of g/L or lb/gal or as a percent relative to oven dry wood. Chemicals are reported on a Na_2O basis in North America.

$$TA = NaOH + Na_2S + Na_2CO_3 + Na_2S_xO_y$$
$$(\text{as } Na_2O) \quad (16\text{-}4)$$

Total titratable alkali (TTA)

TTA is the sum of all of the bases in the white liquor that can be titrated with strong acid. Generally, it is considered as NaOH, Na_2S, and Na_2CO_3 (as Na_2O), although small amounts of Na_2SO_3 and other acids might be present.

$$TTA = NaOH + Na_2S + Na_2CO_3$$
$$(\text{as } Na_2O) \quad (16\text{-}5)$$

Active Alkali (AA)

The sum of the active ingredients in the pulping process is known as active alkali.

$$AA = NaOH + Na_2S \quad (\text{as } Na_2O) \quad (16\text{-}6)$$

Effective alkali (EA)

EA is the sum of sodium chemicals that will produce OH⁻ during kraft pulping. NaOH is completely ionized and for every two sodium atoms of Na_2S, there will be one OH^- produced.

$$EA = NaOH + \tfrac{1}{2} Na_2S \quad (as\ Na_2O) \quad (16\text{-}7)$$

Often AA and EA are given and one needs to determine the concentration of individual species. A very useful relationship to remember is that: $Na_2S = 2\,(AA - EA)$; all species are expressed as Na_2O in this formula.

$$Na_2S = 2\,(AA - EA) \quad (as\ Na_2O) \quad (16\text{-}8)$$

Sulfidity

In the white liquor, sulfidity is the ratio of Na_2S to the active alkali, expressed as a percent. Typically, a mill runs in the vicinity of 25-30% sulfidity, depending largely on the wood species pulped. Sulfidity increases the rate of delignification, which occurs by nucleophilic action of the hydrosulfide anion (HS^-) and appears to protect cellulose against degradation.

$$sulfidity = \frac{Na_2S}{NaOH + Na_2S} \times 100\% \quad (as\ Na_2O) \quad (16\text{-}9)$$

Causticity

The causticity is the ratio of NaOH to active alkali, expressed as a percentage; therefore, causticity + sulfidity = 100%. The term sulfidity is used much more than the term causticity, and both give the same information. It will not be considered further in this text.

$$causticity = \frac{NaOH}{NaOH + Na_2S} \times 100\% \quad (as\ Na_2O) \quad (16\text{-}10)$$

Causticizing efficiency

The causticizing efficiency is the ratio of NaOH to NaOH and Na_2CO_3. This is a measure of how efficient causticizing is; it represents the percentage of the Na_2CO_3 from the recovery boiler that is converted back into useful NaOH cooking chemical. A value of 77-80% is typical.

causticizing efficiency =

$$\frac{NaOH}{NaOH + Na_2CO_3} \times 100\% \quad (Na_2O\ basis) \quad (16\text{-}11)$$

Reduction efficiency

The reduction efficiency is the ratio of Na_2S to the sum of Na_2S and Na_2SO_4 in green liquor expressed as a percentage. This is a measure of the reduction efficiency in the recovery boiler. This value should be high, is usually 95%, and is not routinely measured in the mill. In addition to sodium sulfate, other oxidized forms of sulfur are present, such as sodium sulfite and sodium thiosulfate, that should be considered.

reduction efficiency =

$$\frac{Na_2S}{Na_2S + Na_2SO_4} \times 100\% \quad (Na_2O\ basis) \quad (16\text{-}12)$$

It is convenient to set up a table of conversion factors of use to solve some of the values just given. Table 16-1 gives many of these conversion factors for kraft cooking and chemical recovery. The cooking chemicals of the NSSC process, Na_2SO_3 and Na_2CO_3, are often expressed on an Na_2O basis as well.

Several examples are presented to demonstrate kraft liquor calculations. Example 1 shows the conversion factor for gravimetrically converting NaOH to Na_2O. While Na_2O is a hypothetical species in aqueous solutions and does not occur in aqueous solutions, it is a convenient way of expressing cooking chemicals on a weight basis, but at the same time on an equal molar basis. Example 1 demonstrates how the conversion factors in Table 16-1 are derived.

Example 2 shows how to calculate the actual concentration of chemical species based on cooking liquor parameters. Example 3 is a detailed pulping problem. Example 4 demonstrates the calculation and use of causticizing efficiency values.

There are additional exercises on which to practice these calculations.

Table 16-1. Sodium oxide equivalents and other gravimetric factors for kraft pulping chemicals.

Convert from:	Name	Formula weight	Equival. weight	Convert from Na$_2$O by Multi.	Convert to: by multiplication			
					Na$_2$O	NaOH	Na$_2$S	Na$_2$CO$_3$
White liquor components								
Na$_2$O	Sodium oxide	61.98	31.0	1	1	1.291	1.259	1.710
NaOH	Sodium hydroxide	40.00	40.0	1.291	0.775	1	NA[1]	2.650
Na$_2$S	Sodium sulfide	78.04	39.0	1.259	0.794	NA	1	NA
Na$_2$CO$_3$	Sodium carbonate	105.99	53.0	1.710	0.585	0.377	NA	1
Na$_2$SO$_4$	Sodium sulfate	142.04	71.0	2.291	0.436	NA	0.549	NA
Na$^+$	Sodium ion[2]	22.99	23.0	0.742	1.348	1.740	NA	2.305
NaHS	Sodium hydrogen sulfide	56.06	56.1	1.808	0.553	NA	NA	NA
Na$_2$S$_2$O$_3$	Sodium thiosulfate	158.10	79.0	2.551	0.392	NA	NA	NA
Na$_2$SO$_3$	Sodium sulfite	126.04	63.0	2.034	0.492	NA	0.619	NA
NaHSO$_3$	Sodium bisulfite	104.06	104.1	3.358	0.298	NA	NA	NA
Recovery chemicals[3]								
CaO	Calcium oxide	56.08	28.0	0.905	1.105	1.427	NA	1.890
Ca(OH)$_2$	Calcium hydroxide	74.10	37.0	1.196	0.836	1.080	NA	1.430
CaCO$_3$	Calcium carbonate	100.09	50.0	1.615	0.619	0.799	NA	1.059

[1]NA indicates these chemicals are not directly interconvertible in the kraft process. While this is true of the conversion of all sulfur containing compounds to Na$_2$O, here the weight containing (or reacting with) one mole of sodium ion is used since it is the standard of the industry.

[2]Sodium associated with organic chemicals such as sodium carboxylates or phenolates but not associated with halides such as sodium chloride.

[3]These values can be used to calculate recovery chemical demands; however, recausticizing is typically less than 83% so recovery demand is actually about 20% higher.

1 g/L = 0.008345 lb/(U.S. gal) = 0.06243 lb/ft^3

EXAMPLE 1.   Derive the conversion factor of 0.775 used to express the weight of NaOH on an Na$_2$O basis.

SOLUTION.   First, the molar relationship between these two species is expressed.

$$2 \text{ NaOH} + \rightleftarrows \text{Na}_2\text{O} + \text{H}_2\text{O}$$

From this relationship the gravimetric factor is determined as follows:

$$1 \text{ g NaOH} \times \frac{1 \text{ mol NaOH}}{40 \text{ g NaOH}} \times \frac{1 \text{ mol Na}_2\text{O}}{2 \text{ mol NaOH}} \times \frac{62 \text{ g Na}_2\text{O}}{1 \text{ mol Na}_2\text{O}} = 0.775 \text{ g Na}_2\text{O}$$

EXAMPLE 2.   Given the following information for TTA, AA, EA, and TA (all as Na_2O), determine the sulfidity as a percentage and concentration of the actual species $NaOH$, Na_2S, Na_2CO_3, and Na_2SO_4 in grams per liter. Assume these are the only sodium-containing species present.

TTA = 120 g/L; AA = 104 g/L; EA = 88 g/L; and TA = 128 g/L

SOLUTION: Using Eq. 16-8, $Na_2S = 2(AA-EA)$, the following is obtained:

$$\text{sulfidity} = \frac{Na_2S}{NaOH+Na_2S} \times 100\% = \frac{2(AA-EA)}{AA} \times 100\% = \frac{100\times2(104-88)}{104} = 30.8\%$$

Chemical composition:

$Na_2S = 2(AA-EA) = 2(104-88) = 32$ g/L (as Na_2O);

32 g/L $Na_2S \times 1.259 = 40.3$ g/L Na_2S

$NaOH = AA-Na_2S = 104-32 = 72$ g/L (as Na_2O);

72 g/L $NaOH \times 1.291 = 93$ g/L $NaOH$

$Na_2CO_3 = TTA-AA = 120-104 = 16$ g/L (as Na_2O);

16 g/L $\times 1.710 = 27.4$ g/L Na_2CO_3

$Na_2SO_4 = TA-TTA = 128-120 = 8$ g/L (as Na_2O);

8 g/L $\times 2.294 = 18.3$ g/L Na_2SO_4

Total active chemical = 40.3 + 93 = 133.3 g/L

Total inactive chemical = 27.4 + 18.3 = 45.7 g/L

PROBLEM: What is the causticizing efficiency of this liquor? Answer: 85.3%.

EXAMPLE 3.   Given: 1 metric ton of wood (oven-dry basis). The moisture content of the chips is 50% on the green weight basis. The active alkali (AA) of the white liquor is 20% on oven-dry wood, and the sulfidity is 25%. The liquor:wood ratio is 4:1 (oven-dry wood), not including the water in the chips. Assume the specific gravities of all liquors are 1.00; in fact, they are closer to 1.05 g/ml. In this problem consider only the active chemical species $NaOH$ and Na_2S, ignoring the presence of Na_2CO_3, Na_2SO_4 and other chemicals normally present. Calculate:

1. $NaOH$ per ton as Na_2O and as $NaOH$
2. Na_2S per ton as Na_2O and Na_2S
3. Total white liquor mass per ton (do not include the mass of water in the chips here).
4. A) Mass of water in the original liquor; B) mass of water in chips; C) total mass of water in digester
5. A) Concentration of $NaOH$ and Na_2S in the white liquor (as Na_2O); B) in the diluted liquor.

SOLUTION: Wood mass × chemical charge = mass of chemical (active alkali).

1000 kg × 20% = 200 kg active alkali (as Na_2O).

1.   Active alkali × sulfidity = Na_2S.

200 kg AA × 25% = 50 kg Na_2S (as Na_2O);

50 kg Na_2S × 1.259 = 63.0 kg Na_2S (as Na_2S).

2.   Active alkali - Na_2S = NaOH   (all of these can be on a mass basis, as Na_2O)

200 kg AA - 50 kg Na_2S = 150 kg NaOH;

150 kg NaOH × 1.291 = 193.5 kg NaOH (as NaOH).

3.   Oven-dry wood mass × L:W ratio = total mass of added liquor.

1000 kg × 4 (kg liquor)/(kg wood) = 4000 kg of liquor.

4A. Total liquor mass - mass of chemicals = mass of water in the liquor.

4000 kg - (63 kg + 193.5 kg) = 3743.6 kg of water in the added white liquor.

4B. MC_{GR} = mass of water in chips/(mass of water chips + oven-dry mass of chips)

0.500 = $x/(x + 1000$ kg dry wood);

x = mass of water per 1000 kg dry chips = 1000 kg

4C. Water in liquor + water in chips = total water in cook

3743.6 + 1000 = 4743.6 g (or kg) of total water.

5.   Mass of dry chemical/volume of solution = concentration

A.   150 kg/4000 L = 37.5 g/L NaOH (as Na_2O).

50 kg/4000 L = 12.5 g/L Na_2S (as Na_2O).   Therefore, AA = 50 g/L.

B.   150 kg/5000 L = 30.0 g/L NaOH (as Na_2O).

50 kg/4000 L = 10.0 g/L Na_2S (as Na_2O).   Therefore, AA = 40 g/L.

EXAMPLE 4.   Your boss tells you that the white liquor fresh from the recovery cycle contains 85 g/L NaOH (as NaOH) and 35 g/L Na_2CO_3 (as Na_2CO_3). He asks you if everything is all right. Use the definition of causticity and (knowing typical mill values for this) indicate whether or not you think there is a problem.

SOLUTION: 85 g/L NaOH × (62 g Na_2O/80 g NaOH) = 65.9 g/L NaOH (as Na_2O)
35 g/L Na_2CO_3 × (62 g Na_2O/106 g Na_2CO_3) = 20.5 g/L Na_2CO_3 (as Na_2O)
Causticity = 100% × 65.9/(65.9 + 20.5) = 76.3%; this indicates that only 76.3% of the sodium carbonate is being converted to the active species sodium hydroxide. The conversion should be 77-80%, so you tell your boss there is a problem.

16.3 KRAFT LIQUOR--CHEMICAL ANALYSIS

The relative proportions of NaOH, Na_2S, and Na_2CO_3 are determined by titration with HCl in so-called "ABC titrations." The procedures are detailed in TAPPI Standard T 624 for kraft and soda white and green liquors and in T 625 for kraft and soda black liquors. The pH versus the

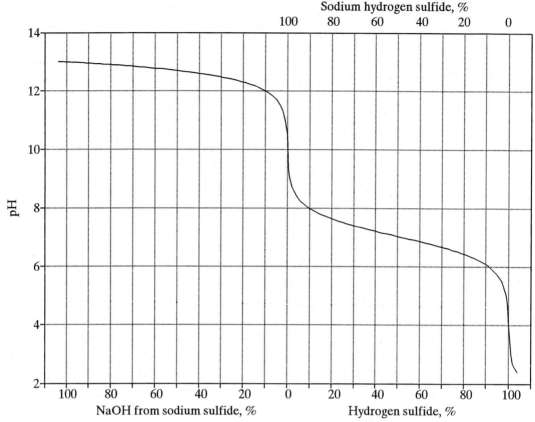

Fig. 16-1. Titration curve of 0.1 *M* Na₂S with two moles of HCl.

forms of Na₂S and Na₂CO₃ are very useful to examine to understand the titration method. The titration curve for NaOH was presented in Fig. 14-1. Fig. 16-1 shows the pH as a function of the species Na₂S, HS⁻, and H₂S. Fig. 16-2 shows this for Na₂CO₃. These graphs are valid for dilute solutions at 25°C. There is appreciable variation in the acid dissociation constants from various sources, which may alter the pH from 0.1 to 0.5 or more units. Titration curves are approximately accurate, but not exact. For acids such as H₂CO₃ and H₂SO₃ which exist in equilibrium with gases, CO₂ and SO₂, respectively, the total concentration of both the acid and the gas are included in the equilibrium constant.

The titration curve of Na₂S (sodium sulfide) represents the titration of a weak base in the presence of a strong base. Na₂S reacts in water essentially to completion to give NaHS (sodium hydrogen sulfide) and NaOH. NaHS, however, is a weak base with a pK_b of 6.96. The titration curve

is that of a strong base-strong acid titration until all of the initially formed NaOH is neutralized, then it is a weak base-strong acid titration until all of the NaHS is H₂S (hydrogen sulfide).

For the titration curve of H₂CO₃, it is assumed that CO₂ is kept in the solution, even though it is not very soluble. The dashed line shows the titration curve if CO₂ is removed from solution; this is commonly done by boiling the CO₂ off (as it is not soluble in water at elevated temperatures) in order to sharpen the endpoint. As CO₂ is boiled off, all that remains is NaHCO₃ so the pH is approximately constant at $(pK_1 \cdot pK_2)^{0.5}$ = $(6.37 \cdot 10.32)^{0.5}$ = 8.345. A similar method involves titrating beyond the endpoint with HCl, swirling the solution to drive off the CO₂, and then back titrating to the endpoint with standardized NaOH solution. The *x*-axis represents the addition of two moles of HCl. As the first mole is added the carbonate anion decreases from 100% to 0%, while the bicarbonate anion increases from 0% to

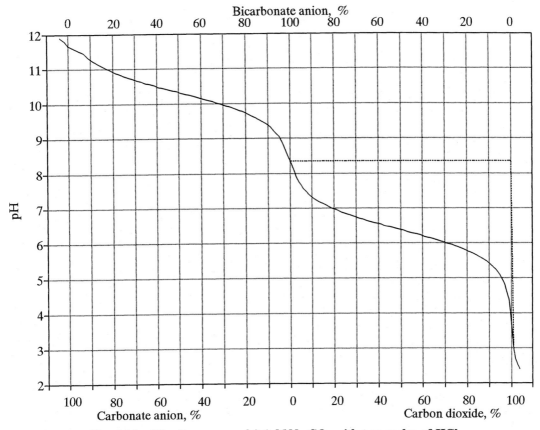

Fig. 16-2.   Titration curve of 0.1 M Na$_2$CO$_3$ with two moles of HCl.

100%. As the second mole of HCl is added the bicarbonate anion decreases from 100% to 0%, while the carbon dioxide increases from 0 to 100%. The first endpoint is titrated to a phenolphthalein endpoint and the second endpoint is titrated to a methyl orange endpoint. A graph in this form gives the relative concentration of two species at any pH.

An ABC titration (shown in Fig. 16-3) is carried out with three titrations. Before the first titration, excess BaCl$_2$ is added to the liquor to precipitate carbonate anion according to the reaction:

$$Ba^{2+} + CO_3^{2-} \rightleftarrows BaCO_3\!\downarrow \quad K_{sp} = 8.1 \times 10^{-9}$$

The sample is then titrated with acid to the endpoint at 10.5 (phenolphthalein), the "A" titration. This corresponds to all of the NaOH and half of the Na$_2$S, since all of the Na$_2$S is converted to NaHS in water with the liberation of NaOH. The amount of acid corresponds to the effective alkali.

The titration curve up to this point is identical to that of Na$_2$S with HCl (Fig. 16-1). In the second titration, excess formaldehyde is added to the same sample to convert the NaHS to NaOH by the reaction:

$$HS^- + CH_2O + H_2O \rightleftarrows CH_2(OH)SH + OH^-$$

The pH rises as NaOH is liberated upon addition of formaldehyde and the solution is titrated again to the phenolphthalein endpoint in the "B" titration. This corresponds to half of the original Na$_2$S as all of the NaHS is converted to the complex (avoiding H$_2$S fumes that would be emitted without formaldehyde addition). The acid consumed from A to B corresponds to AA - EA. The solution is then titrated with HCl to a methyl orange endpoint of pH 4, the "C" titration. The acid consumed from B to C corresponds to Na$_2$CO$_3$. At this endpoint the carbonate has redissolved and been converted to CO$_2$ according to the reaction:

$$BaCO_3(s) + 2HCl \rightleftarrows BaCl_2 + H_2O + CO_2\uparrow$$

To summarize, the significance of A, B, and C are as follows:

EA	$= A$	$NaOH$	$=$	$2A - B$
AA	$= B$	H_2S	$=$	$2(B-A)$
TTA	$= C$	Na_2CO_3	$=$	$C - B$

Black liquor is titrated in a similar fashion if one is interested in residual concentrations of these species after the pulping process. One important difference is that after the $BaCl_2$ is added, the sample is centrifuged to collect the precipitate at the bottom of a test tube. An aliquot is taken from the supernatant for titration. Otherwise, the $BaCO_3$ interferes with the first two endpoints as it does not precipitate well. The "C" endpoint must then be determined on a separate aliquot that has not been treated with $BaCl_2$. This brings up the important point that during an ABC titration of white liquor, it is important to avoid high concentrations of acid that might redissolve the $BaCO_3$; it would probably be better to handle it like black liquor.

EXAMPLE 5. Given: $0.500\ N\ H_2SO_4$ is used to titrate 5.00 ml of a white liquor sample. The titration values were A = 30.6 ml, B = 34.2 ml, and C = 38.7 ml. Calculate AA, EA, and TTA.

SOLUTION: Each ml of H_2SO_4 is equivalent to 0.500 meq/ml × 31 mg Na_2O/meq = 15.5 mg Na_2O. 15.5 mg Na_2O/5 ml liquor = 3.1 mg Na_2O/(ml white liquor)/(ml titrant) = 3.1 g/L Na_2O. Therefore,

$$EA = 30.6 \times 3.1 = 94.9\ \text{g/L}$$
$$AA = 34.2 \times 3.1 = 106\ \text{g/L}$$
$$TTA = 38.7 \times 3.1 = 120\ \text{g/L}$$

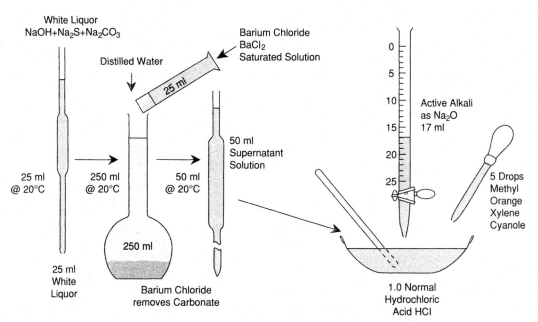

Fig. 16-3. Titration of white liquor. Redrawn from J. Ainsworth, *Papermaking,* ©1957 Thilmany Paper Co., with permission.

16.4 SPECIFIC GRAVITY AND VIS-COSITY OF KRAFT LIQUORS

The specific gravity of white liquor at room temperature may be *estimated* as follows:

specific gravity = 1.0 + (% solids/100%)

The specific gravities and Baumé values as a function of southern pine kraft black liquor solids content is given in CPPA data sheet C-5 (issued in 1943, reissued in 1952, but now apparently abandoned; it was at one time TAPPI DATA SHEET 58A). It should be used with care with other species; however, it is a good approximation lacking any other information. Fig. 16-4 is a graphical presentation of the specific gravity versus solids content. The data sheet gives Baumé values for black liquor at 80 to 210°F in 10° intervals. Work in my laboratory with commercial kraft black liquor of Douglas-fir (from a brown paper

mill) at 72°F gave a specific gravity of 1.300 for 51.0% solids which agrees well with the earlier results. Equations are easily determined from this table to calculate °Bé (which can be converted to specific gravity) as a function of solids content for various temperatures.

$$°Bé = 2/3 × (solids, \%) \qquad at \ 60°F$$

$$°Bé = -2.296 + 0.6524 × (solids, \%) \quad 140°F$$

$$°Bé = -4.288 + 0.6384 × (solids, \%) \quad 210°F$$

specific gravity = 145/(145-°Bé);   sp gr > 1.0

The viscosity of black liquors depends on several factors, particularly the temperature and solids content, as shown in Fig. 16-5. The source of the black liquor is also important, with hardwood black liquors generally having lower viscosities than softwood black liquors.

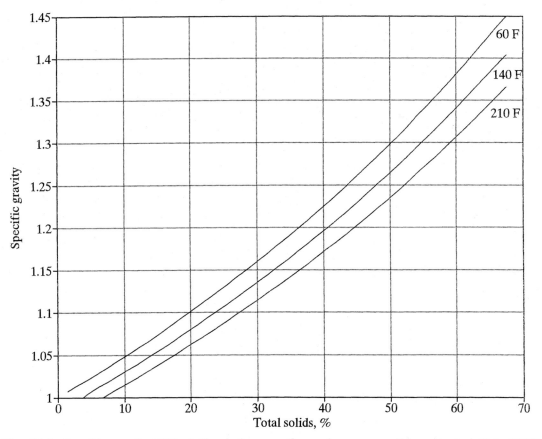

Fig. 16-4. Specific gravity of black liquor from southern pine versus solids content at 60 to 210°F.

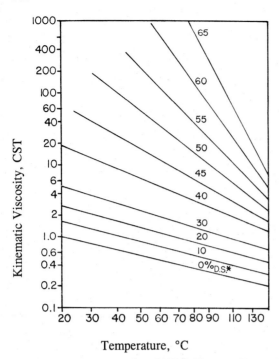

Temperature, °C

Fig. 16-5. The viscosity of black liquor. From U.S. EPA. *DS is dissolved solids.

16.5 IMPORTANCE OF BLACK LIQUOR pH

If the black liquor pH decreases below 12, lignin begins to precipitate. This is due to conversion of phenolate anions to phenol. The salt form is much more polar and water soluble. The pK_a values of phenols representing structures of lignin (model compounds) are shown in Table 16-2.

16.6 KRAFT H-FACTOR AND OTHER PROCESS CONTROL EQUATIONS

The H-factor is a single variable used in the kraft cooking process to combine the variables of temperature and time into a single variable representing the extent of the cook. It is the integral of the *relative reaction rate* with respect to time. It is an extremely important operating parameter in modern process control. The H factor combines only time and temperature, not active alkali, sulfidity, and other variables assumed to be held constant. The original publication on the H-factor is Vroom (1957).

The H-factor is based on the assumption that delignification is a single reaction. While this model holds well, lignin is not actually a "pure"

Table 16-2. The pK_a values for various phenols.

Compound	pK_1	pK_2
phenol	9.99	
4-methylphenol	10.26	
2-methoxyphenol	9.99	
1,2-dihydroxybenzene	9.36	12.98

compound undergoing a single chemical reaction to achieve delignification. Rather lignin is a complex molecule with many types of reactions occurring during the delignification process.

In 1889 Arrhenius observed that the rate of reaction, k, for most chemical reactions fits Eq. 16-13, where E_a is the *Arrhenius activation energy*, A is the *pre-exponential factor*, T is the temperature in Kelvin, and R is the gas constant. E_a can be determined within about 1 kcal/mole, and A within a factor of 3. Most chemical reactions increase by a factor of 2 to 3 with a 10°C increase in temperature; this is attributed to the increased number of collisions of reactants with sufficient energy to achieve the activation energy of the transition species in order for the reaction to proceed. The rate of delignification during kraft pulping increases by a factor of about 2 for an increase in temperature of 8°C. Taking the natural log, ln, of both sides of Eq. 16-13 leads to Eq. 16-14. For a *relative reaction rate*, which is what the H-factor is, Eq. 16-14 is simplified to Eq. 16-15, where B and C are constants.

$$k = Ae^{-E_a/RT} \qquad (16\text{-}13)$$

$$\ln k = \ln A - E_a/(RT) \qquad (16\text{-}14)$$

$$\ln k = B - C/T \qquad (16\text{-}15)$$

Eq. 16-15 was solved by Vroom using $C = 16113$ (based on data in the literature) and arbitrarily setting a relative reaction rate $= 1$ at 100°C (373.15 K). This leads to Eq. 16-16 which gives the relative reaction rate (which I will call the relative *H-factor rate*) as a function of temperature. The H-factor is determined by the area under a cooking curve corresponding to the H-factor rate at a given temperature and the time (in hours) at each temperature. The H-factor is

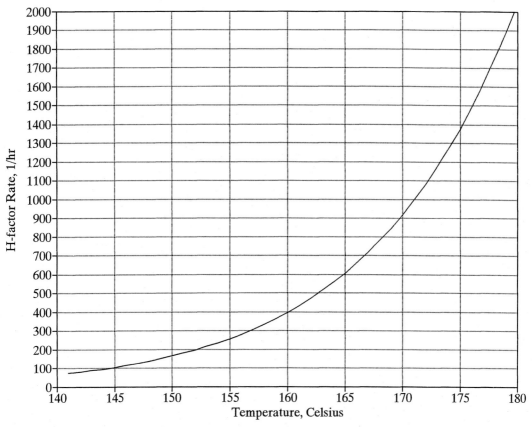

Fig. 16-6. The relative H-factor rate/hr as a function of temperature.

therefore the integral of the H-factor rate with respect to time expressed in hours. The relative H-factor rate is shown as a function of temperature in Fig. 16-6. Fig 16-7 shows the H-factor rate for a linear heating temperature ramp from 80-170°C. In this case the H-factor may be estimated by using a suitable approximation method, such as the one explained below. Temperatures below 130°C (265°F) contribute very little to the overall extent of the cook.

$$\ln k = 43.181 - 16113/T \qquad (16\text{-}16)$$

There are slight discrepancies between calculated values and values given in Vroom's paper, but this is due to the available calculation methods of the time that Vroom did his work. Also, Vroom used 100°C = 373 K (obtaining a B of 43.20) instead of 373.15 K. For 180°C and using B = 43.20, the H-factor is 2060/hr at 453 K and 2084/hr at 453.15 K (a 1.2% difference) although Vroom reported it as 2042/hr. I suggest using Eq. 16-16 and adding 273.15 to the temperature in Celsius for reasonable agreement with Vroom's published values and to promote a consistent theoretical basis. At 170°C this modification gives 917/hr, Vroom's equation gives 923/hr, and Vroom reported 927/hr. At 160°C this gives 396/hr, Vroom's equation gives 398/hr, and Vroom reported 401/hr. These differences are insignificant and presented only to alleviate concern an individual might have when calculating values that differ from published values.

The H-factor alone cannot be used to predict the yield or other properties of the pulp. However, once the H-factor versus yield relationship (or other pulping parameter relating to lignin content such as kappa number) is determined for a particular set of pulping conditions, the yield for a given H-factor will be known (presumably) for any time-temperature combination used to achieve that H-factor.

EXAMPLE 6. Sawdust is pulped in a continuous digester at a constant temperature (zero heat-up time) of 180°C for 45 minutes to a kappa number of 30. If the digester temperature is increased by 5°C and the other pulping conditions are held constant, solve the following:

a) What pulping time is required to reach the same kappa number?

b) What is the percent increase in digester output?

Note: the H-factor rate at 180°C = 2042/hr and at 185°C = 3054/hr

SOLUTION. a) The H-factor of the cook at 180°C is 2042/hr × 0.75 hr = 1532. To cook to the same H-factor at 185°C, solve for time as: $t = 1532/3054$ hr^{-1} = 0.50 hr = 30 min. Cooking at the higher temperature gives 3 batches/90 min or a 50% increase in production.

Although Eq. 16-16 enjoys widespread use in industry, there is no reason to assume that the activation energy of the Arrhenius equation is the same for hardwoods and softwoods or even among species of either group. Furthermore, the composition of cooking chemical may influence the activation energy somewhat. Taking the derivative of the Arrhenius equation and solving for the activation energy gives Eq. 16-17.

$$E_a = RT^2 \, d\ln k/dT$$

$$\approx RT_1T_2(\Delta T)^{-1} \ln(k_{T2}/k_{T1}) \qquad (16\text{-}17)$$

EXAMPLE 7. Solve for E_a based on the H-factor relative reaction rates at 165°C and 175°C.

SOLUTION. $(0.001987$ kcal mol^{-1} K$^{-1}) \times (438$ K$)(448$ K$)(10$ K$)^{-1} \times \ln(1387/610) = 32.0$ kcal. Also, comparison of Eq. 16-14 to Eq. 16-16 shows that $16113 = E_a/R$; therefore, $E_a = 32.0$ kcal/mole and is independent of temperature.

Often linear temperature ramps are used or assumed during digester heating. It would be ideal to integrate Eq. 16-16 with respect to time using a linear temperature ramp to solve for the H-factor during the heat-up time. By integrating from absolute zero the form of the equation could be written as follows, where T_r is the rate of temperature increase.

$$\text{H-factor} = \int k(t)dt = -e^{43.181} \int e^{-16113/(Tr \cdot t)} \, t^2 \, t^2 \, dt$$

Let $a = -16113/T_r$ and $x = 1/t$ so, $dx = -t^2 dt$; the integral is now in this form:

$$\text{H-factor} = C \int e^{ax}/x^2 \, dx$$

Unfortunately this integral leads to a series that requires several hundred terms before diverging, due to the large value of a, involving numbers on the order of 100! (factorial), making it more difficult to solve than "manual" methods. Below is a short program listed in BASIC which can be used to solve the H-factor for linear temperature ramps (or any ramp with suitable modification). The program is set up for a temperature ramp of 80-170°C during 90 minutes. Running the program as is (with 6 time intervals of 15 minutes gives an H-factor of approximately 165.4 for the time period. Fig. 16-7 is a graphic depiction of what the program accomplishes. The program works by dividing the temperature interval defined by the final temperature, TF, and initial temperature, TI, into rectangles of time. The reaction rate is determined at the temperature corresponding to the center of the time interval and applied to the entire time interval to give an H-factor for that rectangle, TOTAL. The more rectangles used, the closer the approximation. Using 10 time intervals gives an H-factor of 171.6, while using 900 time intervals gives 175.2; obviously only a few intervals are needed for sufficient accuracy.

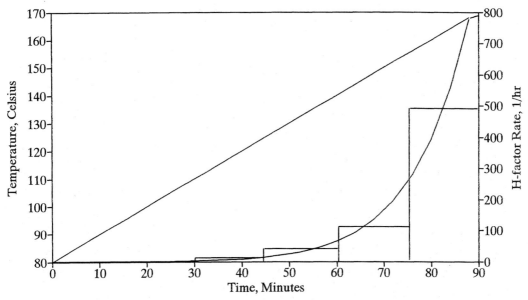

Fig. 16-7. A linear heating ramp from 80-170°C with an approximation of H-factor by the area under the curve as rectangles.

```
10   TI = 80
20   TF = 170
25   TIME = 90
30   REM REMARK STP IS THE WIDTH OF
     THE RECTANGLES IN DEGREES
     (WHICH IS CONVERTED TO TIME),
     AND INCR IS THE AREA OF EACH
     RECTANGLE.  BE SURE TO THAT THE
     NUMBER OF DEGREES IN THE TEM-
     PERATURE RAMP IS EVENLY DIVISI-
     BLE BY STP.
40   STP = 15
50   FOR I = TI TO (TF-STP) STEP STP
60   INCR = EXP(43.181-(16113/(273.15 + I +
     0.5 * STP)))
70   TOTAL  =  TOTAL  +  (INCR*
     (STP/((TF-TI)/(TIME/60))))
80   NEXT I
90   PRINT TOTAL
```

Empirical equations have been published which have found uses in process control. Hatton (1973, 1976) discusses many of these and presents his own model for pulp yield and kappa number. His equation of pulp yield is in the form of the equation: Yield = $A - B[(\log H)(EA)^n]$, where EA is the effective alkali and H is the H-factor.

The coefficients were solved for a variety of different species of both hardwoods and softwoods. Generally, the softwoods had much higher correlation coefficients.

Edwards and Norberg (1973) developed an extension of the H-factor called the τ (tau) factor, which is a single variable combining the effects of time, temperature, initial EA, and liquor to wood ratio. Earlier work had established the τ factor concept for hypochlorite and chlorine dioxide bleaching data. At constant sulfidity, $\tau = (EA/L:W)^2 H$. Lin et al. (1978) calculated the kappa number as a function of H-factor, liquor to wood ratio, and alkali to wood ratio, obtaining a single equation for hardwoods of Taiwan and the Ivory Coast. Tasman (1981) calculated the yield as a function of the liquor sulfidity, effective alkali, and the H-factor for several species of hardwoods and softwoods in his equations. Recently, Paulonis and Krishnagopalan (1991) discussed a sophisticated process control system for kraft batch digesters that uses liquid density, electrical conductivity, and other techniques in addition to the usual parameters in order to predict kappa number despite variations in wood chip supply and digester operation.

Kubes et al. (1983) have developed a G-

Table 16-3. The H-factor rate, G-factor rate, and ratio for various temperatures.

Temp., °C	H-fact.	G-fact.	H/G ratio
160	400	2960	0.135
165	610	5220	0.117
170	927	9100	0.102
175	1387	15600	0.089
180	2042	29600	0.069

factor that is analogous to the H-factor except that remaining cellulose viscosity is correlated to cooking temperature and time under constant conditions for any alkaline cooking process. The method works for any alkaline method since cellulose viscosity loss is due to random alkaline cleavage, and there are no known additives that decrease this reaction. While one may not expect viscosity to be correlated to extent of reaction, the authors' model fits the data fairly well. The authors obtain the value of 175-180 kJ/mol for this reaction.

Table 16-3 lists the H-factor rate, G-factor rate, and ratio of H-factor to G-factor rates at various temperatures. The higher this ratio, the less cellulose viscosity decrease is predicted when cooking to a specified kappa number. Table 16-3 indicates that cooking at 160°C would give pulp with twice the viscosity as pulp cooked at 180°C.

16.7 SULFITE LIQUOR CALCULATIONS

The concentrations of chemicals in sulfite pulping are reported on an SO_2 basis. This allows the various forms of sulfurous acid and its salts to be compared on an equal molar basis, although the concentrations are reported on a weight basis. (This is the same principle as using Na_2O as the basis for kraft pulping and recovery chemicals.) The concentration is reported as a percent or g/100 ml, both of which are identical units if the specific gravity is unity. Full chemical sulfite methods all use approximately 20% chemical (SO_2 basis) on a dry wood basis. Table 16-4 gives conversion factors to and from the *actual* chemical

concentrations and the SO_2 basis. See Example 8 to see how these conversion factors are derived.

The various salts of sulfurous acid are of central importance to the various sulfite-based processes. The chemistry of sulfurous acid as a function of pH is very important. For the titration curve of H_2SO_3, the first proton is a moderately strong acid and titrates along with any strong acids present. Therefore no initial "endpoint" below 0 ml base (corresponding to excess acid) is present as there is for a weaker acid such as acetic acid. The Henderson-Hasselbalch equation cannot be used for this part of the titration curve since the H_2SO_3 ionizes appreciably, thereby reducing the concentration of H_2SO_3 and changing the ratio of acid to conjugate base significantly. The quadratic equation must be used to solve $[H^+]$ for the initial part of the titration curve. After about 40 ml of NaOH is added, the pH is calculated accurately by this equation.

It is assumed that SO_2 is kept in the solution, even though it is only partially soluble at room temperature. The x-axis represents the addition of two moles of NaOH. As the first mole is added the SO_2 decreases from 100% to 0% while the bisulfite anion increases from 0% to 100%. As the second mole of NaOH is added the bisulfite anion decreases from 100% to 0% while the sulfite ion increases from 0 to 100%.

The ionization of sulfurous acid is temperature dependent. Rydholm, in *Pulping Processes,* p. 456, gives the pK_a as 1.8 at 25°C, 2.3 at 70°, 2.6 at 100°, 3.0 at 120°, and 3.3 at 140°C.

Sulfurous acid is a moderately strong acid. It will react completely with any of the five common hydroxides used in sulfite cooking. The form of the reaction is independent of the base and occurs in two steps if enough base is present.

$$MOH + H_2SO_3 \rightleftarrows MHSO_3 + H_2O$$

$$MOH + MHSO_3 \rightleftarrows M_2SO_3 + H_2O$$

M is $\frac{1}{2}Mg^{2+}$, $\frac{1}{2}Ca^{2+}$, Na^+, NH_4^+, or K^+

Since $CaSO_3$ has limited solubility ($K_{sp} = 6.8 \times 10^{-8}$ at 25°C), calcium based sulfite cooking has to be carried out at a pH of 1-2 to maintain SO_2 as the soluble HSO_3^-. Similarly, since $MgSO_3$ has a limited solubility ($K_{sp} = 3.2 \times 10^{-3}$ at 25°C), magnesium based sulfite cooking is carried out at

Table 16-4. **Gravimetric factors for chemicals involved in sulfite pulping.**

Convert From:	Name	Formula weight	Convert to SO_2 by multiplication	Convert from SO_2 by multiplication
SO_2	Sulfur dioxide	64.06	1	1
SO_3	Sulfite ion	80.06	0.800	1.250
HSO_3^-	Bisulfite ion	81.07	0.790	1.266
H_2SO_3	Sulfurous acid	82.07	0.780	1.282
Na_2SO_3	Sodium sulfite	126.04	0.508	1.968
$NaHSO_3$	Sodium bisulfite	104.06	0.616	1.624
$MgSO_3$	Magnesium sulfite	104.36	0.614	1.629
$Mg(HSO_3)_2$	Magnesium bisulfite	186.43	0.687	1.455
$CaSO_3$	Calcium sulfite	120.14	0.533	1.875
$Ca(HSO_3)_2$	Calcium bisulfite	202.21	0.634	1.578
$(NH_4)_2SO_3$	Ammonium sulfite	116.13	0.552	1.813
$(NH_4)HSO_3$	Ammonium bisulfite	99.10	0.646	1.547

a pH below 5. At the lower pH values the sulfite exists as HSO_3^-.

Fig. 16-8 shows clearly that H_2SO_3 and SO_3^{2-} do not exist in solution together in any appreciable amount. Since the standard terminology of free and combined SO_2 are used, the following equation is important.

$$2HSO_3^- \rightleftarrows SO_3^{2-} + H_2SO_3 \qquad (16\text{-}18)$$

Again, the equilibrium for the above reaction lies far to the left. When free and combined forms of SO_2 are added in solution they will react together until one of them is exhausted as the reaction proceeds to the left. On the other hand, if the solution is of known composition and the free and combined equivalents are to be determined, one imagines the reaction to proceed to the right until all the bisulfite is exhausted.

EXAMPLE 8.   Derive the gravimetric factor for converting $Ca(HSO_3)_2$ to SO_2.

SOLUTION.   One mole of $Ca(HSO_3)_2$ gives two moles of SO_2: $Ca(HSO_3)_2 \rightarrow 2SO_2 + Ca(OH)_2$

Therefore,

$$1 \text{ g } Ca(HSO_3)_2 \times \frac{2 \text{ mol } SO_2}{1 \text{ mol } Ca(HSO_3)_2} \times \frac{1 \text{ mol } Ca(HSO_3)_2}{202.21 \text{ g } Ca(HSO_3)_2} \times \frac{64.06 \text{ g } SO_2}{1 \text{ mol } SO_2} = 0.634 \text{ g } SO_2$$

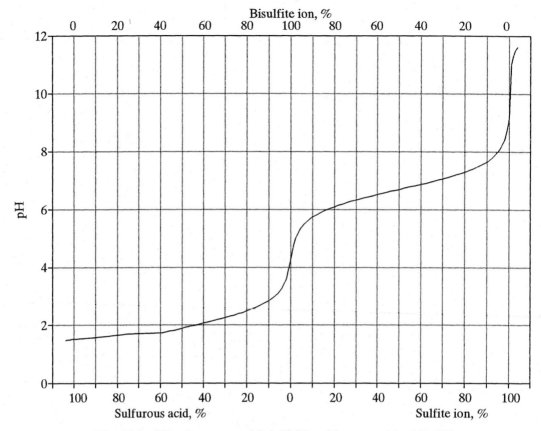

Fig. 16-8. Titration curve of 0.1 M SO$_2$ with two moles of NaOH.

EXAMPLE 9. Case 1. A cooking liquor consists of 6% free SO$_2$. What is the form of SO$_2$? Answer. All of the SO$_2$ is in the form of H$_2$SO$_3$ and the pH is about 1.2 at 25°C. This is only a hypothetical cooking liquor.

Case 2. A cooking liquor consists of 4.2% free SO$_2$ and 1.8% combined SO$_2$. What is the form of the SO$_2$? Answer. The limiting chemical is combined SO$_2$, so 1.8% combined SO$_2$ reacts with 1.8% free SO$_2$ (leaving 2.4% free SO$_2$ as H$_2$SO$_3$) to produce 3.6% MHSO$_3$, pH $\approx$ 1.85.

Case 3. A cooking liquor contains 2.5% free SO$_2$ and 2.5% combined SO$_2$. What is the form of the SO$_2$? Answer. All of the free reacts with all of the combined SO$_2$ to give 5% MHSO$_3$. The pH is 4.3, but there is little buffering capacity. This is a *square* liquor.

Case 4. A cooking liquor contains 1% free SO$_2$ and 5% combined SO$_2$. What is its actual composition? Answer. 1% free SO$_2$ reacts with 1% combined SO$_2$ (leaving 4% M$_2$SO$_3$) to produce 2% MHSO$_3$. The pH is about 5.8.

Case 5. NSSC is carried out with sulfite at pH 8-9. Fig. 16-8 shows that all of the SO$_2$ is in the combined form. Notice also there is no buffer capacity in this region. NSSC cooking is carried out with 10-15% carbonate to supply some buffering action.

EXAMPLE 10.   A solution is 3% Na_2SO_3 (as SO_2) and 2% $NaHSO_3$ (as SO_2). What is the free and combined SO_2? What is the actual concentration of $NaHSO_3$?

SOLUTION.   2% $NaHSO_3$ → 1% H_2SO_3 + 1%Na_2SO_3. Therefore, this corresponds to 4% combined SO_2 and 1% free SO_2. From Table 16-4 the gravimetric factor of 1.624 is obtained to convert SO_2 to $NaHSO_3$. 2% $NaHSO_3$ (SO_2 basis) × 1.624 = 3.25% $NaHSO_3$ = 32.5 g/L $NaHSO_3$.

EXAMPLE 11.   Given: A digest charge with 1 kg of dry wood, a calcium liquor containing 5% free SO_2 and 1% combined SO_2 and a liquor:wood ratio of 4:1 (including the water in the wood).

Calculate: the concentrations and amounts of the <u>actual</u> chemical species. Assume the specific gravity of the liquor is 1.00.

SOLUTION.   The total SO_2 is equal to the concentration in the liquor times the liquor to wood ratio, or 24% on wood, which is 240 grams SO_2 for the charge. The 1% combined reacts with 1% free to give 2% (60 g on wood) as bisulfite leaving 4% (160 g on wood) as sulfurous acid. From Table 16-4 80 grams bisulfite (SO_2 basis) corresponds to 80 × 1.578 = 126.4 g $Ca(HSO_3)_2$ on wood or 31.6 g/L. In a similar fashion 160 g free SO_2 corresponds to 205.2 g H_2SO_3 on wood or 51.3 g/L.

PROBLEM.   For this problem, what would the concentrations be for Mg based cooking? Answer: 29.1 g/L $Mg(HSO_3)_2$.

16.8 SULFITE LIQUOR ANALYSIS

Sulfite pulping liquors could be titrated with NaOH to each endpoint to determine the free SO_2 and combined SO_2. However the first endpoint is not very sharp, and even the second endpoint may not be very sharp in sulfite liquors. Palmrose (1935) developed a method where, under acidic conditions, all of the SO_2 is converted to SO_4^{-2} by periodate ion as shown in the following two chemical equations. Periodate thus measures the *total SO_2*.

$$KIO_3 + 3H_2SO_3 \rightarrow KI + 3H_2SO_4$$

$$2KIO_3 + 3(HSO_3^-)_2 \rightarrow 2KI + 3H_2SO_4 + 3SO_4^{2-}$$

I of KIO_3 is reduced from +5 to -1 while each sulfur of SO_3^{2-} is oxidized from +4 to +6. The equivalent weight of KIO_3 is 1/6 the molecular weight, and the equivalent weight of H_2SO_3 is one-half the molecular weight. These reactions are actually fairly slow and the endpoint would easily be overrun. Small amounts of KI (from the indicator solution), however, allow the following two reactions, which are rapid, to occur:

$$IO_3^- + 3H_2SO_3 + 5I^- \rightarrow 3SO_4^{2-} + 3I_2 + 3H_2O$$

$$3SO_4^{2-} + 3I_2 + 3H_2O + 3H_2SO_3 \rightarrow 6I^- + 6H_2SO_4$$

The titration is carried out with 0.1 N potassium iodate to blue endpoint using KI/starch indicator. The excess I_2 at the end of the reaction reacts with starch to give a characteristic blue color, a well known reaction used as an indicator for starch. The slight excess of KIO_3 at the end of the reaction is reacted with thiosulfate.

The liberated acid is then titrated with 0.1 M NaOH to a methyl red endpoint and represents the *free SO_2*. Notice that the latter titration is a strong acid-strong base titration with a well-resolved endpoint. When this method was first developed calcium was the only base used for sulfite pulping and the liquors were necessarily acidic. With other bases where the cooking liquor is above pH of 4-5, a known amount of sulfuric acid is added before the iodate titration. The liberated acid is then titrated but the additional amount of sulfuric acid added is subtracted from the free SO_2 value obtained in the second titration. This procedure is described in TAPPI Standard T 604.

EXAMPLE 12. A sulfite liquor was diluted 1:10. A 10.00 ml aliquot of the diluted solution was titrated by the Palmrose method. Calculate the total and combined SO_2 in the original solution based on the following amounts of titrants: 12.18 ml of 0.2060 N KIO_3 and 10.78 ml of 0.0946 N NaOH.

SOLUTION: The total SO_2 in $N = 10 \times (12.18$ ml $\times 0.2060$ $N)/10$ ml $= 2.51$ N.

The total $SO_2 = 2.51$ $N \times 32$ g/eq $SO_2 = 80.32$ g/L total SO_2 or 8.03% total SO_2.

The free $SO_2 = 10 \times (10.78$ ml $\times 0.0946$ $N)/10$ ml $= 1.020$ N.

The free $SO_2 = 1.02$ $N \times 32$ g/eq $SO_2 = 32.64$ g/L free SO_2 or 3.26% free SO_2.

16.9 THE CHEMISTRY OF SULFUR

Elemental sulfur

Elemental sulfur occurs in many complex forms. The chemistry of elemental sulfur presented here is a simplification of its complex chemistry but will be useful to explain its properties. Crystalline sulfur contains sulfur rings with 6 to 20 sulfur atoms or chains of sulfur atoms called catenasulfur, S_∞. The most common form is cyclooctasulfur, S_8, which has two common allotropes: orthorhombic sulfur, S_α, and monoclinic sulfur, S_β. S_α is, thermodynamically, the most stable of the S_8 forms. The structures are:

Cyclooctasulfur   Catenasulfur

Above 95°C S_α slowly converts to S_β. With rapid heating, the melting point of S_α (113°C) is obtained. S_β melts at 119°C. S_β crystallizes from sulfur melts and over the course of months converts to S_α. Liquid S_8 sulfur becomes increasing viscous above 160°C as it is converted to catenasulfur. Above 200°C the viscosity of the catenasulfur decreases as its maximum degree of polymerization is at 200°, where the formula weight is above 200,000. The boiling point of sulfur is 445°C. If catenasulfur of high viscosity is quenched by pouring into ice water, a plastic solid results. The solid catenasulfur slowly reverts to S_8 over time. S_8 is soluble in CS_2, S_∞ is not soluble.

Sulfur is recovered in large amounts from H_2S in natural gas by the reaction:

$$2H_2S + SO_2 \rightarrow 3S + H_2O$$

In Europe large amounts of sulfur are used from iron pyrite, FeS (the substance called fool's gold because of its similarity to real gold), which is a solid. Sulfur is also a by-product of copper production from CuS. Sulfur is also mined in large amounts in its elemental form from volcanic deposits by the *Frasch* process, where steam is injected into the ground to heat the sulfur and the molten sulfur is pumped from the ground.

Sulfur combustion to produce SO_2

Elemental sulfur is burned to produce sulfur dioxide.

$$S + O_2 \rightarrow SO_2 \text{ (gas)}$$

Above 1000°C (1830°F) no sulfur trioxide is produced; however, some might be produced in the process of cooling the gases. Sulfur trioxide, which produces sulfuric acid upon reaction with water, is very undesirable in pulping reactions and is removed during the cooling/scrubbing process (by counter current flow of water and sulfur dioxide).

$$SO_3 + H_2O \rightarrow H_2SO_4$$

The SO_2 forms H_2SO_3 in water which in turn is reacted with metal bases to produce sulfite pulping liquors, as described in other sections. Because H_2SO_3 is much more acidic than H_2CO_3, salts of carbonate can be used to form salts of sulfite. For example, wet SO_2 is traditionally formed into calcium bisulfite by calcium carbonate (limestone) by the following equation:

$$CaCO_3 + 2H_2SO_3 \rightarrow Ca(HSO_3)_2 + CO_2 + H_2O$$

Total sulfur by gravimetric analysis

Often the total content of organically bound and inorganic sulfur is desired. This is accomplished by treating the sample with a strong oxidant under alkaline conditions to convert the sulfur to sulfate. For example, when sodium peroxide is used, sodium sulfate is formed and the organic chemicals are largely converted to carbon dioxide and water. After the sample is acidified with HCl, the SO_4^{2-} is precipitated with excess $BaCl_2$. The precipitate is washed with small amounts of water and then dried in a muffle furnace. The precipitate is weighed and converted to a sulfur equivalent with the appropriate gravimetric factor.

16.10 CALCINING EQUATIONS

Two equations are used to characterize calcining of lime mud to produce fresh lime. The *specific energy consumption* is an indication of how much fuel is required to process the lime mud and is often reported as Btu per ton of lime.

$$\text{specific energy consumption} = \frac{\text{fuel to kiln}}{\text{CaO output}}$$

The *lime availability* is an indication of the purity of the lime in terms of available CaO divided by the amount of lime product.

$$\text{lime availability} = \frac{\text{CaO}}{\text{lime}} \quad \text{as mass ratio}$$

16.10 ANNOTATED BIBLIOGRAPHY

H-factor and process control equations
1. Edwards, L. and S.-E. Norberg, Alkaline delignification kinetics, A general model applied to oxygen bleaching and kraft pulping, *Tappi J.* 56(11):108-111(1973).

2. Kubes, G.J., B.I. Fleming, J.M. MacLeod, and H.I. Bolker, Viscosities of unbleached alkaline pulps. II. The G-factor, *J. Wood Chem. Tech.* 3(3):313-333(1983).

3. Hatton, J.V., Development of yield prediction equations in kraft pulping, *Tappi J.* 56(7):97-100(1973).

4. Hatton, J.V., The potential of process control in kraft pulping of hardwoods relative to softwoods, *Tappi J.* 59(8):48-50(1973).

5. Lin, C.P., W.Y. Mao, and Jane, C.Y., Development of a kappa number predictive equation in kraft pulping for all types of hardwood, *Tappi J.* 61(2)72(1978).

6. Paulonis, M.A. and A. Krishnagopalan, Adaptive inferential control of kraft batch digesters as based on pulping liquor analysis, *Tappi J.* 74(6):169-175(1991).

7. Tasman, J.E., Kraft delignification models, *Tappi J.* 64(3)175-176(1981).

8. Vroom, K.E., The "H" factor: A means of expressing cooking times and temperatures as a single variable, *Pulp Paper Mag. Can.* 58(3):228-231(1957).

Sulfite liquor analysis
9. Palmrose, G.V., A mill test for the exact determination of combined sulphur dioxide, *Tech. Assn. Papers*, XVIII:309-310(1935); the same article is reprinted as *ibid*, *Paper Trade J.* 100(3)38-39(1935).

Chemistry of sulfur
10. Cotton, F.A. and G. Wilkinson, *Advanced Inorganic Chemistry*, 4th ed., Wiley and Sons, New York, 1980. 1396 p.

EXERCISES

Kraft liquor-chemical calculations

1. Derive the gravimetric conversion factor to express Na_2CO_3 as Na_2O.

2. White liquor has the following composition of the <u>actual</u> chemical species. Calculate TTA, AA, EA, TA, causticizing efficiency and sulfidity.

   NaOH = 145 g/L Na_2S = 55 g/L

   Na_2CO_3 = 17 g/L Na_2SO_4 = 14 g/L

Kraft liquor-chemical analysis

3. The effective alkali concentration of a white liquor is 93.7 g Na_2O/L. The sulfidity is 37.5% and the causticizing efficiency is 77.9%. Calculate the following.

   a) The A, B, C values in ml for a 10 ml aliquot of liquor titrated with 1.017 N HCl.

   b) The molar concentrations of the chemical species NaOH, Na_2S, and Na_2CO_3 in this liquor.

   c) The active and effective alkali charges on wood (as a percent) if 200 ml of this liquor is used to pulp 200 g Douglas-fir chips at 50% MC_{gr} (with some additional dilution water).

H-factor

4. What is the H-factor for a cook which is held at 165°C for 1.43 hours? If this H-factor is satisfactory, how long should one cook if the temperature is increased to 170°C for a different cook?

Sulfite liquor calculations

5. By comparing sulfite cooking chemicals all on a SO_2 basis, one is really comparing them on an equal (molar or weight) basis. Circle the correct choice.

6. Analysis of a sulfite cooking liquor indicates a total SO_2 of 7% and 4% free SO_2.

   The combined SO_2 is ____%.

   The SO_2 as sulfurous acid is ____%.

   The SO_2 in the form of the bisulfite ion (HSO_3^-) is ____%, and the SO_2 in the form of sulfite ion (SO_3^{2-}) is ____%.

7. Analysis of a sulfite cooking liquor indicates a total SO_2 of 8%, and a combined SO_2 of 2%. Solve for unknown free SO_2, as well as the concentration of the sulfurous acid, bisulfite ion, and sulfite ion species (all as % SO_2 which equals g/100 ml), and the pH of the liquor.

8. A solution is made by weighing 1.1 grams sulfurous acid (as sulfurous acid) and 10.3 grams sodium sulfite (as sodium sulfite) to make 0.193 liters of solution. Solve for unknown free SO_2, combined SO_2, total SO_2, as well as the concentration of the sulfurous acid, bisulfite ion, and sulfite ion species (all as % SO_2 which equals g/100 ml), and the pH of the liquor.

Sulfite liquor analysis

9. Analysis of a pulping liquor with iodometric titration gives a SO_2 value of 10%. Titration with NaOH gives 6.5% SO_2. Solve for unknown free SO_2, combined SO_2, total SO_2, as well as the concentration of the sulfurous acid, bisulfite ion, and sulfite ion species (all as % SO_2 which equals g/100 ml), and the pH of the liquor. If sodium is the base, how many grams of sulfurous acid, sodium bisulfite, and/or sodium sulfite are needed to make one liter of cooking liquor?

17

BLEACHING AND PULP PROPERTIES CALCULATIONS

17.1 DILUTION WATER CALCULATIONS

The water per ton of pulp ratio (V) is solved from the consistency (c) as $V = (100-c)/c$.

▧▧▧▧▧▧▧▧▧▧▧▧▧▧▧▧▧▧▧▧▧▧▧▧▧▧▧▧▧

EXAMPLE 1.   After the brown stock washers, the consistency of unbleached pulp is 11.2%. Calculate the volume of water (in m^3/t oven-dry pulp) required to dilute the slurry to 3% consistency for chlorination.

SOLUTION.   At 11% consistency there are 89 t water/(11 t pulp) = 8.09 t water/t pulp. At 3% consistency there are 97 t water/(3 t pulp) = 32.33 t/t. Therefore 32.33 - 8.09 = 24.24 t water/t pulp to be added. Since 1 t = 1000 kg = 1 m^3 of water, this is 24.24 m^3 of water per metric ton of pulp.

▧▧▧▧▧▧▧▧▧▧▧▧▧▧▧▧▧▧▧▧▧▧▧▧▧▧▧▧▧

17.2 CHLORINE BLEACHING

The pH is of utmost importance in chlorine bleaching for several reasons. First the form of chlorine in solution is dependent on the pH according to the following equilibrium reactions.

$$Cl_2 + H_2O \rightleftarrows H^+ + Cl^- + HOCl$$

$$K_a = 3.9 \times 10^{-4} \text{ at } 25°C \qquad (17\text{-}1)$$

$$HOCl \rightleftarrows H^+ + OCl^-$$

$$K_a = 3.5 \times 10^{-8} \text{ at } 18°C \qquad (17\text{-}2)$$

From these equilibria, Fig. 17-1 can be constructed in the manner of Giertz (1951). This is very similar to other diprotic acids plotted in Chapter 16, except the three species are plotted together. Determining the concentration of the actual species at low pH is tricky because chlorine is a moderately strong acid (and so its ionization alters the pH

significantly) and three species are formed by the ionization instead of two.

First, the concentration of chlorine in water at 25°C is 0.091 M at saturation (i.e., one atmosphere pressure of Cl_2). (Like the solubility of CO_2, the solubility of Cl_2 is highly pH dependent.) The pH of saturated chlorine water is calculated as follows:

$$K_a = [HOCl][H^+][Cl^-]/[Cl_2] = \frac{[H^+]^3}{(0.091 - [H^+])}$$

Solving for $[H^+]$ gives $[H^+] = 0.029\ M$. Therefore, a saturated chlorine solution in water, without addition of acid or base, has a pH of -log 0.029 or 1.54, an actual chlorine concentration, $[Cl_2]$, of 0.062 M, a hydrochloric acid concentration of 0.029 M, and a hypochlorous acid concentration of 0.029 M.

The concentration of chlorine species will be considered using 0.05 M Cl_2 initially, since this will not exceed the saturated pressure of chlorine at any pH, as a function of pH. Acid or base can be hypothetically added without dilution to achieve the pH's shown in Fig. 17-1. Initially $[Cl_2]$ = 0.028, $[H^+]$ = 0.022, and the pH is 1.66.

The ratio of Cl_2 to HOCl as a function of pH (where the source of the acid is immaterial) is solved in the following manner, where x = $[HOCl]$ = $[Cl^-]$:

$$K_1 = [H^+]\, x^2/(0.05 - x)$$

$$[H^+]\, x^2 + K_1\, x - 0.05\, K_1 = 0$$

This equation is easily solved at various pH's by use of the quadratic equation (see page 322). (At pH = 0, x = 0.00423 M. At pH 1.66, x = 0.0222 in agreement with the calculation above.)

When 0.05 M NaOH is added, all of the Cl_2 is converted to HOCl and the pH is of the weak acid system HOCl. HOCl is easily handled as any other weak acid. The pH is then calculated from Eq. 13-5 as -log($K_a \times C)^{\frac{1}{2}}$ = 4.38. When 0.10 M

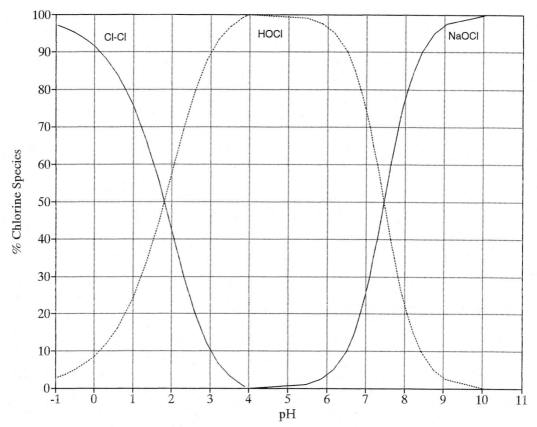

Fig. 17-1. Chlorine species as a function of pH in saturated chlorine water (0.05 M Cl$_2$) at 25°C. Obviously the total solubility is pH dependent with NaOCl being very soluble at the higher pH's.

NaOH is present the chemical is essentially all OCl⁻ and the pH is given as:

$$(14 - (K_b \times [OCl^-])^{1/2}) = 10.08.$$

The pH at intermediate amounts of base is given in the form of the Henderson-Hasselbalch equation (see Eq. 14-2 for this equation); consequently, the ratio of HOCl to OCl⁻ is not concentration dependent (except for the effect of activity coefficients) from pH 5 to 9.

$$\text{pH} = pK_a + \log(C_A/C_A)$$

$$= 7.46 + \log([OCl^-]/[HOCl])$$

Because the actual concentration of chlorine species is pH dependent, pulp bleaching with chlorine is also pH dependent. The desired reaction of substitution occurs most rapidly at the lowest pH values; however, acid hydrolysis of cellulose and hemicellulose also increases with decreasing pH. (Acid hydrolysis leads to a decrease in the cellulose viscosity.) The hydrolysis of chlorine is also temperature dependent. Rydholm in *Pulping Processes* (p. 920) gives the hydrolysis constants of Cl$_2$ as a function of temperature; they are 1.45×10^{-4} at 0°C and 9.75×10^{-4} at 60°C.

17.3 CHLORINE DIOXIDE

The action of ClO_2 has been studied in detail by Schmidt (1923). He found ClO_2 to be highly reactive with aromatics and phenolics but not reactive with carbohydrates. Today ClO_2 is used in the later bleaching stages since it is very selective for lignin removal. It is also being used in early stages for environmental reasons. Chlorine dioxide can be produced from chlorite by oxidation with hypochlorite or reduction with suitable reducing agents (Giertz, 1951). Chlorine dioxide formation by reduction of sodium chlorate in acidic aqueous solutions occurs with different reducing agents that give rise to many processes as shown in Table 17-1. Chlorine is reduced from a valence of $+5$ in chlorate to a valence of $+4$ in chlorine dioxide. Chlorate is produced by the disproportionation reaction as follows:

$$3ClO^- \rightarrow 2\ Cl^- + ClO_3^-$$
$$+1 \qquad -1 \quad\ +5$$

The overall reaction is:

$$NaCl + 3H_2O + electricity \rightarrow NaClO_3 + 3H_2$$

Chlorine dioxide is also produced by the *Lurgi* process from Cl_2, H_2O, and electricity. The chlorine dioxide is produced by the same reaction as the Rapson process. The NaCl and Cl_2 products of this reaction are used to generate additional chlorine dioxide. Also $NaClO_3$ is generated by electrolysis. Fig. 17-2 shows the process. The process is electrical intensive and uses 8-9,000 kWh/t of ClO_2.

17.4 CHEMICAL ANALYSIS OF BLEACHING LIQUORS AND CHLORINE EQUIVALENCY

The determination of active bleaching agents is performed by iodometric titrations. The bleaching solution is added to an aqueous solution containing excess potassium iodide, KI. The KI is oxidized to iodine while the bleaching agent is reduced. The amount of iodine liberated is measured by titration with thiosulfate using a starch indicator to observe the final disappearance of iodine. In the case of chlorine the reactions are as follows:

$$Cl_2 + 2I^- \rightarrow 2Cl^- + I_2 \tag{17-3}$$

$$I_2 + 2Na_2S_2O_3 \rightarrow 2NaI + Na_2S_4O_6 \tag{17-4}$$

Bleaching chemicals are normally expressed as amount of "available chlorine." This allows bleaching chemicals to be considered in terms of their equivalent weights. Table 17-2 shows the conversions for several bleaching agents. Keep in mind that in Eq. 17-3 one mole of KI reacts with one mole of electrons; therefore, the equivalent

Table 17-1. Formation of ClO_2 by reduction of $NaClO_3$.

Name of Process	Reducing Agent		Chemical Equation
	Species	Change of Valence	
Mathieson	SO_2	S: $+4 \rightarrow +6$	$2NaClO_3 + SO_2 + H_2SO_4 \rightarrow 2ClO_2 + 2NaHSO_4$
Solvay	CH_3OH	C: $-2 \rightarrow +4$	$6NaClO_3 + CH_3OH + 6H_2SO_4 \rightarrow 6ClO_2 + CO_2 + 6NaHSO_4 + 5H_2O$
Rapson	HCl	Cl: $-1 \rightarrow 0$	$2NaClO_3 + 4HCl \rightarrow 2ClO_2 + 2NaCl + Cl_2 + H_2O$
R-2, R-3, SVP	NaCl	Cl: $-1 \rightarrow 0$	$NaClO_3 + NaCl + H_2SO_4 \rightarrow ClO_2 + \frac{1}{2}Cl_2 + Na_2SO_4 + H_2O$

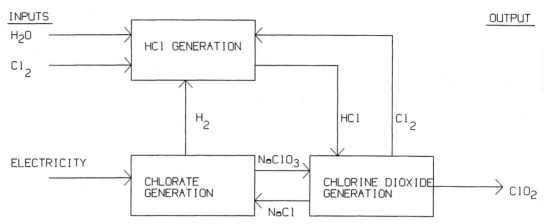

INPUTS   OUTPUT

Fig. 17-2. Chlorine dioxide generation from Cl$_2$, H$_2$O, and electricity by the Lurgi process.

weight of KI is equal to its formula weight. On the other hand, one mole of I$_2$ reacts with two moles of electrons in Eq. 15-4, and the equivalent weight of I$_2$ is one-half of its formula weight. It should be clear that the equivalent weight of Cl$_2$ is one-half of its formula weight.

Table 17-2. Equivalent weights of bleaching chemicals.

Bleaching agent	Chemical equation	Equivalent weight		Gravimetric factors		$E°$, V
		as formula weight	g/equiv	to Cl$_2$	from Cl$_2$	
Chlorine	Cl$_2$ + 2e → 2Cl⁻	FW/2	35.5	1.000	1.000	1.36
Hypochlorite	NaOCl + H$_2$O + 2e → NaCl + 2OH⁻	FW/2	37.25	0.953	1.049	0.89
Hypochlorous acid	HOCl + H⁺ +e → ½Cl$_2$ + H$_2$O; then Cl$_2$ rxn. as above	FW/2	26.25	1.352	0.739	1.63 1.36
Chlorine dioxide	ClO$_2$ + 4H⁺ + 5e → Cl⁻ + 2H$_2$O	FW/5	13.5	2.630	0.380	1.27
Chlorine monoxide	Cl$_2$O + 4e → Cl⁻	FW/4	21.75	1.632	0.613	
Sodium chlorate	NaClO$_3$ + 6e → Cl⁻	FW/6	17.75	2.029	0.493	
Oxygen	O$_2$ + 2H$_2$O + 4e → 4OH⁻	FW/4	8.00	4.438	0.225	0.41[1]
Iodide (sodium)	I$_2$ + 2Na⁺ + 2e → 2NaI	FW	149.89	0.237	4.222	0.620

[1]In 1 N NaOH.

EXAMPLE 2. Write the balanced reaction of chlorine dioxide with iodide.

SOLUTION: $2ClO_2 + 10\ I^- + 8\ H^+ \rightarrow 2Cl^- + 5I_2 + 4H_2O$

EXAMPLE 3. Suppose a chemical loading of 2% NaClO on pulp (as chlorine) is desired. What is the actual amount of NaClO that should be used?

SOLUTION: 2% as $Cl_2 \times (37.25$ g NaClO)/(35.5 g Cl_2) = 2.1% NaClO as NaClO.

PROBLEM: Suppose 2% ClO_2 as chlorine is desired. What should the actual concentration be? Answer: 0.76% ClO_2.

EXAMPLE 4. Given the data in the following table, fill in the blank spaces.

Stage	C	E	H	D
Chemical input on pulp, % as Cl_2	6	3	3	2
Chemical input on pulp, % true	?	?	?	?
Yield per stage, %	99	96	97	98
Cumulative yield, %	?	?	?	?
Pulp input, kg	1000	?	?	?
Pulp output, kg	?	?	?	?
Chemical input, kg	?	?	?	?

SOLUTION: Chlorine and caustic are expressed "as is", so the chlorine input would be 6% of the pulp charged and the caustic input would be 3% of the charge. Hypochlorite and chlorine dioxide charges are given in terms of equivalent chlorine and must be converted as explained.

Hypochlorite: 3% (as Cl_2) × 1.049 = 3.15% (as NaClO) based on pulp.

Chlorine dioxide: 2% (as Cl_2) × 0.380 = 0.76% (as ClO_2) based on pulp.

The yield per stage is the ratio of output per stage to the input per stage, and the cumulative yield (for the entire bleaching sequence) is the ratio of the final output to the input of the first stage of bleaching. It is also the sequential product of the yields per stage:

Cumulative yield stage 1: 100% x 0.99 = 99%

Cumulative yield stage 2: $100\% \times 0.99 \times 0.96 = 95.0\%$ ... and so forth for each stage.

The pulp input and pulp output per stage can be calculated by using the above percentage figures. What comes out of one stage must be the input for the following stage. The chemical input per stage is figured from the pulp input times the chemical charge (percentage) per stage:

C stage: $1000 \text{ kg} \times 6\% = 60 \text{ kg Cl}_2$ as Cl_2.

E stage: $990 \text{ kg} \times 3\% = 29.7 \text{ kg NaOH}$ as NaOH.

H stage: $950 \times 3.15\% = 29.9 \text{ kg NaClO}$ as NaClO.

D stage: $922 \times 0.76\% = 7.0 \text{ kg ClO}_2$.

Final pulp yield: $1000 \times 0.99 \times 0.96 \times 0.97 \times 0.98 = 903 \text{ kg}$ bleached pulp.

The completed chart appears as follows:

Stage	C	E	H	D
Chemical input on pulp, % as Cl_2	6	3	3	2
Chemical input on pulp, % true	6.0	3.0	3.15	0.76
Yield per stage, %	99	96	97	98
Cumulative yield, %	99	95	92.2	90.3
Pulp input, kg	1000	990	950	922
Pulp output, kg	990	950	922	903
Chemical input, kg	60	29.7	29.9	7.0

PROBLEM: Complete the following table.

Stage	C	E	H	D
Chemical input on pulp, % as Cl_2	5	2.5	3.7	1.5
Chemical input on pulp, % true	?	?	?	?
Yield per stage, %	98.5	97	98	99
Cumulative yield, %	?	?	?	?
Pulp input, kg	1000	?	?	?
Pulp output, kg	?	?	?	?
Chemical input, kg	?	?	?	?

Table 17-3. CSF consistency correction table. From Eq. 17-5.

CSF	Consistency, %				
---	0.2	0.25	0.30	0.35	0.40
50	-23	-12	0	12	23
80	-34	-17	0	13	26
100	-36	-19	0	19	36
150	-49	-24	0	23	45
200	-58	-28	0	26	51
250	-66	-31	0	29	55
300	-72	-34	0	30	57
350	-76	-36	0	32	59
400	-80	-37	0	32	59
450	-82	-38	0	32	58
500	-82	-38	0	32	57
550	-81	-37	0	30	54
600	-79	-36	0	29	51
650	-74	-34	0	27	46
700	-68	-31	0	24	41

Table 17-4. CSF temperature correction table. From Eq. 17-6

CSF	Temperature, °F				
---	10	15	20	25	30
50	11	6	0	- 6	-11
80	20	10	0	-10	-20
100	25	12	0	-12	-25
150	33	16	0	-16	-33
200	38	19	0	-19	-38
250	42	21	0	-21	-42
300	44	22	0	-22	-44
350	46	23	0	-23	-46
400	46	23	0	-23	-46
450	46	23	0	-23	-46
500	44	22	0	-22	-44
550	42	21	0	-21	-42
600	40	20	0	-20	-40
650	37	18	0	-18	-37
700	33	16	0	-16	-33

17.5 CSF CORRECTION EQUATIONS

The Canadian Standard freeness is a commonly used indicator of the level of refining. The test calls for determining freeness at 20°C and 0.30% consistency; however, tabular corrections are given for temperatures from 10-30°C and 0.20-0.40% consistency. These tables are time-consuming to use and are unnecessary with the use of computerized electronic spreadsheets.

Two equations (Biermann and Hull, 1992) have been formulated to correct freeness values above 50 CSF; one for consistency, Eq. 17-5, and one for temperature, Eq. 17-6. While these equations are awkward, once entered into the spreadsheet they can be used to automatically correct freeness. Tables 17-3 and 17-4 are abbreviated versions of the correction tables generated from Eq. 17-5 and Eq. 17-6, respectively.

$$\text{CSF corr} = (\text{cons} - 0.30) \times 590 \times$$
$$(1 + \frac{0.4 - \text{cons}}{0.2} \times \frac{\text{CSF}}{1000}) \times (1 - \frac{(\text{CSF} - 390)^2}{\text{CSF}^{0.2} \times 87000}) \quad (17\text{-}5)$$

$$\text{CSF corr} = (20 - t) \times 4.6 \times (1 - \frac{(400 - \text{CSF})^2}{\text{CSF}^{0.25} \times 61000}) \quad (17\text{-}6)$$

Recent work (Sundrani et al., 1993) shows that the freeness tables supplied with the TAPPI and CPPA methods give corrections with an absolute value below what they should be for kraft pulps. The correction tables were developed with stone groundwood and, for the higher freeness values, sulfite chemical pulps in the 1930s. There is no original data in the literature and there is no indication of the algorithm used to generate the tables. Since the viscosity of water decreases exponentially with temperature, there is no reason to believe the temperature correction table should be symmetrical as it is given in these standards.

17.6 COMPUTER SIMULATION OF FIBER CLEANING SYSTEMS

When evaluating the overall cleaning efficiency of a system of vortex cleaners containing primary, secondary, tertiary, and, perhaps, higher order banks before the paper machine, one soon discovers that the calculations are tedious and that evaluating numerous configurations or flow rates is laborious. It is very easy to simulate the overall performance using simple computer programs. BASIC will be used here since many personal computers are equipped with this tool.

The philosophy of evaluating these systems is this: Do not try to solve the problem; simply define the system and put a specified number of fibers or dirt particles through the system, solve the approximate accept and reject rates, run the system enough times so that the solution converges to the desired answer within a tolerance limit, and count where the particles come out. It is convenient to have 100 particles exit the system so the various exit points can be calculated as percentages. Two programs must be made: one for the dirt particles and one for the fibers since they behave differently.

An example will be used to show the principle. We will simulate the operation of a system containing primary, secondary, and tertiary cleaners. In each of these sections 75% of the dirt particles are rejected. The fiber reject rate is 10% for the primary cleaners, 15% for the secondary cleaners, and 20% for the tertiary cleaners. The primary accepts leave the cleaning system and go to the paper machine. The primary rejects go to the secondary inlet, the secondary rejects go to the tertiary inlet, and the tertiary rejects are sewered. The secondary accepts go to the primary inlet. Initially we will study a system where the tertiary accepts go to the secondary inlet. (We will then reinvestigate the system with the tertiary accepts going to the primary inlet.)

Table 17-5 is the program with the results of the run to simulate the fibers. Table 17-6 shows the results of the same system from the point of view of the dirt particles. The outputs of each cleaner bank are shown when fibers are used, since this reflects the size of each bank that will be required.

When the tertiary rejects go to the secondary inlet the overall fiber reject rate is 0.377% and the overall dirt reject rate is 67.50%. To simulate the system where the tertiary accepts go to the primary inlet, changes must be made in three lines of both programs. The changes are identical in both programs. The corrected lines read:

```
130  SEC.IN = PRI.REJ
210  IF ABS((DIRT + SEC.ACC +
     TER.ACC)-PRI.IN) < TOLER GOTO 240
220  PRI.IN = DIRT + SEC.ACC + TER.ACC
```

When the modified programs are run, the fiber reject rate is 0.332% and the dirt reject is 62.79%. In the first system the selectivity (the dirt reject rate divided by the fiber reject rate) is 179, and in the second system the selectivity is 189. When primary accepts are not treated further in a cleaning system with tertiary cleaners, the tertiary rejects should go to the secondary cleaners not the primary cleaners since 67.5% of the dirt will be removed compared to 62.8%, while the fiber reject rate changes very little. When the secondary accepts are returned to the primary cleaners the tertiary accepts should always go to the secondary inlet; the reason is that the dirt particles must then go through two sets of cleaners before being accepted rather than just going to the primary cleaners where they have a relatively good chance of being accepted.

Once one has learned how this technique is used, it is easily modified to accomplish numerous chores. For example, one might try altering the flow rates or reject rates in different banks of cleaners. The model can be made more complex by actually defining a distribution of dirt or fiber particle sizes with different reject rates for each particle size range. The technique can also be adapted to screening, cleaning, deinking, and other systems. The limits are defined only by the ingenuity of the programmer. If systems get too complicated, one should consider using commercial simulation programs.

17.7 PAPER MACHINE CALCULATIONS

The relationships between basis weight, consistency, paper machine width and speed, and production rates are easily derived. It is assumed that the speed of the stock exiting the headbox is approximately equal to the speed of the machine wire. The volumetric flow rate may be expressed as a function of the three primary paper machine dimensions or in terms of the dry fiber rate and pulp consistency at the headbox.

Table 17-5. The simulation of a cleaning system with fibers.

10	REM THIS PROGRAM HAS PRIMARY, SECONDARY, AND TERTIARY CLEANERS FOR FIBER. THE TERTIARY REJECTS GO TO THE SECONDARY CLEANERS.
20	FIBER = 100: REM 100 FIBERS MEANS OUTPUTS WILL BE PERCENTAGES
30	PRI.RATE = 0.9: REM THE FIBER ACCEPT RATE OF THE PRIMARY CLEANERS
40	SEC.RATE = 0.85: REM THE FIBER ACCEPT RATE OF THE SECONDARY CLEANERS
50	TER.RATE = 0.8: REM THE FIBER ACCEPT RATE OF THE TERTIARY CLEANERS
60	PRI.IN = FIBER
70	TOLER = .005:REM THE ERROR BETWEEN ITERATIONS SHOULD BE LESS THAN 0.005% BEFORE STOPPING THE PROGRAM
80	ITERATE = ITERATE + 1: REM THE PROGRAM WILL RETURN HERE BE-TWEEN ITERATIONS
90	PRINT "ITERATION = ";:PRINT USING "##";ITERATE;:PRINT "; FIBERS IN = ";:PRINT USING "####.##";PRI.IN;
100	PRI.ACC = PRI.IN * PRI.RATE
110	PRI.REJ = PRI.IN - PRI.ACC
120	ACCEPTS = PRI.ACC
130	SEC.IN = PRI.REJ + TER.ACC
140	SEC.ACC = SEC.IN * SEC.RATE
150	SEC.REJ = SEC.IN - SEC.ACC
160	TER.IN = SEC.REJ
170	TER.ACC = TER.IN * TER.RATE
180	TER.REJ = TER.IN - TER.ACC
190	REJECTS = TER.REJ
200	PRINT "; ACCEPTS = ";:PRINT USING "###.##";ACCEPTS;:PRINT "%; REJECTS = ";:PRINT USING "###.##";REJECTS;:PRINT"%"
210	IF ABS((FIBER + SEC.ACC)-PRI.IN)<TOLER GOTO 240
220	PRI.IN = FIBER + SEC.ACC
230	GOTO 80
240	PRINT "PRIMARY ACCEPTS = ";PRI.ACC;"%; ";"PRIMARY REJECTS = ";PRI.REJ;"%; ","SECONDARY ACCEPTS = ";SEC.ACC;"%; ";"SECONDARY REJECTS = "; SEC.REJ;"%; ","TERTIARY ACCEPTS = ";TER.ACC;"%; ";"TERTIA-RY REJECTS = ";TER.REJ;"%; "

RUN

ITERATION = 1;	FIBERS IN = 100.00;	ACCEPTS = 90.00%;	REJECTS = 0.30%
ITERATION = 2;	FIBERS IN = 108.50;	ACCEPTS = 97.65%;	REJECTS = 0.36%
ITERATION = 3;	FIBERS IN = 110.24;	ACCEPTS = 99.22%;	REJECTS = 0.37%
ITERATION = 4;	FIBERS IN = 110.60;	ACCEPTS = 99.54%;	REJECTS = 0.38%
ITERATION = 5;	FIBERS IN = 110.67;	ACCEPTS = 99.61%;	REJECTS = 0.38%
ITERATION = 6;	FIBERS IN = 110.69;	ACCEPTS = 99.62%;	REJECTS = 0.38%

PRIMARY ACCEPTS	= 99.61916 %;	PRIMARY REJECTS = 11.0688 %	
SECONDARY ACCEPTS	= 10.69104 %;	SECONDARY REJECTS = 1.886653 %	
TERTIARY ACCEPTS	= 1.509322 %;	TERTIARY REJECTS = .3773306 %	

Ok

Table 17-6. The simulation of a cleaning system with dirt particles.

```
10      REM THIS PROGRAM HAS PRIMARY, SECONDARY, AND TERTIARY CLEANERS
        FOR DIRT.  THE TERTIARY REJECTS GO TO THE SECONDARY CLEANERS.
20      DIRT = 100: REM 100 DIRT PARTICLES MEANS OUTPUTS WILL BE PERCENTAGES
30      PRI.RATE = .25:  REM THE DIRT ACCEPT RATE OF THE PRIMARY CLEANERS
40      SEC.RATE = .25:  REM THE DIRT ACCEPT RATE OF THE SECONDARY CLEANERS
50      TER.RATE = .25:  REM THE DIRT ACCEPT RATE OF THE TERTIARY CLEANERS
60      PRI.IN = DIRT
70      TOLER = .005:    REM THE ERROR BETWEEN ITERATIONS SHOULD BE LESS
                        THAN 0.005% BEFORE THE PROGRAM ENDS
80      ITERATE = ITERATE + 1: REM THE PROGRAM RETURNS HERE EACH ITERATION
90      PRINT "ITERATION = ";:PRINT USING "##";ITERATE;:PRINT ";  DIRT IN =
        ";:PRINT USING "####.##";PRI.IN;
100     PRI.ACC = PRI.IN * PRI.RATE
110     PRI.REJ = PRI.IN - PRI.ACC
120     ACCEPTS = PRI.ACC
130     SEC.IN = PRI.REJ + TER.ACC
140     SEC.ACC = SEC.IN * SEC.RATE
150     SEC.REJ = SEC.IN - SEC.ACC
160     TER.IN = SEC.REJ
170     TER.ACC = TER.IN * TER.RATE
180     TER.REJ = TER.IN - TER.ACC
190     REJECTS = TER.REJ
200     PRINT ";  ACCEPTS = ";:PRINT USING "###.##";ACCEPTS;:PRINT "%;  REJECTS =
        ";:PRINT USING "###.##";REJECTS;:PRINT"%"
210     IF ABS((DIRT + SEC.ACC)-PRI.IN)< TOLER GOTO 240
220     PRI.IN = DIRT + SEC.ACC
230     GOTO 80
240     END

RUN

ITERATION =  1;   DIRT IN =  100.00;    ACCEPTS =  25.00%;   REJECTS =  42.19%
ITERATION =  2;   DIRT IN =  118.75;    ACCEPTS =  29.69%;   REJECTS =  58.01%
ITERATION =  3;   DIRT IN =  125.78;    ACCEPTS =  31.45%;   REJECTS =  63.94%
ITERATION =  4;   DIRT IN =  128.42;    ACCEPTS =  32.10%;   REJECTS =  66.17%
ITERATION =  5;   DIRT IN =  129.41;    ACCEPTS =  32.35%;   REJECTS =  67.00%
ITERATION =  6;   DIRT IN =  129.78;    ACCEPTS =  32.44%;   REJECTS =  67.31%
ITERATION =  7;   DIRT IN =  129.92;    ACCEPTS =  32.48%;   REJECTS =  67.43%
ITERATION =  8;   DIRT IN =  129.97;    ACCEPTS =  32.49%;   REJECTS =  67.47%
ITERATION =  9;   DIRT IN =  129.99;    ACCEPTS =  32.50%;   REJECTS =  67.49%
ITERATION = 10;   DIRT IN =  130.00;    ACCEPTS =  32.50%;   REJECTS =  67.50%

Ok
```

volumetric flow rate =

PM speed × PM width × slice height (17-7)

mass flow rate =

(dry fiber rate)/(consistency) (17-8)

In Eq. 17-8 one must report the consistency in units of mass per volume (such as kg/m^3) in order to obtain the volumetric flow rate in units of volume. Often the consistency is reported as a mass % such as 0.4% (implying 0.4 kg dry pulp per 100 kg pulp slurry). If one is not careful, errors in calculations will occur since $10\ kg/m^3$ is 1% consistency! The dry fiber rate is easily given in terms of basis weight and machine production in area (Eq. 17-9). Eq. 17-10 shows the relationship between slice height and consistency.

dry fiber rate =

basis weight × PM speed × PM width (17-9)

basis weight =

slice height × consistency (17-10)

As with any equation, when solving these equations one must use care to insure that consistent units are used throughout the solution. In Eq. 17-10 the slice height is conveniently expressed in meters and the consistency in kg/m^3. The basis weight is solved in terms of kg/m^2, which is easily converted to g/m^2.

The speed of the stock exiting the headbox is usually within 5% of the machine wire speed. As shown in Section 9.3, the pressure in the headbox is dependent on the wire speed. Some speed is lost due to friction. The amount of this loss is dependent on the slice geometry.

EXAMPLE 5. What is the speed of a paper machine producing 500 MT per day of paper that is 50 g/m^2 and makes paper 10 m wide?

SOLUTION: The dry fiber rate is:

500,000 kg·day^{-1}/(86,400 s·day^{-1})

= 5.79 kg/s.

5.79 kg·s^{-1}/(0.050 kg·m^{-2}) = 115.8 m^2/s. Since the width = 10 m, the speed = 11.6 m/s.

PROBLEM: What is the slice opening for this machine if the consistency is 0.5%? Answer: 0.01 m, which is the basis weight divided by the consistency.

EXAMPLE 6. Derive the relationship between slice velocity and machine speed assuming no frictional losses.

SOLUTION: The potential energy of a given mass of pulp slurry is mgh, where m is the mass, g is the acceleration due to gravity, and h is the height of the headbox (or a pressure which can be converted to an equivalent height). The potential energy is converted to kinetic energy that is given as $\frac{1}{2}mv^2$, where v is the velocity of the slice which is approximately equal to the velocity of the machine. The slice velocity is realistically given with a term for energy loss, C_q. C_q is approximately 0.85-0.90 for a tapered slice and 0.65-0.75 for a slice with an abrupt opening (as viewed from the inside). Therefore:

$$\frac{1}{2}mv^2 = mgh$$

$$v = \sqrt{2gh}$$ (17-11)

actually, $v \approx C_q\sqrt{2gh}$

EXAMPLE 7. The pressure of a headbox corresponds to a height of 10 m. If 50% of the potential energy is converted to kinetic energy, what is the velocity of the slice?

SOLUTION.

$$\frac{1}{2}mv^2 = 0.5 \times mgh$$

$$v = \sqrt{0.5} \times \sqrt{2gh}$$

$$v = 9.9 \text{ m/s}$$

17.8 PROPERTIES OF DILUTE PULP SLURRIES, FLOCCULATION

The behavior of dilute slurries of pulp (0.1 to 3% or higher) is critical to paper formation on the paper machine. (It is also important in pulp pumping, mixing, and screening, but receives less attention in these applications.) This area is complex and there are few resources that give a good overview to it. The results of some studies are given in this section, but this is only a starting point of a complex area.

One important aspect of dilute pulp slurries is the tendency for pulp fibers to flocculate, which leads to poor paper formation on the paper machine. Flocculation is a nonuniform fiber distribution, or clustering, in the slurry. Kerekes (1983) gives a review with 32 references on fiber flocculation with particular attention to the presence of decaying turbulence, which is applicable to many areas of pulp processing such as flow beyond the holey roll in the headbox, at impeller tips of mixers and pumps, and perforated screens.

Mason (1954) showed that the critical concentration (c_o) for flocculation is inversely dependent upon the square of the axis ratio (the length divided by the width) of the fiber. For example, the calculated c_o on a volume basis for a L/w of 100 is 0.0015% and for a L/w of 60 is 0.0042%.

Jokinen and Ebeling (1985) list some ways of decreasing flocculation. These include reducing the pulp consistency (which is why headbox consistencies are about 0.5%), lowering the temperature of the pulp suspension, reducing fiber length (such as with increased refining or using hardwood instead of softwood), increasing the pH, using anionic polymers, and using high stock velocities and turbulent flow (such as holey rolls in headboxes). The authors experimentally evaluated the relative importance of these factors and concluded that fibers flocculate for mechanical reasons. Flocculation is most effectively decreased by decreasing the consistency, using shorter fibers (but this decreases paper strength), and adding anionic polymers of high molecular weight such as polyacrylamide.

Kerekes and Schell (1992) indicate that *uniformity* of fiber distribution and *mobility* of fibers in suspension both contribute to good formation. *Superposition*, the piling of fibers on the wire during sheet formation, also affects formation. The authors develop a mathematical model called the crowding factor (the number of fibers within the spherical volume formed by the diameter of a fiber) to describe fiber flocculation. They tested their model with a variety of fiber types and include some stunning pictures of fibers in slurries.

Gorres et al. (1989) used a simulation model to predict paper formation based on fibrous floc characteristics. Smith (1986) compared handsheets as a way to rank formation potentials.

17.9 STRENGTH OF WET FIBER MATS

Introduction

In their classic study, Lyne and Gallay (1954) measured the wet strength of pulp webs and various types of glass fibers to determine the relative effect of surface tension and hydrogen bonding on the wet web strength. Their results are shown in Fig. 17-3 for groundwood and sulfite pulps and Fig. 17-4 for glass fiber webs.

Based on their results, the authors conclude that up to 20-25% solids, the fibers are held together by surface tension forces. Surface tension forces decrease with increasing solids beyond this point (as observed by the unmodified glass fibers), but hydrogen bonding begins and the strength increases as the water is removed.

The authors also demonstrate that papers formed from liquids of lower surface tension than water do not achieve nearly the strengths of paper formed from water. It would be ideal if paper could be dry-formed like most fiberboards, but this is not possible when hydrogen bonding provides the interfiber bonding in the final product. However, pulps are sometimes dried from acetone in the laboratory so that they remain bulky for subsequent work. Even just lowering the surface tension of water to 33 dyn/cm (from 73 dyn/cm at 18°C) by the addition of 0.1% surfactant decreases the ultimate breaking length of a groundwood paper from 300 meters to 95 meters. Making the surface of pulps hydrophobic with mineral oil also drastically decreased the strength of paper made from them. (Freeze-drying of pulp slurries also gives a very weak fiber mat.)

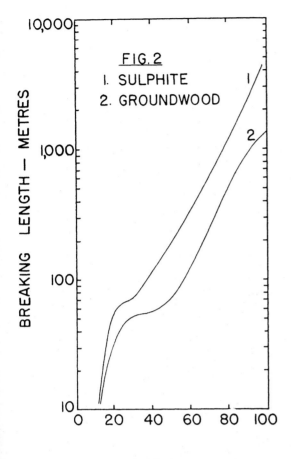

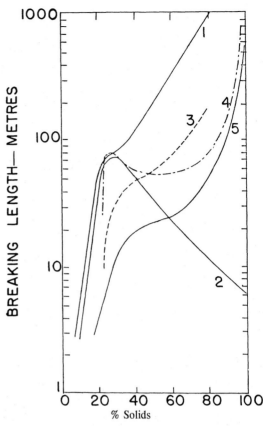

1) sulfite pulp, 2) 2.5 μm untreated glass fibers, 3) HF treated glass fibers, 4) glass fibers with sodium silicate adhesive, and 5) as 3 with gelatine adhesive.

Fig. 17-3. Strength of pulp webs with increasing dryness. ©1954 TAPPI. Reprinted from Lyne and Gallay (1954) with permission.

Fig. 17-4. Strength of fiber webs with increasing dryness. ©1954. TAPPI. Reprinted from Lyne and Gallay (1954) with permission.

By the same token, normal paper formed from water does not lose much strength when exposed to liquids of low surface tension and polarity, such as toluene (28 dyn/cm) or hexane (18.5 dyn/cm). Paper loses moderate amounts of its strength in solvents capable of hydrogen bonding, such as methanol (23 dyn/cm), even if the surface tension is not particularly high. Paper loses most of its strength in water (unless treated with special wet-strength agents).

Surface tension effects

Obviously the surface tension of the liquid from which the fiber mat is formed is of central importance to the papermaking process. The pulp fibers must approach each other within a few tenths of a nanometer in order to form hydrogen bonds. The fiber surfaces approach each other when dried from liquids of high surface tension. Tremendous forces from capillary action and hydrogen bonding bring the fiber surfaces very close together.

Swanson (1961) discusses some of the surface tension forces in wet webs. Fig. 17-5 shows how some of these forces act. The three diagrams are in order of decreasing water content and increasing forces: In part (a) the web is about 8% solids, or the consistency on the fourdrinier wire in the hivacs. The force in (b) with a surface tension of 75 dynes/cm for water and a fiber diameter of 0.03 mm would be about 5 kPa; with fibrillation the diameter may be effectively 0.003 mm, giving

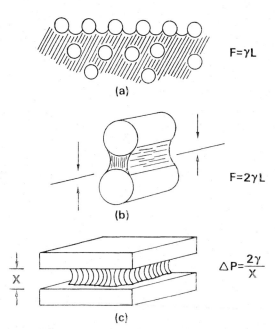

Fig. 17-5. Surface tension forces during paper drying. ©1961 TAPPI. Reprinted from Swanson (1961) with permission.

a force of 50 kPa (7 psi). In the region of 20-25% solids the water becomes discontinuous and the surface tension forces decrease and (c) applies. As the film thickness decreases, the pressure may reach one or two hundred atmospheres.

17.10 FIBER PHYSICS

Introduction

Some basic aspects of wood and fiber physics important to understanding paper were presented in Section 2.7.

The strength of individual fibers

Often the strength differences in the machine direction and cross-machine direction of paper are attributed to fiber alignment. While this accounts for some of the difference, another factor is the fact that paper is dried under tension in the machine direction. The ratio of the zero-span tensile strength has been used as an indicator of fiber alignment, but this method is flawed since the strength of individual fibers increases when dried under a stress. Methods relying on the speed of sound waves do not measure fiber alignment. The

speed of sound waves is proportional to the stiffness and the square root of density. While fibers are stiffer in the longitudinal direction, fibers that dry under stress have a higher stiffness too. The effect of each variable can not be ascertained by a single number.

Work by Jentzen (1964) shows that the strength and stiffness of individual fibers increases when they are dried under tension. Others have verified this effect. However, Kim et al. (1975) found that certain fibers would not increase in strength when dried under tension, including pre-dried kraft fibers and fibers where the hemicellulose has been removed by alkali extraction. The workers hypothesized that fibers would increase in strength if the fibrils of the fibers could flow when shear forces are applied. In short, individual papermaking fibers can be expected to increase in strength in the machine direction.

Effect of fiber strength and bonding on paper properties

Van den Akker et al. (1958) studied the effect of fiber strength on sheet strength. They showed that if the orientation of fibers is random in the sheet, the zero-span tensile strength is 3/8 (37.5%) of the load if all of the fibers were aligned in the direction of loading. Their laboratory work verified that the zero-span tensile strength is an important measure of fiber strength. This fact is useful to determine the amount of fiber strength loss that occurs during various methods of refining. The ideal refining method should cause fibrillation with a minimum of fiber strength loss. For example, high consistency refining in disk refiners does this more effectively than low consistency refining in Jordan refiners.

Investigators have studied fiber bonding. For example, Skowronski and Bichard (1987) developed a technique to measure bond strength in papers. The method is based on the energy of delamination.

17.11 ANNOTATED BIBLIOGRAPHY

Bleaching

1. Giertz, H.W., Delignification with bleaching agents, in E. Hägglund, Ed., *Chemistry of Wood*, Academic Press, New York, 1951. 631 p.

2. Giertz, H.W., *Svensk Papperstidn* 46:152(1943).

3. Schmidt, E., *Ber.* 54:1860(1921); ibid., 56:25(1923).

Freeness
4. Biermann, C.J. and J.L. Hull, Replacement of the Canadian Standard freeness temperature and consistency correction tables with equations suited to computer use, *Tappi J.* 75(10):245-246(1992).

5. Sundrani, R.S., J.L. Hull, and C.J. Biermann, Consistency and temperature correction for Canadian Standard freeness with chemical pulps, *Tappi J.*, In press. (1993).

Fiber flocculation
6. Kerekes, R.J., Pulp flocculation in decaying turbulence: A literature review, *J. Pulp Paper Sci.* 9:(3):TR86-TR91(1983).

7. Mason, S.G., Fibre motions and flocculation, *Pulp Paper Mag. Can.* 55(13, Dec.):96-102(1954).

8. Jokinen, O. and K. Ebeling, Flocculation tendency of papermaking fibres, *Paperi ja Puu- Papper och Trä* 67(5):317-325(1985).

9. Kerekes, R.J. and C.J. Schell Characterization of fibre flocculation regimes by a crowding factor, *J. Pulp Paper Sci.* 18(1):J32-J38(1992).

10. Gorres, J., T. Cresson, and P. Luner, Sheet formation from flocculated structures, *J. Pulp Paper Sci.* 15(2):J55-J59(1989).

11. Smith, M.K., Formation potential of west coast kraft pulps, *Pulp Paper Can.* 87(10):T387-T394(1986).

Forces in wet fiber mats
12. Lyne, L.M. and W. Gallay, Studies in fundamentals of wet web strength, *TAPPI* 37(12):698-704(1954).

13. Swanson, J.A., The science of chemical additives in papermaking, *Tappi* 44(1):142A-181A(1961).

Fiber physics
14. Jentzen, C.A., The effect of stress applied during drying on some of the properties of individual pulp fibers, *Tappi* 47(7):412-418(1964). The article was reprinted in *For. Prod. J.* (9):387-392(1967).

15. Kim, C.Y., D.H. Page, F. El-Hosseiny, and A.P.S. Lancaster, The mechanical properties of single wood pulp fibers. III. The effect of drying stress on strength, *J. Appl. Polym. Sci.* 19:1549-1561(1975).

16. Van den Akker, J.A., A.L. Lathrop, M.H. Voelker, and L.R. Dearth, Importance of fiber strength to sheet strength, *TAPPI* 41(8):416-425(1958).

17. Skowronski, J. and W. Bichard, Fibre-to-fibre bonds in paper. Part I. Measurement of bond strength and specific bond strength, *J. Pulp Paper Sci.* 13(5):J165-J169(1987).

General interest on pulp slurries
18. Wahren, D. (Bonano, E.J., Ed.), *Paper Technology, Part 1: Fundamentals,* Institute of Paper Chemistry, Appleton, Wisconsin, 1980, pp. 199-250. This work includes mathematical development of flow through pipes, properties of fiber suspensions, critical volume concentrations for fiber flocculation, network flow, turbulence, and colloidal flocculation.

EXERCISES

Bleaching
1. From the point of view of pollution abatement, why is ClO_2 preferable to Cl_2? What are the number of electrons transferred in the redox reactions per mole of Cl in each?

2. Write the reaction of hypochlorite with iodide.

3. A 50.0 ml aliquot of ClO_2 solution consumed 19.96 ml of 0.152 N thiosulfate solution. What is the concentration of ClO_2 in g/L?

4. In the Solvay process for making ClO_2, how much methanol is theoretically required per ton of sodium chlorate?

5. A softwood, unbleached kraft pulp has a kappa number of 34. During the bleaching process, all of the lignin and 2% of the carbohydrates are removed, what is the pulp yield of the bleach plant? If the yield from the pulp plant is 46%, what is the overall bleached pulp yield from wood?

6. Describe how to prepare and standardize 5 gallons of $KMnO_4$ solution of 0.1000 N to be used in the pulp mill for quality control. The laboratory has 2 M H_2SO_4, 1 N H_2SO_4, 2 M KI, 0.1000 N $Na_2S_2O_4$, and starch indicator solution.

Paper machine calculations

7. A paper machine with a width of 2.5 m has a speed of 10 m/s and produces a paper with a basis weight of 50 g/m^2. Calculate the weight of paper produced per second. From this, and assuming a consistency of 0.5% in the headbox, calculate the volumetric flow rate necessary through the headbox. Since one knows the width of the slice, and the length per second going through the slice, you should be able to easily solve the slice height. Assume this is an open headbox with 100% efficiency of conversion of potential energy to kinetic energy (unlikely in the first case; impossible in the second case). Calculate the height of the water above the slice from the equations for potential energy (PE = mgh, g = 9.8 m/s^2) and kinetic energy (KE = 0.5 mv^2) to obtain the equation.

8. A paper machine operates at 20 m/sec and is 10 m wide. Calculate the required equivalent height of water in the headbox assuming all of the potential energy of water is converted to kinetic energy. For paper of basis weight 50 g/m^2, what is the required slice height? (The headbox consistency is 0.5%.) What is the maximum annual production in tons?

9. A paper machine has a speed of 5 m/s with a basis weight of 100 g/m^2 and stock consistency of 0.6% at the headbox. What should the slice height be?

Strength of wet fiber mats

10. In Fig. 17-6, where are the strongest bonds formed between fibers?

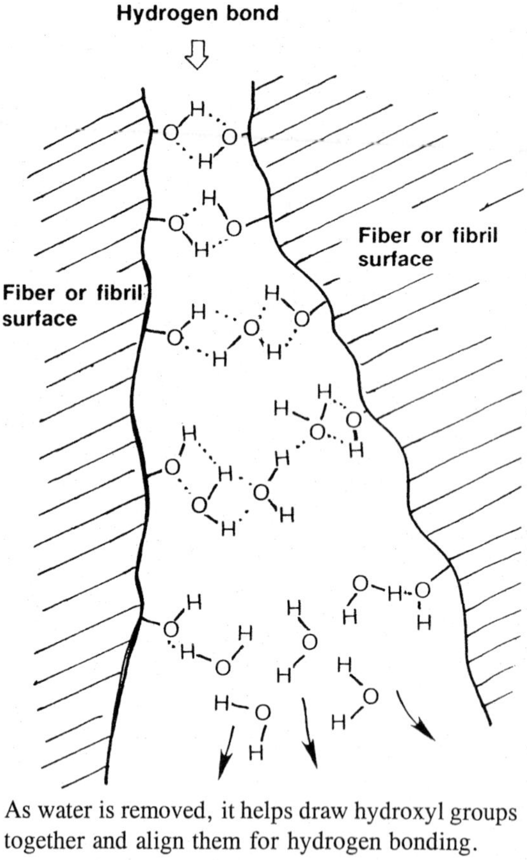

Hydrogen bond

As water is removed, it helps draw hydroxyl groups together and align them for hydrogen bonding.

Fig. 17-6. Hydrogen bonding and water removal. ©1991 James E. Kline. **Reprinted from** *Paper and Paperboard* **with permission.**

18
POLYMER CHEMISTRY

18.1 INTRODUCTION AND TYPES OF POLYMERS

Introduction

This chapter is an introduction to polymer chemistry with emphasis on topics related to pulp and paper. Polymers are very important to many aspects of pulp and paper, including wet end chemistry, surface sizing, and coating. Of course, the principal constituents of pulp fibers are all polymers. Like other chapters in this book, many of the introductory details are not included; for further information, one should consult a textbook on polymer science.

History of polymers

One of the earliest industrial developments was the use of natural, soft rubber (poly-*cis*-isoprene) in the early 1800s; the development of the vulcanization process (reaction of the carbon double bonds with sulfur to form crosslinked chains) by Goodyear in 1839 led to hard rubber.

The first human-made plastic may be considered to be cellulose nitrate, which was discovered in 1846 by Schönbein of Switzerland by the action of a mixture of nitric and sulfuric acids on cotton. Cellulose nitrate was used as a propellant/explosive by the Austrian army in 1852 (now called guncotton or smokeless powder), with amyl acetate solvent as the first modern lacquer in the United States in 1882, and in early photographic films (plasticized with camphor to form Celluloid, and as the first artificial silk by Count de Chardonnet in France in 1884. Cellulose acetate (discovered in 1865), being much safer since it is less flammable, soon replaced cellulose nitrate for most uses. Regenerated cellulose (see cellulose xanthate) was invented by Cross and Bevan in 1892; this is the viscose rayon process and was used by Brandenberger to make cellophane, marking the beginning of modern packaging with transparent, plastic films in 1924 when the first cellophane plant started operation in Buffalo, New York. Indeed the cellulose-based plastics domi-nated the synthetic plastics field for about 50 years.

The development of purely synthetic polymers began with the discovery of the phenol-formaldehyde resins by Baekland with the trade name of Bakelite, small scale production of which began in 1907. Other developments included the use of styrene in synthetic rubbers in the 1930s, the appearance of nylon (invented by Carothers) in 1939, and the appearance of Teflon in 1941.

Polymers

Polymers are high molecular weight chemicals made of repeating units, called *monomers*, which are linked by covalent bonds. The physical properties of polymers depend on 1) the chemical composition of the monomeric units, 2) stereochemistry, if present, between the monomeric units, 3) the mechanical configuration of the polymer chain (that is, is it coiled or linear), and 4) the chain length of the polymer, that is, the number of monomers, known as the *degree of polymerization* (DP), of the polymer. Typical DP values of commercial polymers range from several hundred to many tens of thousands.

Polymers may be grouped according to their component monomers. The simplest type of polymer is the *homopolymer*, a polymer containing only one type of monomer, Fig. 18-1. For convenience, letters of the alphabet are assigned to different types of monomers so that a homopolymer may be described as one having the configuration of: A-A-A-A-A-A-A-A-A-A. This can also be written: A-$(A)_n$-A; if $n = 8$ the DP of these polymers is 10. A simple example of this is polyethylene, made from the ethylene monomer $CH_2=CH_2$; the structure of the polymer is CH_3-$(CH_2$-$CH_2)_n$-CH_3. *Copolymers* are polymers containing two types of monomer, and *terpolymers* contain three types of monomers. A copolymer with an *alternating* structure is of the form ABABABABABABABABABABABA; a copolymer with a *random* structure is ABBAAABAABBABBBABABAABBBABA; a

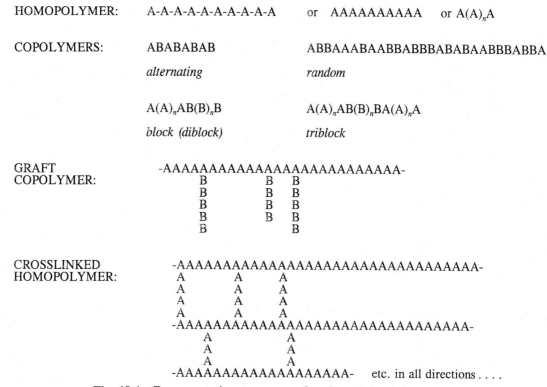

HOMOPOLYMER: A-A-A-A-A-A-A-A-A-A   or AAAAAAAAAA   or A(A)$_n$A

COPOLYMERS: ABABABAB   ABBAAABAABBABBBABABAABBBABBA

alternating   *random*

A(A)$_n$AB(B)$_n$B   A(A)$_n$AB(B)$_n$BA(A)$_n$A

block (diblock)   *triblock*

GRAFT
COPOLYMER: -AAAAAAAAAAAAAAAAAAAAAAAAAAA-
                         B   B   B
                         B   B   B
                         B   B   B
                         B   B   B
                         B   B

CROSSLINKED
HOMOPOLYMER: -AAAAAAAAAAAAAAAAAAAAAAAAAAAAAAAAAA-
                A   A   A
                A   A   A
                A   A   A
                A   A   A
          -AAAAAAAAAAAAAAAAAAAAAAAAAAAAAAAAAA-
                 A   A
                 A   A
                 A   A
         -AAAAAAAAAAAAAAAAAA-   etc. in all directions . . . .

Fig. 18-1. Representative structures of various classes of polymers.

block copolymer has the structure A(A)$_n$AB(B)$_n$B if it is *diblock* or A(A)$_n$AB(B)$_n$BA(A)$_n$A if it is *triblock*. Block copolymers are made by anionic polymerization techniques since monomers can be added consecutively with this method as the anion is stable indefinitely. The structure of a *graft copolymer* is shown in Fig. 18-1.

A *branched homopolymer* is like a graft copolymer except all of the monomer units are identical. Polymers may be *crosslinked* to give additional strength and rigidity. A crosslinked homopolymer is depicted in Fig. 18-1.

Many plastics or adhesives form crosslinks during their manufacture or curing phase. These plastics are called *thermosets* because once set in position they will not soften with heating. Epoxy resins, phenol formaldehyde resin, and polystyrene crosslinked with divinyl benzene are several examples of thermosetting materials. Other polymers consisting of linear chains or linear chains with little or no branching are called *thermoplastics*, which soften when heated and harden when cooled to ambient temperatures.

18.2 ADDITION POLYMERS

There are two categories for types of polymerization reactions used to form polymers: *condensation* and *addition mechanisms* to form *condensation* and *addition polymers*, respectively.

The addition mechanism is used to make polymers from monomers with ring structures or double bonds by a chain reaction. The "extra" bond of the monomer is used to form the bond between monomers; this means that no molecules are lost during polymerization, that is, there is no change in the molecular weight of the monomer incorporated into the polymer. These polymers are usually formed by free radical reactions; however, anionic and cationic mechanisms may also be used, but require special solvents and reaction conditions. Free radical reactions are started using *initiators*. Initiators are compounds that form free radicals, such as peroxides, to start the reaction. The free radical is always carried by the terminal carbon atom between propagation steps. Initiation may also be carried out with high energy radiation, photolysis, or thermal energy. The reaction of a vinyl monomer, which has the form $CH_2=CHR$, is shown in Fig. 18-2.

Monomers almost invariably add head to head, $-CH_2CHR-CH_2CHR-$, as opposed to head to tail, which is $-CH_2CHR-CHRCH_2-$. Head to head addition can form three types of polymers which differ in rotation around a carbon-carbon single bond: in *atactic* polymers the R groups are randomly oriented around the longitudinal axis of the polymer; in *isotactic* polymers the R groups branch out on one side of the polymer chain; and in *syndiotactic* polymers the R groups alternate on one side of the axis to the opposite side of the axis. Atactic polymers are usually amorphous and less dense than their often crystalline syndiotactic or isotactic counterparts. Their structures are:

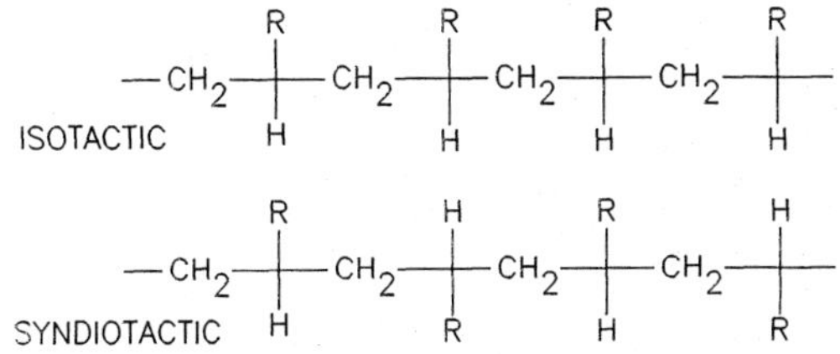

Addition polymers are characterized by high molecular weight averages, rubbery or brittle solids, and, usually, amorphous structures. The chain reaction goes very quickly and the entire polymer is built in a matter of seconds. It should be noted that polymerization proceeds until all of the monomer is removed or the free radicals are all terminated. It is quite possible that residual monomer will be present after polymerization. Fig. 18-3 shows a variety of vinyl diene structures, Fig. 18-4 shows a variety of vinyl polymers, and Fig. 18-5 shows a variety of acrylic polymers all of which are usually formed by the addition mechanism. Two other polymers include $-CH_2CCl_2-$ [poly(vinylidene chloride), Saran] and $-CF_2CF_2-$ [poly(tetrafluoroethylene), Teflon]. The diene polymers are very elastic because the double bonds cause the backbone to assume a kinked, random coil shape; these are rubber materials. Tensile forces allow these coils to be partially straightened; relaxation of the forces allow the original structure to be assumed.

One other type of acrylic polymer is polyacrylic acid where R and R' in Fig. 18-5 both consist of H atoms. The so-called superabsorbent

ISOTACTIC

SYNDIOTACTIC

STEP	EXAMPLE REACTION	DESCRIPTION
Initiation:	$I_2 \rightarrow 2I\cdot$	The initiator forms two free radicals.
	$I\cdot + H_2C=CHR \rightarrow ICH_2\overset{R}{\underset{H}{C}}\cdot$	The free radical reacts with the first monomer.
Propagation:	$I(CH_2CHR)_nCH_2CHR\cdot + H_2C=CHR$ $\rightarrow I(CH_2CHR)_{n+1}CH_2CHR\cdot$	Monomers are added to the growing chain.
Chain transfer:	$-CH_2CHR\cdot + H_2C=CHR'$ $\rightarrow -CH=CHR + CH_3\overset{H}{\underset{R'}{C}}\cdot$	One possibility.
Termination:	$2\ -CH_2CHR\cdot \rightarrow -CH_2CHRCHRCH_2-$	Termination by *coupling* or *combination*.
	$2\ -CH_2CHR\cdot \rightarrow -CH_2CH_2R + -CH=CHR$	Termination by *disproportionation*.

Fig. 18-2. The steps involved in polymerization by the addition mechanism.

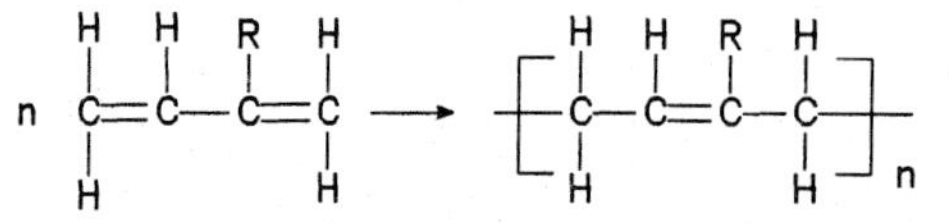

R	Polymer Name
H	Poly(butadiene) (synthetic rubber)
CH$_3$	Poly(isoprene) (natural rubber, cis config·)
Cl	Poly(chloroprene) (Neoprene)

Fig. 18-3. Polymerization of dienes. The position of the double bond is variable; it may be a side group.

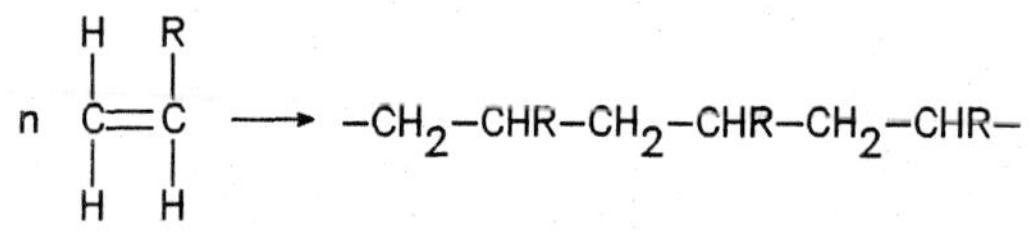

R	Polymer Name
H	Polyethylene, PE
CH$_3$	Poly(propylene), PPE
Cl	Poly(vinyl chloride), PVC
OH	Poly(vinyl alcohol), PVA
OAc	Poly(vinyl acetate), PVAc
C≡N	Poly(acrylonitrile), PAN
CONH$_2$	Poly(acrylamide), PAM
COOH	Poly(acrylic acid)
 ⬡	Polystyrene, PS

Fig. 18-4. Vinyl polymer structures.

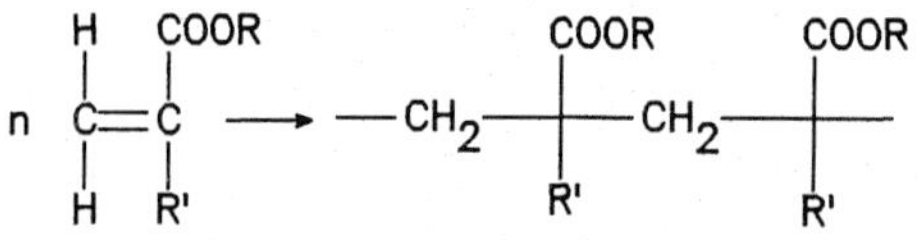

R'	R	Polymer Name
H	CH$_3$	Poly(methacrylate)
H	CH$_2$CH$_3$	Poly(ethacrylate)
CH$_3$	CH$_3$	Poly(methyl methacrylate) (Lucite, Plexiglas)

Fig. 18-5. Acrylic ester structures.

polymers that are used in diapers and incontinence products are based on polyacrylic acid, where some of the carboxylic acids occur as the sodium salt. Sometimes the polyacrylic acid polymers are reacted with starch to form the superabsorbent polymers.

Another type of addition polymerization is the *ring opening* reaction. One example is the polymerization of ethylene oxide used to form polyethylene glycols and ethoxylated molecules used as surfactants.

$$n \; \overset{\displaystyle O}{\overset{/ \; \backslash}{CH_2CH_2}} \; \overset{H^+}{\rightarrow} \; HOCH_2CH_2OCH_2CH_2{}^+{}-$$

18.3 CONDENSATION POLYMERS

The condensation mechanism is used to make polymers from monomers with two (or more for crosslinked polymers) reactive functional groups by a step-wise reaction. Each reaction step between two monomers produces a simple molecule, often water, but sometimes NH_3 or another molecule, as a side product. These reactions are common, well known organic chemistry reactions. They are easily controlled reactions, unlike free radical reactions. The most common reactions of synthetic polymers are esterification (formation of ester linkages) and amidation (formation of amide linkages). Fig. 18-6 gives an example of each. In the case of nylon 6,6 (6 is the number of carbon atoms in each of the two monomers if a dicarboxylic acid and diamine were used to form it) it is possible to use a single compound to make the polymer, and that is $H_2N(CH_2)_5COOH$.

It is apparent that monomers will react together initially and small polymers (*oligomers*) will react later. This means that after a short time essentially all of the monomers will be gone and oligomers will exist. These oligomers will then react to form short chain polymers, then the short chained polymers will react to form moderately long chain polymers. The number average $\overline{DP}_n$ can be predicted as a function of the *extent of reaction*. The extent of reaction is the fraction of functional groups which have participated in the polymerization reaction. For bifunctional reactants present in equal proportions, such as in Fig.

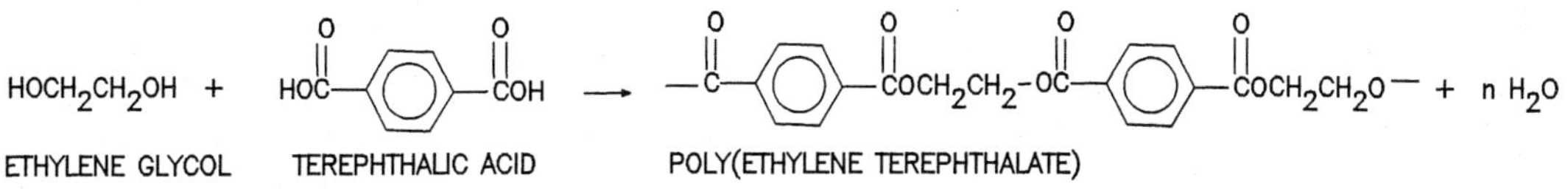

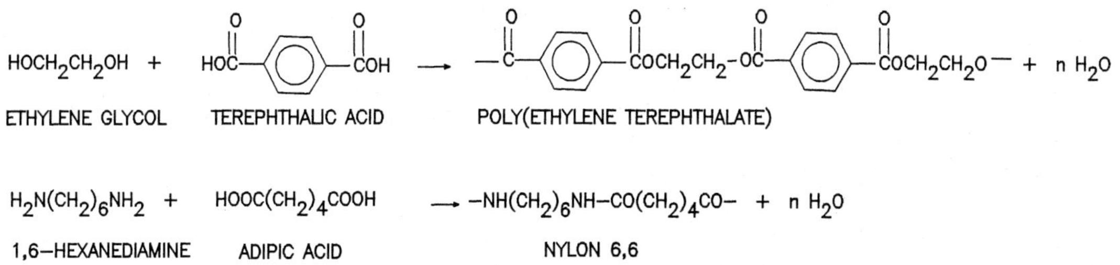

Fig. 18-6. Formation of polyester and polyamide polymers.

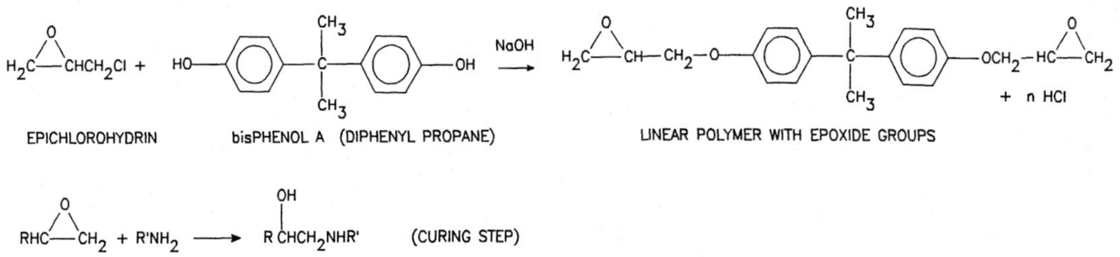

Fig. 18-7. Formation of epoxy resins.

18-6, the $\overline{DP}_n = 1/(1-p)$. For example, if 50% of the functional groups reacted $p = 0.5$ and $\overline{DP}_n = 2$; this would occur if each monomer reacted with one other monomer so that only dimers were left. If p = 75% (if these dimers reacted to form tetramers), then $\overline{DP}_n = 4$. It is apparent that if $p = 99\%$, $\overline{DP}_n = 100$; if $p = 99.9\%$, $\overline{DP}_n = 1000$; and if $p = 99.99\%$, $\overline{DP}_n = 10,000$. It is very difficult to obtain such high extent of reactions; therefore, condensation polymers tend to have relatively low molecular weights unless the monomers have, on the average, more than two functional groups each. Two other useful relationships include $\overline{DP}_w = (1+p)/(1-p)$ and $\overline{DP}_w/\overline{DP}_n = 1+p$; consequently, for high extents of reaction the polydispersity approaches 2.

Condensation polymers are characterized by polar linkages, relatively low molecular number averages, unless crosslinking agents are used, and, often, crystalline polymers. Other examples of condensation polymers include polysaccharides (Chapter 19), proteins, and polyurethanes. Proteins are made from amino acids, which have the structure HOOC-CHR-NH₂ and are joined by amide linkages to form:

-COCHRNH-COCHRNH-COCHRNH-

Polyurethanes are formed by the reaction of diisocyanates (O=C=N-R-N=C=O) and glycols with the linkage -NH-CO-O- by the reaction:

$$OCNCH_2CH_2CNO + HOCH_2CH_2OH \rightarrow$$
$$-[OCONHCH_2CH_2NHCOOCH_2CH_2-]_n)$$

Epoxies are polyethers but retain the name of one of the reactants. The most common epoxies involve reaction of excess epichlorohydrin with bisphenol A; the excess epoxide groups are cured by reacting with polyamines as shown in Fig. 18-7. Epichlorohydrin is used in numerous resins supplied to the pulp and paper industry.

18.4 MOLECULAR WEIGHTS OF POLYMERS

Synthetic polymers consist of individual polymers with varying molecular weights. The properties of the polymers depend not only on the average molecular weight but on the distribution of molecular weights as well. An example of a molecular weight distribution typical of free radical reactions is shown in Fig. 18-8. Many of the methods used to characterize polymers give average molecular weights that depend on the molecular weight distribution as well as the number average molecular weight. It becomes useful, as will be demonstrated shortly, to define several types of molecular weight distributions.

The first is a simple average where each molecule contributes equally to the average (a number average molecular weight depicted as $\overline{M}_n$). Let N_i equal the number of the molecules (or moles or other measure of number of molecules) with mass M_i (or a measure of mass such as DP). Using summation terminology for each fraction from $i = 1$ to ∞, abbreviated here as Σ, the number of molecules in each fraction multiplied by the molecular weight (or other measure of molecular weight such as DP) for each fraction are summed and the entire sum is divided by the total number of molecules as shown in Eq. 18-1.

$$(18\text{-}1)$$

$$\overline{M}_n = \frac{\Sigma N_i \times M_i}{\Sigma N_i}; \quad \overline{DP}_n = \frac{\Sigma N_i \times DP_i}{\Sigma N_i}$$

A number average molecular weight is obtained by methods that depend on the colligative properties of polymer solutions. One such method is vapor phase osmometry that indirectly measures vapor pressures of polymer solutions. Such methods are limited to low molecular weight polymers (several thousand grams per mole or less). This method has the drawback that it does not represent the behavior of polymer mixtures very well.

Consider an extreme, hypothetical example; suppose during kraft pulping that 10 individual glucose units are cleaved from a cellulose chain with a DP of 1000. We now have 11 "polymers", one with a DP of 990 and 10 with a DP of 1. The number average DP is $(1 \times 990 + 10 \times 1)/(10+1) = 90.9$; in reality, however, the only effect one would notice is a slight loss of yield (1%) with identical strength properties. In fact, if every cellulose chain had a DP of 91, the pulp would be useless for structural papers.

A second type of average gets around these difficulties; it is the weight average, $\overline{M}_w$ where the weight of each molecule is considered in importance not the number. We introduce the term W_i, which is the weight of the fraction of with molecular weight M_i. The weight of fraction i can be substituted as follows: $W_i = N_i \times M_i$.

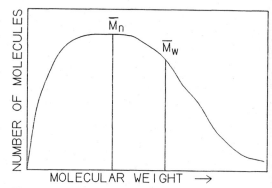

Fig. 18-8. Typical distribution of molecular weights in polymers produced by free radical polymerization.

Using this substitution, the following relationships are obtained:

$$(18\text{-}2)$$

$$\overline{M}_w = \frac{\Sigma W_i \times M_i}{\Sigma W_i} = \frac{\Sigma N_i \times M_i^2}{\Sigma N_i \times M_i}$$

$$\overline{DP}_w = \frac{\Sigma N_i \times DP_i^2}{\Sigma N_i \times DP_i}$$

A weight average molecular weight is determined by light scattering methods but is limited to relatively high molecular weights, above several tens of thousand g/mol. One can calculate the weight average DP in the example above where the number average DP was 91. The $\overline{DP}_w = (1 \times 990^2 + 10 \times 1^2)/(1000) = 980.1$; $\overline{M}_w/\overline{M}_n$ is a measure of the *polydispersity*, that is, how disperse the molecular weight distribution is. Since the weight average molecular weight is always higher than the number average molecular weight, except for the special case when they are equal if all of the polymers have the exact same molecular weight (a *monodisperse* sample), the ratio is greater than one. The polydispersity of this hypothetical example is extremely high. It is $980.1/91 = 10.8$

EXAMPLE 1. Imagine a cellulose chain which has a DP of 5000. In case "A" it is split into two equal parts of 2500 each. In case "B" it is split into one part of 1000 DP and one part of 4000 DP. Calculate $\overline{DP}_n$, $\overline{DP}_w$, and $\overline{DP}_w/\overline{DP}_n$ in each of the two cases.

SOLUTION: Case A: $\overline{DP}_n = (2 \times 2500)/2 = 2500$
$\overline{DP}_w = (2 \times 2500^2)/(2 \times 2500) = 2500$
$\overline{DP}_w/\overline{DP}_n = 2500/2500 = 1$

Case B: $\overline{DP}_n = (1000 + 4000)/2 = 2500$
$\overline{DP}_w = (1 \times 1000^2 + 1 \times 4000^2)/(1000 + 4000) = 3400$
$\overline{DP}_w/\overline{DP}_n = 3400/2500 = 1.36$

EXAMPLE 2. What is $\overline{M}_n$ and $\overline{M}_w$ for a polymer mixture containing 0.5 moles of each of three polymers where $M = 20,000$; $40,000$; and $50,000$; respectively.

SOLUTION: $\overline{M}_n = (0.5 \times 20,000 + 0.5 \times 40,000 + 0.5 \times 50,000)/(0.5 + 0.5 + 0.5) = 36,700$
$\overline{M}_w = (0.5 \times 20,000^2 + 0.5 \times 40,000^2 + 0.5 \times 50,000^2)/(0.5 \times 20,000 +$
$0.5 \times 40,000 + 0.5 \times 50,000) = 37,500$

EXAMPLE 3. Repeat Example 2, except instead of 0.5 moles each assume you have 0.5 parts by weight of each component.

SOLUTION: Use Eq. 18-2 directly and find:

$$\overline{M}_w = (0.5 \times 20,000 + 0.5 \times 40,000 + 0.5 \times 50,000)/(0.5 + 0.5 + 0.5) = 36,700$$

In order to solve for the number average we need a *relative* measure of moles of the different materials. Assume we have 0.5 kg of each polymer: 500 g/20,000 g/mol = 0.025 mol; 500 g/40,000 g/mol = 0.0125 mol; and 500 g/50,000 g/mol = 0.01 mol. Now:

$$\overline{M}_n = (0.025 \times 20,000 + 0.0125 \times 40,000 + 0.01 \times 50,000)/0.0475 \text{ mol} = 31,600$$

A molecular weight profile of soluble polymers may be obtained by a chromatographic method known as size exclusion chromatography (SEC), which is also known as gel permeation chromatography (GPC). The chromatography column consists of a "gel" such as polystyrene crosslinked with divinyl benzene. The gel effectively has "tunnels" of various sizes on the order of the size of the molecules to be separated. Large molecules cannot enter these tunnels, whereas small molecules can. Therefore the volume of the chromatography column is effectively larger for the small molecules. The large molecules come off the column in the solvent first, and the smallest molecules come off last. The fact that large molecules of similar structure come off the column first is unique to this method that separates by size. In all other forms of chromatography, large compounds elute later than small molecules of similar structure.

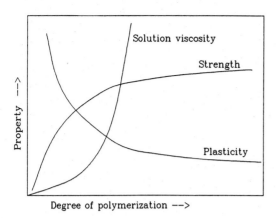

Fig. 18-9. Properties of polymers as a function of average DP.

The mechanical properties of polymers and polymer solutions are greatly dependent on their molecular weights. Fig. 18-9 shows the effect of the DP on plasticity, strength properties, and solution viscosity.

EXERCISES

1. Describe the mechanism whereby GPC separates compounds and indicate the unique characteristic, in regards to the order of elution of similar compounds, of GPC.

2. A polymer contains 1 mole with DP 200 and 1 mole with DP 400. What are the number average and weight average DPs. If the polymer is polystyrene, what is the number average molecular weight?

3. What are some differences between addition polymers and condensation polymers?

4. Draw the general structure of vinyl polymers and show the structures of three.

5. A polymer has an average molecular weight of 6×10^6 and is monodisperse. How was it made?

19
CARBOHYDRATE CHEMISTRY

19.1 INTRODUCTION

The field of carbohydrate chemistry has seen many advances in the past thirty years. The interest in carbohydrate chemistry is not surprising considering the importance of carbohydrates in most aspects of our environment. It is only in the last few decades that the significance of carbohydrates, in their many forms, has truly been recognized. Carbohydrates have long been recognized for their roles as structural materials and sources of energy in the biological world, but their role as informational molecules has only relatively recently been understood, and we have much more to learn. Some interesting facts regarding carbohydrates only hint at their remarkable properties.

1. Cellulose is the most abundant organic chemical on the face of the earth.

2. Some 400 billion tons of carbohydrates are produced annually by photosynthesis.

3. The typical diet consists of more than 60% (dry weight) of carbohydrates.

4. The major blood group types are determined by sequences and branching of carbohydrates.

5. Clearance of erythrocytes from the blood stream by the spleen is determined by the structure of oligosaccharides on the erythrocyte membrane.

Generally, carbohydrates may be defined as polyhydroxy aldehydes or polyhydroxy ketones, occurring in their open chain forms or in their heterocyclic ring forms (the acetal or ketal forms). The simplest types of carbohydrates are called neutral *sugars*. Sugars may be chemically modified to form derivatives. Some examples include, reduction of the carbonyl group to give alditols, oxidation of either terminal carbon to form sugar acids, acetylation or methylation of hydroxyl groups. The chemistry of carbohydrates is presented here to the extent that it is of importance to the understanding of cellulose and hemicelluloses during pulping and papermaking. For further information, introductory biochemistry, wood chemistry, and organic chemistry texts should be consulted. Many of these texts contain chapters devoted to carbohydrate chemistry.

19.2 NOMENCLATURE

The *monosaccharides*, simple sugars that cannot be easily hydrolyzed into smaller units, are classified according to the number of carbon atoms in the molecule. This classification is used for carbohydrates with three to seven carbon atoms; that is, with trioses, tetroses, *pentoses*, *hexoses*, and heptoses. *Aldoses* are monosaccharides that have an aldehyde when in the acyclic form (in the absence of the hemiacetal form); ketoses are monosaccharides with a ketone when in the acyclic form (absence of the hemiketal bond). Glucose is an example of an aldohexose, and fructose is an example of a ketohexose or hexulose, a six-carbon ketose, as shown in Fig. 19-1.

If the terminal R-CH$_2$OH (at the C-6 position) of an aldose is oxidized to a carboxylic acid, then the monosaccharide is known as an *uronic acid*; if the aldehyde is oxidized to a carboxylic acid, the compound is referred to as an aldonic acid; and if both terminal carbon atoms are oxidized to carboxylic acids, the compound is referred to as an aldaric acid. Monosaccharide constituents of particular importance in woody plant cell wall polysaccharides are the pentoses arabinose and xylose; the hexoses glucose, mannose, and galactose; and the uronic acid (4-*O*-methyl) glucuronic acid; these structures are shown with the hemicelluloses in Section 2.6.

19.3 FORMS OF MONOSACCHARIDES

Monosaccharides may be represented, as in Fig. 19-1, by the Fischer projection, introduced by Emil Fischer in the late 19th century. The Fischer projection and the ball-and-stick model of D-glucose are presented in Fig. 19-2, as well as the Fischer projection of L-glucose. Horizontal lines of the Fischer projection represent H and OH

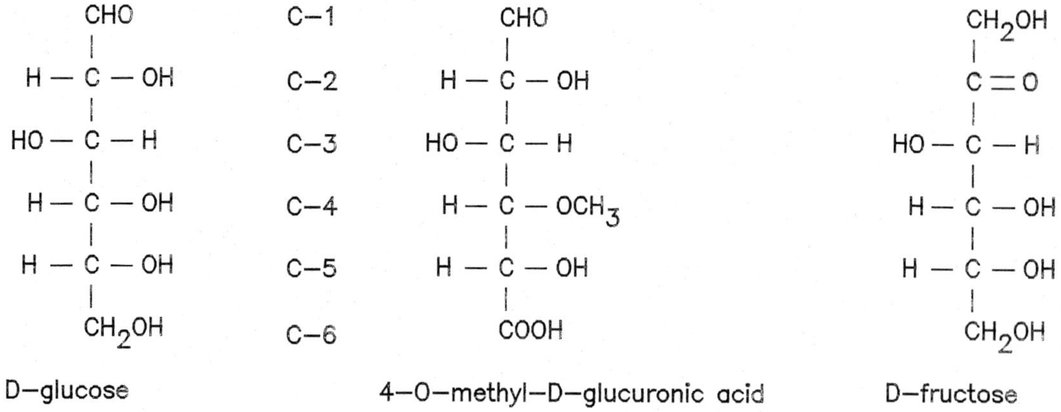

CHO   C-1   CHO   CH$_2$OH

H — C — OH   C-2   H — C — OH   C = O

HO — C — H   C-3   HO — C — H   HO — C — H

H — C — OH   C-4   H — C — OCH$_3$ H — C — OH

H — C — OH   C-5   H — C — OH   H — C — OH

CH$_2$OH   C-6   COOH   CH$_2$OH

D—glucose   4—O—methyl—D—glucuronic acid   D—fructose

Fig. 19-1. Some examples of hexoses.

groups coming towards the viewer. Vertical lines represent bonds going into the plane of the paper. Although these models force the carbon backbone to be curved, it is projected onto the plane of the paper as flat. Carbon atoms are numbered consecutively from the top down. When *n* is the number of carbon atoms, aldoses have *n* minus 2 and ketoses have *n* minus 3 *chiral* (asymmetric) substituted carbon atoms. In glucose these are carbon atoms C-2, C-3, C-4, and C-5. The stereochemistry of the chiral carbon farthest from the anomeric carbon determines whether or not a monosaccharide is designated as D or L; in the case of glucose this is the C-5. If the hydroxyl group on this carbon is on the right of the Fischer projection, then it represents a D-series; if the hydroxyl group is on the left, it represents an L-series. Most naturally occurring monosaccharides are members of the D-series. Notable exceptions

are arabinose, fucose, rhamnose, and iduronic acid. Since aldohexoses have four chiral carbons there are 2^4 (or 16) possible neutral monosaccharides, eight of the D-series and eight of the L-series. With ketohexoses there are 2^3 (or 8) possibilities.

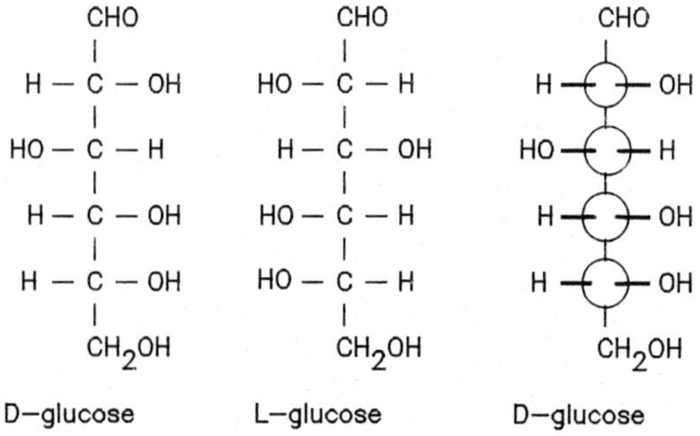

In solution, monosaccharides do not remain as aldehydes and ketones, but form hemiacetal and hemiketal bonds, respectively, as shown in Fig. 19-3. In the case of aldoses, one of the hydroxyl groups will attack the carbon atom of the carbonyl group to give a C-O bond (a hemiacetal); the carbon oxygen double bond of the aldehyde becomes a C-OH group. (The mechanism is protonation of the oxygen of the carbonyl group; attack of the carbonyl carbon atom by a hydroxyl group which transfers the positive charge to the attacking hydroxyl group; loss of a proton by the attacking hydroxyl group leading to the cyclic form.)

Usually a five-member ring (a *furanose*) or a six-member ring (a *pyranose*) is formed, which means that the hydroxyl group of the C-4 or the C-5, respectively, will react with the aldehyde group. The pyranose form is more common than the furanose form, especially in polysaccharides. With some monosaccharides, both pyranose and furanose forms exist in equilibrium.

Since stereochemistry is introduced at the aldehyde carbon atom (i.e., it becomes asymmetric), there are two possible products, which are called *anomers*, for the pyranose or furanose. The carbon atom of the aldehyde or ketone is termed the *anomeric carbon atom*. The equilibrium

CHO   CHO   CHO

H — C — OH   HO — C — H   H —⊖— OH

HO — C — H   H — C — OH   HO —⊖— H

H — C — OH   HO — C — H   H —⊖— OH

H — C — OH   HO — C — H   H —⊖— OH

CH$_2$OH   CH$_2$OH   CH$_2$OH

D—glucose   L—glucose   D—glucose

Fig. 19-2. Fischer projections and ball-and-stick form of D-glucose.

between these possible forms depends on the stereochemistry of the monosaccharide (i.e., glucose is different from mannose, etc.), the composition of the solution and solvent, and the temperature of the solution.

The cyclic forms of carbohydrates are often represented by *Haworth* projections. In fact, pyranoses actually exist in a shape somewhat more complicated than the Haworth projection indicates. These shapes are *chair* conformations of which there are two possibilities. The chair form with the most bulky groups in the equatorial position is greatly favored. The Haworth projection is a convenient approximation of the chair form to use. Fig. 19-3 shows the Fisher projections, the corresponding Haworth projections, and the favored chair conformations of the two anomers of glucopyranose. α-Glucopyranoside is represented with the anomeric OH group down in the Haworth projection (trans to the terminal CH_2OH), and ß-glucopyranoside is represented with the anomeric OH group in the down position. Any group that is on the right in the Fischer projection is projected down in the Haworth projection.

The acid- or based-catalyzed conversion of one anomer into its equilibrium mixture of anomers is called *mutarotation* and is accompanied by change in the optical rotation. (Carbohydrates, like any compound containing asymmetric atoms, cause plane polarized light to rotate when passed through a solution of the carbohydrate; while this phenomena provides an important tool to carbohydrate chemists, it is not of particular significance to the pulp and paper field.) The equilibrium mixture of glucose in dilute aqueous solution at 20°C is about one-third in the α-glucopyranose, two-thirds in the ß-glucopyranose form, and much less than 1% in the acyclic and furanose forms.

More complex carbohydrates occur in nature when two or more simple sugars are linked together. In nature, the linkages are glycosidic linkages, that is, acetal or ketal bonds involving the anomeric carbon of at least one of the monosaccharides involved. These bonds are known as *glycosidic linkages* and allow the formation of dimers, trimers, tetramers, and so forth of monosaccharides termed *disaccharides, trisaccharides, tetrasaccharides*, etc. In general, any compound containing a glycosidic linkage may be termed a

Fischer projections

Haworth projections

Chair conformations

α–D–glucopyranose   ß–D–glucopyranose

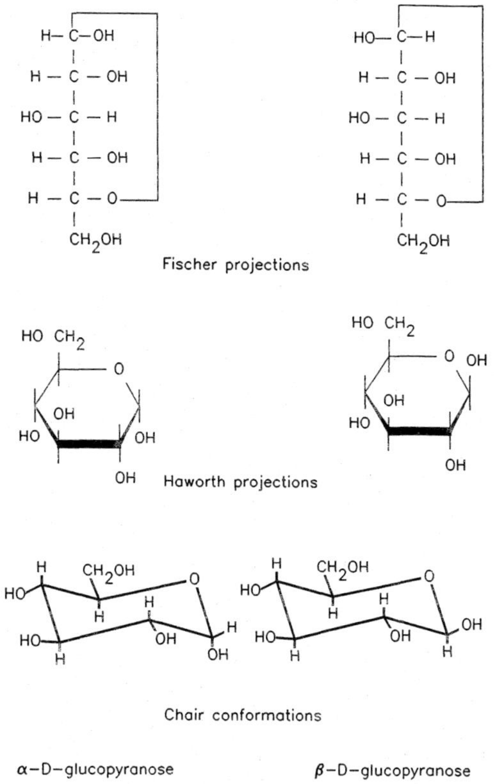

Fig. 19-3. The Fischer and Haworth projections and conformations of the two glucopyranoses.

glycoside. Each glycosidic linkage is formed with removal of a water molecule between the two monosaccharides; this makes polysaccharides *condensation polymers*.

Glycosidic linkages of compounds in aqueous solutions are subject to *hydrolysis* in the presence of acid at reflux temperatures. Hydrolysis is the breaking of a bond by the addition of a water molecule across it. For example, hydrolysis of a disaccharide would yield the original monosaccharides. Carbohydrates with a degree of polymerization (DP, number of monosaccharide units) of more than eight units are considered polysaccharides. Sucrose (table sugar) and maltose are examples of disaccharides; cellulose and starch are examples of polysaccharides.

A glycosidic linkage involving the anomeric OH of an aldose is called an acetal linkage; if the anomeric OH of a ketose is involved, the bond is

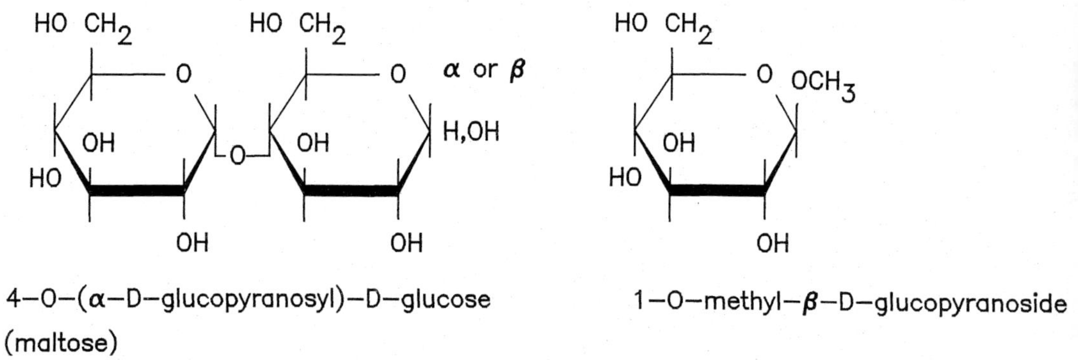

4−O−(α−D−glucopyranosyl)−D−glucose
(maltose)

1−O−methyl−β−D−glucopyranoside

Fig. 19-4. Two examples of glycosidic linkages.

a ketal. The anomeric OH group may also react with an ROH group in general. The reaction with methanol produces a methyl glycoside, the reaction with phenol produces a phenyl glycoside, and so forth. Once a glycosidic linkage is formed it is usually fairly stable; however, glycosidic linkages of glycosides in aqueous solutions may often be cleaved quantitatively by acid-catalyzed hydrolysis at reflux temperatures or by the use of specific enzymes known as glycosidases. Fig. 19-4 shows some examples of glycosidic linkages.

19.4 SELECTED REACTIONS OF CARBOHYDRATES

Acid-catalyzed hydrolysis was discussed in Section 19.3. Anthraquinone (Section 3.6) causes oxidation of cellulose reducing end groups and thereby protects it against the alkaline peeling reaction.

Acid hydrolysis

Acid hydrolysis becomes important in the pulp and paper industry whenever we process wood fibers below pH 2 or so at elevated temperatures. Let us consider some examples of acid hydrolysis. If one treats cellulose (which has been swelled in 72 % w/w sulfuric acid at room temperature) with 6 % w/w aqueous sulfuric acid under reflux, glucose is obtained in high yield with little secondary decomposition (Fig. 19-5).

If polysaccharides are treated with strong acid such as 20% sulfuric acid at high temperatures, they are first hydrolyzed to the component monosaccharides. The monosaccharides then undergo decomposition reactions. Under these conditions, furfural can be produced near-quantitatively from pentoses or *pentosans* (polymers of pentoses such as xylan, a polymer of xylose) if it is distilled from the solution as it is formed. (The furfural can be measured colorimetrically with addition of suitable reagents, and this was an important method for determination of pentoses and pentosans in the older literature.) The reaction of hexoses, or polymers of hexoses such as cellulose, starch, and glucomannans, produces 5-(hydroxymethyl)-2-furfural that is not volatile and undergoes decomposition to levulinic acid and numerous other compounds. These reactions are summarized in Fig. 19-6.

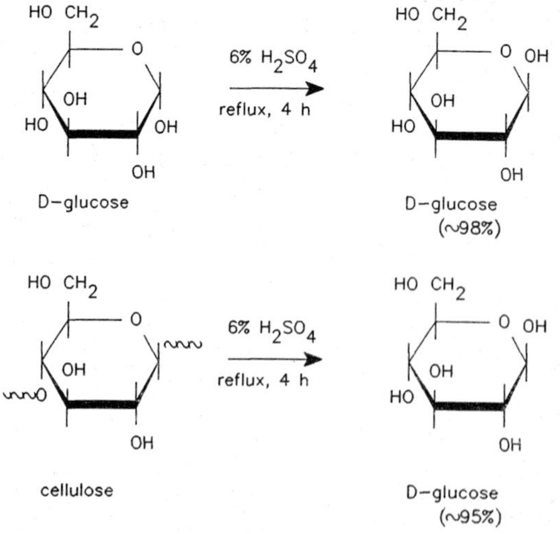

Fig. 19-5. Dilute acid hydrolysis of glycosidic linkages.

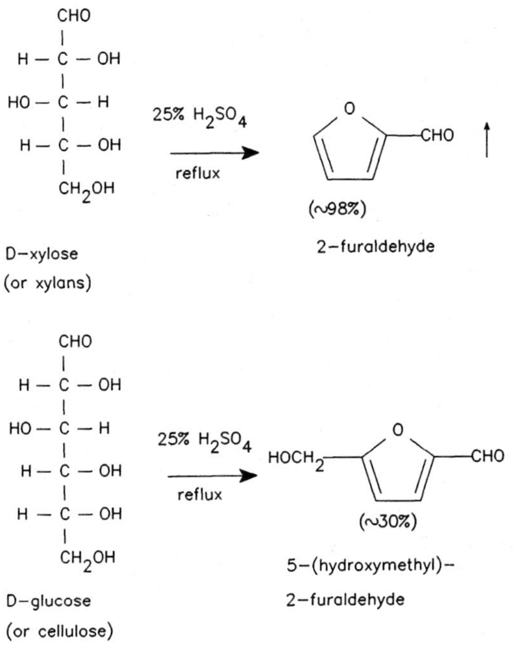

CHO
|
H — C — OH
|
HO — C — H   25% H_2SO_4
| —————→
H — C — OH   reflux
|
CH₂OH

D—xylose
(or xylans)

2—furaldehyde (~98%)

CHO
|
H — C — OH
|
HO — C — H   25% H_2SO_4
| —————→
H — C — OH   reflux
|
H — C — OH
|
CH₂OH

D—glucose
(or cellulose)

5—(hydroxymethyl)—
2—furaldehyde (~30%)

Fig. 19-6. Decomposition of carbohydrates by strong acids.

CHO   CH₂OH
| |
H — C — OH   H — C — OH
| |
HO — C — H   NaBH₄   HO — C — H
| —————→ |
H — C — OR   OH⁻   H — C — OR
| |
H — C — OH   H — C — OH
| |
CH₂OH   CH₂OH

D—glucose (R=H)   D—glucitol or
Cellulose (R=polyglucose)   Protected end group

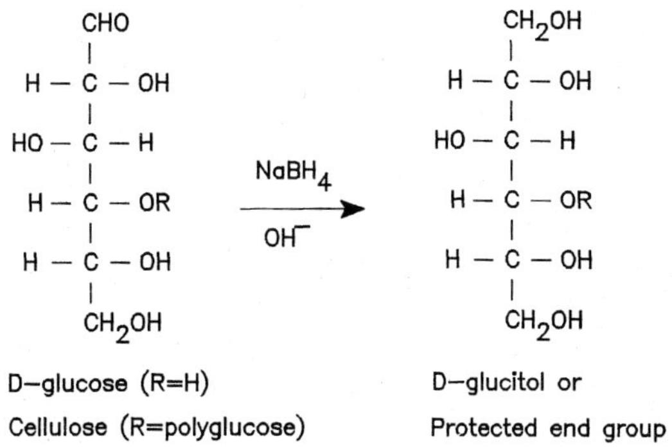

Fig. 19-7. Reduction of the reducing end of a carbohydrate.

Reduction

One can reduce the reducing end of mono- or polysaccharides by using sodium borohydride (NaBH₄) under alkaline conditions at room temperature. NaBH₄ must be used under alkaline conditions as it decomposes by reacting with protons. This reaction is important because it stops the alkaline peeling reaction of cellulose. While NaBH₄ is too expensive to use commercially, it is has been an important tool to investigate the alkaline peeling reaction. The reduction of glucose and cellulose is shown in Fig. 19-7.

Reactions with alkali

Under conditions of dilute alkali (perhaps 0.1 *M* NaOH at 100°C) monosaccharides (and the reducing group of polysaccharides) will slowly undergo a reaction that causes C-2 epimerization (a change in configuration of the second carbon atom) as shown in Fig. 19-8. The reaction is not of any significance to the pulp and paper industry.

Under conditions of fairly concentrated alkali (≈1 *M* NaOH above ≈80°C), polysaccharides undergo several important degradation reactions. These reactions are important during kraft pulping, alkaline sulfite pulping, alkali extraction during bleaching, and other pulping and bleaching operations conducted at elevated temperatures and pH because they degrade cellulose and hemicelluloses.

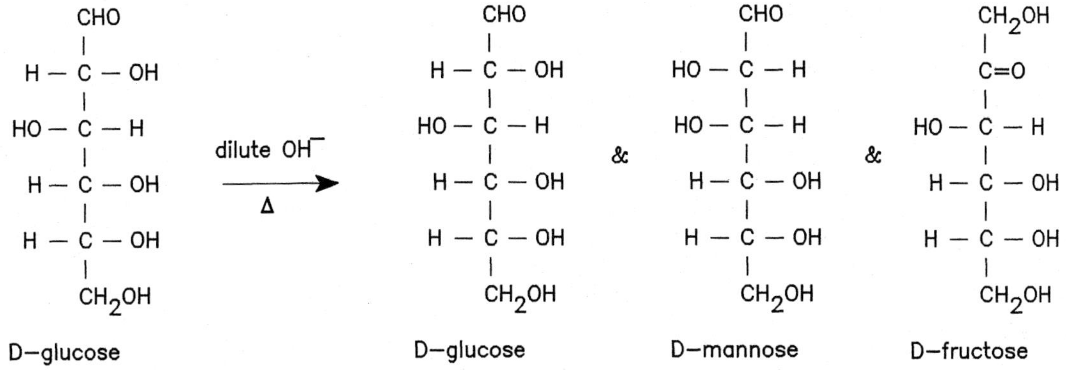

CHO   CHO   CHO   CH₂OH
| | | |
H — C — OH   H — C — OH   HO — C — H   C=O
| | | |
HO — C — H dilute OH⁻ HO — C — H   HO — C — H   HO — C — H
| —————→ | & | & |
H — C — OH   Δ   H — C — OH   H — C — OH   H — C — OH
| | | |
H — C — OH   H — C — OH   H — C — OH   H — C — OH
| | | |
CH₂OH   CH₂OH   CH₂OH   CH₂OH

D—glucose   D—glucose   D—mannose   D—fructose

Fig. 19-8. C-2 epimerization of glucose with dilute alkali.

The first reaction is random cleavage where some glycosidic linkages anywhere along the chain are broken. Alkali cleavage is not nearly as effective as acid hydrolysis at reducing the chain length; however, a relatively few breaks in the cellulose chain greatly decreases its average degree of polymerization. This, in turn, greatly diminishes fiber strength. Oxygen increases this reaction so air should be excluded from the digester; however, magnesium ion helps protect cellulose during oxygen bleaching.

The second reaction is the *alkaline peeling reaction*. The reaction is summarized in Fig. 19-9 and the mechanism is given in Fig. 19-10. The peeling reaction will typically cleave 50 to 100 glucose units from the reducing end of cellulose until another reaction occurs, the *stopping reaction*, which leaves the reducing end of cellulose as a carboxylic acid not subject to the alkaline peeling reaction.

Each time a RO⁻ acts as a leaving group it quickly abstracts a proton from solution to give ROH, another reducing end which is capable of undergoing the peeling reaction. When the stopping reaction occurs, that particular reducing end is protected from further cleavage of glucose units; however, random alkali cleavage does expose new reducing end groups.

The peeling reaction decreases the yield of fiber from kraft cooks but does not affect the cellulose viscosity to an appreciable extent. Imagine a cellulose chain containing at least 1000 glucose molecules. If 50 of these are removed as isosaccharinic acid, the DP has gone from 1000 to 950, a relatively small change for chain length. However, the yield will decrease by 50/1000 or 5%, which is 50 tons per day for a mill that produces 1000 tons per day; this corresponds to something on the order of $25,000 a day or $9 million per year. Also, the carboxylic acids that are formed deplete the alkali in the pulping liquor increasing the amount of alkali required and put a higher load on the recovery system.

Anthraquinone (described at the end of Section 3.6) is used to decrease the peeling reac-

D—glucose (R = H)   or
Cellulose (R = glucose chain)

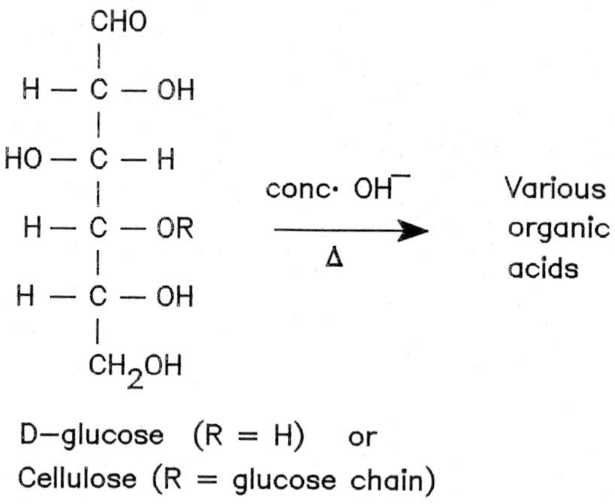

Fig. 19-9. Conversion of D-glucose (R = H) or cellulose (R = glucose chain) to organic acids with strong alkali.

tion by oxidation of the reducing group of cellulose. Polysulfide pulping (page 90) can be used to oxidize the reducing end of cellulose too. Both of these methods increase the pulp yield from 1 to 3%; for example, from 50% to 51-53% yield.

Summary of reactions of carbohydrates

Table 19-1 is a summary of acid and base degradation reactions of carbohydrates. It includes the reaction of $NaBH_4$ that protects against the alkaline peeling reaction.

Holocellulose isolation

Holocellulose from wood or high yield pulps is isolated from lignin in the laboratory by the reaction of chlorine dioxide. Chlorine dioxide is produced directly in the pulp slurry by the following reaction:

$$NaClO_3 + HAc \rightarrow ClO_2$$

This procedure allows the viscosity of cellulose in high yield pulps to be measured where lignin would otherwise interfere. It is also an effective tool for isolating holocellulose for further studies since the holocellulose is not degraded appreciably during the isolation procedure.

$$\overline{M}_n = \frac{\Sigma N_i \times M_i}{\Sigma N_i}; \quad \overline{DP}_n = \frac{\Sigma N_i \times DP_i}{\Sigma N_i} \qquad (18\text{-}1)$$

A number average molecular weight is obtained by methods that depend on the colligative properties of polymer solutions. One such method is vapor phase osmometry that indirectly measures vapor pressures of polymer solutions. Such methods are limited to low molecular weight polymers (several thousand grams per mole or less). This method has the drawback that it does not represent the behavior of polymer mixtures very well.

Consider an extreme, hypothetical example; suppose during kraft pulping that 10 individual glucose units are cleaved from a cellulose chain with a DP of 1000. We now have 11 "polymers", one with a DP of 990 and 10 with a DP of 1. The number average DP is $(1 \times 990 + 10 \times 1)/(10+1) = 90.9$; in reality, however, the only effect one would notice is a slight loss of yield (1%) with identical strength properties. In fact, if every cellulose chain had a DP of 91, the pulp would be useless for structural papers.

A second type of average gets around these difficulties; it is the weight average, $\overline{M}_w$ where the weight of each molecule is considered in importance not the number. We introduce the term W_i, which is the weight of the fraction of with molecular weight M_i. The weight of fraction i can be substituted as follows: $W_i = N_i \times M_i$.

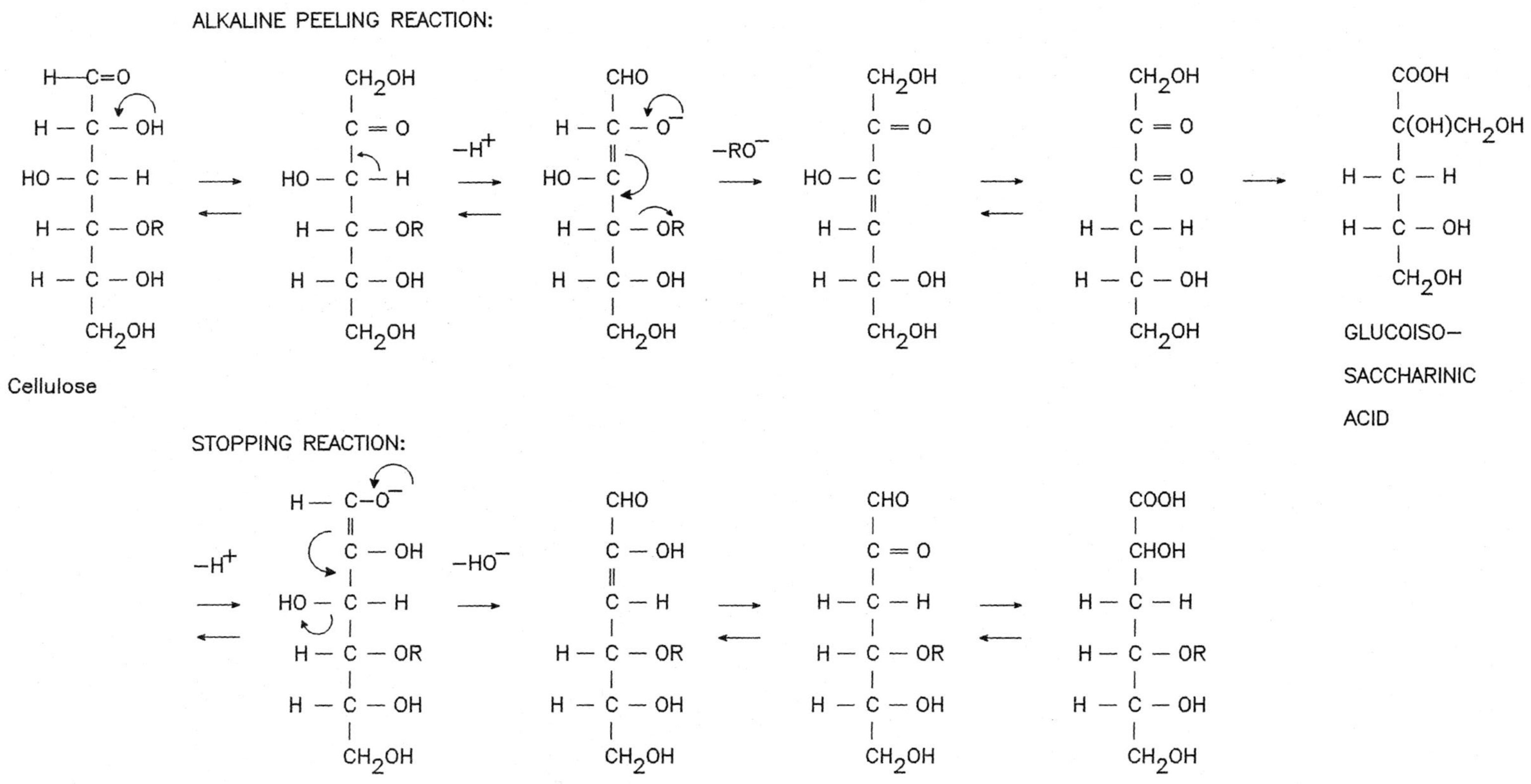

Fig. 19-10. The mechanism of the alkaline peeling reaction and stopping reaction of cellulose.

Table 19-1. Summary of products when model carbohydrate compounds are treated under various conditions.

sugar	strong OH⁻	weak OH⁻	weak H⁺	strong H⁺	NaBH4
xylose	acids	C-2 epimers	no reaction	furfural	xylitol
glucose	acids	C-2 epimers	no reaction	HMF, etc.	glucitol
maltose	acids	C-2 epimers on reducing end	glucose	HMF, etc.	4-O-gluco-glucitol
phenyl-D-glucoside	no reaction	no reaction	glucose, phenol	phenol, HMF, etc.	no reaction

EXERCISES

1. Explain how sodium borohydride stabilizes cellulose and hemicelluloses. (From what does it stabilize these polysaccharides?)

2. What is a hexose? What is a pentose?

3. What is the difference in structure between cellulose and the backbone of starch? How does this affect their physical properties?

4. If you are given a polysaccharide and want to isolate its monosaccharide constituents for subsequent analysis, what sort of treatment will you subject it to?

5. Does one usually measure the viscosity of cellulose in mechanical pulps? Why or why not?

6. How does anthraquinone increase pulp yield during alkaline pulping methods?

20
TOTAL QUALITY MANAGEMENT

20.1 INTRODUCTION TO TQM AND SPC

Introduction

When speaking about quality control it is important to know the background of the key people behind developments in this field. Three people from Bell Telephone Laboratories, W. A. Shewhart, H. F. Dodge, and H. G. Roming, contributed much to statistical quality control (SQC) during its early years in the 1920s.

Dr. W. Edwards Deming is considered to be the originator of total quality management (TQM), which includes management philosophy and style along with statistical process control. By the end of the 1930s Deming (1938) had written a book, helped edit one (Shewhart, 1939) about statistical analysis in regard to quality control, and began working for the U.S. Census Office. During World War II quality control rose to prominence and with it, the work of Deming. In the 1950s his ideas on how companies should manage their people and processes were not widely accepted by American management. He took his ideas to Japan, where he started by assisting with their census, and is credited with helping that country become highly successful in industry. TQM has two important elements to it: a management philosophy (the human aspect) and methods of process control (the process aspect).

Philip B. Crosby has contributed much to the terminology and economic analysis of investments in quality improvement. Dr. Joseph M. Juran contributed to the organization methodology to implement and support project improvement. Juran (1951, 1962) believed that inspection of the final product is not an efficient or successful method of making high quality products. Instead, inspection of materials and methods should be made throughout the process. Quality should not be stamped on the product as the last step but should be incorporated throughout the process.

One can discuss quality at length without defining it, but trying to achieve quality control means specific goals must be kept in mind. Quality has many facets. One of the most obvious

in paper products is product variation. It is always desirable to make one's product in a reliable and consistent manner. It is said that the color of your paper is not as important as making it the same color all the time. Quality is also the suitability of a product for its intended purpose. Another aspect is the cost. One can always make a quality product if money is no object; however, a quality product that few can afford does very little good unless you are contracting for the government.

Total quality management, TQM

Total quality management involves more than just statistical process control, which is not a management style, only a tool. This aspect of TQM is considered separately.

GOAL/QPC, an organization of proponents and practitioners of TQM in the U.S., defines TQM as "a structured system for creating organization-wide participation in the planning and implementation of a continuous improvement process that exceeds the needs of the customer.

Deming (1982) has 14 points of management covering aspects of purpose, quality in products, eliminating waste, error and inefficiency, improving methods, removing fear in the workers, communication between departments, and implementing education and self-improvement programs. Deming points out that you pay for mistakes twice: once to make them and once to correct them. If you have 10% of your workers reworking products, that means 20% of your workforce is making and correcting mistakes. Deming also has a list of 66 questions in Chapter 5 to help managers understand their responsibilities.

Managers should allow workers to work together in teams throughout this process. An important aspect of this is two way communication between all parties with a sincere desire to work together, not just hand down orders.

TQM assumes that, given the chance, workers (or suppliers) will perform best when they know and understand the company or process goals. Productivity should not be enforced by rules, regulations, monitoring, performance standards, and other degrading methods. Arbitrary standards, personnel evaluations, and other humiliating tools should not be used to increase production and efficiency from workers. One might summarize these aspects of TQM as "you catch more flies with honey that with vinegar.

With paper, there are intrinsic tradeoffs in quality. Choices are made about which properties must be decreased in order to improve others. For example, for any type of paper, there is always a choice between tensile and tear strengths. Improving one means decreasing the other. However, once a choice is made, the tools of TQM and SPC can be used to maintain the process, reduce variation, and improve efficiency.

Deming's 14 points for management

Deming (1982) presents 14 points for management in his classic work. They are presented here in concisely with examples relating to the pulp and paper industry, but Deming's work must be used to get the full meaning of them.

1. Constancy of purpose to improve. The purpose of a paper company is to make high quality paper in an efficient manner with little waste. Teams of employees should always be on the lookout to make the process better, especially with communication between and among all levels.

2. New philosophies. The pulp and paper world is a much different place than it was in the past. Management must now compete internationally and use present circumstances.

3. Cease inspection as the means to quality. Quality should not be stamped on the product at the end, but built into the entire system so that quality is insured, not just a matter of chance.

4. Do not consider the lowest price as the ultimate bargain. For example, "Sawmill A" may sell wood chips at $85 per bone dry unit while "Sawmill B" sells wood chips at $90 per ton. "Mill A", however, has 1% more bark, 2% more fines, and 6% more pin chips and actually costs $10 per ton extra to process. Develop long-term relationships of loyalty and trust with suppliers rather than jumping around from supplier to supplier to try to save a dollar.

When selecting a supplier for a part that is part of a larger assembly, it is wise to let the supplier help in the design process, rather than just making specifications that later need to be altered at great expense.

5. Always be looking for methods of improving the process. Process improvement does not necessarily mean buying new equipment. Often the present equipment can be improved or used more efficiently. The relative amounts of variation by each component of a system must be identified to determine if an improvement (such as new equipment) will actually mean a benefit. For example, it may be useful to work with chip suppliers to decrease the amount of overthick chips rather than purchase an additional chip slicer.

It should be clear that putting out fires should be only a minor component of the process engineer's time. In the pulp and paper industry, however, it tends to be a major portion of the engineer's time. If more time were spent on improving the process, then little time would be spent on fighting fires. Unfortunately, the argument is that there is no time to work on the process because of all the fires to fight. This is only short-term thinking and, in the long run, will lead to ruin.

6. Use on the job training. On the job training is very successful since the tools being taught are immediately put to use. Deming says "the greatest waste in America is failure to use the abilities of people.

7. Leadership. The goal of leadership should be to help people and machinery do a better job. To merely demand increased productivity without improving the process will lead nowhere. If management spends its time supervising workers, then there are two people doing the same job.

8. Drive out fear. Setting production goals, other quotas, and demands of zero-defect strongly imply "punishment" if they are not met. If one shows people how to do a good job, or improves the process so a good job can be done, quality and production will automatically improve. Management should not assume the worker is failing, the failure may be attributable to the manager.

9. *Break down barriers between departments internally, and customers and suppliers externally.* Chip suppliers must know what a pulp mill is looking for in wood chips and why it is important to the pulping process. Management must also promote internal teamwork.

10. *Do not use slogans, work targets, and exhortations to try to increase quality and production.* These techniques lead to fear by implied consequences as described in (8) above. Furthermore, if the system was capable of higher productivity as is, why isn't that level of productivity already being achieved?

11. *Do not manage by objectives, but by leadership.* As in (10), to demand more productivity is not as useful as showing people how the process or testing methods can be modified to achieve higher productivity.

12. *Remove barriers that rob people of pride in workmanship.* One barrier is the annual merit rating. Often the letter of the "law" is followed without the spirit of the "law" so that compliance may appear to be occurring, but in fact is not. Workers must not feel that they are only following orders, but that they are in control of the situation and responsible for the final product.

13. *Make education and self-improvement a goal.* People who are challenged and see growth in themselves will enjoy their jobs more and perform better in the long run.

14. *Everyone is involved in the transformation of the company.* Accomplishments are the result of team effort. This reduces duplication of effort and allows different departments to interact more successfully.

20.2 STATISTICAL PROCESS (QUALITY) CONTROL, SPC, SQC

Introduction

Statistical process control (SPC) or statistical quality control (SQC) involves data collection and analysis, modeling of systems, problem solving, and design of experiments. I like to summarize SPC as the application of elementary statistical analysis to control a process. It is the scientific method applied to manufacturing. Shewhart (1939) made the following comparison:

Mass Production	*Scientific Method*
Specification	Hypothesis
Production	Experiment
Inspection	Test of Hypothesis

Statistical analysis

Statistical analysis of the process is a key part of SPC since it is crucial to determine what is random variation and what is nonrandom variation that can be controlled. Anyone who wants to implement SPC must understand elementary statistics, experimental design, and sampling techniques. There are many good books on statistical analysis so there is little point including all this information here, but some basic statistical equations are included later. When using statistical analysis, the underlining assumption is that all of the variation is random. SPC tools must be developed when the process is in control and there are no trends in the data. (This does not mean all of the product will be satisfactory, only that the operator is doing the best that he/she can with the equipment.) This is not always easily done since an apparent trend in the short run could be due to statistical fluctuation. Steps can be taken to decrease the random variation so that "actual" changes can be observed more easily.

Most aspects of statistical analysis in common use assume that statistical deviation follows the *normal distribution*, which is symmetrical and has a bell-shaped curve as shown in Fig. 20-1. Many statistical equations are only applicable to this distribution. Most measurements have variations that follow the normal distribution. Some, however, such as the time between failures, follow other distributions. These other distributions can be predicted, but require tools generally found in advanced simulation or statistical analysis textbooks.

EXAMPLE 1. What is the probability that a sample from a normal distribution will be 1 or more standard deviations less than the mean?

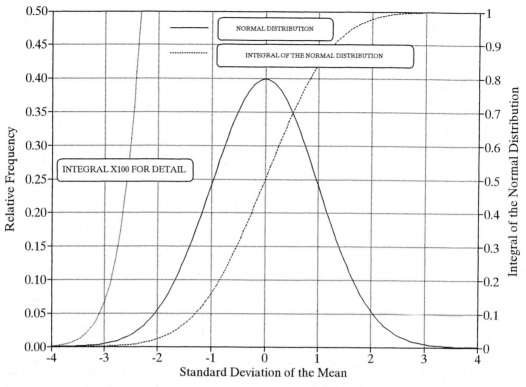

Fig. 20-1. The normal distribution and its integral for determining probabilities.

SOLUTION. The integral in Fig. 20-1 shows that 16% of the values lie below 1 standard deviation. Since the curve is symmetrical, 32% of the values lie outside ± 1 standard deviation of the mean.

EXAMPLE 2. What is the probability that a sample from a standard will be within two standard deviations of the mean?

SOLUTION. Since 2.3% are below 2 standard deviations, 4.6 are outside ± 2 standard deviations (σ) of the mean. Stated conversely, 95.4% of the values lie within $\pm 2\sigma$ of the mean.

The invention of statistical analysis for process control can be credited to W. S. Gosset (1908). He published his work in 1908 under the pseudonym of Student, because he knew the importance of statistics to control processes, and he did not want his competitors to know that he was using this tool. He discovered the t-distribution from a normally distributed population, which is defined as:

$$t = \frac{\bar{X} - \mu}{s/\sqrt{n}}$$

This distribution is still called the *Student's t*. The test is used to make inferences about the mean values of populations based on the measurements of relatively small numbers of samples. The value of t is analogous to z of a normal distribution, but the actual standard deviation is unknown and is approximated as s based on a finite sample size. The distribution is dependent on the sample size and t approaches z for large sample sizes.

Important process variables should be monitored using the concepts of statistical analysis. Process variables are loosely meant to be qualities of the raw materials, important variables in the process itself, and qualities of the product. In kraft pulping the key process variables of the raw material (wood) are chip species, thickness, moisture content, bark content, etc. The key pulping variables are H-factor and liquor characteristics. The key product variables are kappa number and cellulose viscosity.

This is the field of statistical quality control (SQC) or statistical product control (SPC). These techniques should be second nature to any scientist, but they have met with opposition in production facilities where they are treated as one more fad brought down by management. One aspect of this is to reduce product variation. This assumes that all of the important variables are measured in timely manners. It is interesting to speculate on companies that join the SPC bandwagon, indicating that SPC is the best thing they have ever heard; what were these companies doing before they heard of SPC?

The cost versus benefit of SPC

The cost of quality assurance throughout the manufacturing process is an important consideration. With too little quality assurance, high costs will result because there will be a high rejection rate at the end. Too much quality control and the tail wags the dog. The amount of quality control practiced depends on the intended use of the product. Transistors for consumer radios will not be made with the same tight specifications as electronic components designed for space satellites, so the level of SPC is appropriately different in these two products.

Inventions and new processes change the quality control picture dramatically. The best quality control in the design of vacuum tubes will never compete with the transistor, which is inherently of higher quality for most purposes.

Computers and computer networks now make many aspects of quality control essentially automatic to the production worker. Data is collected continuously by sensors, sent to computers, graphed, and presented to operators. Recent trends, warnings of process variables going out of a specified range, and other information is given continuously. This allows the operators to see a potential problem long before the quality of the product is seriously affected. These automatic systems also save money by not requiring much operator time. Much of the drudgery of statistical process control, such as plotting points by hand, should be gone. For other data that is not collected automatically and for teaching purposes, computers with graphic packages should be made available.

20.3 STATISTICAL PROCESS CONTROL TOOLS

Data collection

It is very important to collect useful data before trying to analyze it by charting and statistical analysis. TAPPI Standard Methods cover areas of sampling for paper, chips, and other materials. If the sample one obtains does not represent the material that is actually being used, then data analysis may do more harm than good by throwing people off track. This has been mentioned in various parts of this book already. For example, a bucket sampler that is filled by the first few chips off a chip truck potentially allows chip suppliers to put poor quality material in the truckload without detection. Mention was also made of a neglectful worker who would collect an entire shift worth of eight samples all at the same time, but submit them hourly.

Another point to consider is the sampling frequency. Generally, the more variability in a material, the more often it should be sampled. Wood has a very large amount of variability due to the nature of the material. Wood is not like sodium hydroxide or other chemicals that have relatively little variation between batches.

The accuracy of data is paramount. No amount of statistical analysis will detect errors in pulp kappa numbers that are all too low because a buret could not be filled to the top. Such an error is fixed and not random.

One should also consider what variables should be tested. In the past, kraft mills tested wood chips on the basis of size for pulping. However, thickness is a much more suitable variable, and the industry now saves millions of dollars every year by testing and separating wood on the basis of its thickness.

Graphs and charts

Data for statistical process control can be presented in any form in which data is commonly presented for analysis, presentations, and comparison. These tools include bar charts, histograms, line graphs, pie charts, scatter plots to relate two or more variables, Pareto charts, and control charts.

By graphing the data, *trend analysis* becomes much easier. It is virtually impossible to see

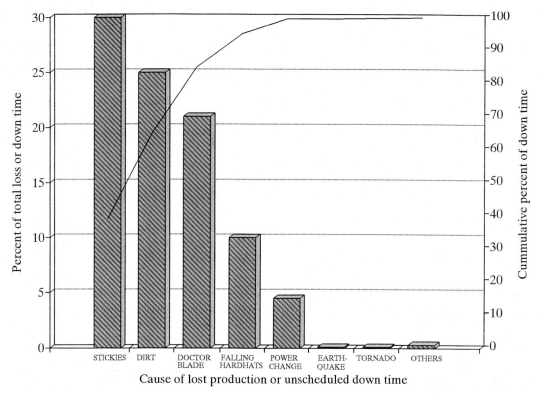

Fig. 20-2. A Pareto chart for hypothetical down time of a tissue paper machine due to web breaks.

subtle trends in data presented tabularly, but they become more apparent when presented graphically. For example, one problem was seen to occur with a cycle of the same length as the bleaching cycle changeover from one species to another. By recognizing this fact, it became obvious that the problem was due to some carryover between two tanks. The cause of the problem could be verified by fiber analysis techniques. Once the source of the problem was recognized, several solutions were suggested in a brain-storming session and the solution implemented with teamwork. One might define SPC as the presentation of useful, accurate information in a timely and useful manner (via the tools of elementary statistics).

Pareto charts

Pareto charts were invented in 1906 by Vilfredo Pareto. He modified Lorenz type-plots that were first used to show unequal distribution of wealth (Juran, 1962). Pareto charts are histograms assigning failure to various causes. Often just a few causes result in most of the failures, just

as a few people control most of the wealth. Fig. 20-2 shows a Pareto chart showing the cause of down time on a paper machine.

Control charts

One of the unique tools of statistical process control is the control chart devised by Dr. Walter A. Shewhart in 1924 (Juran et al., 1962). The control chart is a line graph where the *x*-axis is the independent variable and the *y*-axis is the process control parameter or variable. Fig. 20-3 shows a control chart for the kappa number of pulp going to the bleach plant.

Upper and lower boundaries of the "desired" values of *y* are plotted as lines. These are the upper control limit (UCL) and lower control limit (LCL). The average value ($\overline{X}$, said as "X-bar") is centered between the control limits. The range of the control limits, *R*, depends on the statistical variation of the process, the measurements of the process, and other variations. Usually the range is defined as $\pm$ two standard deviations of the mean so that 95% of the data points should be

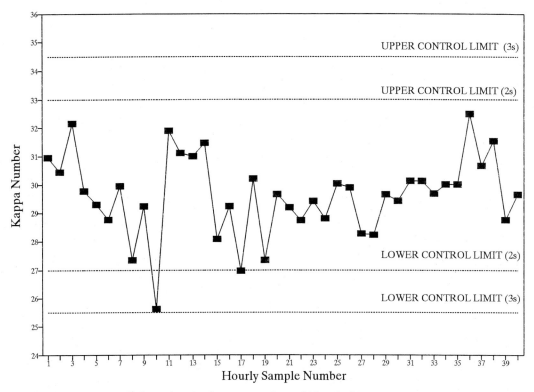

Fig. 20-3. Control chart for the kappa number of bleachable pulp leaving the pulp mill.

within the range if all of the variation is random (not systematic error such as a process upset). If it is defined as plus or minus three standard deviations, it will include 99% of the data points. Control charts must be developed when the process is under statistical control, that is, all of the variation is random. Otherwise, the control limits will be in error.

One can quickly see if the process control variable is within the control area. If the actual y values (kappa number in this example) are consistently within the boundaries, the process is said to be in control. This allows an operator to make changes when appropriate, but not to try to make changes when no action is called for.

The data for the chart in Fig. 20-3 was generated by a computer program (described at the end of the chapter) designed to give an average of 30 and a standard deviation of 1.5. In this case, all of the variation is random (unless, in the unlikely event, the random number generator used to generate the data is faulty). This data set has an actual average of 30.72 and a standard devia-

tion of 1.45. If it is assumed that the kappa number should be between 27 and 33 for this grade of pulp, then the process or measurement of the process must be improved as it is uncontrollable due to the random error. In this case one might try using an automated device to see if variability of measurement is an appreciable portion of the overall variation (in which case the kappa number may not actually be changing, but one would be tempted to control the process without realizing that). If this does not solve the problem then new SPC techniques that are being introduced in the pulp mill would probably decrease process variability of the pulp going to the bleach plant from the pulp mill.

20.4 IMPLEMENTING TQM AND SPC

Commitment of management

One statement about management that has a lot of truth is that an organization take on the attitude of their top manager (i.e., the attitude comes from the top down). In order for these tools to work, the top *management must be com-*

mitted to the techniques of TQM and SPC. Employees should be trained from the top down, with the level of detail higher at each lower level so that decisions can be made at the front line. The techniques and purposes of TQM and SPC should be explained to the satisfaction of the employees. Training and implementation should go hand in hand to reinforce the learning process and so the relevance of the method is apparent. The expectations of each employe must be clear. Too often TQM and SPC are talked about at length, but what they are and how they relate to the process are never mentioned. For many people this is sheer frustration; for too many others, unfortunately, it is normal to discuss things at great length that they do not understand. Management must also be committed to solving problems once they are identified.

Specify objectives and goals clearly

There must be specific goals and objectives in mind so that improvements can be ascertained. Often the goal might be to meet established specifications with less variation or reject product. Quality improvements often increase production by decreasing reject product. Vague goals cannot be measured, and so improvement cannot be ascertained. Suppliers, and all the other parties in the process, should be involved and kept up to date with clear communication. Not surprisingly, if people know what one wants and why it is necessary for one to have this, people will bend over backward to help you. If they do not know why you want something, you have only yourself to blame when you do not get it.

Effective communication and feedback

The key to letting people know what is expected of them is effective communication. Be sure to always let someone know when they are doing things well; do not just indicate to them when they are not doing things well. There must be effective feedback for any person to know how they are doing. One cannot control the basis weight of paper on a paper machine if one does not presently know what the basis weight is and what choices one has to change the basis weight. Also, a person cannot know how well he or she is doing if there is no feedback!

Verification of data

It is important to insure that the collected data is correct. The accuracy of sensors should be checked periodically. One should always consider whether parameters are being measured in the most effective way possible. Other questions like "is this the most effective process?" should be asked. As the systems come under control, the specifications can become tighter.

What to avoid

One thing that should avoided at mills is the writing of data by hand on charts or scraps of paper, especially when this data is to be used by others. Rewriting of the data, with the introduction of errors, is time wasted. Data that is directly entered into computer spreadsheets can be immediately evaluated graphically in many forms and transmitted to others who might make use of it. It is not uncommon to see the paper testing lab write down the test results (when the measuring devices could automatically enter the values into a computer system) on a scrape of paper to be given to the appropriate paper machine operator (when the "courier" happens to come by the paper testing lab) about 30 minutes later. The data is then rewritten on another chart. Later, others must manually look up the data for monthly production reports or other purposes.

Occasionally one sees a small research laboratory in a mill with state-of-the-art equipment such as electronic basis weight scales that are used once or twice a day. Out in the production area, however, paper is still (slowly) weighed on a mechanical balance about 8 times an hour with perhaps 2% or more error. One should think of the mill as a giant research experiment where the gathering and analysis of data are just as critical as in the most serious scientific work.

20.5 MISCELLANEOUS TOPICS

ISO 9000 series certification

The International Organization for Standardization (ISO) developed the ISO 9000 standards in 1987. The standards focus on twenty criteria of a quality control program. There are three levels of ISO 9000 certification. ISO 9001 is the most comprehensive and includes meeting requirements in design, production, installation, and servicing as applicable to the product. ISO 9002 includes meeting requirements during production and installation. ISO 9003 includes meeting requirements in product quality only by final testing and inspection. (ISO 9004 is a guidance document.) The European Communities (EC) will require most industries to have ISO 9000 certification. Rabbitt and Bergh (1992) give a useful summary of ISO 9000 certification.

Just in time concept

This concept means that inventory is kept to a minimum and parts are produced almost as needed. The "just in time" concept originates in post World War II Japan. At that time America was largely using the methods of Henry Ford, who developed the assembly line as a means of manufacturing large numbers of automobiles in an efficient manner. This included having large inventories of each piece that would be brought to the assembly floor in large quantities as needed.

Japan simply could not afford to tie up any resource that would not soon become part of a product to be sold. So methods were developed that coordinated the manufacture of individual parts with assembly of the whole item to limit the inventory at any time. Today it is realized that large inventories tie up large amounts of capital. Interest must be paid on this capital. Large building and storage costs also result.

20.6 EQUATIONS

A good description of elementary statistics is found in SCAN-G2:63 (1962), although any elementary statistics textbook will have similar information with more detail.

$$\text{mean} = \overline{X} = \Sigma x/n = (x_1 + x_2 + \ldots + x_n)/n$$

range $= R$ is the difference between the largest and smallest values

$$\text{standard deviation} = s = (\Sigma(X-\overline{X})^2/(n-1))^{1/2}$$

$$\text{coefficient of variation} = V = s/\overline{X} \times 100\%$$

20.7 ANNOTATED BIBLIOGRAPHY

1. Deming, W.E., *Statistical Analysis of Data*, Wiley, New York, 1938, 1943. 261 p.

2. Shewhart, W.A., (with the editorial assistance of W.E. Deming), *Statistical Method From the Viewpoint of Quality Control*, Washington, The Graduate School, 1939.

3. Juran, J.M., L.A. Seder, and F.M. Gryna, Jr., *Quality Control Handbook*, 2nd ed., McGraw Hill, New York, New York, 1962. (The first edition was published in 1951).

4. W. E. Deming, *Out of the Crisis*, Massachusetts Institute of Technology, Center for Advanced Engineering Study, Cambridge, Mass., 1982, 1986, 507 p. This book is an absolute must in the library of management and anyone interested in implementing the management aspects of TQM!

5. Student, The probable error of a mean, *Biometrika* 6:1-25(1908).

6. Statistical treatment of test results, SCAN-G2:63 (Accepted - November 1962).

ISO 9000 certification
7. Rabbitt, J.T. and P.A. Bergh, The whys and hows of ISO 9001 certification, *Tappi J.* 75(5):81-84(1992).

8. Cox, J., Is mastering the confusion of ISO 9000 the key to the market place?, *Am. Papermaker* 55(6):20-23(1992).

EXERCISES

1. Given the following two sets of kappa numbers, calculate the mean and standard deviation of each set.

   31.0, 32.2, 29.3, 30.0, 29.3, 31.9, 31.0, 28.1, 27.0, 27.3

   29.2, 29.4, 30.0, 28.3, 29.7, 30.1, 29.7, 30.0, 30.7, 28.8.

2. Using Fig. 20-3 as a control chart, plot the following data and indicate where the process may be out of control and what the reason might be. These are kappa numbers taken every two hours of pulp going to the bleach plant.

29.2, 29.6, 30.4, 28.9, 30.5, 31.1, 30.9, 31.4, 32.4, 32.2, 34.6, 33.9, 35.3, 34.8, 35.5, 35.8

3. Draw a Pareto chart of the following failures or downgrades during the month of December for rolls of paper.

low burst strength, 10	low brightness, 4
poor sizing, 3	poorly wound, 18
defects in paper, 2	too narrow, 5

4. Using Fig. 20-1, what percentage of values from a normal distribution lie within 3 standard deviations of the mean? Within 4 means?

APPENDIX--GENERATING RANDOM DATA WITH A NORMAL DISTRIBUTION

The data in Fig. 20-3 was computer generated by the program below written in BASIC. This program can be used to generate random data to simulate a process where all of the variation is random, and not possibly controllable by the operator. Indeed, attempts to control the process only increase the variability. The data is easily plotted by *importing* the file into a program such as QuatroPro or generating it within QuatroPro with its random number generator.

```
5    OPEN "B:OUTPUT" FOR OUTPUT AS #1
10   RANDOMIZE
20   MEAN = 30: REM THIS IS THE DESIRED MEAN
30   S = 1.5: REM THIS IS THE DESIRED STANDARD
     DEVIATION
40   FOR I = 1 TO 40:  REM GENERATES 40 RANDOM
     NUMBERS WITH A NORMAL DISTRIBUTION
50   Y1 = 2*RND-1
60   Y2 = 2*RND-1
70   R = Y1*Y1+Y2*Y2
80   IF (R>1 OR R=1) GOTO 50
90   FAC = SQR(-2*LOG(R)/R)
100  NUM = FAC*Y1:  REM THIS IS THE RANDOM
     NUMBER WITH MEAN OF 0 AND S OF 1
110  VAR = NUM*S+MEAN :REM THIS IS VARIABLE
     WITH THE DESIRED MEAN AND S
120  PRINT VAR,: PRINT#1, VAR: REM PRINTS VARI-
     ABLE  TO  SCREEN  AND  FILE  FOR  GRAPHING
     LATER
130  STOTAL   =   (VAR-MEAN)*(VAR-MEAN)   +
     STOTAL:REM FOR CALCULATING ACTUAL S
140  MEANTOT = TOTAL + VAR: REM USED TO CAL-
     CULATE THE ACTUAL MEAN FOR THIS SET
150  NEXT I
160  PRINT: PRINT "ACTUAL MEAN OF THIS SET =
     ",(MEAN + MEANTOT/I)
170  PRINT "ACTUAL STANDARD DEVIATION OF THIS
     SET = ",SQR(STOTAL/(I-1))
```

Run

Random number seed (-32768 to 32767)? 4

```
30.95114  30.44624  32.16494  29.78161  29.29416  28.76899
29.95173  27.35375  29.23539  25.62650  31.90901  31.10704
30.99176  31.46926  28.06687  29.24346  26.97691  30.21155
27.33691  29.65250  29.20096  28.75179  29.42387  28.79731
30.03629  29.88838  28.27955  28.22769  29.65059  29.44141
30.13715  30.13631  29.69470  30.01222  30.00990  32.50525
30.65970  31.54469  28.77460  29.65343
```

ACTUAL MEAN OF THIS SET = 30.72326
ACTUAL STANDARD DEV. OF THIS SET = 1.454831
Ok

21

COLLOID AND SURFACE CHEMISTRY

21.1 INTRODUCTION

Much of what will be considered in this chapter is *surface chemistry*. Chemically, a *surface layer* (*interfacial* or *interface layer*) is the boundary between two phases. The boundary is a few molecules thick. Two examples are a cellulose fiber in contact with aqueous solution and an oil phase in contact with a water phase in an emulsion. Colloidal chemistry is one branch of surface chemistry. Surface chemistry and physics are important in colligative properties of solutions, capillary action, papermaking chemistry, water treatment, refining behavior, and foaming.

21.2 COLLOID CHEMISTRY

Colloids

A *colloid* is a **stable** combination of particles of one substance that are dissolved or suspended in a second substance; the particles have at least one of the principal dimensions (somewhat arbitrarily) from about 1 nm to 1 μm. (The time frame of stability is also somewhat arbitrary.) The small size of colloidal particles means that a high percentage of the molecules will be on the surface, so surface chemistry is of vital importance to their properties. A colloid may be considered to include the gas, liquid, or solid in which the particles are dispersed in addition to the particles themselves. The *dispersion medium* forms a *continuous phase*; both of these terms are used to describe the principal phase.

Since a gas in a gas does not exist as discrete phases, there are eight ($3 \times 3 - 1$) different types of colloids (Table 21-1). A *sol* is a colloidal system in which the dispersion medium is a gas or liquid. An *aerosol* is a sol with the continuous phase a gas. Fog is an aerosol of water droplets; smoke is an aerosol of solid particles. An *emulsion* is a sol in which the suspended particles are liquid droplets and the continuous phase is also a liquid; the two phases are immiscible, otherwise a solution would form. A *colloidal suspension* is a sol of solid particles suspended in a liquid. *Foam* is a colloidal system of gas bubbles dispersed in a liquid or solid. Styrene-butadiene copolymer and paper are both colloidal solids.

This size range includes soluble polymers of a few thousand g/mol at the small extreme to most fillers and pulp fines at the large extreme; weak chemical bonds can hold objects in this size range together because of the high surface areas.

Colloid surfaces

The surfaces of colloids are of utmost importance since the interaction of the particles with each other and the principal phase is of primary concern. The surface—to—volume ratio of a given particle shape is inversely proportional to the length of the principal axis. For example, in a sphere the volume increases in proportion to r^3 but the surface area increases only in proportion to

Table 21-1. Two phase colloidal systems.

Dispersed phase	Dispersion medium	Name	Examples
Liquid	Gas	Liquid aerosol	Fog, mist, hair spray
Solid	Gas	Solid aerosol	Smoke, dust, allergens
Gas	Liquid	Foam	Head on beer, foaming
Liquid	Liquid	Emulsion	Italian salad dressing, milk
Solid	Liquid	Sol, paste	Ink, mud, clay slurries
Gas	Solid	Solid foam	Foam rubber, styrofoam
Liquid	Solid	Solid emulsion	Pearl, ice cream
Solid	Solid	Solid dispersion	ABS, paper, pigmented plastics

r^2. A cube of 1 m on each side has a surface area of 6 m^2/1000 kg or 0.000006 m^2/g. A cube of 1 cm (1/100 the length of the larger cube) has a total surface area of 6 × 10^{-4} m^2/g or 0.0006 m^2/g (100 times the specific surface area of the large cube). For a small molecule like Al_2O_3, about 1 in 100 molecules will lie on the surface of a 0.1 μm particle and about 1 in 10 of a 0.01 μm particle.

It takes energy to create surface area of particles. The energy required is related to the surface tension of a material. (One speaks of *surface tension* of a liquid against a gas, usually its vapor, and *interfacial tension* of two liquids against each other.) The surface tension (γ) of water at 20°C is about 72 mJ/m^2. The work (W) or *surface excess free energy* (G) required to make new surface area (A) is given as follows:

$$\Delta G = \Delta W = 2\gamma A \qquad (21\text{-}1)$$

Table 21-2 demonstrates this concept with one mole of water at 20°C. On the first line of the table the one mole of water forms a single sphere. As this water is divided to form smaller spheres, the surface area increases, as does the surface excess free energy. Since the surface free energy is a bulk property, it has little meaning for very small particles; however, for the last entry, corresponding to separation into individual molecules, this value (40.3 kJ/mol) corresponds well with the heat of vaporization of water (40.7 kJ/mol at the boiling point and lower at room

temperature). This energy provides a driving force for the coalescence of small water droplets.

Table 21-2 is also a good reference for the approximate specific surface area of particles (based on particle diameter), for a first—order approximation (see the second footnote).

Emulsifying agents act to reduce the surface tension between two dissimilar liquids. Soap emulsifies oil and water by decreasing the interfacial tension of the oil—water interface via fatty acid molecules that have affinity for both phases. (The fatty acid may be called a *compatibilizer*.)

The high proportion of molecules on the surfaces of small particles means that colloidal particles behave much differently than larger materials. For example, the flat surfaces of kaolinite clay tend to be negatively charged, while the edges are positively charged below pH 7, which explains why clay tends to coagulate at low pH [it is described as a *house of cards* structure (Hunter, 1987)]. The edges of clay particles will absorb phosphate ions $(PO_4)^{3-}$ and obtain a negative charge. This negative charge causes them to repel each other in suspension, making the clay much easier to disperse (thinning). Calcium ions have an opposite effect (thickening).

Micelle formation

The classic behavior of soap molecules in water is a good example of a very practical application of colloidal chemistry. (Both individual

Table 21-2. Colloidal properties of one mole of water versus spherical particle size.

Particle radius	Particle number	Particle volume, cm^3	Particle area, cm^2	Total area, cm^2	Total surface[1] energy, J/mol	Specific area[2] m^2/g
1.63 cm	1	18	33.2	33.2	0.00048	1.8·10^{-4}
1.00 mm	4 300	4.19·10^{-3}	0.126	540	0.00619	3.0·10^{-3}
0.10 mm	4.30·10^6	4.19·10^{-6}	1.26·10^{-2}	5 400	0.0619	3.0·10^{-2}
0.01 mm	4.30·10^9	4.19·10^{-9}	1.26·10^{-5}	5.4·10^4	0.619	3.0·10^{-1}
1.00 μm	4.30·10^{12}	4.19·10^{-12}	1.26·10^{-7}	4.3·10^5	6.19	3.0
0.10 μm	4.30·10^{15}	4.19·10^{-15}	1.26·10^{-9}	4.3·10^6	61.9	3.0·10^1
0.01 μm	4.30·10^{18}	4.19·10^{-18}	1.26·10^{-11}	4.3·10^7	619	3.0·10^2
1.00 nm	4.30·10^{21}	4.19·10^{-21}	1.26·10^{-13}	4.3·10^8	6192	3.0·10^3
0.1925 nm	6.02·10^{23}	2.98·10^{-23}	4.66·10^{-15}	2.8·10^9	40300	3.0·10^4

[1] From Eq. 21-1.

[2] This is applicable for spherical particles of specific gravity 1.0; for other materials, divide this number by the specific gravity of the solid material for the given particle radius.

molecules and aggregates of soap molecules are colloids, but with different properties.) Soap molecules are sodium or potassium salts of fatty acids (Fig. 2-31). At very low concentrations soap particles are individually dissolved in the water and contribute to the colligative properties (such as increased osmotic pressure) according to the number of moles present.

As the concentration increases to the *critical micelle concentration, cmc*, soap particles abruptly agglomerate into spherical structures called *micelles*. Micelles are groups of about 20 to several hundred molecules with the nonpolar (hydrophobic) hydrocarbon tail toward the center and the polar (hydrophilic) head toward the surface in contact with water (Fig. 21-1). The cmc is typically 10^{-2} to 10^{-3} mol/dm^3 for ionic surfactants and 10^{-4} mol/dm^3 for nonionic surfactants. Several cmc values are given later in the section on surfactants.

If the solution is diluted, the micelles dissociate into individual molecules; hence they are termed *association colloids*. It is the hydrophobic center that "dissolves" grease and oil when washing with soap. Well above the cmc, micelles begin to change shape and may form liquid crystals and other mesomorphic shapes. The biological double—layer membranes of cells are also related to micelles.

Solution properties of colloids

Since Freundlich's classic work (1926), colloids are classified as *lyophilic (solvent loving) colloids* when the dispersed phase is dissolved in the continuous phase. (If the dispersed phase is a very large or crosslinked polymer, such as some starches or gelatin, then a *gel* may form. In the case of a gel, the dispersed phase is also continuous.) *Lyophobic (solvent hating) colloids* occur when the dispersed phase is suspended in the continuous phase; they are thermodynamically unstable. If the dispersed phase is water, the two terms are *hydrophilic colloids* and *hydrophobic colloids*, respectively. In papermaking systems, the solvent is almost always water.

These terms are arbitrary, but some properties of these colloids can be contrasted after Alexander and Johnson (1949); they are sometimes referred to as reversible and irreversible

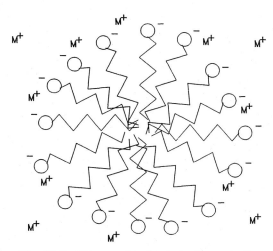

Fig. 21-1. Soap molecule array in a micelle.

colloids, respectively. High concentrations of lyophilic dispersed phase (such as water—soluble proteins) are usually stable and often increase the viscosity of the continuous phase substantially; rehydration after desiccation leads to dispersion (reversible); coagulation requires very high levels of electrolytes and usually leads to a viscous residue. On the other hand, lyophobic dispersed phases (such as clay particles, metal oxides, or silver halides in water) are easily precipitated by low levels of electrolytes, are irreversibly coagulated on drying, tend to give granular precipitates, and have high levels of light scattering (if they are not translucent).

In 1857 Michael Faraday observed and described some of the fundamentals of colloidal systems. He studied colloidal gold (a well-known occurrence at that time) prepared by reduction of gold salts and proved that the solutions contained very small gold particles. The resultant colloid was ruby red or blue in color; it exhibited what was to be later called the *Tyndall effect*, a bluish opalescence that appears when a bright, narrow beam of light passes through the solution (due to light scattering). He showed that the addition of salts caused the particles to increase in size (aggregate) and even to deposit gold in the vessel. He found that the addition of gelatin would prevent the gold from forming large particles with the addition of salts. We will see that similar phenomena occur in papermaking systems with fiber fines or fillers in place of the gold.

Most colloids have charged surfaces from anionic or cationic functional groups, which play central roles in their behaviors. Vegetable fibers have negative charges on their surfaces due to the carboxylic acid groups of hemicellulose and cellulose. Mineral fillers and many polymers used in colloidal systems also have charges. These charges may be positive from cationic groups (from polymers with amine groups or metal cations) or negative (from phosphate groups in potato starch, carboxylate groups in many anionic polymers, sulfate groups, and carbonate groups, to name several examples). The amount and type of surface charges are critical to the stability and solubility properties of colloids. The type and amount of electrolytes (especially the amount and valence of metallic cations), pH, and temperature play critical roles in the solution properties of charged colloids. These factors will be considered in detail presently.

Microscopy of colloids

The resolution of optical microscopy is limited by the wavelength of light. The resolution is about 0.15 μm under the best of conditions and perhaps double this for routine work (about 500 to 1000 magnification since the unaided human eye can distinguish objects about 0.1 mm apart). A technique known as *dark field* microscopy can be used to detect small particles, but with a low level of detail. In this technique, light is directed perpendicular to the direction of viewing. With no particles present, the light does not reach the objective. Lyophobic (insoluble and of different index of refraction as the solution) particles (0.02 μm and larger) can scatter enough light that some of it will be visible. Although detail is lacking, the motion of colloids can be followed to study flocculation, sedimentation, and electrophoretic mobility. Zsigmondy used dark field microscopy extensively in his classic work on colloidal gold for which he received the 1925 Nobel Prize. Light scattering is an important property of colloids and will be considered in detail. Scanning electron microscopy (SEM) can be applied to colloid particles but not colloid solutions.

Ionic interactions and (external) electric fields

There are four electrokinetic phenomena of charged surfaces in solution; the first is much more important than the others. *Electrophoresis* is the movement of charged surfaces (with associated ions and water) in the stationary liquid induced by an external field. *Sedimentation potential* is the charged field generated by charged particles moving in a stationary liquid. *Streaming potential* is the generation of an electric field by movement of the liquid along stationary charged surfaces. Finally, *electroosmosis* is the movement of liquid relative to stationary charged particles induced by an external electric field.

The net charge on particles in aqueous solutions can be measured by the process of electrophoresis, which is the basis of some wet end chemistry sensors. When an electric field is applied to two metal plates across a solution of charged particles, the particles will move in response to the force set up by electrostatic attraction (between oppositely charged objects) or repulsion (between like charged objects) (Fig. 21-2). Negatively charged particles move toward the positive plate (anode), and positively charged particles move toward the negative plate (cathode).

Uncharged particles or particles with no *net* charge do not respond to electric fields. For most charged particles it is possible to adjust the solu-

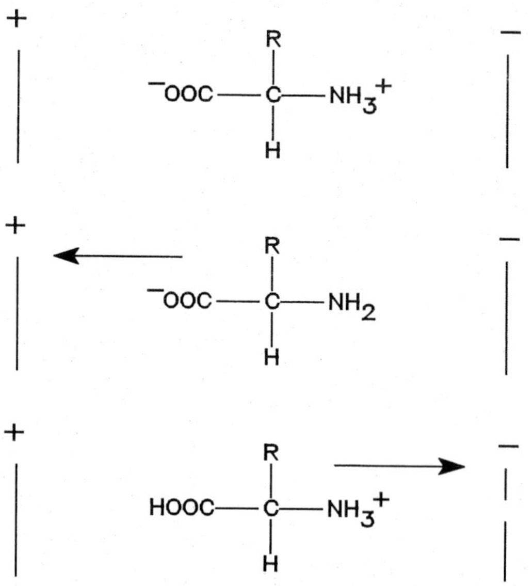

Fig. 21-2. **An amino acid at its isoelectric point (top), as an anion (center) at higher pH, and as a cation at lower pH. The arrows show the direction of movement for the applied charge.**

tion so that the particle has no net charge. This is accomplished by changing the pH or changing the composition of electrolytes in solution. When a particle has no net charge it is at its *isoelectric point* and it will not move in the electric field. Electrophoresis is an important tool of biochemistry used to separate groups of proteins and other molecules; it is also one method of measuring zeta potential (effective surface charge).

Many amino acids are *zwitterions*, dipolar ions or inner salts, and represent structures with no net charge. Figure 21-2 (top) shows a representative amino acid at its isoelectric point (around pH 8). Although it is has two opposite charges, there is no net charge. With increasing pH (middle), the amine spends more and more of its time deprotonated (since this is an equilibrium reaction) and the molecule has a slight to moderate negative charge. With decreasing pH the carboxylate salt is protonated (a hydrogen atom is bound to it) a higher and higher percentage of the time to give the molecule a positive charge. The solubility of proteins (and other materials) is usually lowest at their isoelectric point.

Electrical double layer

Most materials involved in papermaking have charged surfaces. For example, wood fibers in aqueous suspensions have a net negative surface at pH above 3.5 due to the carboxylic acid groups of cellulose and hemicellulose, which exist as their carboxylate salts. We will consider the behavior of such a fiber in the development of surface charge properties.

If metallic salts are added to a suspension of wood fibers, the wood fibers will absorb (bind) a certain portion of them (this is an equilibrium reaction) and become less negatively charged or even positively charged depending on how many are absorbed (Fig. 21-3). The positively charged ions (*counter—ions*, since their charge is counter or opposite to the particle) and water form the *electrical double layer* at the surface.

In 1879, Helmholtz hypothesized that counter—ions would be held firmly to charged surfaces, depending on the type and concentration of counter—ions in the surrounding solution. In 1910, Gouy and, in 1913, Chapman independently concluded that the counter—ions would spread diffusely from the surface due to thermal motion.

In 1924, Stern modified the double layer to include some tightly bound ions of Helmholtz and some loosely bound ions of the Gouy—Chapman model (Fig. 21-3). The model holds that the distribution of counter—ions is more highly concentrated at the surface and less concentrated with increasing distance from the surface to yield a Boltzmann distribution (of statistical mechanics).

The "thickness" of the double layer is given by $1/\kappa$ where κ is the ionic radius from Debye-Hückel theory. $1/\kappa$ is a function of the dielectric constant, ε; the universal gas constant, R; temperature, T; Faraday's constant, F; and the concentration (c, mol/m^3) of each ionic species (i) of valence z as follows:

$$\frac{1}{\kappa} = \left(\frac{\varepsilon RT}{F^2 \sum c_i z_i} \right)^{0.5} \tag{21-2}$$

Zeta potential

The *slipplane* is the plane defined by the distance at which the structure with its chemically bound water and ions moves in bulk through the solution as indicated in Eq. 21-2. It is the plane at which the *zeta potential* is valid. The zeta

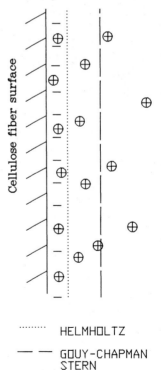

Fig. 21-3. Electrical double layer.

potential is the "modified" or "effective" surface charge. The equation shows that with increasing concentration of electrolyte, the slipplane will contract toward the particle surface and the zeta potential will become less negative, until the isoelectric point is achieved, and may even become positive. Each type of particle (softwood fibers, hardwood fibers, parenchyma, TiO_2 filler, clay filler, etc.) will have its own zeta potential or even a zeta potential distribution.

The valence of the cation is of exponential importance in decreasing the negative charge on anionic surfaces (as determined by precipitation). This was independently observed by Schultze in 1882 and Hardy in 1900 and is referred to as the Schultze—Hardy rule. Later work showed that charge neutralization is proportional to the sixth power of z (more detail is found below). Other factors such as the size of the ion and the size of the hydration layer modify these numbers somewhat. Despite this enormous influence other factors may have strong influences as well, such as the formation of coordinate complexes or insoluble precipitates. For example, Cu^{2+} ions form strong complexes with amines and may impart more of a charge to polyamines than Al^{3+} ions because the copper—amine interaction is much stronger than ionic interactions. In ammonia solution, copper ions exist as $Cu(NH_3)_6^{2+}$. High ionic strength (corresponding to conductivity > 1000 μmho) suppresses the zeta potential and its usefulness as a control parameter in papermaking.

Coagulation

Under many conditions, colloidal particles can clump together to form large particles that are no longer stable to suspension; this is the process of coagulation. Coagulation is analogous to the formation of a liquid from a gas or the precipitation of a solid from solution. In all three cases it is largely a matter of the thermodynamic stability of the products and activation energies to be overcome. The stability of a suspension depends on whether the repulsive forces between particles are larger than the attractive forces between particles and the energy of the various states. In some cases the solution is unstable to begin with, and it is only a matter of time before coagulation occurs. Conversely, an unstable or stable colloid

can sometimes be formed from two continuous phases by a large amount of agitation.

Attractive forces (Levine, 1978) are due to any of the normal secondary chemical forces (as opposed to covalent bonds) that hold materials together. The forces between identical particles are usually induced dipole-dipole interactions such as the weak *van der Waals force* (based on the work of 1873), which is proportional to $1/r^7$; they depend (only to a small degree) on the properties of the solvent. Even a material like argon is subject to van der Waals forces, since these are the attractive forces between atoms of argon in its liquid or solid form. The energy of this interaction is the *London* (based on the work in 1930) or *dispersion* energy and is proportional to $1/r^6$. The magnitude of this interaction at the most favorable distance is typically 0.5 to 2 kcal/mol. These forces are very dependent on distance but act only at short range. Hydrogen bonding may be another important attractive force at higher consistencies with energies of 2—8 kcal/mol.

Two molecules are prevented from being "sucked into each other" by repulsive forces, *Born repulsion*, predicted by the Pauli exclusion principle, which act at shorter range than van der Waals forces and are proportional to $1/r^8$—$1/r^{18}$.

The bottom curve of Fig. 21-4 shows how these forces act in the gas-liquid equilibrium of a pure material. The bottom curve is in the form of the *Lennard—Jones 6—12 intermolecular potential* (for Ar or N_2), where $\nu = 4\varepsilon[(\sigma/r)^{12} - (\sigma/r)^6]$. ($\nu$ is the sum of the very short range repulsive potential and short—range attractive potential, ε is the depth of the minimum potential, and σ is the intermolecular distance where $\nu = 0$.) At a given intermolecular distance two molecules form a (somewhat) stable dimer (or liquid or solid) depending on the temperature (average kinetic energy of the species). At high temperatures the atoms or molecules overcome the binding energy and exist as a gas. Liquids are incompressible because of the high dependence on distance of the short—range Pauli exclusion repulsive forces.

In colloids with charged surfaces, relatively long range repulsive forces have two origins: stabilization of the particles by the solvent (hydrogen bonds and London energy) and double-layer ionic interactions. In the former, if particles have

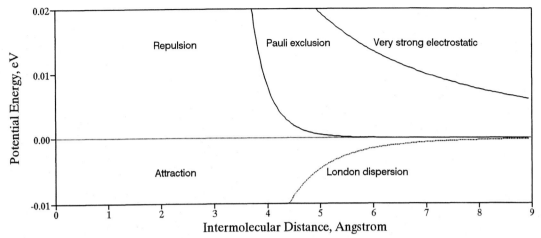

Fig. 21-4A. Individual potentials for Pauli exclusion, London, and electrostatic repulsion.

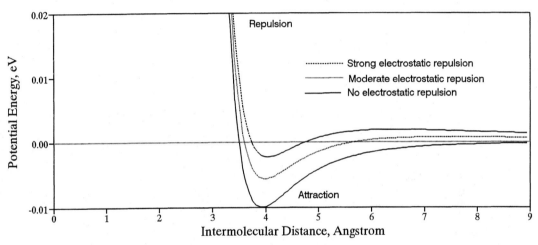

Fig. 21-4B. Lennard-Jones 6-12 intermolecular potential for Ar (bottom curve) and modifications with moderate (middle curve) and strong (top curve) electrostatic repulsion of colloid particles.

large layers of hydration, they cannot approach each other. (In the case of micelles, it has been shown that these colloids form because of the solvent-solvent interaction, which is increased when the hydrocarbon chains of soap molecules are "kept together" which is much larger than the nonpolar—nonpolar interaction inside the micelles.)

The ionic interactions are electrostatic forces that obey Coulomb's law. The forces are proportional to $1/r^2$ and the energy of interaction is proportional to $1/r$. These are relatively weak forces that act over relatively long ranges. The dielectric properties of the solution may be instrumental to the types of reactions that occur; this is

not critical in dilute aqueous solutions but may be very important as the solids content of the sheet reaches 50% and higher in the dryer section. (The exact interaction of electrical double layers of two particles can be considered with various degrees of additional sophistication.)

The additional consideration of electrostatic interactions in Fig. 21-4 is considered in the upper two curves. The electrostatic forces operate over a much longer range than the other forces. With moderate electrostatic charges (middle curve), the long—range repulsive forces can be overcome. With high electrostatic charges (top curve), the long—range repulsive forces cannot be easily overcome. When two particles combine, the electrostatic charges rearrange, because they must occur at the surface of particles. This rearrangement will mean that the entire short—range attractive potential will be realized.

In fact, it is not necessary to consider the Pauli exclusion forces when considering the stability of colloids (since these act only at short ranges after the weak chemical attraction forces have caused coagulation or coalescence). In the DLVO (from Deryaguin, Landau, Verway, and Overbeek) theory, colloid stability is considered in terms of the long—range repulsive forces and the short—range attractive forces.

With increasing temperature, pure materials go from solid to liquid to gas (decreasing particle size). With increasing temperature, colloids often go from solutions to solids (protein coagulation, such as occurs when cooking an egg) or from small particles to large particles. The apparent contradiction lies in the fact that increasing temperature in colloidal reactions is overcoming repulsive forces, whereas increasing temperature in the vaporization of a pure material is overcoming attractive forces. Increased agitation or increased temperature may cause colloids to form from larger particles if the London potential is weak, just as in a liquid.

At great distances there is no interaction between identical colloid particles. As they approach each other, the electrostatic repulsive forces act at relatively long distances. Pushing the particles together causes the van der Waals force to increase until the particles attract each other and combine. (The Pauli exclusion repulsive forces act to prevent the particles from approaching closer than in normal solids and liquids.)

Therefore, coagulation depends on decreasing the electrostatic repulsive forces and/or forcing particles to be in close proximity to each other. Small amounts of polymers can sometimes be used to prevent coagulation. For example, gelatine prevents coagulation of gold particles by *steric stabilization*, whereby particles are prevented from getting close to each other by adsorbed layers of gelatin. Work by Neuman, Berg, and Claesson (1993) indicates that **steric stabilization of cellulosic surfaces (that does not behave according to DLVO theory) may be much more important**

than ionic interactions. The contribution of this phenomenon in papermaking chemistry needs more consideration in future work. This finding explains why the use of the zeta potential is not always helpful in papermaking chemistry and why too much polymer can reverse flocculation.

The electrostatic potential depends on the usual factors of electrolyte concentration and composition, pH, temperature, and degree of agitation. For example, when the zeta potential approaches zero, the electrostatic forces are decreased; even with a zeta potential of zero there may be electrostatic repulsion as the particles tend to polarize each other and the double layers affect each other; therefore, the maximum coagulation often occurs within ± 5 mV of the isoelectric point (i.e., where the zeta potential is zero). A compressed double layer (resulting from higher levels of electrolytes and/or high—valence cations) also decreases electrostatic repulsion; generally the effectiveness toward coagulation of $M^+:M^{2+}:M^{3+}$ is 1:75:625 according to the *Schulze-Hardy rule*. Within a given valence, a large ionic radius is a little more effective toward coagulation than one of small ionic radius.

Overbeek (1952) demonstrated these concepts in his work (reprinted in Hunter, 1987 and Shaw, 1980). For example, the *critical coagulation concentrations, ccc,* for a sol of As_2S_3 is 58 mmol/L of LiCl, 51 mmol/L of NaCl, and 50 mmol/L of KNO_3 for monovalent ions; 0.72 mmol/L of $MgCl_2$ (10% higher for the sulfate salt) and 0.69 mmol/L of $ZnCl_2$ for divalent ions; 0.093 mmol/L of $AlCl_3$ (barely higher for the sulfate salt) and 0.08 mmol/L for $Ce(NO_3)_3$. The same approximate ratios are realized with two other sols as well in this work, with one of the sols positively charged and precipitated by Cl^-, SO_4^{2-}, and other negative ions. [Theoretically the cmc decreases to the sixth power of z (Hunter, 1987); for $z = 1, 2, 3$ this is 1:64:729.] If too much electrolyte is present, charge reversal is possible and coagulation may be reversed. It is interesting to note that aluminum behaves according to this rule despite its complex chemistry with water. One wonders how polyelectrolytes behave since they may have numerous charges per molecule.

Up to this point this section has considered **identical** colloidal particles. With different particles there can be very strong attractive electrostatic forces if some particle are negatively charged and others are positively charged. In this case coagulation is very quick.

Flocculation in the papermaking system

Flocculation in the papermaking system is analogous to coagulation except that a charged polymer is used either to decrease the repulsive forces or to act with several particles at one time (bridging) to achieve aggregation. Consider the use of cationic polymers in a suspension of wood fiber fines with negative surface charges. Without polymers, the fines repel each other due to the like charges. The addition of a low molecular weight cationic polymer will decrease the zeta potential so that fines may flocculate.

Flocculation can also occur by the *bridging* mechanism, whereby small amounts of cationic polymer attach to the particles and create local neutral charges (a *patch*), although the overall particle may have a net negative charge. Two patches, each on a different particle, may form a bridge even though the overall zeta potential is still appreciably negative.

The molecular weights of polymers are important to their use. Usually a minimum molecular weight is required for a particular application. High molecular weights allow direct bridging of particles. Sometimes two component polymer systems are used, with one polymer having a low molecular weight (perhaps a few tens of thousands) to provide suitable charge on a part of the surface and a second, high molecular weight molecule used to provide bridging between patches. High shear forces in solution must be avoided if high molecular weight is required in an application since shear forces will cleave large molecules. The use of branched polymers (for a given overall molecular weight) decreases their effectiveness toward bridging particles.

Like coagulation, flocculation can be reversible with shear forces (mixing or agitation) especially if the flocculation is weak such as when barely enough polymer is added to initiate flocculation. If too much polymer is added, charge reversal or steric stabilization may occur.

One area that has been overlooked by the paper industry, but offers great potential, is the use of diblock and other block polymers. This approach has the potential of allowing fillers to attach directly to fibers and fines to increase opacity. Too, fillers (of different indexes of refraction) could be made to attach to each other to increase light scattering.

[Strictly speaking, coagulation and flocculation are both the clumping of small particles (like or not) together into groups; the terms are usually used interchangeably by colloid chemists, although coagulation implies a stronger interaction that is generally not reversible. (Admittedly, this is contradictory in terms of how these behave to shear forces in the papermaking system.) *Coalescence* is the fusion of two small particles into one large particle where the original particles are indistinguishable. Oil droplets of an oil in water emulsion may coalesce into larger oil droplets. The large materials may settle from solution, if they are more dense, by *sedimentation* or rise to the top, if they are less dense, by *creaming*.]

21.3 POLYELECTROLYTES

Polymers with charged groups may be called *polyelectrolytes*. Anionic groups are usually carboxylate salts and occasionally sulfonate or phosphate salts; cationic groups are usually amines or quaternary ammonium salts (quats). The *charge density* (often measured as percentage of monomers containing a charge) is an important criterion for polyelectrolytes. A polymer with a high charge density (above 30%) tends to be completely "unraveled" in solution.

21.4 SURFACE TENSION AND SIZING

Mechanism of sizing action

The Lucas (1918) equation [also referred to as the Washburn (1921) equation] shows that the liquid penetration (depth of penetration, l) into a material is a function of the size of capillaries or pores in a material, r; the contact angle of the liquid with the solid, θ; the surface tension of the liquid, γ; the viscosity of the liquid, η; and time, t. The equation is:

$$l^2 = rt\gamma \cos \theta / 2\eta \qquad (21-3)$$

Lucas used this equation to calculate pore sizes in paper based on the speed of the rise of

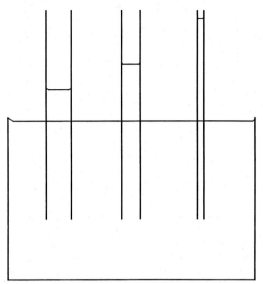

Fig. 21-5. Demonstration of the capillary effect with glass tubes of various diameters placed in water.

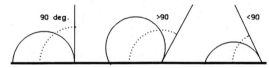

Fig. 21-6. Three contact angles of a droplet on a flat surface.

organic liquids in paper strips. Lucas used organic liquids since they would not change the pore sizes by swelling or hydration. By assuming complete wetting ($\cos \theta = 1$), lower limits of the pore size in paper were determined to be 0.5 to 6 μm. Later investigators used water and obtained pore sizes of about one-tenth of these values.

Capillary action (which is the ultimate depth of penetration) is demonstrated by the height of water in glass tubes of decreasing diameter (Fig. 21-5). It is the same force that draws fibers together during papermaking (Fig. 17-5) and is also relevant to paper drying. Capillary action is the result of a pressure gradient between the water phase and the air with water vapor phase. The pressure gradient (P) is equal to the surface tension (γ) difference between two phases (α and β) divided by the radius (R) of the liquid phase for a spherical interaction. The surface tension of water against air (see Section 33.1) is 73 dyn/cm at 18°C, 66 dyn/cm at 60°, and 59 dyn/cm at 100°.

The capillary equation, derived independently by Young and Laplace in 1805 (Levine, 1978), is:

$$P^\alpha - P^\beta = 2\gamma/R$$

This equation is modified to give the actual capillary rise for water by solving for the force of water and the wettability of the glass by the water. If the water completely wets the glass, then $\theta = 0°$. Here, r is the capillary tube radius, ρ is the density of a phase, g is the acceleration due to gravity, and h is the height of the water.

$$\gamma = \frac{1}{2}(\rho_\alpha - \rho_\beta)ghr/\cos \theta$$

For example, for a water-air interface at 20°C in a capillary tube with an inside diameter of 0.5 mm, the capillary rise will be 6 cm if the glass is completely wetted. The capillary rise is generally the most accurate method for measuring the surface tension of liquids.

The contact angle is a measure of the adhesive force of the liquid with the surface relative to the cohesive forces in the liquid. The following equations indicate this relationship with examples in Fig. 21-6:

if adhesive > cohesive then $0° \leq \theta < 90°$

if adhesive < cohesive then $90° < \theta \leq 180°$

For example, water on a freshly waxed surface "beads up," an example of a low adhesive force between water and the surface (wax).

In the case of water in glass tubes, the adhesive force is higher than the cohesive force (the contact angle is < 90°) and the meniscus is concave; this leads to capillary rise. In the case of mercury in glass tubes, the adhesive force is lower than the cohesive force (the contact angle is > 90°); this leads to capillary depression.

Cellulose is easily wetted by water, and water on cellulose gives a low contact angle that approaches zero; this causes the water to be drawn into the capillaries. The purpose of internal sizing is to present a surface to the water that results in a high contact angle. This is accomplished with nonpolar functional groups. Internal sizing agents are generally *amphipathic* (having polar and nonpolar moieties on the same molecule) materials. The polar portion (more significantly, it is chemically reactive) attaches to the fiber or fines directly through a covalent bond or indirectly through a mordant such as alum. Since the adhesive strength of water to the sized surface is now less than the cohesive strength of water, the water does not penetrate the pores of the fibers. (Surface sizing with starch physically plugs the surface capillaries of paper, which is its chief mechanism of sizing even though starch is hydrophilic.)

Some observations will be considered in light of this theory. Hexane has very little cohesive strength, that is, a low surface tension, and easily penetrates paper, whether sized or unsized. Lowering the surface tension of water (by use of a surfactant) will cause a decreased capillary action (sizing would be observed to improve) directly as indicated in Eq. 21-3; however, this will cause the contact angle to decrease, which has the effect of decreasing the observed sizing. Usually the latter effect induces a larger relative change, so that the use of surfactants will decrease the observed sizing to a modest degree (see Chen and Biermann, 1995, in Chapter 22).

Inverse gas chromatography has been used to characterize surface energy; in pulp and paper this has been accomplished with treated and untreated pulps (Pyda et al., 1993). Another technique that has been used is dynamic contact angle analysis (Huang et al., 1995)

The contact angle of buffer solutions at various pH values on individual wood fibers was studied by Jacob and Berg (1993). Their study indicates that the change in surface ionizable groups brought about by variation of pH affects the wettability of the fibers. More recent work of some workers refutes this approach, however.

Capillary condensation (Stamm, 1962)

Above 90% relative humidity, wood and paper can adsorb small amounts of water vapor as free water due to surface tension effects of water. The relative humidity must reach 99.5% to fill pit chambers and the tips of fibers and 99.9% to fill fiber lumens of wood.

A water droplet with a positive radius (concave surface) has a slightly higher vapor pressure than a flat surface of water. A water droplet with a negative radius (convex) has a lower vapor pressure, which allows condensation to occur. The radius of the pores in which water will condense is calculated by a variation of the Kelvin equation:

$$r = \frac{2\gamma \overline{V}}{RT \ln(p_o/p)}$$

where r is the radius in meters, γ is the surface tension of the liquid (0.073 J/m^2 for water at 20°C), $\overline{V}$ is the molar mass of the liquid (1.8·10^{-5} mol/m^3 for water), R is 8.314 $J/(mol \cdot K)$, T is temperature in Kelvin (293), and p/p_o is the relative vapor pressure of the liquid.

At 20°C and 50% relative humidity the pore size is 1.6·10^{-9} (1.6 nm); at 90% RH the pore size is 1.02·10^{-8} m or 0.01 μm; at 99% RH the pore size is 0.1 μm; at 99.9% RH the pore size is 1 μm. Possibly, pocket ventilation at the last few dryer cans must be particularly effective (to give a sufficiently low vapor pressure of water) to get the last water out of paper during drying.

[There is an equation (derived by Oswald in 1907) that is analogous to the Kelvin equation but applies to the solubility of particles in solution. A small particle actually has a slightly higher solubility than a larger particle. If the solubility is sufficiently high, with aging, small particles decrease in size while larger particles increase in size. This is the basis of aging precipitates in some analytical chemistry experiments. The surface energy of solid-solution interfaces is about 1 J/m^2.]

21.5 SURFACTANTS

Introduction

Surfactants are surface—active agents that aggregate near or have a strong effect on modifying the interface between two materials. This occurs because of their dual nature: hydrophobic and hydrophilic. The hydrophilic moiety may be *anionic* (carboxylate, sulfate, or phosphate), *cationic* (quaternary ammonium salt), *ampholytic* (cationic and anionic), or *nonionic* (Span®, Tween®, Triton®) depending on the type of charge(s) carried, if any. Performance of surfactants depends strongly on their surface activity and micelle formation.

Figure 21-7 shows the effects of some chemicals on the surface tension of water. The effect of NaOH in increasing the surface tension is unusually high among the electrolytes. The hydrophobic moiety must be large enough to "resist" the aqueous phase. For example, 5 g/L of ethanol will lower the surface tension of water at 20°C (73 mJ/m²) by about 2 mJ/m², 5 g/L propanol by 5 mJ/m², 5 g/L butanol by 7 mJ/m², 5 g/L pentanol by 22 mJ/m², and 5 g/L hexanol by 35 mJ/m².

Anionic surfactants are generally less expensive than cationic ones, though the latter sometimes have biocidal action. Span and Tween materials (Fig. 21-8) are formed from hexahydric alcohols (Span, derived from sorbitol), fatty acids, and alkene oxides (Tween). Tween 80 is called Polysorbate 80 when used in pharmaceutical and food products. The Span materials tend to be less water soluble than the Tween materials. The Triton family is shown in Fig. 21-8. Surfactants are often named according to their use. The same material could be a detergent, a wetting agent, an emulsifier, or a dispersant. The majority of all surfactants are detergents. Detergents are often alkyl sulfates. Alkyl aryl sulfates used to have widespread use, but they are not readily biodegradable. Detergents are used with silicates to help bind interfering metal ions, phosphates to

help prevent flocculation (scum formation), and other materials to keep them free-flowing.

Critical micelle concentration, cmc

The cmc of surfactants depends on several factors (Shaw, 1980). The most important factor is the strong water—water interaction (the *hydrophobic effect*) that is allowed when water does not interact with individual fatty acid chains in solution; the nonpolar—nonpolar interaction in the center of the micelle does not provide the thermodynamic energy required for micelle formation. Most ionic surfactants have a cmc that depends on the hydrocarbon chain. In Table 21-3 all of the C-12 ionic surfactants have a similar cmc. Nonionic surfactants usually have lower cmc values.

Longer chain fatty acid portions increase the hydrophobic effect; an example was already given with the lower chain alcohols. After about 18 CH₂ groups, an additional effect is not noted since these would coil on their own in solution by themselves. For example, the cmc of sodium

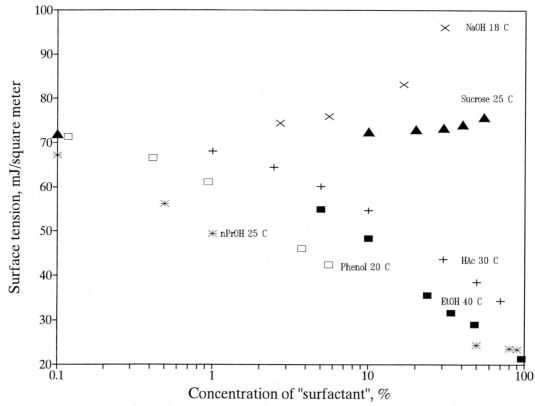

Fig. 21-7. The effect of various chemicals on the surface tension of water. Data from CRC *Handbook of Chemistry and Physics*, 67th ed.

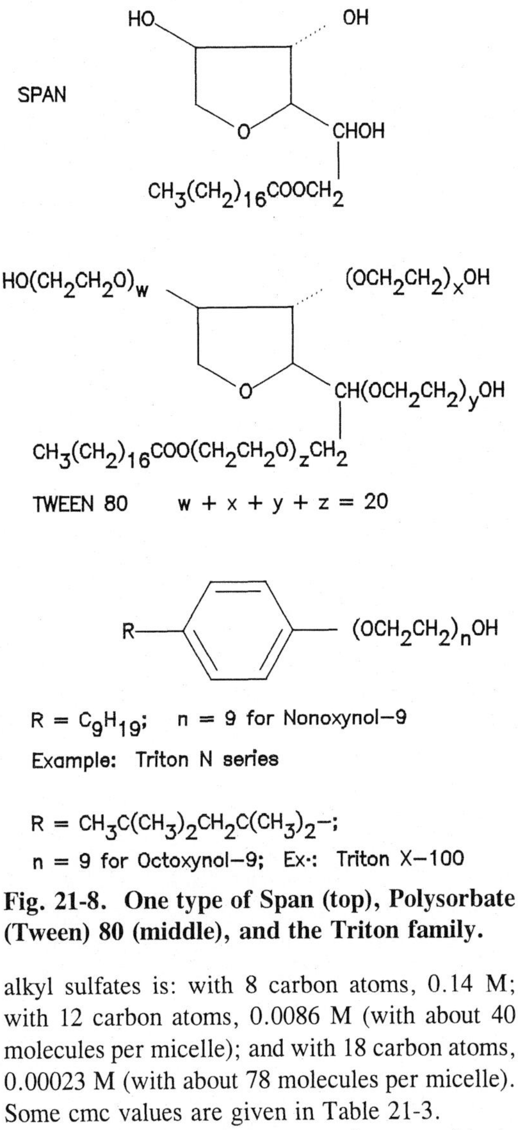

HO. OH

SPAN

$CH_3(CH_2)_{16}COOCH_2$

CHOH

$HO(CH_2CH_2O)_w$ $(OCH_2CH_2)_xOH$

$CH(OCH_2CH_2)_yOH$

$CH_3(CH_2)_{16}COO(CH_2CH_2O)_zCH_2$

TWEEN 80   w + x + y + z = 20

R—⟨ ⟩—$(OCH_2CH_2)_nOH$

R = C_9H_{19};   n = 9 for Nonoxynol–9

Example: Triton N series

R = $CH_3C(CH_3)_2CH_2C(CH_3)_2$—;

n = 9 for Octoxynol–9; Ex·: Triton X–100

Fig. 21-8. One type of Span (top), Polysorbate (Tween) 80 (middle), and the Triton family.

alkyl sulfates is: with 8 carbon atoms, 0.14 M; with 12 carbon atoms, 0.0086 M (with about 40 molecules per micelle); and with 18 carbon atoms, 0.00023 M (with about 78 molecules per micelle). Some cmc values are given in Table 21-3.

Addition of salts decreases the cmc of ionic surfactants because it has a tendency to screen the electrostatic repulsion at the surface of the micelle. In the case of sodium alkyl sulfate with 12 carbon atoms, the cmc decreases to 0.0056 M in 0.01 M NaCl, 0.0015 M in 0.1 M NaCl, and 0.0007 M in 0.3 M NaCl. Increasing the temperature will increase the cmc somewhat due to thermal motion.

Many surfactants that form micelles have a solubility below their cmc. Thus at a certain temperature, known as the *Krafft point*, the solubility increases dramatically since the micelles are quite soluble. For the sodium alkyl sulfates the Krafft temperature increases from 16°C for 12

carbon atoms, to 30°C for 14 carbon atoms, to 56°C for 18 carbon atoms.

Monomolecular films

Some surfactants that have low solubility in water form monomolecular films on the surface of water. Very small amounts of oil form iridescent surfaces on water due to the thin films formed. Usually these are alcohols or carboxylic acids of long—chain fatty acids. These are routinely used to combat mosquitoes in wet areas, decrease evaporation from bodies of water, etc. These concepts have application to internal sizing of paper. The hydrophilic portion is oriented toward the water and the hydrophobic tails form a layer on top, where they stabilize each other.

Sulfated polyoxyethylated alcohols

Schwuger (Rosen, 1984) is a good reference for the structure versus performance of sulfated polyoxyethylated alcohols and was used for the following paragraphs. The use of ethylene oxide (EO) to form alkyl ether sulfates from fatty alcohols provides a surfactant with unique characteristics, including favorable interfacial and application properties especially with mixtures of other surfactants. Although alkyl ether sulfates are expensive anionic surfactants, they are widely used in industry. Alkyl ether sulfates have characteristics of both anionic and noncharged surfactants.

Ether groups in anionic surfactants lower the Krafft temperature. While the Krafft point (in this case defined as the temperature of solubility of a 1% solution of surfactant) of Ca dodecyl sulfate is 50°C, one ethylene oxide group (the anion is $C_{14}H_{29}OCH_2CH_2OSO_3^-$) lowers it to 15°C. A second ethylene oxide unit lowers it to below 0°C.

Metal counter ions also play an important role in the action of a surfactant. Tetradecyl (C_{14}) sulfate has a Krafft point of 21°C as the sodium salt and 67°C as the calcium salt. For this reason, in many applications (such as laundry detergents) ionic surfactants are often used with complexing agents or other methods to bind or remove higher weight metal ions. Since calcium ions are often unavoidable in pulp and paper processing slurries, anionic surfactants often contain ethylene oxide units when used in our industry.

Calcium or magnesium ions can improve the washing effect of alkyl ether sulfates (relative to sodium ions); this effect is attributed to compres-

Table 21-3. Cmc values of various surfactants (10^{-3} mol/L).[1]

Base molecule	Total number of carbon atoms in the fatty acid chain						
	8	10	12	14	16	18	20
Anionic							
$CH_3(CH_2)_n$-COO$^-$ Na$^+$, 25°			12		1.0		
$CH_3(CH_2)_n$-CH$_2$OSO$_3^-$ Na$^+$, 40°C	140	33	8.6	2.2	0.58	0.23	
$CH_3(CH_2)_n$-CH$_2$OSO$_3^-$ Na$^+$, 25°C			8.1				
$CF_3(CF_2)_n$COO$^-$ K$^+$	29						
Cationic							
$CH_3(CH_2)_n$-NH$_3^+$ Cl$^-$, 25°C				13			
$CH_3(CH_2)_n$-N(CH$_3$)$_3^+$ Br$^-$, 25°C		68	20				
$CH_3(CH_2)_n$-N(C$_5$H$_5$)$_3^+$ I$^-$, 25°C		5.6			0.94		
Ampholytic							
$CH_3(CH_2)_n$-NH$_2^+$COO$^-$, 25°C			18	1.8	0.18	0.018	
Nonionic							
$CH_3(CH_2)_n$(OCH$_2$CH$_2$)$_6$OH	9.9	0.99	0.087				
Span materials						15 (in benzene)	
Tweens 20 EO units			8				

[1] From Shaw (1980), Everett (1988), and Hunter (1987).

sion of the electrical double layer at the interface. Tetradecyl 3 EO sulfate is said to be a particularly effective laundry detergent and does not require the use of phosphate to complex Ca^{2+} and Mg^{2+}.

Other aspects of surfactant structure (Rosen, 1984)

With ionic surfactants, the cmc decreases with increasing level of electrolyte concentration. With nonionic ethylene oxide surfactants, electrolyte level has little effect.

In pure nonionic ethylene oxide surfactants, an increase in the number of ethylene oxide units results in an increase in the Krafft point; an increase in surface tension at the cmc also results.

The nature of the aliphatic chain of aliphatic ethylene oxide surfactants is very important. When three short chains are used instead of one long chain for surfactants containing 10 carbon atoms in the aliphatic chain and 7 EO units, the cloud point is 22°C compared to 75°C and the cmc is 6 mM compared to 0.85 mM. This unusual behavior is attributed to the formation of a bilayer micelle in double-chain hydrophobes.

When sodium dodecyl sulfate (SDS) in water at 25°C is subjected to increasing NaCl concentration (0-0.5 M), the aggregation number in the micelles increases from about 70 to 145. In 0.80 M NaCl, the aggregation number is over 1000 and the micellar shape is a rod.

21.6 FOAM

Foams are large air bubbles separated by thin films of liquid with thicknesses in the range of those of colloids. Foams are relatively unstable with lifetimes that increase with increasing viscosity of the liquid phase, which decreases the rate of water drainage at the junction of three gas bubbles, the *plateau borders*.

Foams are important because they are used in froth flotation deinking and because they are detrimental in many other aspects of pulping and papermaking (such as in brown stock washing). Froth flotation of minerals uses a surfactant (the *collector*) to attach to the mineral particles, making them hydrophobic, and a *foaming agent* to stabilize the foam that is formed.

Foam formation agents

Agents used to induce foam are selected empirically, although a few rules can act as guidance. Usually mixtures of surfactants have a synergistic effect toward inducing foam formation. Surfactants used near their cmc are particularly effective since they can form monolayers.

Proteins are sometimes useful in foam formation, as anyone who has beaten an egg white or cream knows. These foams are formed by denaturing of the proteins.

Defoaming agents

Chemical defoamers have been discussed on page 203. Foams can sometimes be destroyed by blowing hot air on them. This lowers the viscosity of the liquid and helps evaporate liquid at the interfaces. Some surface active chemicals that do not stabilize the foam can displace the surfactants that do stabilize the foam. Thus, low molecular weight alcohols or fatty acids (4-6 carbon atoms) can act as defoamers. Ca^{2+} may decrease the foaming tendency of fatty acid soaps by rendering them insoluble, like taking a bath in hard water.

21.7 EMULSIONS

Emulsions easily fall into two categories: an oil in water (O/W) or water in oil (W/O) emulsion, depending on the continuous phase. The type of emulsion that forms depends largely on the volume ratio of the two materials, with the more abundant phase forming the continuous phase. The emulsion can be diluted with the same type of material as the continuous phase. Milk, an O/W emulsion, is diluted with milk; mayonnaise, a W/O emulsion, is diluted with cooking oil.

As mentioned, emulsions are stabilized with surfactants (emulsifying agents) that have affinities for both phases; these decrease the energy required to make new surfaces between the two phases, the interfacial surface tension. Destabilization of an emulsion is an example of coagulation (or coalescence) of colloids.

21.8 LIGHT SCATTERING BY COLLOIDS

Light scattering (Shaw, 1980) is an important method of monitoring and studying colloid suspensions. While the theory may not be necessary for most pulp and paper technologists, some aspects

will be presented here as a foundation for future discussions. (The wavelength of visible light is about 0.45 μm for blue light and 0.7 μm for red.)

There are three recognized types of light scattering of individual particles. Small particles (with each dimension less than about 5% of the wavelength of the light) act as point sources of light and obey Rayleigh scattering, for example, a solution of large polymers such as proteins. Light scattering by larger particles in a dispersion medium of similar index of refraction is Debye scattering. Light scattering by larger particles in a dispersion medium of much different index of refraction, such as in printing papers of high opacity, is Mie scattering.

Rayleigh scattering

Rayleigh described the scattering of light by smoke in 1871. Light of intensity I_o and wavelength λ interacting with a small particle of polarizability α induces oscillating dipoles in the particle. The particle will then emit light of the same wavelength, but in any direction. If the light is not polarized, the light emission, I, is given as a function of the distance from the particle, r, and the angle relative to the incident beam, θ as:

$$\frac{Ir^2}{I_o} = \frac{8\pi^4\alpha^2}{\lambda^4}(1+\cos^2\theta) = R_\theta(1+\cos^2\theta)$$

In reality, there are many particles, and these will cause secondary scattering, so the overall scattering may be more complex in relatively concentrated materials.

Since Rayleigh scattering is proportional to $1/\lambda^4$, blue light is scattered about eight times more than red light. When the incident light is white, high levels of scattering filter the blue light from it, so the remaining incident light is red. However, the scattered light itself appears blue. Thus the sky is blue during the day, but the setting or rising sun appears distinctly red (do not look directly into the sun as this may cause permanent damage to the retina of the eye). Dilute or low—fat milk has a blue appearance, as do the deep footprints in some types of cold, fluffy snow.

This type of light scattering can be used to give absolute molecular weights (Debye, 1947), but we will not consider this now. Suffice it to say that the amount of light scattered (and, therefore, the sensitivity) is proportional to the square of the molecular weight. The lowest molecular weights observable are around 50,000 and the maximum molecular weight can be around 10^7 for spherical particles. Doppler light scattering (where the shift in wavelength is observed) can be used to determine whether or not the particles are in motion.

Scattering by large particles

Light scattering by large particles is very complicated due to the high level of destructive interference. Mie described the physics of light scattering by spherical particles in 1908.

Neutron scattering

Neutron scattering (Everett, 1988) is analogous to light scattering, but the smaller wavelength of neutrons (0.1 to 1 nm) can be applied to concentrated dispersions impossible to study with light scattering, especially in solutions of high opacity.

21.9 ANNOTATED BIBLIOGRAPHY

1. Alexander, A.E. and P. Johnson, *Colloid Science*, Oxford University Press, Oxford, 1949.

2. Debye, P., *J. Phys. Chem.* 51:18(1947).

3. Einstein, A., *Investigations on the Theory of Brownian Movement*, Methuen, 1926; Dover, 1956.

4. Everett, D.H., *Basic Principles of Colloid Science*, Royal Soc. Chem., Letchworth, 1988.

5. Freundlich, H., *Colloid and Capillary Chemistry*, Methuen, London, 1926 (from the 1909 German). Ibid, *Capillary Chemistry* (Ger.), Vol. 2, Akad. Verlag, Leipzig, 1932.

6. Hiemenz, P.C., *Principles of Colloid and Surface Chemistry*, 2nd ed., Marcel Dekker, New York, 1986, 815 p. This is an advanced book that is a very good starting point for serious theoretical considerations of the title topic.

7. Huang, Y., D.J. Gardner, M. Chen, and C.J. Biermann, Surface energetics and acid/base character of sized and unsized paper handsheets, *J. Adhes. Sci. Tech.* 9(11):1403-1411(1995).

8. Hunter, R.J., *Foundations of Colloid Science*, Vol. 1, Clarendon Press, Oxford, 1987; p 95 gives Overbeek's work.

9. Jacob, P.N. and J.C. Berg, Contact angle titrations of pulp fiber furnishes, *Tappi J.* 76(5):133-137(1993). Contact angles are determined as a function of pH to indicate the forms of ionizable groups on the surface of four different types of pulp. (It is not a titration.) These results are a little esoteric; more pragmatic information would be obtained by determining zeta potentials of these fibers as a function of pH.

10. Levine, I.N., *Physical Chemistry*, McGraw-Hill, New York, 1978. Probably any introductory physical chemistry book could be used.

11. Lucas, R., *Kolloid-Z* 23:15-22(1918).

12. Neuman, R.D., J.M. Berg, and P.M. Claesson, Direct measurement of surface forces in papermaking and paper coating systems, *Nordic Pulp Paper Res. J.* 8(1):96—104(1993).

13. Overbeek, J.T.G., in *Colloid Science*, Kruyt, H.R., Ed., Vol. 1, Elsevier, Amsterdam, 1952.

14. Pyda, M., M. Sidqi, S. Keller, and P. Luner, An inverse gas chromatographic study of calcium carbonate treated with alkylketene dimer, *Tappi J.* 76(4):79-85(1993).

15. Rosen, M.J., Ed., *Structure/Performance Relationships in Surfactants, American Chemical Society Symposium Series No. 253*, Washington, DC, 1984. This is very useful.

16. Shaw, D.C., *Introduction to Colloid and Surface Chemistry*, 3rd ed., Butterworths, London, 1980. This work is good mix of theory and explanation. Page 184 has Overbeek's work reprinted.

17. Stamm, A.J., *Wood and cellulose-liquid relationships*, North Carolina Ag. Exp. Station Tech. Bull. No. 150, Sept., 1962, 56 p.

This is a very good overview of the wood and fiber-water relationship by one of the pioneers of the field. It includes adsorption, swelling, capillary flow, and diffusion by wood and paper.

18. Tanford, C., *Physical Chemistry of Macromolecules*, Wiley and Sons, New York, 1961, 710 p. This is the classic reference on the title subject of use to some advanced areas of wet end chemistry.

19. Washburn, E.W., *Phys. Rev.* 17:273-283(1921).

EXERCISES

1. Describe how soap removes grease during washing.

2. What is electrophoresis?

3. What is the significance of charge on wood fibers as related to wet end chemistry?

4. Does the DLVO theory describe steric stabilization of colloid suspensions? Explain.

5. Describe the difference in mechanism of action of internal and external sizes.

6. How does propanol break up foam? How does hot air break up foam?

7. Why is the sky blue?

8. What is zeta potential? Does one value describe the zeta potential for the entire wet end? Explain.

APPENDIX

Sedimentation rate of spherical particles

The sedimentation rate of particles will influence the drainage rate of pulp slurries. The driving force for sedimentation is the mass minus the buoyancy of an object. The rate of sedimentation will depend on the terminal velocity. An object of mass m, with a specific volume v, and density ρ, in gravity field g will have a terminal velocity as follows:

$$m(1 - v\rho)g = f\,dx/dt$$

Stokes' law applies for dilute colloids and gives $f = 6\pi\eta a$ for spherical particles so that:

$$dx/dt = [2a^2(\rho_2-\rho)g]/9\eta$$

Particles with irregular shapes will have slower sedimentation for a given mass (volume) since the surface area increases. Frictional ratios of particles (to the equivalent volume sphere) are given in some locations (Shaw, 1980, p 21).

Equation 21-1 is similar to one of the above, except that the force originates from electrostatic attraction instead of gravitational action.

Brownian motion and radius of gyration

Einstein (1926) studied Brownian motion (named after the botanist who studied this phenomenon), which is the random motion of particles in liquids and is due to thermal energy. By treating it as a "random walk" the displacement of a particle along an axis is, on average, given as a function of its diffusion coefficient, D:

$$\overline{x} = (2Dt)^{0.5}$$

D is related to f as $Df = kT$.

This theory is used to predict the behavior of polymers in solution. Each segment of the polymer is treated as a step of the random walk with an angle of (for atoms with tetrahedral angles between substituents) 109.5°. The end—to—end distance of the molecule is then given as a function of the length between backbone atoms and the number of such atoms as follows:

$$l(2n)^{0.5}$$

Flory (1969) made an approximation for the root mean square of the radius of gyration for many carbon-backbone polymers as $<r^2>^{1/2}$ (in nm) $\approx 0.06\,M^{1/2}$ where M is the molecular weight of the polymer (Flory, P.J., *Statistical Mechanics of Chain Molecules*, Interscience, New York, 1969). As already mentioned, polymers with a high charge density do not behave in this random pattern since the like charges will repel each other.

Miscellaneous

The HLB scale was developed by W.C. Griffin (1954), *J. Soc. Cosmet. Chem.* 5:249.

22
PAPERMAKING CHEMISTRY

22.1 INTRODUCTION

Some practical applications and methods of wet end chemistry have been discussed in Sections 8.4—8.6 and 14.7. Fiber slurries and fiber mats have been considered in Sections 17.8 and 17.9. Chapter 18, Polymer Chemistry, is also of use since many wet end chemicals are polymers. Chapter 21, Colloid Chemistry, is of utmost importance in this area. Some of the practical aspects of papermaking chemistry in this chapter must be taken with a grain of salt since the field is based more on empirical results than theory. In practice, the concepts are often used qualitatively and not quantitatively.

In traditional wet end chemistry one is almost always interested in the interactions of colloids in aqueous solutions. For sizing molecules, wet and dry strength agents, and other functional additives, retention must occur for these molecules to be effective in the final paper. Wet end chemistry has dealt primarily with interactions in aqueous solutions and, by inference, it is often, at least subconsciously if authors and reader are not careful, assumed that these interactions occur in the dry sheet. In fact, rosin does not give sizing with alum in paper until most of the water is removed. Undried paper will not have sizing. If the same interactions are present in the wet end of the paper machine as in the dry sheet, one would expect to have sizing in the wet sheet. But how can a wet sheet be resistant to water? The interactions between pulp fibers and paper additives in the dryer section are completely different from those in the wet end since free water is not present in the dryer section. While the two processes must be considered in terms of separate mechanisms, the two processes are intricately related in the overall papermaking process.

The fact that the term dryer chemistry (even by any other name) does not exist in the pulp and paper literature shows the inherent bias in pulp and paper thinking. An appropriate term to cover both areas with less bias might be "papermaking chemistry." For example, current thinking about sizing with alkylketene dimer (AKD) or alkyl succinic anhydride (ASA) involves two steps. Size molecules must be retained on the sheet in association with fibers (as opposed to fillers) under conditions that are not reactive (to prevent hydrolysis), but they must be reactive in the dryer section so that they form covalent bonds to the fibers. When one understands the contradiction of not being reactive initially but becoming reactive later on, then it may be easier to design overall papermaking chemistry conditions that allow this rather than limiting oneself to wet end considerations.

Even this overly limits papermaking chemistry since papermaking chemistry really starts with the pulping operation. In practice papermaking chemistry considerations should begin at least at the refining stage (see the discussion following references 14 and 15 on page 156.) Also, many factor relate to the physics of the system.

The field of papermaking chemistry is inexact due to its complexity. This problem is magnified by the lack of specific information in much of the literature describing the polymers that are used. Work that does not give detailed information about the polymers used is almost useless! This situation reflects the willingness of mill personnel to rely on manufacturers to run their mills.

Many polymers, especially those with high charge densities, should be used in dilute solutions of 0.5 to 2% to facilitate their dispersion throughout the stock. Polymers with high molecular weights are generally added in locations that avoid high turbulence (for reasons to be discussed).

One final word of advice. Each mill is different and will have different problems and considerations. NSSC and CTMP mills may have a fairly high carryover of pulping chemical and lignin degradation products that must be handled differently than other mills. Any mill that has a buildup of electrolytes, which is likely as systems become more and more closed, will have particular problems with decreased electrostatic interactions where they are desired. Mills using unbleached fiber may be able to take advantage of the lignin of their pulps in papermaking chemistry systems, although this technique needs much development (Zhuang and Biermann, 1993).

Fines

Fines are the materials that pass through a 76 μm (200 mesh) screen. This is essentially the definition, since the Britt jar test and Bauer—McNett schemes both use these screens. Fines include *pulp fines* (parenchyma cells, especially from hardwoods and nonwood pulps, and small fiber fragments), fillers, and most dissolved or colloidal additives. Materials larger than this are fibers and large fiber fragments. Fines typically have five times the surface area of fibers per given unit of weight, but it may be as much as ten times. Fines also tend to swell and adsorb more water than fibers. Some typical fines levels in stock are 40—60% for newsprint, 30—55% for fine papers, and less than 25% for linerboard.

Anionic trash refers to materials that contribute to the anionic charge of the stock, but are not intentionally added components. This might include lignin degradation products, pulping or bleaching chemicals, or materials carried over from secondary fiber, depending on the mill. The cations associated with these materials are most apt to be Na^+ ion, which, as we will see, do not (at low concentrations) have a marked affect on papermaking chemistry—thus, the lack of the term cationic trash.

Fillers

In addition to the fillers discussed in Section 8.4, some manufactures have plastic or plastic—encapsulated fillers. Others include aluminum oxide—based materials.

The shape of fillers is important for several reasons. It affects the light scattering properties (opacity and gloss), the calendering process, the drying rate, and the air permeability of the paper. The specific surface area of fillers ranges from 2 to 40 m^2/g for most fillers and is inversely related to particle size. Small particles, however, have a high tendency to aggregate. Ironically, larger diameter fillers may be more difficult to retain (although they will require lower levels of retention aids) due to their decreased specific surface. On the other hand, furnishes containing fillers of small particles are more difficult to give internal sizing to (in the final sheet) since there is a higher surface area competing for the sizing agent.

Polyacrylic acid or polyphosphates are often used, with anchoring cations, to decrease the zeta potential (described below) of some fillers. This helps disperse the fillers but may interfere with their aggregation (retention) during drainage. It may be useful to characterize each additive in terms of the amount of a standard polymer required to cause flocculation in order to see if the raw materials change with time, as when a problem occurs. Even if changes do not occur in the raw materials, this knowledge will allow one to focus on the real problem.

Kaolin clay

Kaolin clay has an empirical formula of $Al_2O_3 \cdot 2SiO_2 \cdot 2H_2O$. Although there are several forms of the mineral, kaolinite is the most abundant of these. It occurs in three morphologies that are available to the pulp and paper industry; these are nondelaminated (booklet or standard hydrous), delaminated (platelet), and calcined.

Nondelaminated clay occurs in a stack (much like a roll of coins) of hexagonal (more or less) plates (Fig. 22-1). Partially delaminated clay is often used in wet end filler applications.

These stacks are broken into individual plates by the process of delamination (Figs. 22-2 and 22-3). Delaminated clay of ultrafine particle sizes is used in coatings to give high gloss. Fine particle sizes give lower gloss. Coarse particles are used for matte finished sheets.

Calcined clay is made by heating clay (as in a lime kiln) at approximately 1100°C. The structure is changed somewhat to become more porous; it is of lower overall density than standard clays (Fig. 22-4) [the material density (excluding the pores) has changed very little]. The brightness may be as high as 92—94% in premium grades (calcining increases the brightness by 3—4 points). Calcined clay is appreciably more abrasive (due to the formation of some mullite) and expensive than untreated clay. The opacity of calcined clay is higher than that of uncalcined clay due to the increase number of clay—air interfaces.

According to the China Clay Producers Association, the U.S state of Georgia produces

Fig. 22-1. Central Georgia nondelaminated kaolin clay. Courtesy of Evans Clay Co. Bar = 4 μm.

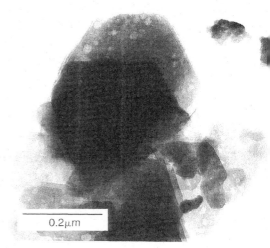

Fig. 22-3. Delaminated clay (Dry Branch Kaolin) slurried from $CaCl_2$ (pH = 5) to promote "house of cards" structure. TEM bar = 0.2 μm.

Fig. 22-2. Central Georgia delaminated kaolin clay. Courtesy of Thiele Kaolin. Bar = 5 μm.

about 60% of the world's kaolin. In 1992 Georgia produced 7.649 million short tons (industry survey) and the total U.S. production was 9.9 million tons (informal number from U.S. Bureau of Mines). Most of this comes from a narrow belt in the center of the state from Columbus to Augusta with particularly high concentrations from Macon to Augusta; it originated from erosion of mica, feldspar, and other minerals of the Piedmont Plateau. In addition to its use in paper, kaolin is used in plastics, rubber, paints, and numerous other products.

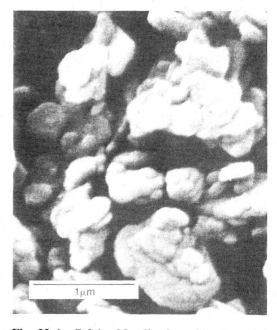

Fig. 22-4. Calcined kaolin clay. Courtesy of J. M. Huber Co. Bar = 1 μm.

22.2 POLYMERIC ADDITIVES

A wide variety of water—soluble polymers are used for papermaking; many were considered in Sections 8.4 and 8.5. Miscellaneous additional aspects of polymers are discussed here. Paper-making polymers can be categorized as naturally occurring polymers (polysaccharides or proteins), modified naturally occurring polymers, and synthetic polymers. They may also be classified as neutral, cationic, anionic, or *zwitterions*.

(The term *amphoteric* is often used to mean a material containing cationic and anionic groups; strictly speaking, however, it means a material that is both acidic and basic. HCO_3^- is amphoteric; it is not a zwitterion. A molecule with both an amine and carboxylic acid is amphoteric and a zwitterion. A molecule with an amine and a sulfonate exists as a zwitterion but is not strictly amphoteric, since the sulfonate is the conjugate base of a strong acid and not appreciably basic.)

Polymers with charged groups may be called *polyelectrolytes*. Anionic groups are usually carboxylates and occasionally sulfonates or phosphates; cationic groups are usually amines or quaternary ammonium salts (quats). The *charge density* (often measured as the percentage of monomers containing a charge) is an important criterion for polyelectrolytes. A high charge density (above 30%) tends to completely "unravel" the normally coiled molecules in solution.

Polyacrylamides

Polyacrylamides (PAMs) are relatively inexpensive polymers that are easily formulated to high molecular weights on the order of several million g/mol. PAM was first used in the paper industry in the mid—1950s. Anionic PAM usually contains about 5% polyacrylic acid groups formed by copolymerization of acrylamide and acrylic acid monomers or by hydrolysis of PAM homopolymer under conditions to convert some of the amide groups to carboxylate salts. PAM with as much as 50% acrylic acid is used. Cationic PAM is made by copolymerization of acrylamide monomers with cationic monomers (described below).

Ionic charges of amines as a function of pH

A wide range of amine functional groups are used in papermaking polymers. Amines or quaternary ammonium salts usually supply the positive charge of cationic polymers. We will consider the behavior of several model compounds to describe the charge on polyamines as a function of pH (Biermann, 1992). Quaternary ammonium compounds always carry a charge regardless of the pH since they do not react with protons.

Acid dissociation of protonated tertiary amines can be written as follows:

$$R_3NH^+ \rightleftarrows R_3N + H^+$$

Acid dissociation reactions of secondary and primary amines are analogous where one or two, respectively, of the R groups are protons. Protonated trimethylamine has a pK_a of 9.80 and triethylamine has a pK_a of 10.72 (Dean, 1985). The pK_a of protonated diethylamine is 10.93 and the pK_a of protonated butylamine is 10.61, which represent secondary and primary amines, respectively. Protonated ammonia has a pK_a of 9.24; its titration curve is shown in Fig. 14-4 on page 339. This indicates that most amines will be well over 90% protonated at pH 8 or above. This assumes the amines are three or more carbon atoms away from each other. This is true for starches made cationic with tertiary amines and most polymers with a charge density below 10%.

Consider the effect of amines' proximity on their protonation. Diprotonated 1,3—propane diamine has a pK_1 of 8.49 and pK_2 of 10.47. Polyethyleneimine (PEI) has some use in papermaking. Its structure is $-(CH_2CH_2NH_2)_2-$ with about 25% primary amines, 50% secondary amines, and 25% tertiary amines. Consider the ionization of the representative model compound pentaprotonated pentaminetetraethane, $NH_2CH_2CH_2(NHCH_2CH_2)_3NH_2$; consecutive pK_a values are 2.98, 4.72, 8.08, 9.10, and 9.67. Therefore, in the pH range of 6—8, only about 50% of the amines of PEI are protonated.

Amide groups ($R-CO-NH-\acute{R}$) are not cationic. The pK_a of acetamide (CH_3CONH_2) is -0.37. At pH 0.63 less than 10% of the acetamide molecules are protonated; at pH 6, less than one in a million are ionized.

This discussion uses model compounds to predict ionization of polyamines as a function of pH. It is useful to do a direct acid titration to determine this information empirically for a particular polymer.

Acetic acid serves as a model compound for the acid ionization behavior of many carboxylic acid groups (Fig. 14-3). Above pH 7, ionization is 100% and materials containing these moieties will have negative charges. At pH 4.75 ionization is about 50%; at pH 2.5 there is little ionization. This is true for cellulose fibers and carboxylated polymers. Sulfuric acid is a strong acid; sulfate groups will be completely ionized above pH 3.

In addition to ionic interactions of polymers with each other and particles, one should consider coordinate bonding and hydrogen bonding possibilities in their charge behavior and in their action in the papermaking system. For example, the free electron pair of amines readily forms coordinate bonds with many coordinating metal ions.

Starch

Potato starch has about 0.1 to 0.3% equivalent PO_4 corresponding to a degree of substitution of 0.002 to 0.005. Potato starch has little protein, while corn starch is about 0.3% protein. Fig. 22-5 shows starch granules from corn and potato. These granules and the starch therein are made soluble by the cooking process.

22.3 COLLOID CHEMISTRY

Classical colloid chemistry has been considered in detail in a separate chapter since this area has many applications in water treatment, emulsions, papermaking, and refining.

In papermaking chemistry, the colloid size range of 1 nm to 1 μm includes soluble polymers of a few thousand g/mol at the small extreme to most fillers and some pulp fines at the large

extreme; weak chemical bonds can hold objects in this size range together in solutions. Colloids can also attach to larger objects, such as fibers; however, two fibers will not be held together by weak chemical bonds in dilute suspensions.

Almost all colloids involved in the papermaking system have charged surfaces from anionic or cationic functional groups. The most obvious example are vegetable fibers that have negative charges on their surfaces due to the carboxylic acid groups of hemicellulose (4—O—methyl—glucuronic acid) and slightly degraded cellulose. Other particles such as fillers or many of the diverse papermaking polymers also have charges. These charges may be positive from cationic groups (of amine—type polymers or aluminum in alum—based systems) or negative (from phosphate groups in potato starch, carboxylate groups in many anionic polymers, sulfate groups, and carbonate groups). The amount and type of surface charge are critical to the stability and solubility properties of colloids. The type and amount of electrolytes (especially the amount and valence of metallic cations), pH, and temperature play critical roles in the solution properties of charges colloids.

Zeta potential

The zeta potential of each species to be retained during sheet formation should be within ± 10 mV of zero to give maximum retention,

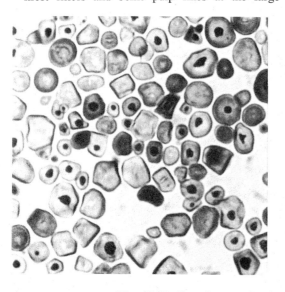
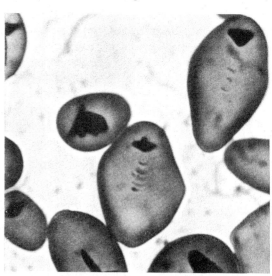

Fig. 22-5. Starch granules from corn (left) and potato. 600×.

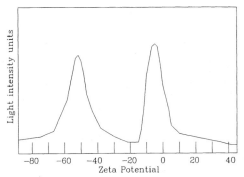

Fig. 22-6. Zeta potential distribution of two liposomes (phosphatidylserine and egg-phosphatidylcholine). Courtesy of Dr. P. Smejtek.

although this is less true when high molecular weight charged polymer systems are used for retention. Often people in the paper industry speak of zeta potential as a single value that applies to the entire system. Zeta potential distributions are demonstrated in Fig. 22-6. This distribution was taken with a device that measures light scattering with particle speed by the Doppler effect so that an entire zeta potential distribution for all of the components can be obtained. The role of zeta potential as a tool for the papermaker has been discussed (Stratton and Swanson, 1981). Again, each type of particle can be expected to have its own zeta potential that depends on pH, ionic strength, composition, and other factors.

Experience at a mill may allow the zeta potential to be used empirically, if not theoretically. Continuous monitoring of the zeta potential of paper machine stock may be useful for spotting upsets. Zeta potentials can also be used in colloidal titrations. Zeta potentials will be of limited use with additives that work by hydrogen bonding, such as nonionic polymers like neutral PAM and in cases where steric stabilization is appreciable. With high molecular weight polymers that work by bridging, only patches of positive areas are required (so it is said), so retention may be high, even though the overall zeta potential is fairly negative. High ionic strength (corresponding to conductivity > 1000 μmho) suppresses the zeta potential and its usefulness as a control parameter.

Flocculation in the papermaking system

Flocculation in papermaking chemistry is analogous to coagulation except that a charged polymer is used to either decrease the repulsive forces or act with several particles at one time (bridging) to achieve aggregation. Consider the use of cationic polymers in a solution of wood fiber fines with negative surface charges. Without polymer, the fines repel each other due to the like charges. The addition of a low molecular weight cationic polymer will decrease the zeta potential so that fines may flocculate.

Flocculation occurs by the *bridging* mechanism whereby small amounts of cationic polymer attach to the particles and create local neutral or positive charges (a *patch*), although the overall fine may have a net negative charge. Two patches, each on different particles, may form a bridge even though the overall zeta potential is still appreciably negative.

The molecular weights of polymers are important to their use. Usually a minimum molecular weight is required for a particular application. High molecular weights allow direct bridging of particles. Sometimes two—component polymer systems are used, with one polymer having a low molecular weight (perhaps a few tens of thousands) to provide a suitable charge on a part of the surface and a second, high molecular weight molecule used to provide bridging between patches. High shear forces in solution must be avoided if high molecular weight is required in an application since shear forces will cleave large molecules. Branching of polymers (for a given overall molecular weight) decreases their effectiveness toward bridging particles.

Like coagulation, flocculation can be reversible with shear forces (mixing or agitation) especially if the flocculation is weak when barely enough polymer is added to initiate flocculation. If too much polymer is added, charge reversal or steric stabilization may occur.

22.4 RETENTION, FORMATION, AND DRAINAGE

Retention occurs by direct mechanical trapping of relatively large particles (*filtration*) or by colloidal interactions that result in the *flocculation* of fines with large fibers or with each other to form large particles that are retained by filtration. Retention is a wet end effect. Overall retention should be 90—95% to keep the process economical. As long as excess white water is minimal and excess white water is treated to remove solid materials, this is easily achieved.

First—pass retention

First—pass (or *one pass*) retention is often separated into various subcategories of total solids including fiber retention, fines retention, and ash (i.e., filler) retention. The mechanisms of retention of these fractions are different, so it is useful to know the retention level of each. The first—pass fiber retention on paper machines using unbleached softwood fibers is usually over 98%. Furnishes using bleached hardwood and softwood fibers have a first—pass retention of about 95%. Occasionally the first—pass fiber retention is below 90%. The retention of fines is much lower, however, and should be the focus of retention strategy. A high first—pass retention means that there will be less solids in the headbox so that the drainage rate will increase. (Smaller particles are the ones less apt to be retained but contribute most to decreased slurry freeness if they build up in the system due to low retention.)

Fines also adsorb a higher percentage of additives such as polymers and sizing agents due to their high surface area. Additives such as ASA or AKD will hydrolyze in the white water and become useless. Hydrolyzed ASA causes pitch problems and actually decreases the level of sizing since it acts as a surfactant. Mills with excessive pitch problems on the paper machine may have to sewer excess white water rather than use it for purposes such as wash water in the bleach plant.

An increased first—pass retention implies that particles are held more firmly to the web. This results in secondary benefits of less sheet two—sidedness (since particles on the wire side will not be selectively removed) and decreased felt filling by fines and fillers. Tanaka et al. (1982) showed that the clay and TiO_2 distribution in the sheet z direction was more uniform with the use of retention aids. Systems with high first—pass retention are often free of slime—producing organisms.

Conversely, high retention values may mean overflocculation of filler particles, which leads to decreased light scattering and lower opacity. Extra filler may have to be added, the point of addition of materials may be changed, or the retention aid changed (by using an alternate product). Ideally, the fillers should flocculate with fibers and not with each other. Usually there is some tradeoff between retention of fillers and opacity. If high levels of fillers are used, if the filler is inexpensive, and if adequate strength can be achieved, this will probably not be a problem (these are the main advantages of alkaline papermaking for printing papers).

First—pass retention measurement

First—pass retention (Ret_{FP}) can be approximated by the relative change in solids content from the headbox (HB) stock to the white water (WW) tray divided by the initial solids content of the headbox. This is done by measuring the consistencies (c) at the headbox and in the white water. This could be done for total solids or fractionated samples. This method suffers from the fact that white water consistency may vary from place to place in the tray. Also, dilution from shower water may be appreciable.

$$Ret_{FP}, \% = (c_{HB} - c_{WW})/c_{HB} \times 100\%$$

A more comprehensive first—pass retention determination is accomplished by looking at the material entering the headbox compared to the sheet coming off the couch roll. Samples are taken from the headbox slurry, the white water slurry, and the couch roll. A couch trim sample can be used if one is confident that its composition represents that of the entire sheet at that point (ignoring moisture content variation). TAPPI Standard 269 pm-85, which uses the Britt jar, and Standard T 261 cm-90 describe this method in detail with example calculations. This method measures the fiber to fines ratio and uses the fibers as an internal standard with a correction for the fibers that appear in the white water. Moisture contents are not critical in the couch sample but the white water fraction must not be contaminated with fibers from any other source.

Colloidal retention

Since retention of the fines fraction is of central importance, the *Britt jar* (or *dynamic drainage retention jar*, page 151 and TAPPI Standard T 232) is a useful tool because it measures retention through a screen with 76 μm holes without formation of a mat; it measures colloidal interactions without interference from filtration mechanisms. (Initially, before filtration of fiber, the hole size in the forming media of the paper machine is about 300 μm.) It provides turbulence and shear forces (via the stirrer) that are always present during the papermaking process. The Britt jar consists of a precision controlled stirrer and a jar with the screen. Three baffles (present in the one shown Fig. 6-21) may or may not be used.

Even though retention might be high in a laboratory sheet former, retention may be low under actual papermaking conditions because of the turbulence and shear. The reason is that under certain conditions *soft flocs* may form that, by definition, break apart under moderate agitation. Conversely, *hard flocs* will withstand a moderately high level of shear and remain intact during the sheet forming process (Unbehend, 1976). Obviously, there is a continuum of floc strengths.

Hard flocs may decrease opacity and the quality of formation in the final sheet. Opacity is decreased by aggregation of mineral fines in large clumps, meaning that light travels through fewer interfaces and is not scattered as much. (However, below particle sizes of 0.3 μm the particles are so small relative to the wavelength of light that the light scattering coefficient decreases with decreasing particle size.)

When soft flocs are dispersed with high turbulence, they reform with the same level of retention in a papermaking system (Fig. 22-7 with alum). Hard flocs will be retained at a lower level after they are dispersed with high turbulence (Fig. 22-7, PAM with and without initial disruptive turbulence); this is hypothesized as occurring because polymer loops that extended in solution are "flattened" onto the surface of the particle and decrease particle—particle interactions. Also, the polymer is cleaved into shorter lengths. **When using high molecular weight polymers, one must be careful to avoid unnecessarily high shear to achieve maximum retention.**

Generally, soft flocs form with inorganic salts and polymers with molecular weights of a few thousand g/mol or less, whereas hard flocs form with high molecular weight polymers. [A few metallic elements (such as Al in polyaluminum chloride, PAC) can form high molecular weight materials.] Flocs that form only by surface patches (using low molecular weight cations without a high molecular weight anionic bridging polymer) are apt to be soft flocs (Unbehend, 1976). This means that high speed paper machines using inorganic fillers will require at least one high molecular weight polymer in the retention system or the use of the microparticle retention system.

The behavior of high molecular weight cationic polyacrylamide has been studied by Tanaka et al. (1993) in an elegant study. These workers found that cationic PAM that is first

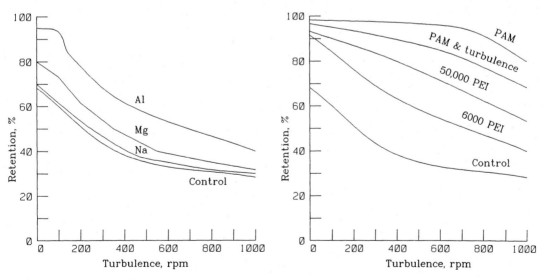

Fig. 22-7. Model floc retention with cations (left) and cationic polymers of decreasing molecular weight (with and without initial disruptive turbulence in the case of PAM) (after Unbehend, 1976).

added to fibers is redistributed to fines in the presence of hydrodynamic shear (they used polystyrene latex particles) within 1—2 min with some cleavage of the PAM. Much of the PAM was cleaved from the original molecular weight of $4 \cdot 10^6$ to about $4 \cdot 10^5$ depending on the amount of shear and mixing times.

Retention of low molecular weight materials such as sizing agents and dyes should probably be considered separately from suspended materials. Sizing agents can be anchored to the surfaces of fibers and fines with metallic salts and polyelectrolytes. Because of their small dimensions, they may continue to adhere to surfaces, even though the particles themselves are separated from each other (i.e., under turbulent conditions).

The Britt jar can be used with a variety of stirring speeds to simulate (somewhat) various degrees of turbulence. Often a corresponding stirrer velocity (in rpm) can be determined for a given commercial paper machine.

The above discussion indicates that retention aids can be classified as inorganic salts or polyelectrolytes. The pH will have an effect on the retention system; it should always be monitored and controlled. These materials and the chemistry of aluminum have been considered in detail in other sections of this book. Other aspects of polymers will be covered later in this chapter.

Carbohydrate—based retention aids

Starches have been described already. Cationic starches have a much higher level of retention (and affinity for cellulosic materials with carboxylate groups) than unmodified neutral starches, which decreases the biological oxygen demand in the excess white water. The addition of tertiary amines often occurs by reaction with compounds of the form $R_2NCH_2CH_2$—Cl (where R is an alkyl such as the ethyl group) where the starch molecule attaches through one of its R'OH groups by displacing the Cl atom. A typical DS is 2 to 5 amines per 100 glucose units (0.02 to 0.05). Too much cationic starch may decrease retention of fines by charge reversal. Even so, the relatively low charge density of cationic starch means they are not as effective retention aids as some highly cationic synthetic polymers. Cationic starches are subject to shear damage and are often added late in the approach system. Commercial anionic starches include only starch phosphates (potato starch), which are used from pH 4.5 to 6 with alum. Their absorption may be reversible unless a cationic promoter (of the quaternary amine type) is used.

Guar gum (from the seed) is a galactomannan (similar in structure to the top diagram of Fig. 2-24). They exist as rigid rods. Unaltered, they are retained by hydrogen bonding; when derivatized, using many of the same approaches used to derivatize starch, they act as retention aids.

Synthetic polymer retention aids

Neutral synthetic polymer retention aids include polyacrylamide (PAM), polyvinyl alcohol, and polyethylene oxide (PEO). PAM is the most widely used of these and adds to the dry strength. PEO forms hard flocs with mechanical pulps or in systems where sulfonated polymers are added.

Polymers are made anionic by carboxylate groups (RCOO⁻), sulfonate groups (R-SO₃⁻) or phosphate groups (RPO₃⁻). The polymers are prepared as metallic salts. The most common anionic polymer is made by introducing acrylic acid groups to high molecular weight PAM as mentioned above. Anionic polymers must be used with cationic intermediates in order to interact with negatively charged fibers and fines. At pH 4—6, this may be alum, which can cause charge reversal on the cellulose fibers. In dual polymer systems the cationic intermediates are small cationic polymers that act by forming patches. In the *microparticle systems* the cationic intermediate is cationic polyacrylamide with anionic bentonite or cationic starch with anionic colloidal silica. A recent method is the use of cationic starch with anionic polyaluminum chloride.

Cationic groups are introduced into polymers with amines or quaternary ammonium ions (tetra—substituted nitrogen atoms). Rarely sulfonium ions (tri—substituted sulfonates) or phosphonium ions (tetra—substituted phosphates) are used. Several monomers (Fig. 22-8) may be incorporated into PAM during its polymerization. The copolymers formed often have a cationic mole percentage (charge density) as high as 10%. DADMAC has the advantage of being a crosslinking agent. This is necessary since allyl monomers are difficult to polymerize (Biermann, 1992). In single polymer systems molecular weights of 1—10 million are used to allow direct bridging of anionic materials, but higher molecular weight polymers are very shear sensitive. If pyridine units are incorporated (Fig. 22-8), they may catalyze acylation reactions, such as the binding of AKD to cellulose.

Cationic/anionic polymers usually have more cationic than anionic functionalities. An example is potato starch with naturally occurring phosphate groups corresponding to a DS of 0.002, which is made cationic with quaternary ammonium salts corresponding to a DS of 0.03. Cationic/anionic polymers will work in a wide variety of systems but are not maximized for any particular system.

Retention systems

Retention by the single additive systems of flocculation and bridging has been mentioned. In *dual polymer systems*, cationic intermediates are small cationic polymers (such as PEI) that are added first to form patches on the surfaces of fibers and fines. Large anionic polymers (PAM) are added second to bridge these patches to induce flocculation. This system also adds to dry strength. This system forms very hard flocs that do not regain their strength fully if they are disrupted by high turbulence.

In the microparticle systems the large anionic polymer is replaced by bentonite or another colloidal silica with a negative charge. This system is a recent advancement in papermaking, although silica was used for AKD retention by Weisgerber (1958). It is said to result in relatively small, reversible flocs so that good formation is obtained; it is discussed in more detail in the section on practical aspects of alkaline sizing agents.

Formation

Formation is a function of fiber length, retention aids employed, and turbulence and is controlled by the size of the flocs. *Microflocs* are flocs with a size on the order of 100 μm; *macroflocs* are flocs that are visible to the naked eye. Macroflocs are a problem because they can lead to large local differences in the basis weight of the final paper. They occur when fibers begin to clump together. The reader will not be surprised to find that conditions that give good microflocs (desirable) will also lead to macroflocs. These conditions include the use of high molecular weight polymers in bridging systems.

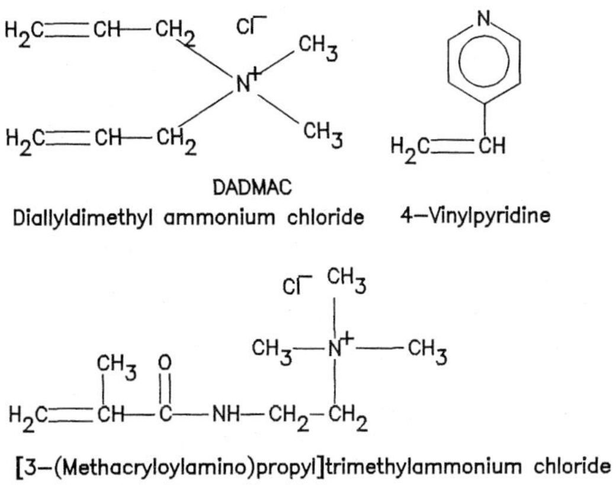

DADMAC
Diallyldimethyl ammonium chloride

4—Vinylpyridine

[3—(Methacryloylamino)propyl]trimethylammonium chloride

Fig. 22-8. Cationic monomers used to form cationic PAM copolymers.

Control of the floc size can be achieved through the use of controlled shear forces. This is accomplished, not by changing the paper machine, but by controlling where, how, and what polymers are added to the system. If polymer addition near the headbox gives large flocs, one may consider polymer addition at the secondary fan pump or some other place where **moderate** turbulence decreases flocculation.

Large polymers probably contribute to formation by limiting fiber—fiber contacts. A moderate charge density on the polymer prevents the polymer from coiling in solution. The argument has been given that large polymers increase the viscosity of the solution so that after turbulence is induced fibers cannot regroup as quickly.

Drainage, filtration, web consolidation

The mechanics of drainage on the paper machine have been discussed in detail in Chapter 9. Some additional aspects related to papermaking chemistry will be mentioned here. The freeness of wood fibers will increase with the addition of cationic polyacrylamide polymer to the isoelectric point. Flocculation of fines will also increase the freeness. Both of these effects will increase the drainage rate observed on the paper machine. (Drainage implies the removal of water; another term, *consolidation*, views the subject from the point of view of the fibers.)

High drainage rates mean faster paper machine operation and/or energy savings at the dryer section. Since increased solids in the web means higher web strength, paper machine speed can be increased without additional web breaks. Faster drainage also means the sheet will be set more quickly, before fiber flocculation can take place.

Water removal from a fiber suspension may occur by *filtration* or *thickening* (Parker, 1972). In the process of filtration, water is removed from the bottom of the suspension and the fibers pile on top of the forming material. A sharp boundary forms and the undrained suspension on top remains at the same original consistency. In *thickening* water removal occurs from the entire suspension and the entire suspension thickens at about the same rate with no sharp boundary. Thickening occurs if the fibers interlock such as by entanglement. The principal drainage mechanism on the paper machine is initially filtration. (However, the application of a pressure gradient in the suction boxes and presses causes drainage by thickening.)

Water removal by filtration results in the layered nature of paper. (Filtration theory is discussed in *Perry's* 7th edition starting on pages 19-65 and 20-98), The D'Arcy equation models flow through a porous media. The equation is:

$$U = \frac{1}{\eta R} \frac{dp}{dw}$$

where the filtration velocity, U, is a function of the specific filtration resistance, R, the viscosity of the liquid, η, and the pressure gradient, dp/dw. R is given in the next equation.

Filtration of paper webs has been developed by Campbell (1947) and Ingmanson (1954, 1959, 1963) based on treatment of the Kozeny—Carman (1937) equation, which gives R as follows:

$$R = kS^2 \frac{(1-E)V}{E^3}$$

where k is a constant based on the mat composition, S is the specific surface area of the solids per unit volume, E is the mat porosity, and V is the volume fraction of the web occupied by solids. Information on drainage is found in Rance (1980).

Drainage can be increased by other factors as well. An increased stock temperature gives a reduced water viscosity and has a similar effect as in pressing (page 237). The advantages of increased first—pass retention have been discussed. Retention aids, however, prevent fine material

from clogging the mat of fibers and may help draw cellulose microfibrils together, decreasing their hydration. **Drainage, then, goes hand in hand with retention**. Surfactants also increase drainage, but they can cause foaming and interfere with fiber—fiber bonding and sizing. Entrained air should be avoided.

An increase in chemically bound water (hydrogen—bonded, as opposed to free water) decreases drainage and may result by increased refining, which also generates fines, and the use of hygroscopic additives. The low level of chemically bound water in newsprint may partially explain why newsprint machines that use a furnish of 100 CSF can operate at speeds in excess of 15 m/s (3000 ft/min), but machines using mixtures of bleached softwood and hardwood (which have higher levels of chemically bound water) with much higher freeness are limited to much slower speeds. An extreme example of the problems of hydration would be drainage from a thick gel.

Sheet density is also a factor in drainage. A very dense sheet like glassine will presumably have smaller pores in which the water can flow. Sheet density is in turn a function of fiber flexibility, fines content, and other factors that influence drainage. Too, thick sheets (high basis weight) will drain more slowly as the water has a thicker mat through which it must percolate. The high bulk of mats of mechanical pulp fibers also helps to explain why newsprint machines can operate at high speed with a low CSF pulp.

Drainage in vacuum boxes (hivacs) is somewhat different from drainage by gravity. The reason is that uniform formation increases water removal since higher vacuums can be achieved. Holes or areas of low sheet density allow air to pass easily through the sheet. Therefore, too much retention aid will give poor drainage on the hivacs in addition to poor formation. Modern slotted vacuum boxes work much like foil blades in causing a pressure gradient and doctoring water that is pulled away so that rewetting does not occur. In the early 1960s, Attwood showed this by comparing dewatering on the paper machine with his own laboratory dynamic drainage tester.

Sometimes improvements in sheet formation will result in improved water removal on the vacuum boxes, shifting the dry line toward the headbox, while decreasing the drainage before and after the wet boxes. To a point, refining assists with formation by making more flexible fibers and fines, but increases chemically bound water and can even make fibers gelatinous. (Good formation may also hinder initial drainage.) Some additives also have this effect. **A delicate balance of effective retention additives and induced shear is required to balance good retention with appropriate formation and drainage**.

Drainage in the press section depends on capillary action so that water removal is proportional to the square root of the ratio of surface tension to viscosity of water according to Nissan (1954). This value increases about 40% from 38° to 99°C (100° to 210°F).

Drainage aids

Retention of fines to each other and directly to larger fibers (flocculation) will result in faster drainage since pores in the fiber mat will not become clogged. Most retention aids act as drainage aids; however, cationic materials (alum and polymers) seem to be particularly effective at increasing drainage. Excessive use of cationic materials will cause charge reversal and may decrease the drainage rate.

From the above discussion, it is not surprising that maximum drainage occurs for a particular type of particle near its isoelectric point. As with retention, the flocs must be strong enough to withstand the shear forces they will encounter in the paper machine. For example, while alum may appreciably increase the observed Canadian Standard freeness, alum may have a much smaller effect on drainage and retention on a high speed paper machine.

22.5 INTERNAL SIZING

Sizing is the ability of paper to resist wetting by a given fluid. Usually an aqueous fluid is considered. Internal sizing is important to the operation of the size press, coating operations, printing operations, and paper in service. Davidson (1975) reviewed the topic in detail.

Surface sizing is also used to resist wetting but has other purposes that internal sizing cannot achieve. These include controlling the surface porosity for printing, increasing the surface strength to avoid linting and to improve the picking resistance, and improving strength properties. This explains why many machines use both methods of sizing.

Inks have wide ranges of compositions and solids content. Their surface tension can be quite low relative to water; therefore, paper with internal sizing that is effective against water may not have suitable size properties for some inks.

The pulp and paper literature is replete with unproven dogma (as well as material that contradicts established chemical principles) on internal sizing. Hypotheses usually come after observations, and these are not specifically tested by additional experiments that may refute them.

Investigators often say that the purpose of alum in rosin sizing is to orient the rosin with the hydrophobic tail outward; true, but if one sprays a sheet of paper with rosin (to insure retention) and then dries it, there will be no sizing. Rosin orientation is a consequence of anchoring; no anchoring, no sizing. (Besides, without anchoring, the carboxylic acid groups would prefer to hydrogen bond to the surface rather than not.) It is beneficial to try to think of situations (or experiments) that argue or test the accepted hypotheses. Unfortunately, interesting stories are accepted too easily by the pulp and paper industry as fact. Often what one perceives to be "too complicated" or "outside of one's understanding" is really outside of anyone's understanding.

Critical thinking, the difference in knowing when you are right from when you do not know, is essential to the scientific process. Whatever the case, experimental results from any study (while ignoring the discussion) may be useful.

Attachment of rosin size molecules to fiber

The mechanism of attachment of the rosin size molecule to the paper is an important, poorly understood, aspect of sizing. It was said that rosin size could not be used above pH 7 until it was demonstrated that certain polyamines were highly effective mordants up to pH 10 (Biermann, 1992). It used to be said that mechanical pulps were more difficult to size than other pulps, but Zhuang and Biermann (1993) show that the use of iron—based mordants is highly effective with mechanical pulps but ineffective with bleached chemical pulps.

Ligands (of coordinate chemistry) in the liquid against which sizing is determined (or paper is subjected to in service) will play an important role in the stability of the linkage.

The formic acid used in the Hercules size test (HST) test has been said (in several publications) to reduce HST values by lowering the ink surface tension relative to neutral ink. This is an example of unproven dogma that is easily tested. Work by Chen and Biermann (1995) refutes this claim. Formic acid lowers the surface tension of water from 73 to 71 mJ/m^2. The use of a small amount of hexanol (0.25%) lowers the surface tension much more than 1% formic acid, yet does not decrease the HST time to the same extent that 1% formic acid does. Furthermore, formic acid and neutral ink give about the same values for handsheets sized with AKD (no alum present). In some cases, formic acid actually takes longer to penetrate than neutral ink (Chen and Biermann, 1995).

Formic acid was used in the HST test before ASA and AKD sizing methods were developed. Early inks used HCl in most formulations. Formic acid (a fairly strong acid) was used to simulate the HCl in ink formulations while avoiding the problems of the more volatile HCl that would make the acidity of ink change with time. The acid adversely affects the stability of the rosin—alum linkage in the size complex.

Sizing is not just a property of the paper, but of the liquid against which sizing is tested. Gess (1981) has a very useful paper demonstrating this point. This also presents an opportunity for studying the size—mordant—fiber interaction by using a variety of different solutions to test sizing. For example, fluoride (a strong ligand that reacts with aluminum) in water destroys alum—rosin sizing quickly, but has little effect on ASA or AKD size (Chen and Biermann, 1995).

Practical aspects of rosin sizing

Rosin sizing with dispersed rosin requires the use of cationic starch to help with its retention. Dispersed size is more expensive than soap size, but it is said to be more effective and does not decrease the strength of paper to the same extent that soap size does. Dispersed rosin size is less sensitive to the presence of calcium ions than soap

size. At stock temperatures above 40—45°C, soap sizes may be more effective than dispersed sizes.

Neutral rosin sizing with dispersed rosin size and PAC should be used with headbox stock at pH 7 ± 0.2. The overall charge (as measured by colloidal titration) should be near zero and is controlled by the addition of cationic polymers. Retention above 85% is ideal.

Rosin fortified with fumaric acid (*trans*-isomer) has better chelating properties than that fortified with maleic acid (*cis*-isomer), and most rosin is fortified with fumaric acid. Traditionally fumaric acid has been cheaper than maleic acid.

Casein is often used with (anionic or neutral) dispersed rosin sizes (e.g., Neuphor) to stabilize the rosin product; the casein itself may improve the final sheet (as a useful binder) but probably plays a minor role in the size anchoring process in the final sheet. Cationic dispersed rosin formulations (e.g., Hi—pHase) are emulsified with various cationic polymers. In either case the rosin is not chemically linked to the casein or the cationic polymer.

Some mills add rosin and alum prior to refining. The discussion on page 156 indicates that alum interferes with refining.

Practical aspects of alkaline sizing

Some of the advantages of AKD and ASA are the low melting temperatures of 40—45° that make them easy to emulsify. The high melting temperature of stearic anhydride and its relatively high reactivity make it much harder to use than ASA or AKD. ASA is more reactive than AKD, which partially explains why it develops sizing on machine rather than off machine. There is not a good theoretical explanation as to how AKD is able to give improved sizing with time.

One application of ASA size is on newsprint machines that operate below pH 7. As many newspaper presses switch to offset printing, some degree of sizing in the newsprint is required. Many mills find using alum with mechanical pulps causes pitch to deposit (by precipitation). Since ASA is more reactive than AKD, and since lower pH results in a slower reaction rate for transesterification reactions, ASA size proves to be more effective than AKD under these conditions as AKD takes too long to develop sizing.

Williams (1991) gives one mill's experience with conversion to alkaline papermaking with AKD sizing. AKD slipperiness gave poor runnability on the winder and converting operations even though paper gives a higher coefficient of friction. (Perhaps the coefficient of static friction is higher, but the coefficient of dynamic friction is lower; see Tappi Standard T 549 pm-90). AKD caused static electricity to build up on the paper. Size reversion over days and weeks was also a problem depending on the PCC used, with larger PCC particles causing fewer problems but giving lower opacity. A bentonite microparticle retention system worked better here than the dual polymer retention aid system of anionic polyacrylamide with cationic promoter. Higher retention was achieved with a high level of retention control without overflocculation; improved drainage and drying are also noted.

The use of alum with AKD provides a retention aid and detackifying agent for hydrolyzed AKD to decrease press picking. Improvements without problems should be possible with addition up to 2.5 kg/t (5 lb/ton) (on pulp). The alum can be added to the thin stock when AKD is added to thick stock. It seems that more mills use alum with AKD than do not, especially with the dual polymer retention system.

In alkaline papermaking one will get improved sizing by using higher levels of high DS cationic starch. Starch at 30 lb/ton will give much better sizing than 10 lb/ton and 0.35% N will give much better results than 0.18% N in the starch. Also, higher drainage rates are realized. A high cationic charge also helps with the starch used to emulsify the ASA or AKD. The molecular weight of the starch used to emulsify the ASA or AKD is not critical if adequate emulsification is achieved. Hexanedioic (adipic) acid (1 to 3% on starch) has been used to help stabilize ASA emulsions. Chen and Biermann (1995) show the mechanism of anchoring for AKD and rosin/alum sizing.

Practical aspects of retention systems

The dual retention aid system usually works with a high molecular weight anionic polyacrylamide that is added as a flocculent to the stock after the pressure screens and a low molecular weight cationic polymer that is added in the thin stock as a charge neutralizer/coagulation agent.

A bentonite microparticle retention system with AKD size (Dixit et al., 1991) works with a medium to high molecular weight cationic polymer (starch or polyacrylamide) added early to the thinstock system. A high surface area anionic microparticle [2.5 kg/t (5 lb/ton) bentonite] is injected near the headbox so that microflocs form. About 0.75 kg/t (1.5 lb/ton) high molecular weight cationic polymer and 7.5 kg/t (15 lb/ton) cationic starch with tertiary amines was an effective system on a machine making 50 g/m^2 (34 lb/ream) paper at 610 m/min (2000 ft/min).

Optimization of this retention system on a machine making 44 g/m^2 (30 lb/ream) paper with 11.25% UFGC and 2.5% PCC at 600 m/min used 1 kg/t (2.1 lb/ton) HMW cationic polymer added at the primary fan pump, 9 kg/t (18 lb/ton) cationic starch with tertiary amines, and 2.5 kg/t (5 lb/ton) modified bentonite. The headbox consistency was 0.42%, the FPR was 71% and the FPAR was 35%. In mills using clay, one wonders if clay could be modified to act like bentonite in the microparticle retention system.

22.6 WET AND DRY STRENGTH

Sections 17.8 and 17.9 describe how fibers flocculate or bond as a function of consistency in water. Chapter 6 (aspects of refining) is also relevant since paper strength properties are developed by refining. Possible mechanisms for fiber to fiber bonds include mechanical entanglement, covalent bond formation, and secondary chemical bonding including ionic, hydrogen, and van der Waals bonding.

In most paper, hydrogen bonding is the most important of all of these. Hydrogen bonds are stronger than the other weak chemical bonds. The facts that starch and hemicellulose increase the strength of paper and that paper loses it strength when wet both indicate hydrogen bonding. Also, paper made from nonpolar materials has very little strength. Dry strength additives promote hydrogen bonding between fibers (through a polymeric adhesive such as starch).

Wet strength implies a sheet will maintain a high level of its strength (usually between 20% to 40%) even when it is saturated with water. Wet strength additives work by forming covalent bonds between fibers (through an adhesive) or by forming their own crosslinked network of covalent bonds. They increase the dry strength as well. Since these materials are necessarily resistant to water, they are not used on tissue and other papers that must be repulped or be water dispersible.

Even small molecules like butane tetracarboxylic acid (useful in durable or permanent press treatments for textiles) can give a wet tensile strength that is 67% of the dry value, although the fold strength is lost (Horie and Biermann, 1994).

Dry strength additives

Reynolds (1974) showed that dry strength additives have a more pronounced effect on internal bonding in paper at relatively low degrees of refining. With increased refining, pulp fibers effectively "make" their own bonding agents by increased hydration of fibers and increased fiber flexibility. Dry strength additives can help make bulky, porous sheets since lower levels of refining are required. Dry strength aids may allow the use of higher levels of recycled or hardwood fibers.

Unmodified starch is difficult to retain on the paper, so cationic starches (with DS of 0.02 to 0.04) have replaced them as the most common dry strength additive now used. It is typically used at 2 to 4% on pulp.

Polyacrylamide for dry strength development can be of molecular weight 100,000 to 500,000 (usually with a charge density of 5—10%), while that used for retention is one to five million. Consequently, this parameter provides the papermaker with a degree of control over the product. Linke (1962) showed 500,000 to be ideal (for this study's particular conditions), with higher weights actually decreasing the strength somewhat. This work also showed that a charge density of 10—12% (if acrylic acid is used) was ideal for strength development. PAM had a high effect on the fold strength.

Anionic PAM is used with 1% alum at a pH of 4 to 4.8. Once PAM is added to the stock refining must be avoided if alum is present, otherwise moderate refining is possible. PAM is not very effective with furnishes containing 50% or more of mechanical pulps.

Cationic dry strength resins that are *self-substantive* (do not need an alum or polymer promoter for retention) work over a pH of 4 to 8.5 depending on the particular formulation. If rosin is used it should be set with alum prior to the addition of the cationic dry strength resin.

Fillers interfere with dry strength resins. Dry strength resins may not be necessary if wet strength resins are also used.

Wet strength resins

Anionic PAM dry strength resin can be used with a cationic polymer (such as a polyamide-polyamine that is added first) to make wet strength resins up to a pH near 7. The cationic polymer decreases the alum requirements.

Wet strength resins have been discussed. In some systems a partially polymerized material is added to the wet end during papermaking. (A second material, such as monomer, may be used to saturate the paper.) Curing in the dryer section produces the final polymer. This system, *aminoplasts*, is used with urea— or melamine—formaldehyde (UF or MF), and epoxidized polyamine-polyamide resins (Unbehend, 1987).

UF resins have been modified by replacing some of the urea with amine groups that become cationic. These are self—substantive. The curing can continue off—machine for several weeks. Low pH ($\approx$ 4.5) helps UF resins cure more quickly. Like UF, MF must be used under acidic papermaking conditions. MF tends to impart a high level of sizing to paper. The dry paper is also stiffer and stronger in tensile strength. Paper with MF resin is repulped at pH 4.5 or lower at elevated temperature.

Epoxidized polyamine—polyamide (EPP or PPE) resins work from pH 4 to 10 but are best suited for use at pH 6 to 8. Their cationic nature means they are well retained without the use of retention aids. They are more effective than MF per given level of addition and do not impede water absorption appreciably. The dry paper is not much stiffer or stronger. Repulping is done at pH 10 or above at ambient temperatures. The epoxy groups of this and other molecules that use it (such as materials used to prepare cationic starch) are stored in a stable form and made reactive before using them (Fig. 22-9). Kymene is a widely used wet strength resin that is formulated from diethylenetriamine, adipic acid, and epichlorohydrin to help anchor it to the fibers.

22.7 MONITORING WET END CHEMISTRY

The properties of all of the materials used in wet end chemistry should be known as appropriate. Therefore, most of the quality control tests and methods discussed in other areas of this book are applicable. Methods of particular interest to wet end chemistry will be explored in this section. They are summarized in Table 22-1.

The large number of tests in this table indicates that only a few of all of the possible tests can be done routinely. If one deals with only one filler, then an ash content is all that is required to follow the filler. Some tests might be done to determine if a certain material is causing a problem, but is then no longer used. Much ingenuity must be provided as each mill is different and no standard outlines exist.

It is well known (Koethe and Scott, 1993) that cationic polymers cause maximum flocculation shortly (within a few seconds) after their addition to fiber slurries. With time (on the order of 10 to 100 or more minutes) zeta potential decreases. It is thought the cationic polymers diffuse into the fibers since the decrease in zeta potential is smaller with higher molecular weight polymers. Therefore, cationic polymers should be added relatively late into the papermaking system.

The composition of the stock includes dissolved and suspended solids. Dissolved solids include polymers, inorganic electrolytes, and contaminants such as carryover from recycled fiber and pulping operations. Suspended solids include fillers and the fibers and fiber fines.

The types of pulps (hardwood or softwood; mechanical or chemical; CTMP or TMP; size distribution of materials) are also important. CSF should be used to monitor refining, but has little application to papermaking chemistry; CSF is dominated by fines content, which does not adversely affect paper machine speed to a large degree. One of the drainage tests should be used instead. (Industry standardization in this area would be helpful.) Other additives are also included. First—pass retention of the three main groups (fibers, fines, and ash) is important.

The properties of the electrolytes include pH, buffer capacity (often a measure of alum content), types of ions present (such as aluminum, sulfate, and water hardness from calcium and magnesium),

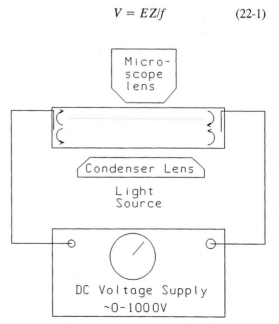

Stable form   Reactive form

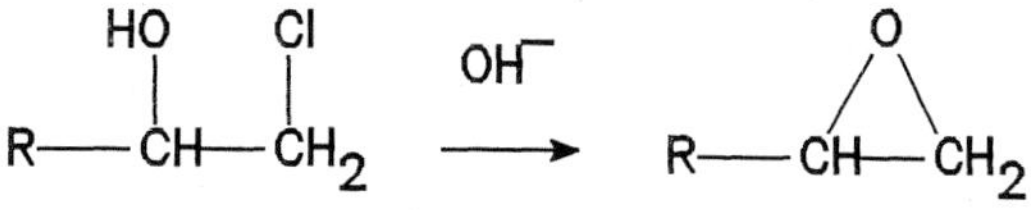

Fig. 22-9. Formation of epoxide group.

overall conductivity, and zeta potential of each type of particle. The properties of each additive should potentially be monitored separately to verify that there is no change from batch to batch. At least some viscosities and colloidal titrations should be considered.

Zeta potential measurement

The zeta potential can be measured by electrophoresis (Fig. 22-10). The electric field causes movement of the water, so the measurements must be made where it is motionless. A microscope must be used to view the particles and determine their velocity. The velocity (V) of migration of a molecule is proportional to the electric field (E, which is the volts of the field divided by the distance between the plates, also referred to as the voltage drop per unit length, V/m)) and the charge of the species (Z) and is indirectly proportional to the frictional coefficient, f. The velocity can be related to zeta potential.

$$V = EZ/f \qquad (22\text{-}1)$$

Fig. 22-10. Measurement of zeta potential.

Table 22-1. Some TAPPI and other methods of potential interest to papermaking chemistry slurries.

Test	Purpose	Methodology	Test number and/or notes[1]
Ash, %	Inorganic filler, etc.	Ashing at 900°C	T 413 om-85 (T 240 first)[2]
Ash, %	Fast analysis, 5 min	Ashing with oxygen	UM 496[2]
Chlorine	Biocide	Iodometry or ISE	T 611 om-89 (UM 612)
Consistency, %	Solids content	Filtration, drying	T 240 om-88
Consistency, %	Fines retention	Turbidity of white water	Use with caution![3]
Dissolved inorganic	Corrosion, pulp washing	Specific conductance[4]	T 252 om-90, ionic interference
Dissolved organic	Determine retention, etc.	Concentration, extraction	GPC, NMR, FTIR, GLC, N, etc.
Dyes	Retention	Color absorption curve	T 442 om-88 is loosely related
Filler determination	Qualitative analysis	Microscopy, chemistry	T 421 om-91, tedious[5]
Talc		Spectroscopy, etc.	T 665 om-89
Fines content	Particles < 70 μm	Clark or Bauer-McNett	T 233 cm-82
Fines content	Particles < 70 μm	Britt jar	T 261 cm-90
pH	Fundamental	Paper strip, pH electrode	T 252, 435, 509, 667, UM 214
Hardness, Ca & Mg	Scale, soap precipitation	Chelant (EDTA) titration	
Inorganic anions[6]	Liquor carryover, etc.	Ion chromatography	T 699 om-87, T 700 om-87
Cl^-	Corrosion, buildups	Mercuric nitrate	T 256 cm-85 (UM 254, ISE)
SO_4^{2-}	Buildup interferes rosin	Precipitation with Ba^{2+}	UM 254, (also T 255 om-89)
H_2SO_3	Added in bleach plant	Iodometry	T 604 cm-85
Inorganic cations[7]	Corrosion, scale, etc.	Atomic absorption (AA)	T 266 om-88
Al	Numerous	ISE with F	*Tappi* 62(2):43(1979)
Ca in pulp	Dissolving pulp	Ashing/EDTA titration	T 247 cm-83
Fe	Discoloration	Acid soluble/colorimetric	T 434 cm-83 (T 242 cm-83)
Na	Poor pulp washing	ISE	*Tappi* 54(12):2051(1971)
Zn, Cd	Fillers and pigments	Polarography	T 438 cm-90
Melamine resin	Wet strength papers	HCl liberation/colorimetry	T 493 cm-84
Nitrogen (total)	From PAM, amines, etc.	Kjeldahl digestion	T 418 om-90 (NH_3 ISE detect.)
Paper			
Air permeability	Formation, floc sizes	Caliper & density interact	T 251 cm-85, 460, 536, 547
Internal bond	Fundamental	Mechanical test	T 541 om-89
Opacity	Formation, floc size	Optical	T 425 om-91, T 519 om-91
Sizing	Fundamental	Usual tests	pages 171-174
Wet tensile	Wet strength papers	Wet paper tensile test	T 456 om-87
Pitch	Common problem	Extraction/spectroscopy	T 204, FTIR, NMR, GPC, GLC
Pulp			
Carboxyl groups	Fiber anchor points	Acid-base titration	T 237 om-88
Copper number	Oxidized cellulose	Copper reduction	T 430 om-88 (oxidized starch too)
External surface	Surface for retention	Clark's silver method	T 226 cm-86
		Nitrogen BET absorption	Dehydration may decrease area
Fiber length dist.	Fundamental	Optical methods	Kajaani device (Also T 232 cm-85)
Hemicellulose	Fiber bonding, colloids	Alkali solubility	T 235 cm-85 (10-21% NaOH)
Solubility	Colloidal contribution	1% NaOH solubility	T 212 om-88
Water retention	Dewatering behavior	Centrifuge/moisture	UM 256
Retention, total	Fundamental	Headbox, tray, couch bal.	T 269 pm-85
Rosin	Retention, buildup	Various	T 408 om-88, T 601, T 628
Starch	Cationic should retain	Colorimetric	T 419 om-91 (ok for low DS)
TiO_2	Quantitative analysis	Reduction/titration	T 627; UM 408, 508 by colorimetry[5]

Some newer methods measure scattered light with the Doppler effect (laser doppler velocimetry) during electrophoresis to measure the particle speeds without direct visual observation. These devices give spectra (Fig. 22-6) of zeta potentials for various particles in the suspension. Study of the isolated components under a variety of pH conditions and their interactions with other components should yield a lot of useful information on practical aspects of wet end chemistry for a given paper machine.

Fibers are too large to easily measure their zeta potential. Often the assumption is made that the zeta potential behavior of fiber fines is the same as that of the fibers. The validity of this assumption is an area of controversy. Zeta potential is useful for determining the cationic demand, the amount of cation necessary to cause particles to have a zeta potential of zero.

Problems in zeta potential measurement occur if the water has a relatively high ionic strength since heating will occur, which induces convection currents. Decreasing the applied voltage or measurement time will be necessary. Samples must be diluted with a solution of similar inorganic electrolyte composition to see individual particles. Samples should be analyzed without undue storage time, if possible.

Details of this and other procedures may be obtained from the manufacturers of these instruments.

Polyelectrolyte titrations

McDermott (1992) gives a good recipe for polyelectrolyte titrations, sometimes called colloidal titrations. Consider an anionic polymer (A, usually the potassium salt of polyvinylsulfate) that is used to titrate a cationic polymer (C) that is a component of a pulp furnish. The anionic polymer has an affinity for a dye indicator (I, such as *o*-toluidine blue):

$$A + I \rightleftarrows AI$$

At the same time the anionic polymer should have an even stronger affinity for the cationic polymer than for the indicator. To increase the stability of the complex, the titration must be done in solutions of low ionic strength; Lindström in Roberts (1991) gives limiting values of $0.05\ M\ Na^+$, $0.005\ M\ Ca^{2+}$, and $0.0005\ M\ Al^{3+}$ for this system.

In direct titration methods, the indicator is added to the solution containing the cationic polymer. Anionic polymer is added by a buret. Initially it complexes completely with the cationic polymer. After the cationic polymer is complexed

---

Notes for Table 22-1. 1) cm denotes that superior methods have superseded these. Many of these methods would require adaptation to pulp slurries. 2) High—temperature ashing for fillers requires the use of gravimetric factors (Unbehend, 1987). For example, $CaCO_3$ is recovered as CaO so multiply the ash by 1.785 to convert to filler. Uncalcined clay requires a conversion factor of about 1.15 for constitutional water loss. ZnS, 1.2; TiO_2, 1.008; various silicates, 1.05-1.10; these can be determined by ashing pure samples of filler. 3) Changes in flocculation change the turbidity per unit mass; use to spot machine upsets. 4) Measured in μmho (backwards ohm) or mS/m; do not directly compare slurries to filtrates. Report as ppm (mg/L). 5) Other tools for filler qualitative and quantitative analyses include SEM-EDAX, X-ray fluorescence, electron microprobe, and X-ray diffraction. 6) Inorganic anions can be analyzed by ion chromatography. 7) Several of the listed methods may have to be modified for wet end solutions. See a major scientific supply catalog for the many available ISE electrodes and their limitations (CO_2, Cl_2, O_2, NH_3, F^-, Cl^-, I^-, CN^-, NO_3^-, S^{2-}, Ca^{2+}, Cu^{2+}, K^+, Na^+, Pb^{2+}). Some electrodes allow indirect determination of other species, like F for Al. AA works for virtually any metal element and has very good sensitivity (below 1 ppm) and specificity. Ion chromatography is another possible method. 5-7) Many of the wet chemical methods listed can be done by other methods if the more complex equipment is not available (as described above).

Other tests have generic interest such as T 620 cm-83, Analysis of industrial process water; T 656 cm-83, Measuring, sampling, and analyzing white waters; UM 470, Sulfates, barium, titanium, aluminum, zinc, calcium, and magnesium in coated paper (qualitative). Carbonates can be detected by adding HCl and observing for bubbles (UM 531). Sulfides can be detected by HCl with smell or darkening of paper saturated with lead acetate solution. Um 634 is used for oil absorption of bentonite.

with the titrant, excess titrant causes a color change (from blue to pink) in the indicator.

In an indirect titration to characterize surface charges of solids, excess anionic polymer is added, the solution is filtered, and the excess anionic polymer is back—titrated with another cationic polymer such as 1,5—dimethyl—1—diazaundecmethylene polymethobromide, hexadimethrine bromide (DDPM) (McDermott, 1990) from a pink to blue endpoint. In a second set of titrations, DDPM is used to titrate anionic polymer in the furnish. The original solution can be filtered of solids and directly titrated to subtract ionic charges due to water—soluble components.

Net colloidal titrations

Halibisky (1977) developed a method where the amount of standard cationic and anionic polymer that absorbs onto the headbox furnish is measured. The amount is measured by titration with a colorimetric indicator. The ratio of anionic polymer used to cationic polymer used in these titrations is known as the *colloid titration ratio* or the *ionic ratio* (CTR). If more anionic polymer is absorbed than cationic polymer, the ratio is greater than one and the surfaces presumably have a net positive charge. Alum causes some interference. Hubbe (1979) describes a modification of this.

Four titrations are used, titrating the filtered solution to subtract its effect and titration of the solids by indirect titration each with anionic and cationic titrants.

Colloidal titrations have several disadvantages. It is hard to reproduce the method from operator to operator without automation of the system. The endpoint determination is not very sharp to the unaided eye. The use of a spectroscopic instrument to help determine the endpoint is useful, but adds to the complexity and cost of the method. Many mill people feel this technique is best left to research facilities.

Simulating the paper machine in the laboratory

Paper machine production personnel will not let academics recreationally test ideas when the costs can be tens of thousands of dollars an hour. Thus, any information one can glean from laboratory investigations is essential.

The papermaking system can be simulated off—line with handsheet formation, the Britt jar

test, pilot plant paper machines, or other tools as appropriate. Handsheet formation has the limitations of very dilute solutions (typically 0.017% consistency for Tappi Standard handsheets of 60 g/m^2 in the British sheet mold), low turbulence during formation, and air drying of the paper. Sheet dryers are available, but the sample does not dry under restraint. Often, all of the chemicals can be added when the stock is at about 1% consistency with dilution occurring only during the sheet forming phase. A higher basis weight sheet can be formed from lower amounts of water to increase the consistency.

Static settling tests (*jar test*) are often used in the water treatment industry. Various levels of retention aid are added to stock solution in each of several beakers. The stock is mixed and allowed to settle. Materials that promote flocculation will cause settling more quickly than in the controls. After a given period of time, the turbidity of the supernatant is determined.

Because of the low turbulence in the British sheet mold and jar test, any retention aid that does not work in the laboratory will not work on the paper machine; however, the converse is not necessarily true. One would at least like relative results, but under static conditions, soft flocs may behave similarly to hard flocs.

The application of the Britt jar test to retention has been discussed. This device allows one to determine the effect of a wide range of shear forces and turbulence on retention. It is said that one should be able to determine first—pass retention to within 3 or 4%. The stirring time should be kept relatively short after addition of the additives.

Zunker (1982) described modifications to the Britt jar that allow it to be used as a drainage tester. He added vanes (which approximately doubled the turbulence and decreased the vortex formation), an electrically activated valve at the bottom, and the use of vacuum. This apparatus allowed the measurement of a wide variety of wet end phenomena.

Water retention value (WRV) has been used to predict strength and dewatering of pulp on the paper machine. The technique has been applied to newsprint by LeBel et al. (1979). The WRV newsprint pulp was fairly similar to press consistency in commercial newsprint machines. This is probably a good routine quality control measure on a paper machine's thick stock since it is said to correlate well with paper machine breaks.

The WRV procedure involves centrifuging a 50 mL sample of $\approx 3\%$ consistency in a cup with 100 mesh screen at the bottom. The centrifuge is operated at 840 g (g is acceleration of gravity) for 12 minutes. The moisture content of the pad after centrifugation corresponds to the WRV, which is a percentage. A procedure is given in UM 256 (900 g for 30 min).

22.8 PAPERMAKING CHEMISTRY CONTROL

It is always desirable to optimize the performance of the wet end. This is particularly true with expensive paper machines where one desires to use high speeds but must not suffer web breaks. Too, the extremely high cost of many papermaking additives must not be taken lightly. The goal is to do more with less, but there are always tradeoffs. Chapter 20 on total quality management should be consulted for the implementation of papermaking chemistry control.

Control implies feedback. Feedback consists of a few parameters of product quality at the very minimum. The ideal would be numerous product quality parameters in addition to the results of a battery of continuous on—line tests and immediate access to all mill operating data.

Variability occurs in pulp preparation, stock refining, raw materials used in the wet end, mill upsets, and so on and on. Variability must always be considered! The properties of each additive are ideally monitored separately to verify that there is no change from batch to batch. At least some viscosities and colloidal titrations could be performed. (You say your supplier does that for you and your supplier never has variation in the material sent to you.)

The entire process should be viewed for what it is, a scientific experiment. The critical step is the development of a hypothesis; this is, unfortunately, often the last step after analysis of results. A protocol is established to test the hypothesis, the determination of the necessary data is made, data is collected, verification of data collection and analysis is made, and the hypothesis is determined

to be true or false. New hypotheses usually present themselves in well—designed experiments.

It is often helpful to make a series of graphs relating various parameters to each other. These graphs may show correlations between variables. The correlations will have to be determined as direct cause—and—effect or indirect relationships; this may require additional experiments. (Cigarette manufacturers have long argued that the correlation between smoking and disease does not prove cigarettes cause disease; they argue that the type of people who choose to smoke are predisposed to disease.) These graphs also show which variables are easy to control and which have high levels of variation or obvious trends.

On—line sensors

On—line sensors are superior to off—line measurements that take operator time and data is available immediately to control the process. Measurements of pH, conductivity, temperature, flow rates, zeta potential, certain ions, and flow rates are well established and do not require operator time to collect. Flow rates are determined with ultrasonic Doppler (reflective for solutions containing particles or bubbles or clamp-on for solutions) or magnetic devices. On—line sensors should be routinely calibrated. Kortelainen (1992) discusses on—line sensors.

Off—line sensors

Off—line sensors usually require manual sampling of process streams. It is critical to obtain a representative sample. This is not easy in the case of paper machine white water; in this case sampling of more than one region may be required, and each sample must be representative of the area from which it is collected. A small study needs to be done just to determine that a representative sample is being obtained. Samples are generally collected with equal time intervals between them. Grab samples are generally useful only if the process is in control (operating under near steady-state conditions).

Flow rates can be measured by bucket and stopwatch, chemical dilution (TIS 0410-09, the old 018-3), and flow balance methods.

Material balance of a paper machine wet end

One can monitor a particular flow (perhaps

tray 1 white water) for routine quality control of drainage. If new drainage aids are used, however, one will want to do a complete balance around the paper machine to determine where drainage is occurring. The reason is that drainage aids work by different mechanisms and affect components of drainage differently. As already mentioned, improved drainage at the flat boxes does not mean the consistency at the couch will be improved.

Complete balances can be solved with computer—based steady—state balance programs such as GEMS or MAPPS if the data is available. Problems and errors in material balances (such as first—pass ash retention) occur because small variations is headbox and tray consistencies and ash content determinations are amplified in the calculations. Digital analytical balances with a precision of 1 mg or (preferably) 0.1 mg should be used in solids determinations where the weights (with tared vessels) are below 100 grams. (Solution volumes over 100 mL are easily measured by weight with a precision of 0.01 g.)

22.9 OTHER CONSIDERATIONS

Wood pitch

Pitch has been discussed on page 264. Pulp mills that use nonalkaline processes often have trouble with wood pitch. Wood pitch consists of over 50% saponifiable materials that are hydrolyzed and solubilized under strongly alkaline conditions. Pitch is usually anionic if it has any charge and may contribute to anionic trash. Calcium ions form precipitates with wood pitch.

Wood pitch is concentrated in parenchyma cells and may be squeezed out in refining, paper pressing, and other operations. Pitch may contribute to foam formation when the pitch molecules attach to air bubbles. Wood pitch in bleached sulfite pulps may be relatively abundant and of high molecular weight. Analytical techniques for pitch analysis are considered elsewhere (p. 434).

Cationic retention aids such as alum may help bind wood and other pitch materials to the stock to prevent their concentration in the system. Pitch may be absorbed on talc if the pitch is dispersed. Dispersants are used to keep the pitch from forming aggregates, but may contribute to poor retention and foaming.

Foam

In addition to surfactants, finely divided solids can contribute to foam problems. These solids include fillers, alumina, fines, and other colloids that can surround air bubbles to decrease the surface tension. Because of their high surface areas, foams are thermodynamically unstable, but may be quite stable kinetically.

Dyes

Basic dyes, unlike other dyes, are soluble in moderately polar to nonpolar materials like the lower alcohols and oil. They have a tendency to work unevenly and may give a surface that appears somewhat like granite.

Because titanium dioxide and many fillers absorb or reflect UV light, they mitigate the effect of fluorescent brightening agents. Light color shades tend to be direct dyes, while deep colors tend to be basic dyes. Biocides in the white water, such as chlorine compounds, may oxidize dyes to change (bleach) their colors. Like phenolphthalein, many dyes are pH sensitive.

Refining

The level of refining can be followed with Simons (1950) stain (TAPPI UM 17) of fibers observed through a microscope. Even without staining, refining can be followed qualitatively with microscopic observation. Pulps with low cellulose viscosity refine faster, so the refiner operator should be aware of changes in the pulp viscosity.

Areas of more research

The areas of colloid chemistry and electrochemistry are very important to papermaking chemistry, as is coordinate chemistry, but these areas do not receive the attention they deserve. One area of research is the use of di— or triblock polymers with affinities for different components of the furnish. This would be useful, for example, for allowing fillers to attach directly to fiber (with the improvement in opacity mentioned above), although one cannot help wondering what this may do to the strength properties of the paper.

Battan (1995) shows that ammonium salts of partially saponified styrene—maleic anhydride copolymers are better at surface sizing than sodium salts. He attributes this to the fact that ammonia is removed in the dryer section, leaving the carboxylic acid to coordinate with cationic species in the paper. Some rosin formulations use ammonium salts.

Additional comments on internal sizing

It is interesting to consider that hydrolyzed ASA (Hodgson, 1994) and unreacted rosin are surfactants which, as shown by Chen and Biermann (1993) and others, lower the level of sizing. Therefore, using additional ASA or rosin to try to improve sizing may have the opposite effect.

Shimada (1996) ivestigated the use of various carboxylic acids (in small amounts of ethanol to help disperse them in water) with alum in the sizing of laboratory handsheets. He found that primary fatty acids gave good sizing when the chain length was 12—18 carbon atoms with stearic acid giving the best results. Larger fatty acids gave poor results. Rosin gave better results than stearic acid presumably because it is locked into shape instead of having a chain that can coil to various degrees. Oleic acid was not as useful as stearic acid against formic acid ink because the *cis* carbon—carbon double bond kinks the hydrocarbon tail giving a less dense surface (Shaw, 1980, in the previous chapter).

A six carbon isocyanate gave no sizing while an 18 carbon isocyanate was better than stearic acid showing that a well bound six carbon moiety does not impart sizing.

Some parting thoughts

There are too many hard and fast rules in alkaline papermaking chemistry. There are so many variables in fiber types, filler types, polymers among various manufacturers, and so forth. Only with detailed characterization is it possible that the effects (and intereffects) of all of these parameters can be sorted out at least empirically if not theoretically.

Therefore, authors of publications must characterize their polymers and other additives to as high a degree as possible. It is necessary for papermakers to have detailed specifications of the products of various manufactures to really be able to maximize the attributes of their paper and paper

machine; however, manufacturers may have a different point of view. A comprehensive, industry—wide survey of who is doing what, and with what additives, would be one of the most productive undertakings to advance papermaking chemistry to the level it should be at. This would include a detailed chemical characterization of the additives (with or without the help of the manufacturers).

It is unfortunate that the industry takes the attitude that the supplier will solve all of their problems. Suppliers sell complete systems that were optimized in **their** laboratories or on **someone else's** machine at some other time. There is no reason to assume their particular system is optimized for your product on your machine using your pulp.

We have not even touched on the possible effect of molecular weight distribution (see Chapter 18). Most papermaking polymers are made by free radical polymerization techniques. This means that a given molecular weight, without describing the method used to determine it, is misleading. A great deal more care must be paid to detail in the papermaking chemistry literature.

I especially welcome comments, suggestions, pointing out of erroneous statements, and other correspondence from the reader in this area.

22.10 ANNOTATED BIBLIOGRAPHY

In addition to the references below, Vol. 8, No. 1 (1993) *Nordic Pulp and Paper Research Journal* contains 40 of the 70 papers presented at the June, 1992 Paper and Coating Chemistry Symposium, Stockholm, in the areas of polymer adsorption, interactions at interfaces, flocculation, rheology, and applied paper chemistry (234 pages). There is a lot of new work and ideas in this reference, but it is overwhelming on the first reading. Similarly, Vol. 4, No. 2 (1989) issue included 17 of the 37 papers of the September, 1988, Paper Chemistry Symposium, Stockholm.

1. Battan, G.L., The effects of SMA surface sizes on paper end—use properties, *Tappi J.* 78(1):142-146(1995).

2. Biermann, C.J., Rosin sizing with polyamine mordants from pH 3 to 10, *Tappi J.*,

75(5):166-171(1992). See Subrahmanyam, S. and C.J. Biermann, Generalized rosin soap sizing with coordinating elements, *Tappi J.* 75(3):223-228 for the role of coordinate chemistry in aluminum and papermaking chemistry.

3. Britt, K.W., Mechanisms of retention during paper formation, *Tappi* 56(10):46-50(1973). This paper describes Britt's dynamic retention jar and its application to retention studies with several cationic polymers.

4. Britt, K.W. and J.E. Unbehend, Electrophoresis in paper stock suspensions as measured by mass transport analysis, *Tappi* 57(12):81-84(1974).

5. Campbell, W.B., *Pulp Paper Mag. Can* 38(2):189(1937); 48(2):104(1947); *Paper Trade J.* 95(8):29(1932). Campbell worked out much of the theory on drainage and forces developed between fibers. In Canadian Dept. Interior, *Forest Service Bull.* 84(1933), he points out that the capillary forces between fibers of 30 μm diameter is 87 psi; of fibrils 3 μm in diameter, 540 psi; and of 0.3 μm fibrils, 2480 psi.

6. Campbell, W.B., The physics of water removal, *Pulp Paper Res. Can.* 103-109, 122, Convention Issue, 1947. One of the earliest looks at filtration of water in the forming web. This work also includes water removal by compression of the fiber mat (pressing).

7. Carman, *Trans. Inst. Chem. Eng. (London)*, 15:150(1937).

8. Chen, M. and C. J. Biermann, Investigation of the mechanism of paper sizing through a desizing approach, *Tappi J.* 78(8):120-126(1995). A variety of inks containing surfactants, acid, and aluminum complexing species such as fluoride were used to test the chemical properties of size anchoring bonds of rosin—alum and AKD size. Rosin sizing behaved as a coordinate bond through aluminum, and AKD behaved as a direct bond.

Calcium carbonate papers should not be tested with inks containing formic acid. A surfactant may be used to accelerate testing.

9. Davison, R.W., The sizing of paper, *Tappi* 58(3):48-57(1975); this has 94 references.

10. Dean, A., *Lange's Handbook of Chemistry and Physics*, 13th ed., McGraw—Hill, New York, 1985, Chapter 5.

11. Dixit, M.K., C.W. Jackowski, and T.A. Maleike, Retention strategies for alkaline fine papermaking with secondary fiber at Appleton Papers, Inc., West Carrollton, Ohio; a case review, *TAPPI 1991 Papermaker's Conference*, 5 p. (Reprinted in *Alkaline Papermaking 1982-1992*, TAPPI Anthology, pp. 13-18.)

12. Forsberg, S. and G. Ström, The effect of contact time between cationic polymers and furnish on retention and drainage, *J. Pulp Paper Sci.* 20(3):J71-J76(1994). The authors present data indicating that maximum retention usually occurs when cationic polyacrylamide or starch is added 10-20 seconds before drainage. Longer contact times decrease retention and drainage rates.

13. Gess, J.M., The measurement of sizing in paper, *Tappi* 64(1):35-39(1981). This paper compares several sizing methods and the use of different sizing test solutions.

14. Halibisky, D.D., Wet—end control for the effective use of cationic starch, *Tappi* 60(12):125-127(1977).

15. Hodgson, K.T., A review of paper sizing using alkyl ketene dimer versus alkenyl succinic anhydride, *Appita* 27(5):402-406 (1994). This is a very useful review and comparison of the two sizing methods.

16. Horie, D. and C.J. Biermann, The application of durable press treatment to bleached softwood kraft handsheets, *Tappi J.* 77(8):135-140(1994).

17. Hubbe, M.A., A modified reporting procedure for polyelectrolyte titrations, *Tappi* 62(8):120-121(1979).

18. Ingmanson, W.L., et al., *Tappi* 37(11):523-534(1954); 42(6):449-454(1959); 42(10)840-849(1959); 46(3):150-155(1963). This work investigates drainage in terms of filtration theory.

19. Koethe, J.L and W.E. Scott, Polyelectrolyte interactions with papermaking fibers: the mechanism of surface—charge decay, *Tappi J.* 76(12):123-133(1993). The authors contend that cationic polymers have a decreasing effect on zeta potential with time as they diffuse into the fibers; therefore, high molecular weight polymers are more effective at having a sustained effect with relatively long periods of time.

20. Kortelainen, H., Tools for successful wet—end chemistry control, *Tappi J.* 75(12):112-117(1992). The author describes on—line sensors of interest to wet end chemistry and suggests a control system (38 references).

21. LeBel, R.G., G.C. Nobleza, and R. Paquet, Water retention value indicates machine runnability of pulp, *Pulp Paper Can.* 80(5):T135-T140(1979).

22. Linke, W.F., Polyacrylamide as a stock additive, *Tappi* 45(4):326-333(1962). This useful paper describes the effect of PAM on the papermaking process.

23. McDermott, D., Wet end starch, alum, anionic retention aids used in alkaline papermaking, *1990 TAPPI Neutral—Alkaline Papermaking Short Course Notes*, 9 p. Reprinted in *Alkaline Papermaking 1982-1992*, TAPPI Anthology, pp 304-312.

24. Nissan, A.H., *Tappi* 37:640(1954).

25. Parker, J.D., *The sheet forming process*, Monograph No. 9, 1972, TAPPI Press, Atlanta.

26. Rance, H.F., Ed., *Handbook of Paper Science*, Elsevier, Amsterdam, Vol. 1, 298 p., 1980; Vol. 2, 288 p., 1982. This is a fundamental review of the paper physics literature that is very well referenced and easy to read. Many of the topics are directly or indirectly related to papermaking physics. Volume 1 chapter titles are Cellulose—water interactions, The fibrous raw materials of paper, Preparation of pulp fibres for papermaking, Forming the web of paper, and Consolidation of the paper web; Vol. 2 titles are The porosity of paper (including transport properties), Mechanical properties of paper, Optical properties of paper, and The structure of paper.

27. Reynolds, W.F., New aspects of internal bonding strength of paper, *Tappi* 57(3):116-120(1974). This includes a lot of well-graphed experimental results (no references).

28. Roberts, J.C., Ed., *Paper Chemistry*, Blackie, London, 1991, 234 p. This work gives details of papermaking additives, chemistry, and control.

29. Shimada, K., Unpublished data, 1996.

30. Simons, F.L., A stain for use in microscopy of beaten fibers, *Tappi* 33(7):312-314(1950). Two aniline dyes, direct blue and direct orange, have differential affinity for cellulose fibers with beaten pulp turning orange and unbeaten pulp turning blue. This method was used by Blanchette et al. in their study of biomechanical pulps [*Tappi J.* 75(11):121-124(1992) with four color photographs.]

31. Stratton, R.A. and J.W. Swanson, Electrokinetics in papermaking, *Tappi* 64(1):79-83(1981). This is a useful paper for people thinking about using zeta potential for control of their paper machine.

32. Tanaka, H., P. Luner, and W. Côté, How retention aids change the distribution of filler in paper, *Tappi* 65(4):95-99(1982).

33. Tanaka, H., A. Awerin, and L. Ödberg,

transfer of cationic retention aid from fibers to fine particles and cleavage of polymer chains under wet—end papermaking conditions, *Tappi* 76(5):157-163(1993). Also see Forsberg and Ström (1994).

34. Unbehend, J.E., Mechanism of "soft" and "hard" floc formation in dynamic retention measurement, *Tappi* 59(10):74-77(1976). This paper demonstrates laboratory retention with varying turbulence using retention systems of metallic ions with varying valences, low and high molecular weight cationic polymers, and high molecular weight polymers with and without disruptive turbulence before drainage.

35. Unbehend, J., *Introduction to Wet—end Chemistry*, Home Study Library Vol. 14, 1987, TAPPI Press, Atlanta. This is a useful introduction with well—referenced text (79 pages) of tables and figures and accompanying audiotapes. It contains very little information since 1982, which makes it dated in the area of alkaline papermaking. The author is not unbiased regarding the equipment he sells.

36. Washburn, E.W., *Phys. Rev.* 17:273-283(1921).

37. Weisgerber, C.A., Ketene dimer sizing composition and process for sizing paper therwith. U.S. Patent 2,865,743 (1958).

38. Williams, M., An alkaline experience at Boise Cascade—Wallula, *TAPPI 1991 Papermakers Conf.*, 2 p.

39. Zhuang, J. and C.J. Biermann, Rosin soap sizing with ferrous and ferric ions as mordants, *Tappi J.* 76(12):141-147(1993). The title ions were exceedingly effective in sizing mechanical pulps with rosin (though the paper

darkened) while they were completely ineffective at sizing bleached chemical pulps.

40. Zunker, D.W., *1982 TAPPI Papermakers Conference Proceedings*, pp 197-204.

EXERCISES

1. Why would CTMP pulp produced with sulfite be expected to behave differently than TMP pulp in terms of papermaking chemistry?

2. What are the advantages of high first pass retention levels? Give at least three advantages.

3. A particular papermaking polymer has a very high cationic charge even at pH approaching 11. What functional group probably makes this material cationic?

4. Why do fines require more wet end additives than fibers?

5. One is using a dual polymer retention system. What is the concern over high levels of hydrodynamic shear?

6. How would one prove or disprove the hypothesis that formic acid decreases the surface tension of water as its principlal means of lowering HST values relative to neutral ink?

7. Why is surface sizing used even in papers with internal sizing? Why is internal sizing used in papers with surface sizing?

8. Why are high molecular weight polymers required in high—speed paper machines for retention?

23

PRINTING AND THE GRAPHIC ARTS

23.1 INTRODUCTION

Papermakers must serve their customers and know about their processes. Printing is one major use of paper. Also, many paper converting operations include printing.

International Paper (1983) is a useful (but somewhat dated) reference on printing and provided some of the information included in this chapter, although a 1989 edition exists. Glassman (1985) is another excellent source of information.

In 1990, there were over 60,000 printing and publishing establishments with 1.54 million employees of whom 818,000 were production workers. The annual payroll was $38.8 billion. The total value of the publications was $157 billion. Almost two—thirds of this was value added, a very high ratio for any industry.

Historical context

Printing is often defined as "the use of ink to transfer an image on a substrate such as paper"; it implies the generation of a "large" number of reproductions. The term "printing" was synonymous for the letterpress method, which was predominant until about 1940; now very little letterpress printing is practiced. The use of computers and instrumentation to generate material must modify these traditional dcfinitions and implications.

Early letterpress plates were prepared by carving wood plates corresponding to entire pages, and these were of no use once the printing was complete. The invention of movable type in the western world is credited to Johannes Gutenberg in Mainz, Germany in 1455; but a primitive movable type made from metal was used in Korea, China, and Japan several hundred years earlier.

Movable type consisted of individual letters and characters that could be arranged to form an individual line. Once a mold was made for a particular letter, it could be filled with lead to give the letter, cooled and separated, and filled again; in this manner individual characters could be mass produced. Lines of type were assembled to give a page. When the printing job was completed, the characters could easily be recovered for reuse. The invention of Gutenberg's movable type was of paramount importance to the development of civilization, as the availability of printed material increased by several orders of magnitude within a relatively short period of time.

Another major advance was the invention of linotype machines by Ottmar Mergenthaler, which appeared in 1886. These machines automatically assemble lines of type as the characters are typed in at a keyboard. Typesetting was common until the late 1960s. The Linotype—Hell Company licenses many fonts used in printing including some used in computer printers.

Introduction

The various methods of printing high numbers of reproductions center on the *plate* which carries the image to be printed. Plates may carry type (text), pictures, or both (the copy material). The purpose of the plate is to transfer the ink (from the *ink—fountain*) in the form of an image to the paper (perhaps through a blanket) to give a clear copy of the image of interest. Three possible geometries are available and define the basic processes (Fig. 23-1) of *letterpress*, raised printing surface; *lithography*, printing and nonprinting surfaces that are in the same plane; and *gravure* (recessed printing surface). The fourth main geometry is used in *screen* printing, which uses a stencil or porous plate.

When an image is transferred by direct contact, the image on the plate must be a mirror image of the desired result. If one ink transfer is used, the plate uses a positive image.

Many advances in printing have caused common terms to be redefined. While the definition of the term was apparent at one time, the present definition is not necessarily straightforward. A good example of this is the term *offset*. When letterpress printing was common, offset referred to the undesirable transfer of ink to the blanket. In the early 1900s, it was found by accident that when offset came in contact with paper, a superior image was produced. With that discovery, offset printing (printing to a blanket and then to the paper) was developed. Now most offset printing is practiced as lithography so <u>the term "offset" is now synonymous with "lithography"</u>.

Not every possible variation of printing that has ever been used (especially in regard to sheet handling) will be considered here so that the common methods of printing used today may be emphasized. Many subtle aspects of printing that are not critical to paper use are not presented either. Some practices that are no longer in common usage are presented because they are necessary for comparison or historical development. For example, letterpress printing has negligible commercial use now but must be considered in any chapter on printing.

23.2 LETTERPRESS

Introduction

Letterpress uses raised surfaces (*relief* printing) upon which ink is transferred. An even layer of ink is applied to the raised surfaces by inked rollers. When paper is pressed against the plate, the raised surfaces contact the paper and leave the ink impression. This is the only method whereby replaceable type can be used; it is useful for text information that may be changed often. Later, most letterpress plates were made by photographic methods (with plates coated with egg albumin and ammonium dichromate). Since the ink may exude around the characters during the printing operation, a ring of ink may form at the edges of printed areas (visible under magnification).

Variations of letterpress

Flexography uses rotary web letterpress with flexible rubber plates and fast—drying inks. The rubber plates are wound around a printing cylinder. Flexography has widespread use in packaging and has grown rapidly with the packaging industry. For example, it is widely used to print corrugated boxes, plastic films, and foils. Photopolymers are used to photographically prepare the plates. In modern flexography, the ink is applied to the plate by *anilox* rolls, which are discussed in the appendix of this chapter.

Thermography is used for printing raised letters (*embossed*) for wedding invitations, greeting cards, and especially high—value printing jobs. An adhesive is applied (which remains on the surface), a powder is applied which is retained on the areas with adhesive, and the excess powder is removed by an air stream or vacuum.

23.3 LITHOGRAPHY

Introduction

Lithography (stone printing) was invented by Alois Senefelder in 1798. The original method used sandstone rock upon which an image was made using a thick, greasy material. During printing, the surface was treated with a water solution of gum arabic wetting only the nonimage areas. The surface was then treated with a nonpolar ink (oil, wax, soap, and carbon soot) that soaked into the image area, but avoided the water-treated, nonimage area. The image was then transferred to paper. The same basic technique is

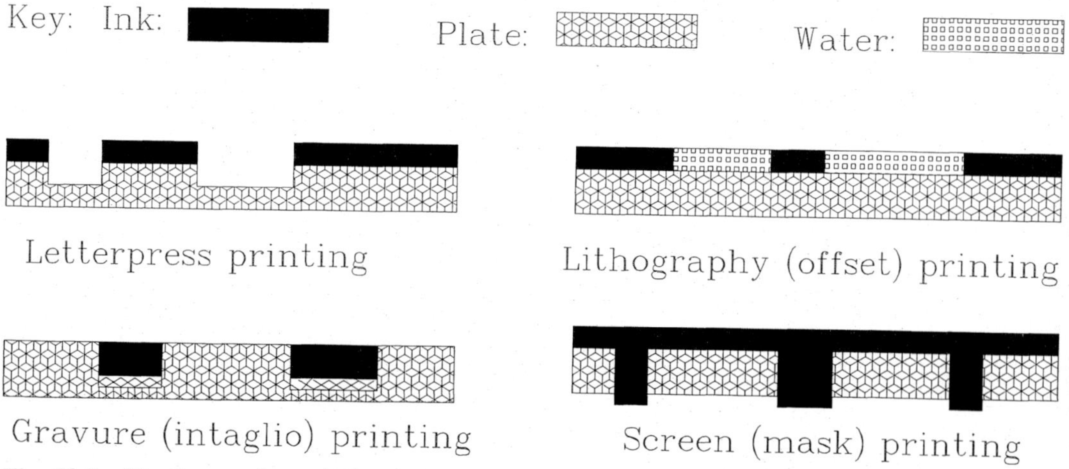

Fig. 23-1. The four major printing processes as seen through a side view of a plate cross section.

still used today. The water solution used to prevent printing in the nonprint areas is called the *fountain solution*. This method is sometimes called *planographic* printing since the image and nonimage areas are in the same plane.

The system used to apply the water solution to the plate surface is called the *dampening* system. It makes the nonimage area hydrophilic and leaves the image area hydrophobic [water hating, but nonpolar ink loving (oleophilic)]. The dampening solution usually contains some isopropanol (though it is used less for environmental and health reasons) which decreases its surface tension and increases its viscosity.

Lithographic plates

The plate surface is often a thin aluminum plate wrapped tightly around a cylinder. Every time the cylinder rotates, a copy can be made. Bimetallic plates are made using two metals of different properties. Oleophilic materials include certain resins, copper, and brass (a copper alloy with up to 40% zinc). Hydrophilic metals include aluminum, chromium, and stainless steel. The hydrophilic surface is kept hydrophilic by treatment with materials such as gum arabic (a water—soluble polysaccharide). Platemaking can also be done by use of computer—controlled laser light on plates with a Mylar (polyester) coating. This method has become very popular as new Mylar coatings allow 100,000 or more impressions to be made from the plate.

Offset printing

Most lithography is carried out as *offset* printing; the image on the ink plate is transferred (offset) to a rubber sheet (the *blanket*) and from the blanket to the paper (as opposed to direct printing). In this way the original plate is not a mirror image of the desired print. Offset printing gives smooth edges on printed areas (as observed with magnification). [Offset printing is possible with letterpress or gravure printing; however, most lithography is done with offset, and as already mentioned the term *offset printing* is essentially synonymous with lithography.] Modern lithography prints a thin layer, about 3–5 μm, so complete opacity is very difficult to obtain.

The rubber blanket usually contains air pockets (Fig. 23-2), if it is the common compressible type and not the conventional type, in a layer

below the surface layer to give it good compressibility so that printing on relatively rough surfaces is improved. This makes paper compression a relatively unimportant factor, but has the disadvantage that some print distortions (such as dot gain or paper distortion) may be realized. In most cases, compressible blankets give higher print quality, better wear, and are more forgiving of paper wrinkles than conventional blankets; they are used for over 90% of offset printing on paper.

Offset printing has several advantages. The rubber blanket gives a smooth image on relatively rough pieces of paper. Since the plate only contacts a rubber offset blanket, plate wear is minimized for long plate life. (Paper can be a very abrasive material and a material over which the printer does not have direct control.) Since there are two transfers, the original plate is a positive of the image.

Offset lithography is the most widely used printing method for commercial printing, newspapers, books, and magazines. There are three major types of this: sheet—fed, non—heat—set web, and heat—set web. The nonheat—setting process requires an absorbent paper and is used for newsprint and book papers. Heat—set presses are used for higher quality publications, often on coated papers.

Sheet—fed presses run as fast as 12,000 copies per hour, while web presses may be double

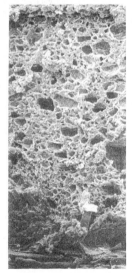

Fig. 23-2. Cross section of a compressible blanket (left, 30×) and air pockets (200×).

this. However, the highest quality papers are often available in the sheet form only.

Offset *duplicators* (such as A.B. Dick Co. products) are very similar to presses, but there are some important differences including their lower cost. Duplicators use a higher pressure on the paper to the blanket that results in a higher dot gain and possible smearing on some coated papers; however, duplicators can often use paper with lower smoothness. Press blankets are said to just *kiss*, i.e., lightly contact, the paper.

Offset press configurations

Figure 23-3 shows a schematic diagram of the dampening and ink rollers in relation to the plate, blanket, and impression rolls or cylinders; it includes many of the features common to most offset presses, but presses have many more rolls and many configurations possible. Many two—color presses operating in a single unit will use a double—sized impression cylinder with two blanket rolls. (Each blanket roll has its own plate roll with separate ink and dampening systems.) Here, the registration is very good since the paper is held by a single gripper for both impressions.

Waterless printing

The ink/water balance in offset lithographic printing is critical. Also, the use of the dampening system complicates the press and its operation and adds to its initial cost. The 3M company developed and introduced waterless printing in the late 1960s; the product was soon abandoned. Toray Industries further developed the process in the 1970s. The process was not widely used until the mid—1990s. It offers the potential to greatly affect the printing industry. This method is a bit of a competitor to stochastic printing (below).

The Toray plate uses a silicone rubber layer (2 μm) on top of a aluminum plate coated with a light—sensitive photopolymer. The silicone rubber is removed from the print areas on the plates by exposure to light. In fact, the printing area of the plate is slightly recessed, but the ink rises above the surrounding silicon rubber area.

Inks must be designed especially for this system to be effective; the principal difference is that the ink vehicle that is selected must give suitable rheological properties such as a fairly high viscosity. The temperature of the plate during

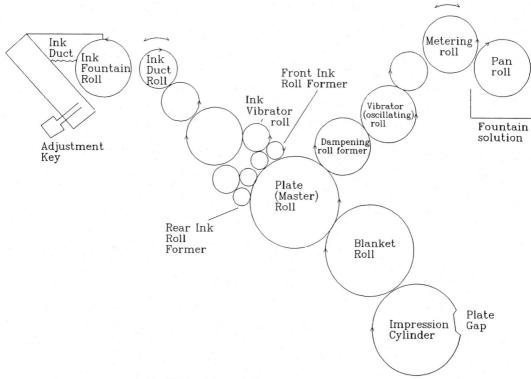

Fig. 23-3. General diagram of an offset press.

operation is extremely critical and should be kept within a 2°C (5°F) tolerance. This is done using a cooling system to control the temperature of the ink vibrator rollers. The exact temperature used is variable, depending on the ink that is used. If the ambient temperature and relative humidity of the press area are high, water may condense onto the ink with an unsatisfactory result.

Most offset presses can be converted to a waterless printing by adding an ink cooling system and using the appropriate plates and inks. Dot gain in the midtones is said to be about 30% lower than with conventional lithography.

Waterless inks should be used with papers having a high surface strength (picking resistance) because waterless inks tend to have higher tack when pulling away from the paper (*release factor*); these inks have no dampening solution with which to emulsify, which would otherwise tend to decrease their tack at this point. Press rooms should be as clean as possible, and paper that does not lint should be used since the dampening solution that usually lubricates the system is absent.

23.4 GRAVURE

Introduction

Gravure printing (also called *intaglio*, pronounced with a silent g) uses an indented area of the plate from which to print. An excess of fluid is applied to the plate, and the excess is removed. This leaves the raised areas free of ink, and the ink is transferred from the *depressed* areas. The edges of printed areas have a serrated appearance (under magnification). Web printing with the gravure process is called *rotogravure*.

Gravure printing is usually carried out as a web process on lightweight (below 40 lb/ream) coated or uncoated groundwood paper with minimal pressure on the plate. Therefore, paper smoothness is very important. Compressible paper and good ink receptivity help print quality.

The major components of a rotogravure press include the engraved cylinder (the plate), the ink-fountain, a doctor blade to remove the ink from the nonprinting area, and the impression roller.

Gravure plates

The ink is held in the plate by a series of cells or wells etched in a metal surface (often copper). The cells are filled by rolling the plate cylinder in a bath of ink. Ink is removed from the nonprinting areas by a doctor blade (usually made of flexible steel).

Gravure is the preferred method for high—quality pictures (especially color), but the high cost of plate preparation limits its use to medium to long press runs. The resolution of gravure is 150 or more lines per inch.

Conventional gravure uses fixed area and fixed depth cells; it is used for short—run, high—quality reproductions of photographs. Variable area gravure is used for packaging, and variable area and depth is used for newspaper supplements and catalogs. The latter method allows some direct reproduction of shades of gray by controlling the amount of ink that is transferred.

Plate preparation

The printing plate is prepared (Blasche, 1993) on a copper plated [25—75 μm (1—3 mil) thick] steel roll. The surface is engraved by one of several methods: the majority are engraved by electromechanical methods, but carbon tissue and direct transfer methods are also employed.

Electromechanical methods use a computer—controlled oscillating (in and out) diamond that penetrates variable distances (in 256 steps) to cut the copper plate, which results in pyramid—shaped cells of variable depth and variable area. Carbon tissue (gelatin light—cured surface followed by ferric chloride etching) methods give variable depth cells. Direct transfer (photoresist surface followed by chemically etching with ferric chloride) results in variable area cells. After engraving, a chrome surface is applied by electroplating to give a hard, long—lasting surface that can withstand millions of impressions on paper. Ideally, plates can be reused by removal of the engraving and application of a new layer of copper.

Steel—die engraving

This method uses a mechanically engraved or chemically etched copper (for one—time use) or steel plate. Ink is applied to the plate and doctored off. Slightly moistened paper is pressed with high pressure against the plate. The result is a characteristic embossed printed surface and a slightly indented impression on the backside of the paper. This method is capable of very sharp images and is used for printing currency and certificates with fine details.

23.5 SCREEN PRINTING

Introduction

Screen printing uses a mask (*stencil* or *porous* plate) to control the location of the ink. (Perhaps you have used pieces of heavy paper with the area of a letter cut away and a can of spray paint to "print" a sign.) Silk screen printing is the only commonly used stencil printing method used. It uses a fine—mesh screen (silk, hard plastic, or stainless steel) mounted to a frame. The screen supports the prepared stencil (the plate). The stencil is held against the surface to be printed, and ink is forced through the stencil (and supporting screen) with use of a squeegee. A thick layer of ink that gives a good impression on virtually any material is obtained. Mimeographing is the other common method of stencil printing, although it is seldom used anymore.

23.6 OTHER PRINTING METHODS

Introduction

Other types of printing methods should be mentioned because one often encounters them on a daily basis and they account for an appreciable use of paper because so many people use them, although individually they are not high—volume reproduction methods. The geometry can usually be compared to one of the four main methods in most cases. (Some people may not wish to call these processes printing, but they still consume lots of paper and make lots of images.) Many of these methods are used with computers or other electronic instrumentation to produce the output.

Electrophotocopy, laser printing

The *electrophotocopy* (photocopy) process was introduced by the Xerox company in 1960. An image is produced on a charged drum (a flat surface, making it a planographic method) by means of a light source that scans the surface of the document and, by means of optics, reflects the light to the charged drum. Dark areas on the original correspond to dark areas on the reflected image. The light neutralizes the charge on the

drum where it strikes. The drum then picks up dry toner particles on the charged surfaces. These particles are then transferred to the paper and fused onto the paper surface by heat or other mechanisms (TIS 108-03 has four pages of definitions relevant to this process).

The original drums used a surface of selenium (Xerography) or cadmium sulfide. Both of these materials are toxic and most manufacturers have used organic polymer coatings on the drum surfaces since the early 1980s.

Laser printers use this same process, except the light source is a laser that is controlled by a computer to give the computer—generated output without an original. Lasers allow a more precise (higher resolution) image to be formed. Photocopy machines are available that include scanners and image—processing computers and use a laser to generate the output; these have 600 dots per inch resolution or higher.

Color copiers or color laser computer printers use the same process as black and white copiers, but use four different color toners to produce a wide range of colors.

Photocopy methods are efficient for somewhere around 5,000 copies or less. The printed product is less durable, does not fold well due to toner separation from the paper, and has poor halftone properties. However, photocopiers are fast (no intermediate steps from the original to the product), inexpensive for small jobs, and allow the operator flexibility with little training.

Ink jet and related printers

Ink—jet printers spray the ink directly through a series of holes onto the surface of paper as the printhead scans back and forth across the paper. Several mechanisms are available to effect the transfer. Piezoelectric crystals surrounded by ink vibrate as electric current activates them, causing a droplet to form and pass through a fine nozzle; very small electric heaters boil the ink to transfer it to the paper. Four printheads (or portions of) are used for color printing.

Dye sublimation printers use a dye from a page—size ribbon that vaporizes and condenses on the paper. (Four ribbons are used for color printing.) This method allows a continuous tone of ink intensities to be produced; it produces high—quality outputs for modest capital cost. It is the only computer printer (at this time) suitable for color printing of photographs. *Thermal wax* printers use an ink that is a solid wax at room temperature. The wax is purchased as an inexpensive stick that is easy to load into the printer. The wax is melted by heating the entire wax delivery system and applied to the paper. Four waxes are used for color printing.

Thermal printers

Thermal printers are generally reserved for instrumentation. For example, EKG machines and other "chart recorders," many FAX receivers, some computer printers, and other devices use this method. A special ink is in the paper that, when heated, turns dark (such as black or blue). Printing occurs by a scanning needle or printhead that heats appropriately as it scans. This method has the advantage that ink will not clog in equipment that gets intermittent use so that the reliability is greatly improved.

23.7 HALFTONE PRINTING

Halftone printing

A particular area of paper is printed either as black or white by most printing methods. This is useful for text and line art, artwork characterized by relatively large areas of black or white, but not gray. A photograph is the opposite of line art and is *continuous tone* with various shades of gray. However, areas of various shades of gray cannot be printed directly as gray. In 1886, Frederick Ives developed tonal reproduction by use of the halftone screen; it is based on the optical illusion that areas of very small black dots appear as gray, since the eye cannot distinguish the individual dots without magnification. The sizes of the dots are such that on the order of 50 (newspapers) to 200 (high quality reproduction of photographs) occur per linear inch. The first edition of this book used 150 dots per inch (22,500 dots per square inch). The screen is aligned so that the rows and columns are at a 45° angle in the printed product.

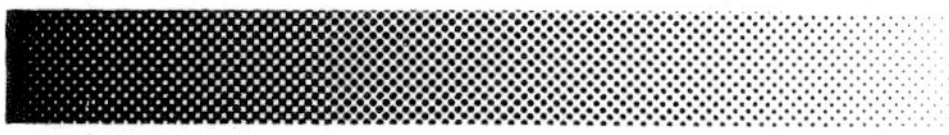

These patterns of dots are prepared with half—tone screens, special photographic techniques, and other methods. Figure 23-4 shows a very low resolution halftone that gives continuous shades of gray. Figure 23-5 shows several techniques to reproduce a photograph; the area shown is the pupil of an eye. For additional examples it is recommended that the reader secure a dissecting microscope and examine several books with reproduced photographs (for example, 50% tones are often produced by a checkerboard pattern of black and white squares). Perhaps some original photographs can also be examined for contrast.

The overall *tone* of a halftone is defined as the relative area that is inked or the overall area that is inked multiplied by the transmittance of the printed area. Stochastic printing is a modification of angled, conventional halftone printing and is discussed in the next section.

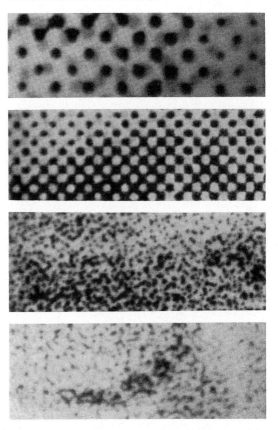

Fig. 23-5. A portion of a printed photograph that demonstrates halftones (top two) and stochastic screening (bottom two) with fixed size variable spaced dots at two resolutions each.

Fig. 23-4. CorelDraw shades of gray for 40 dots per inch.

23.8 COLOR PRINTING

Introduction

Multicolor presses use printing units in tandem to apply several colors (Fig. 23-6), often two (a single color and black) or four colors (three colors and black for "full color" printing). Four—color printing will be considered here. (For relatively small runs, a single color press can be used by a skilled craftsman to produce multicolor prints by using several passes through the press with a change of ink between passes. It is also common to carry out four—color printing with two passes through a two—color press.)

Most of the colors can be adequately reproduced with translucent yellow, magenta, and cyan inks. Since these inks do not block (absorb) all light, black ink is used to give the overall shade to a particular color. The four *process colors* are known as the CYMK color system, where K indicates black (if it were designated B it may be confused with blue) or key. These inks produce various colors by subtractive mixing to obtain red, blue and green (Plate 41). Color theory is presented in Chapter 24.

The four process colors have the same names globally. The shades are uniform in North America, but a different set of shades is used in Europe, and other sets are used in other parts of the world; therefore, be careful when printing outside your own country (Beach, 1993). The 1992 worldwide color printing market was about $150 billion.

Color separation

Color separation is the process of making four halftone negatives for color pictures. Until the mid—1980s, it was accomplished by taking consecutive pictures of the colored object with red, green, and blue filters, respectively. Color is usually evaluated with a light source corresponding to blackbody radiation at a temperature of 5000 K.

If a colored object is photographed through a red lens, red light is transmitted, while blue and green colors are blocked. Since this photograph is a negative, it best represents the blue and green areas of the object. The red negative will be printed with cyan (a combination of blue and green). In an analogous fashion, the green negative is printed with magenta and the blue with yellow. Color corrections must usually be applied, and this is accomplished in any of a variety of methods. The entire process can now be accomplished with the use of computer—controlled scanners, but the method of collecting the information does not change the theory of the color printing process.

While color separation is now always carried out with scanners, the principle is still the same. When four—color printing is used, *trapping* is often employed in platemaking. This is the process of overlapping color ink (on the order of 0.01 in.) to insure faithful color reproduction and paper register. A color dot that is made smaller is *choked*, while one that is made bigger is *spread*.

Fig. 23-6. A five color offset printing press.

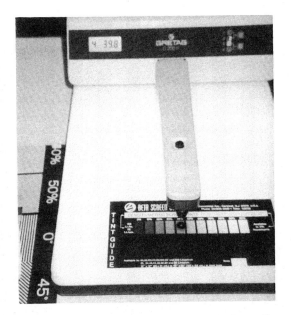

Fig. 23-7. A densitometer measuring the darkness of a film.

Register is a term that is used to describe alignment of the color negatives during color separation or alignment of the paper between printings of various colors. Misregister indicates poor alignment of the colors. Register of color negatives is accomplished by *register pin* systems.

Color printing on the press

Register on the multicolor printing press is the degree to which the paper remains in position between the printing operations for the various colors. Press manufactures have systems that maintain register. One method is by the use of register marks that are optically scanned.

Misregister indicates poor alignment of the colors. The tolerance may be from 0.005 in. to 0.05 in., depending on the nature of the print, i.e., a folding box print versus a high—quality photographic reproduction. *Commercial* register is within one dot, while *hairline* (tight or close) register must be within about one—half dot. One can often see the color register marks in obscure areas of packaging papers or boxes. These marks are printed across the entire width of the press so that the color balance across the press can be adjusted to make it uniform.

The darkness of the printed areas is measured by a *densitometer* (as in Fig. 23-7), and the ink

flow can be adjusted by the *ink keys* across the width of the printing press where the ink is metered (much like the slice adjustment on a headbox). Figure 23-8 shows a densitometer checking the print density across the width of a printed page where the numbers correspond to the ink keys of the ink fountain. Transmission densitometers are used for photographic films, and reflectance densitometers are used for printed surfaces. Color filters may be used to evaluate each of the four (or more) colors used in the print process. Polarized light may be used to mitigate the effect of gloss on shiny paper or wet ink.

The halftone patterns are usually printed at 30° angles to each other (Plate 42), except yellow, since it is a light color, which is printed at a 15° relative to two of the colors, to avoid what are called *moiré* patterns. A common set of angles is black, 45°; magenta, 75°; yellow, 90°; and cyan, 105°. (This is the same as yellow, 0°; cyan, 15°; black, 45°; and magenta, 75°.) Figure 23-9 demonstrates moiré patterns when two dot grids are offset by 4° (left) and the lack of these patterns when they are offset by 30°.

The first ink alters the surface properties of paper, so succeeding inks in multicolor printing usually have lower tack and/or surface tension in order to achieve uniform results. The usual order for color printing is yellow, magenta, cyan, and black. The yellow, magenta, and cyan inks are usually transparent to show the color of inks underneath, though opaque inks are occasionally used for specialized purposes. Paper color, of course, strongly affects the product color.

Stochastic screening

The traditional method of color printing uses a grid of halftone dots for each color (angled to avoid moiré patterns). A new method of color printing has the generic term *stochastic printing* (Fig. 23-5). It is also called frequency modulation or random dot. Two systems are possible: fixed dot sizes with variable dot spacing or variable dot sizes with variable dot spacing. This system is now possible because computers can make the calculations required to set up the dot patterns.

Agfa uses the trademark CristalRaster. It has a different structure for gain than conventional halftone dot printing. Another trade name is Diamond Screen of Linotype—Hell.

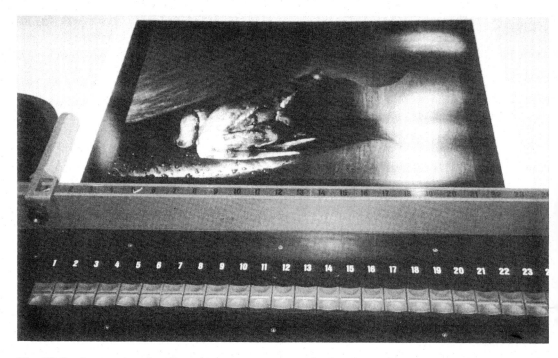

Fig. 23-8. A scanning densitometer on a press console that is used to check the print intensity across the printed page. Adjustments can be make at the ink keys of the ink fountain for each color using the register marks.

Other aspects of color printing

It is not uncommon to have a five or six color press. The extra press may be used for a variety of purposes. A protecting topcoat (perhaps containing a UV absorbing material to limit color fading) may be used in one press. A special color of ink may be used that is impossible to achieve by mixing the primaries, such as vibrant red colors. If this color is used only in certain parts of the print without mixing with other colors, it is

Fig. 23-9. Two dot grids offset by 4° (left) showing the moiré pattern and offset by 30° (right) showing a more uniform dot distribution.

termed a *spot color* or a *flat color*. When flat colors are combined with other colors, one is said to be *building colors*.

A common name and brand of spot color is *Pantone Matching System* or preferably *Pantone Color*. Other manufacturers of spot colors are Focoltone and Trumatch. Swatch books containing samples of colors identify the color by number (not name), the industry practice.

Eight color presses are being used increasingly to produce a wider color range; this is sometimes called hi—fi (high—fidelity) color. When used with other printing techniques such as stochastic printing or variable dot size, photograph—like quality can be achieved and certain effects not available in photography are possible. The use of eight colors on each side of a page puts extra demands on the paper substrate.

23.9 COMMON ASPECTS OF PRINTING

There are some other important considerations in printing. *Dot gain* represents the increase in dot size from the plate to the printed image. For example, a dot that is 50 μm on the plate may print as 75 μm on the paper. Dot gain is a function of the size of the dot to be printed and often is greatest in the midtones; this function is called a *dot gain curve*. *Dot distortion* refers to a round dot being printed unevenly. These are a function of the amount of ink used to print, paper characteristics, and other factors.

Other terms

A *perfecting press* prints on both sides of the sheet with one pass through the printing press. Printing of individual sheets (*sheet—fed press*) used to be quite common, but the use of a continuous web (*roll—fed press*) is much faster and has replaced individual sheet printing in many applications. The paper is then sheeted and folded at the end of the press.

23.10 INK AND INK APPLICATION

Introduction

Early inks were all pretty similar and consisted primarily of boiled linseed oil (from flax) and carbon black (soot or lampblack). Some milestones are as follow: in the middle 1700s, ink becomes available in more colors than just black;

1850s, coal tar dyes developed; 1920s, introduction of synthetic resins; 1978, acrylic resins used in inks for coated papers.

In 1990 the U.S. printing ink industry had about $3 billion in sales. Gravure inks are produced in the highest mass, but lithographic inks are produced with the highest sales (value). About 200 companies produce ink at nearly 500 plants. Inks cost about $5 to $40 per pound.

Ink components

Ink consists of pigments or dyes, a vehicle (binder) to attach the pigment to the paper, and other components. Other components include solvents for faster drying and ink additives.

Ink pigments

Ink consists of pigments, such as carbon black or titanium dioxide, to supply color and opacity and a vehicle to carry the pigment and bind it to the paper. Pigments should be insoluble in water and the ink formulation to prevent running. Some pigments are inorganic materials such as alumina, chrome yellow, or iron blues. Many pigments are insoluble variants of organic dyes. Heat transfer pigments are solids that sublime upon heating (to fabrics).

Ink additives

Ink additives have numerous functions; some of the more common ones are considered here. Ink driers are often heavy metal salts of (unsaturated fatty acid) soaps that promote oxidation and/or polymerization. The metals include cobalt, manganese, cerium, and others (these are often used as free radical initiators in polymer chemistry). Wax makes ink more slippery, increases abrasion resistance, decreases setoff, and decreases tackiness with minimal effect on other properties. Small amounts of antioxidants prevent ink from drying in bulk but allow it to dry on the substrate.

Ink vehicles and ink drying or setting

The type of resin depends on the type of ink and how it attaches to the paper. Inks may "set" on the paper by one of several mechanisms. Glossy inks use high molecular weight resins that decrease the absorption of ink into the paper.

Absorption of the hydrocarbon vehicle into the paper substrate occurs by capillary action.

This type of ink (such as soy bean oil vehicle) is used in newsprint, but tends to smudge.

Evaporation of the ink vehicle leaves the solid ink with rosin esters and metal binders behind; this is used in magazine and catalog grades using letterpress or heat—set web offset printing. To keep heating costs relatively low, solvents with low boiling points, but not so low as to be unsafe, are used. The ink further sets by cooling to ambient temperatures.

Oxidation of a drying oil with multifunctional carboxylic acids and alcohols left on the paper surface after the vehicle is absorbed into the paper may promote *polymerization*. Heavy metals (cobalt, manganese, lead) may promote oxidation with atmospheric oxygen.

UV or electron beam radiation curing causes monomers or prepolymers in the vehicle to form polymers; it is useful with acrylics. This method does not have much use with paper, but is often used for metal and plastic substrates to avoid the use of large ovens.

Infrared hardening (heat—setting), *precipitation* of binders, *gelation*, and *cooling* of hot thermoplastic inks used in electrostatic printing (photocopy and laser printers) are other mechanisms used less often, but are increasing in use.

The use of volatile organic solvents improves the speed of ink dying; however, more and more environmental restrictions are being placed on the use of these materials. Vapors must be trapped and recovered and incinerated. Also, these materials have the potential to explode if the vapors are concentrated.

Ink mileage or coverage

Ink mileage is the amount of paper that can be covered by a certain mass of ink to achieve a specified print density under operating conditions of a press. It is measured in English units of square inches per pound of ink. It is a measure of the ink requirement and varies with the type of paper. For example, one manufacturer's black ink for lithographic work covers 275,000 (square inches per pound) for uncoated paper; 375,000 for dull coated; 380,000 for litho coated; and 425,000 for enamel coated. On the other extreme, metallic or opaque white inks have only about one—half of this coverage.

Critical ink properties

Some of the properties of interest of inks include viscosity, ink tack, the length of ink, pigment particle size distributions, avoidance of particle clumping, opacity, color and color strength, gloss, and drying rate. Some of these parameters depend on colloid chemistry, general opacity and filler theory, color theory, and other topics discussed elsewhere in this book in regard to paper. These analogies should not be ignored.

Ink viscosity, ink yield point, and thixotropic properties are all important properties for ink flow and transfer rates which are dictated by the end use. Nondrip paints containing silicic acid are good examples of a *thixotropic material*. The paint will flow when applied with a brush or roller, but without this shear force the paint does not flow. The shear forces break the weak bonds between particles so that the viscosity decreases with even minor shear forces.

Ink tack is a measure of the force or energy required to split a film of ink. In this regard the ink is like a contact adhesive. If a high force is required to split a film of ink, it may cause picking of the paper. (Think of the Dennison wax pick test of paper.)

The length of ink refers to the length ink can be drawn out before breaking. (Hot mozzarella cheese or taffy behaves like a long ink; on the other hand, a short ink is said to have a buttery consistency.) Ink length is a function of the cohesive forces of the ink and other properties.

An example of the suitable properties of an ink for a particular application is given by Williams (1985a) for lithographic inks. (As with paper, tradeoffs must be made between properties since many contradict each other.)
The requirements are:

1. Ink in an open container should not form a skin at the surface.
2. Ink should remain a liquid in bulk but dry quickly when on a substrate (paper).
3. Ink must be able to absorb (dissolve) a small amount of water without emulsifying.
4. Ink must dry with a good gloss while being subject to many processing operations.
5. Ink should be compatible with a wide range of papers and other substrates so that ink

changes are not required and compatibility problems do not arise.

6. High ink holdout gives better gloss and decreases the amount required.
7. Suitable tack for widespread use. Low tack reduces picking, but high tack gives sharper detail.
8. Ink color must be permanent and consistent from batch to batch.

Flexography, gravure, and silk screen require inks of low viscosity where drying is by vehicle evaporation.

23.11 PAPER PROPERTIES AND PRINTING

General considerations

Standard sizes for printing (whether sheet—fed or web—fed) include the form web (trimmed to 8½ in. by 11 in.), mini web (trimmed to 11 in. by 17 in.), half web (trimmed to 17 in. by 22 in.), and full web (trimmed to 23 in. by 35 in.). If the machine direction (MD) is parallel to the long direction, the paper is said to be *grain long*; if the MD is parallel to the short direction the paper is termed *grain short*.

Paper properties are important for *runnability* and *print quality*. Runnability is usually more of a consideration in offset processes than with the other processes. For example, a high level of water—soluble components in paper will require relatively frequent washing of the blanket. Also, the use of tacky inks and lightweight magazine papers in web offset printing requires a considerable tension on the paper to pull it through the press, with the potential for web breaks.

It seems many printers learn what paper works well and what paper works not so well by trial and error. Once some suitable suppliers are found, they tend to stick with them. A little extra cost of paper is not a large factor compared to

Wavy edges   Tight edges

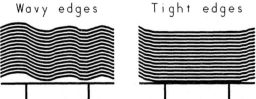

Fig. 23-10. Improper conditioning may cause wavy or tight edges.

press costs. (Government institutions, however, are often forced to look at cost of materials directly and not the overall cost of using a material. This is a good example of how total quality management (TQM), if it were practiced, could increase the efficiency and quality of an operation.)

Some aspects of this section cover material presented in TIS 108-01. High paper stiffness is generally an attribute to printing; stiffness is often lost in papers with high levels of fillers. The importance of the surface of the paper to resisting picking has already been discussed, especially in regard to inks with a high degree of tack (such as those used in offset or letterpress gloss printing).

Moisture and dimensional stability

If a stack of paper is of appreciably lower equilibrium moisture content (EMC) than it would be in the ambient atmosphere, moisture will be adsorbed, causing *wavy edges*. On the other hand, if moisture is given off by the paper, the result is *tight edges* (Fig. 23-10).

Dimensional stability is an indication of the ability to resist strain caused by stress or changes in moisture content. It should be considered in both the machine direction and cross direction. Contact with moisture, heating of the sheet to dry ink, or high stresses required to handle paper during printing may cause appreciable changes in the size of paper. Uneven application of these factors may cause the paper to cockle. Paper may be kept in sealed containers prior to use to prevent some dimensional stability problems.

Curl

Curl has been discussed on page 171. Lack of curl in paper is important, especially in multiple printing operations such as four color printing. Changes in moisture content (between conditioned EMC and EMC conditions in the press room) often cause paper to curl (*atmospheric curling*), although paper manufactured on twin—wire machines tends to be less prone to curling. Curl tends to occur around the axis of a cylinder, and the direction of curl is defined as the concave side. On single—wire machines, atmospheric curling tends to occur around an axis parallel to the machine direction. The direction is toward the wire side as water is lost and toward the felt side as water is gained. *Mechanical curl* is cause by stress whose origin is not from moisture content fluctuations.

Paper smoothness

Gravure, because of the low viscosity of ink used in this process, can work well with reasonably rough paper surfaces. Even with this process, however, as the resolution of halftones increases (more points per inch) paper smoothness becomes more important so that each halftone dot can contact the surface of the paper to give uniform printing. The rubber blanket of offset printing can conform to surface imperfections to some degree. Paper smoothness and other properties in regard to printing quality are often best approached by making actual prints and using standardized techniques to evaluate those prints. Related to paper smoothness and print quality is paper compressibility.

Papers with some compressibility tend to give better prints than paper without compressibility, especially with letterpress. A very smooth uncalendered sheet gives better printing results than the same sheet that has been heavily supercalendered.

Ink receptivity

Ink receptivity is a measure of the amount of ink that is transferred from the plate to the paper. The ink receptivity must be high enough to leave a clear, even print on the paper. A number of factors (just as in paper sizing and a few additional ones such as the rate of ink drying) are important. Smooth paper usually requires less ink for a given print quality than rough paper due to a lower surface area and possible lower degree of ink migration below the surface. Related to roughness is paper *levelness*, which refers to evenness of caliper on a larger scale, above 0.1 in. (TIS 108-01). Paper that is somewhat compressible gives a more even distribution of ink.

The ink—paper sizing relationship is very important to ink receptivity. As pointed out in other areas of this book, sizing of paper must always be considered in terms of the liquid against which paper is sized. Since the sizing characteristics of paper in terms of ink penetrations are very important, sizing should be considered in terms of the components and properties of ink.

Ink holdout refers to the tendency of paper to resist the penetration of ink inwardly. A high level of ink holdout means the ink remains on the surface of the paper. When the vehicle of the ink (which causes a decrease in paper opacity by filling air pockets that otherwise increase light scattering) is absorbed too deeply in the paper *strike—through* (*soakiness* is severe strike—through) occurs; i.e., the ink is visible from the other side of the paper. In papers where the vehicle takes a long time to set, strike—through may appear several hours after printing.

Printing with a high degree of ink holdout usually results in a print with higher gloss in the printed areas. If the ink holdout or receptivity is uneven, a *mottled* (uneven) result may be obtained; this is often caused by uneven binder migration in coated papers (Engström, 1994).

Lithography and paper properties

Chalking occurs in lithographic printing when the ink does not bind well to a coating. It may be especially profound in the early stages of drying the ink where the ink can be rubbed off with minimal effort. *Snowflaking* occurs when water is present in the printing nip; small (0.0005 to 0.005 in.) white specks may appear in printed areas.

Tinting occurs when small ink particles appear on unprinted areas in an (otherwise) properly operating press. One factor that may cause this is paper coating ingredients that may alter the surface tension properties of the printing plate. Emulsification of ink in the press water may also cause this. *Scumming* is the same defect with a cause traceable to the printing process such as mechanical or chemical degradation of the plates, improper preparation of the plates, or improper ink—water balance during printing.

Water sensitivity refers to the weakening of paper due to press moisture. This may lead to picking at the surface or paper curling. The ink—water balance, therefore, is critical in web offset (lithographic) printing.

Paper coating, binders, and printing

Grades in order of decreasing quality include ultra—premium, premium, #1, #2, #3, #4, and #5. The latter two are available in roll form only for use in catalogs and magazines of short life and may contain mechanical pulp. The highest quality grades are available in sheet form only.

Engström (1994) discusses mottle as a defect in offset printing of coated paper; this is a common problem. The coating is usually applied as a slurry of 45—60% solids containing 10—15 parts binder per 100 parts pigment (mostly kaolin clay). The final dry coat is about 10 g/m^2 (7 μm thick). Mottle, nonuniform print density, can be caused by uneven distribution of the coating binder on the surface of the paper due to *binder migration* during the coating and drying operation.

Polyvinyl acetate binder in coated paper is useful for web offset coatings because it is somewhat porous and imparts stiffness in the paper (useful for improving the stiffness in the cross machine direction); these properties decrease paper blistering during printing. Synthetic binders often have poorer ink receptivity than starch or protein binders, but give enhanced ink gloss due to improved ink holdout.

Miscellaneous considerations

Paper used in books and magazines should be printed such that the binding edge is parallel to the machine direction on the paper machine so that pages turn easily and lie flat. The binder used in coatings decreases the gloss and brightness, so as little as possible is used. Gravure and letterpress put much less stress on the coating adhesive than offset so paper made for gravure will probably give picking problems when used for offset.

23.12 SPECIFIC PRODUCTS

Glassman (1985) provides technical information on this subject, and the U.S. Dept. of Commerce (1992) provides statistical information.

Employment and advertising

In the U.S. in 1990, 1.6 million people were employed in printing and publishing (compared to 700,000 in paper and allied products), a substantial increase over the 1975 employment of 1.08 million. The value of all printed materials was $157 billion. In commercial printing operations (which excludes newspapers, periodicals, and books), 560,000 people earned $14.3 billion and produced $52.9 billion in shipments. Advertising

expenditures were $32.3 billion in newspapers, $6.8 billion in magazines, $28.4 billion in television, and $23.4 in direct mail.

Of course, letterpress was the dominant method of printing for virtually all products at one time, but this method has been replaced in most applications.

Newspaper

Most newsprint is printed by web offset (lithography). The ink is usually carbon black in mineral or plant oil (soybean oil). (The triglycerides of plant oils are much more easily removed during deinking of recycled fiber than mineral oils that are straight chain hydrocarbons.) Offset has the problem of lint from the paper accumulating on the blanket. The use of twin—wire formers for newsprint has greatly decreased this problem. The introduction of special inks and fountain solutions have also helped mitigate this problem.

A few newspapers with large circulations, especially on the west coast of the U.S., use the flexographic process with water—soluble inks. This process has the disadvantage that the newsprint is very difficult to recycle. The problem is that the ink vehicle allows the carbon particles to disperse easily in water giving a very dark suspension. This water is difficult to clean or to treat as effluent.

In the U.S. in 1990, 443,000 people were paid $10.41 billion to produce $36.6 billion in newspapers using $8.09 billion of materials. In 1992 there were 11,339 newspapers (of which 1755 were daily with a total circulation of 62 million). Most newspapers are published weekly.

Magazines and periodicals

Magazine papers are often machine coated with clay and possibly small amounts of calcium carbonate and titanium dioxide and supercalendered. Increasing amounts of mechanical pulp are used in these papers. Lightweight coated papers are being used to larger degrees due to high postal costs. Brightness, opacity, blister resistance, and folding properties are very important in these papers.

Web offset is widely used for magazine papers. Gravure accounts for a little under half of magazine production. It is used for the largest print runs (like the *National Enquirer*) or moderate sized jobs where high—quality color reproduction is desired. In 1994, R.R. Donnelley and Sons, Inc. controlled about 45% of the magazine gravure market, which accounted for about 20% of magazines printed. Gravure has a long tradition of use in Europe, but this is changing.

In the U.S. in 1990, 115,000 people were paid $3.66 billion to produce $20.4 billion in periodicals using $6.58 billion of materials. In 1992 there were 11,143 periodicals of which 466 were weekly and 4326 were monthly.

Books

Books represent an area where factors outside printing influence printing methods. (Too, with periodicals postage costs and regulations play an important role in how products are printed.) Many book publishers print short runs because they do not want large inventories of books on hand. (This is a trend of TQM, to limit inventory, and, in some cases, inventories are taxed by governments.) Most books are printed by offset, but flexographic printing is commonly used for paperback books.

Every year about 40 to 50 thousand new books or new editions of books are published in the U.S., not including government publications and certain other categories. In 1990, the overall average retail price of a hardcover book was $39.91. Mass market paperbacks averaged $4.47, while trade paperbacks averaged $16.60.

In the U.S. in 1990, 74,000 people were paid $2.30 billion to produce $15.3 billion in books using $4.47 billion of materials.

Packaging

Much of the flexographic printing that is carried out is applied to packaging. It is applied to bags, milk cartons, and corrugated boxes. Lithography is used for short press runs requiring high—quality pictures or print, while gravure is used in long press runs for the same purpose.

Sheet—fed paper for offset printing

Sheet—fed paper for offset printing is the most difficult grade of paper to print. The demands on this grade of paper are quite high.

Plain paper copiers

Plain paper refers to the fact that special coatings are not required as in earlier forms of office copying methods. The paper for this process must be able to withstand the high heating required to set the toner without curling, which would otherwise cause paper jamming.

Other papers

Other papers that are printed have special uses but low levels of production. For example, consider butter wrappers that must be greaseproof or soap wrappers that must be alkali resistant.

23.13 COMPUTERS AND PRINTING

With the advent of personal computers, many people are able to do their own printing by "desktop" or "micro" publishing. The publishing industry tends to use Apple computer products (especially Macintoshes) more than IBM compatible types. The former have been designed for graphical interfaces from the beginning. Many of the programs are available for either environment.

Computers are involved in every aspect of printing from prepress to press control. They offer the ability to decrease *turnaround* time from when the customer submits a job to when the job is completed. Inventories of printed materials can be greatly decreased or entirely eliminated in certain cases. Printing of regional editions of documents is much more easily accomplished and new stochastic imaging techniques are now made possible by computers. The best—selling products (in 1994) are mentioned as examples.

Most people now use word processing programs (such as WordPerfect or Microsoft Word) for routine typing and moderately high level page layout. (WordPerfect 5.1 was used to produce the final copy of this book.) Page layout programs (such as Aldus PageMaker or QuarkXPress) represent the next step of sophistication for page production; they are designed to bring text and graphics (such as those produced on the programs mentioned below) together on the page(s).

Illustrations are made with programs such as Adobe Illustrator, Aldus FreeHand, or Corel Draw. Images (photographs) can be manipulated with programs such as Adobe Photoshop (about 90% of the market) or Aldus PhotoStyler. Aldus Persuasion is used to make slide presentations.

The computer can now take the digitalized layout and directly make a negative for a plate.

This bypasses the tedious steps of making films, assembling films so that the plate can be made (stripping), and blue—lining the film proofs. Also the chance of transcription mistakes is greatly decreased. Trapping of a document can be accomplished by Island Trapwise or Aldus TrapWise. Imposition software (Aldus PressWise, Impostrip, or QuarkXTensions) allows one to lay out pages from one or more documents to form signatures (groups of pages to be printed together on a single sheet as described briefly in the section on binding). In the future, complex printing plates will be made directly from the computer.

Possible pitfalls

The use of computers does come with some potential pitfalls. There is a tendency to over-design one's product. For example, the wide choice of fonts occasionally detracts from the product. Consider the examples in Fig. 23-11. More information on fonts and types is contained in this chapter's appendix. The use of numerous fonts often makes the final work look like the format used in "ransom notes" (kidnappers sometimes use words removed from a wide variety of printed matter to "write" ransom notes).

Furthermore, it is easy to design many special effects on computer screens, but one should still have some knowledge of printing, or work with a printer, to produce the final product. For example, if small text is placed on top of metallic paint, the text may rub off the final product because ink does not adhere well to metallic paint.

Unless one is using an entire system by a single manufacturer (read proprietary and very expensive) designed for accurate color reproduction, the color results one observes on one's screen will not be the same as the final output. Each computer monitor is different and subject to color alignment and calibration. Computer monitors use the RGB color system, and printing uses the CYMK color system; the sets of colors these two methods can achieve are different.

PostScript

PostScript is a computer language (as are Basic, Fortran, C, and Pascal) that is used for page description to control computer printers. It is another example of a big company making a big mistake because they thought "they knew the market and nobody was interested in this technology." A group of people interested in this technology defected from Xerox, who did not have interest in this technology, in 1982 to begin development at Adobe Systems. In 1985, the first printer to use this was commercially available.

Since PostScript is a set of protocols, it can be run on any printer that supports the protocol. Its operation becomes transparent to the user; that is, one does not need to know anything about it in order to achieve good results with it. PostScript files use the file extension filename.eps to designate their purpose; eps is an acronym for Encapsulated PostScript. A printer controller, a raster image processor (RIP), converts the PostScript file into the final product, whether a printed piece of paper from a laser printer or an imagesetter. An imagesetter is a very high quality printer capable of 1200—5000 dots per inch. They use inks, not toner particles as in laser printers. They are available at prepress service bureaus.

PostScript has the advantages that it provides high text and graphics quality. Many typeface choices are available, and it is device (printer) independent. PostScript uses the CIE color model.

Digital imaging technologies

Digital imaging technology not only increases the speed of many production steps, it allows many new capabilities. The data for a print job is entered into the computer, the data is processed, and the results are printed to photographic film, paper, or to a printing plate.

Text data is entered directly from a keyboard or storage device. *Scanners* are devices that allow images to be stored in digital form. These images can be photographs or artwork, for example.

IT IS SMART TO BE
THRIFTY

It is *SMART* to be
THRIFTY

Fig. 23-11. The careful use of fonts adds to the message, the overuse of fonts detracts from it.

Because of the large amounts of data involved, storage devices such as compact disks (CDs) are often used to store the digital information. A compact disk stores 600 Mbytes (a byte is 8 bits and a bit is the smallest unit of storage consisting of a variable that is 0 or 1). An image with 256 shades of color on a 1000 by 800 pixel matrix (SVGA computer screen) uses 25.6 Mbytes of data and begins to provide the quality of a small photograph. Photographic CD systems developed by Kodak use 18 Mbyte images that (using computer algorithms) compress images to 6 Mbytes in digital form, allowing about 100 images per CD.

Color images should be scanned with a resolution of 300—1200 dots per inch, while black and white material should be scanned at 1000—1200 dots per inch. The human eye—brain system gets more of certain types of information from black and white images than from color images. For example, night vision is black and white.

Scanners are also used to "read" printed material so that manual keyboard entry of already printed text is not required. The digital image is converted to text data by the use of *optical character recognition* (OCR). Various OCR software programs have various levels of reliability and can handle certain fonts and font sizes.

Once the data is entered into the computer, the data is processed. The main categories of data processing include manipulation of individual images and page layout. For example, an object in a photograph can be separated from its background, put on a new background (or foreground), and modified to remove imperfections in ways that were essentially impossible a few years ago.

The results must be printed. It is desirable to be able to do this to check the layout, color rendition, and other aspects of the job, but high-quality color printing still requires an expensive printer.

Examples of good communication

When Hewlett—Packard was developing and marketing ink jet printers, they provided about 20 paper companies (after signing secrecy agreements) samples of their ink in bulk so that companies could develop papers that would work well in their printers. This allowed the paper manufacturers to tailor their products to meet this niche. The result is that paper companies sold more paper and Hewlett—Packard sold more printers.

Examples of poor communication

When Perfect Binding was introduced in the late 1970s, the manufacturer of the binding machine suggested that print shops leave the glue pot heated so that binding could be immediately carried out as the need arose. If only occasional binding was done, decomposition of the adhesive occurred and a binding job which did not last long resulted. After several years, the manufacturer of the adhesive learned of this problem and indicated that of course this would happen and one should not leave the adhesive pot heated unless the binding machine is being used. Since that time the overall quality of Perfect Binding has increased.

When many fine—paper mills went to alkaline papermaking with calcium carbonate fillers, the characteristics of their papers changed drastically. For instance, many offset fountain solutions are considerably acidic and the alkaline paper can change the carefully balanced chemistry of the ink—water balance of an offset press. Many paper vendors did not know or did not care that their products were suddenly different, and many printers had serious problems because this change was not communicated to them ahead of time.

23.14 BINDERY OPERATIONS

Bindery operations include anything that must be done after printing in order to complete the printing job. It includes perforation, scoring and folding (Fig. 23-12), signature formation (Fig. 23-13), cutting and trimming, collating, drilling or punching holes, binding, packing, counting, storage, and transportation. These operations may be in—house or out—of—house. Many business specialize in binding operations.

Scoring

Scoring is the formation of a crease by folding paper around a rounded edge before the actual fold. It is carried out when precision folding is required, when folding stock thicker than 5 mil, or when folding coated stock.

Perforation

Perforation of straight edges in one direction can be accomplished off press using a knife that alternates with sections of sharp edges and blunt portions. More complicated perforations are carried out on the press.

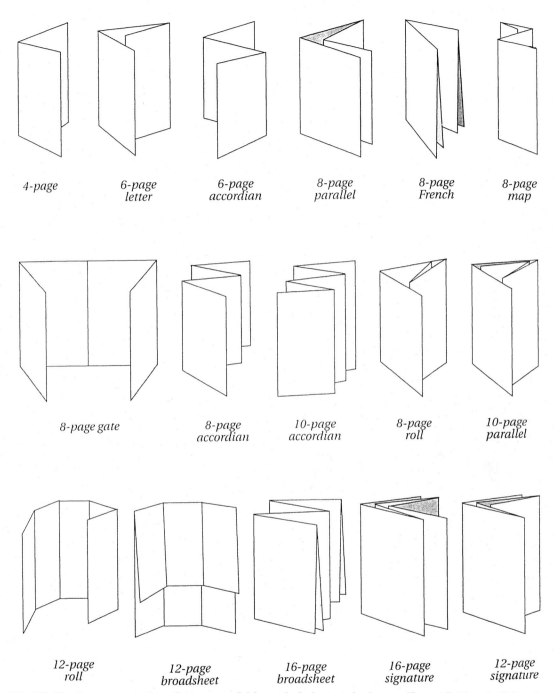

4-page 6-page letter 6-page accordian 8-page parallel 8-page French 8-page map

8-page gate 8-page accordian 10-page accordian 8-page roll 10-page parallel

12-page roll 12-page broadsheet 16-page broadsheet 16-page signature 12-page signature

Fig. 23-12. Some examples of common folds and their terminology. From *Getting It Printed*, revised edition ©1986, 1993 by Mark Beach. Used by permission of North Light Books.

Signatures

Signatures are groups of pages that are printed together on a single, flat sheet; they are assembled together for binding. The signature is formed by cutting, folding, and trimming operations. Signatures of 16 pages are common in book publishing, but those of 4, 8, 12, 24, and 32 are often used too (Fig. 23-13). Signatures can be assembled in nested signatures or gathered signatures, depending upon the desired binding method; Obviously the placement of pages during prepress operations must take this into account. Nested signatures are used in binding by saddle stitching; otherwise, it is not common.

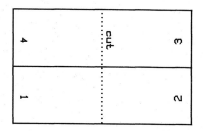

Cutting

Cutting of a stack of paper, the *lift*, is done by a sharp, guillotine blade that may cut through three inches of paper. It is important for the lift to be help down tightly while it is cut. Even so, there is some movement, called *draw*, that leads to variation of the cut size.

Binding

A wide variety of bindings are available for attaching sheets or signatures to form the product (Beach, 1993). They differ in appearance, durability, equipment requirements, and cost. (Fig. 23-14 shows a binding test machine that turns a page back and forth to test its anchoring in the binding.) The main binding methods are summarized in Fig. 23-15.

Case binding

This is considered to be the most durable and visually pleasing of the binding methods (Fig. 23-16). Case binding may be square back or round back and uses a hard cover (hardbound). The paper MD should be parallel to the spline for optimal results.

Signatures are sewn together (Smythe is the major manufacturer of the sewing machines), trimmed on three sides, and placed inside the case of binder boards covered with paper, cloth, plastic, or leather.

Mechanical binding

Mechanical binding uses wire or plastic through a series of small holes to hold pages together as in spiral notebooks. A plastic comb (GBC is a trade name) and double wire (Wire—O is a trade name) are two other examples. They are commonly used in technical manuals, cookbooks, and in—house publications because the book stays open when set down and the process can be done almost anywhere with simple equipment.

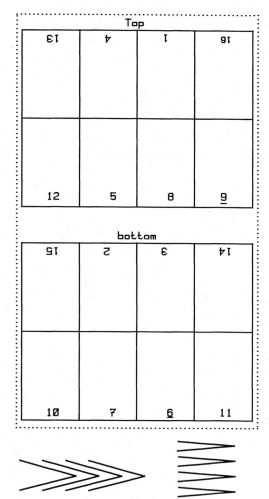

Nested Signatures   Gathered Signatures

Fig. 23-13. A 4 page—signature printed on one side only (top) and a 16—page signature printed with 8 pages on each side.

Perfect binding

Gathered signatures can be bound by perfect binding. The spine is trimmed so that the edge of each page is exposed. Adhesive is then applied to the edges and this is glued to the binding. The remaining three edges are trimmed after the adhesive has set. This method is used in paperback books. The MD of paper should be parallel to the binding. Some binders recommend against the use of C2S (coated two side paper) if this type of bind is to be used.

Traditional binding adhesives for perfect binding are thermal (hot melt adhesives) or cold glues (polyurethane resin, PUR). Polyurethane resin is said to offer the advantages of increased strength, increased flexibility (so books do not snap shut), and increased recycling efficiency (the polyurethane does not break up and is easily removed when the paper is recycled). PUR is recommended for *lay flat* perfect binding. Polyurethane is moisture resistant and it mitigates gutter ripple; it requires a 24 hour cure time.

Saddle—stitch binding

This method of binding usually uses a staple that goes from the top to the bottom of the center of the open work. It requires at least 8 pages and is expandable in units of 4 pages. It is a quick and inexpensive binding commonly used for small magazines, catalogs, brochures, and pamphlets.

Padded binding

In padding, the edges of loose sheets are held together by glue as in note pads.

23.15 ANNOTATED BIBLIOGRAPHY

1. Beach, M., *Getting it Printed*, Revised Edition, North Light Books, Cincinnati, 1986, 1993 (199 p.) "How to work with printers graphic art services to assure quality, stay on schedule, and control costs." This and other books by the author are easy to read, well illustrated, reasonably priced, and highly recommended by many people in the trade.

2. Blasche, D.W., Methods of cylinder engraving for gravure printing, *Tappi J.* 76(8):169-174(1993) (no references).

3. Engström, G., Formation and consolidation of a coating layer and the effect on offset-print mottle, *Tappi J.* 77(4):160-172(1994), (63 references).

4. Glassman, A., Ed., *Printing Fundamentals*, Tappi Press, Atlanta, 1985, 388 p. This book gives much detail about the characteristics of various papers used in printing.

Fig. 23-14. Binding test machine.

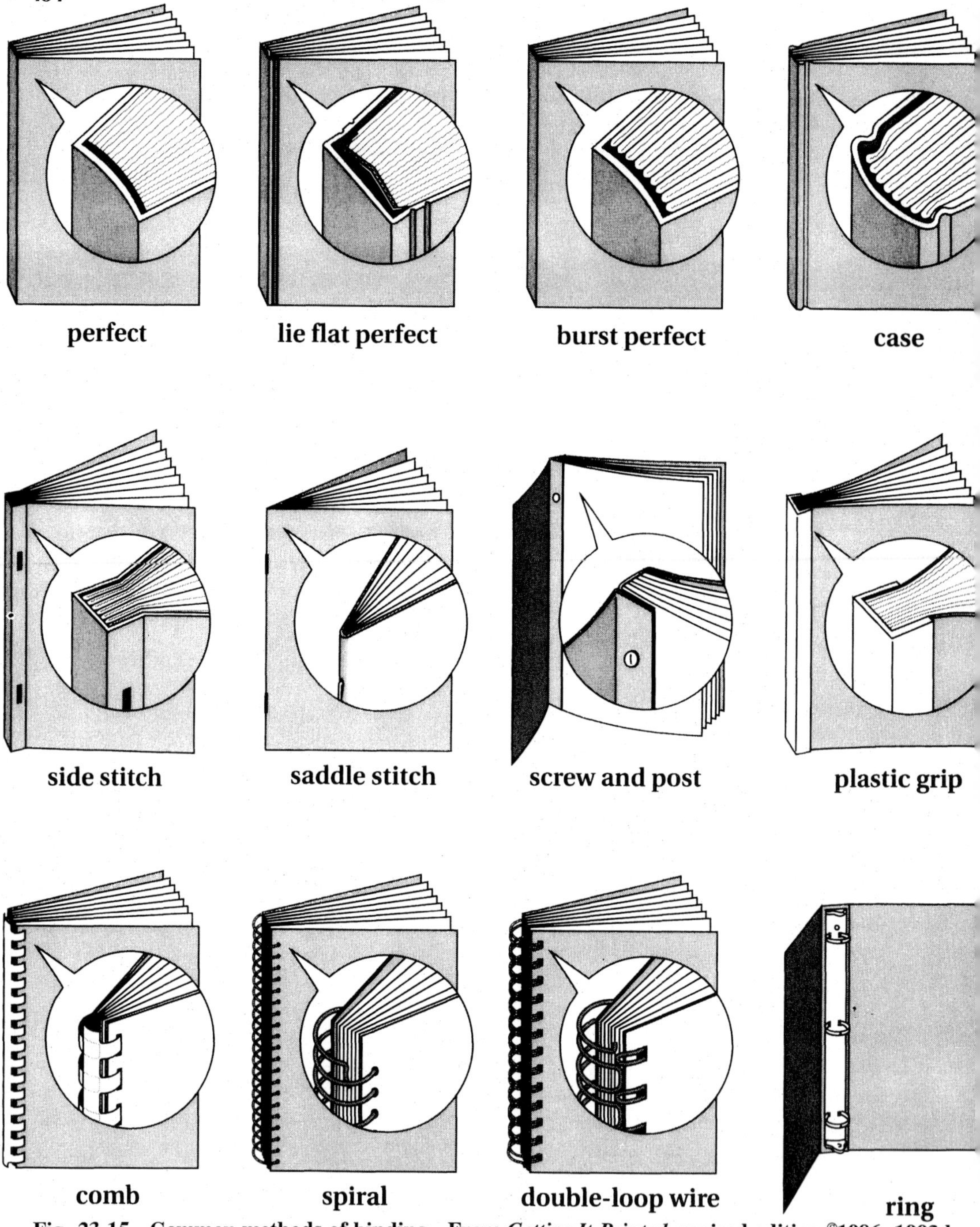

| perfect | lie flat perfect | burst perfect | case |

| side stitch | saddle stitch | screw and post | plastic grip |

| comb | spiral | double-loop wire | ring |

Fig. 23-15. Common methods of binding. From *Getting It Printed*, revised edition ©1986, 1993 by Mark Beach. Used by permission of North Light Books.

5. International Paper Co., *Pocket Pal, A Graphic Arts Production Handbook*, 13th ed., International Paper, New York, 1983 (216 p. with several color plates). The first edition was published in 1934. It is called pocket pal due to its 4.2 by 7.2 in. size. This is a concise, easy to read book on the topic.

6. Kline, J.E., *Paper and Paperboard, Manufacturing and Converting Fundamentals*, 2nd

Square back

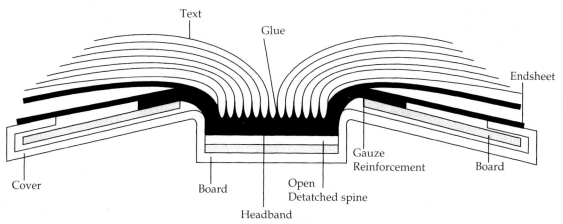

Text

Glue

Endsheet

Cover

Board

Open
Detatched spine

Gauze
Reinforcement

Board

Headband

Round back

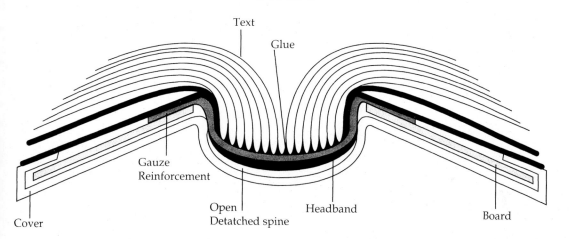

Text

Glue

Gauze
Reinforcement

Open
Detatched spine

Headband

Board

Cover

Fig. 23-16. Case binding. ©1994 Lincoln and Allen Bindery, Portland, Oregon (with permission).

ed., Miller Freeman Pub. Inc., San Francisco, 1991. One chapter (pp. 162-183) contains information on printing and printing grades of paper.

7. U.S. Dept. of Commerce, *Statistical Abstract of the United States*, 112th ed., 1992. There is a section on the printing industry.

8. Williams, R.L., *Paper & Ink Relationships*, Mennonite Press, Inc., Newton, Kansas, 1985, 150 p. a) p. 44. This is an easy to read, perhaps more practical reference on printing methods than the ones listed above; it is a useful addition to one's library.

Tappi Technical Information Sheets

Information sheets exist for the properties of aqueous slurries of various pigments and fillers including calcium carbonate (TIS 106-01), hydrous clay (106-02), and anatase (106-03). Pigment indexes for these materials are 106-05 to 106-07, respectively. Graphic art terms (somewhat dated in terms of twin-wire formers since the last version is 1964) are defined in 108-01.

TIS 0305- (14-19, 38) and 0306-01 to 0306-09 contain methods for printing corrugated containers. These, like many of the TIS sheets for corrugated containers, are fairly generic (of widespread use) in their treatment of the subject matter. Examples include 0305-16 (Flexographic ink), 0305-38 (Printing section, troubleshooting), 0306-05 (Glossary of flexographic terms), and 0306-03 (Photopolymer plate update).

EXERCISES

1. Describe the geometries of the four major types of printing processes.

2. What is offset printing? To what process does this usually refer?

3. What is halftone printing? Why is it used?

4. What is stochastic printing? Give two major advantages of this process over traditional screen printing of halftones. One advantage is especially applicable to color printing.

5. What is four—color process printing? What is its theoretical basis?

6. What is an important measurement for quality control on a four—color printing press?

7. What printing process requires a paper surface that is very resistant to surface picking?

APPENDIX

Anilox, ink transfer rolls

The ink is applied to the plate by a series of ink rollers. Traditionally, soft roll applicators were used, but many flexographic applications now use metal rolls having a mechanically engraved surface. These are termed *anilox rolls*, which were developed in the 1930s (the term is related to the fact that flexography was initially called aniline printing, since aniline—based coal—tar dyes were used as inks). The engraved surface allows metering of the ink (which is otherwise difficult to control in flexography) and can carry more ink that flat—surfaced rolls. The engraved surface is chemically and physically protected by chrome plating or ceramic (Al_2O_3) spray. A ceramic—coated roll may be engraved with a laser.

TIS 0305-17 (1982) states that for flexographic printing on corrugated boxes, the industry standard for mechanical surface engraving of anilox rolls is the 165—pyramid cell (42 μm deep) when the ink—fountain contains a doctor blade or the 140—quadragravure cell (64 μm deep) when presses use a doctor blade wiper (Fig. 23-17). (Cells are listed as number per linear inch.) Laser engraved ceramic—covered rolls can be engraved with much finer patters (250 is common, 550 is specialty grade) to give better printing resolution TIS 0306-1 (from 1992). Engraving angles are typically 45° for corrugated box applications, though other angles are available for more viscous ink. Since most of the ink is carried at the surface of the roll, it is very important to maintain ink capacity by decreasing wear. This is done by running the system with ink (a lubricant) and keeping the rolls clean.

Various types and sizes of fonts (Beach, 1993)

Point sizes vary from 6 to 72 and refer to the height of the line of type in 1/72 in. increments (in North America and the United Kingdom) from the bottom of a letter with a *descender* (like y or p) to the top of a letter like T or W; the common 10 point type with 2 point spacing is 6 lines per inch. A *pica* is 12 points, and 6 picas make an inch.

Fonts of the *serif type* have bases on letters such as A, F, H, and f. Examples include Times Roman and Garamond Antiqua, which are used for use of text and captions. Since *sans* means without, *sans serif fonts* are solid and include Antique Olive and Arial. These fonts are used in headlines, posters, callouts, and a short index.

This book is set in CG times proportional spaced font. Some examples of other fonts are presented here. This is fixed spaced courier, like a typewriter. Notice the width of the i and w are the same. *This is the same typeface as italics* **and bold italics**. This is letter gothic, *and with italics*. This is new times roman.

Gamma function

A *gamma function* is a function used to alter the shades of darkness of a printed surface. For example, it is used to enhance darkness (since the eye is more sensitive to dark areas) so that the eye can more easily recognize images.

165—Pyramid cell, 42 um high, 300,000 um^3 ink/cell

140—Quadragravure cell, 64 um cell depth

Fig. 23-17. Engraving of anilox rolls.

24

OPTICAL PROPERTIES OF PAPER

24.1 INTRODUCTION

Some aspects of optics and dyes have been included in Chapter 8 on additives. The optical properties of many additives are of central importance to their use. Optical properties of paper have been briefly discussed on pages 185 and 186.

Electromagnetic waves and light

Some of the basic properties of light will be presented in this chapter. The theoretical development is fairly complicated, but is of little use to papermakers. Most introductory physics textbooks at the college level cover these topics in various levels of mathematical detail.

Light is an electromagnetic wave and has properties of waves and particles. It consists of individual photons that have properties of waves. The properties and applications of electromagnetic waves depend on their frequency (hertz, Hz, "cycles per second") or wavelength. Both units have common uses depending on the area in which they are used; for example, light is often reported in wavelengths while radio waves are reported as either depending on the band. Frequency and wavelength are related to the speed of light, c, (which is 2.9979246×10^8 m·s^{-1} in a vacuum) as:

$$c = f\lambda \qquad (24\text{-}1)$$

EXAMPLE 1. Shortwave radio bands are often described in terms of their wavelengths. What is the approximate frequency of the 40 meter band?

SOLUTION: Rearrangement of Eq. 24-1 gives $f = 3.00 \times 10^8$ m·s^{-1}/40 m = 7.50 MHz

PROBLEM: The wavelength of light used in the TAPPI standard brightness test is 457 nm. What is its frequency?

Answer: 6.56×10^{14} Hz.

Electromagnetic waves (in increasing order of energy and frequency) include radio waves, microwaves (used in satellite dish communications and microwave ovens), infrared light, visible light, ultraviolet light, x—rays, and gamma rays. Table 24-1 shows the electromagnetic spectrum with landmarks inserted in nearly correct positions.

The energy of a light particle, E, is usually given as a function of frequency, f, and *Planck's constant, h*, as:

$$E = hf \qquad (24\text{-}2)$$

Planck's constant has the value of 6.62×10^{-34} J·s. Since photons are involved in chemical reactions, such as color reversion in mechanical pulps, it is useful to consider the energy of a mole of light particles. The energy of light at 400 nm is approximately 300 kJ/mol (72 kcal/mol), which is below that of most chemical bonds (see page 315). UV light at 200 nm has 600 kJ/mol, which is sufficient to break many chemical bonds.

Table 24-1. The electromagnetic spectrum.

Wavelength	Landmarks	Frequency (Hz)
higher energy ↑		
0.01 nm	- gamma ray; cosmic ray	- 3×10^{19}
0.01 nm	- X-ray; gamma ray	- 3×10^{19}
0.1 nm	- X-ray	- 3×10^{18}
1 nm	- far UV; X-ray	- 3×10^{17}
10 nm	- near UV	- 3×10^{16}
100 nm	- visible light 400-700 nm	- 3×10^{15}
1 μm	- near infrared light	- 3×10^{14}
10 μm	- far infrared light	- 3×10^{13}
100 μm	-	- 3×10^{12}
1 mm	-	- 3×10^{11}
10 mm	- radar, microwave	- 3×10^{10}
100 mm	- radar, UHF TV	- 3×10^{9}
1 m	- TV chan. 7-13	- 3×10^{8}
10 m	- FM 88-108 MHz	- 3×10^{7}
100 m	- shortwave radio	- 3×10^{6}
1 km	- AM 550-1600 kHz	- 3×10^{5}
10 km	- long radio waves; VLF	- 3×10^{4}
100 km	-	- 3×10^{3}
lower energy ↓		

The overall relative sensitivity to light by the human eye (i.e., the standard luminosity function of the standard observer as will be given in $\tilde{y}$ of Fig. 24-10) is about 20% at 470 nm, 40% at 490 nm, 60% at 500 nm, 80% at 510 nm, 100% at 550 nm, 80% at 590 nm, 60% at 610 nm, 40% at 630 nm, and 20% at 650 nm.

However, the brain applies compensation in that the appearance of brightness is a logarithmic function of the light level (as is the perception of sound level by our ears). For example, a surface that reflects 10% of its light contrasted to one that reflects 20% has an equal apparent brightness difference when compared (at the same time to avoid dilation of the pupil as a variable) to the difference in a surface that reflects 20% of its light contrasted to one that reflects 40%.

Index of refraction, reflection, and absorption

The speed of light in matter is always smaller than that in a vacuum. The ratio of the speed of light in a vacuum to that in a given material is known as its *index of refraction, n*. The index of refraction of a material is usually a function of the wavelength of light. Values in the literature for various materials correspond to n determined with the yellow light of a sodium light (589 nm) near the center of the visible spectrum.

In practice the index of refraction is important not because the speed of light in a material is of consequence, but because light "bends" (unless it is normal to the surface) when it travels across

an interface of two materials of differing indexes of refraction. This is what causes the image over a hot road or heater to appear to ripple; differences in the density of air cause differences in its index of refraction. The index of refraction of air is about 1.0003 and is usually treated as unity.

24.2 BEHAVIOR OF LIGHT RAYS

If a beam of light traveling in a vacuum (air is close enough) strikes a flat surface of a transparent material at an angle of incidence ϕ_a relative to normal (Fig. 24-1), some of the light will be reflected at angle ϕ_r and some of the light will be transmitted to the object at angle ϕ_b. It is known that the angle of reflection from normal is equal to the angle of the incident light from normal ($\phi_a = \phi_r$). It is also known that the $\sin \phi_a / \sin \phi_b = n$ of the transparent material. The incident, reflected, and refracted rays and the normal line are all in the same plane. If the refracted light were to travel the same path backward the same relationship would exist, that is $\sin \phi_a / \sin \phi_b = n$.

Refraction means that when light travels from a material of low index of refraction to a material of higher index of refraction, the light is "bent" toward the normal. The converse is also true. Fig. 24-1 can be generalized for any two materials as $n_a \sin \phi_a = n_b \sin \phi_b$.

With total internal reflection, it is possible that a light ray will be reflected back into a material when striking at a high angle. It is this phenomenon that keeps light inside glass fiber cables so that it can be transmitted large distances with relatively little loss of intensity.

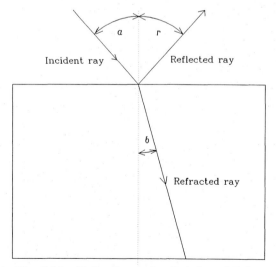

Fig. 24-1. Reflection and refraction of light.

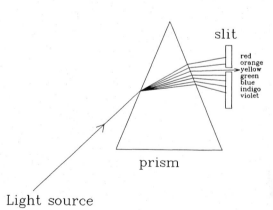

Fig. 24-2. Use of a prism and slit to produce monochromatic light.

Dispersion uses the fact that the index of refraction of a material is a function of the wavelength of the light ray. The index of refraction of a material is often higher for blue light than for red light. White light can be separated into its component colors by the use of a prism as demonstrated in Fig. 24-2 or as observed in a rainbow.

Light may also be absorbed as it is transmitted through a material (see Section 14.6). Reflection, refraction, and absorption of light are all very important in determining and predicting the optical properties of paper, as we will see.

Absorption, transmittance, and reflection in matter

Of course the interaction of light with paper and other materials is much more complex than the simple example just considered. In fact, the surfaces of the particles are in random orientations and a beam of light is usually scattered numerous times (multiple scattering), as shown in Fig. 24-3, and the scattered light goes in haphazard directions in all three dimensions. We refer to diffuse reflectance (as opposed to specular reflection as in a mirror and Fig. 24-1) and transmittance.

It is interesting to consider some overall generalizations. Imagine a sheet of paper or powder that is infinitely thick in terms of light penetration (> 1 cm). If no light is absorbed, then all light would be scattered out the top surface; the light will continue to scatter indefinitely until it reaches the top surface where nothing can send it back into the sheet. (See the average end—to—end distance of a polymer for the reasoning on page 437.) Reference standards such as MgO and $BaSO_4$ have low absorption coefficients with high scattering coefficients and, therefore, are

nearly perfectly white (because they reflect nearly all visible light that shines on them).

Black paper has very high absorbance across the spectrum and so little light is reflected or transmitted. Gray paper has some absorbance which is approximately constant across the visible light spectrum. Gray or white is said to be *achromatic* (without color) and has a flat spectral reflectance curve.

Glassine paper scatters light very little and has relatively low absorbance so that most light is transmitted through it, unless it is very thick. An extreme example of this case is a clear, plastic, solid, film, such as cellophane (cellulose), which behaves as in Fig. 24-1, mostly transmitting light through the material with essentially no scattering.

Fillers and scattering coefficients

It is known that the maximum amount of scattering results when the minimum particle dimension is about one—half the wavelength of the light being scattered. This corresponds to about 0.25 μm, which is a common size for certain fillers. While some fillers have smaller particle dimensions than this, they often tend to agglomerate into particles of larger effective size. More information is contained in the chapters on colloid and wet end chemistry.

24.3 COLOR

Color

Visible light consists of electromagnetic radiation from about 380 nm to 700 nm, though some people may be able to detect light below 300 nm and above 800 nm. Monochromatic light ideally consists of photons all having exactly the same wavelength. Lasers approach this very closely and give pure colors within 0.1 nm. High—quality color filters typically have 90% of the light within about ± 25 nm. The colors (the order is ROY G. BIV) one perceives have the following approximate wavelengths in vacuum:

violet	380—430 nm
indigo	430—450 nm
blue	450—500 nm
green	500—560 nm
yellow	560—590 nm
orange	590—630 nm
red	630—720 nm

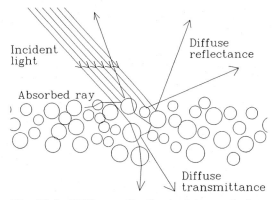

Fig. 24-3. Diffuse reflection and transmission.

Human response to color, three—color theory

The light—sensing components of the human eye consist of rods and cones, named on the basis of their shapes. Rods respond to all light (with an absorption peak at 500 nm) and confer black and white vision, the primary source of visual information. Because rods are much more sensitive to light than cones, night vision is based largely on rods, but does not allow for color perception.

The three—color theory of light perception was proposed by the English physician Thomas Young in 1801. The German physicist H.L.F. Helmholtz published his hypotheses in 1852 and 1856 regarding the three channel nature of color vision. He hypothesized that three types of sensors in the eye have their own spectral response curves and send signals to the brain when triggered. The brain takes the three signals for a given area and converts them into a single color. All combinations of light consist of various levels of response from these three types of sensors in the eye. Furthermore, it should be possible to define color in terms of the equivalent amount of the three monochromatic components of red, blue, and green. We now know these three channels correspond to the three types of cones in the eye.

Three types of cones differ in their photoreceptive pigments. (Light must be absorbed in order to cause the chemical reactions that lead to its perception.) Each of these pigments has a different absorption spectrum (absorption spectra will be described below and are demonstrated in Fig. 24-8 for various color papers) as a function of wavelength (color).

Fig. 24-4 shows the approximate relative response of the three types of cones (see Levine, 1985, page 338 or Judd and Wyszecki, 1975, page 84). The three cones have absorption spectra with peaks at about 440 nm (perceived as indigo), 535 nm (perceived as green), and 565 nm (perceived as green—yellow) and are distinguished as the short—, medium—, and long—wave (*S, M,* and *L,* referring to wavelength of light) absorbing cones, respectively (Levine, 1985). There is a high degree of overlap in the absorption spectra of the latter two pigments. It is important to note that the long—wave type cones are best stimulated (with a minimum level of stimulation of the medium wave absorbing cones) by using red light. All

shades of color correspond to the relative intensities of these three types of cones. Conversely, it should be possible to simulate any color by appropriately stimulating the three types of cones (by using three suitable colors). Much of color theory is logical if one understands these concepts.

When one speaks of color, one is referring to the way the eye responds to the light spectra that are emitted from objects. Hunter and Harold (1987) point out three visual conditions: the *illuminant mode,* where the stimulus one sees is the source of light itself, such as a light bulb; the *object mode,* where one see the stimulus as being illuminated by a light source; and the *aperture mode,* where one sees the stimulus simply such as light, as light coming through a door that is opened just a crack.

(It is estimated that most people can distinguish well over 100,000 different shades of color.) Therefore, it is relevant to consider how the human eye works. With this background, color measurement methodologies become logical.

Additive colors

A practical demonstration of the three—component nature of color lies in CRT

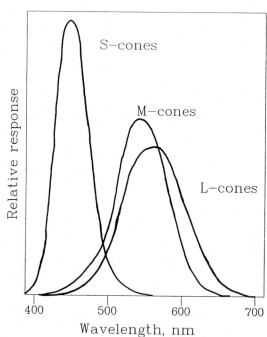

Fig. 24-4. The spectral response of the three cones of the human eye.

(computer or television) screens that have separate red, blue, and green light sources that are reasonably monochromatic. If each of the three light sources selectively stimulated one type of cone, then all possible colors that are perceived could be achieved by adding the light outputs in various proportions. In practice, the system is somewhat limited in response due to the overlap in response of the three types of cones. Nevertheless, blue, green, and red are called the *additive primaries*.

Red, blue, and green are the primary colors. Black (what do you see in a "pitch black" room?) is the absence of any color (light). As shown in Fig. 24-5 and depicted in Plate 43 (although the shades are not quite correct due to color reproduction in the printing process), equal parts of green and red light added together give yellow; equal parts of red and blue give magenta; and equal parts of green and blue give cyan. Approximately equal parts of red, blue, and green give white light. A *complementary color* of a color is the color that is added such that the overall product is white. From the argument above yellow and blue equal white, so blue is the complementary color of yellow (and the converse is also true).

Subtractive mixing

Dyes and pigments behave differently because they absorb light that is not part of the given color. This is called *subtractive* mixing.

Color printing uses yellow, magenta, and cyan inks to produce various colors by subtractive mixing to give red, blue, and green. (Each color ink should be thought of as transmitting two of the additive primaries, which is the same as absorbing one of the additive primaries.) These inks are transparent so where they overlap they each subtract one of the main additive primary colors. As Fig. 24-6 shows, yellow ink transmits red and green; it is yellow by preferentially absorbing blue light (leaving the green and red components). Cyan ink selectively absorbs red light. Where cyan and yellow are printed together, blue and red are selectively absorbed to leave green light to be reflected. Thus the subtractive mixing of cyan and yellow ink gives the appearance of green. By using the inks in various combination, most shades of color can be achieved.

Subtractive mixing of these colors is shown in Fig. 24-5 and is depicted in Plate 41 (again with poor color reproduction due to printing limitations). Imagine the three color filters in front of a white light source where these filters partially overlap. Theoretically, where the three filters overlap, all light is absorbed. This diagram only shows mixing of equal parts; the entire rainbow is more or less available by mixing (printing together) various proportions of each dye.

Mixing green and red dyes or pigments gives brown or a mud color because most light is ab-

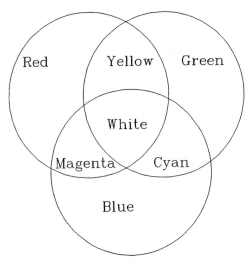

Additive colors

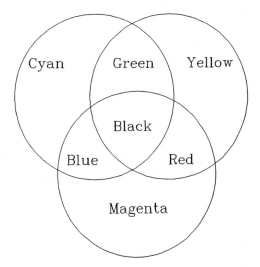

Subtractive colors (for printing)

Fig. 24-5. Additive mixture of light and subtractive mixing of inks used in printing.

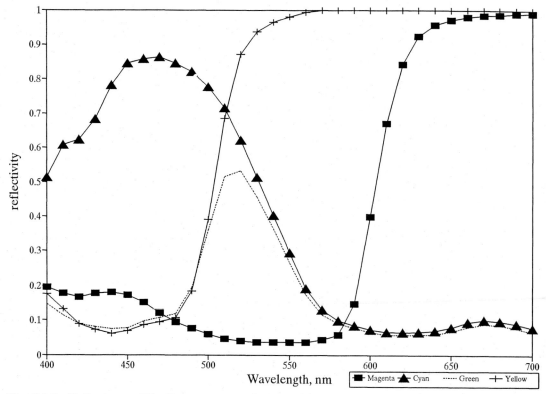

Fig. 24-6. Reflectance of the three process colors and green produced by printing cyan and yellow together. (Using printed color from *Pocket Pal*, 1983, page 94.)

sorbed and little light is reflected. (Brown is the perceived color of dark yellow or dark orange.) A similar example is the use of two filters of different colors (in series or partially overlapping) between a source of white light and the observer.

Paper whiteness

Since yellow light and blue light give white light, it is arguable that white light minus blue light should equal yellow light. Thus, a material that absorbs blue light leaves yellow light. Paper that is somewhat yellow appears to be dingy (drab). That is why we measure paper brightness in the blue region—to insure that we are not taking any of the blue light away. (If blue is being selectively removed, we add a blue dye which selectively removes all colors except blue. This balances the reflectance spectrum of the paper.) On the other hand, paper that we want to be yellow will absorb blue light from the spectrum, but very little of other colors.

Laboratory measurement of color

Spectrophotometers. A reliable way to measure color involves the recording spectrophotometer. Traditionally spectrophotometers consist of a light source that [via a prism (Fig. 24-2) or diffraction grating] produces reasonably monochromatic light (the *bandwidth* is often as low as 2-4 nm at any given point) as it scans across the visible spectrum and measures the intensity of that light as it reflects off the surface of paper (or other object) for each point in the spectrum relative to the level of lighting which is measured by a reference photocell. Any type or model of spectrophotometer should give the same result for a given sample.

In the so—called reverse spectrophotometer, one shines a white light on paper and then applies a diffraction grating (or filters) to analyze the relative reflectance across the visible spectrum. Differences may result in these two methods if fluorescent dyes or other agents are used or are present (most discussions of color in this chapter

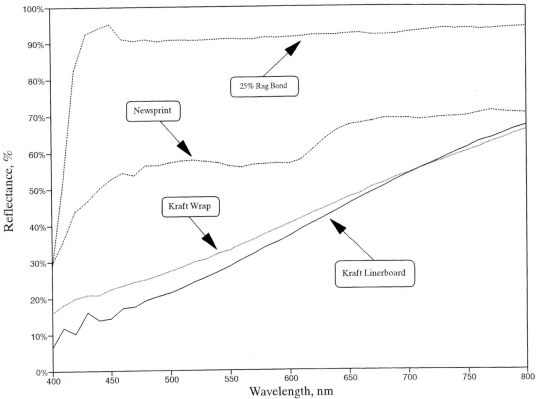

Fig. 24-7. Spectral responses of several types of undyed papers relative to a white standard.

ignore this possibility). The reverse spectrophotometer was used to generate the curves shown in Fig. 24-6 and Fig. 24-7 for various papers. The small shoulder in the blue region of the 25% rag bond curve in Fig. 24-7 (which could theoretically go above 100% with a suitable light source) is most likely due to fluorescent brighteners. If a conventional spectrophotometer is used, one would expect this shoulder to disappear.

This latter point demonstrates how the source of the light affects light reflected from an object. Another example comes from photography. When film designed for outdoor use is used indoors under incandescent (tungsten) light without corrective filters, the pictures appear to be yellowish since tungsten light bulbs emit much more light in the yellow region of the spectrum than the blue region, while sunlight is fairly uniform in intensity across the visible light spectrum.

(One's eye automatically compensates for the source of light, which is why the yellowishness of incandescent lights is not as noticeable to the eye as to the camera. *Color constancy* is the ability of

the eye (and brain) to see a color as (essentially) constant, even when the light source changes considerably. However, this also makes the human eye an unreliable judge in some cases. For example, gray against a green background appears slightly reddish, while gray against a red background appears slightly greenish.)

In summary, color perception depends on the reflectance curve and fluorescence of an object, the nature of the light shining on the object, and the spectral response of the human eye. A consequence of this statement is that two objects may appear to be the same color under one set of lighting conditions, but different colors under a second set of lighting conditions; this phenomenon is called *metamerism*. Metamerism is avoided (or at least determined in the laboratory before it becomes a problem) by viewing the two objects under two light sources of widely differing characteristics. [Metamerism could not happen to two objects with the same spectral reflectance curve; a color will always appear the same as itself under the same light source. It occurs when two objects have the same tristimulus values (described in the following section) under one set of lighting conditions, but different tristimulus values under a second set of lighting conditions.]

It is impossible to know how another person perceives color, but there are some known phenomena such as yellowing of the lens with age and certain types of color blindness which affect the way a person perceives color. Furthermore, one's experiences alter one's perception of color.

Fig. 24-7 shows that dark brown paper (unbleached kraft paper) has a low absorbance in the blue range and a fairly high absorbance in the yellow range. (Remember, however, that TAPPI brightness is measured in the blue region at 457 nm.) This is typical of brown materials. Gray objects, however, would have a fairly uniform absorbance across the spectrum. Dark gray has a uniformly high absorbance (low reflectance or transmittance), and light gray has a low absorbance (high reflectance or transmittance). Yellowish papers, such as newsprint, fall between gray and brown in their spectral curve.

Fig. 24-8 shows the reflectance curves (obtained with a reverse spectrophotometer) of various dyed papers (top). [The absorbance curve (see page 342 for a mathematical development) is included for interest because the eye has a logarithmic response that is more like the absorbance curve.] One observes that red construction paper reflects a high percentage of the red light (80%), but very little of the blue light (about 10%). The opposite is true of blue paper. Because the shapes of these curves are much different, they are said to have different *hues*, or shades. Red, orange, yellow, green, yellowish green, blue, indigo, violet, and reddishorange are examples of different hues. Purple (and related colors such as mauve) is a hue that does not exist in the spectrum and corresponds to blue and red added together.

Saturation refers to the "spectral purity" of a color. A pure blue light mixed with some white light still appears to be blue (a somewhat different blue), but has lost some of its saturation. Monochromatic light has perfect saturation. Hue and saturation define the *chromaticity* of an object.

One also observes that the shape of the blue bond is the same as that of the blue report cover, but the blue report cover has a much deeper blue appearance. These two samples differ in their *strength* or *brightness* values. Brightness is the least critical of the three definitions of color since it is easily changed by the level of incident lighting. (In other words, brightness is always changing anyway, and the eye—brain system compensates for various levels of lighting automatically.)

Colorimeters. Because the spectral curves in the visible region are relatively smooth, it is not necessary to scan the entire spectrum. A series of filters can be used across the visible spectrum. For example, the Elrepho (photoelectric reflectance photometer made by Carl Zeiss Co.) has a series of seven filters that have dominant wavelengths of 420, 460, 490, 530, 570, 620, and 680 nm; it is an example of an *abridged spectrophotometer.* Special filters can also be used to measure the tristimulus values described below.

24.4 TRISTIMULUS SYSTEMS

Introduction

Since there are three types of color sensors in the human eye, it should be possible to define color by three coefficients corresponding to the primary colors. The exact wavelengths of these primary colors must be agreed upon and are the basis of the coefficients. Any color of any brightness could then be plotted on a three—dimensional coordinate system (sometimes called a color space). If one ignores brightness (since humans tend to perceive color independent of the level of lighting), then a color plane can be defined that intersects through the space; such a diagram is called a chromaticity chart and takes into account hue and saturation. The CIE chromaticity diagram was developed in 1931.

[A certain color stimulus (what a spectrophotometer sees from a surface) absolutely defines a certain color sensation (in a standard observer), but a certain color sensation can be caused by an infinite number of possible color stimuli.] It stands to reason that much more information is contained in a spectral curve of (potentially) an infinite number of points than in three numbers of a tristimulus system.

Munsell

The Munsell system (1929) is a nonquantitative three—dimensional system for plotting colors. It was specifically derived such that there is uniform spacing of perception. In other words, the distance required such that the observer can recognize two separate shades is

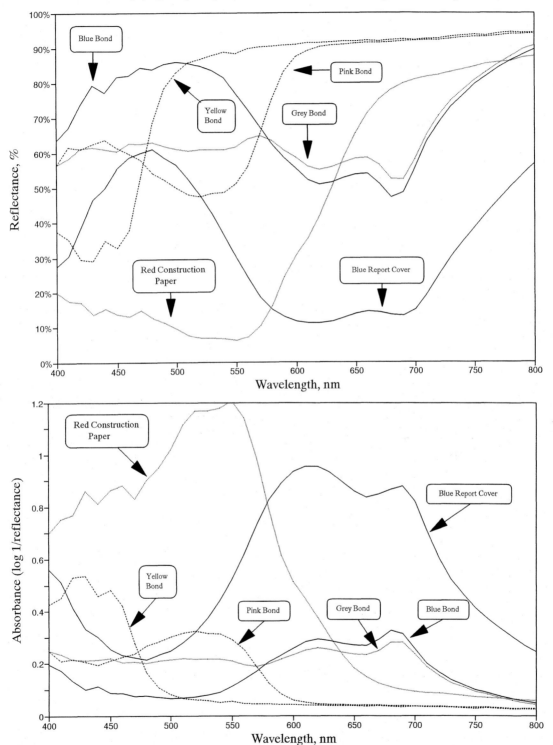

Fig. 24-8. Spectral responses of dyed papers as reflectance curves (top) and absorbance curves.

about the same throughout the system. The *Munsell value* corresponds to brightness from black (0) to perfect diffuse reflectance (10). The *Munsell hue* and *chroma* are the other axes.

CIE system

The first quantitative tristimulus system was developed by the CIE (Commission Internationale de l'Éclairage), which defines the internationally accepted standard observer (since perception of color is observer—dependent), used in the other systems. (Older literature calls this the ICI system.) The 1931 observer is based on the average 2° vision of 17 observers, while this was expanded to 76 observers in 1964. The sample is illuminated at a 45° angle with observations at 0°. (A 45° illumination with 45° observation is an observation of incident light, which has a higher proportion of the color source characteristics.)

Since the color of an object is dependent upon the source of the illumination, several standard illuminants were also specified (Section 24-5). Illuminant A corresponds to a tungsten light operating at a temperature of 2854 K. Illuminant B corresponds to direct sunlight, and Illuminant C, perhaps the most common reference illuminant, corresponds to north sky (in the northern hemisphere) average diffuse daylight (as on an overcast day); obviously these illuminants are simulated in the laboratory with a specified arrangement. Illuminant D_{65} is also daylight but specifies the UV and visible region for use in fluorescence; it is also widely used as the reference for color television monitors. Illuminant E is a theoretical source of pure white light.

CIE tristimulus values are obtained for a given set of lighting conditions (for expected further illumination of the object, such as paper expected to be viewed under diffuse sunlight) since, as has been said, two objects that appear the same color under one set of lighting conditions may appear different under a second set of lighting conditions. The light source used when the CIE values are determined must also be specified since the light shining on the object partly determines the spectral reflectance of the object. Usually the light source is Illuminant E, perfectly white light, (unless fluorescence is being considered in which case Illuminant D is specified), since a spectro-photometer operates as if the light source is pure white light, because the light reflected off a sample is always compared to the light incident on the sample for each wavelength under consideration.

For example, an object with Illuminant E may have several different sets of tristimulus values corresponding to a set for use in tungsten light illumination, direct sunlight illumination, or diffuse sunlight illumination. If one has a reflectance curve then the tristimulus values can be obtained with a computer that automatically factors in the effect of any light source used to generate the reflectance curve.

The 1931 CIE RGB system was defined with primaries of red, R, at 700.0 nm; green, G, at 546.1 nm; and blue, B, at 435.8 nm. While the exact selection of wavelengths is arbitrary, they were selected from regions of the primary colors, according to Grassman's (published in 1853) laws of color mixing (Brunner et al., 1990). When the relative amounts of each of these colors were added together to match monochromatic light of each wavelength across the spectrum (based on the agreement of a group of standard observers), the result in Fig. 24-9 was obtained, which is a plot of the tristimulus values. However, this necessarily led to negative values of the red primary in order to match some colors (the mathematical equivalent of having to add red primary to the sample to obtain perceived agreement).

In order to avoid the use of negative values, primaries (color matching functions) that had no physical meaning were selected in the CIE XYZ system so that positive values of each could be combined to make any color of the spectrum (the first criterion). (It was more important that certain criteria be met rather than have color matching functions that could be physically reproduced.) The coefficients of the three color matching functions at any particular wavelength are designated $\tilde{x}$, $\tilde{y}$, and $\tilde{z}$, respectively. The green primary function, $\tilde{y}$, was chosen to match the standard luminosity function (sometimes called visual efficiency in older literature) of the standard observer (the second criterion). The blue primary function, $\tilde{z}$, was chosen to be zero over as much of the visible spectrum as possible (the third criterion). The tristimulus values of each of the three *XYZ* primaries are given in Fig. 24-10 at 10

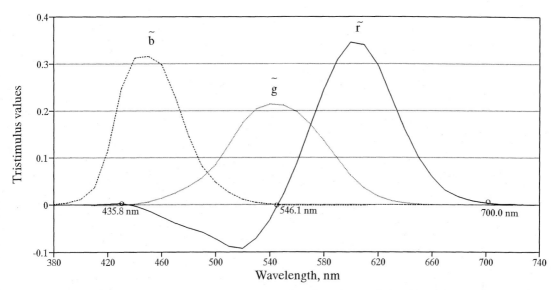

Fig. 24-9. Tristimulus values (color matching functions) of the CIE RGB system.

nm intervals. The tabular values will be used presently and are included in Table 24-2. [The standard specifies the values from 360 nm to 830 nm at 1 nm increments to five decimal places, but such precision and range is seldom required; these values are available in Wyszecki and Stiles (1967).] Separate integration (for each tristimulus function over the visible spectrum) of the energy of the light source (E_λ) multiplied by the spectral reflectance from the object expressed from 0 to 1 (R_λ) multiplied by tristimulus values $(\widetilde{x}, \widetilde{y}, \text{ or } \widetilde{z})$ gives X, Y, and Z as follow:

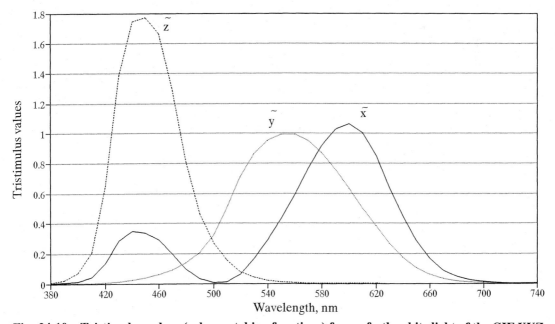

Fig. 24-10. Tristimulus values (color matching functions) for perfectly white light of the CIE XYZ system .

$$X = C \int E_\lambda \, \bar{x} \, R_\lambda \, d\lambda$$

$$Y = C \int E_\lambda \, \bar{y} \, R_\lambda \, d\lambda$$

$$Z = C \int E_\lambda \, \bar{z} \, R_\lambda \, d\lambda$$

where $\qquad C = 100 / (\int E_\lambda \, \bar{y} \, d\lambda)$

(By these definitions, Y will always be 100 for perfectly white light.) The integration is approximated by a summation, such as by taking values at every 10 nm from 380 nm to 720 nm ($d\lambda \approx \Delta\lambda = 10$ nm); high precision, of course, requires a smaller increment. High precision is obtained by taking smaller increments where the functions have high values in the *selected ordinate integration method*. Tables that give intensities for each wavelength (or every 10 nm) for a particular standard illuminant simplify the calculations. Many computerized spectrophotometers have the software to makes these calculations automatically.

The fourth criterion of the three color functions is that white light corresponds to $X = Y = Z$. The three *chromaticity coordinates* are the ratio of each of the tristimulus values divided by the sum of all three tristimulus values. This gives the ratios of the three tristimulus values with loss of the overall magnitude information. The three equations are:

$$x = X/(X + Y + Z)$$
$$y = Y/(X + Y + Z)$$
$$z = Z/(X + Y + Z)$$

Any two of these specify the third since $x + y + z = 1$. The use of x and y alone in a plot (since the value of z is implicit in $x + y$) allows all colors to be plotted in a two—dimensional system where the overall brightness is ignored. (The selection of x and y is arbitrary; a plot of any two of the variables would have worked.)

Therefore, the *chromaticity diagram* is a plot of x and y values as shown in Fig. 24-11 [with the data obtained from Wyszecki and Stiles (1967)]. The data are plotted every 10 nm from 380 nm to 720 nm; plotting every 1 nm would have made the curve smoother. Let us examine how the points on a chromaticity chart are determined and what they represent.

The simplest example is monochromatic light. For example, let us determine the point for blue light at 450 nm. Since overall brightness is not considered, consider all light to be in units from 0 to 1 transmittance. Monochromatic light at 450 nm corresponds to 0 at all wavelengths and 1 at 450 nm. From Table 24-2, X, Y, and Z are calculated as 0.3362, 0.0380, and 1.7721, respectively. The solutions of x and y are then 0.3362/2.1463 and 0.0380/2.1463 or 0.1566 and 0.0177, respectively. Fig. 24-11 shows the location of this point at those coordinates. Thus, points on the outer "circle" represents monochromatic light.

White light is another useful example, but takes a little more work. One takes the light output for each wavelength and multiplies by the factors in Table 24-2. It is important that the spacing of the wavelengths be equal so that one wavelength does not have more weight than another in the calculations (unless one wants to divide each increment by the wavelength interval). But using a relative light output at each wavelength of 1, a simple sum is taken as shown in Table 24-2. Since all of the sums are equal, it follows that $x = y = z = 0.3333$.

The lower line connects monochromatic blue and red lights in various proportions to obtain various shades of purple. Obviously this does not occur when light is separated with a prism, but can nevertheless occur to make colors.

If monochromatic green light of wavelength 520 nm is mixed with various parts of white light the line is obtained that goes from 520 nm to white (achromatic) light. This line corresponds to various shades (various excitation purities) of the same hue (dominant wavelength). A color, therefore, can be described by its dominant wavelength and excitation purity or its x and y values.

The reduced CIE Y, x, y system uses x and y and retains Y for a measurement of brightness.

A problem of the system demonstrated in Fig. 24-11 is that the sensitivity to changes in color by the human eye varies greatly in different regions of the plot. A perception of color change in the green area corresponds to a much larger distance than in the red area. This should not be surprising since the CIE XYZ space is not a uniform color space as is the Munsell system. [See McAdam, *J. Opt. Soc. Am.* 32:247(1942).] Also, there is no area that corresponds to black.

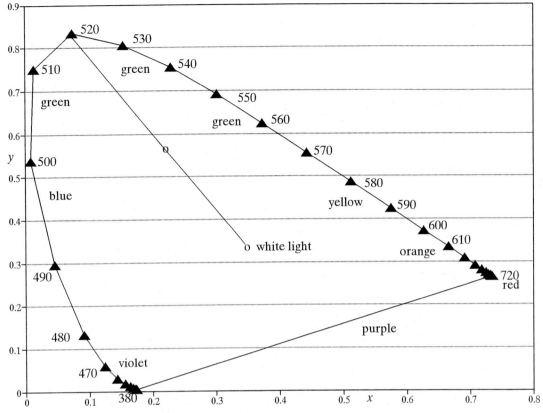

Fig. 24-11. The 1931 chromaticity chart.

L, a, b systems

Several systems use *L, a,* and *b* variables. These systems are designed to have a relatively uniform color space, which makes them useful for determining perceptual color differences. The variables have similar physical meaning in all of these systems, but somewhat different numerical values. It is an example of an opponent color scale. Adams coordinates is one system that uses this notation. Two others are discussed presently.

As shown in Fig. 24-12, *L* has values of 0 (corresponding to black) to 100 (perfect whiteness); positive values of *a* correspond to redness, while negative values correspond to greenness; and positive values of *b* correspond to yellowness, while negative values correspond to blueness. When *a* and *b* are both zero, the sample is achromatic (black, gray, or white).

The CIE *L*a*b** (1976) system (pronounced as "sealab") defines the variable as follows for Illuminant C (*X, Y,* and *Z* are each > 0.01):

$$L^* = 116\ Y^{1/3} - 16$$
$$a^* = 500\ [(1.02\ X)^{1/3} - Y^{1/3}]$$
$$b^* = 200\ [Y^{1/3} - (0.847\ Z)^{1/3}]$$

For use in perfectly white light the coefficients 1.02 and 0.847 are changed to 1.

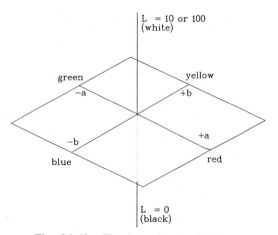

Fig. 24-12. The *L, a, b* color systems.

Table 24-2. Tristimulus values (color matching functions) for perfectly white light of the CIE XYZ system.

nm	$\bar{x}$	$\bar{y}$	$\bar{z}$
380	0.0014	0.0000	0.0065
390	0.0042	0.0001	0.0201
400	0.0143	0.0004	0.0679
410	0.0435	0.0012	0.2074
420	0.1344	0.0040	0.6456
430	0.2839	0.0116	1.3856
440	0.3483	0.0230	1.7471
450	0.3362	0.0380	1.7721
460	0.2908	0.0600	1.6692
470	0.1954	0.0910	1.2876
480	0.0956	0.1390	0.8130
490	0.0322	0.2080	0.4652
500	0.0049	0.3230	0.2720
510	0.0093	0.5030	0.1582
520	0.0633	0.7100	0.0782
530	0.1655	0.8620	0.0422
540	0.2904	0.9540	0.0203
550	0.4334	0.9950	0.0087
560	0.5945	0.9950	0.0039
570	0.7621	0.9520	0.0021
580	0.9163	0.8700	0.0017
590	1.0263	0.7570	0.0011
600	1.0622	0.6310	0.0008
610	1.0026	0.5030	0.0003
620	0.8544	0.3810	0.0002
630	0.6424	0.2650	0.0000
640	0.4479	0.1750	
650	0.2835	0.1070	
660	0.1649	0.0610	
670	0.0874	0.0320	
680	0.0468	0.0170	
690	0.0227	0.0082	
700	0.0114	0.0041	
710	0.0058	0.0021	
720	0.0029	0.0010	
730	0.0014	0.0005	
740	0.0007	0.0003	
750	0.0003	0.0001	
760	0.0002	0.0001	
770	0.0001	0.0000	
Totals =	10.68	10.69	10.68

The Hunter (1958) *L, a, b* values are related to the CIE *XYZ* tristimulus values as follow:

$$L = 10\ Y^{0.5}$$
$$a = 17.5\ (1.02\ X - Y)/Y^{0.5}$$
$$b = 7.0\ (Y - 0.847\ Z)/Y^{0.5}$$

Other color systems

There are numerous color systems that have been developed for various purposes. TIS 0804-04 gives formulae for interconversion of many color systems.

Determination of tristimulus values

Tristimulus values can be obtained by the comparison method where the color is matched by mixing three primary standards until visual observation gives a match, spectrophotometric analysis, or use of trichromatic methods employing analysis with three separate filters.

Video color systems

Video color—based systems (such as broadcast television) are tristimulus systems. Video based systems, however, have additional constraints (Brunner et al., 1990). For example, these systems have to use efficient encoding of information because their transmissions have limited bandwidth during broadcast. Also, they must be compatible with black and white systems.

This point is mentioned because the use of video based color is important in computer vision and other techniques, which will have more importance to the industry in the future. For example, broadcast color television can reproduce only about 50% of the area of the CIE chromaticity chart (Fig. 24-11).

24.5 BLACKBODY RADIATION AND OTHER LIGHT SOURCES

Since the source of light is very important in the appearance of colored objects, blackbody radiation will be considered. A blackbody is an object that absorbs all electromagnetic radiation that falls on it. The rate of radiation emitted from a blackbody is a function of temperature only and is independent of the type of material. The energy of the emitted radiation at a given wavelength $R(v)$ was derived by Planck in 1900 and is given as a function of frequency (v), Planck's constant (h), the speed of light in a vacuum (c), temperature (T), and Boltzmann's constant (k):

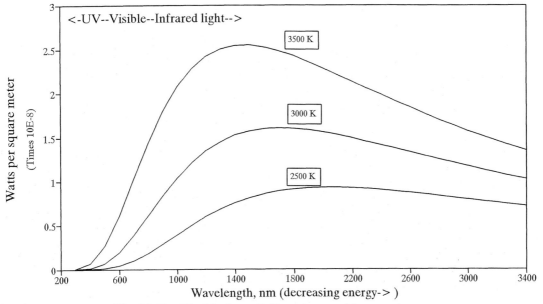

Fig. 24-13. Blackbody radiation at several temperatures.

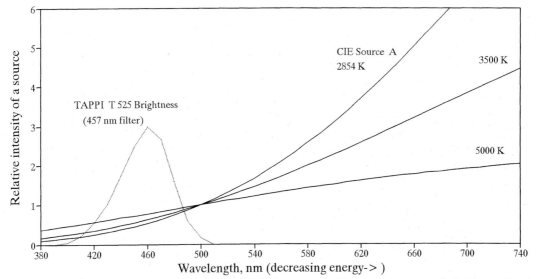

Fig. 24-14. Blackbody radiation in the visible region at several temperatures and the normalized transmittance of the blue filter used in paper brightness determinations.

$$R(v) = \frac{2\pi h v^3}{c^2} \frac{1}{e^{hv/kT} - 1} \qquad (24\text{-}3)$$

Fig. 24-13 shows the energy distribution as a function of wavelength over a broad range for several temperatures. Fig. 24-14 considers only the visible region of the electromagnetic spectrum.

The number of photons per wavelength drops off faster at lower wavelengths since each photon has more energy so fewer photons "carry" the energy. One observes that with increasing temperature more energy is given off and the frequency of maximal energy increases. As one heats a piece of metal it begins to glow faint red, then brighter yellow, then white hot, then blue—white. A tungsten incandescent light typically operates at 2800 K; one can see that most of the energy is given off as heat in the infrared region (i.e., it is an inefficient producer of light) and most of the visible light is toward the red end of the spectrum. Light sources are often described by their *color temperature*, i.e., the spectral distribution of a blackbody radiator of a specified temperature that most closely matches the light source. CIE Illuminant B is approximated by 4870 K and Illuminant C by 6740 K.

The use of halogen gases allows the temperature at which a tungsten light operates to increase. This gives a whiter distribution of light and increases the efficiency of the light. Halogen lights operate in the temperature range of 2800—3400 K.

Light can be produced by the excitation of molecules or atoms (often by an electric charge, but also by the processes of fluorescence and combustion) in addition to blackbody radiation. This mechanism gives different energy distributions. For example, arc lights tend to give a high percentage of their energy at a few selected wavelengths. Sodium lights give out a lot of light in the yellow region of the spectrum at 589 nm. They may be approximated by a color temperature, but this is often not satisfactory.

Light is modified by filters to achieve certain purposes. The blue filter centered at 457 nm for paper brightness is well known. Heat filters are also used to remove the IR energy from blackbody radiation sources. UV filters are used for a similar purpose, especially with high—temperature blackbody radiation, where large amounts of UV light (which can cause eye and skin damage) are produced. The UV c band is defined as 375—405 nm, UV b is defined as 345—375 nm, and the UV a band is defined as 300—345 nm (presented in order of increasing energy).

24.6 KUBELKA—MUNK THEORY

Introduction

The optical properties of paper are extremely important for many of its uses, especially when paper is used for printing. The important optical properties include reflectance (color and brightness, especially diffuse reflectance) and opacity. It is important to know how processing and

variations in the composition of paper will affect the optical properties. A generalized approach to the optical properties of paper allows one to predict the properties of paper to avoid a labor—intensive trial—and—error work approach. The concepts of brightness and color of paper have already been considered in this chapter.

A few notes on common conventions used in this chapter. Reflectance values that are determined relative to MgO should be reported as a percentage. Absolute reflectances can be calculated if the reflectance of the MgO standard is known or the photometer is calibrated as such. Absolute reflectance is reported from 0 to 1. Paper with 85.0% reflectance relative to a certain MgO standard that is known to have an absolute reflectance of 0.980 corresponds to an absolute reflectance of 0.833 (calculated from 0.850×0.980). Most equations use absolute reflectance values.

The relationships expressed in this section strictly apply to only a single wavelength (or a relatively narrow band of wavelengths) of light (and preferably measurements taken with the same light source and geometry). Difficulty may arise in nonwhite (or nongray) papers since TAPPI opacity is determined with the CIE Y function as a weighing factor (centered near 567 nm) while brightness is determined at 457 nm. This is a common occurrence that is usually ignored, but (usually unwittingly) implies that k and s are equal under the two conditions!

There are many other reflectance models in the literature with various underlying assumptions. The dichromatic reflection model is one, but it does not work well with paper at present. The Kubelka—Munk theory is relatively simple in terms of the number of constants involved, works very well for many papers, and is well documented for use by the pulp and paper industry.

Opacity

Opacity is the property of a material that indicates the ability to hide what is behind it. Opacity is usually evaluated as a *contrast ratio*, C, such as TAPPI opacity, which is the reflectance of a single sheet with a perfectly black backing (R_0) divided by the reflectance with a white (or nearly white) backing (such as 89% reflectance, $R_{0.89}$). [See Judd (1934) below to see how the value of

89% reflectance accidentally came to be for TAPPI opacity.] The reflectance of an (effectively infinite) thick pad of paper is R_∞ and is known as *reflectivity*. The reflectance of a single sheet with a perfectly white backing is given as R_1, and the corresponding opacity is called the ideal opacity.

$$\text{TAPPI opacity} = C_{0.89} = \Omega_T = R_0/R_{0.89}$$

$$\text{Printing opacity} = C_\infty = \Omega = R_0/R_\infty$$

Judd (1937, 1938) describes (based on Kubelka—Munk theory) how any opacity value may be converted to any other opacity value by formula (see Table 24-4 for TAPPI opacity from any other opacity) or graphical aid. If a 0.8900 backing is not available, TAPPI opacity could be determined by using another backing (of known reflectance) near 0.89 and applying a correction.

R is the reflectance of a layer (e.g., of paper or coating on paper) on top of a background with reflectance of R_g. For example, R_0 corresponds to a black background where $R_g = 0$. The use of a general form of R allows the theory to have much more general application, such as the effect of a coating layer on paper.

Uses of Kubelka—Munk Theory

Kubelka—Munk theory is used in papermaking in several ways. If one knows the optical properties of each pulp, filler, and dye used in papermaking, then the optical properties of a paper made with any combination of the materials can be predicted. If one knows the contrast ratio and reflectivity of a paper, the changes in these properties with a change in basis weight can be predicted. If any two variables of R_0, $R_{0.89}$, and R_∞ are known, the third can be determined. Thus, one can calculate R_∞ from a single sheet of paper by measuring $R_{0.89}$ and R_0. (However, if one uses $R_{0.89}$ and R_∞, and if these are close in value, then the error in determining R_0 will be large.) Similarly, for nearly opaque papers it is difficult to determine other parameters with high precision.

Historical context

Judd (1937) documents much of the early work on the relationships between incident, reflected, and emergent light from a thin, homogeneous layer of absorbing and scattering materials

such as that of G.G. Stokes in 1860; Channon, Renwick, and Storr in 1918; and Silberstein in 1927. Judd points out that many of these methods have ignored light that escapes from the sides of these layers because of the added complexity.

The Kubelka—Munk (1931) theory of light allows one to predict quantitatively the behavior of light in coatings such as paint for which it was developed; this work has become the basis of much of this field. Steele (1935) reproduced many of the derivations of equations from the Kubelka and Munk derivation (for an English language audience), gave a discussion of their work, and provided additional work for routine application of this theory to paper. Kubelka (1948) points out that other derivations are available that independently lead to many of the same equations presented in this section.

The variables k and s

Two parameters are the basis of the theory: S, the scattering coefficient, and K, the absorption coefficient, each for an infinitely thin layer of material (dX). These variables, then, are intensive properties and do not depend on the amount of material (such as thickness of the sheet).

[Coffee has a high value of K and a low value of S. Conversely, cream has a low value of K and a high value of S. Therefore, when one puts creamer in coffee it turns lighter in color; S of the coffee is increased dramatically, K increases only negligibly (or decreases with dilution), and S/K increases dramatically.]

This theory was developed to define properties based on the thickness of a material (X), but in the case of porous materials like paper, workers quickly started using basis weight. Later, Van den Akker (1949) proved that basis weight (W) can be used with many advantages; he recommended that the lowercase variables s and k be used if the thickness of material is reported as a basis weight (the practice used in this chapter). These substitutions are mathematically equivalent to the original theory, and all derived relationships remain the same. For example, SX, scattering power, is dimensionless and equal in magnitude to sW; therefore, $s/k = S/K$, etc.

While these variables are often reported for a particular pulp, processing variables such as refining (increases s)and pressing (decreases s by increasing fiber—to—fiber bonding and decreasing air—fiber interfaces) are certain to affect scattering and supercalendering will affect gloss (gloss is not diffuse reflectance required by the theory). Nevertheless, the theory is useful if used with care.

These variables vary as a function of wavelength, so the wavelength should be specified. Also, these variables are specified according to basis weight so their values depend upon the basis of the basis weight measurement (i.e., g/m^2 or the various ream—based systems give different values to these variables); if the basis weight system in not specified without ambiguity, then the values given for s and k are meaningless. One advantage to using sW is that it is unitless.

In white or gray papers k varies very little with wavelength, but in brown or colored papers the variation is appreciable (as shown in Fig. 24-7). Most printing papers, however, are white or gray. Hemicellulose and cellulose have very little absorption of light in the visible region, but lignin has a high absorbance, so most light absorption is due to lignin. This has been considered in the chapter on pulp bleaching and earlier in this chapter with paper reflectance.

The use of soluble dyes in paper allows k to increase with very little change in the value of s, thereby allowing opacity to be increased.

Expected deviations from theory

Deviations from this theory are expected with glossy papers, but polarized filters can be used to remove the effect of gloss (light reflected according to Fresnel's law which has a degree of polarization). Deviations might also occur in papers where fillers are nearly absent on the wire side; in this case one should determine if optical properties (such as R_0) are the same when the sheet is turned over.

If large deviations from optical theory result, it often indicates that nonrandom distribution of components is occurring, almost always leading to less scattering. This often happens with filler loading above 10% because the filler begins to agglomerate. On the other hand, a second filler is often used with titanium dioxide to improve its opacity by keeping the titanium dioxide particles separated from each other (so they can act inde-pendently to a larger degree); the effectiveness of such strategies can be determined.

Another source of deviation arises from the fact that paper is not perfectly uniform in terms of its formation. A paper with poor formation will have the same brightness but a lower opacity than paper with good formation. Consider an extreme case where with perfect formation the opacity is about 99%. With poor formation one would expect areas where more light goes through and the opacity would be lower. This is an especially large problem with papers of low basis weight.

Kubelka—Munk theory development

The development of the theory will rely on Kubelka (1948) and i will denote the light flux moving downward and j will denote the light flux moving upward. X is positive upward and negative downward. (A detailed derivation is not presented here; it is anticipated that this audience mainly desires to use the resulting equations.) The derivation is valid for a diffuse light source or light source 60° from normal to the surface for perfectly dull surfaces (Kubelka, 1948) of homogeneous (each layer is identical) materials.

Refer to Fig. 24-3 to observe that when light travels from above through a thin layer parallel to the surface (dx) of translucent, diffusing material (e.g., paper) distance x from the bottom of the paper of total thickness X, some light will be scattered above the layer, some light will be absorbed by the layer, and most of the light will continue through the layer.

Consider a light source entering the top of the paper with intensity I_0. Let i be the light entering element dx from the top due to transmittance through the layers above and j be the light entering the bottom of dx due to reflectance from the layers below. In passing through dx by definition of S and K:

i decreases by $i(S + K)dx$
j decreases by $j(S + K)dx$
i increases by $j\,S\,dx$
j increases by $i\,S\,dx$

so $-di = -(S + K)i\,dx + Sj\,dx$
$dj = -(S + K)j\,dx + Si\,dx$

These equations are divided by i and j, respectively, then added together to give:

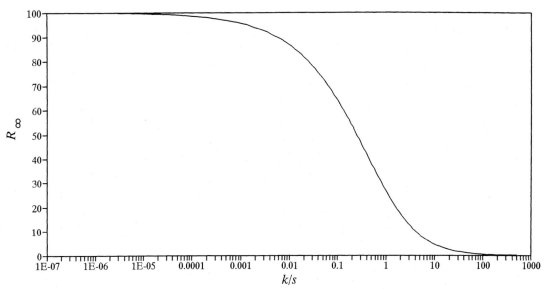

Fig. 24-15. R_∞ (as percentage) versus k/s.

$$\frac{dj}{j} - \frac{di}{i} = d\ln\frac{j}{i} = -2(S + K)\,dx + S\left(\frac{i}{j} + \frac{j}{i}\right)dx$$

since $j/i = R$ (reflectance) the equation can be expressed as an integral as follows:

$$\int_{R'}^{R} \frac{dR}{R^2 - \dfrac{2(S + K)r}{S} + 1} = S\int^{K}_{\cdot} dx$$

This equation is integrated in a variety of ways with a variety of substitutions to yield various equations. The most general exponential solution to the equation was reported by Kubelka and Munk (1931) as (using variables already introduced in this chapter):

$$R = \frac{\dfrac{1}{R_\infty}(R_g - R_\infty) - R_\infty(R_g - 1/R_\infty)\,e^{SX(1/R_\infty - R_\infty)}}{(R_g - R_\infty) - (R_g - 1/R_\infty)\,e^{SX(1/R_\infty - R_\infty)}}$$

It is possible to solve the integrated equations in terms of hyperbolic functions to facilitate their practical use by manual calculation with tables (Steele, 1935; Kubelka, 1948). Ironically, computers can often solve the exponential solutions faster than the hyperbolic solutions. (See the appendix of this chapter for details on hyperbolic functions.) Simplifying variables such as a and b (Table 24-4) are introduced. One equation is:

$$R = \frac{1 - Rg(a - b\coth bsW)}{a + b\coth bsW - Rg}$$

Specific solutions are obtained for certain cases. For example, by setting X as (effectively) infinity (∞) so that $R = R_\infty = R_g$ one obtains a fundamental relationship between brightness and the Kubelka—Munk variables as:

$$R_\infty = 1 + k/s - [(k/s)^2 + (2k/s)]^{0.5}$$

which is rewritten as:

$$\frac{k}{s} = \frac{(1 - R_\infty)^2}{2R_\infty}$$

The hyperbolic functions were used to solve enough points so that Judd could produce a graph that allows most solutions to be done graphically, since these equations are so complex. It is not surprising that the Judd opacity chart has been reproduced in many locations within the literature. Numerous equations appear in the various references of this chapter (most of which originate from Kubelka, 1948) and many are presented in Table 24-4 in the appendix.

When R_∞ is evaluated with the blue filter at 457 nm it is the TAPPI brightness (absolute) and can be used to determine k/s at that wavelength. Tables of R_∞ versus k/s exist [e.g., TIS 0606-18; Casey (1981), Vol. 3, pp. 1860—1864 (see p. 9

for the citation); MacLaurin and Alfenzer (1954)], but are unnecessary with the availability of computational equipment. The relationship is shown as a graph in Fig. 24-15. It is apparent that when brightness above 90% must be reached, k must be made very small or s must be very large. The value of s of papermaking fibers does not change (with refining, calendering, pressing, etc.) nearly as much as k changes by bleaching operations; however, fillers like titanium dioxide have much higher scattering coefficients than wood fiber.

In order to solve for k or s separately, and not just as a ratio, another parameter is necessary. This is usually provided from R_0 or a contrast ratio that is a function of R_0. Calculations using mathematical relationships (Table 24-4) or charts (Fig. 24-16 or TIS 0804-03) allow s and k to be determined individually. Usually sW is determined, and then k can be solved from R_∞ using Eq. 24-2. Once k and s are each known, many optical properties of the paper (at any basis weight) can be determined.

Additive values of k and s for mixtures

When one works with any mixture of several types of pulps and fillers, the overall k and s values can be determined if one knows their values for each of the components. If C_i is the mass ratio of component i, such that $\Sigma C_i = 1$, then:

$$\begin{aligned} (24-4) \\ s = C_1 s_1 + C_2 s_2 + C_3 s_3 + \cdots + C_n s_n \\ k = C_1 k_1 + C_2 k_2 + C_3 k_3 + \cdots + C_n k_n \end{aligned}$$

In a similar manner for both variables:

$$ sW = s_1 W_1 + s_2 W_2 + s_3 W_3 + \cdots + s_n W_n $$

The previous two equations assume that the components behave independently, that there is no interaction between them. This is usually not the case, especially for s. The k and s values can be determined for the filler in the paper at a given level so compensation for the interactions occurs. If the k and s are determined for paper without filler, and if k and s are determined for paper containing this paper with a known proportion of filler, then the effective k and s values for this filler in this type of paper (at the specified concentration) can be calculated. Certainly this information is useful for determining changes in the

optical properties of the paper with modest changes in the concentration of the filler. Steele (1937) gave the calculation to determine the scattering coefficient of the pigment filler (S_p) in the filled paper as a function of the scattering power of the fiber [(S_f), determined separately with unfilled paper], the scattering coefficient of the filled paper (S), and the fraction of filler in the filled sheet (y):

$$ S_p = \frac{S - (1-y)S_f}{y} \qquad (24-5) $$

One may wish to determine the s and k values as a function of filler loading so that more accurate calculations can be made. Steele (1937) pointed out that this calculation may have more value as an indicator of how effective the pigment is in a particular sample.

Applications to coating

Coating of one side of paper represents the reflectance of a (thin) homogeneous layer backed with a material of known reflectance R_g. R_g is, of course, the reflectance of the paper which could be one sheet with a black backing, R_0 of the paper, a relatively thick pad of paper with reflectance of R_∞, or other term. Under these conditions Eq. 24-9 may be of use. (This approach does not usually work with size press coatings of low solids because they do not stay on top of the sheet.) Often two sets of variables must be distinguished, one for the coated paper and one for the uncoated paper. The optical properties of the coating material can be determined by coating two sheets of different brightness with a known amount of coating formulation. Robinson (1975) gives the solution of R_∞ for the coating where R and R' are the reflectances of the coating over papers of reflectance R_g and $R_{g'}$, respectively:

$$ R_\infty = (e - (e^2 - 4f^2)^{1/2})/(2f) $$

$$ f = a - b $$

$$ e = ac - c - bd + d $$

$$ a = RR_{g'} $$

$$ b = R'R_g $$

$$ c = R' + R_g $$

$$ d = R + R_{g'} $$

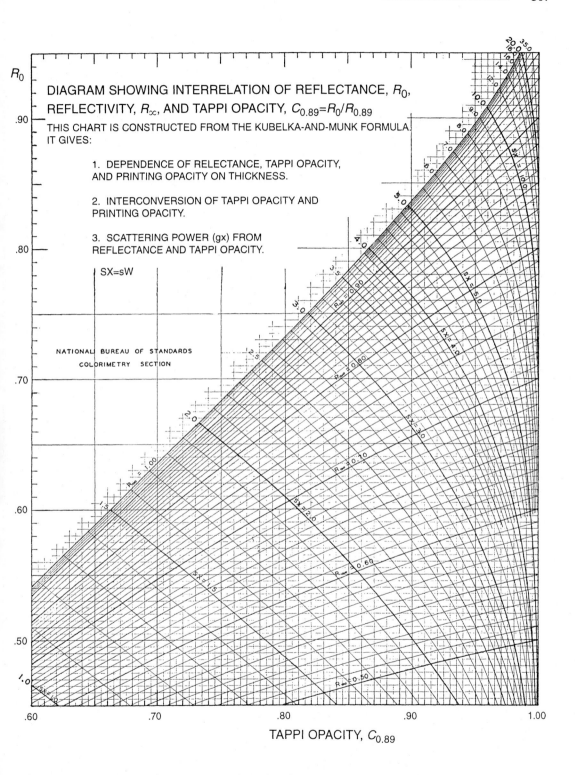

R_0

DIAGRAM SHOWING INTERRELATION OF REFLECTANCE, R_0,
REFLECTIVITY, R_∞, AND TAPPI OPACITY, $C_{0.89}=R_0/R_{0.89}$

THIS CHART IS CONSTRUCTED FROM THE KUBELKA-AND-MUNK FORMULA.
IT GIVES:

    1. DEPENDENCE OF RELECTANCE, TAPPI OPACITY,
    AND PRINTING OPACITY ON THICKNESS.

    2. INTERCONVERSION OF TAPPI OPACITY AND
    PRINTING OPACITY.

    3. SCATTERING POWER (gx) FROM
    REFLECTANCE AND TAPPI OPACITY.

SX=sW

NATIONAL BUREAU OF STANDARDS
COLORIMETRY SECTION

TAPPI OPACITY, $C_{0.89}$

Fig. 24-16. Interrelation of reflectance R_0, reflectivity R_∞, contrast ratio $C_{0.89}$, and sW. From Judd (1937, 1938) and sometimes called a Judd opacity chart.

The problems with coating calculations are analogous to those with using fillers at various levels of loading; that is, the behavior of the coating pigments depends on how they are used. The scattering coefficient, s, is a function the coating formulation, the level of application of coating to the base stock, and the degree of calendering. The binder in coating formulations will decrease opacity just as wax decreases the opacity of waxed paper, and binder migration is a consideration. In one case s in m^2/g was 0.30 for the pigment, 0.15 in the coating before calendering, and 0.10 after calendering. Thus, calendering can decrease brightness by decreasing light scattering.

The gloss of coated paper may increase to the level where appreciable deviation from Kubelka—Munk theory occurs. Nevertheless, Kubelka—Munk theory gives good results with the diffuse component of light. Useful references on this subject include Robinson (1975), Robinson and Linke (1963), and Clark and Ramsay (1965).

Problemsolving with Kubelka—Munk equations

A number of important applications exist for Kubelka—Munk theory. Some of the more common applications will be demonstrated in this section. Some other uses are considered in the appendix of this chapter.

Changes in basis weight

Often one wants to know how optical properties of paper will change with a change in basis weight of the paper.

EXAMPLE 2. Given: a paper of basis weight of 65 g/m^2 with a brightness (R_∞) of 0.81 and TAPPI opacity of 0.92. What is the opacity if the basis weight is dropped to 60 g/m^2?

SOLUTION: Using the Judd opacity chart with brightness and TAPPI opacity, sW is determined as 4.3. Solve s as 0.0662 m^2/g and the new sW as 4.0. Since brightness does not change with basis weight, reuse of the chart gives the new TAPPI opacity as 0.91.

One can use Eq. 24-28 to solve sW as 4.282 (after solving a as —0.0276, b as 0.1688, c as 0.00640, and y as 6.160). Using bright-

ness and the new sW of 3.95, Eq. 24-9 gives $R_{0.89}$ as 0.828, Eq. 24-13 gives R_0 as 0.751, and TAPPI opacity is 0.907.

PROBLEM: What is the assumption that is made in this question if TAPPI Standard tests were used to obtain the original brightness and TAPPI opacity?

Mixtures of pulps or pulp and filler

It is possible to solve the optical properties of paper without making paper if one knows the properties of the furnish. In this problem a clay with properties listed in Table 24-3 will be mixed with the pulp from Example 2 to make paper. The optical properties will be determined with the additive rule.

EXAMPLE 3. What is the printing opacity of paper made with 0.9 parts (by mass) of the pulp in Example 2 with 0.1 part of clay (R_∞ = 0.883, s = 0.11 m^2/g, and k = 0.00085 m^2/g) and basis weight of 65 g/m^2?

SOLUTION: The value of k of the first pulp is determined from Eq. 24-8 to be 0.001475; so:

s_{mix} = 0.9(0.0662) + 0.1(0.11) = 0.07058
k_{mix} = 0.9(0.001475) + 0.1(0.00085) =
= 0.00141
sW = 4.588
R_∞ (Eq. 24-15) is 0.817; R_0 (Eq. 24-13) is 0.771; $R_{0.89}$ (Eq. 24-9) is 0.832; and, by its definition, $C_{0.89}$ is 0.927.

PROBLEM: How does the TAPPI opacity of this paper compare to that of Example 2, and why does this occur?

24.7 ANNOTATED BIBLIOGRAPHY

Color theory

1. Billmeyer, F.W. Jr. and M. Saltzman, *Principles of Color Technology*, 2nd ed., Wiley, New York, 1981, 240 pp. This is a useful introduction to the field. Its presentation with numerous figures helps convey difficult concepts in an easy to understand manner. It has a thorough annotated bibliography in

addition to the bibliography. It includes some information on Kubelka—Munk theory.

2. Brunner, C.C., G.B. Shaw, D.A. Butler, and J.W. Funck, Using color in machine vision systems for wood processing, *Wood & Fiber Sci.* 22(4):413-428(1990). A good resource on general color theory and the title subject with application to vision systems for paper with 60 references.

3. Hunter, R.S., Photoelectric color difference meter, *J. Opt. Soc. Amer.* 48:985-995(1958).

4. Hunter, R.S. and R.W. Harold, *The Measurement of Appearance*, 2nd ed., Wiley—Interscience, New York, 1987. (1st ed., 1975). This highly readable book explains the methods used to study color and optical perception. Many different color systems are explained in this resource. The authors make some interesting points about perception and appearance we take for granted based on experience. When one drives down a road, one does not look for glossy surfaces (which is what one actually observes) that may spell trouble, one looks for water or ice. When a hand casts a shadow on a piece of white paper, one does not say the hand paints the paper black, one says that certain parts of the white paper are in the shadows.

5. Judd, D.B. and G. Wyszecki, *Color in Business, Science, and Industry*, 3rd ed., Wiley, New York, 1975, 553 p. This is commonly cited in the applied color literature and has many references. The 2nd ed. is from 1963.

6. Levine, M.D., *Vision in Man and Machine*, McGraw—Hill, New York, 1985, 574 p. This is a well referenced book with lots of mathematical detail (the author is an electrical engineer) and covers color vision in biological, historical, and theoretical manners in addition to machine vision.

7. Munsell Color Company, *Munsell Book of Color*, Baltimore, Maryland, 1929.

8. Wyszecki, G. and W.S. Stiles, *Color Science (Concepts and Methods, Quantitative Data and Formulas)*, John Wiley and Sons, New York, 1967, 628 p. The 2nd edition was

published in 1982 and is 950 pages. This is the best reference for a theoretical basis of color theory. It has solutions to the Kubelka—Munk equation (pp. 188-189).

Kubelka—Munk, scattering properties

9. Biermann, C.J., Construction of opacity charts, *Tappi J.* 78(1)238-239(1995).

10. Clark, H.B. and H.L. Ramsay, Predicting optical properties of coated papers, *Tappi* 48(11):609-612(1965). This is useful for learning the optical properties of coatings.

11. Hillend, W. J. Opacity problems in printing papers: Kubelka—Munk theory gives good, quick, answers, *Tappi* 49(7):41A-47A(1966).

12. Judd, D.B., Opacity standards, *J. Res. Nat. Bur. Standards* 13:281-291(1934). In this work Judd points out that the results of the Tappi opacity of the time were dependent upon each machine and each working background. For his particular machine and standard he stated, "It is found that the Tappi opacity corresponds to a reflectance of white backing in contact with the sample of about 0.89 reflectance." Tappi then defined its opacity to fit this one result.

According to earlier work of Judd [Sources of error in measuring opacity of paper by the contrast—ratio method, *J. Res. Nat. Bur. Standards* 12:345-351(1934)] the original Tappi opacity, which called for "The white surface standard must not touch the surface of the test specimen but must be so near it that a further decrease in distance will not affect the test results", was designed to have as high a white background as was practical. In fact, it did not work out that way, but we religiously cling to this arbitrary method instead of printing opacity, which is amenable to Kubelka—Munk calculations.

13. Judd, D.B., Optical specifications of light-scattering materials, *J. Res. Nat. Bur. Standards* 19(3):287-317(1937). This reference includes a large foldout chart of Fig. 24-16 that is reproduced in many places in the pulp and paper literature. A condensed version of this paper was published as *ibid., Tech.*

Assoc. Papers 21:474-481(1938) and identically as *ibid., Paper Trade J.* 106(1):39-46(Jan. 6, 1938).

14. Koller, E.B. and S.M. Chapman, Tables of light scattering coefficients from reflectance tables, *Svensk Papperstidn.* 64(4):113-124(1961). sW is given as a function of the matrix of R_0 and R_∞.

15. Knox, J.M. and D. Wahren, Determination of light scattering coefficient of dark and heavy sheets, *Tappi J.* 67(8):82-85(1984). This is a discussion of the usefulness of a device for transmission measurements in papers of high opacity.

16. Kubelka, P., New contributions to the optics of intensely light—scattering materials. Part I, *J. Opt. Soc. Am.* 38(5):448-456(1948) with Errata *ibid.* 38(12)1067(1948). This is the one paper to read on Kubelka—Munk theory and includes numerous equations that reappear in other literature. He also includes many approximate solutions, but these are often unnecessary (unless a variable is removed from the equation) with modern computational tools. *Ibid.*, Part II, Nonhomogeneous layers, *J. Opt. Soc. Am.* 44(4):330-335(1954).

17. Kubelka, P. and F. Munk, Ein Beitrag zur Optik der Farbanstriche (A contribution to the optics of paint), *Zeitschrift für Techn. Physik* 12:593-601(1931).

18. MacLaurin, D.J. and F.A. Aflenzer, Predicting the brightness of groundwood—sulfite pulp mixtures, *Tappi* 37(9):388-393(1954). This contains a listing of k/s as a function of R_∞ from 0 to 1.000 in 0.001 increments (two ratios are incorrect: replace 7.5048 for 0.059 and 3.72640 for 0.107, but these errors are minor). Values of s and k for a groundwood pulp bleached with various amount of H_2O_2 and a sulfite pulp bleached with $Ca(ClO)_2$ are given and listed here in Table 24-3.

This paper also contains data evaluating W. J. Foote's approximation [Simple method for predicting brightness of mixed pulp furnishes, *Tech., Assoc. Papers* 29:180-182(1946)

Table 24-3. Some values of s and k for various materials from the literature (for use as rough values and trends in processing only).

Material	s, m^2/g	k, m^2/g	Ref.[1], R_∞
Pulps			
Unbeaten sw sulfite	0.0310	0.00128	3, 0.751
Uunbleached sulfite		0.0254	1
Bleached kraft	0.0495		1
Groundwood			3
hw, 100 mL CSF	0.0830	0.00451	0.720
hw, 100 mL CSF	0.0830	0.01102	0.600
Spruce groundwood			2
	0.0670	0.00924	0.595
	0.0709	0.00798	0.625
	0.0738	0.00623	0.665
	0.0854	0.00545	0.701
	0.1165	0.00441	0.760
Spruce sulfite			2
	0.0649	0.00848	0.603
	0.0750	0.00692	0.653
	0.0846	0.00539	0.701
	0.0923	0.00377	0.752
	0.0936	0.00221	0.805
	0.0941	0.00123	0.851
Softwood kraft			3
Unbeaten	0.0350	0.00149	0.748
400 mL CSF	0.0275	0.00149	0.720
100 mL CSF	0.0250	0.00149	0.716
Hardwood kraft			3
Unbeaten	0.0442	0.00193	0.745
400 mL CSF	0.0355	0.00193	0.720
100 mL CSF	0.0328	0.00193	0.715
Fillers			
Clay	0.186		1
filler	0.11	0.00085	3, 0.883
No. 1 coating	0.1578	0.00129	3, 0.879
No. 2 coating	0.2100	0.00073	3, 0.920
TiO_2			3
anatase	0.42	0.00019	0.970
rutile	0.55-0.75		0.985
Black pigment dye	0.9500	21.	3, 0.16

References are 1, Steele (1937); 2, MacLaurin and Aflenzer (1954); and 3, Hillend (1966).

ibid. Paper Trade J. 122(8):35-37(Feb. 21, 1946)]. I personally do not feel that there is much advantage to Foote's approximation since measuring opacity is simple (thereby giving values to both k and s); it leads to increased error.

19. Robinson, J.V., A summary of reflectance equations for application of the Kubelka—Munk theory to optical properties of paper, *Tappi* 58(10):152-153(1975).

20. Robinson, J.V. and E.G. Linke, Theory of the opacity of films of coating pigment and adhesive: A method for calculating the opacity of coatings, *Tappi* 46(6):384-390(1963).

21. Steele, F.A., The optical characteristics of paper. I. The mathematical relationships between basis weight, reflectance, contrast ratio and other optical properties, *Tech. Assoc. Papers* 18:299-304(1935) and identically as *ibid., Paper Trade J.* 100(12):37-42, TS151-156(Mar. 21, 1935). This work reproduces many mathematical derivations of Kubelka and Munk (1931) in English.

22. Steele, F.A., The optical characteristics of paper. III. The opacifying powers of fibers and fillers, *Tech. Assoc. Papers* 20:37-39, 151-152(1937) and identically as *ibid., Paper Trade J.* 104(8):157-158(Feb. 25, 1937). See also Steele, F.A., *Paper Trade J.* 101(17):31(1935).

23. Van den Akker, J.A., Scattering and absorption of light in paper and other diffusing materials, *Tappi* 32(11):498-501(1949).

Tappi Standards
24. T 218, Forming handsheets for reflectance tests of pulp. This method allows "the preparation of a smooth and reproducible surface for reflectance measurements with a minimum of washing or contamination of the sample."

25. T 442, Spectral reflectance factor, transmittance, and color of paper and pulp (polychromatic illumination).

26. T 524, *L, a, b,* 45°, 0° colorimetry of white and near—white paper and paperboard.

27. T 527, Color of paper and paperboard in CIE *Y; x, y,* or *Y,* dominant wavelength and excitation purity.

Tappi Technical Information Sheets
28. TIS 0606-18, The Kubelka—Munk relationship between absorption/scattering coefficients and reflectance. This reference is easily replaced by a programmable calculator or computer. Do not use the equation listed for k/s for pulp mixtures without reading the papers by Foote (1946) and MacLaurin and Aflenzer (1954).

29. TIS 0804-01, Light sources for evaluating papers including those containing fluorescent whitening agents. This is an excellent, thorough reference.

30. TIS 0804-02, Optical measurements terminology (related to appearance evaluation of paper). This is a 12 page glossary of several hundred terms.

31. TIS 0804-03, Interrelation of reflectance, R_0; reflectivity, R_∞; TAPPI opacity, $C_{0.89}$; scattering, s, and absorption, k. These charts are similar to the Judd opacity chart [they allow SX (or sW) to be determined when R_0 and $C_{0.89}$ are known] but have higher resolution in certain regions and cover regions not covered by Judd.

32. TIS 0804-04, The determination of instrumental color differences. This reference includes mathematical formulae for numerous color systems. There are many color systems in use for specialized purposes.

33. TIS 0804-05, Indices for whiteness, yellowness, blue reflectance factor, and luminous (green) reflectance factor. Mathematical formulae of many color systems are included.

34. TIS 0804-06, Photometric linearity of optical properties instruments.

35. TIS 0804-07, Calibration of reflectance standards for hemispherical geometry.

EXERCISES

1. What are "optical brighteners" and how do they work?

2. Why does one sometimes add blue dye to pulp to make the final paper appear brighter? Is the paper really brighter?

3. What is the correct method of expressing the true, fundamental color of a material such as a sheet of colored paper? Why is this expression independent of the light source illuminating the paper?

4. Remember your optical friend's name, ROY G. BIV, and fill in the blanks. A red piece of paper will reflect mostly _____ colors of the spectrum. Blue light contains mostly _____ color light. Thus, if one looks at a red piece of paper under a blue light, it will appear to be _____. The worst color light in which to grow green plants is _____.

5. What are the dominant wavelength and saturation of the color corresponding to $X = 82$, $Y = 100$, and $Z = 53$?

6. What are two advantages of halogen incandescent lights over conventional incandescent lights?

7. When I was doing some touch—up painting on the exterior of a house, I noticed (after the paint had dried) that the paint matched perfectly when I viewed it straight on. But when I looked at the newly painted spots at an angle, they looked brighter than the old paint. Give a possible explanation.

8. What is the advantage to using fillers with high indices of refraction? What property of paper does this improve?

9. Why is oil—saturated paper much more translucent (of lower opacity) than unsaturated paper?

10. A bleached kraft softwood pulp has a scattering coefficient, s, of 0.05 m^2/g. Determine k for a brightness of 0.87 and for 0.95.

11. What is the Tappi opacity, $C_{0.89}$ for a sheet that is 40 g/m^2 using the kraft pulp of 0.87 brightness in Exercise 9?

12. One might argue that refining of bleached softwood pulp will increase its opacity in paper by generating fines. What is wrong with this argument?

13. Generate your own opacity chart for a given grade of paper. See Biermann, C.J., Construction of opacity charts, *Tappi J.* 78(1):238-239(1995).

APPENDIX

Hyperbolic functions

The function coth is the hyperbolic cotangent of the argument and coth^{-1} is the arc hyperbolic cotangent of the argument (abbreviated as ctgh in much of the optical literature). They have many analogies to the geometric functions of sin, cos, tan, etc. The functions are defined as follows:

$$\coth u = (\cosh x/)(\sinh u) = (e^u + 1/e^u)/(e^u - 1/e^u)$$

$$\sinh u = (e^u - 1/e^u)/2$$

$$\cosh u = (e^u + 1/e^u)/2$$

For the inverse functions, with coth as an example, if $v = \coth u$ then $u = \coth^{-1} v$. Other functions are defined as simple reciprocals so that $\tanh u = 1/(\coth u)$. These functions are discussed in most elementary calculus books.

The inverse hyperbolic functions can be solved directly for use in computer programs that do not support these functions. The solutions are:

$$\sinh^{-1} u = \ln [x + (x^2 + 1)^2$$

$$\cosh^{-1} u = \ln [x - (x^2 - 1)^2 \qquad u \leq 0$$
$$\cosh^{-1} u = \ln [x + (x^2 - 1)^2 \qquad u \geq 0$$

$$\coth^{-1} u = \tfrac{1}{2} \ln[(u + 1)/(u - 1)]$$
$$\qquad u < -1 \text{ or } u > 1$$

$$\tanh^{-1} u = \tfrac{1}{2} \ln[(1 + u)/(1 - u)]$$
$$\qquad -1 > u > 1$$

Light transmission versus opacity

Since opacity is a measure of the degree of resistance to light transmission, increased accuracy is obtained by measuring the transmittance of paper directly as a measure of opacity instead of measuring its opacity. Transmittance is measured with a beam incident on one surface (although a diffuse light source is theoretically preferable, there is often little difference in practice with reasonably opaque sheets) with collection of the transmitted light with an integrating sphere on the other side. When R_0 approaches 1.00, T can be measured to give a higher level of precision and a small change in opacity causes a large change in transmittance as shown in Fig. 24-17. This subject is discussed in Knox and Wahren (1984) and UM 550 (but this method is not included in the 1991 issue of Useful Methods).

Since opacity is a measure of the degree of resistance to light transmission, it should not be surprising that a darker paper generally has a higher opacity than a brighter paper and dyes may be added to paper to increase its opacity.

Miscellaneous

Other interesting facts. In 1989 the cost per ton per point increase in brightness was $0.11 at 70% brightness, $0.30 at 80%, $0.71 at 85%, and $3.01 at 90%.

It is interesting to consider the optical properties of snow and a clear ice cube.

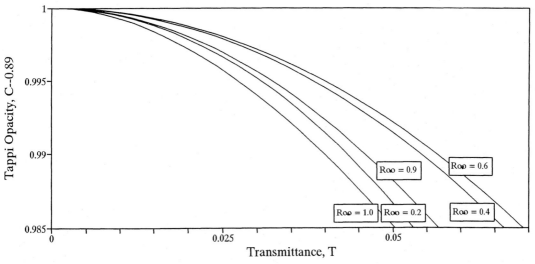

Fig. 24-17. Tappi opacity versus transmittance shows that low transmittance values can be used to calculate opacity with extremely high precision.

Table 24-4. Exact mathematical relationships of the Kubelka—Munk theory.[1]

VARIABLES AND CONSTANTS

Primary variables

R_∞ reflectivity of thick pad of paper
R_0 reflectance of a single sheet of paper backed by a blackbody
$R_{0.89}$ reflectance of a single sheet of paper backed by material with reflectance of 0.89
R reflectance with backing R_g
$C_{0.89}$ Tappi opacity $= \Omega_T = R_0/R_{0.89}$
C_∞ printing opacity $= \Omega = R_0/R_\infty$

W basis weight (of paper, filler, etc.)
s scattering coefficient
k absorption coefficient
T transmittance of a single sheet

Secondary constants, a and b

$$a = \frac{s+k}{s} = 1 + \frac{k}{s} = \frac{1}{2}(1/R_\infty + R_\infty) \tag{24-6}$$

$$b = (a^2 - 1)^{0.5} = \frac{1}{2}(1/R_\infty - R_\infty) \tag{24-7}$$

MATHEMATICAL RELATIONSHIPS

Miscellaneous solutions

$$\frac{k}{s} = \frac{(1-R_\infty)^2}{2R_\infty} \tag{24-8}$$

Solutions of R (24-9)

$$R = \frac{\frac{1}{R_\infty}(R_g - R_\infty) - R_\infty(R_g - 1/R_\infty)e^{sW(1/R_\infty - R_\infty)}}{(R_g - R_\infty) - (R_g - 1/R_\infty)e^{sW(1/R_\infty - R_\infty)}}$$

$$R = \frac{1 - R_g(a - b\coth bsW)}{a - R_g + b\coth bsW} \tag{24-10}$$

$$R = \frac{R_0 - R_g(2aR_0 - 1)}{1 - R_gR_0} \tag{24-11}$$

Solutions of R_0 (24-12)

$$R_0 = \frac{1}{a + b\coth bsW} = \frac{\sinh bsW}{a\sinh bsW + b\cosh bsW}$$

$$R_0 = \frac{1 - e^{sW(1/R_\infty - R_\infty)}}{R_\infty - \frac{1}{R_\infty}e^{sW(1/R_\infty - R_\infty)}} \tag{24-13}$$

$$R_0 = \frac{R - R_g}{1 - R_g(2a - R)} \tag{24-14}$$

Solutions of R_∞

$$R_\infty = 1 + k/s - [(k/s)^2 + (2k/s)]^{0.5} \tag{24-15}$$

$$a = \frac{1}{2}(R + \frac{R_0 - R + R_g}{R_0 R_g}) \tag{24-16}$$

$$R_\infty = a - (a^2 - 1)^{1/2} \tag{24-17}$$

Solution of Rg (24-18)

$$R_g = \frac{-1 + R_\infty R + \frac{1}{R_\infty}(R_\infty - R)e^{sW(1/R_\infty - R_\infty)}}{R - \frac{1}{R_\infty} + (R_\infty - R)e^{sW(1/R_\infty - R_\infty)}}$$

Formulas involving R_1

$$R = \frac{R_0(1 - R_g) + R_1 R_g(1 - R_0)}{1 - R_g R_0} \tag{24-19}$$

$$R_1 = \frac{R - R_0 + R_0 R_g(1 - R)}{R_g(1 - R_0)} \tag{24-20}$$

Table 24-4. Exact mathematical relationships of the Kubelka—Munk theory, Continued.[1]

Solutions of sW (24-21)

$$sW = \frac{1}{1/R_\infty - R_\infty} \ln \frac{RR_g R_\infty - RR_\infty^2 - R_g + R_\infty}{RR_g R_\infty - R - R_g R_\infty^2 + R_\infty}$$

(24-22)

$$sW = \frac{1}{b}[\coth^{-1}((a-R)/b) - \coth^{-1}((a-R_g)/b)]$$

special cases where $R = R_0$:

$$sW = \frac{1}{1/R_\infty - R_\infty} \ln \frac{-R_0 R_\infty^2 + R_\infty}{-R_0 + R_\infty}$$ (24-23)

$$sW = \frac{1}{b}\coth^{-1}[(1 - aR_0)/(bR_0)]$$ (24-24)

Eq. 24-23 is rearranged in terms of R_0/R_∞, useful for constructing charts based on printing opacity.

$$sW = \frac{1}{1/R_\infty - R_\infty} \ln \frac{1 - R_\infty^2}{1 - R_0/R_\infty}$$ (24-25)

Formulas involving Tappi opacity

To convert opacity $C_g = R_0/R_g$ to $C_{0.89}$: (24-26)

$$C_{0.89} = \frac{1 - 0.89R_0}{0.11 + \dfrac{0.89(1 - R_g R_0 - C_g + R_g C_g)}{R_g C_g}}$$

$$C_{0.89} = \frac{1}{1 + \dfrac{0.89}{R_0(1 - 0.89R_0)}T^2}$$ (24-27)

R_∞ as $f(R_0, C_{0.89})$: $R_{0.89} = \dfrac{R_0}{C_{0.89}}$; →Eq. (24–23)

The following equation uses *a, b, c,* and *y,* which are unrelated to their use elsewhere:

$$sw = \frac{\ln y}{(1/R_\infty - R_\infty)}$$ (24-28)

$$where:\ y = \frac{-b - (b^2 - 4ac)^{1/2}}{(2a)}$$

$$a = (0.89 - 1/R_\infty)(1 - C_{0.89})$$

$$b = (0.89 C_{0.89}(R_\infty^2 + 1/R_\infty^2) +$$

$$(R_\infty + 1/R_\infty)(1 - C_{0.89}) - 1.78$$

$$c = (0.89 - R_\infty)(1 - C_{0.89})$$

Formulas involving transmittance, T (24-29)

$$sW = \frac{1}{b}[\sinh^{-1}(b/T) - \sinh^{-1} b]$$

$$= \frac{1}{b}(\sinh^{-1}(b/T) + \ln R_\infty)$$

$$T = \frac{b}{a \sinh bsW + b \cosh bsW}$$ (24-30)

$$R_0 = a - (T^2 + b^2)^{1/2}$$ (24-31)

$$T = ((a - R_0)^2 - b^2)^{1/2}$$ (24-32)

$$a = \frac{1 + R_0^2 - T^2}{2R_0}$$ (24-33)

$$R = R_0 + \frac{T^2 + R_g}{1 - R_0 R_g}$$ (24-34)

$$T = [(R - R_0)(1/R_g - R_0)]^{1/2}$$ (24-35)

[1]See the appendix of this chapter for more information on the hyperbolic functions coth and coth⁻¹. Most equations are from Kubelka (1948); exceptions include: the solutions of sW as $f[\ln(u)]$ and R_g Robinson (1975), the conversion of opacity to Tappi opacity (Judd, 1937, 1938), and Tappi opacity as $f(R_0, T)$ [Å.S. Stenius, *Tappi J.* 56(11):157(1973)]. **Be creative**; although sW as $f(R_0, R, R_g)$ does not appear in this table, R_∞ can be solved first as $f(R_0, R, R_g)$ followed by sW as $f(R_\infty, R, R_g)$ or $f(R_\infty, R_0)$.

25

WOOD AND FIBER—GROWTH AND ANATOMY

25.1 INTRODUCTION

It is interesting to detail the development of the entire tree, or woody plants in general, but this information is of little use to the practicing pulp and paper scientist. Unfortunately, the terms to describe various plant tissues are somewhat arbitrary in this limited context; nevertheless, the features of cell types can be used to distinguish various species without the need to understand the entire woody plant as a single organism.

This chapter presents an overview of the anatomy of the wood of temperate softwoods and hardwoods. The growth characteristics of tropical species are somewhat different, especially in regard to growth rings, and a few examples are included as examples of this vast resource. Most tropical woods are hardwoods, but a few softwoods are commercially important. Softwoods are "designed" to resist the drought of winter where water exists as ice and is not available.

Detailed anatomies of softwoods, hardwoods, and some nonwood fiber sources are each considered in their own separate chapters. An overview of the growth of wood is considered in this chapter to give some context to wood anatomy. The anatomy of various woods is important for several reasons. For example, the characteristics of papermaking fibers depend much upon their anatomy. The pulping characteristics of various fiber sources are dependent upon their species and growing conditions.

One should review the fundamental concepts presented in Chapter 2 if the terminology is unclear in this chapter; Chapter 2 also has tables of wood properties and chemical compositions.

Flat or plain sawn wood has a tangential surface as the widest surface. Quartersawn boards have a radial surface as the widest surface. Letters are often used to indicate which plane of the wood to observe to see a particular feature including x for cross sectional, r for radial, and t for tangential surfaces. Some features are observed in individual fibers.

Traditionally wood has been used locally in manufacturing processes even if the final products have been transported some distance. More and more, however, wood is being bought and sold on the international market as local supplies dwindle in many locations. Because of this trend, much information on global supply and properties of many internationally important woods is included in this chapter. As wood becomes a more valuable material, its properties become even more important. Wood quality and the manipulation of wood quality are becoming increasingly important.

Woody plants

There are over 500 species of softwood and over 12,000 species of hardwoods (dicotyledons, plants containing seeds with two leaves, that are woody) worldwide. In the U.S. there are about 30 softwood and 50 hardwood species of commercial importance. This number is increasing as wood becomes in higher demand. For example, some *Populus* species were not used commercially until it was realized that they had good papermaking qualities. Even red gum, *Liquidambar styraciflua*, was considered to be a weed tree before 1900 until methods were developed to properly kiln dry it. About two—thirds of the commercial species used for lumber have appreciable use in pulp and paper, but others are also considered in this chapter as common commercial species which may enter the pulp mill.

All woody plants are perennial, i.e., live for a number of years. They have stems consisting of xylem and phloem tissues that conduct water and nutrients. The xylem is lignified and constitutes the wood. Their stems live and thicken by growing outward from the cambium each year. Although most paper has more softwood fiber than hardwood fiber, the anatomy and number of hardwoods often make them more difficult to identify than softwood. Woody plants can also be classified as trees, shrubs, or liana (climbing vines). *Dendrology* is the study of woody plants. Some herbaceous (nonwoody) plants (such as cereal straws and bagasse) are important sources of fiber for paper products in various countries.

Global and U.S. distribution of trees

Of the world's 3.70 billion hectares of forest land, one—third (1.22 billion) consists of conifers and two—thirds (2.49 billion) of broadleafs (FAO, 1966). Most of the conifers are in North America (36%) and the (former) U.S.S.R. (45%). Africa and the Pacific have negligible amounts of conifers. Due to the importance of the long fibers of conifers, the relatively untapped resources of Eastern Europe will play an important role in the source of fiber in the future. Table 25-1 shows harvest rates for 1973 by region of the world. Table 25-2 shows this in more detail for 1990.

Most of the world's broadleaf acreage is in Central and South America (34% of the total), Africa (27%), Asia (16%), North America (10%), and the (former) U.S.S.R. (7%).

Figs. 25-1 and 25-2 show the composition of U.S. forests. Most of the commercial hardwoods of the U.S. grow east of the Great Plains and overall account for less than 20% of the total timber resource. Most of the commercial softwood resource is west of the Great Plains with high concentrations in western Oregon and Washington (Douglas—fir) and in the South Atlantic and Southern States (southern pines).

In 1990 the world used 298 million m^3 of conifer pulpwood and 142 million m^3 of hardwood pulpwood (FAO, 1992). In 1990, the U.S. pulp-

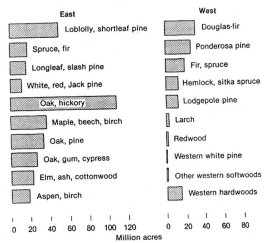

Fig. 25-1. U.S. commercial timberland by area, 1977. USDA For. Res. Rep. No. 23 (1982).

wood consumption was about 100 million cords (16%, West; 36%, South Central; 32%, South Atlantic; 7%, North Central; and 7%, North East). Softwood was 70% of the resource and hardwood was 30%. The U.S. pulp and paper industry had sales of $131 billion in 1991.

Botanical classification of wood (Jane, 1956)

All living organisms are classified according to their similarities. Because of the hundreds of thousands of organisms, methodical titles must be given in order to assure that one is talking about

Table 25-1. Global harvest of roundwood and growing stock in 1973.[1]

Region or country	Softwoods			Hardwoods		
	Total Harvest 10^9 ft^3	Percentage softwood growing stock	Percentage of harvest that is industrial	Total harvest 10^9 ft^3	Percentage hardwood growing stock	Percentage of harvest that is industrial
North America	13.4	1.6	99	3.5	1.1	86
Latin America	1.6	2.0	50	8.1	0.3	14
Europe	7.5	2.5	95	3.8	2.0	66
Africa, N. East	0.4	13.3	50	10.6	0.5	12
Asia-Pacific	4.5	2.5	40	19.9	1.9	23
Japan	0.9	2.5	100	0.6	2.0	100
U.S.S.R.	11.3	0.5	82	2.3	0.5	52
Total	39.6	1.1	84	48.8	0.7	30

[1] From *An Atlas of the Timber Situation in the United States 1952-2030*, USDA, For. Res. Rep. 23, 1982.

Table 25-2. 1990 wood production (FAO, 1992).

Region, Country	Total Harvest 10^6 m^3	
	Softwood	Hardwood
Africa	15.9	433.5
Cameroon	--	13.1
Ethiopia	2.8	38.3
Kenya	1.8	22.6
Nigeria	--	98.8
South Africa	6.5	12.8
Tanzania	--	32.9
Zaire	--	36.1
N. & C. America	490.6	221.9
Canada	140.0	15.5
Guatemala	6.4	1.4
United States	329.7	168.3
South America	59.3	237.7
Argentina	2.0	7.4
Brazil	45.6	178.1
Chile	11.4	7.0
Columbia	0.1	16.1
Asia	185.6	867.9
Bangladesh	--	30.9
China	132.0	145.0
India	9.9	252.8
Indonesia	0.7	170.1
Japan	19.6	10.0
Malaysia	--	48.6
Philippines	--	38.4
Thailand	--	34.0
Turkey	9.2	6.3
Vietnam	0.5	28.7
Europe	280.5	115.9
Austria	14.8	2.5
Czechoslovak	14.1	3.7
Finland	33.7	8.0
France	22.9	21.2
Germany	70.9	13.8
Norway	10.7	1.1
Poland	15.4	4.0
Romania	6.4	10.2
Spain	10.8	6.6
Sweden	47.4	8.4
Oceania	18.9	22.4
Australia	7.2	12.9
New Zealand	11.5	0.5
USSR, the former	300.3	64.3
Total	1351	1964

the same thing as one's colleagues who are working around the world. Common names may vary from one region to another. Often misleading names are assigned to wood from various species to help market them; there are distinctions made between a tree and the wood from the tree.

The botanical system of nomenclature is complicated and uses scientific (Latin) names. This system has been under development for hundred of years and undergoes changes with time so that each name should have an author name with it in the most technical and formal literature. This practice has not been followed in this work since commercial species are well known.

The most obvious division is that of the Kingdoms: plants and animals. Additional divisions are made as follows: phylum, class, order, family, genus, species. (The genus and higher levels always begin with a capital letter; the species always begins with a lowercase letter. Each type of plant has a unique combination of genus and species, but the species *rubra*, for example, might be a red oak, *Quercus rubra*, or red alder, *Alnus rubra*.) These have subgroupings such as subphyla; however, a given species of plant or animal may not have classifications at each of these levels.

For example, all seed plants (members of the phylum Spermaphyta) are divided into two subphyla: Gymnospermae (gymnosperms, those with naked seeds) and Angiospermae (angiosperms, those with seeds enclosed within the ovary of the flower). The Gymnospermae are not a group with many members so it is divided next into orders. The commercial timber species are all in the order Coniferales, the wood from which are the softwoods. From here they are divided into families, genera, and species. Fig. 25-3 demonstrates the classification of one hardwood and one softwood.

Softwoods are divided into seven botanical families. Four of these each contain commercial species important in the U.S. Table 25-3 shows the genera in each of these families that contain commercial species.

Angiospermae are arranged into two groups: commonly referred to as *monocots* (*monocotyledons*) and *dicots* (*dicotyledons*). Monocotyledoneae have one leaf in the seed and include mostly herbaceous plants such as grasses and corn.

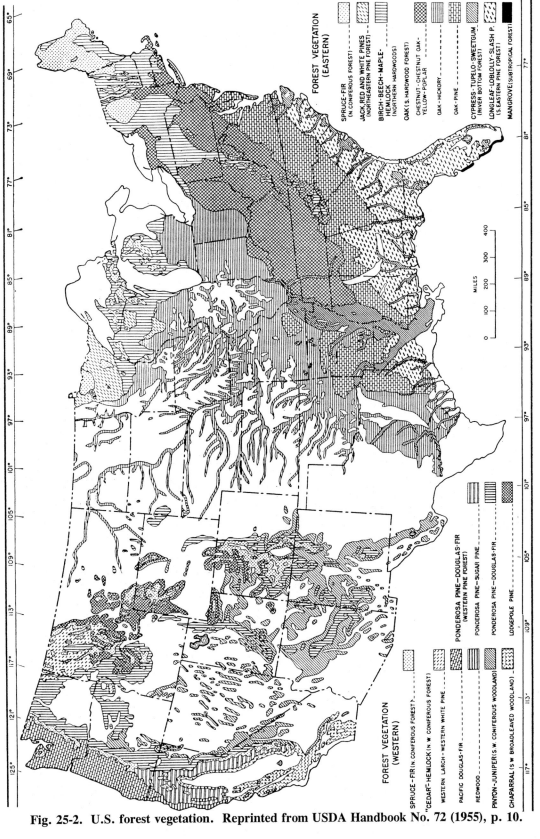

Fig. 25-2. U.S. forest vegetation. Reprinted from USDA Handbook No. 72 (1955), p. 10.

FOREST VEGETATION
(EASTERN)

SPRUCE-FIR (N. CONIFEROUS FOREST)
JACK, RED AND WHITE PINES (NORTHEASTERN PINE FOREST)
BIRCH-BEECH-MAPLE-HEMLOCK (NORTHERN HARDWOODS)
OAK (S. HARDWOOD FOREST)
CHESTNUT-CHESTNUT OAK-YELLOW-POPLAR
OAK-HICKORY
OAK-PINE
CYPRESS-TUPELO-SWEETGUM (RIVER BOTTOM FOREST)
LONGLEAF-LOBLOLLY-SLASH P. (S. EASTERN PINE FOREST)
MANGROVE (SUBTROPICAL FOREST)

FOREST VEGETATION
(WESTERN)

SPRUCE-FIR (N. CONIFEROUS FOREST)
"CEDAR"-HEMLOCK (N.W. CONIFEROUS FOREST)
WESTERN LARCH-WESTERN WHITE PINE
PACIFIC DOUGLAS-FIR
REDWOOD
PINYON-JUNIPER (S.W. CONIFEROUS WOODLAND)
CHAPARRAL (S.W. BROADLEAVED WOODLAND)
PONDEROSA PINE-DOUGLAS-FIR (WESTERN PINE FOREST)
PONDEROSA PINE-SUGAR PINE
PONDEROSA PINE-DOUGLAS-FIR
LODGEPOLE PINE

MILES
0   100   200   300   400

Kingdom:	PLANTS	PLANTS	
Phylum:	SPERMAPHYTA	SPERMAPHYTA	
Sub-Phylum:	GYMNOSPERMAE	ANGIOSPERMAE	
Class:		DICOTYLEDONEAE	
Sub-Class:		SYMPETALAE	
Order:	CONIFERALES	CONTORTAE	
Sub-Order:		OLEINEAE	
Family:	PINACEAE	OLEACEAE	
Genus:	PINUS	FRAXINUS	
Species:	STROBUS L.	AMERICANA L.	

Fig. 25-3. The full classification of a softwood and a hardwood.

A few woody plants such as bamboo, palm, and other "nonwood" fiber sources used in pulp and paper are members of this class.

Dicotyledoneae have two leaves in the seed and include peanuts, string beans, and many other plants including the hardwoods. The numerous Dicotyledoneae are further divided into subclasses, orders, suborder, families, and some subfamilies. The anatomy of monocot and dicot stems will be considered in Chapter 29 along with pulping on nonwood fibers.

Hardwoods are divided into scores and scores of families. A relatively few contain all of the species used for papermaking in the U.S. Table 25-4 shows the genera for some of the major families of the U.S. with commercial species and Table 25-5 lists several genera with commercial species of global importance.

Table 25-3. Genera of softwood families with commercial species in North America.

Family	Genus	Common name	Notes (N. or S. hemisphere)
Pinaceae	*Pinus*	Pine	N.
	Pseudotsuga	Douglas-fir	North America, Eastern Asia
	Picea	Spruce	N.; temperate
	Abies	Firs, true, silver	N.; temperate
	Larix	Larch	N.
	Tsuga	Hemlock	N.; Eastern Asia
	Cedrus[1]	True cedars	North Africa, Asia Minor
Cupressaceae	*Chamaecyparis*	False cypress	North America; East Asia
	Juniperous	Juniper, pencil cedar	N.
	Libocedrus	Incense-cedar	North America
	Thuja	Cedar	North America; Eastern Asia
	Cupressus	True cypresses	N.
	Widdringtonia[1]		South and East Africa
	Fitzroya[1]		Also *Callitris*[1], *Tetraclinis*[1]
Taxodiaceae	*Taxodium*	Swamp cypresses	
	Sequoia	Sequoia, Redwood	Western North America
	Cryptomeria[1]		Eastern Asia
Taxaceae	*Taxus*	Yew	N.; temperate

Other genera of global importance (are not native to North America)

Podocarpaceae	*Dacrydium*		S.; temperate, esp. New Zealand
	Podocarpus	Podocarp	S.; esp New Zealand
Araucariaceae	*Araucaria*		S.; not Africa; also *Phyllocladus*
	Agathis	Kauris	S.; Southeast Asia, Australasia

[1] Not native to North America, but an important genus globally.

Table 25-4. Genera of hardwood families with commercial species in North America (Jane, 1956).

Order	Family	Genus	Common name or example	Notes (N. or S. hemisphere, etc.)
Subclass Archichlamydeae				
Salicales	Salicaceae	*Populus*	Cottonwood, aspen	N.; temperate
		Salix	Willow	N.; temperate
Juglandales	Juglandaceae	*Carya*	Hickory, pecan	N.Amer.; Asia
		Juglans	Walnut, butternut	N.
Fagales	Betulaceae	*Alnus*	Alder	N. mostly
		Betula	Birch	N.
		Carpinas	Hornbeam	N.
		Ostrya	Hophornbeam	N.
	Fagaceae	*Castanea*	Chestnut	N.; temperate
		Castanopsis	Chinkapin	N. Amer.
		Fagus	Beech	N.; temperate
		Lithocarpus	Tanoak	N. Amer.
		Quercus	True oak	N.
		Nothofagus[1]	Beech	S. Amer; Austral
Urticales	Ulmaceae	*Celtis*	Hackberry	N.; South Africa
		Ulmus	Elm	N.;
		Phyllostylon[1]		Tropical Amer.
Moraceae	Moraceae	*Maclura*	Osage-orange	N.
		Morus	Mulberry	N.
		Chlorophora[1], *Artocarpus*[1], *Antiaris*[1], *Brosimum*[1], *Piratinera*[1]		
Ranales	Magnoliaceae	*Liriodendron*	Yellow-poplar	N. Amer; China
		Magnolia	Magnolia	N. Amer; Asia
	Lauraceae	*Sassafras*	Sassafras	
		Umbellularia	Myrtle	
		Phoebe[1], *Ocotea*[1], *Eusideroxylon*[1], *Beilschmiedia*[1], *Endiandra*[1]		
	Cercidiphyllaceae[1]	*Cercidiphyllum*	*C. japonicum*	Japan, China
	Anonaceae[1]	*Oxandra*	*O. lanceolata*	Tropical Amer.
	Myristicaceae[1]	*Virola*, *Pycnanthus*		Trop. Amer, Africa
	Monimiaceae[1]	*Laurelia*, *Persea*		Chile, N. Zealand
Rosales	Hamamelidaceae	*Liquidambar*	Sweet gum	N.
	Platanaceae	*Platanus*	Sycamore	N.; temperate
	Rosaceae	*Prunus*, *Malus*	Cherry, apple	N.; temperate
	Leguminosae[2]			
	-Mimosoideae[1]	*Acacia*, *Xylia*, *Piptadenia*, *Cylicodiscus*, *Erythrophleum*		
	-Caesalpinioideae[1]	*Mora*, *Copaifera*, *Peltogyne*, *Intsia*, *Berlinia*, *Daniellia*, etc.		
	-Papilionatae	*Gleditsiack*	Honeylocust	N.
		Robinia	Black locust	North America
		Laburnum[1], *Byra*[1], *Dalbergia*, *Pterocarpus*, *Andira*, etc.		

[1] Not native (or minimal) to North America, but of importance globally.
[2] An extremely complicated family of thousands in three subfamilies

Table 25-4. Genera of hardwood families with commercial species in North America (Cont'd).

Order	Family	Genus	Common name or example	Notes (N. or S. hemisphere, etc.)
Subclass Archichlamydeae				
Sapindales	Aquifoliaceae	*Ilex*	Holly	
	Aceraceae	*Acer*	Maple	N.; temperate
	Hippocastanaceae	*Aesculus*	Buckeye	N.temperate; S.
	Rhamnaceae	*Rhamnus, Maesopsis*[1]		
	Buxaceae[1]	*Buxus*	Box	Eurasia, C. Amer.
	Anacardiaceae[1]	*Dracontomelum, Campnosperma, Astronium, Schinopsis*		
	Celastraceae[1]	*Euonymus, Goupia*		
	Sapindaceae[1]	*Harpullia*	*H. pendula*	
Malvales	Tiliaceae	*Tilia*	Basswood	N.; temperate
		Pentace[1], *Cistanthera*[1]		
	Gonystylaceae[1]	*Gonystylus*	(Melawis, ramin)	Malaya
	Triplochitonaceae[1]	*Triplochiton*	(Obeche, wawa, ayous)	West Africa
	Bom Bacaceae[1]	*Bombax, Ceiba, Ochroma*		Tropical species
	Steruliaceae[1]	*Sterculia, Tarrietia*		Mostly tropical
Myrtiflorae	Nyssaceae	*Nyssa*	Tupelo	N. Amer, Asia
	Myrtaceae	*Eucalyptus*		Common
		Eugenia[1], *Syncarpia*[1], *Tristania*[1]		
	Lythraceae[1]	*Lagerstroemia*		Tropical
	Lecythidaceae[1]	*Cariniana*		Tropical America
	Rhizophoraceae[1]	*Bruguiera, Carallia, Anisophyllea* the mangroves		
	Combretaceae[1]	*Terminalia, Anogeissus*		Tropical
Umbelliforae	Cornaceae	*Cornus*	Dogwood	N.; temperate
	Araliaceae[1]	*Acanthopanax*		Asia
Subclass Sympetalae				
Ericales	Ericaceae	*Arbutus*	Madrone	North America
		Erica		
Ebenales	Sapotaceae	*Palaquium, Achras, Mimusops*		
	Ebenaceae	*Diospyros*	Persimmon	U.S., Africa
		Oxydendron	Sourwood	
Contortae	Oleaceae	*Fraxinus*	Ash	N.
		Olea[1]	Olive	Warm climates
	Apocynaceae[1]	*Gonioma, Alstonia, Dyera, Aspidosperma*		
Tubiflorae	Bignoniaceae	*Catalpa*	Catalpa	America, East Asia
		Tabebuia[1], *Paratecoma*[1]		Tropical America
	Scrophulariaceae	*Paulownia*		East Asia, now U.S.
	Boraginaceae[1]	*Cordia*		Tropical America
	Verbenaceae[1]	*Tectonia, Gmelina*	Teak, gray teak	Asia, Australia

[1] Not native (or minimal) to North America, but of importance globally.

Table 25-5. Genera of hardwood families with commercial species globally (Jane, 1956).

Order	Family	Genus	Common name (or example)	Notes (N. or S. hemisphere, etc.)
Subclass Archichlamydeae				
Verticillatae	Casuarinaceae	*Casuarina*	She or Forest oak	Australia, Polyn.
Proteales	Proteaceae	*Grevillea, Knightia, Cardwellia*		Australia
Santalales	Santalaceae	*Santalum*	Sandalwood	Indomalaya
Geraniales	Erythroxylaceae	*Erythroxylum*	*E. mannii*	West Africa
	Zygophyllaceae	*Guaiacum, Bulnesia*		
	Rutaceae	*Fagara, Flindersia, Chloroxylon, Amyris*		
	Simarubaceae	*Ailanthus*	Tree of heaven	Asia
	Burseraceae	*Canarium, Aucoumea*		Trop. Asia Africa
	Meliaceae	*Cedrela, Khaya, Carapa, Guarea, Lovoa, Sandoricum*, etc.		
	Vochysiaceae	*Vochysia, Qualea*		Tropic America
	Euphorbiaceae	*Aextoxicon, Croton, Hura, Ricinodendron*		Most tropical
Parietales	Ochnaceae	*Lophira*		Tropical Africa
	Theaceae	*Tetramerista*		Malaya
	Guttiferae	*Cratoxylon, Mesua, Calophyllum, Symphonia*		
	Dipterocarpaceae	*Dipterocarpus, Hopea, Shorea, Pentacme*, etc.		Eastern Tropics
	Flacourtiaceae	*Scottellia, Gossypiospermum*		
Subclass Sympetalae				
Primulales	Myrsinaceae	*Rapanea*		Africa, Asia
Rubiales	Rubiaceae	*Calycophyllum, Adina, Mitragyna, Gardenia, Canthium*		

Common names of woods that do not really belong to suggested genera are hyphenated or spelled as one word. Douglas—fir is not a true fir, tanoak is not a true oak, and parana—pine is not a true pine.

Plant growth tissues

The growth tissues of plants are meristems; they have rapidly dividing cells that enlarge (grow) to provide the bulk of the plant. Examples of several meristematic tissues will be considered here: the root tip (radical), the terminal bud (apical meristem), and the *cambium*. The first two provide growth with an increase in length; the latter provides growth in diameter and is of the most interest to wood scientists. A perfunctory discussion is given here.

Primary growth results in plant elongation at the root and bud apical meristems. Fig. 25-4 shows a root tip from onion. (Onion was selected because it shows many features unusually well.) The root shows the division of the cell nucleus; individual chromosomes can be seen. Fig. 25-5 shows a shoot apex (bud) for maple (hardwood) and pine (softwood). Growth originates from the tiny mass of close—packed cells known as the apical meristem. As the apical meristem grows upward, the center cells remain parenchymatous. The epidermis is formed from the outermost cells, and the cortex is adjacent to the epidermis. The vascular tissue forms between the cortex and pith and consists of the xylem (which will expand and grow into wood) and phloem (bast, which will grow into the inner bark).

The primary vascular tissue exists in small bundles. In monocots these bundles are scattered in the stem, whereas in dicots the bundles are all about the same distance from the center and form

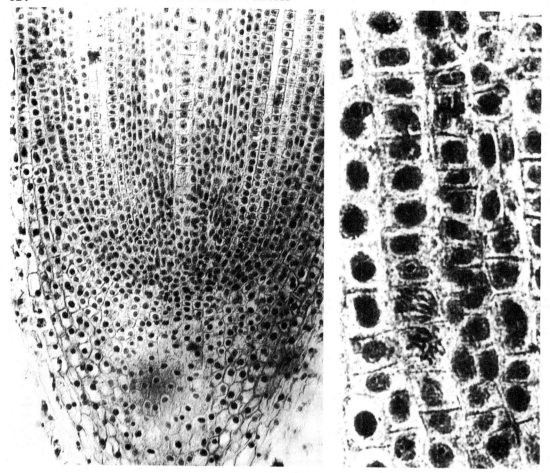

Fig. 25-4. Onion root tip magnified 150× and 600×. The individual chromosomes can be seen in cells undergoing division.

a ring of bundles. The morphology of the vascular tissue of various of these plant categories is detailed in Chapter 29. The scattering of the xylem remains in those monocots that undergo secondary growth.

The vascular cambium (Plate 44) of dicots and coniferous species will form between the xylem and phloem and result in *secondary growth*; it increases the diameter of the stem but not the length and is called a *lateral meristem*. The cambium has two types of cells: *fusiform initials*, which are vertically elongated and divide into the longitudinal cells, and *ray initials*, which are more cubical in shape and divide into the ray cells. These cells divide by forming tangential walls (periclinal division) and give rise to daughter cells. One of the daughter cells, called the mother cell,

retains the ability to divide. The daughter cells differentiate into the xylem (if interior to the mother cell) or phloem (if exterior to the mother cell). Some hardwoods have cell division in the radial direction (anticlinal) for an increase diameter. Softwoods and other hardwoods have a diagonal division and differential growth forms two side—by—side cells for an increase in diameter.

Parenchyma cells of the xylem will usually live until heartwood formation; the prosenchyma cells of the xylem differentiate into their final form within a few days or weeks and die. The anatomy of the xylem (wood) will be considered in detail.

Phloem

Phloem typically grows at only one—sixth to one—tenth of the rate of xylem. Phloem consists

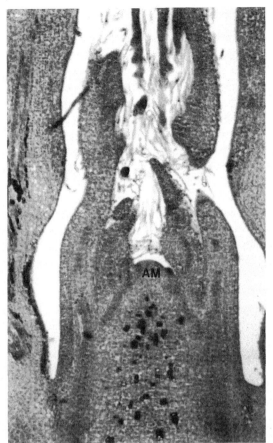 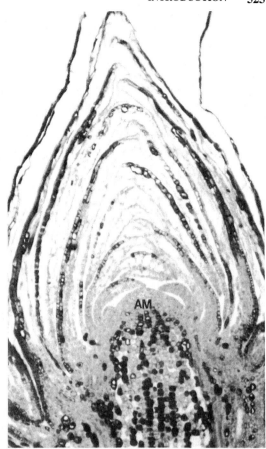

Fig. 25-5. Shoot apices with apical meristems (marked AM) of maple (left) and pine (55×).

of sieve tubes, companion cells, parenchyma, and bast fibers (*sclerenchyma*). Bast fibers are formed directly from cambial initials of modified parenchyma. The latter are called *sclereids* and may have highly branched, unusual shapes. Companion cells are so named because each sieve tube has at least one companion cell associated with it. Bast fibers are lignified and do not easily crush.

The walls of sieve tubes and companion cells do not become lignified. The sieve tubes usually function for one year and then die and collapse. The collapse provides some additional space for new growing tissue to the interior. Tangential growth in the bark is sometimes provided by flaring of the rays.

Bark

Initially a twig is protected from the environment (against desiccation and mechanical injury) by the thin epidermis. When the cambium becomes active, the increase in diameter will cause the epidermis to rupture. By this time, however, the *periderm* (also called *rhytidome* or *outer bark*) has begun to form. It consists of the *phellogen* or cork cambium, the tissue outward of the phellogen that is called the *phellum* and is composed cork cells with much suberin, and the tissue inner to the phellogen that is called the *phelloderm* and is composed of a thin layer of thin—walled tissue. Thick—walled fibers (sclerenchyma) may also be present in the periderm. Sclerenchyma cells from the bark are called *stone cells* by the pulp and paper industry, but the botanical use of the term stone cells is much more limited.

The periderm is eventually cut off as dead phloem underneath is shed. Secondary periderms form in arcs from phloem parenchyma cells to continue the growth of the outer bark (page 13). This accounts for the layered nature of bark and the deep fissures that may form. Detailed bark anatomy is considered by Chang (1954).

Rays

Most cells of wood are oriented vertically to conduct nutrients and water up and down the bole. Some cells provide conduction from the pith outward to the bark. They vary in height from a few cells in some species to several inches in the oaks; in the oaks they cause the *ray flecks* or silver grain observed on the radial surfaces. The rays of softwoods generally account for less than 5% of the volume of the wood. The rays serve to store food and transport it horizontally in the stem.

General types of cells in wood

(The term cell is correctly used for living cells only, but its use to previously living cells of wood is in common practice.) There are two broad categories of cells in mature wood: *prosenchyma* and *parenchyma*. They may each be oriented longitudinally and, in the case of ray cells, horizontally. The term prosenchyma is not used commonly in the pulp and paper literature, but it includes fibers (tracheids of softwoods and hardwoods), libriform fibers, and vessel elements. The latter two occur in hardwoods only. Prosenchyma are dead in the mature xylem.

Parenchyma cells are rectangular with an aspect (length to width ratio) of about 2—4 compared with 50—100 that is typical of elongated fibers. They are small, thin—walled cells and function in the wood by storage, secretion, and

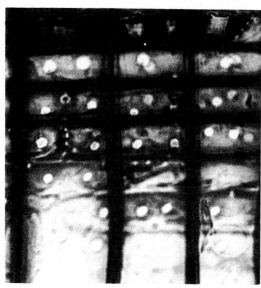

Fig. 25-6. Crystals (top) of the marginal ray cells of *Abies concolor*.

wound healing. They may occur longitudinally or radially with the latter outnumbering the former in most species. They may store calcium oxalate as crystals, tannins, starch, fats, or proteins. Parenchyma cells are often living in sapwood. Calcium oxalate crystals are observed in some true firs, sitka spruce, incense cedar, and black walnut. Rhomboidal and rectangular crystals are found in marginal parenchyma cells of some *Abies* (Fig. 25-6). Specific crystal types occur in certain tropical species (Brown and Panshin, 1940, p. 214).

Softwood cell types

The volume of parenchyma cells in softwoods is typically 5% in spruce and pine to 10% in larch and fir; however, their mass is only 1—2.5% of the total mass. They may occur as strand or epithelial (around resin canals) cells in the longitudinal or horizontal (as part of the rays) directions. Likewise tracheids may be longitudinal or horizontal. All softwoods have at least longitudinal tracheids and ray parenchyma. The non-parenchyma cells of softwoods are all tracheids and fibers. Fig. 25-7 shows the structure of a representative softwood.

Hardwood cell types

The amount of parenchyma cells in hardwoods is much higher and more variable with a volume percentage of 10—35% and mass percentage of 5% or more. They may occur as strand, fusiform, or epithelial cells (such as those found in longitudinal gum canals in injured sweetgum) if they are oriented longitudinally (named axial) and as ray (*upright* or *procumbent*) or epithelial parenchyma if oriented horizontally. (Some of these parenchyma types are rare in U.S. species.) Fusiform parenchyma occurs as marginal parenchyma of red maple.

Fibers include *vessel elements* and fiber cells. The fiber cells include the ubiquitous *libriform fibers*. The other fiber cells include *fiber tracheids*; *vasicentric tracheids* found in the genus *Fraxinus* and the family Fagaceae; and *vascular tracheids* found in the Ulmaceae family of elms and hackberry. Fig. 25-8 shows the structure of a representative hardwood.

The cell composition of hardwoods and other properties such as fiber length are very important due to their higher complexity compared to soft-

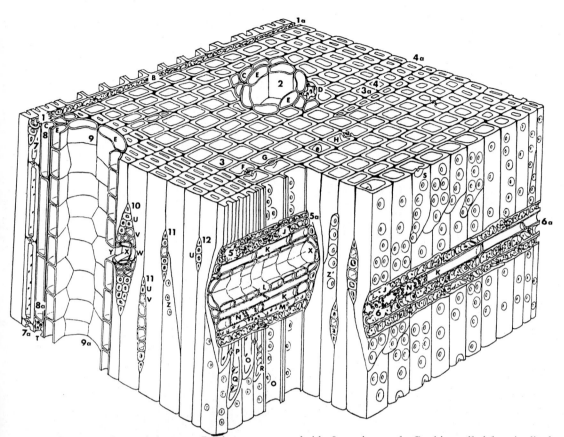

Key: **Cross section:** 1-1a, ray; B, dentate ray tracheid; 2, resin canal; C, thin-walled longitudinal parenchyma; D, thick-walled longitudinal parenchyma; E, epithelial cells; F, radial bordered pit pair cut through torus and pit apertures; G, pit pair cut below apertures; H, tangential pit pair; 4-4a, latewood.

Radial view: 5-5a, sectioned fusiform ray; J, dentate ray tracheid; K, thin-walled ray parenchyma; L, epithelial cells; M, unsectioned ray tracheid; N, thick-walled parenchyma; O, latewood radial pit (inner aperture); O', earlywood radial pit (inner aperture); P, tangential bordered pit; Q, callitroid-like thickenings; R, spiral thickenings; S, radial bordered pits (middle lamella removed); 6-6a, sectioned uniseriate, heterocellular ray.

Tangential view: 7-7a, strand tracheids; 8-8a, longitudinal parenchyma (thin walled); T, thick-walled parenchyma; 9-9a, longitudinal resin canal; 10, fusiform ray; U, ray tracheids; V, ray parenchyma; W, horizontal epithelial cells; X, horizontal resin canal; Y, opening between horizontal and vertical resin canals; 11, uniseriate, heterocellular rays; 12, uniseriate, homocellular ray; Z, small tangential pits in latewood; Z', large tangential pits in earlywood. Reprinted from Howard and Manwiller (1969).

Fig. 25-7. The three—dimensional structure of southern pine, a softwood.

woods. The high amount of parenchyma cells in many species is detrimental to papermaking. Table 25-6 gives some values for U.S. southern hardwoods growing on softwood sites. This table also shows that average fiber length is quite variable among species, position in the tree, and other factors.

In hardwood pulp the parenchyma cells are often selectively removed with fines during processing. Because of their small size they do not contribute to paper strength but they contribute appreciably to low freeness in pulp.

The vessel element content of hardwoods is about 10% in birches, 40% in oaks, and over 55%

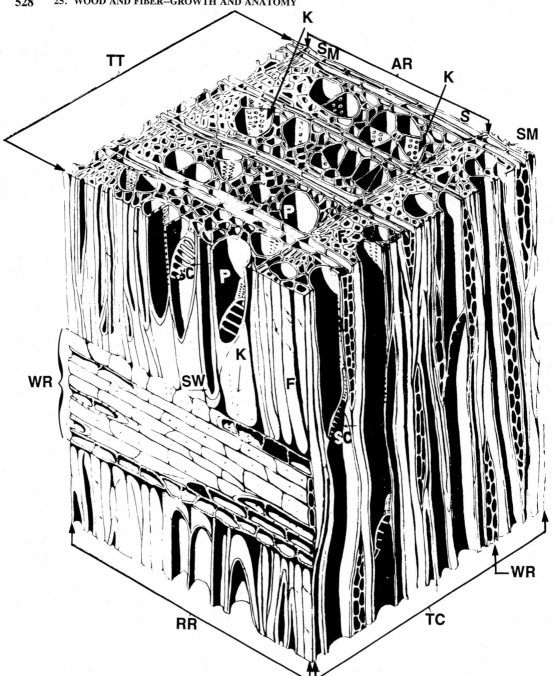

Key: Surfaces are given as TT for the cross section, RR for the radial surface, and TC for the tangential surface. Structures include WR for wood ray and AR for annual growth ring with S for earlywood and SM for latewood. P is a vessel with SC for sclariform plate. F denotes libriform fiber. K is cell pit. Approx. 125×. Reprinted from USDA For. Ser. Tech. Note No. 210, 1952.

Fig. 25-8. The three—dimensional structure of yellow poplar, a hardwood.

in red gum by volume. In the latter two cases, the amount by mass is much smaller since most of the volume is empty space. The length of vessel elements ranges from 0.1 to 2 mm (but 0.5 to 1

Table 25-6. Proportions of cells in 6 in. diameter hardwoods growing on southern pine sites.[1]

| Species | - % composition by volume of stems- | | | | --Mean cell length, millimeter-- | | Vessel[2] Elements | Mean tree age (Years) |
	Vessels	Fiber	Rays	Longitud. Parenchy.	Stem	Branch		
Ash, green	14.5	57.3	14.2	14.0	1.16	0.84	0.26	37
Ash, white	14.5	57.0	14.0	14.4	1.22	0.87	0.29	40
Elm, American	31.1	42.2	12.0	14.8	1.30	0.99	0.22	40
Elm, winged	28.2	40.0	15.5	16.2	1.25	0.93		47
Hackberry	20.1	50.9	13.6	15.4	1.12	0.86	0.26	28
Hickory, true, *Carya*	14.7	53.2	18.1	14.0	1.29	0.98	0.44	54
Maple, red	23.2	50.6	12.8	13.4	0.83	0.66	0.42	40
Oak, black	14.7	37.1	20.8	27.4	1.30	0.98	0.43	39
Oak, northern red	15.3	41.3	19.0	24.5	1.27	0.90	0.42	40
Oak, white	14.7	41.2	20.7	23.5	1.22	0.88	0.40	41
Sweetbay (Magnolia)	30.1	50.3	15.2	4.4	1.24	0.97		39
Sweetgum	46.2	37.6	14.0	2.1	1.54	1.20	1.32	29
Tupelo, black	38.4	39.8	17.2	4.6	1.76	1.40	1.33	59
Yellow poplar	43.3	42.6	11.4	2.7	1.39[3]	0.97	0.89	27
Average, stemwood	20.5	44.4	17.0	18.1	1.27			
Aver., branchwood	19.7	45.6	15.9	18.8		0.96		

[1] From Koch (1985). Averages of woods of 10 trees. [2] From Panshin and de Zeeuw as listed in Koch (1985). [3] Other works show values on the order of 1.7 to 1.8 mm in mature wood.

mm is common) depending on the species; some values are given in Table 25-6. Libriform fibers have small, simple pits. Fiber tracheids have bordered pits smaller in size than the intervessel pits. Tracheids have numerous pits of about the same size as the intervessel pits. All of these fiber types will be discussed in more detail in the appropriate chapters.

Cell pits

Pits (page 17) are openings in the fiber wall that allow conduction of materials between cells. A pit of one cell is almost always associated with a pit of a second wall so the term *pit pair* is often used. There are three types of pit pairs as shown in Fig. 25-9. Each half of a pit pair may be simple (with a uniform hole, as in parenchyma cells) or bordered (with a thickening in the wall around it, as in most tracheids and vessels). Two simple pits form a simple pit pair, one simple pit with a bordered pit form a half—bordered pit pair, and two bordered pits form a full bordered pit (Fig. 25-9). The pits of hardwoods lack the

thickening in the middle, the *torus* (*tori* is the plural form), and are much more variable than those of softwoods.

Miscellaneous considerations

The wood first laid down in an annual growth ring of temperate woods is called the earlywood. It is especially suited for conduction. The remainder of the growth ring, the latewood, is especially suited for strength. These purposes will become more clear as the anatomy of wood is considered.

25.2 SAMPLE PREPARATION FOR IDENTIFICATION OR MICROSCOPY

Preparation of wood samples

Details on sample preparation and microscopy of a wide variety of plant tissues are found in any of the several classic microtechnique books such as Johansen (1940) or Sass (1940). Details on the preparation of wood samples are given in many of the wood anatomy resources listed below. If you are serious about wood identification with

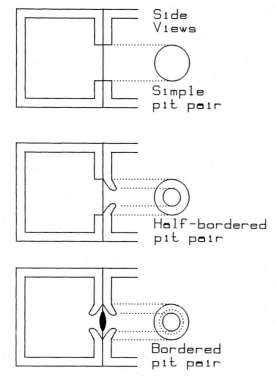

Fig. 25-9. The three types of pits: simple, half-bordered, and bordered.

a hand lens and knife, see Hoadley, 1990; this is an excellent and reasonably priced resource.

For routine analysis of large numbers of wood samples in a mill, where often only a few species must be distinguished, freehand cutting with a very sharp knife or single—edged razor blade may be suitable. A 10× hand lens may provide the necessary resolution of the transverse section (and others, if necessary) for identification. Moistening the freshly cut surface often helps see detail. In most cases, one will probably want to use a single—edged razor blade to prepare samples for examination under the microscope. Double—edged platinum—chrome blades will give very good results (for safety, the edge not being used can be wrapped in cardboard and taped); also, they may be cut in half lengthwise with two pairs of pliers and mounted in a pencil—type hobby knife holder with split collet.

The preparation of wood and fiber samples for microscopy is in many regards easier than for other plant tissues since wood is relatively inert and strong. It is, however, difficult to cut on the microtome. Wood sections for high—quality photomicrographs (i.e., those presented in this chapter) are prepared on a microtome to give sections about 10—25 μm thick. The magnification of a microscope is about 35—600×, although an oil immersion lens brings this to 1000×. This corresponds to a resolution of about 0.3 μm for routine work. Dissection microscopes have a range in magnification from about 5× to 50×.

For high—quality sectioning (photomicroscopy), small cubes (about 1 cm per side) are prepared. Soft species of wood (white pine, aspen, and spruce) are merely boiled until they sink prior to cutting. Other species are softened with 2—4% KOH in refluxing ethanol for one hour or longer, followed by washing. Acetic acid/hydrogen peroxide or any of a multitude of reported methods have also been used. Tropical species with high levels of SiO_2 (several percent) may require treatment with HF (special precautions must be used with this material) to remove it as SiF_4. Fig. 25-10 shows the standard arrangement of wood sections on a microscope slide. An alternate arrangement is to have the cross and radial sections on top and the tangential section beneath the radial section.

Wood and fiber samples are often stained with safranin O (0.5—1% in water or 50% ethanol), a red stain selective for cellulose. Fast green is used with safranin on many other botanical samples as a counterstain (for example, to show phloem tissue).

Analysis of wood chips

Many mills purchase wood chips from a wide variety of suppliers. Often they wish to confirm the identity of the wood species occurring from a supplier. A random sample should be obtained that represents the population of interest. In the laboratory highly skilled anatomists can identify about one wood chip each minute (60/hr). About

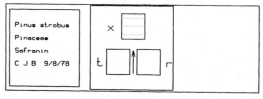

Fig. 25-10. One arrangement of wood samples on a slide (↑ is longitudinal direction).

50 or 100 chips might be identified from the representative sample depending on the requirements of the study. Since the chips have already darkened, microscopic examination is usually required for confidence in the species determination.

It may be difficult to determine the orientation of the wood chip, but snapping it in half usually presents a fresh radial surface. Decay in wood chip piles occurs faster with hardwoods than with softwoods. Hardwoods have more parenchyma (that initially heat the pile by respiration of food) and the heartwood of species used in pulping is generally not as decay resistant as that of the softwoods. Hemlocks, lodgepole pine, and most spruces have only moderate decay resistance in the heartwood. True firs are the only U.S. softwoods with minimal decay resistance in the heartwood.

Macerations

Wood samples are macerated to study the liberated fibers, such as for fiber length studies. Maceration is a highly selective pulping operation. Wood is first cut into pieces about 2 mm square by 25 mm in the longitudinal direction. Air is removed by soaking (perhaps with vacuum) or boiling in water.

The most gentle maceration technique is the use of sodium chlorite acidified with acetic acid [Spearin and Isenberg, *Science*, 105(2721):214 (1947)]. Caution: impurities can make sodium chlorite explosive. In one method, 35 mL of water is used with 0.6 g sodium chlorite and 5 drops of glacial acetic acid for 1 hour at 90°C in a hood. Alternately 2 grams of sodium chlorite with 12 drops of acetic acid can be used for 3 hours at 85°C in a hood.

The method of Franklin is also gentle. Wood pieces are treated with a solution consisting of equal parts of glacial acetic acid and 6% hydrogen peroxide at 60°C for 48 hours. Wilson [*Pulp Paper Mag. Can.* 55(7):127-129(1954)] has reviewed maceration techniques.

After the chemical treatment the wood chips are washed in gently flowing water overnight so the fibers do not separate. The pieces are then teased apart and stained as appropriate. Temporary slides are made with Karo. Permanent slides are made by dehydration using an ethanol series of 15%—30%—50%—70%—85%—95%—abs—abs—50% abs + 50% xylene—xylene.

Preparation of paper samples

A small sample of paper can be boiled in 1% NaOH to hydrolyze the sizing and help disperse the fibers. The sample is then washed and agitated to disperse the fibers.

Mounting of sections and macerated fibers

Mounting media have an index of refraction similar to that of the glass microscope slide. Corn syrup (Karo) is useful for temporary slides; it is often diluted with water (about 80:20 syrup:water) to give a good working solution where bubbles do not interfere. Tappi T 263 suggests the use of 50:50 solution of glycerin and 95% ethanol with heating to the boiling point (after the cover glass is applied) to drive off air. These do not require dehydration of the sample. (Dehydration involves replacing water with ethanol and ethanol with xylene so that nonpolar materials can be used.)

Canadian balsam (from the pitch pockets of the bark of balsam fir) is the traditional mounting medium for permanent slides. Synthetic resins are also available. A coverslip can be held in place with two clothespins when waiting for the mounting medium to dry when making permanent slides.

The use of keys

Dichotomous keys involve a series of separations of possible wood species based on identifying characteristics. Each selection is a choice of two possibilities (such as the presence or absence of a feature). A key using hand lens identification is included to demonstrate their use. For example, a wood that has vessels must be a hardwood so all softwood species are eliminated. If it is ring porous then many of the hardwood species are eliminated. Identification of additional features eventually results in its species determination.

An alternate to the key is the use of end—punched cards where each number corresponds to (the presence or absence of) a single feature (hardwoods: Brazier and Franklin, 1961; softwoods: Phillips, 1948). A hole away from the edge is used if the feature is absent and a notch (where the hole is enlarged to include the edge) is used if the feature is present. By using a wire in a location corresponding to a given observed feature in the unknown, appropriate cards can be made to fall from the deck. The use of a second feature narrows the determination to a smaller group. Eventually the correct species is determined. The observation of a few key features (such as spiral thickening in tracheids) allow the correct determination very quickly. This system has been computerized for faster use (North Carolina State University). The use of keys for softwoods and hardwoods is considered in their respective chapters.

25.3 WOOD VARIATION

Reaction wood consists of compression wood in softwoods and tension wood in hardwoods. A variety of grain distortions are also considered in this section.

Compression wood

Pillow and Luxford (1937) summarize compression wood as an abnormal type of wood occurring as a rule on the lower side of nonvertical trunks and branches among all coniferous species of trees. An increase in the amount of deviation of trunks from a vertical position or in the rate of diameter increment of individual trees increases the formation of compression wood.

Compression wood is identified by markedly eccentric annual growth rings (Fig. 25-11). In the wide portion of the growth ring there is a higher than normal level of latewood that has a more gradual transition from the earlywood to latewood than in normal wood.

Under a microscope the latewood tracheids of compression wood appear to be nearly circular in cross section whereas those of normal wood are more or less rectangular (Fig. 25-12). The fibrils of the secondary cell walls in compression wood make a higher angle in relation to the longest axis of the cells than do the fibrils of normal wood and these walls contain microscopic checks. The spiral thickenings that normally occur in Douglas—fir and a few other coniferous species are fewer in number and less distinct in compression wood than in normal wood and may be confined to earlywood fibers in pronounced compression wood.

The length of tracheids in compression wood is generally 20—30% less than that in normal wood. The fibril angle of the secondary cell wall layer of softwood tracheids is typically about 20—24° in earlywood and below 10° in latewood; in pronounced compression wood the angles are about 35—38° and 24—30°, respectively.

The lignin content of compression wood is about 5% higher and the cellulose content is about 8% lower than normal wood in spruce and redwood. The density of pronounced compression wood is 15—40% greater than normal wood. The longitudinal shrinkage of compression wood from

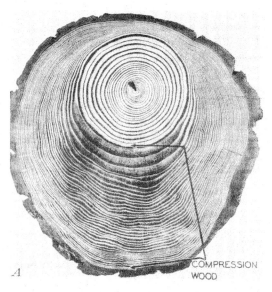

Fig. 25-11. Cross section of southern pine log with conspicuous compression wood. From Pillow and Luxford (1937).

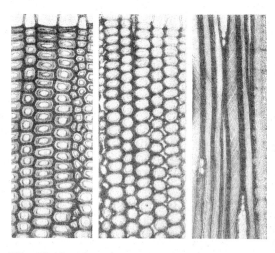

Fig. 25-12. Normal wood of Douglas-fir with rectangular cells (left) and compression wood with rounded cells (center), and checks in fibers with low S-2 fibril orientation (right) (200×).

green to oven—dry condition varies from 0.3 to 2.5%, whereas normal wood has a shrinkage of 0.1 to 0.2%. The transverse shrinkage of compression wood is less than that of normal wood.

When adjustments are made for differences in density, compression wood is lower in practically all strength properties as compared to normal wood (although it is quite hard) and is highly correlated to the differences in the slope of fibrils in compression wood. It is brittle and fails in clean pieces without the splintering normal wood has during failure. Compression wood accounts for much bowing and twisting in manufactured lumber. Pulp from compression wood tends to fragment rather than fibrillate and generally leads to very weak pulps with poor bonding potential.

Tension wood

Tension wood (Perem, 1964) occurs on the underside of branches and leaning stems of angiosperms (hardwoods) and is not easily observed in freshlyfelled timber. Upon aging (by weathering or exposure to strong UV light) it becomes visible by its light color or higher luster compared to normal wood (Fig. 25-13). Tension wood often occurs without the eccentric growth ring pattern of

compression wood. It is less obvious and considered to be much less of a problem than compression wood and, therefore, is not as well studied.

Tension wood has much higher longitudinal shrinkage (0.3 to 0.7% compared to 0.0 to 0.25%) and toughness and lower modulus of rupture than normal wood. It has a strong tendency to give a fuzzy surface on green wood and to cause warping and other distortions in lumber products during drying. It may be slightly higher, slightly lower, or of no significantly different specific gravity than normal wood depending on the species.

Most species (basswood is an exception) have gelatinous fibers in the tension wood (Fig. 25-14). It is the layer inside the S—2 layer that is gelatinous and this often separates from the rest of the fiber cell wall during sample cutting for microscopy, especially in thin—walled fibers. This layer is almost entirely free of lignin. On average, these fibers are about 5% lower in lignin and corre-

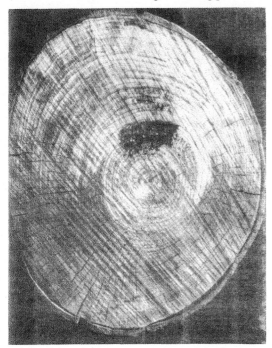

Fig. 25-13. Tension wood (top) of weathered trembling aspen. From Perem (1964).

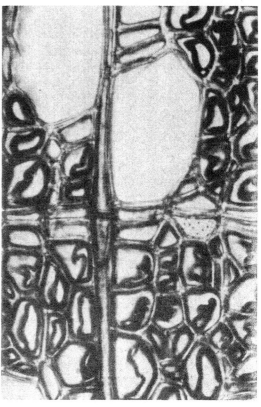

Fig. 25-14. Fibers with gelatinous layers (separated from the rest of the cell wall) of tension wood from trembling aspen at 500×. From Perem (1964).

spondingly higher in cellulose than normal fibers. Staining with safranin and fast green shows normal fibers as red and gelatinous fibers as green. The number of vessels is decreased in tension wood. The microfibril angle is lower than in normal wood, but the numerous incipient tension failures in the fiber cell walls give a weak wood or pulp.

Elm is an example of a hardwood with much branching (bifurcating trunk). It is bound to have large areas of tension wood due to its normal growth form. Tension wood from the genus *Eucalyptus* is associated with permanent collapse during drying.

Pith, juvenile wood

The pith and the first five to ten growth rings, corresponding to a core of several inches of wood, have properties that vary greatly from the rest of the wood (Besley in Anon., 1962). The juvenile wood is generally much weaker than normal wood and often shows leaf scars. The wood has lower specific gravity, shorter fibers, and lower cellulose content than normal wood. In loblolly pine the juvenile wood averages 7 growth rings with a diameter of about 3.5 in. and has a density of about 80% of that of normal wood. Trees (such as southern pine) growing in the open have more knot wood and juvenile wood of lower quality than trees growing under some cover (Paul, 1963).

False growth rings

False growth rings are an additional growth region (visible in the cross section of wood) that is produced by a pronounced decrease in the growth rate followed by resumption of growth within a single season. This may be a result of defoliation followed by foliage growth or temporary adverse conditions of temperature or moisture (too much or too little). False growth rings are relatively common in baldcypress.

Interlocking grain

Interlocking grain occurs in wood when the fibers are inclined in one direction through several growth rings, slowly return to perpendicular in succeeding growth rings, and then gradually reverse to become inclined in the opposite direction for several growth rings, and reverse again to repeat the process. Interlocked grain gives a ribbon or stripe figure on radial surfaces.

Spiral grain

Spiral grain is a type of growth where the fibers take a spiral course around the center of the tree instead of being longitudinal. The spiral may be clockwise or counterclockwise.

25.4 SILVICULTURE AND WOOD QUALITY

Wood is a variable material. Its properties depend on the species, individual genetics, location within a tree, and growing conditions. There are many well—known relationships among the actual properties such as fiber length and microfibril angle with position within a growth ring or tree, but other relationships vary with species. All of these factors contribute to wood quality. There is much talk about growing "super" trees that meet specific purposes. For example, fast—growing trees with long fibers and low lignin contents are ideal for pulp and paper. However, genetic selection of trees has been largely limited to growth rates and specific gravities of wood. Wood quality is in the domain of the forest scientist, but the wood scientist must work with the forest scientist to define wood quality.

Wood quality

Wood quality might be defined in a number of ways from the practical to the esoteric. It may be defined as the "suitability for its purpose" or merely "the economic value of the wood." Like wood chip quality, wood quality assumes someone is actually monitoring it; i.e., there is no impetus such as financial reward (or even recovery of investment) for increased wood quality if the wood user is not quantifying it.

Many properties contribute to wood quality, and these depend on the end use of the wood. Specific gravity, cellulose content, number of knots, uniformity, and microfibril angle of the S—2 cell wall layer are several important factors. Specific gravity is, in turn, a function of the percentage of latewood and cell diameters and wall thicknesses. These are functions of the species and growing conditions (see pages 41 on fibril angle and 153 on latewood versus earlywood). It is no coincidence that the two major U.S. structural woods, Douglas—fir and southern yellow pine, have thick, dense latewood.

There is some contradiction of the desired properties of wood. For various paper types one might want more or less latewood. There are some common qualities that most wood users agree on, however, including a higher level of uniformity. To a degree, wood quality depends on technology. New processes allow some woods to be used in products where they could not be used previously. Too, as technology changes, the ideal requirements of the wood may change.

Wood variability within a tree

Panshin and de Zeeuw (1980) is a useful, general resource for this topic. Fig. 25-15 shows the variation in fiber length for spruce pine as an example of a softwood. It is well known that the average fiber length in trees increases from the pith outward to a nearly constant average fiber length after about 30—50 years growth from the pith. Also the fiber length in latewood is appreciably higher than in the earlywood of softwoods.

Fig. 25-16 demonstrates the variability of fiber length in a hardwood, southern red oak. The same trend (as in softwoods) in average fiber length is shown.

With other factors equal (like growth rate), the specific gravity in a tree is variable, decreasing from the center outward in oak (Paul, 1963); other hardwoods have increasing specific gravity from the pith outward. Softwoods generally have

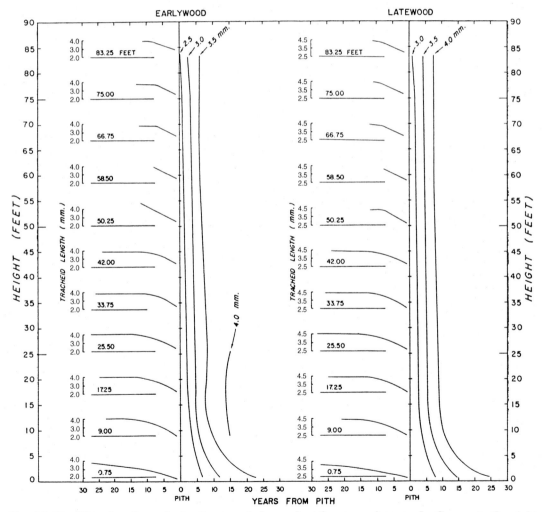

Fig. 25-15. Fiber length analysis of macerated samples of spruce pine wood. Curves to the right of the pith are tracheid length contour lines. From Manwiller (1972).

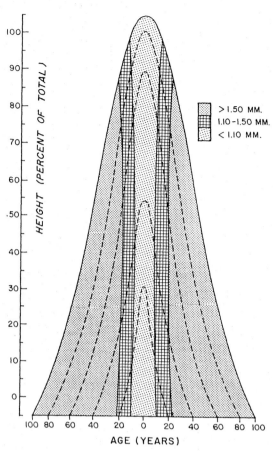

Fig. 25-16. Variation in fiber length of southern red oak. Reprinted from Koch (1985).

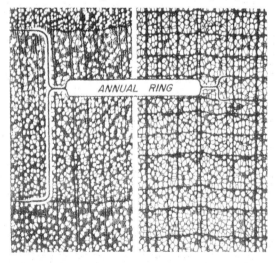

Fig. 25-17. Fast (left, specific gravity of 0.41) and slow (sp. gr. of 0.31) growing yellow-poplar from a single tree. Reprinted from Paul (1963).

increasing specific gravity from the pith outward to the bark. *Chamaecyparis* and *Thuja* are exceptions where the opposite is true.

Silviculture and wood specific gravity

Latewood is much more dense than earlywood in softwoods. Since slow—growing softwoods have a higher proportion of latewood, they also have a slightly higher density. However, the percentage of latewood accounts for only about 45—75% of the variation in wood specific gravity since the specific gravity of both latewood and earlywood is variable. The amount of moisture in the soil and the size of the tree crown are key variables that are inversely related to the percentage of latewood.

Other factors such as the age of the growth ring and position within a tree also contribute to the variation. In softwoods, the wood density generally increases from the pith outward and levels off (at age 20—50) until the tree is very old (100 years old), when it may decrease.

Paul (1963) summarized his work of four decades on the application of silviculture to control specific gravity of wood and gave a good review of the topic. Conifers have an ideal growth rate with at least 10 growth rings/in., although 15 is ideal. One worker suggested 12 ring/in. for Douglas—fir in Germany.

With hardwoods, however, a slow growth rate usually leads to a wood of low density. Fig. 25-17 demonstrates the effect of growth rate on specific gravity of yellow—poplar (diffuse—porous) and Fig. 25-18 for white ash (ring—porous). Contrasted this to Fig. 25-19, where slow growth gives somewhat higher density in softwood.

Genetics plays a role in wood quality. Selection of trees for fast growth, high—density wood, and other desirable characteristics improves wood quantity and quality. Tree selection, however, tends to reduce the genetic variability within the species. Large tracts of land are planted with a single species of the same age in monoculture that is susceptible to massive damage by disease.

Many other factors contribute to wood quality. Do the logging operations involve clearcutting of large tracts of land or is the logging limited to thinning and cutting rows? The proper selection of fungi from the *Rhizobium* genus in the growth of leguminous plants under various conditions can

greatly improve their productivity. Should labor—intensive pruning and tree planting be used?
How can one predict the economic return, which
may not occur for 40 years? Are economic
considerations important only for the relatively
short term when predictions can be made?

Silviculture and compression wood

Proper forestry can minimize compression
wood formation in young, second—growth stands
under a relatively high degree of management
(Pillow and Luxford, 1937). In such even—aged
stands thinnings should be made to remove over-
topped trees if they are at an incline, twin trees
with two stems from a single stump, and defective
or crooked trees. The dominant trees tend to be
straighter than suppressed trees that may grow at
an angle trying to obtain light. Other trees should
be removed to obtain the proper stocking to give
the proper spacing between trees.

However, in partially cut stands, wind action
and increased vigor may contribute to increased
compression wood formation. In partial cuttings
made to improve a stand, the formation of compression wood will be decreased if large, irregular
openings in the crown canopy are avoided and the
action of violent winds thereby reduced.

Wood use patterns and wood quality

In areas that use residual chips from sawmill
operations, one would expect very good wood
quality for packaging papers since most of the
wood comes from slabs containing wood of the
outer growth rings. On the other hand, wood that
comes from precommercial thinnings will have a
very high percentage of juvenile wood. Usually
it costs more to handle small wood than large
wood. Questions remain as to the long—term
effects of plantation grown wood of short rotation
compared to longer rotations on soil properties.

25.5 ANNOTATED BIBLIOGRAPHY

General wood anatomy

1. Brazier, J.D. and G.L. Franklin, Identifica-
    tion of hardwoods, *Forest Prod. Res. Bull.*
    46, London, 1961.

2. Core, H.A., W.A. Côté, and A.C. Day,
    Wood Structure and Identification, 2nd ed.,
    Syracuse University Press, 1979, 138 p. and
    appendix. Numerous optical and scanning

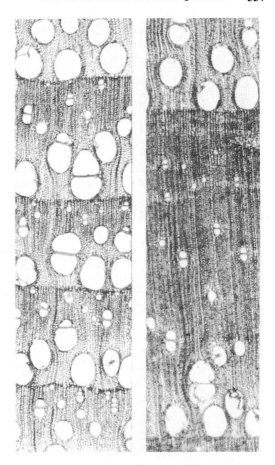

Fig. 25-18. Slow (left, specific gravity of 0.48)
and fast (sp. gr. of 0.65) growing white ash
from a single tree. Reprinted from Paul (1963).

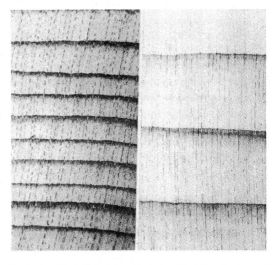

Fig. 25-19. Old growth (left, 24 rings/in.) and
second growth, 12 rings/in. redwood (Paul, 1963).

electron micrographs describe the important features of wood; identification keys are included.

3.   Harrar, E.S. and J.E. Lodewick, Identification and microscopy of woods and wood fibers used in the manufacture of pulp, *The Paper Ind.* February:630-637; *ibid.*, May:103-111; *ibid.*, Aug.:327-335(1934). A summary of wood and anatomy of fiber for pulp and paper technologists. Coniferous woods are considered in the first article, hardwoods in the second, and fibers in the third.

4.   Hoadley, R.B., *Identifying Wood*, Taunton Press, Newtown, Connecticut, 1990, 223 p. This book has a high circulation and is of modest cost. It includes many color plates and a long bibliography; it is a practical book on wood identification with a hand lens and includes sources of authentic wood samples, prepared slides, equipment, etc.

5.   Ilic, J., *CSIRO Atlas of Hardwoods*, CSIRO, Australia, 1991. Text is minimal and photographs are maximal: 69 pages show the crosssections of 1284 species (1.6 by 1.7 in., in color) organized by family and genus at 6.5×; 402 pages show microscopic images of the three (*x, r, t*) sections (25×), and vessel—ray pitting (100×).

6.   Jane, F.W., *The Structure of Wood*, A. and C. Black, Ltd., London, 1956, 427 p. (2nd ed., 1970). This resource has a more detailed discussion of many aspects of wood anatomy than other references. It is global in perspective.

7.   Koehler, A., Guidebook for the identification of woods used for ties and timbers, USDA Forest Service, Misc. RL-1, 1917, 79 p. and 31 plates showing 62 cross sections at 15×. Numerous maps are included showing the distribution of many species. The photographs were made by passing light through thin specimens, which gives better detail than photographing the cross sections of large pieces.

8.   Panshin, A.J., and C. de Zeeuw, *Textbook of Wood Technology*, 4th ed., McGraw-Hill Book Company, New York, N.Y., 1980, 722 p. "Structure, identification, uses, and properties for the commercial woods of the U. S. and Canada" with many micrographs and descriptions by wood species. Keys for identification are included for hardwoods and softwoods. The 4th ed. has minor changes over the 1970 3rd ed. (705 p.) This work goes back before the 1st ed. of 1949 to the 1934 work *Identification of Commercial Timbers of the United States* of H.P. Brown and A.J. Panshin. [*Commercial timbers of the United States* was published in 1940 by the same authors and **has key references on wood identification that are omitted in the later editions**, which is the origin of much of the material; the quality of photographs in the 1940 ed. with coated paper is higher than the 1980 and 1970 eds. on uncoated paper.]

9.   Phillips, E.W.J., Identification of softwoods by their microscopic structure, *For. Prod. Res. Bull.* 22:1-56(1948), London.

Staining of wood

10.   Kutscha, N.P. and I.B. Sachs, Color tests for differentiating heartwood and sapwood in certain softwood species, USDA FPL Rep. No. 2246, 1962, 13 p. Includes 21 solutions for 23 species.

11.   Parham, R.A., *Tappi* 65(4):122(1982). It is very useful to determine the hardwood fraction of chips in chip mixtures using a reaction which specifically stains hardwoods. A very useful, well—known reaction is the Mäule reaction. Chips in a plastic mesh (for ease of transfer) are treated at room temperature in a fume hood with 1% $KMnO_4$ for 10 min., brief rinse, 6 N HCl for 1 min, brief rinse, 10 % NH_3 for 1 min. The chips are dried in the hood. The weight of the red (hardwood) portion as a percentage of the total is determined.

Plant microtechnique

12.   Johansen, D.A., *Plant Microtechnique*, McGraw—Hill, New York, 1940, 523 p.

13. Sass, J.E., *Elements of Botanical Microtechnique*, McGraw—Hill, New York, 1940, 222 p. The 3rd edition of this book appeared in 1958. The book was rewritten by G.P. Berlyn and J.P. Miksche as *Botanical Microtechnique and Cytochemistry* and published by Iowa State University Press (Ames, Iowa) in 1976.

Fiber and wood microscopy

14. Carpenter, C.H., et al, *Papermaking Fibers*, Tech. Pub, 74, State University College of Forestry, Syracuse, New York, 1963. This is a photomicrographic atlas of woody and other fibers with 77 plates. Entire isolated fibers of numerous species are featured. It is an update of the 1952 work; the original version of 1932 is the *Atlas of Papermaking Fibers* by C. H. Carpenter with 56 plates.

15. Graff, J.H., *A Color Atlas for Fiber Identification*, Inst. Paper Chem., Appleton, Wisc., 1940, 27 p. + v color plates. Three classes of stains are described; the five color plates are used to compare the results of the tests.

16. Isenberg, I.H., *Pulp and Paper Microscopy*, 3rd ed., Inst. Paper Chem., Appleton, Wisc., 1967, 395 p. This work is an expansion of Graff (1949) above with much detail, but without color plates. It includes a foundation in optics and microscopy, fiber dimension measurements by microscopy, descriptive fiber morphologies of wood and nonwood plant fibers, animal fibers, and mineral fibers with numerous pictures of actual fibers, and discussions on numerous color stains useful for fiber characterizations. The color stains can often be used directly on paper to characterize its properties. This is a comb—bound notebook.

17. Parham, R.A. and R.L. Gray, *The Practical Identification of Wood Pulp Fibers*, TAPPI Press 1982, 1990, 212 p. This comb—bound book has very high quality paper that helps reproduce the photographs. This work is very useful for wood fiber identification. Optical microscopy and SEM photos are included with each species. There are no keys, and the format is much like an atlas, although there is moderate text included. Numerous fibers from each species indicate the variability of the fiber morphology.

18. Strelis, I. and R.W. Kennedy, *Identification of North American Commercial Pulpwoods and Pulp Fibres*, University of Toronto Press, 1967, 117 p. This work is especially useful for fiber analysis; photomicrographs of animal, mineral, and wood and nonwoody vegetable fibers are included. Techniques for fiber identification by solubility analysis and staining are included. (Of course, FT—IR analysis of individual fibers as a useful tool for fiber analysis was not commonly available at the time of this work.)

19. Tappi Standard T 401 om-82, Fiber analysis of paper and paperboard, 14 p., 32 references. This standard discusses fiber morphology and many staining techniques for quantitative analysis of fiber types.

20. Tappi Standard T 263 om-82, Identification of wood and fibers from conifers, 13 p., 3 references. It is a reprint of B.F. Kukachka's article in *Tappi* 43(11):887-896(1960) with some added information; *Araucaria—Agathis* is listed as having spirals (should be pits 2—4 alternate).

Reaction wood

21. Perem, E., Tension wood in Canadian hardwoods, Can. Dept. of For. Pub. No. 1057, 1964, 38 p.

22. Pillow, M.Y and R.F. Luxford, Structure, occurrence, and properties of compression wood, USDA Tech. Bull. No. 546, 1937, 32 p.

Silviculture and wood quality

23. Anon., *Wood Quality, Proceeding of three symposia*, Dept. of Lands and Forests Res. Rep. No. 46, Ontario Research Foundation Rep. No. 601, 1962. Published in *Pulp Paper Mag. Can.* (2nd symp. in Aug., 1960 Woodland's Section, but in the nontechnical

pages omitted in many libraries) and *Timber Mag. Can.* (3rd symp., issue unknown).

24. Kellison, R.C. and R.G. Hitchings, Harvesting more southern pines will require pulp mill changes, *Pulp & Paper* 59(7):53-56 (1985). This article summarizes problems associated with using young trees as an increasing percentage of raw material.

25. Paul, B.H., The application of silviculture in controlling the specific gravity of wood, USDA Tech. Bull. No. 1288, 1963, 97 p. This is an superb reference on the topic; it was first published in 1930 as No. 168.

Bark anatomy
26. Chang, Y.-P., *Bark Structure of North American Conifers*, USDA Tech. Bull. No. 1095, 1954, 86 p. Keys are included to identify the origin of bark samples.

27. Chang, Y.-P., *Anatomy of Common North American Pulpwood Barks*, Tappi Monograph Series 14, 1954, 249 p.

Pulpwoods
28. Isenberg, I.H., Ed. (revised by Harder, M.L., and L. Louden), *Pulpwoods of the United States and Canada*, 3rd edition, Institute of Paper Chemistry, Appleton, Wisconsin, 1981. This two—volume set (Volume 1, conifers; Volume 2, hardwoods) lists tree species, their silvics, wood properties, and general pulping characteristics with references. It includes the commercial species of North America.

Statistical information on wood industry
29. F.A.O., *World Forest Inventory 1963*, Rome, 1966.

30. FAO, *Yearbook, Forest Products 1990*, Rome, 1992.

31. U.S. Bureau of the Census, *Statistical Abstract of the United States*, 111th ed., 1991, Washington, D.C., $28, pp 675-683 for wood products industry.

Miscellaneous
32. Bamber, R.K. and J. Burley, *The properties of radiata pine*, Commonwealth Agric. Bureaux, England, 1983, 84 p.; large bibliography.

33. Chattaway, M.M., The development of tyloses and secretion of gum in the heartwood formation, *Aust. J. Sci. Res. B.* 2:227-240(1949).

34. Howard, E.T. and F.G. Manwiller, Anatomical characteristics of southern pine stemwood, *Wood Sci.* 2(2):77-86(1969).

35. Koch, P., *Utilization of Hardwoods Growing on Southern Pine Sites*, USDA Agric. Handbook No. 605, 1985.

36. Kukachka, B.F., Properties of imported tropical woods, USDA For. Ser. Res. Paper FPL 125, March, 1970, 67 p.

EXERCISES

1. Write a one—page essay discussing some of the major differences between softwood anatomy and hardwood anatomy.

2. Would you say compression wood generally has advantages in pulping and papermaking (compared to normal wood)?

3. Describe how softwood of the same species could vary drastically in two different plantations.

4. Which are more prevalent globally: hardwoods or softwoods?

26
SOFTWOOD ANATOMY

26.1 GROSS ANATOMY OF SOFTWOODS

Introduction

This chapter presents the detailed anatomy of the wood of temperate softwoods. The anatomy of various woods is important for several reasons. The characteristics of papermaking fibers depend much upon their anatomy. The pulping characteristics of various fiber sources are dependent upon their species and growing conditions. Some species are lumped together as an indistinguishable lumber commodity, but may have greatly differing pulping and papermaking characteristics. For example, southern pine is really four different species of pine. For these reasons it is important to be able to verify the species types of wood chips or even pulp mixtures. One may wish to examine the alterations of individual fibers by processes such as refining. Identification of trees in forests allows the use of flora, bark, twigs, and other features; this topic is not discussed here.

The gross description and uses of many of the wood species contains information from the USDA *Wood Handbook* (Ag. Handbook No. 52, 1935, 1955, 1974, 1987) with minor alterations. Some of the hardwoods contain anatomy information from Koch (1985) with minor alterations. Ag. Handbook No. 101 (1956) was of some use, and Koehler (1917) was used extensively. These resources are in the public domain. Some details have been added from the references listed in Chapter 25. The wood uses are somewhat dated as plastics and other materials have replaced some of the uses; nevertheless, the listed uses say much about the wood properties in manufacturing and use. Other photographs and information are from OSU forest products courses courtesy of Drs. R.L. Krahmer and A.T. Van Vliet.

Softwoods are distinguished from hardwoods by the absence of vessels. The characteristic features of softwoods that are visible with a hand lens will be demonstrated. In the next section microscopic features of softwood will be considered. Then, in the section after that, the anatomy of many individual softwood species is given.

Resin canals

The most pronounced feature of softwoods is resin canals (shown by microscope in Fig. 26-1) in those species which have them. Resin canals are not formed by cells, but are voids surrounded by epithelium cells. (In contrast, hardwood vessels are made up of individual cells.) Tyloses may occasionally form in the resin canals of heartwood. The pines, spruces, larches, and Douglas—fir genera of Pinaceae contain normal resin canals in the longitudinal and radial directions. The radial canals are part of *fusiform* rays and are smaller than longitudinal resin canals. The resin canals in pines are particularly large and numerous and occur throughout the growth ring. In the other three genera, they are small, less numerous, appear to be missing from some growth rings, and may be grouped in small, tangential rows. Brown and Panshin (1940) give northern white pine longitudinal resin canals as 135—150 μm average diameter but as high as 200 μm; radial canals average 80 μm in diameter. They may be difficult to find in eastern spruce samples (Hoadley, 1990).

The large resin canals of pines appear as short dark lines on the longitudinal surfaces (*r, t*) due to the pitch or dirt trapped by the pitch. Epithelium cells, which secrete resin, surround both the longitudinal and ray resin canals. In pines they are thin—walled and easily torn away when free—hand sectioning (or even often when microtoming an embedded sample as in Fig. 26-1). In the other three genera they are thick—walled and usually remain intact with free—hand sectioning.

True firs, hemlocks, and cedars may form *traumatic resin canals* in response to injury in the tree. These small resin canals occur in relatively long rows in the tangential direction when the wood cross section is viewed, but do not occur enough to cause difficulty in wood identification. These genera only have traumatic longitudinal resin canals; radial resin canals (fusiform rays) are always absent. North American hardwoods normally lack resin canals but can form traumatic resin canals as well (Fig. 27-3).

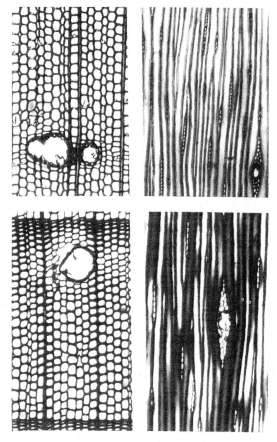

Fig. 26-1. Longitudinal (top left) and radial (top right) resin canals from *Picea sitchensis* with thick-walled epithelium and from *Pinus strobus* (bottom) with thin-walled epithelium.

Earlywood latewood transition

Another easily observed trait is the transition between the earlywood and latewood. It is characterized as abrupt or gradual (shown through a microscope in Fig. 26-2). For example, in the soft pines it is gradual but in the hard pines it is abrupt. Engelmann spruce has a more abrupt transition than the other domestic spruces.

Color

Sapwood always has a relatively light color from near white to tan or yellow. The heartwood of some species may be light or dark or of distinctive color. Redwood heartwood is a distinctive brown—red color. Junipers have a deep purple—red heartwood. Douglas—fir usually has a bright reddish hue. A reference collection for color comparison is very useful.

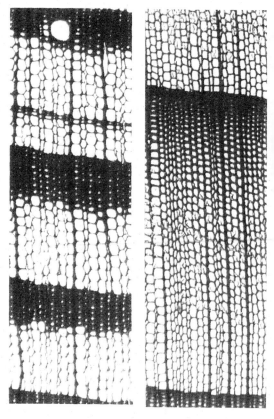

Fig. 26-2. Abrupt transition between earlywood and latewood of *Larix occidentalis* (left) and gradual transition of *Abies concolor*.

Odor

Odor is particularly useful in green wood of freshly cut surfaces. One must learn odors on known samples of the wood; do not try to recognize them by their descriptions. Kiln—dried wood will not exhibit as strong an odor as green wood. The pine genera all share a characteristic odor. Cedars, *Thuja plicata* and *Libocedrus decurrens* have aromatic smells which may be accented by moisture and/or mild heat. Douglas—fir also has a unique odor.

Dimpled surfaces

Some species have a dimpled grain [lodgepole (pronounced) and ponderosa pines and sitka spruce] on the tangential surface (*t*) that is distinctive; this is caused by crystals in the bark that press against the cambium, which deforms it. The growth rings reflect this by being uneven when viewed with magnification.

Greasy feel, etc. (Hoadley, 1990)

Taxodium species have a greasy or tacky feel while redwood feels dry. Tamarack feels waxy (less greasy than baldcypress). Figure in wood, the appearance of the tangential surface (such as on flat—sawn wood) is not overly relevant to pulp and paper, though it can be used in wood identification. Figure is most pronounced on woods with an abrupt latewood transition such as with Douglas—fir and southern pines or ring porous hardwoods such as oak and hickory. Special figures also occur by grain deviations such as at the crotch of trunk splits and at burls (outgrowths of the trunk) especially with hardwoods. Spruces are separated from white pines or firs by their glossy (having luster) longitudinal surfaces.

The volatile material of Jeffrey pine contains a high percentage of *n*—heptane while those of ponderosa pine are high in pinenes. For some mixtures of species machines are being developed to sort logs or boards by, more or less, smelling each one. This technique is similar to that used at airports to detect plastic explosives in luggage. There is some literature on *chemotaxonomy*, taxonomy based on the chemical components of wood and other tissues.

Hand lens key for softwoods

Table 26-1 is a dichotomous hand lens key designed for major U.S. commercial softwoods of the western U.S. It is presented courtesy of R.L. Krahmer and A.C. Van Vliet. Each number is a step where one must decide between a pair of mutually exclusive choices. The format is typical of many keys used to identify species of wood. Microscopic examination of wood is often needed for conclusive identification of most woods.

26.2 MICROSCOPIC ANATOMY OF SOFTWOODS

Rays

Most softwood rays are uniseriate (one cell wide) unless they contain resin canals; fusiform rays, when present, constitute about 5% of the rays. A few species (redwood, incense cedar) are

biseriate for at least a portion of the ray, but many species can be sporadically biseriate. The height of the ray (number of cells) is a useful tool for softwood identification. Cedars tend to have low rays (Port—Orford—cedar and eastern redcedar are generally 6 or less high) while the true firs and baldcypress tend to have high rays (25 to over 40 high and occasionally almost 1 mm high).

Ray tracheids

Ray tracheids (Fig. 26-3) are easily found in many species, especially the hard pines. In other species they are found only with great difficulty and should be counted as absent. Ray tracheids are differentiated from ray parenchyma by their small bordered pits in the cross—field. In larches and hemlocks they occur at the upper and lower margins and are called *marginal ray tracheids*. In pines they occur marginally and throughout the height of the ray. In coniferous species rays containing both tracheids and parenchyma are termed *heterocellular* while rays with only one type of cell are termed *homocellular*. Alaska

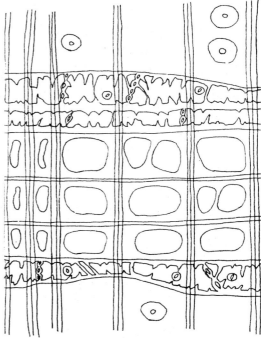

Fig. 26-3. Sketch of *Pinus resinosa* (*r*) showing two rows of dentate ray tracheids on the upper margin of a wood ray and one on the lower margin. The three layers of ray parenchyma show 1-2 simple cross-field pitting. From Kukachka (1960).

Table 26-1. Dichotomous hand lens key of major U.S. commercial softwoods.

1. Wood with normal resin ducts 2
1. Wood without normal resin canals 8
  2. Resin ducts numerous, evenly distributed in outer portion of every ring; generally visible to naked eye, easily seen with hand lens; resinous odor 3
  2. Resin ducts sparse, unevenly distributed and sometimes absent in some rings; occasionally appear as 2-many in a tangential row, barely visible to the naked eye as light or dark flecks 5
3. Transition from early to latewood gradual; wood soft, light, quite even in density; coarse to medium textured
Pinus lambertiana, P. monticola
3. Transition from early to latewood abrupt; wood medium hard to hard 4
  4. Wood medium hard; resin canals confined mostly to outer (latewood) portions of growth rings
Pinus ponderosa, P. contorta
  4. Wood often heavy and hard to cut; resin ducts distributed throughout center and outer portions of growth rings
Pinus spp.
5. Transition from early to latewood gradual 6
5. Transition from early to latewood abrupt 7
  6. Wood light pinkish yellow to pale brown, often somewhat lustrous
Picea sitchensis
  6. Wood yellowish to orange red; characteristic odor
Pseudotsuga menziesii
7. Wood yellowish to orange red or deep red, not oily, characteristic odor on fresh cut surface; rings often wavy in slow growth; resin canals often in tangential lines of 2-3
Pseudotsuga menziesii
7. Wood brownish cast, somewhat oily appearance; no characteristic odor; in slow growth, ring contours generally smooth; resin canals in groups of 2-5, but not in tangential lines
Larix occidentalis
  8. Wood without distinct odor 9
  8. Wood with distinct odor 12
9. Wood medium to coarse textured 10
9. Wood fine textured, dense, fairly heavy; heartwood light orange to rose red
Taxus brevifolia
  10. Wood whitish to pale brown sometimes with purplish tinge; gradual transition from early to latewood; medium textured
Tsuga heterophylla; Abies spp.
  10. Wood red to deep reddish brown or yellowish to dark reddish brown to almost black; coarse textured; abrupt transition 11
11. Wood light cherry red to deep reddish brown; light to moderately lightweight; usually uniform rate of growth
Sequoia sempervirens
11. Wood variable in color, yellowish to light or dark brown, reddish brown to almost black; wood often greasy or oily; cuts like hard rubber with knife; usually irregular rate of growth and discontinuous growth rings
Taxodium distichum
  12. Wood with fragrant, "cedar-like" or sweetish odor 13
  12. Wood with ill-scented, or with rancid odor 17
13. Wood reddish brown to dull brown 14
13. Wood yellowish white to pale yellowish brown 16
  14. Wood dull reddish to pinkish brown, often with streaks of included sapwood; fine textured, moderately hard and heavy; characteristic mild spicy odor
Juniperus occidentalis
  14. Wood reddish brown to dull brown; medium to coarse textured 15
15. Wood sometimes with lavender tinge; Frequently with pecky rot; texture within growth rings even, cuts smoothly across growth rings; distinct odor like pencils and with distinct acrid taste
Libocedrus decurrens
15. Texture with growth rings uneven, cuts brashly across growth rings because latewood considerably harder than earlywood; never with pecky rot; sweet "cedar" odor
Thuja plicata
  16. Wood yellowish white to pale yellowish brown; distinct pungent, ginger-like odor; medium to coarse textured
Chamaecyparis lawsoniana
  16. Wood bright, clear yellow, darkens upon exposure; odor resembles raw potatoes; fine to medium texture
Chamaecyparis nootkatensis
17. Wood yellowish white to pale brown; uniform rate of growth
Chamaecyparis lawsoniana
17. Wood variable in color, yellowish to light or dark brown, reddish brown to almost black; irregular rate of growth
Taxodium distichum

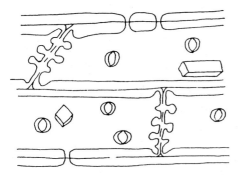

Fig. 26-4. Sketch of *Abies magnifica* (r) showing nodular end walls and indentures. Two crystals also appear. **From Kukachka (1960).**

yellow cedar is unique in that it has homocellular rays composed of either parenchyma or tracheids and very few heterocellular rays. The low rays of hard pines are often homocellular with tracheids.

Ray tracheids may be *dentate*, when tooth—like projections on the horizontal walls are present. In the spruces these are small and observed in the latewood. Hard and red pines have dentate ray tracheids, whereas soft pines have nondentate ray tracheids. These are depicted in Fig. 26-3.

Ray parenchyma end walls may be smooth (e.g., *Thuja*) or nodular with bead—like projections (e.g., *Abies*). In some species the corners of the ray parenchyma are observed to have pit—like depressions, called *indentures*, where the vertical and horizontal walls meet. These features are depicted in Fig. 26-4.

Ray cross—field pits

Cross—field pits occur between ray parenchyma and longitudinal tracheids. Phillips (1948) described five types of cross—field pits that are of importance in identification of softwoods. Use cross—fields of the first few earlywood tracheids (if possible) of a growth ring as those of the latewood tracheids vary in structure. Fenestriform and pinoid pittings are easy to distinguish; the

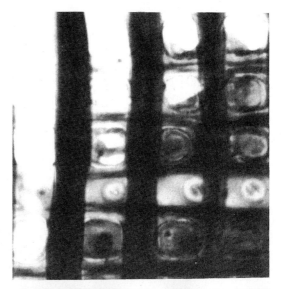

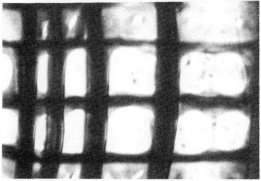

Fig. 26-6. Fenestriform pitting in *Pinus strobus* (top, with a row of interspersed ray tracheids) and *P. monticola*, approximately 600×.

latter generally occur with dentate ray tracheids, although red pine has dentate ray tracheids with fenestriform pitting. Piceoid, cupressoid, and taxodioid (especially the latter two) are difficult to discern since there is overlap between these types. Hoadley, page 26, and Jane, page 95, have excellent diagrams of these. Fig. 26-5 shows diagrams of these pits.

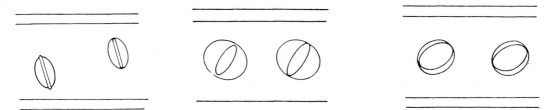

Fig. 26-5. Cross-field pitting: piceoid (left), cupressoid (center), and taxodioid (right).

Fenestriform pits are large, borderless rectangular "window—like" pits with usually 1—2 pits per cross—field (Fig. 26-6). Most other types have 2 to 6 pits per cross—field. These distinctive pits are characteristic of the soft (white, scotch, etc.) and red pines. Sugar pine has 2—4 fenestriform pits per cross—field.

Pinoid are elliptical to irregular polygons in shape (Fig. 26-7). They vary in size. The borders may or may not be visible around the aper-

tures. When they are visible they are narrow and usually wider on one side than the other. These are characteristic of the hard pines (except red pine) and occur in species with dentate ray tracheids.

Piceoid have slit—like apertures that may extend beyond the visible border (Fig. 26-8). The width of the aperture is less than that of the aperture to the border. They occur in spruce, larch, and Douglas—fir.

Cupressoid are similar to piceoid, but the apertures are contained within the boundaries of the border and their width is slightly less than that

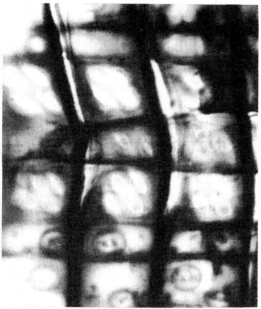

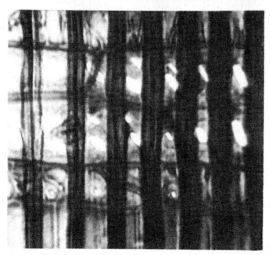

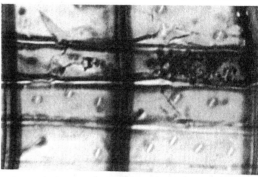

Fig. 26-7. Pinoid pitting in *Pinus banksiana* (top, with shallowly dentate ray tracheid) and *P. ponderosa* (prominently dentate ray tracheids), approximately 600×.

Fig. 26-8. Piceoid pitting in *Picea sitchensis* (top, note nodular end wall), *Larix occidentalis*, and *Pseudotsuga menziesii* (note indentures), approximately 600×.

of the border on either side (Fig. 26-9). They occur in most of the genera of Cupressaceae (except in *Thuja*) and in *Tsuga*.

Taxodioid pits have a regular elliptical shape with apertures that are larger than those of the previous two types (Fig. 26-10). The aperture is wider than the border on either side of the aperture. They occur in *Sequoia, Taxodium, Thuja,* and *Abies* genera.

Number of intertracheid pits across the fiber

The intertracheid, bordered pitting of tracheids pits (*r*) is usually *uniseriate* (generally only one pit across the width of the tracheid, although two may occur where tracheids overlap). Sugar pine may have four consecutive pit pairs in the first few fibers of the earlywood. In some species there are two or more columns of pits that may be alternate or opposite. Opposite multiseriate pitting occurs in *Sequoia* and *Taxodium* (Fig. 26-11). Alternate multiseriate pitting is characteristic of only *Araucaria* and *Agathis* amongst all conifers. Intertracheid pitting is usually confined to the radial surface, although its presence on the tangential surface in the latewood is common. The lack of intertracheid pitting on the radial walls of the latewood of southern pines is used to distinguish them from the soft pines.

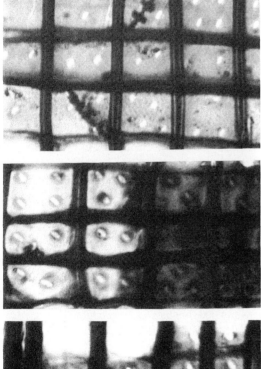

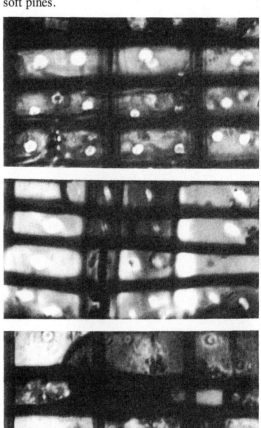

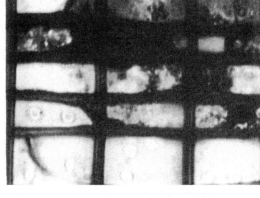

Fig. 26-9. Cupressoid pitting in *Librocedrus decurrens* (top, see nodular end walls), *Juniperus virginiana*, and *Taxus brevifolia*, ~ 600×.

Fig. 26-10. Taxodioid pitting in *Abies concolor* (top, note crystals in marginal parenchyma), *Sequoia sempervirens*, and *Thuja plicada* approximately 600×.

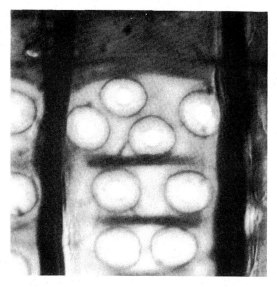

Fig. 26-11. Opposite, paired border pits of *Taxodium distichum* **with crassulae, ~600×.**

In coniferous species, *crassulae*, thickenings in the shape of an arc around the pit, as observed in the radial section, are common (Fig. 22-11).

Spiral thickening of tracheids (Fig. 26-12)

Spiral thickening of tracheids occurs in only a few species (Douglas—fir, Pacific yew, and the relatively rare *Torreya*) so this is an important characteristic, if present. Douglas—fir may not have spiral thickening in the latewood tracheids. Spiral thickening occasionally occurs in the latewood tracheids of southern pines (loblolly) and *larix*, but these species may also have seasoning checks. Spiral thickening occurs at a smaller angle to the longitudinal axis (steeper spirals) in fibers with thin lumens as in the latewood.

Fig. 26-12. Spiral thickening in Douglas—fir tracheids, ~600×.

Compression wood leads to seasoning checks parallel to the S—2 microfibril angle (Fig. 25-12). Checks usually form an angle <45° with the longitudinal axis whereas spiral thickening usually forms an angle >45°.

Tracheid diameter

Table 26-4 shows the diameters of tracheids. This can be used to separate some species from each other. Redwood is very coarse while eastern redcedar has a very fine texture; compare Figs. 26-52 and 26-44.

Longitudinal parenchyma

Strand (axial or vertical) parenchyma are never as abundant in the softwoods as in some of the hardwoods. They are moderately abundant throughout the growth ring in the families Taxodiaceae and Cupressaceae and other genera. These are *diffuse* parenchyma when the strands occur singularly as in redwood (Fig. 26-13) or baldcypress. When strands are grouped they are *zonate parenchyma* which may be *banded* (the old term is metatracheal, as in eastern redcedar or incense cedar) or in *clusters* as in the hardwoods. In the coarse—textured species redwood and baldcypress, they can be barely seen with a hand lens as streaks on the radial and tangential surfaces due to their extractives. Strand parenchyma at the boundaries of the growth rings (*marginal* or *terminal* parenchyma) occur in *Abies* and *Tsuga*. In the heartwood these cells are often filled with extractives. Longitudinal parenchyma are lacking in *Pinus* and *Taxus*. They occur in the root wood of *Picea*, but not in the wood above ground.

The end walls of longitudinal parenchyma of softwoods are usually *nodular*, but they are inconspicuous in redwood and absent (smooth walls) in southern pines and Atlantic white—cedar and help identify these species.

Intercellular spaces

Eastern redcedar has intercellular spaces between the tracheids where they meet at the corners. This feature must not be confused with compression wood where the same effect is noted. The pronounced roundness and spiral checks of compression wood cells help distinguish this from normal intercellular spaces of eastern redcedar.

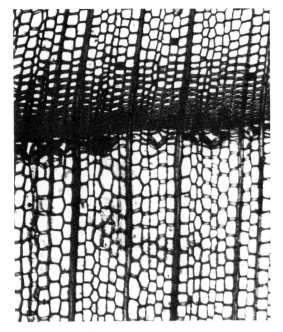

Fig. 26-13. Diffuse strand parenchyma of *Sequoia sempervirens* show up as dark squares due to the pitch contents. Resin crystals are also present at the margin of the growth ring.

26.3 ANATOMY OF SOFTWOOD SPECIES

Table 26-2 helps separate softwoods by genera. Table 26-3 gives a summary of softwood features useful in their identification. Table 26-4 gives properties of tracheids.

The *uses* section includes the properties related to the uses of wood. The mechanical properties are of relevance to wood chipping. The *color* section includes the overall appearance. Many sapwoods are stained with blue stain, a type of fungi, that changes the color.

Macroscopic features can be seen with the unaided eye or with a $10\times$ lens. The terms pores and vessels are used synonymously. The presence of tyloses often tends to make the vessels difficult to see on the cross section with the unaided eye, so the appearance and the actual sizes of vessels are not the same. Parenchyma descriptions in this section refer to longitudinal parenchyma. Most of the macroscopic cross sections are courtesy of R.L. Krahmer and A.C. Van Vliet; they are all the same magnification, about $15-20\times$. While the taste of some woods is mentioned, it is not a good idea to apply this for routine sample identification of large numbers of samples.

Softwoods are conveniently classified by their taxonomy in the order presented in Table 26-3. Most softwood do not have longitudinal parenchyma observable with the unaided eye or even a lens. Softwoods with normal resin canals will have two ray sizes, the fusiform rays that may be visible with the unaided eye and the normal rays that may or may not be visible.

PINACEAE
PINUS

Pines are easily divided into two groups: the soft pines and the hard, yellow, or pitch pines.

Table 26-2. Separation of softwoods into genera by microscopic features.

1.	Axial resin canals normally present	2	4.	Gradual transition between earlywood and latewood; ray tracheids of latewood with minute denticulations *Picea*
1.	Axial resin canals absent, unless traumatic type in tangential rows or groups	5		
	2. Epithelial cells of resin canals are thin walled. Ray tracheids present *Pinus*		5.	Axial parenchyma present and conspicuous. End walls of ray cells not conspicuously pitted. *Sequoia*
	2. Epithelial cells of resin canals are thick-walled	3		
3.	Tracheids with helical thickenings, especially in earlywood. Ray tracheids present. *Pseudotsuga*		5.	Axial parenchyma absent or sparse. 6
3.	Tracheids rarely with helical thickenings.	4		6. End walls of ray parenchyma conspicuously pitted, i.e., nodular end walls. *Abies*
	4. Abrupt transition between earlywood and latewood; ray tracheid walls of latewood rarely with minute denticulations. . . *Larix*			6. End walls of ray parenchyma are not conspicuously pitted. *Thuja*

Table 26-3. Anatomical characteristics of coniferous woods useful in their identification.

Column reference (header of the table):

Col.	Page no.	Scientific name	Common name
1	577	Taxodium distichum	Baldcypress
			Cedar, (not true):
2	574	Libocedrus decurrens	Incense-
3	571	Chamaecyparis nootkatensis	Alaska-
4	571	Chamaecyparis thyoides	Atlantic white-
5	571	Chamaecyparis lawsoniana	Port-Orford-
6	571	Juniperus virginiana	Eastern redcedar
7	576	Thuja plicata	Western redcedar
8	577	Thuja occidentalis	Northern whitecedar
9	561	Pseudotsuga menziesii	Douglas-fir, coastal
			Fir:
10	566	Abies balsamea	Balsam
11	566	A. magnifica	California red
12	566	A. grandis	Grand
13	566	A. procera	Noble
14	566	A. amabilis	Pacific silver
15	566	A. lasiocarpa	Subalpine
16	566	A. concolor	White
			Hemlock:
17	569	Tsuga canadensis	Eastern
18	570	T. heterophylla	Western
			Larch:
19	568	Larix laricina	Tamarack
20	567	L. occidentalis	Western

Characteristic matrix (columns 1–20 as above):

Characteristic	1	2	3	4	5	6	7	8	9	10	11	12	13	14	15	16	17	18	19	20
Use this for alignment	1	2	3	4	5	6	7	8	9	0	1	2	3	4	5	6	7	8	9	0
Resin canals:																				
Present									X										X	X
Epithelium thin-walled																				
Latewood transition abrupt	X						X	X	X									X	X	X
Tracheids																				
Helical thickening (s in LW only)									X										s	s
Bordered pits 2-4 opposite	X																	X	X	X
Bordered pits 2-4 alternate																				
Rays:																				
Homocellular of tracheids present		X																		
Tracheids present		X							X								X	X	X	X
Tracheids dentate																				
Indenture		V	V	V	X	X	X	X		X	X	X	X	X	X	X	X	X	X	X
Parenchyma end walls nodular	s	X	V		X				X	X	X	X	X	X	X	X	X	X	X	X
Marginal parenchyma with crystals										X	X	X			X					
Cross—field pitting:																				
Fenestriform																				
Pinoid																				
Piceoid									X								X	X	X	X
Cupressoid	V	X	X	X	X	X											V	V		
Taxodioid	X						X	X		X	X	X	X	X	X	X				
Longitudinal parenchyma:																				
Present	X	X	X	X	X		X	X	X	X	X	X	X	X	X	X	X	X		
End walls smooth				X																
End walls nodular	X	X	X		X	X	X	X		X	X	X	X	X	X	X	X	X	V	
Miscellaneous:																				
Dimpled tangential surface																				
Heartwood distinct	X		X		X	X	X	X											X	X
Greasy feel	X																			
Odor	X	X	X	X	X	X	X	X	X						X					
Use this for alignment	1	2	3	4	5	6	7	8	9	0	1	2	3	4	5	6	7	8	9	0

Key: X indicates characteristic is pronounced; x indicates characteristic is present but not pronounced (when used with X, it is the less likely of the two choices); V indicates variable according to sample or sources disagree; s indicates sporadic and may not be of diagnostic value.

Table 26-3. Anatomical characteristics of coniferous woods useful in their identification, cont'd.

Columns (Common name — *Scientific name* — Page No.):

1. Pine, soft or white: Jack — *Pinus banksiana* — 554
2. Lodgepole — *P. contorta* — 555
3. Monterey — *P. radiata* — 581
4. Red — *P. resinosa* — 558
5. Sugar — *P. lambertiana* — 558
6. White, northern — *P. strobus* — 559
7. White, western — *P. monticola* — 560
8. Pine, hard or yellow: Ponderosa — *P. ponderosa* — 556
9. Southern U.S., Loblolly etc. — *P. spp* — 553
10. Redwood — *Sequoia sempervirens* — 578
11. Spruce: White, etc. — *P. spp* — 562
12. Sitka — *P. sitchensis* — 563
13. Yew, Pacific — *Taxus brevifolia* — 580
14. Other non U.S., Alerce — *Fitzroya cupressoides* — 582
15. Almaciga (sakar) — *Agathis philippinesis* — 582
16. Klinki pine — *Araucaria klinkii* — 582
17. Parana pine — *A. angustifolia* — 582

Characteristic	1	2	3	4	5	6	7	8	9	10	11	12	13	14	15	16	17
Use this for alignment	1	2	3	4	5	6	7	8	9	0	1	2	3	4	5	6	7…
Resin canals:																	
Present	X	X	X	X	X	X	X	X	X		X	X					
Epithelium thin-walled	X	X	X	X	X	X	X	X	X								
Latewood transition abrupt				X					X		X	X				X	
Tracheids																	
Helical thickening (s in LW only)											s		X				
Bordered pits 2-4 opposite									X		X						
Bordered pits 2-4 alternate															X	X	X
Rays:																	
Homocellular of tracheids present								X	X								
Tracheids present	X	X	X	X	X	X	X	X	X		X	X					
Tracheids dentate		X	X					X	X								
Indenture											X	X	V			X	
Parenchyma end walls nodular					X	V	X				X	X				X	
Marginal parenchyma with crystals																	
Cross—field pitting:																	
Fenestriform				X	X	X	X										
Pinoid	X	X	X					X	X								
Piceoid											X	X					
Cupressoid													X	X	X	X	X
Taxodioid										X							
Longitudinal parenchyma:																	
Present										X							
End walls smooth													X				
End walls nodular															X		
Miscellaneous:																	
Dimpled tangential surface		x									x			x			
Heartwood distinct	X	X	X	X	X	X	X		X		X	X		X		X	
Greasy feel																	
Identifying odor										X							
Use this for alignment	1	2	3	4	5	6	7	8	9	0	1	2	3	4	5	6	7…

Key: X indicates characteristic is pronounced. x indicates characteristic is present but not pronounced (when used with X, it is the less likely of the two choices). V indicates variable according to sample or sources disagree; s indicates sporadic and may not be of diagnostic value.

Table 26-4. Properties of softwood tracheids.

Common name	Scientific name	Diameter, μm[1] Max.	Ave.	Length[2,3] mm	L/D$_{ave}$	Coarseness[3] mg/100 m
Baldcypress	Taxodium distichum	70	52	6.0, 6.2	117	
Cedar, true: Incense	Libocedrus decurrens	50	38	2.0, 3.6	53-95	
Cedar, others						
Alaska—	Chamaecyparis nootkatensis	40	30			
Atlantic white—	Chamaecyparis thyoides	40	30	2.1, 2.1	70	
Port—Orford—	Chamaecyparis lawsoniana	50	40	2.6	65	
Eastern redcedar	Juniperus virginiana	35	25	2.8, 2.8	112	
Western redcedar	Thuja plicata	45	35	3.8, 3.5	105	15.4
Northern whitecedar	Thuja occidentalis	35	25	—, 2.2	88	
Douglas—fir, coastal	Pseudotsuga menziesii	55	40	4.5, 3.9	105	26-31
Fir:						
Balsam	Abies balsamea	50	35	3.5, 3.5	100	
California red	A. magnifica	60	40	3.3	83	
Grand	A. grandis	60	40	5.0	125	
Noble	A. procera	60	40	4.0	100	
Pacific silver	A. amabilis	60	45	3.6, 3.4	78	
Subalpine	A. lasiocarpa	45	38	3.2, 3.0	82	
White	A. concolor	60	40	3.5, 3.4	86	24.0
Hemlock:						
Eastern	Tsuga canadensis	45	34	3.5, 3.0	96	
Western	T. heterophylla	50	35	4.0, 4.2	120	29.0
Larch:						
Tamarack	Larix laricina	45	32	3.5, 3.6	111	
Western	L. occidentalis	60	44	5.0, 5.0	114	32.5
Pine, soft or white:						
Jack	Pinus banksiana			—, 3.5		18.0
Lodgepole	P. contorta	55	40	3.5, 3.1	83	23.0
Monterey	P. radiata					
Red	P. resinosa	45	35	3.7, 3.4	101	21.4
Sugar	P. lambertiana	65	45	4.1, 5.9	90-130	
White, northern	P. strobus	45	30	3.7, 3.0	100-123	19.8
White, western	P. monticola	60	40	4.4, 2.9	73-110	24.0
Pine, hard or yellow:						
Ponderosa	P. ponderosa	60	40	3.6, 3.6	90	26.0
Southern U.S.						
Loblolly	P. taeda	60	40	4.0, 3.6	90-100	23.5
Redwood	Sequois sempervirens	80	58	7.0, 7.0	121	26.8
Spruce:						
White, etc.	Picea spp.	35	28	3.5, 3.3	122	
Sitka	P. sitchensis	55	40	5.5, 5.6	139	
Yew, Pacific	Taxus brevifolia	25	18			
Other non U.S.,						
Alerce	Fitzroya cupressoides					
Almaciga (sakar)	Agathis philippinesis					
Klinki pine	Araucaria klinkii					
Parana pine	A. angustifolia					

[1]Brown and Panshin (1940) p. 112; Panshin and de Zeeuw (1980) p. 133; and Hoadley (1990) p. 17.
[2]USDA FPL-031 (1923, 1964); [3]Isenberg (1981) p. 212.

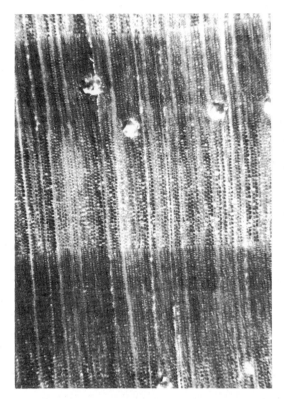

Fig. 26-14. Southern pine, *Pinus* spp.

Soft pines have 5 needles per bundle; hard pines are 2— or 3—needled. Soft pines have a gradual transition to latewood while hard pines have an abrupt transition. Slow growth ponderosa pine

may appear like white pine due to its narrow latewood. The lack of intertracheid pitting on the radial walls of the latewood of southern pines is used to distinguish them from the soft pines. Hard and red pines have dentate ray tracheids, whereas soft pines have nondentate ray tracheids.

The resin canals of *P. lambertina* and *P. ponderosa* are large, while those of *P. contorta* and *P. resinosa* are small for pine resin canals.

Southern pines (Figs. 26-14 and 26-15)

There are a number of species included in the group marketed as southern pine lumber. The most important, and their growth ranges, include: (1) Longleaf pine (*Pinus palustris*), which grows from eastern North Carolina southward into Florida and westward into eastern Texas. (2) Shortleaf pine (*P. echinata*), which grows from southeastern New York and New Jersey southward to northern Florida and westward into eastern Texas and Oklahoma. (3) Loblolly pine (*P. taeda*), which grows from Maryland southward through the Atlantic Coastal Plain and Piedmont Plateau into Florida and westward into eastern Texas. (4) Slash pine (*P. elliotii*), which grows in Florida and the southern parts of South Carolina, Georgia, Alabama, Mississippi, and Louisiana east of the Mississippi River.

Grading standards classify lumber from any one or from any mixture of two or more of these

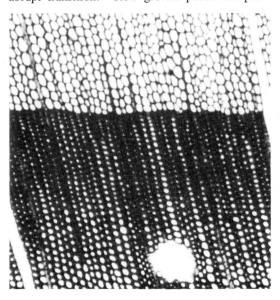

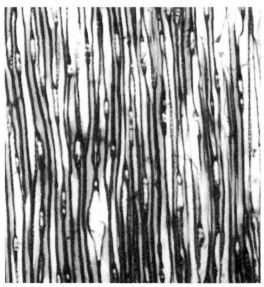

Fig. 26-15. Loblolly pine (*Pinus taeda*) (50×).

species as southern pine and lumber that is produced from longleaf and slash pine species as longleaf pine if conforming to the growth—ring and latewood requirements. The lumber that is classified as longleaf in domestic trade is known also as pitch pine in the export trade. Three southern pines, pitch, pond, and Virginia pine, are designated in published grading rules as "minor species," to distinguish them from the four principal species. Southern pine lumber comes mainly from the Southern and South Atlantic States: from North Carolina to Arkansas and Louisiana.

Uses. Longleaf and slash pines are classed as heavy, strong, stiff, hard, and moderately high in shock resistance. Shortleaf and loblolly pines are usually somewhat lighter in weight than longleaf. All the southern pines have moderately large shrinkage but are stable when properly seasoned. The pith has been used to distinguish longleaf, shortleaf, and loblolly pines (Koehler, 1917). To obtain heavy, strong wood of the southern pines for structural purposes, a density rule has been written that specifies certain visual characteristics for structural timbers.

Dense southern pine is used extensively in construction of factories, warehouses, bridges, trestles, and docks in the form of stringers, beams, posts, joists, piles, and plywood. Lumber of lower density and strength finds many uses for building material, such as interior finish, sheathing, subflooring, and joists, and for boxes, pallets, and crates. Southern pine is also used for tight and slack cooperage. When used for railroad ties, piles, poles, and mine timbers, it is usually treated with preservatives.

The southern pines are widely used in pulp and paper products. They compete with Douglas—fir for most linerboard and packaging papers where high strength is required. Like Douglas—fir, they form coarse papers.

Color. The wood of the various southern pines is quite similar in appearance. The sapwood is yellowish white and is usually wide in second—growth stands. The heartwood is reddish or orangish brown and begins to form when the tree is about 20 years old. In old, slow—growth trees, sapwood may be only 1 to 2 in. wide.

Macroscopic structure. The wood has an abrupt latewood transition with a variable width latewood. Resin canals are scattered throughout the growth ring and are small for pine; the openings are rarely visible without a lens (visible to the unaided eye sometimes in longleaf pine). Nonfusiform rays are fine and distinct with a lens.

Pitch, pond, and Virginia pines

These three species are the "minor" southern pines in the southern pine grading rules; see the previous entry for their structures. Pitch pine (*Pinus rigida*) grows from Maine along the mountains to eastern Tennessee and northern Georgia. The heartwood is brownish red and resinous; the sapwood is thick and light yellow. The wood of pitch pine is medium heavy to heavy, medium strong, medium stiff, medium hard, and medium high in shock resistance. Its shrinkage is medium small to medium large. It is used for lumber, fuel, and pulpwood.

Pond pine (*Pinus serotina*) grows in the coast region from New Jersey to Florida. It occurs in small groups or singly, mixed with other pines on low flats. The wood is heavy, coarse—grained, and resinous, with dark, orange—colored heartwood and thick, pale yellow sapwood. Shrinkage is moderately large. The wood is moderately strong, stiff, medium hard, and medium high in shock resistance. It is used for construction, railway ties, posts, and poles.

Virginia pine (*Pinus virginiana*), known also as Jersey pine and scrub pine, grows from New Jersey and Virginia throughout the Appalachian region to Georgia and the Ohio Valley. The heartwood is orange and the sapwood nearly white and relatively thick. The wood is rated moderately in weight, strength, hardness, and stiffness. It has moderately large shrinkage and high shock resistance. It is used for lumber, railroad ties, mine props, pulpwood, and fuel.

Jack pine (Fig. 26-16)

Jack pine (*Pinus banksiana*), sometimes known as scrub pine, gray pine, or black pine in the U.S., grows naturally in the Lake States and in a few scattered areas in New England and northern New York. In lumber, jack pine is not separated from the other pines with which it grows, including red pine and eastern white pine.

Uses. It is moderately light in weight, moderately low in bending strength and compres-

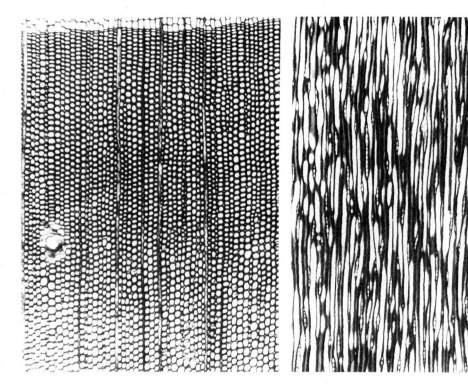

Fig. 26-16. Jack pine (*Pinus banksiana*) (50×).

sive strength, moderately low in shock resistance, and low in stiffness. It also has moderately small shrinkage. Lumber from jack pine is generally knotty. Jack pine is used for pulpwood, box lumber, pallets, and fuel. Less important uses include railroad tie, mine timber, slack cooperage, poles, and posts.

Color. The sapwood of jack pine is nearly white; the heartwood is light brown to orange. The sapwood may make up 50% or more of the volume of a tree. The wood has a rather coarse texture and is somewhat resinous.

Lodgepole pine (Figs. 26-17 and 26-18)

Lodgepole pine (*Pinus contorta*), also known as knotty pine, black pine, spruce pine, and jack pine, grows in the Rocky Mountain and Pacific coast regions as far northward as Alaska. Harvesting occurs largely from the central Rocky Mountain States; other producing regions are Idaho, Montana, Oregon, and Washington.

Uses. The wood is generally straight grained with narrow growth rings (20—30/in.). The wood is moderately light in weight, fairly easy to work,

and has moderately large shrinkage. Lodgepole pine rates as moderately low in strength, moderately soft, moderately stiff, and moderately low in shock resistance. It is used for lumber, mine timbers, railroad ties, and poles. Less important uses include posts and fuel. Other uses include for framing, siding, finish, and flooring.

Color. The heartwood of lodgepole pine varies from light yellow to light yellow—brown and is only slightly darker than the sapwood; often it is not clearly defined from the sapwood. The sapwood is yellow or nearly white, usually 1 in. wide, but sometimes over 2 in. wide.

Macroscopic features. The wood has a fairly abrupt transition to the narrow latewood. The resin ducts are small and numerous. They are not distinctly visible without a lens, although the sap is sometimes visible. The rays are very fine, except for the fusiform rays, which are quite distinct under a lens. The tangential surface is slightly dimpled. The wood has a distinct resinous odor that is stronger than that of most pines.

Similar woods. Ponderosa pine resembles lodgepole pine, but ponderosa pine has darker and more distinct heartwood, wider sapwood, larger growth rings, larger resin ducts, and rarely has the pronounced dimpled surface (*t*) of lodgepole pine.

Ponderosa pine (Figs. 26-19 and 26-20)

Ponderosa pine (*Pinus ponderosa*) is known as ponderosa pine, western soft pine, western pine, California white pine, bull pine, and black jack. Jeffrey pine (*P. jeffreyi*), which grows in close association with ponderosa pine in California and Oregon, is usually marketed with ponderosa pine and sold under that name. Major producing areas are in Oregon, Washington, and California; moderate amounts come from Idaho and Montana; and small amounts come from Wyoming and the Black Hills of South Dakota.

Uses. The wood of the outer portions of ponderosa pine of sawtimber size is generally moderately light in weight, moderately low in strength, moderately soft, moderately stiff, and moderately low in shock resistance. It is generally straight grained and has moderately small shrinkage. It is quite uniform in texture and has little tendency to warp and twist.

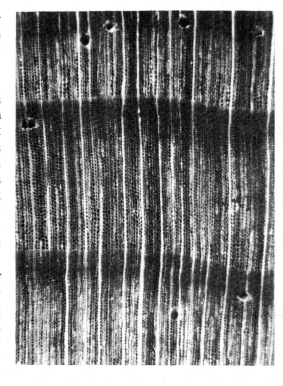

Fig. 26-17.   *Pinus contorta.*

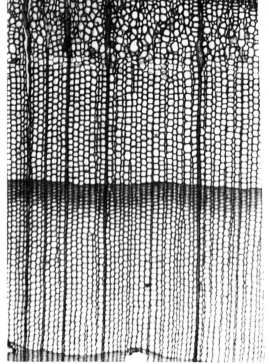

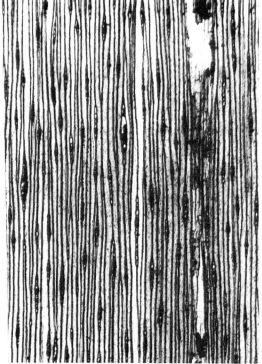

Fig. 26-18.   Lodgepole pine (*Pinus contorta*) (50×).

Ponderosa pine is used mainly for lumber, particleboard, and pulp chips and to a lesser extent for piles, poles, posts, mine timbers, veneer, and ties. The clear wood goes into sash, doors, blinds, moldings, paneling, mantels, trim, and built—in cases and cabinets. Knotty pine is also used in paneling. Lower grade lumber is used for boxes and crates.

Color. Ponderosa pine belongs to the yellow pine group. Much of the wood, however, is somewhat similar to the white pines in appearance and properties. The heartwood is light reddish or orangish brown and clearly defined from the sapwood. The wide sapwood (2.5—4 in.) is nearly white to pale yellow.

Macroscopic structure. The transition to latewood is abrupt as in the southern pines, but the latewood bands are narrow though variable. The tangential surface is often dimpled as in lodgepole pine, but in the latter they are small and more abundant. The numerous resin canals are larger than those of lodgepole pine. The heartwood of lodgepole pine is lighter in color than that of ponderosa pine.

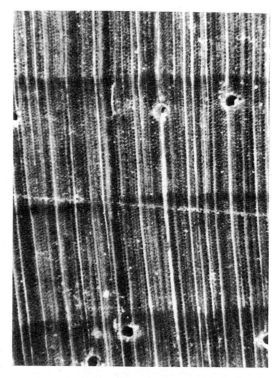

Fig. 26-19. *Pinus ponderosa.*

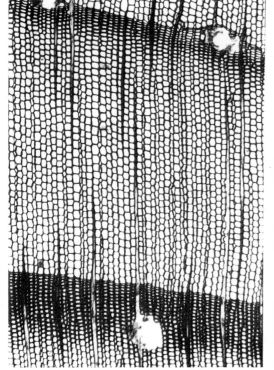

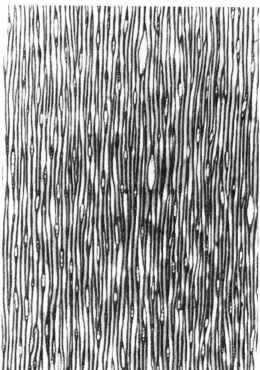

Fig. 26-20. Ponderosa pine (*Pinus ponderosa*) (50×).

Microscopic features. Ponderosa pine may be distinguished from Jack pine because the dentations of the ray tracheids are more pronounced than those of Jack pine.

Red pine

Red pine (*Pinus resinosa*) is frequently called Norway pine. It is occasionally known as hard pine and pitch pine. This species grows in the New England States, New York, Pennsylvania, and the Lake States. In the past, lumber from red pine has been marketed with white pine without distinction as to species.

Uses. Red pine is moderately heavy, moderately strong and stiff, moderately soft, and moderately high in shock resistance. The wood is the lightest of the hard pines east of the Mississippi River. It is generally straight grained, not so uniform in texture as eastern white pine, and somewhat resinous. The wood has moderately large shrinkage but is not difficult to dry and stays in place well when seasoned.

Red pine is used principally for lumber and to a lesser extent for piles, poles, cabin logs, posts, pulpwood, and fuel. The wood is used for many of the purposes for which eastern white pine is used. It goes mostly into building construction, siding, flooring, sash, doors, blinds, general millwork, and boxes, pallets, and crates.

Color. The heartwood of red pine varies from pale red to reddish brown. The sapwood is nearly white with a yellowish tinge, and is generally from 2 to 4 inches wide. The wood resembles the lighter weight wood of southern pine. Latewood is distinct in the growth rings.

Macroscopic structure. The wood has a somewhat abrupt transition to latewood. The resin ducts tend to be concentrated in the outer half of growth rings with openings that are not visible without a lens. Nonfusiform rays are fine. The annual rings are fairly wide.

Similar woods. Red pine resembles the southern pines; it is quickly separated by its fenestriform pitting in the cross—fields.

Spruce pine

Spruce pine (*Pinus glabra*), also known as cedar pine, poor pine, Walter pine, and bottom white pine, grows on low moist lands of the coastal regions of southeastern South Carolina, Georgia, Alabama, Mississippi, and Louisiana, and northern and northwestern Florida. The heartwood is light brown, and the wide sapwood zone is nearly white.

Uses. Spruce pine wood is lower in most strength values than the major southern pines. It compares favorably with white fir in bending properties, in crushing strength perpendicular and parallel to the grain, and in hardness. It is similar to the denser species such as coast Douglas—fir and loblolly pine in shear parallel to the grain.

The principal uses of spruce pine were locally for lumber, and for pulpwood and fuelwood. The lumber, which is classified as one of the minor southern pine species, reportedly was used for sash, doors, and interior finish because of its lower specific gravity and less marked distinction between earlywood and latewood. It has qualified for use in plywood.

Sugar pine (Fig. 26-21)

Sugar pine (*Pinus lambertiana*) is sometimes called California sugar pine. Most of the sugar pine lumber is produced in California and the remainder in southwestern Oregon.

Uses. The wood is straight grained, fairly uniform in texture, and easy to work with tools. It has very small shrinkage, is readily seasoned without warping or checking, and stays in place well. This species is light in weight, moderately low in strength, moderately soft, low in shock resistance, and low in stiffness.

Sugar pine is used almost entirely for lumber products. The largest amounts are used in crates, sash, doors, frames, blinds, general millwork, building construction, and foundry patterns. Like eastern white pine, sugar pine is suitable for use in nearly every part of a house because of the ease with which it can be cut, its ability to stay in place, and its good nailing properties.

Color. The heartwood of sugar pine is buff or light brown, sometimes tinged with red; it is lighter than that of *P. strobus* or *P. monticola*. The sapwood is creamy white.

Macrostructure. Resin canals are abundant and commonly stain the longitudinal surfaces. Transition from earlywood to latewood is gradual, and the growth rings are not prominent on flatsawn surfaces. The wood is coarse textured.

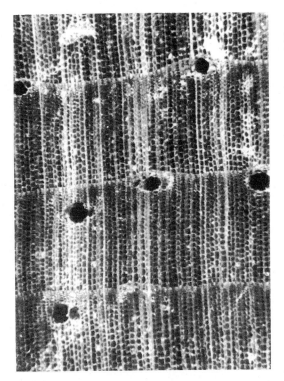

Fig. 26-21. *Pinus lambertiana.*

Microscopic structure. The intertracheid pitting may be biseriate for four or more consecutive pit pairs in the first part of the earlywood.

Eastern white pine (Fig. 26-22)

Eastern white pine (*Pinus strobus*) grows from Maine to northern Georgia and in the Lake States. It is also known as white pine, northern white pine, Weymouth pine, and soft pine. About 50% of the lumber is produced in the New England States, about 30% in the Lake States, and most of the remainder in the Middle Atlantic and South Atlantic States.

Uses. It is easily kiln—dried, has small shrinkage, and ranks high in stability. The wood has comparatively uniform texture and is straight grained. It is also easy to work and can be readily glued. It is light in weight, moderately soft, moderately low in strength, and low in resistance to shock. Almost all eastern white pine is converted into lumber; a large proportion, which is mostly second-growth knotty lumber of the lower grades, goes into container and packaging applications. High-grade lumber goes into patterns for

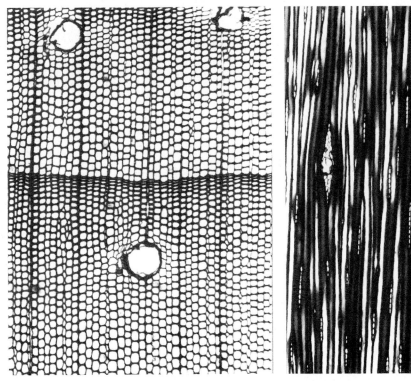

Fig. 26-22. Eastern white pine (*Pinus strobus*) (50×).

castings. Other uses are sash, door, furniture, trim, knotty paneling, finish, caskets and burial boxes, shade and map rollers, toys, and dairy and poultry supplies.

Color. The heartwood of eastern white pine is light brown, often with a reddish tinge. It turns considerably darker on exposure.

Macrostructure. The wood has a gradual transition to latewood. The latewood is narrow and darker than the earlywood.

Western white pine (Figs. 26-23 and 26-24)

Western white pine (*Pinus monticola*) is also known as Idaho white pine or white pine. About four—fifths of the cut comes from Idaho with the remainder mostly from Washington; small amounts are cut in Montana and Oregon.

Uses. The wood is straight grained, easy to work, easily kiln—dried, and stable after seasoning. This species is moderately light in weight, moderately low in strength, moderately soft, moderately stiff, moderately low in shock resistance, and has moderately large shrinkage.

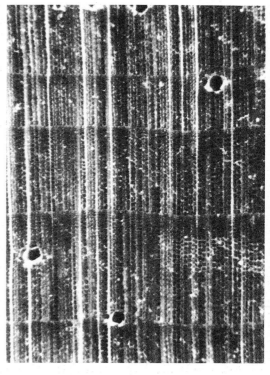

Fig. 26-23. *Pinus monticola.*

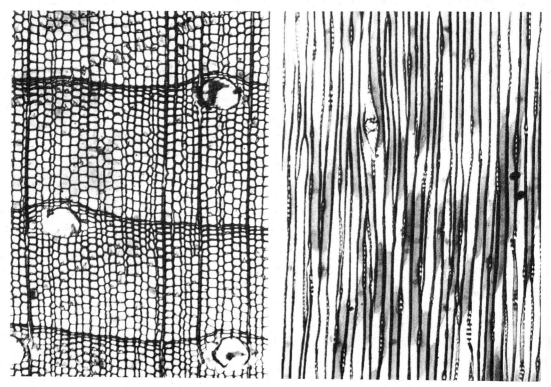

Fig. 26-24. Western white pine (*Pinus monticola*) (50×).

Practically all western white pine is sawed into lumber and used for building construction, matches, boxes, patterns, and millwork products, such as sash, frames, doors, and blinds. In construction, boards of the lower grades are used for sheathing, knotty paneling, subflooring, and roof strips. High—grade material is made into siding of various kinds, exterior and interior trim, and finish. It has practically the same uses as eastern white pine and sugar pine.

Color. Heartwood of western white pine is cream colored to light reddish brown and darkens on exposure. The sapwood is yellowish white and generally from 1 to 2 (occasionally 3) in. wide.

Macroscopic features. The openings of the resin canals are plainly visible with a lens, and sometimes without a lens. The canals are abundant throughout the growth ring. The rays are mostly fine, although the fusiform rays are more pronounced. The annual rings are distinct and moderately wide. The wood has gradual transition to the narrow latewood.

Similar woods. It is separated from sugar pine by the larger resin canals in sugar pine and microscopic features. It is difficult to separate from *P. strobus*. Limber pine is very similar, but it is usually heavier.

Scotch pine

Scotch pine (*Pinus sylvestris*) is native to Europe and Asia, but now grows in southeastern Canada and northeastern U.S. A British term for construction lumber from this species (or softwoods in general) is *deal*.

PSEUDOTSUGA

Douglas—fir (Figs. 26-25 and 26-26)

Douglas—fir (*Pseudotsuga menziesii* and var. *glauca*) is also known locally as red fir, Douglas—spruce, and yellow fir. (The older literature refers to it as *P. taxifola*.) There are two varieties, the coastal and the mountain types. The range of Douglas—fir extends from the Rocky Mountains to the Pacific coast and from Mexico to central British Columbia. The Douglas—fir production comes from Oregon, Washington, California, and the Rocky Mountain States.

Uses. The wood of Douglas—fir varies widely in weight and strength. When structural lumber of high strength is required, selection can be improved by applying the density rule, which has latewood percentage and growth rate criteria.

Douglas—fir is used mostly for building and construction purposes in the form of lumber, timbers, piling, and plywood. Considerable quantities go into railroad ties, cooperage stock, mine timbers, poles, and fencing. Douglas—fir lumber is used in the manufacture of various products, including sash, doors, general millwork, railroad—car construction, boxes, pallets, and crates. Small amounts are used for flooring, furniture, ship and boat construction, wood pipe, and tanks. Douglas—fir is widely used in pulp and paper products, especially in linerboard and sack papers, where it competes with southern pine, where high strength is required. It forms coarse papers and obtaining good formation is difficult.

Color. Sapwood of Douglas—fir is narrow in old—growth trees but may be as much as 3 inches wide in second—growth trees of commercial size. Young trees of moderate to rapid growth have reddish heartwood and are called red fir. Very narrow—ringed wood of old trees may be yellowish brown and is known on the market as yellow fir.

Macroscopic structure. The resin canals may form tangential rows of 2 to 20 (traumatic resin canals would have much longer tangential rows). The resin canals are much smaller than those of the pines, and Douglas—fir normally has a distinctive odor. The bark of Douglas—fir (Fig. 2-1, left) has layers that distinguish it from other species with which it may be confused.

Microscopic structure. Spiral thickening (yew has this but no resin canals) in the tracheids is a characteristic feature of Douglas—fir. Marginal longitudinal parenchyma are sometimes observed.

PICEA

With the exception of sitka spruce, the wood of spruces are not readily assigned to individual species without additional information since their anatomies are very similar. Engelmann spruce, however, tends to have an abrupt transition to latewood since it grows at high elevation in the Rocky Mountains.

Eastern spruce

The term "eastern spruce" includes three species, red (*Picea rubens*), white (*P. glauca*), and black (*P. mariana*) (Fig. 26-27). White spruce and black spruce mainly grow in the Lake States and New England, and red spruce in New England and the Appalachian Mountains. All three species have about the same properties, and in commerce no distinction is made between them.

Fig. 26-25. *Pseudotsuga menziesii.*

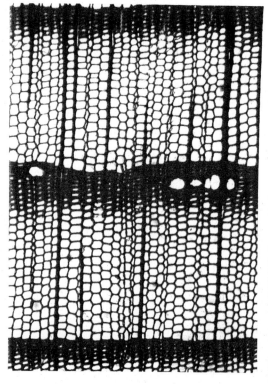

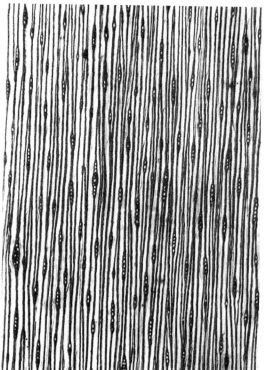

Fig. 26-26. Douglas-fir (*Pseudotsuga menziesii*) (50×).

Uses. The wood dries easily and is stable after drying, is moderately light in weight and easily worked, has moderate shrinkage, and is moderately strong, stiff, tough, and hard. The largest use of eastern spruce is for pulpwood. It is also used for framing material, general millwork, boxes and crates, ladder rails, scaffold planks, and piano sounding boards. Eastern spruce is sold with balsam fir as spruce—fir, but the latter lacks resin canals. Balsam fir is heavier than eastern spruce when wet due to its higher moisture content, but is lighter when dry.

Color. The wood is light in color, and there is little difference between the heartwood and sapwood colors. The high luster of eastern spruce is used to help distinguish it from northern white pine.

Macroscopic structure. The transition to the very narrow latewood is gradual. The resin canals are small and appear as white dots [in the cross section (*x*)]. The rays are visible with a lens on wet surfaces (*x*). Fusiform rays occur randomly and are widely spaced.

Sitka spruce (Figs. 26-28 and 26-29)

Sitka spruce (*Picea sitchensis*) is a tree of large size growing along the Pacific coast from California to Alaska. Local names include yellow spruce, tideland spruce, western spruce, and silver spruce. About 2/3 of Sitka spruce lumber comes from Washington and 1/3 from Oregon.

Uses. It is moderately light in weight, moderately low in bending and compressive strength, moderately stiff, moderately soft, and moderately low in resistance to shock. It has moderately small shrinkage. On the basis of weight, it rates high in strength properties and can be obtained in clear, straight—grained pieces.

Sitka spruce is used principally for lumber, pulpwood, and cooperage. It ranks high as a pulpwood because of its long, strong fibers and ease of pulping. Boxes and crates account for a

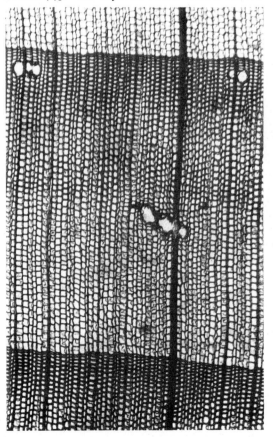
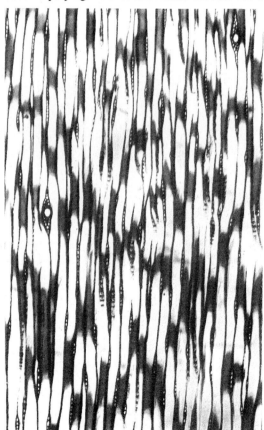

Fig. 26-27. Black spruce (*Picea mariana*) (50×).

considerable amount of the remanufactured lumber. Other important uses are furniture, planing mill products, sash, doors, blinds, millwork, and boats. Sitka spruce has been by far the most important wood for aircraft construction. Other specialty uses are ladder rails and sounding boards for pianos.

Color. The heartwood of Sitka spruce is a light brown with a pinkish or purple cast and is darker than that of eastern spruce. The sapwood is creamy white and shades gradually into the heartwood; it may be 3 to 6 in. wide or even wider in young trees. The wood has a comparatively fine, uniform texture, generally straight grain, and no distinct taste or odor.

Macrostructure. The dimpled grain, if present, its darker and pinkish heartwood, and its somewhat larger resin canals help identify it. It may be mistaken for soft pine.

Microstructure. This species is separated from other spruce by the occurrence of biseriate pitting in longitudinal tracheids. Its ray cells may contain darker (brownish) deposits than other species and crystals.

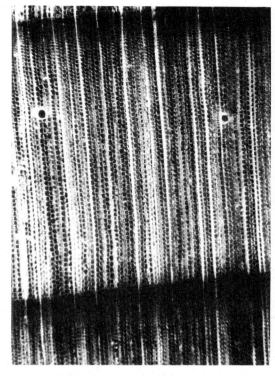

Fig. 26-28. *Picea sitchensis*.

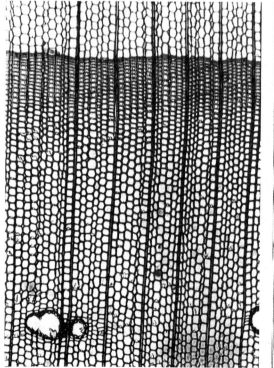

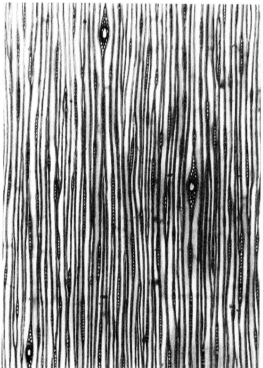

Fig. 26-29. Sitka spruce (*Picea sitchensis*) (50×).

Engelmann spruce (Fig. 26-30)

Engelmann spruce (*Picea engelmannii*) grows at high elevations in the Rocky Mountain region of the United States. It is known by other names, such as white spruce, mountain spruce, Arizona spruce, silver spruce, and balsam. About 60% of the lumber comes from the southern Rocky Mountain States, with most of the remainder from the northern Rocky Mountain States and Oregon.

Uses. The wood has medium to fine texture. It is generally straight grained. Engelmann spruce is rated as light in weight. It is low in strength as a beam or post. It is limber, soft, low in shock resistance, and has moderately small shrinkage. The lumber typically contains numerous small knots. Engelmann spruce is used for lumber, mine timbers, railroad ties, and poles. It is used also in building construction in the form of dimension stock, flooring, sheathing, and studding. It has excellent properties for pulp and papermaking.

Color. The heartwood of Engelmann spruce is nearly white with a slight tinge of red. The sapwood is 3/4 to 2 in. wide and is often difficult to distinguish from heartwood, except for its high moisture content in green timbers.

Macroscopic features. The transition from earlywood to latewood is somewhat more abrupt than in other spruces. Resin canals are present, comparatively few, and are often difficult to find. They often appear as white dots in the latewood. The rays are very fine; the fusiform rays occur sporadically and are widely spaced.

Similar woods. Engelmann spruce is easily confused with the white (true) firs and lodgepole pine, which have almost white heartwoods. The true firs lack resin canals. Engelmann spruce lacks the dimpled surface of lodgepole pine.

Norway spruce

Norway spruce (*P. abies*) is native to Europe. It has been planted in the northeast U.S.

ABIES

Uses. The white firs are very useful in pulp products and produces strong, high—quality paper.

Color. The wood of the true firs (*Abies*) is creamy white to pale brown. The heartwood and sapwood are usually indistinguishable. The similarity of wood structure in the true firs makes it

Fig. 26-30. *Picea engelmannii* (Koehler, 1917) 15×.

impossible to distinguish the species by an examination of the wood alone. Firs tend to be intolerant (of shade, when growing) and are, therefore, often fast growing with wide growth rings having especially gradual transition to latewood.

Fig. 26-31. *Abies* spp.

Microscopic structure. Resin canals and ray tracheids are absent, end walls of ray parenchyma are nodular, and cross—field pits are nodular. *A. lasiocarpa* may be foul smelling with knots that are a distinct yellow color. Crystals may be found in all species, but especially in *A. magnifica* and *A. concolor*. In the eastern species and *A. lasiocarpa* the ray parenchyma contents are yellow to clear, while those of the other species are reddishbrown. Generally the true firs have taller rays than the hemlocks.

Hoadley (1990) describes the use of Ehrich's reagent (4 parts *p*—dimethyl amino benzaldehyde dissolved in 92 parts ethanol made acidic by adding 8 parts concentrated HCl) to distinguish subalpine fir (which turns purple) from Pacific silver fir (which remains clear or pale green).

Eastern species (Figs. 26-31 and 26-32)

Balsam fir (*Abies balsamea*) grows principally in New England, New York, Pennsylvania, and the Lake States; it is the only eastern *Abies* of commercial importance. Fraser fir (*A. fraseri*) has limited distribution in the Appalachian Mountains of Virginia, North Carolina, and Tennessee; this species sometimes has resinous tracheids (near parenchyma) in the heartwood that are filled with reddish brown to black amorphous deposits.

Balsam fir is light in weight when dry, low in bending and compressive strength, moderately limber, soft, and low in resistance to shock. The eastern firs are used mainly for pulpwood, although some lumber is produced from them, especially in New England and the Lake States. Balsam fir latewood often has a light purple cast to it. See eastern spruce for additional information.

Western species (Fig. 26-33)

Six commercial species make up the western true firs: subalpine fir (*Abies lasiocarpa*), California red fir (*A. magnifica*), grand fir (*A. grandis*), noble fir (*A. procera*), Pacific silver fir (*A. amabilis*), and white fir (*A. concolor*).

The western firs are light in weight, but, with the exception of subalpine fir, have somewhat higher strength properties than balsam fir. Wood

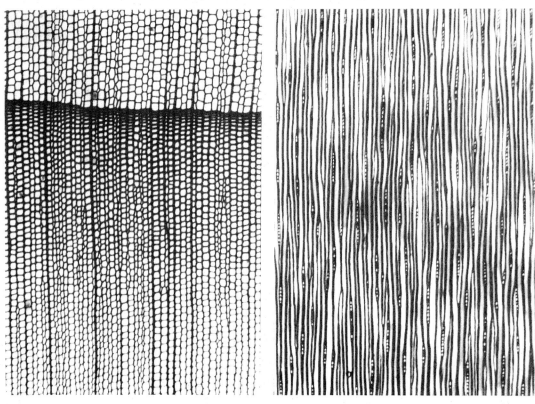

Fig. 26-32. Balsam fir (*Abies balsamea*) (50×).

shrinkage is small to moderately large. The sapwood is not distinguishable from the heartwood in some of these species.

The western true firs are largely cut for lumber in Washington, Oregon, California, western Montana, and northern Idaho and marketed as white fir throughout the U.S. Lumber of the western true firs goes into building construction, boxes and crates, and general millwork. In house construction, the lumber is used for framing, subflooring, and sheathing. Some western true fir lumber goes into boxes and crates. High—grade lumber from noble fir is used mainly for interior finish, moldings, siding, and sash and door stock. The best material is used for aircraft construction and ladder rails.

Macroscopic structure. Resin canals are normally absents, but occasionally traumatic resin canals are present in the outer latewood in tangential rows extending 1/8 in. or more. *A. grandis* sometimes has resinous tracheids (near parenchyma) in the heartwood that are filled with reddish brown to black amorphous deposits.

Similar woods. Western hemlock is very similar, but microscopic features can be used to distinguish these two woods. Western fir has taxodioid cross—field pits and lacks ray tracheids.

LARIX

Larix species have resin canals with thick—walled epithelium cells. There is an abrupt transition from earlywood to latewood. The bordered pitting of vertical tracheids is often biseriate. Eastern larch has lighter color heartwood (more yellow than brown), a lower degree of biseriate bordered pitting, and a finer texture than western larch.

Western larch (Figs. 26-34 and 26-35)

Western larch (*Larix occidentalis*) grows in western Montana, northern Idaho, northeastern Oregon, and the eastern slope of the Cascade Mountains in Washington. About 2/3 of the lumber of this species is produced in Idaho and Montana and 1/3 in Oregon and Washington.

Uses. The wood is stiff, moderately strong

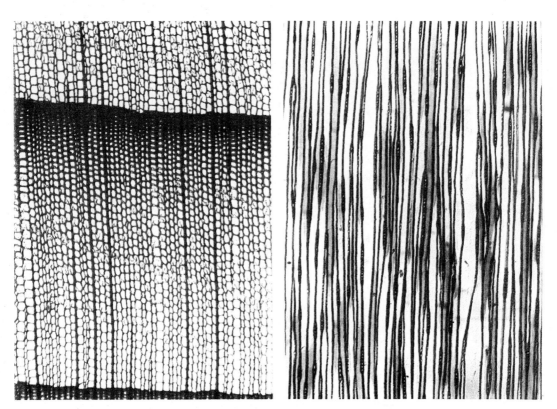

Fig 26-33. White fir (*Abies concolor*) (50×).

and hard, moderately high in shock resistance, and moderately heavy. It has moderately large shrinkage. The wood is usually straight grained, splits easily, and is subject to ring shake. Knots are common but small and tight. Western larch is used mainly in building construction for rough dimension, small timbers, planks and boards, and for railroad tie and mine timbers. It is used also for piles, poles, and posts. Some high—grade material is manufactured into interior finish, flooring, sash, and doors.

Color. The heartwood of western larch is yellowish brown and the sapwood is yellowish white. The sapwood is generally 0.5—0.75 in. and not more than 1 in. thick.

Macrostructure. The wood has a very abrupt transition to the narrow, conspicuous latewood. The resin canals occur sporadically and may be in tangential groups. *Microstructure.* Marginal parenchyma are occasionally present.

Eastern larch, tamarack (Fig. 26-36)

Tamarack (*Larix laricina*) is a small—to medium—sized tree with a straight, round slightly

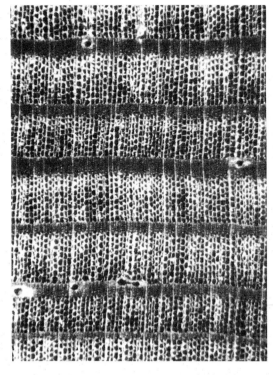

Fig. 26-34. *Larix occidentalis.*

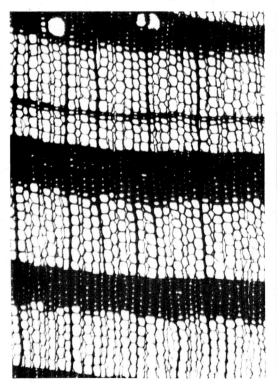

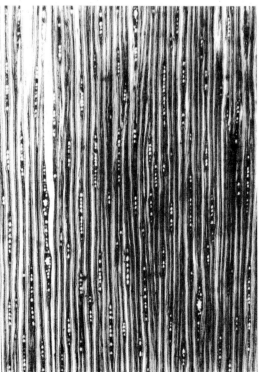

Fig. 26-35. Western larch (*Larix occidentalis*) (50×).

tapered trunk. In the U.S. it grows from Maine to Minnesota, with the bulk of the stands in the Lake States. It was formerly used in considerable quantity for lumber, but in recent years production for that purpose has been small.

The heartwood of tamarack is yellowish brown to russet brown. The sapwood is whitish, generally less than an inch wide. The wood is coarse in texture, without odor or taste, and the transition from earlywood to latewood is abrupt. The wood is intermediate in weight and in most mechanical properties.

Tamarack is used principally for pulpwood, lumber, railroad ties, mine timbers, fuel, fenceposts, and poles. Lumber goes into framing material, tank construction, boxes, and pallets.

European larch

European larch (*Larix decidua*), native to Europe, is planted in Canada and northeast U.S.

TSUGA

The presence of ray tracheids is useful in distinguishing the hemlocks, especially western hemlock, from the true firs (*Abies*). The growth rate of hemlock (especially eastern hemlock) can vary greatly even in a relatively small sample. A smooth cut is difficult to make on the end grain of eastern hemlock (using a knife), but is easily made on western hemlock. Western hemlock wood may have traumatic resin canals, is less splintery, and is usually free of cup shake.

Eastern hemlock

Eastern hemlock (*Tsuga canadensis*) grows from New England to northern Alabama and Georgia, and in the Lake States. Other names are Canadian hemlock and hemlock spruce. The production of hemlock lumber is divided fairly evenly between the New England States, the Middle Atlantic States, and the Lake States.

Uses. The wood is moderately light in weight, moderately hard, moderately low in strength, moderately limber, and moderately low in shock resistance. It is used principally for lumber and pulpwood. The lumber is used largely in building construction for framing, sheathing, subflooring, and roof boards, and in the manufacture of boxes, pallets, and crates.

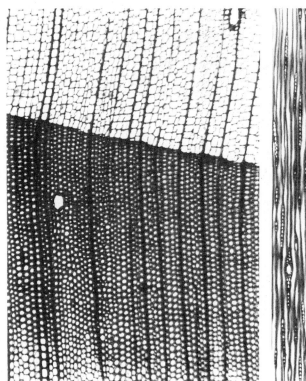
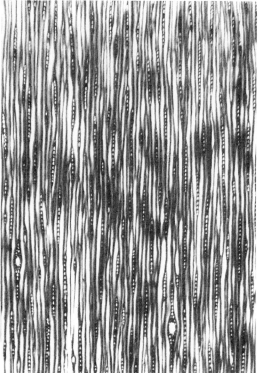

Fig. 26-36. Tamarack, eastern larch (*Larix laricina*) (50×).

Color. The heartwood of eastern hemlock is pale brown with a reddish hue. The sapwood is not distinctly separated from the heartwood but may be lighter in color. The wood is coarse and uneven in texture (old trees tend to have considerable shake).

Similar woods. Spruce is lighter in color, weighs less, does not tend to splinter, and has resin canals. Most of the firs are of light color, especially in the earlywood, and do not tend to splinter. Noble fir should be distinguished with a microscope.

Western hemlock (Figs. 26-37 and 26-38)

Western hemlock (*Tsuga heterophylla*) (alias west coast hemlock, hemlock spruce, western hemlock spruce, western hemlock fir, Prince Albert fir, gray fir, silver fir, and Alaska pine) grows along the Pacific coast of Oregon and Washington and in the northern Rocky Mountains, north to Canada and Alaska. Mountain hemlock, *T. mertensiana*, inhabits mountainous country from central California to Alaska; it is treated as a separate species in assigning lumber properties.

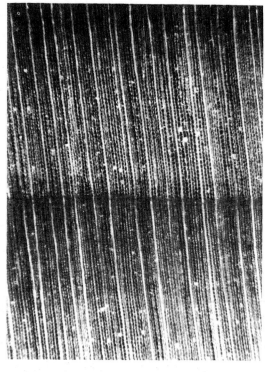

Fig. 26-37. *Tsuga heterophylla.*

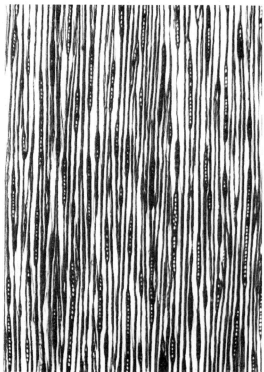

Fig. 26-38. Western hemlock (*Tsuga heterophylla*) (50×).

Uses. Western hemlock is moderately light in weight and moderate in strength. It is moderate in its hardness, stiffness, and shock resistance. It has moderately large shrinkage, about the same as Douglas—fir. Green hemlock lumber contains considerably more water than Douglas—fir and requires longer kiln drying time. Western hemlock is used mainly for pulpwood, lumber, and plywood. The lumber goes largely into building material, such as sheathing, siding, subflooring, joists, studding, planking, and rafters. Major quantities are used in the manufacture of pallets, crates, and flooring, and smaller amounts for furniture and ladders.

Color. The heartwood and sapwood of western hemlock are almost white with a purplish tinge. The sapwood, which is sometimes lighter in color, is generally not more than 1 inch thick. The wood contains small, sound, black knots that are usually tight and stay in place. Dark streaks often found in the lumber and caused by hemlock bark maggots as a rule do not reduce strength.

CUPRESSACEAE

CHAMAECYPARIS (False cypresses)

Alaska—cedar (Fig. 26-39)

Alaska—cedar (*Chamaecyparis nootkatensis*) grows in the Pacific coast region from southeastern Alaska to southern Oregon.

Uses. The wood is fine textured and generally straight grained. It is moderately heavy, moderately strong and stiff, moderately hard, and moderately high in resistance to shock. Alaska—cedar shrinks little in drying, is stable in use after seasoning, and the heartwood is very resistant to decay. The wood has a mild, unpleasant, characteristic odor that resembles that of raw potatoes. Alaska—cedar is used for interior finish, furniture, small boats, cabinetwork, and novelties.

Color. The heartwood of Alaska—cedar is bright, clear yellow; it tends to darken somewhat with exposure to air. The sapwood is narrow, white to yellowish, and hardly distinguishable from the heartwood. The growth rings tend to be very narrow and inconspicuous.

Microscopic structure. The ray tracheids may be large. Some rays have no parenchyma. Ray and longitudinal parenchyma end walls are nodular with indentures.

Fig. 26-39. *Chamaecyparis nootkatensis.*

Port—Orford—cedar (Figs. 26-40 and 26-41)

Port—Orford—cedar (*Chamaecyparis lawsoniana*) (alias Lawson cypress, Oregon cedar, and white cedar) grows along the Pacific coast from Coos Bay, Oregon, southward to California. It does not extend more than 40 miles inland.

Uses. It is moderately light in weight, stiff, moderately strong and hard, and moderately resistant to shock. Port—Orford—cedar heartwood is highly resistant to decay. The wood shrinks moderately, has little tendency to warp, and is stable after seasoning. Some high—grade Port—Orford—cedar was used in the manufacture of battery separators. Other uses are venetian-blind slats, mothproof boxes, archery supplies, sash and door construction, stadium seats, flooring, interior finish, furniture, and boatbuilding.

Color. The heartwood of Port—Orford—cedar is light yellow to pale brown in color and has a strong, pungent odor. The sapwood is narrow (1—3 in.) and hard to distinguish from the heartwood. The wood has a fine, even texture, generally straight grain, and a pleasant spicy (ginger—like) odor and a bitter, spicy taste.

Macroscopic structure. Longitudinal parenchyma are usually visible (wetting helps) with a lens as reddish lines or zones.

Microscopic structure. Ray tracheids are lacking, ray parenchyma have smooth end walls, and the transverse walls of the vertical parenchyma are nodular.

Atlantic white—cedar

This species is strictly a swamp tree and is very similar in properties and uses to northern red cedar; both grow in the eastern U.S. *Chamaecyparis thyoids* is also known as juniper, southern white—cedar, swamp cedar, and boat cedar; it grows near the Atlantic coast from Maine to northern Florida and westward along the gulf coast to Louisiana. Commercial production centers in North Carolina and along the gulf coast. White—cedar lumber is used principally where a high degree of durability is needed, as in tanks and boats, and for woodenware.

Color. The heartwood is light brown with a reddish cast that is sharply demarcated from the sapwood. It is aromatic and has a similar odor to that of northern white—cedar.

Fig. 26-40. *Chamaecyparis lawsoniana.*

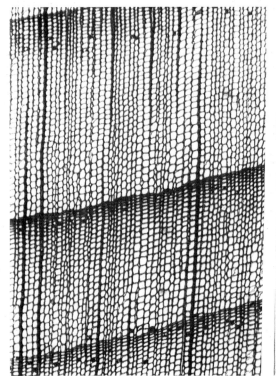
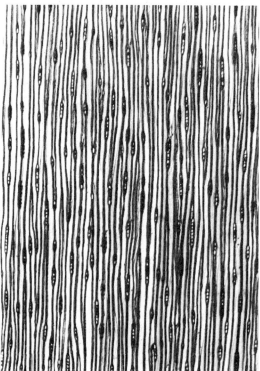

Fig. 26-41. Port-Orford cedar (*Chamaecyparis lawsoniana*) (50×).

Microscopic structure. Ray tracheids are lacking, ray parenchyma have smooth end walls, and the transverse walls of the vertical parenchyma are smooth.

JUNIPERUS

The heartwood of trees of this genus are all aromatic with a pencil or cedar chest smell. Eastern and southern species have deep red heartwood; the western species are more brownish in color. Western juniper is *J. occidentalis* (Figs. 26-42 and 26-43).

Microscopic features. This genus is characterized by large amounts of vertical parenchyma with dark colored contents, nodular end walls, and indenture. Intercellular spaces occur between vertical tracheids.

Eastern redcedar (Fig. 26-44)

Eastern redcedar (*Juniperus virginiana*) grows throughout the eastern half of the United States, except in Maine, Florida, and a narrow strip along the gulf coast, and at the higher elevations in the Appalachian Mountain Range. Com-

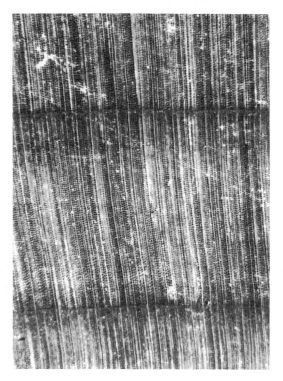

Fig. 26-42. *Juniperus occidentalis.*

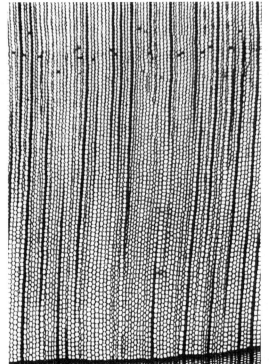

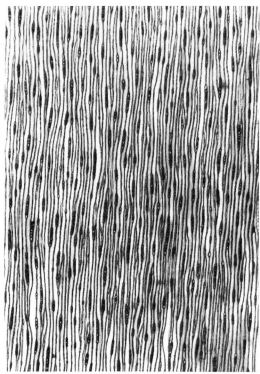

Fig. 26-43. Western juniper (*Juniperus occidentalis*) (50×).

mercial production is principally in the southern Appalachian and Cumberland Mountain regions. Another species, southern redcedar (*J. silicicola*), grows over a limited area in the South Atlantic and Gulf Coastal Plains.

Uses. The wood is moderately heavy, moderately low in strength, hard, and high in shock resistance, but low in stiffness. It has very small shrinkage and stays in place well after seasoning. The texture is fine and uniform. Grain is usually straight, except where deflected by knots, which are numerous. Eastern redcedar heartwood is very resistant to decay. The greatest quantity of eastern redcedar is used for fenceposts. Lumber is manufactured into chests, wardrobes, and closet lining. Other uses are flooring, novelties, pencils, and small boats. Southern redcedar is used for the same purposes.

Color. The heartwood of redcedar is bright to dull red, and the thin sapwood is nearly white.

Macroscopic structure. The longitudinal parenchyma are visible as several dark lines, particularly in the sapwood.

LIBOCEDRUS

Many species in this family are similar in appearance. Heartwood color and odor help to distinguish them.

Incense cedar (Figs. 26-45 and 26-46)

Incense—cedar (*Libocedrus decurrens*) grows in California and southwestern Oregon and a little in Nevada. Most incense—cedar lumber comes from the northern half of California and the remainder from southern Oregon.

Uses. Incense—cedar is light in weight, moderately low in strength, soft, low in shock resistance, and low in stiffness. It has small shrinkage and is easy to season with little checking or warping. Incense—cedar is used for lumber and fenceposts. Nearly all the high—grade lumber is used for pencils and venetian blinds. Some is used for chests and toys. Much of the incense—cedar lumber is more or less pecky; that is, it contains pockets or areas of disintegrated wood caused by advanced states of localized decay

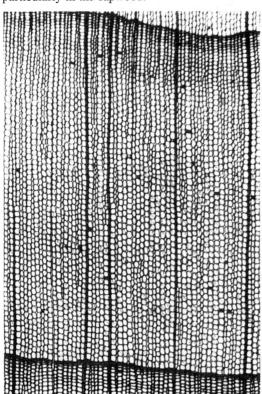 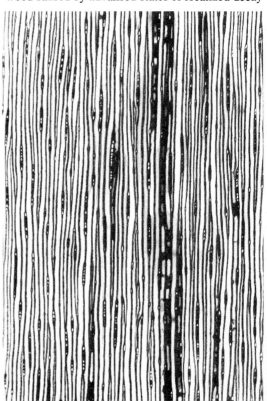

Fig. 26-44.   Eastern redcedar (*Juniperus virginiana*) (50×).

in the living tree. Peck development is arrested once the lumber is seasoned. This lumber is used locally for rough construction where cheapness and decay resistance are important. Because of its resistance to decay, incense—cedar is well suited for fenceposts. Other products are railroad ties, poles, and split shingles.

Color. The sapwood of incense—cedar is white or cream colored, and the heartwood is light brown, often tinged with red or pale purple. The wood has a fine, uniform texture, a spicy odor, and acrid taste.

Macroscopic features. The transition from earlywood to latewood is more or less abrupt, and the growth rings are prominent on flatsawn surfaces. Incense—cedar is similar to western redcedar, but the latter lacks the acrid taste. Longitudinal parenchyma are visible in the latewood as dark reddish lines or zones.

Microscopic structure. Incense—cedar lacks indentures in the ray parenchyma, which distinguishes it from other species with nodular ray cell end walls on transverse surfaces of longitudinal parenchyma.

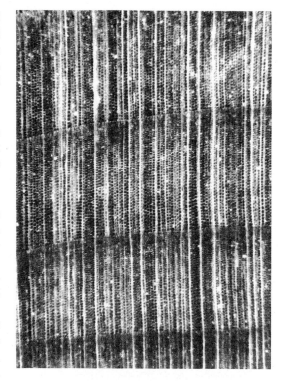

Fig. 26-45. *Libocedrus decurrens.*

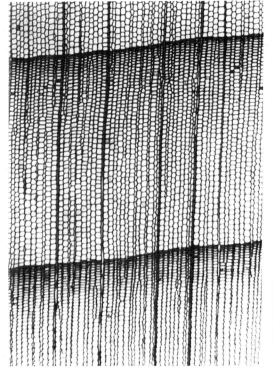

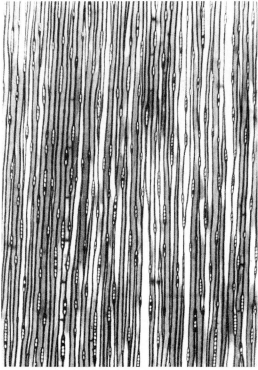

Fig. 26-46. Incense-cedar (*Libocedrus decurrens*) (50×).

THUJA

Western redcedar (Figs. 26-47 and 26-48)

Western redcedar (*Thuja plicata*) grows in the Pacific Northwest and along the Pacific coast to Alaska. Western redcedar is also called canoe cedar, giant arborvitae, shinglewood, and Pacific redcedar. Western redcedar lumber is produced principally in Washington, followed by Oregon, Idaho, and Montana.

The wood is generally straight grained and has a uniform but rather coarse texture and very small shrinkage. It is light in weight, moderately soft, low in strength when used as a beam or post, and low in shock resistance. It has a higher density in the pith that decreases outward. It has a sweet odor that some say is like chocolate. Western redcedar is used principally for shingles, lumber, poles, posts, and piles. The lumber is used for exterior siding, interior finish, greenhouse construction, ship and boat building, boxes and crates, sash, doors, and millwork.

Color. The heartwood of western redcedar is reddish or pinkish brown to dull brown and very

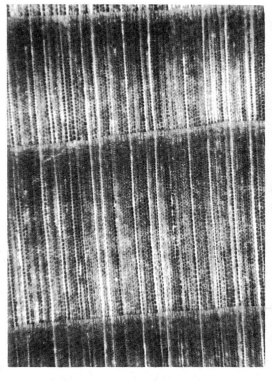

Fig. 26-47. *Thuja plicata.*

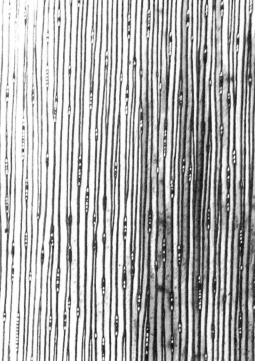

Fig. 26-48. Western redcedar (*Thuja plicata*) (50×).

resistant to decay. The sapwood is nearly white and narrow, often 1/4—1 in. wide.

Similar woods. Western redcedar closely resembles northern white—cedar but the latter is of slower growth and has a plain light brown color usually without a reddish tinge. Dark grades resemble redwood but redwood lacks the characteristic odor and taste.

Northern white—cedar

This species is very similar in properties and uses to Atlantic white—cedar; both grow in the eastern U.S. The two species are used for similar purposes, mostly for poles, ties, lumber, posts, and decorative fencing. Northern white—cedar (*Thuja occidentalis*) is also known as arborvitae, or simply cedar. It grows from Maine along the Appalachian Mountain Range and westward through the northern part of the Lake States. Production of northern white—cedar lumber is probably greatest in Maine and the lake States.

Uses. The wood is light in weight, rather soft and low in strength, and low in shock resistance. It shrinks little in drying. It is easily worked, holds paint well, and the heartwood is highly resistant to decay.

Color. The heartwood of white—cedar is light brown, and the sapwood is white or nearly so. The sapwood is usually thin, not over 1 in. except in very fast growth.

Similar woods. *C. thyoides* is similar but has a more pinkish hue, a spicy odor, and, as a rule, wider growth rings. Western redcedar generally has wider growth rings and a reddish hue. Bald cypress has a more dingy color and a rancid odor that is much different from that of cedar.

CUPRESSUS

Arizona cypress

Arizona cypress (*C. arizonica*) grows in the southwest U.S. The heartwoods merges into the sapwood and both are yellowish in color.

Many of the microscopic features resemble those of *Juniperus*, although the parenchyma end walls are not as obviously nodular and the rays are not as high.

TAXODIACEAE

TAXODIUM

Baldcypress (Figs. 26-49 and 26-50)

Baldcypress (*Taxodium distichum*) is commonly known as cypress, also as southern cypress, red cypress, yellow cypress, and white cypress. Commercially, the terms tidewater red cypress, gulf cypress, red cypress (coast type), and yellow cypress (inland type) are often used. About one—half of the cypress lumber comes from the Southern States and one—fourth from the South Atlantic States, although the supply has decreased.

Uses. The wood is moderately heavy, moderately strong, and moderately hard. The heartwood of old—growth timber is one of our most decay—resistant woods. Shrinkage is moderately small: greater than that of cedar and less than that of southern pine. Often the wood of certain cypress trees contains pockets or localized areas that have been attacked by a fungus. Such wood is known as pecky cypress. The decay caused by this fungus is arrested when the wood is cut into lumber and dried. Pecky cypress is durable and useful where watertightness is unnecessary and appearance is not important or a novel effect is desired. Cypress has been used principally for building construction, especially where resistance to decay is required. It was used for beams, posts, and other members in docks, warehouses, factories, bridges, and heavy construction.

Color. The sapwood of baldcypress is narrow (1—2 in., gradually merging into heartwood) and nearly white. The color of the heartwood varies widely, ranging from light yellowish brown to dark brownish red, brown, or chocolate. The wood is dull, lacking sheen or luster.

Macroscopic features. The darker heartwood has a more or less rancid odor; the longitudinal surfaces feel distinctly greasy or waxy. The wood lacks resin canals and has an abrupt transition from earlywood to latewood. The growth rings are distinct and irregular in outline and width. Longitudinal parenchyma are often visible with a lens on longitudinal surfaces as strands.

Similar woods. Cypress has a high degree of variability. It resembles the cedars, but the odor is entirely different. Some species of *Cupressus* spp. of the western U.S. are called cypress, but are much different from bald cypress.

SEQUOIA

Redwood (Figs. 26-51 and 26-52)

Redwood (*Sequoia sempervirens*) is a very large tree growing on the coast of northern California. Another sequoia, giant sequoia (*Sequoia gigantea*), grows in a limited area in the Sierra Nevada of California, but is used in very limited quantities. Other names for redwood are coast redwood, California redwood, and sequoia. Production of redwood lumber is limited to California, but a nationwide market exists.

Uses. Typical old—growth redwood is moderately light in weight, moderately strong and stiff, and moderately hard. The wood is easy to work, generally straight grained, and shrinks and swells comparatively little. The heartwood has high decay resistance. Redwood lumber is used for building. It is remanufactured extensively into

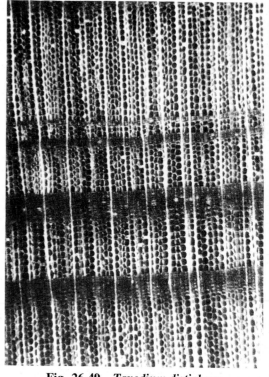

Fig. 26-49. *Taxodium distichum.*

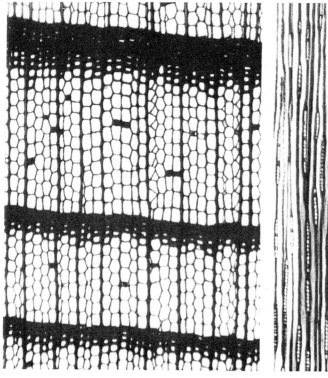

Fig. 26-50. Baldcypress (*Taxodium distichum*) (50×).

siding, sash, doors, blinds, finish, casket stock, and containers. Because of its durability, it is useful for cooling towers, tanks, silos, wood—stave pipe, and outdoor furniture. It is used in agriculture for buildings and equipment. Its use as timbers and large dimension in bridges and trestles is relatively minor. The wood splits readily and the manufacture of split products (posts and fence material) is an important business in the redwood area. Some redwood decorative veneer is used.

Color. The unique color of redwood heartwood varies from a light cherry to a dark mahogany and distinguishes it from baldcypress. The narrow sapwood (1—3 in.) is almost white.

Macroscopic features. The wood is without normal resin canals and has no distinctive odor, taste, or feel. The wood is coarse—textured. The transition to the narrow, dark latewood is abrupt. Western redcedar heartwood may approach redwood in color, but the odor of the former is distinct. The longitudinal parenchyma are visible on longitudinal surfaces with a lens as strands.

Microscopic features. Bi— or triseriate

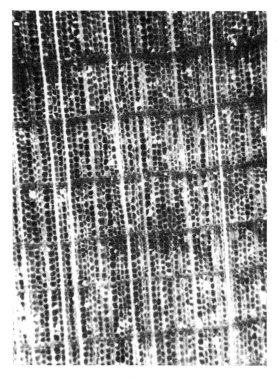

Fig. 26-51. *Sequoia sempervirens.*

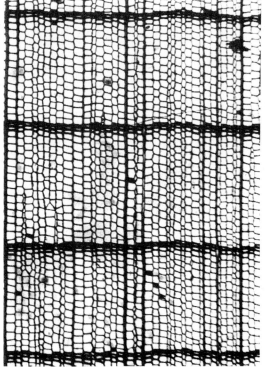

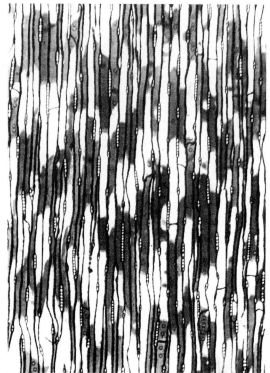

Fig. 26-52. Coast redwood (*Sequoia sempervirens*) (50×).

pitting of tracheids, ray cell smooth end walls, and taxodioid pitting on the cross—fields are present.

TAXACEAE

TAXUS

Yew (Figs. 26-53 and 26-54)

The yew is a genus of minor wood importance in the U.S. Florida yew (*T. floridana*) is rare and confined to a few counties in northwestern Florida. Pacific yew (*T. brevifolia*) is native to the U.S. A compound called taxol, which is extracted from the bark and needles, has been used in the last few years to treat certain types of cancer. Taxol will be synthesized in the laboratory and the yew tree will not be harvested for this drug in the future. Pacific yew grows from southeast Alaska to northern California

Microscopic features. The spiral thickening of yew tracheids and the lack of resin canals are shared with the rare *Torreya*, but yew has cross—field pits that are 8—10 μm, larger than those of *Torreya* (6 μm). The spiral thickenings of *Torreya* are less numerous and steep than those of yew.

Fig. 26-53. *Taxus brevifoia.*

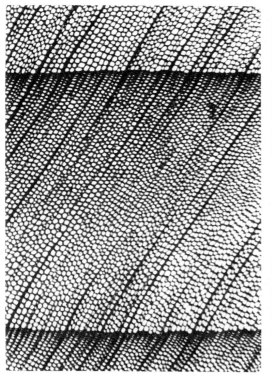

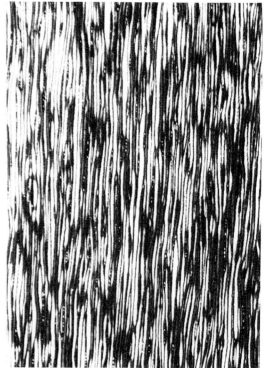

Fig. 26-54. Pacific yew (*Taxus brevifolia*) (50×).

26.4 NON—AMERICAN SPECIES

Information on non—North American species is available in Kukachka (1970).

PINACEAE

PINUS

Pine, Radiata

Radiata or Monterey pine (*P. radiata*) is native to a small area of California but is planted extensively in the southern hemisphere, mainly in Chile, New Zealand, Australia, and South Africa. Plantation—grown trees may reach a height of 80 to 90 feet in 20 years. The species is not commercially significant in its native U.S. in part due to its large number of knots.

Plantation—grown heartwood is variable in color among trees and is light brown to pinkish brown. It is distinct from the paler creamy sapwood in older trees (more than 40 years old). Growth rings are mostly wide and distinct. False growth rings may be common. Texture is moderately even and fine, and the grain is usually straight.

Plantation—grown radiata pine averages about 30 lb/ft^3 at 12% MC_{OD} (specific gravity of 0.48). The strength is comparable to red pine (*P. resinosa*) although there may be considerable variation due to location and growth rates. The sapwood is prone to attack by stain fungi and vulnerable to boring insects. Radiata pine can be used for the same purposes as other pines of the United States.

Bamber and Burley (1983) review the properties of this species. It had been introduced to Australia in the 1850s. In 1979 there were 430,000 hectares in Australia, 800,000 in New Zealand, and 600,000 in Chile. It is mostly three—needled, with markedly dentate walls in the ray tracheids and 1—4 pinoid cross—field pits of small size. The uniseriate rays have marginal tracheids. The fusiform rays may be 3 to 5—seriate and 15—20 cells high. The earlywood longitudinal tracheids average 3.0 mm long, 0.045 mm in diameter, and 0.003 mm in thickness. The latewood longitudinal tracheids average 3.5 mm long, 0.013 mm in diameter, and 0.005 mm in thickness. The cellulose content in mature wood is over 50% and the lignin content is about 26%.

Caribbean pine

Caribbean pine (*P. caribaea*) occurs along the Caribbean side of Central America from Belize to northeastern Nicaragua. It is also native to the Bahamas and Cuba. It is also called slash pine and British Honduras pitch pine. This tree of the low elevations is widely introduced as a plantation species throughout the world tropics.

The heartwood is golden brown to red brown and distinct from the sapwood, which is 1 to 2 inches in thickness and a light yellow. The softwood species has a strong resinous odor and a greasy feel. The weight varies considerably and may vary from 24 to 51 lb/ft^3 at 12% MC_{OD}, although the higher density is much more common.

Caribbean pine may be appreciably heavier than slash pine (*P. elliotti*) but the mechanical properties of these two species are rather similar. The durability and resistance to insect attack vary with resin content but generally the heartwood is rated moderately durable. The sapwood is highly permeable. Caribbean pine is used for the same purposes as the southern pines of the U.S.

Scots pine

Scots pine (*P. sylvestris*), also called red deal, yellow deal, or redwood, grows in Europe.

PICEA

Norway spruce

Norway spruce (*P. abies*) is also called white deal or whitewood and grows in Europe.

LARIX

European larch

European larch (*L. decidua*) is also called common larch and grows in Europe.

CUPRESSACEAE

CYPRESSUS

Bentham cypress

Bentham cypress (*C. lusitanica*) grows in Mexico and Guatemala. The heartwood is distinct from the sapwood and light reddish brown. The wood is aromatic.

The end walls of the ray parenchyma are not nodular and indenture is lacking or not obvious.

FITZROYA

Alerce

Alerce (*F. cupressoides*) is native to southern Chile. The heartwood is reddish brown and sharply defined from the narrow and very light—colored sapwood. The density is 26 lb/ft³. Alerce is much like redwood in most regards and used as redwood is used, but redwood rays are two or three times taller. Alerce has indenture and nodular end walls in ray cells. The rays are mostly under 12 cells high.

ARAUCARIACEAE

All three of the species listed in this section lack growth rings and have vertical tracheids with 2—4 bordered pits across their widths in alternate arrangement that are distinctive for this family.

AGATIIIS

Almaciga, sakar

Almaciga or sakar (*Agathis philippinesis*) is from the Philippine Islands. *Agathis* lacks the leaf scars that cause the pin knots. Klinki pine lacks the reddish streaks.

ARAUCARIA

A unique characteristic of *Araucaria* and *Agathis* amongst all other conifers is the 2—4—seriate, alternate intertracheid pitting.

Resinous tracheids (near parenchyma) occur in the heartwood of *Agathis* and *Araucaria* of Araucariaceae. They are filled with reddish brown to black amorphous deposits.

Parana—pine

Parana—pine (*A. angustifolia*) is not a true pine. It is a softwood that comes from southeastern Brazil and adjacent areas of Paraguay and Argentina. It is available in large clear boards of uniform texture. The small pinhead knots that appear on flat—sawn surfaces (caused by leaf scars)

and the light brown or reddish brown heartwood, which is frequently streaked with red, provide desirable figured effects for matching in paneling and interior finishes. The growth rings are fairly distinct and more nearly like those of white pine (*P. strobus*) rather than those of the yellow pines. The wood has relatively straight grain, takes paint well, glues easily, and is free from resin ducts, pitch pockets, and streaks.

The sapwood is yellowish and not always distinct from the heartwood, which is a very light brown and frequently shows bright streaks of red.

The strength is similar to that of other U.S. softwoods of similar density (loblolly and shortleaf pine), except for its deficiency in compression strength across the grain. Splitting and warping of the kiln—dried wood is due to the presence of compression wood. The principal uses include framing lumber, interior trim, furniture, case goods, and veneer.

Klinki—pine

Klinki pine of hoop—pine (*Araucaria klinkii*) is from Australia, New Guinea, New Caledonia, and Norfolk Island. It typically grows to 150 ft with a trunk diameter of 3 ft. The heartwood is light yellowish brown, not sharply demarcated from the straw—colored sapwood. The basic specific gravity is 0.42.

PODOCARPACEAE

PODOCARPUS

Podocarp

A variety of podocarp species (*P. spp*) have limited use as pulpwood. Common names include podo, yellow wood, and manio. The anatomy of the species varies greatly. Some have resin canals, others do not. The cross—field pitting can be of almost any type, except pinoid. *P. amarus* grows in New Zealand, South Africa, and Chile and looks somewhat like eastern white pine.

References for this chapter are found on pages 537-540 of Chapter 25.

HARDWOOD ANATOMY

27.1 GROSS ANATOMY OF HARDWOODS

Introduction

The references for this chapter are found on pages 537-540 of Chapter 25.

Hardwoods are identified by the presence of vessels (pores), although a few (commercially unimportant) species lack them. *Vessels* are the long tubes (which may extend the length of the tree) for conduction of water made up of individual *vessel elements* (or members) attached end to end. Vessels may be barely visible with a lens in some species and quite obvious in others.

Vessel size distribution (seen in cross section, *x*)

Ring—porous woods (oaks, ash, elm, hickory, hackberry) have large—diameter vessels in the earlywood that soon decrease in diameter toward the latewood. *Semi—ring—porous* woods have a gradual decrease in vessel diameter through the width of the growth ring, with the earlywood vessels of moderately larger diameter than the latewood. *Diffuse—porous* woods have vessels that do not change much in diameter throughout the growth ring (maple, yellow—poplar, sweetgum). Obviously the distinction is somewhat arbitrary, but Fig. 27-1 shows representative

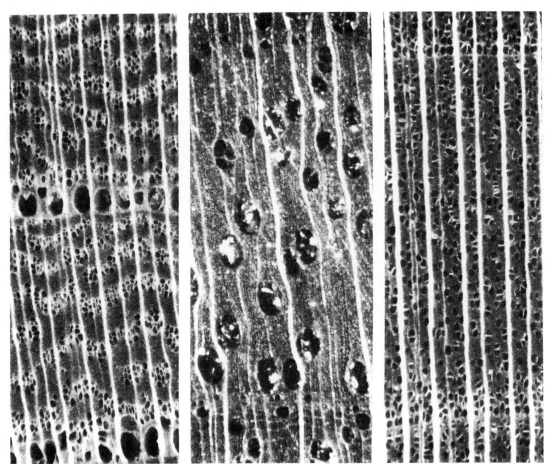

Fig. 27-1. Ring—porous (*Ulmus americana*), semi—ring—porous (*Juglans nigra*), and diffuse—porous (*Prunus serotina*) woods. Tyloses fill many of the vessels of the first two examples.

samples. The diameter of vessels increases with increasing growth ring width and distance from the pith. Tropical woods tend to be diffuse porous.

The vessel diameters can be very small, below 30 μm; small, 30—100 μm; medium, 100—300 μm; and large, greater than 300 μm.

Vessel position arrangement, grouping (Fig. 27-2)

In some species, the vessels are isolated and independent of each other (*solitary* as in some maples). The vessels of other species (maple, cottonwood, and alder) form radial rows or multiples of 2 to 4 or more (*pore multiples*). Some species form radial chains with numerous pores (*pore chains* as in holly). Vessels may also form oblique (diagonal) lines, while still independent of each other. *Ulmiform* latewood vessels occur in wavy, concentric, tangential (parallel to the growth ring) bands in the latewood. Vessels may also cluster (*nested* vessels as in Kentucky coffeetree and black locust) and these clusters may have a characteristic arrangement.

Tyloses

The vessel elements of some species become occluded with tyloses originating from outgrowths of adjacent (often ray, but sometimes longitudinal) parenchyma (Fig. 27-1). This is common during heartwood formation, except in red maple and red oaks. White oaks have large amounts of tyloses, which greatly decreases their permeability.

A classic work (Chattaway, 1932) hypothesizes that if pits between vessel and adjacent parenchyma are 8—10 μm or larger, tyloses are able to form; otherwise occlusion may be by gum formation as in black cherry or maple.

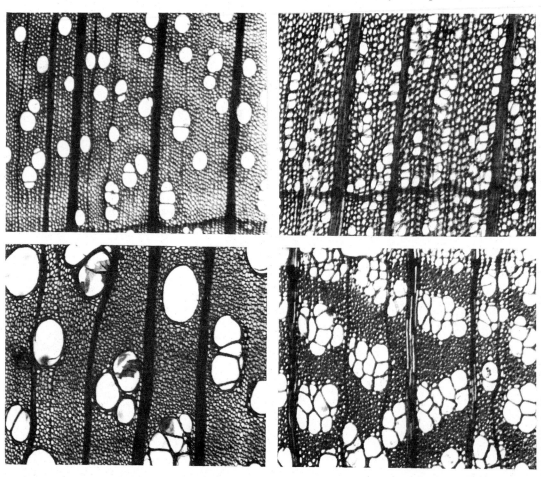

Fig. 27-2. Solitary and pore multiples in *Acer saccharum* (top left), pore chains in *Ilex opaca* (top right), nested pores in *Nyssa silvatica* (bottom left), and ulmiform pores in *Ulmus americana* (60×).

Parenchyma cell arrangement (x)

Parenchyma cells are arranged longitudinally (*axial*) or radially (*ray*). Axial parenchyma may be rare or absent, but are generally present. The longitudinal arrangement has been described by Jane (1956, p. 115) based on work by Kribs in 1950. Parenchyma cells are often visible as white bands in hardwoods. *Marginal* (boundary) parenchyma are found in bands near the boundary of the growth ring Plate 45-a. They may be *terminal* if occurring in the latewood or *initial* if occurring in the earlywood. *Epithelial* parenchyma surround longitudinal gum canals in traumatic gum canals (for example, in wounded sweetgum).

Apotracheal parenchyma are arranged independently of the vessels. They may be *diffuse* or *banded* (forming groups of 2—4 or more cells wide parallel to the growth rings). Lines of banded parenchyma may form a net—like appearance with rays known as *reticulate parenchyma* as in hickory and persimmon (Plate 45-b).

If the parenchyma cells are arranged around the vessels they are called *paratracheal*. If they partially surround the vessels they are *scanty*; if they form a partial band around the vessel all on one side they are *unilateral*. They may totally encircle the vessel (*vasicentric*, Plate 45-c) and have lateral wings (*aliform*) or long lateral wings that merge with each other (*confluent*).

Ray appearance

Oaks, maples, and beeches contain rays that are readily visible in all three views. *Aggregate rays* consist of radial and longitudinal cells that are intermixed and occur in addition to smaller rays (red alder and American hornbeam). *Storied rays* are a series of rays that each have the same height and occur at the same height (in the tangential view, *t*) (American mahogany). Some species have rays that are not observed with a hand lens.

The ray width relative to vessel diameter, the number of rays over a given area, and the percentage of cross—sectional area covered by rays are three items that may be important. Some rays expand at the growth ring boundary (x) in *noded rays* [yellow poplar (Plate 45-a), beech, and sycamore]. Large rays show as *ray flecks* (r).

Color

The heartwood of many hardwood species is characteristic. Black walnut has a dark brown heartwood; the color of black cherry varies from light to deep brown with high luster; ebony is almost black; yellow poplar heartwood has a greenish cast with patches or streaks from red to dark brown; the small black heartwood of persimmon resembles ebony, to which it is related; red gum has a dingy, reddish brown color; buckeye has a uniform creamy yellow color. Osage orange is very similar to black locust, but the former gives off a water—soluble yellowish dye when extracted with water. A reference collection of known species is useful for color comparisons.

Odor

Teak, sassafras, and Oregon myrtle have distinctive odors. Basswood (said to be like raw potatoes) and catalpa (said to be like kerosene) have odors that can be helpful in their identification especially in freshly cut, green wood.

Fluorescence

Hoadley (1990) describes the fluorescence of wood using longwave UV lamp sources. Among U.S. woods, this techniques is useful with members of the Leguminosae family including black locust (to separate it from osage—orange), honeylocust, Kentucky coffeetree, and acacia. Wood from *Rhus* and *Ilex* genera also fluoresce.

Resin canals

Longitudinal resin canals may occur in hardwoods, but they are rare in U.S. species. The *schizogenous* type (Fig. 27-3) occur by separation of cells and are lined with epithelium cells (as in traumatic resin canals of softwoods); they often occur in a single tangential row. The *lysigenous*

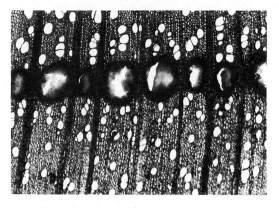

Fig. 27-3. Traumatic resin canals from *Prunus serotina* filled with a gummy mass.

type occur by lysis (dissolution) of cells and are not lined with epithelium cells.

27.2 MICROSCOPIC ANATOMY OF HARDWOODS

Spiral thickening of vessels

Some vessel elements (Fig. 27-4) have spiral thickening (*Tilia americana*, sweetbay, maples). This is limited to the elongated tips (*ligules*) of the vessel elements in sweetgum and black tupelo and to latewood vessels and vascular tracheids in elm and hackberry. It is rare in tropical species.

Vessel element perforation plates (radial view, *r*)

The perforation plates of vessel elements occur where two vessel elements join end to end. *Simple* perforation plates have a single opening (without bars) as in the maples. *Sclariform* perforation plates have a series of bars across the open-

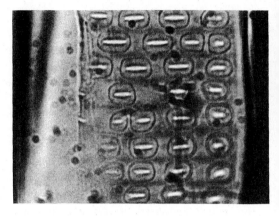

Fig. 27-5. Opposite intervessel pitting of *Liriodendron tulipifera*, 600×.

ing. The number of bars is an important identifying attribute and varies from a few in yellow—poplar to many in sweetgum (Fig. 27-4).

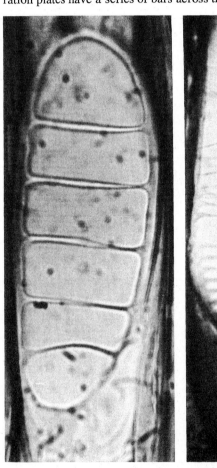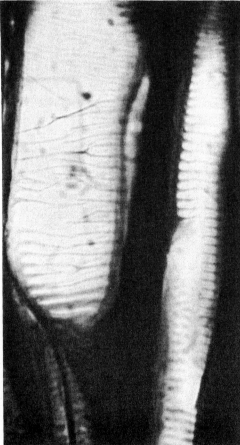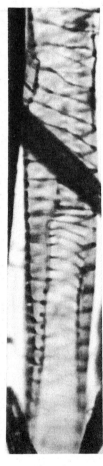

Fig. 27-4. Sclariform perforation plates of vessels in *Liriodendron tulipifera* (left), *Nyssa salvatica* (center, two examples), and *Ilex opaca* (right, at bottom with spiral thichenings as well) (600×).

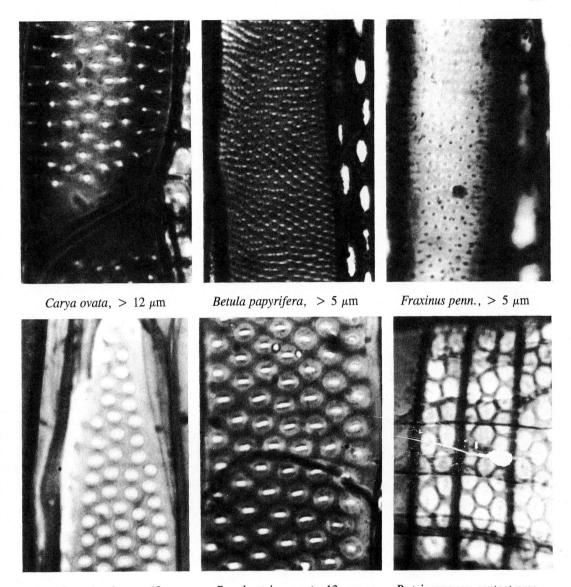

Carya ovata, > 12 μm   *Betula papyrifera,* > 5 μm   *Fraxinus penn.,* > 5 μm

Aesculus octandra, > 12 μm   *Populus tricarpa,* > 12 μm   *P. tricarpa* ray contact area

Fig. 27-6. Examples of alternate, intervessel pitting, 600×.

Intervessel pitting

The pitting between side—by—side vessels (Figs. 27-5 to 27-7) is usually observed on the tangential face, but it may be more visible on the radial face when near the ends of vessels when vessels are isolated from each other.

Intervessel pitting may be opposite (Fig. 27-5, forming horizontal rows) or, more commonly, alternate (forming diagonal rows, Fig. 27-6). The pits of the various species differ in size and shape. This is a key feature for the identification of hardwood fibers. Pits that are long across the vessel diameter and narrow in the longitudinal axis are called *linear* or *sclariform* (Fig. 27-7). The pit outline is also of relevance (angular, rounded). Intervessel pitting is rare in oaks and sweetgum.

The pit pairs from vessels to ray parenchyma may be bordered, half—bordered, or simple (Brown and Panshin, 1940). The ray to vessel pitting is compared to intervessel pitting (similar or dissimilar; rounded, oval, elongated); see ray to vessel pitting, below.

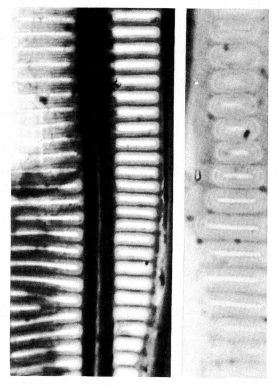

Fig. 27-7. Sclariform intervessel pitting of *Magnolia acuminata* (left) and less pronounced in *Liquidambar styraciflua* (right), 600×.

Woods in the Leguminosae family and many tropical species have *vestured* intervessel pitting. This is characterized by outgrowths in the pit chamber to give it a jagged appearance, though it is difficult to see by light microscopy.

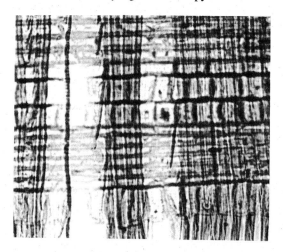

Fig. 27-8. Heterocellular ray from *Nyssa sylvatica* (150×).

Ray appearance

Hardwoods do not have ray tracheids. *Homocellular* rays consist of *procumbent* cells only (cells with the long axis parallel to the radial direction). *Heterocellular* rays have two types of cells: cells with the long axis arranged longitudinally (*upright*) and procumbent cells (Fig. 27-8). Upright cells are often restricted to the margins (top and bottom) of the rays.

Uniseriate rays are one cell wide and may range from 1–3 cells to over 12 cells in height. Rays may also be multiseriate and very wide in oaks and maples. *Aggregate* (occasionally called false) rays are closely spaced rays that form a large structure that is visible with the hand lens.

Some procumbent cells of a ray may be subdivided into many small erect empty cells (*tile cells*), which occur in some families of the Malvales order. Some hardwoods (Sassafras and Oregon Myrtle of the Lauraceae family) contain enlarged upright ray cells known as *oil* cells.

Ray to vessel pitting

Ray to vessel pitting can be similar to or different from intervessel pitting. Fig. 27-6 shows an example where the ray to vessel pitting is different from the intervessel pitting. If it is similar, its appearance is known by the intervessel pitting. If it is different is may be essentially simple and elongated or it may be simple to bordered. It is unique in *Populus* and *Salix*.

Fiber tracheids and libriform fibers

There are two types of fibers that may intergrade to a degree (Fig. 27-9). Fiber tracheids have bordered pits and tend to have relatively large lumens. Libriform fibers have simple pits and tend to have relatively small diameter lumens. Libriform fibers can be stiff due to the thick cell wall. These differences may be difficult to see by light microscopy.

Tracheids

Two relatively rare (in U.S. species) types of tracheids occur. These are not required for identification of large wood samples, but may be useful in the identification of slivers and pulp fibers. *Vascular* tracheids of hardwoods may appear to be like the latewood vessel elements, but have pitting at the end walls instead of perforation plates (elm, hackberry).

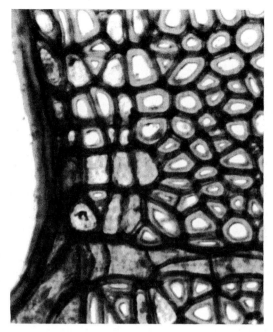 

Fig. 27-9. Libriform fibers of *Fraxinus* (left) and fiber tracheids of *Betula papyrifera* (600×).

Vasicentric tracheids (oaks) are shorter than libriform fibers but of larger diameter, have an irregular outline, distinctive pitting, and *crassulae*, thickenings of the cell wall. They occur in horizontal clusters in association with vessel elements, especially in the earlywood. They are found in the oaks, tanoak, and eucalypts.

Future research

An interesting value for hardwoods would be the parenchyma value by mass (or at least the useful fiber content). This could be obtained by an acid chlorite treatment of wood shavings followed by fiber fractionation (a 200 μm screen is said to isolate fines and parenchyma cells). Each fraction should be examined microscopically to determine the efficiency of the separation. This would give an indication of the useful amount of fiber in hardwood.

Summary

A summary of features for many hardwoods is given in Table 27-1.

27.3 ANATOMY OF HARDWOOD SPECIES

Hardwoods are listed alphabetically in this section (usually ignoring geographical, color, and

other prefixes) because of the numerous families, genera, and variety that occurs. For a listing by family and order see Table 25-4.

The *uses* section includes the properties related to the uses of wood. The mechanical properties are of relevance to wood chipping. The *color* section includes the overall appearance. Many sapwoods are stained with blue stain, a type of fungi, that modifies the natural color of the sapwood.

Macroscopic features refer to those that can be seen with the unaided eye or with a 10× lens. The words pores and vessels are used synonymously throughout. The presence of tyloses often tends to make the vessels difficult to see on the cross section with the unaided eye, so the appearance and the actual sizes of vessels are not the same. Parenchyma descriptions in this section refer to longitudinal parenchyma. Most of the macroscopic cross sections are courtesy of R.L. Krahmer and A.C. VanVliet; they are all the same magnification, about 15—20×.

While the taste of some woods is mentioned, it is not a good idea to apply this for routine sample identification of large numbers of samples.

Table 27-1. Anatomical characteristics of deciduous woods useful in their identification.

Use this for alignment	1	2	3	4	5	6	7	8	9	0	1	2	3	4	5	6	7	8	9	0	1	2	3	4	5	6
Page	592	592	594	595	597	598	599	602	601	602	605	605	605	605	606	608	609	609	609		609			610		611
Scientific name	Ailanthus altissima	Alnus rubra	Fraxinus spp.	Tilia americana	Fagus grandifolia	Betula papyrifera	Aesculus glabra	Juglans cinerea	Catalpa spp.	Prunus serotina	Castanea dentata	Magnolia acuminata	Cornus florida etc.	Ulmus americana	Ulmus rubra	U. thomasii	Celtis occidentalis	Carya spp.	C. spp.	Ilex opaca	Gleditsia triacanthos	Ostrya viginiana	Carpinus caroliniana	Robinia pseudoacacia	Arbutus menziesii	Magnolia grandiflora
Common name	Ailanthus	Alder, red	Ash	Basswood, American	Beech, American	Birch, paper etc.	Buckeye, Ohio etc.	Butternut	Catalpa	Cherry, black	Chestnut, American	Cucumber tree	Dogwood	Elm, American	Elm, red (slippery)	Elm, rock (hard)	Hackberry, etc.	Hickory, pecan	Hickory, true	Holly, American	Honeylocust	Hophornbeam	Hornbeam, American	Locust, black	Madrone, Pacific	Magnolia, south. etc.
Vessels																										
Ring porous	X		X								X			X	X	X	X	x	X		X			X		
Semi-ring porous								X	X	X												X	X			
Diffuse porous		X		X	X	X	X					X	X							X					X	X
Spiral thickening				A				S		L		A		L	L	L	L			F	L	A	A	L	F	A
Arrangement-mostly solitary	-	S	S	X						S	X														S	S
-of clusters	-	M	M	M	M	M	M	M	N	M	M	M	M	W	W	W	W	M	M	C	N	M	M	N	M	M
Perforation plates-simple	-	X	X					X	X	X	X			X	X	X	X	X	X		X	X	X	X		
-scalariform (T some >10 bars)	-			T	S	S	T					S	S							T					X	X
Intervessel pitting																										
Opposite	-	S											X													
Alternate	-	X	X	X	X	X	X	X	X	X	X	X		X	X	X	X	X	X	X	X	X	X	X	X	
Scalariform	-												X													X
Linear	-												X													X
Fibers																										
Libriform	-	X			X		X		X					X	X	X	X	X	X		X	X		X		
Fiber tracheids	-	X		X	X	X	S	X			X	X	X	X						X	X	X		X	X	X
Ray composition																										
heterocellular					?		?	S	S	S		X	X				X	X	X	X		S	X	S	X	X
homocellular	X	X	X	X	X	X	X	X	X	X	X	X		X	X	X					X	X		X		X
Ray-vessel pitting (1, 2, 3)	3	1	1		3	1	2	1	1		3	3	1	3	3	3	2	1	1	1	1	3	3	3	3	3
Tracheids present	-		C										C	R	R	R										
Use this for alignment	1	2	3	4	5	6	7	8	9	0	1	2	3	4	5	6	7	8	9	0	1	2	3	4	5	6

Key: X indicates characteristic is pronounced. x indicates characteristic is present but not pronounced (when used with X, it is the less likely of the two choices). S indicates a feature is sporadic, ? indicates sources disagree. Spiral thickening [Core et al, 1979, P. 78] S) sporadic, A) all vessels, L) latewood vessels only, F) fibers and vessels, T) vessel tips only. Vessel arrangement of clusters M) multiples, C) chains, W) wavy bands, N) nested. Ray vessel pitting [Core et al, 1979, P. 68] 1) essentially like intervessel pitting, 2) essentially simple, 3) simple to bordered. Tracheids present C) vascicentric, R) vascular. Source: Panshin and de Zeeuw (1980); Hoadley (1990); and others.

Table 27-1. Anatomical characteristics of deciduous woods useful in their identification, cont'd.

Common name	Maple, hard	sugar (and black)	Maple, soft red	silver	Mulberry, red	Oak, red group	Oak, white group	Oak, live	Oregon-myrtle	Osage orange	Persimmon	Poplar family:	aspens	cottonwoods	Sassafras	Sugarberry	Sweetgum	Sycamore, American	Tanoak	Tupelo	Walnut, black	Willow, black	Yellow-poplar	Non U.S. species	Eucalypts	Gmelina		
Scientific name		*Acer saccharum*	*A. rubrum*	*A. saccharinum*	*Morus rubra*	*Q. rubra* etc.	*Q. alba* etc.	*Q. viginiana*	*Umbellularia californica*	*Maclura pomifera*	*Diospyros virginiana*		*Populus spp.*	*P. spp.*	*Sassafras albidum*	*Celtis laevigata*	*Liquidambar styraciflua*	*Platanus occidentalis*	*Lithocarpus densiflorus*	*Nyssa spp.*	*Juglans nigra*	*Salix nigra*	*Liriodendron tulipifera*		*Eucalyptus* spp.	*Gmelina arborea*		
Page		612	612	612	614	615	616	616		617	618		594	603	618		619	619	622	623	624	625	625		627	627		
Use this for alignment	1	2	3	4	5	6	7	8	9	0	1	2	3	4	5	6	7	8	9	0	1	2	3	4	5	6	7	8
Vessels																												
Ring porous					X	X	X			X					X	X					X							
Semi-ring porous								X			X		X	X					X		X	X						
Diffuse porous		X	X	X					X				X	X				X		X		X	X					
Spiral thickening		A	A	A	L					L						L	T			T								
Arrangement-mostly solitary		X	X	X					S						S			X	X	S			S		-	-		
-of clusters		M	M	M	N	M	M	M	M	N	M		M	M	M	W	M	M		M	M	M	M		-	-		
Perforation plates-simple		X	X	X	X	X	X	X	S	X	X		X	X	X	X		X	X		X	X			X	X		
-scalariform (T >10 bars)									X						S		T	S		T			T					
Intervessel pitting																												
Opposite																	X			X			X					
Alternate	X	X	X	X	X	X	X	X	X	X	X		X	X	X	X		X	X		X	X			-	-		
Scalariform																	S						S					
Linear																	X			S			S		-	-		
Fibers																												
Libriform		X	X	X	X	X	X	X	X	X	X		X	X	X	X			X			X	X		-	-		
Fiber tracheids		X	X	X		X	S	X							X		X	X	X	X	X		X		-	-		
Ray composition																												
heterocellular					X				X	X	?				X	X	X			X	?	?	X		X	X		
homocellular		X	X	X	X	X	X	X			X		X	X	X		X	X	X		X		X					
Ray-vessel pitting (1, 2, 3)		1	1	1	3	3	3	3	2	2	1		3	3	3	2	3	1		1	1	2	3					
Tracheids present						C	C	C								R			C						C			
Use this for alignment	1	2	3	4	5	6	7	8	9	0	1	2	3	4	5	6	7	8	9	0	1	2	3	4	5	6	7	8

Key: X indicates characteristic is pronounced. x indicates characteristic is present but not pronounced (when used with X, it is the less likely of the two choices). S indicates a feature is sporadic, ? indicates sources disagree. Spiral thickening [Core et al, 1979, P. 78] S) sporadic, A) all vessels, L) latewood vessels only, F) fibers and vessels, T) vessel tips only. Vessel arrangement of clusters M) multiples, C) chains, W) wavy bands, N) nested. Ray vessel pitting [Core et al, 1979, P. 68] 1) essentially like intervessel pitting, 2) essentially simple, 3) simple to bordered. Tracheids present C) vascicentric, R) vascular. Source: Panshin and de Zeeuw (1980); Hoadley (1990); and others.

Red alder (Fig. 27-10)

Red alder (*Alnus rubra*) grows along the Pacific coast between Alaska and California. It is used commercially along the coasts of Oregon and Washington and is the most abundant commercial hardwood species in these States. European black alder (*A. glutinosa*) is native to Europe, North Africa, and Asia; it now grows in the northeast to northcentral U.S. into Canada. Since alder is a legume, the wood has high levels of nitrogen and is a rapidly decaying wood. Vessel elements of aged alder contain an orange—red pigment that may survive pulping and bleaching [Karchesy et. al., *J. Chem. Soc. Chem. Comm.* 649-650(1974)].

Uses. The main use is for furniture, but it is also used for sash, doors, panel stock, and mill-work. It has appreciable use in pulp and paper.

Color. The wood of red alder varies from almost white to pale pinkish brown and has no visible boundary between heartwood and sapwood. It is moderately light in weight, intermediate in most strength properties, but low in shock resistance. Red alder has relatively low shrinkage.

Macrostructure. The wood is diffuse porous,

and the growth rings are delineated by a whitish or brownish line. The latewood vessels are numerous, small, and occur in small clusters or multiples. Longitudinal parenchyma are not visible. The wide, aggregate rays are distinctive.

Similar woods. Red alder is sometimes confused with birch, but birch is harder and does not have the wide, aggregate rays.

Ash (Figs. 27-11 and 27-12)

Important species include white ash (*Fraxinus americana*), green ash (*F. pennsylvanica*), blue ash (*F. quadrangulata*), black ash (*F. nigra*), pumpkin ash (*F. profunda*), and Oregon ash (*F. latifolia*). Another 12 species account for only a few percent of the commercial cut. The first five of these species grow in the eastern half of the U.S. and are grouped together commercially. Oregon ash grows along the Pacific coast. Commercial white ash is a group of species that consists mostly of white ash and green ash, although blue ash is also included.

Uses. Second—growth white ash is sought since it is especially heavy, strong, hard, stiff, and

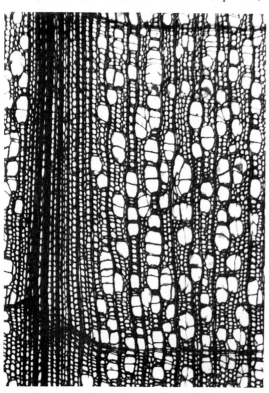
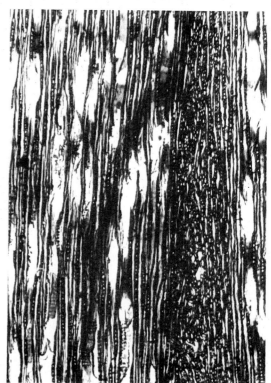

Fig. 27-10. Red alder (*Alnus rubra*) (50×).

has high resistance to shock; it is used principally for handles, oars, vehicle parts, baseball bats, and other sporting and athletic goods. The best handle specifications require 5 to 17 growth rings per inch and a weight of at least 43 lb/ft³ at 12% MC.

Ash wood of lighter weight, including black ash, is sold as cabinet ash and is suitable for cooperage, furniture, and shipping containers. Some ash is cut into veneer for furniture, paneling, and wire—bound boxes.

Color. The heartwood is grayish brown, sometimes with a reddish tinge; the sapwood is light colored or nearly white and several inches wide. Second—growth trees have a large proportion of sapwood. The annual rings are very distinct and mostly of moderate width.

Macroscopic structure. Green ash and white ash are ring porous woods that are difficult to distinguish anatomically. The earlywood vessels form a ring that is 2 or 3 (4 in wide rings) pores wide; they usually contain tyloses, except in the outer sapwood. Latewood vessels are very small and appear as white specks that are not numerous and are scattered singly or in groups of two. The

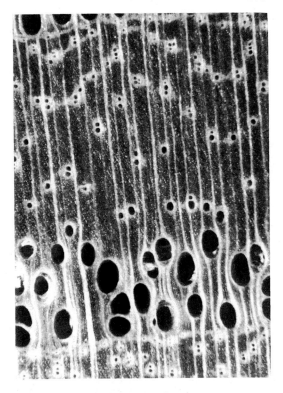

Fig. 27-11. *Fraxinus* **spp.**

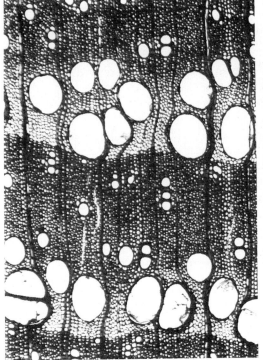

Fig. 27-12. Green ash (*Fraxinus pennsylvanica* Marsh.) (50×).

latewood pores are arranged in short, radial multiples that are sparsely distributed.

The rays are small, normally spaced, and uniform in size; they are not distinct without a lens. The longitudinal parenchyma are paratracheal, being associated with the latewood vessels; they form wavy, tangential lines visible without a lens, especially in the outer latewood of wide growth rings. The tangential bands of parenchyma are wider in green ash and more narrow in black ash than in white ash. White ash has more luster, darker heartwood, and usually has wider growth rings than brown (black) ash, which may be very slow growing. The latewood pores of white ash are more prominent than those of black ash.

Microscopic structure. Intervessel pitting of white ash is alternate with rounded pits in outline.

Similar woods. Hickory is distinguished from ash by the presence of abundant apotracheal parenchyma in the latewood, while white ash has paratracheal and paratracheal confluent parenchyma. The earlywood pores of hickory are more distinctive than in ash. In chestnut, the vessels of the latewood are numerous and in radial bands. In catalpa and sassafras the latewood vessels are numerous but in tangential bands, while in the ashes they are few and scattered and surrounded by parenchyma so as to appear in tangential bands. Sassafras is readily distinguished by its spicy odor.

Aspen

Aspen is applied to bigtooth aspen (*Populus grandidentata*) and to quaking aspen (*P. tremuloides*), but not balsam poplar (*P. balsamifera*) and the species of *Populus* that make up the group of cottonwoods (see Figs. 27-28 and 27-29). The lumber of aspens and cottonwood may be mixed in trade and sold either as poplar ("Popple") or cottonwood. The name poplar should not be confused with yellow—poplar (*Liriodendron tulipifera*), also known in the trade as poplar. Aspen lumber is produced mainly in the Northeastern and Lake States. There is some production in the Rocky Mountain States.

Uses. The wood of aspen is lightweight and soft. It is low in strength, moderately stiff, moderately low in resistance to shock, and has a moderately high shrinkage. Aspen is cut for lumber, pallets and crates, pulpwood, particleboard, matches, veneer, and miscellaneous turned articles.

Color. The heartwood of aspen is grayish white to light grayish brown. The sapwood is lighter colored and generally merges gradually into heartwood without being clearly marked. Aspen wood is usually straight grained with a fine, uniform texture. It is easily worked.

Macroscopic structure. Aspen is semi—ring-porous to diffuse—porous. The pores are small and usually not visible to the unaided eye. Growth rings are usually faint. Rays are small, uniform in height, and visible only on radial surfaces.

Basswood (Figs. 27-13 and 27-14)

American basswood (*Tilia americana*), is the most important of the several native basswood species; next in importance is white basswood (*T. heterophylla*). Other species occur only in very small quantities. No attempt is made to distinguish them in lumber form. Other common names of basswood are linden, linn, and beetree.

Basswood grows in the eastern half of the United States from the Canadian provinces southward. Most basswood lumber comes from the Lake, Middle Atlantic, and Central States. In commercial usage, "white basswood" is used to specify white wood or sapwood of either species.

Uses. When dry, the wood is without odor or taste. It is soft and light in weight, has fine, even texture, and is straight grained and easy to work with tools. Shrinkage in width and thickness during drying is rated as large but basswood seldom warps in use. Basswood is used for crates, lumber for furniture, millwork, and core material in plywood.

Color. The heartwood of basswood is pale yellowish brown with occasional darker streaks. Basswood has wide (several inches), creamywhite or pale brown sapwood that merges gradually into the heartwood. The annual rings are fairly distinct and defined by a change in vessel size and a white line of marginal parenchyma.

Macroscopic structure. Basswood has small pores (distinct with a lens) that may form groups or chains radially, diagonally, or tangentially. The rays of basswood are more pronounced (broader and higher) than those of aspen, are fairly distinct without a lens, and form high, scattered flecks on the radial surfaces. The rays

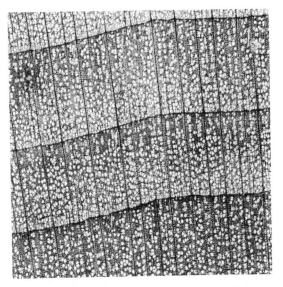

Fig. 27-13. *Tilia americana* 15× (Koehler, 1917).

of basswood vary in height, whereas those of aspen are more uniform in height. Parenchyma may or may not be visible with a lens as very fine tangential lines (apotracheal). The wood has a slight, characteristic odor.

Similar woods. See yellow buckeye for differentiation of this species.

American beech (Figs. 27-15 and 27-16)

Only one species of beech, Americana beech (*Fagus grandifolia*), is native to the United States. It grows in the eastern one—third of the United States and adjacent Canadian provinces. The highest production of beech lumber is in the Central and Middle Atlantic States.

Uses. The wood of beech is classed as heavy, hard, strong, high in resistance to shock, and highly adaptable for steam bending. Beech has large shrinkage and requires careful drying. It machines smoothly, is an excellent wood for turning, wears well, and is rather easily treated with preservatives.

Large amounts of beech go into flooring, furniture, brush blocks, handles, veneer woodenware, containers, and cooperage. When treated, it is suitable for railway ties.

Color. Beech wood varies in color from nearly white sapwood to reddishbrown heartwood in some trees. Sometimes there is no clear line of

Fig. 27-14. American basswood (*Tilia americana* L.) (50×).

demarcation between heartwood and sapwood. The sapwood may be 3 to 5 in. thick. The annual rings are distinct. The wood has little figure and is of close, uniform texture. It has no characteristic taste or odor.

Macroscopic features. Beech is diffuse—porous. The vessels are very small and not visible to the unaided eye. They are crowded in the earlywood but gradually decrease in number and size through the growth ring, forming a darker and denser band of latewood. Tyloses are present.

The parenchyma form very fine, irregular, light colored, tangential lines in the latewood that usually cannot be seen distinctly with a lens.

The rays are of two sizes with the broad ones visible on all surfaces. They appear uniformly spaced on the cross section and of uniform height on the radial surface. The cross section shows 5—10% of the rays to be conspicuously broad, fully twice as wide as the widest vessels. These rays appear as "silver grain" on the radial surfaces. The other rays are narrower than the vessels.

Similar woods. Beech is distinguished from other native species by its weight, conspicuous

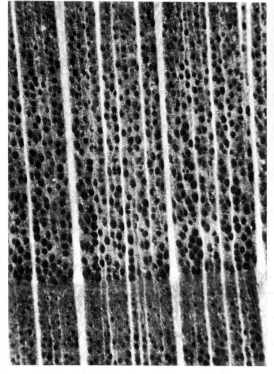

Fig. 27-15. *Fagus grandifolia.*

Fig. 27-16. American beech (*Fagus grandifolia***) (50×).**

rays, and tiny pores. The tangential surface shows the wide rays to be less numerous than in sycamore, which has only wide rays.

Birch (Figs. 27-17 to 27-19)

The important commercial species of birch are yellow birch (*Betula alleghaniensis*), sweet birch (*B. lenta*), and paper birch (*B. papyrifera*). Other birches of some commercial importance are river birch (*B. nigra*), gray birch (*B. Populifolia*), and western paper birch (*B. papyrifera* var. *commutata*). Yellow birch, sweet birch, and paper birch grow principally in the Northeastern and Lake States. Yellow and sweet birch also grow along the Appalachian Mountains to northern Georgia. They are the source of most birch lumber and veneer. The inner bark of yellow and sweet birch has a slight wintergreen flavor; these species are difficult to discern unless the outer bark is present.

Uses. Birch wood is used in furniture, boxes, baskets, crates, woodenware, cooperage, interior finish, doors, and pulp. Birch veneer goes into plywood used for flush doors, furniture, paneling, cabinets, aircraft, and other specialty

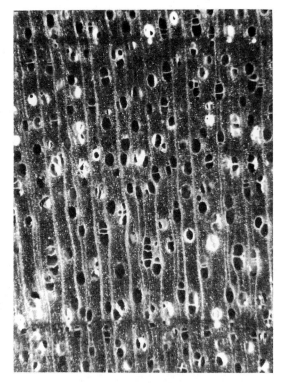

Fig. 27-17. *Betula* spp.

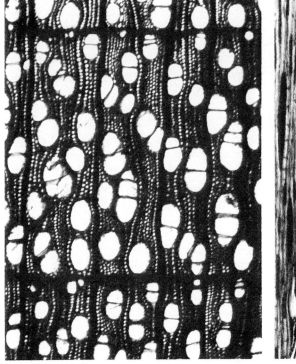

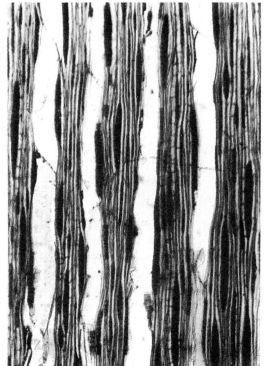

Fig. 27-18. River birch (*Betula nigra*) (50×).

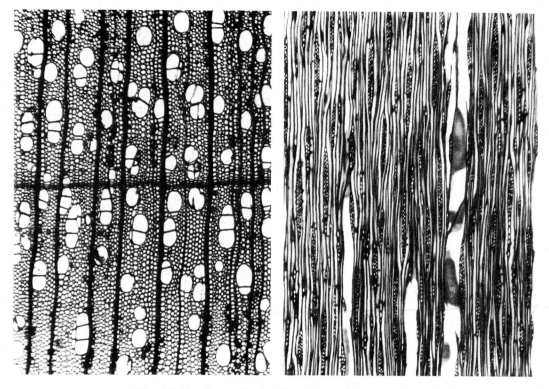

Fig. 27-19. Paper birch (*Betula papyrifera*) (50×).

uses. Paper birch is used for turned products, including spools, small handles, and toys.

Color. Yellow birch has white sapwood (several inches wide, but variable) and light reddishbrown heartwood. Sweet birch has light—colored sapwood and dark brown heartwood tinged with red. The annual rings are not distinct to the unaided eye; they are defined by a faint line and a small change in pore size. (Pith flecks are usually absent in yellow birch, but abundant in river birch.) Wood of yellow birch and sweet birch is heavy, hard, strong, and has good shock—resisting ability. The wood is fine and uniform in texture. Paper birch is lower in weight, softer, and lower in strength than yellow and sweet birch. Birch shrinks considerably during drying.

Macroscopic features. Birch is diffuse—porous. Yellow birch vessels may be barely visible to the unaided eye (x); they occupy less than one—third of the area between rays, except in very narrow growth rings. They may form radial rows of 2-4. Tyloses are absent. The vessels form small groves on the longitudinal surfaces. Longitudinal parenchyma are not visible.

Rays are visible only on the radial surface without magnification as very fine reddish brown "flakes". They appear to be one size and of uniform height along the grain. The rays are narrower than the vessels (x). Growth rings are moderately distinct on tangential surfaces. Paper birch has smaller pores, is lighter in weight, and softer than yellow and sweet birch.

Microscopic features. The simple vessel perforation plates of maple distinguish it from birches, which have sclariform perforation plates.

Similar woods. The maples have smaller pores and wider rays than the birches.

Buckeye (Figs. 27-20 and 27-21)

Buckeye consists of two species, yellow buckeye (*Aesculus octandra*) and Ohio buckeye (*A. glabra*). They range from the Appalachians of Pennsylvania, Virginia, and North Carolina westward to Kansas, Oklahoma, and Texas. Buckeye is not customarily separated from other species when manufactured into lumber and can be utilized for the same purposes as aspen, basswood, and sap yellow poplar.

Uses. The wood is uniform in texture, usually straight—grained, light in weight, weak in beam use, soft, and low in shock resistance. It is rated low on machinability such as shaping, steam bending, boring, and turning. Buckeye is suitable for pulping for paper; its lumber is used principally for furniture, crates, food containers, woodenware, novelties, and planning mill products.

Color. The white sapwood of buckeye merges gradually into the creamy or yellowish white heartwood. The annual rings are not always distinct and are defined by light—colored lines of marginal parenchyma.

Macroscopic features. The pores are very small, uniform in size, and distributed evenly throughout the growth ring (diffuse—porous). The rays are very fine (uniseriate), even under a lens. They are storied (arranged in tiers), producing very fine bands or ripple marks running across the tangential surface (not as pronounced in *A. glabra*, but made more visible by wetting its surface). Marginal parenchyma are seen with a lens.

Similar woods. Basswood is differentiated from buckeye by the conspicuous rays of bass-

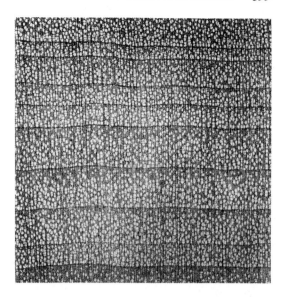

Fig. 27-20. *Aesculus octandra* 15× (Koehler, 1917).

wood. The fine lines of buckeye on the tangential surface also distinguish it from basswood.

Butternut (Figs. 27-22 and 27-23)

Butternut (*Juglans cinerea*) is also called

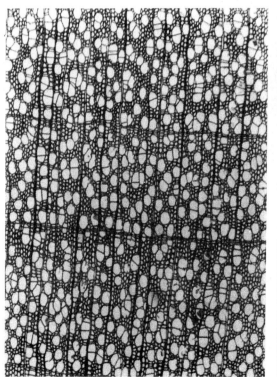

Fig. 27-21. Yellow buckeye (*Aesculus octandra*) (50×).

white walnut, American white walnut, and oilnut. It is grown from southern New Brunswick west to Minnesota. It extends southward into northeastern Arkansas and eastward to western North Carolina.

Uses. The wood is moderately light in weight, about the same as eastern white pine, rather coarse—textured, moderately weak in bending and endwise compression, relatively low in stiffness, moderately soft, and moderately high in shock resistance. Butternut machines easily and finishes well. In many ways it resembles black walnut, but it does not have the strength or hardness. Principal uses are for lumber and veneer, which are further manufactured into furniture, cabinets, paneling, trim, and assorted rough items.

Color. The sapwood (rarely over 1 in. wide) is nearly white or grayish brown; the heartwood is light brown, frequently modified by pinkish tones or darker brown streaks. The annual rings are distinct, marked by the abrupt change in pore size.

Macroscopic features. The wood is semi—ring—porous. The pores are relatively large and easy to see, especially in the earlywood. Tyloses are present, but not dense. They are not crowded

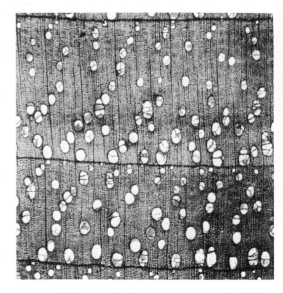

Fig. 27-22. *Juglans cinera*, 15× (Koehler, 1917).

but occasionally form radial rows of 2 to 6. Rays are very fine, inconspicuous, and indistinct without a lens. The apotracheal parenchyma are visible as fine, tangential lines on the cross section, especially in the outer portion of the growth ring.

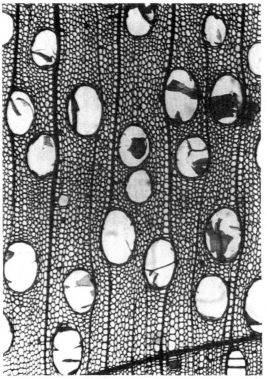

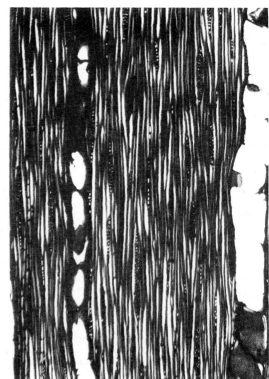

Fig. 27-23. Butternut (*Juglans cinera*) (50×).

Black cherry (Figs. 27-24 and 27-25)

Black cherry (*Prunus serotina*) is also known as cherry, wild black cherry, wild cherry, or chokecherry. It is the only native species of the genus *Prunus* of commercial importance for lumber. It is scattered from southeastern Canada throughout the eastern U.S. Production is centered in the Middle Atlantic States.

Uses. The wood has a fairly uniform texture and very satisfactory machining properties. It is moderately heavy. Black cherry is strong, stiff, moderately hard, and has high shock resistance and moderately large shrinkage. After seasoning, it has good dimensional stability. Black cherry is used principally for furniture, fine veneer panels, and architectural woodwork.

Color. The heartwood of black cherry varies from light to dark reddish brown and has a distinctive luster. The sapwood is narrow (less than 1 in.) in old trees and nearly white. The annual rings are fairly distinct and defined by the change in pore size.

Macroscopic features. Individual pores are not visible to the unaided eye, but they form

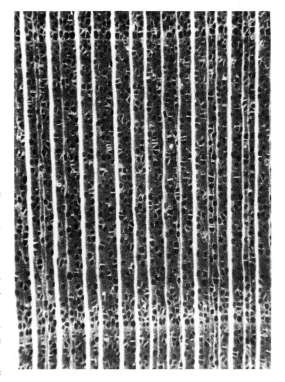

Fig. 27-24. *Prunus serotina.*

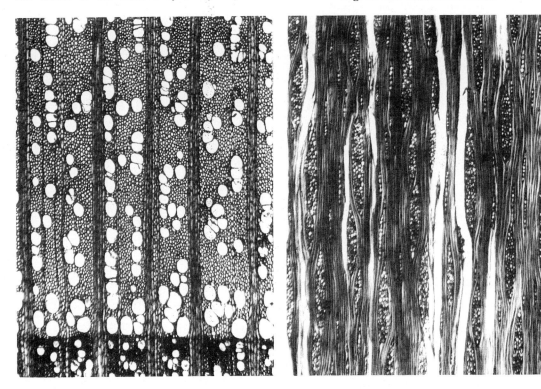

Fig. 27-25. *Prunus avium* (50×).

tangential lines. The pores occupy one—third or slightly more of the area between rays. Tyloses are absent, but gum may be present in some pores. The rays are barely visible to the unaided eye on the cross section (as wide or almost as wide as the widest pores) and tend to produce a distinctive flake pattern on radial surfaces (the identifying characteristic) that are more pronounced than those of black walnut. Parenchyma are not visible. Traumatic resin canals may be present and appear as dark lines on longitudinal surfaces.

Catalpa (Fig. 27-26)

Catalpa (*Catalpa speciosa*), alias northern catalpa, western catalpa, cigartree, and Indian bean, is widely planted outside its natural range of central to southeastern U.S. A minor species is southern catalpa or common catalpa (*Catalpa bignonioides*).

Color. The sapwood is very narrow, rarely over ½ in. wide. The heartwood is grayish brown with an occasional tinge of light purple. The wood is moderately light, straight grained, and without a unique odor or taste. Sometimes it has a faint odor resembling kerosene. The growth rings are usually fairly wide and often variable in width, but slow growth is possible.

Macroscopic features. The earlywood vessels are large and distinct and filled with glistening tyloses. The pores in the middle of the growth ring are isolated, in groups of 2—5, or more often in wavy, tangential bands. The pores in the latewood near the end of the growth ring are very small and always in wavy, tangential bands. The normally spaced rays are not distinct, nor are the paratracheal parenchyma distinct from the latewood vessels without a lens.

Sassafras and black ash resemble catalpa in color and somewhat in structure. Sassafras is separated by its spicy odor. Black ash is separated by its higher density and relatively few vessels in the latewood. The pith of catalpa is often three—sided, whereas in black ash it is round.

American chestnut (Fig. 27-27)

American chestnut (*Castanea dentata*) is known also as sweet chestnut. It grew in commercial quantities from New England to northern Georgia, but practically all standing chestnut has been killed by blight.

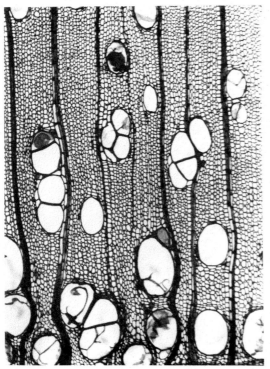 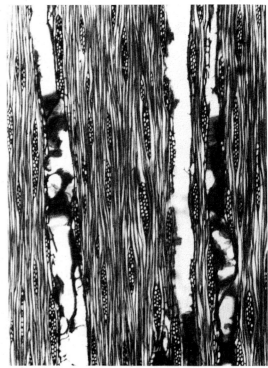

Fig. 27-26. Catalpa (*Catalpa speciosa*) (50×).

Uses. Chestnut was used for poles, railway ties, furniture, caskets, boxes, crates, and core stock for veneer panels. It appears as "wormy chestnut" for paneling, trim, and picture frames, while a small amount is still used in rustic fences.

Color. The heartwood of chestnut is fairly uniform grayish brown or brown and darkens with age. The sapwood is very narrow (rarely over 0.5 in. wide) and almost white. Chestnut wood is moderately light in weight. It is moderately hard, moderately low in strength, moderately low in resistance to shock, and low in stiffness. It seasons well and is easy to work with tools.

Macroscopic features. The wood is ring—porous and coarse in texture. The growth rings are made conspicuous by several rows of large, distinct vessels (almost entirely plugged with tyloses) at the beginning of each year's growth. The latewood vessels are very small, numerous, and arranged in irregular branched radial bands. The rays are barely distinct with a hand lens. The apotracheal and paratracheal parenchyma are scattered among the pores and are not distinct.

Similar woods. The radial bands of vessels are found in no other native timber species except oaks, but oaks have large rays. Black ash resembles chestnut, but the vessels of black ash latewood are few, isolated, and never arranged in radial bands.

Kentucky coffeetree

Kentucky coffeetree (*Gymnocladus dioicus*) grows in the central section of the eastern U.S. The sapwood is narrow, usually less than 0.5 in. but up to 1 in. wide. The heartwood is bright cherry red to reddish brown. The annual rings are distinct and highly variable in width. The wood has the same weight and physical properties of honey locust. Its color and structure are similar to those of honey locust. See honey locust for structure and distinguishing characteristics.

Cottonwood (Figs. 27-28 and 27-29)

Cottonwood includes several species of the genus *Populus*. Most important are eastern cottonwood (*P. deltoides* and varieties), also known as Carolina poplar and whitewood; swamp cottonwood (*P. heterophylla*), also known as river

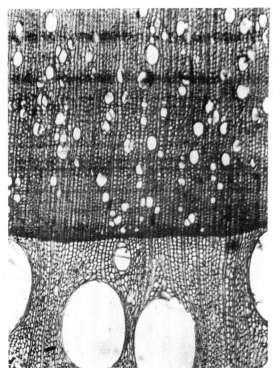
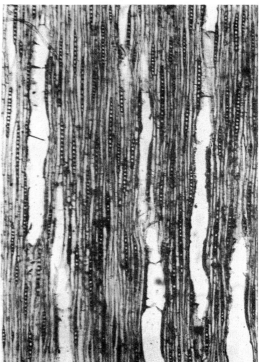

Fig. 27-27. American chestnut (*Castanea dentata*) (50×).

cottonwood, cottonwood, and swamp poplar; and black cottonwood (*P. trichocarpa*).

Eastern cottonwood and swamp cottonwood grow throughout the eastern half of the United States. The greatest production of lumber is in the Southern and Central States. Black cottonwood grows in the West Coast States and in western Montana, northern Idaho, and western Nevada. Balsam poplar grows from Alaska across Canada and in the northern Great Lake states.

Eastern cottonwood is moderately low in bending and compressive strength, moderately limber, moderately soft, and moderately low in ability to resist shock. Black cottonwood is slightly below eastern cottonwood in most strength properties; both have moderately large shrinkage. Some cottonwood is difficult to work with tools because of fuzzy surfaces. Tension wood is largely responsible for this characteristic.

Uses. Cottonwood is used principally for lumber, veneer, pulpwood, particleboard, and fuel. The lumber and veneer go largely into boxes, crates, baskets, and pallets.

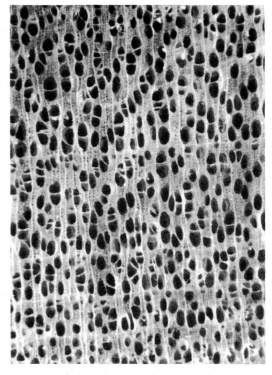

Fig. 27-28. *Populus tricocarpa.*

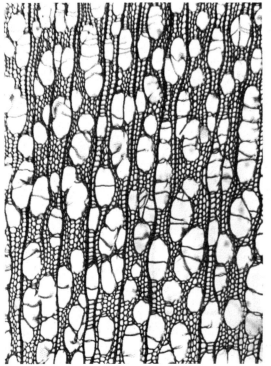

Fig. 27-29. Black cottonwood (*Populus tricocarpa*) (50×).

Color. The heartwood of the three cotton-woods is grayish white to light brown. The sapwood is whitish and merges gradually with the heartwood. The wood is comparatively uniform in texture and generally straight grained. It is odorless when well seasoned.

Macroscopic structure. The annual rings are semi—ring—porous (approaching diffuse—porous in some samples), distinct, but not conspicuous, and defined by a slight difference in pore size. The vessels are usually visible to the unaided eye. They decrease slightly in size from the earlywood to the latewood. They are numerous and occupy one—half of the area between the rays, except in the widest rings. They are scattered singly or occasionally in radial rows of 2—4. The rays are barely distinct with a lens and normally spaced. Marginal parenchyma are not distinct, but form a narrow, light—colored line.

Similar woods. Cottonwood is very similar in appearance to black willow. The heartwood color of willow is more brown to reddish. Microscopic examination of sapwood also distinguishes them. Cottonwood pores are larger than those of aspen and are barely visible without magnification.

Cottonwood may be confused with other light woods like tupelo, yellow poplar, basswood, and buckeye. However, the pores of the earlywood of cottonwood are larger than those of the other species, being nearly visible with the unaided eye.

Dogwood

Dogwood does not usually grow to very large sizes. Dogwood appears to be similar to maple. Its grain tends to be interlocked. Its sapwood has a pinkish cast. The heartwood (of large samples) is narrow and dark brown. It is diffuse—porous with indistinct annual rings. The vessels are mostly solitary or form short radial multiples in the latewood. The apotracheal banded parenchyma are not visible. The rays are of two sizes and occupy nearly half the cross section. Pacific dogwood (*Cornus nuttallii*) has rays up to about 40 cells high whereas those of flowering dogwood (*C. florida*) may be up to 80 cells high.

Similar woods. The tendency toward radial multiples of pores separates it from redgum. Swamp tupelo is similar in appearance.

Elm (Figs. 27-30 and 27-31)

Six species of elm grow in the eastern United States: American elm (*Ulmus americana*), slippery elm (*U. rubra*), rock elm (*U. Thomasaii*), winged elm (*U. alata*), cedar elm (*U. crassifolia*), and September elm (*U. serotina*). American elm is also known as white elm, water elm, and gray elm; slippery elm as red elm; rock elm as cork elm or hickory elm; winged elm as wahoo; cedar elm as red elm or basket elm; and September elm as red elm. More than 80% of rock elm comes from Wisconsin and Michigan. American elm has been almost eradicated by two diseases, Dutch elm and phloem necrosis. Slippery elm is distinguished by the mucilaginous nature of its inner bark, if it is available.

Uses. The elms may be divided into two general classes, hard elm and soft elm, based on the weight and strength of the wood. Hard elm includes rock elm, winged elm, cedar elm, and September elm. American elm and slippery elm are the soft elms. Soft elm is moderately heavy, has high shock resistance, and is moderately hard and stiff. Elm has excellent bending qualities and is used in the bent portion of chairs. Some hard elms tend to have interlocked grain.

Elm lumber is used principally in boxes, baskets, crates, slack barrels, furniture, and caskets. For some uses the hard elms are preferred. Elm veneer is used for furniture, fruit, vegetable and cheese boxes, baskets, and decorative panels.

Color. The sapwood of the elms is nearly white (1—3 in. wide) and the heartwood light brown, often tinged with red.

Macroscopic features. American elm is a ring—porous wood, and the earlywood pores are easily observed in one row (or two to three rows in very wide growth rings) rarely with tyloses. Winged elm usually has a more interrupted, uniseriate row of earlywood vessels at the start of the growth ring. (In rock elm earlywood vessels are observed only with magnification and tend to have tyloses.) The latewood vessels are arranged in an ulmiform pattern (wavy, tangential bands) characteristic of elms and hackberry. The rays are fine and not distinct without a hand lens. The longitudinal parenchyma are paratracheal but not seen without a hand lens. The wetted heartwood of *U. rubra* has a disagreeable odor.

Microscopic features. Intervessel pitting of American elm is alternate, angular, and crowded. Rays are normally spaced and 3—4 seriate.

Similar woods. See hackberry (which has more distinct rays, paler heartwood with a tinge of green, and wider sapwood) for separation from this species.

Hackberry (Fig. 27-32)

Hackberry (*Celtis occidentalis*) and sugarberry (*C. laevigata*) supply the lumber known in the trade as hackberry. Hackberry grows east of the Great Plains from Alabama, Georgia, Arkansas, and Oklahoma northward, except along the Canadian boundary. Sugarberry overlaps the southern part of the range of hackberry and grows throughout the Southern and South Atlantic States.

The sapwood of both species varies from pale yellow to greenish or grayish yellow and may be over 3 in. wide; it is subject to blue stain. The heartwood is commonly darker and yellowish or greenish gray. The wood resembles elm in structure. Hackberry lumber is moderately heavy. It

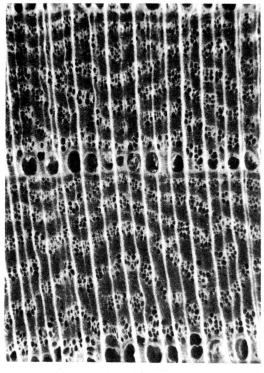

Fig. 27-30. *Ulmus americana.*

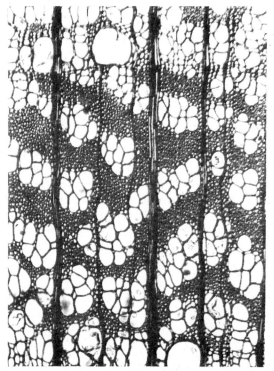

Fig. 27-31. Rock elm (*Ulmus thomasii*) (50×).

Fig. 27-32. Hackberry (*Celtis occidentalis*) (50×).

is moderately strong in bending, moderately weak in compression parallel to the grain, moderately hard to hard, high in shock resistance, but low in stiffness. It has moderately large to large shrinkage. Hackberry lumber keeps its shape well during seasoning.

Hackberry is cut into lumber, with small amounts going into dimension stock and veneer. It is used for furniture and some for containers.

Macroscopic features. Hackberry is a ring—porous wood; the earlywood vessels form a ring that is 2—3 pores wide, except in narrow growth rings. The earlywood pores of the heartwood are partly occluded with tyloses. The latewood vessels are in a ulmiform arrangement. The rays are normally spaced and relatively conspicuous. Longitudinal parenchyma are paratracheal, being intermixed with latewood vessels and vascular tracheids, and visible only with a hand lens.

Hackberry is similar to elm but the large earlywood vessels are in more than one row, the ulmiform pattern of latewood vessels is less pronounced, the rays are more distinct, and the heartwood may have a greenish tinge.

Hickories (Figs. 27-33 and 27-34)

The hickories are divided into two groups, the pecan hickories and the true hickories, based on the twigs, leaves, and fruit. It is difficult, even with a microscope, to distinguish among the species, or even the groups, except for the water hickory. The hickories range throughout the eastern U.S. The wood is very heavy and exceedingly hard and tough.

Color. The sapwood is 1—2 in., occasionally to 3 in., wide. The heartwood is brown to reddish brown. It has no characteristic odor or taste.

Macroscopic features. Individual pores are visible, but the zone of large pores is not sharply outlined as in the oaks and the ashes. The earlywood has a ring of large vessels of 1 to 2 (or 3 pores in wide rings) width. The pores decrease in size somewhat gradually toward the latewood. In the latewood, the pores are not numerous, isolated, or in radial rows of 2—3.

The rays are very small and seen without magnification on radial surfaces only. The apotracheal, banded parenchyma are distinct with a lens and form numerous fine, light—colored, tangential bands that do not encircle the vessels (as in ash), but usually extending between the vessels or passing around them to one side. The parenchyma form a net—like appearance with the rays.

Similar woods. The hickories are distinguished from the other ring—porous woods by the numerous fine, tangential lines of parenchyma and the presence of few pores in the latewood. Persimmon is similar but has finer tangential lines of parenchyma and its rays are in horizontal rows that are plainly visible on tangential surfaces as very fine bands across the grain.

Pecan hickories

Species of the pecan group include bitternut hickory (*Carya cordiformis*), pecan (*C. illinoensis*), water hickory (*C. aquatica*), and nutmeg hickory (*C. myristicaeformis*). Bitternut hickory grows throughout the eastern half of the United States. Pecan hickory grows from central Texas and Louisiana to Missouri and Indiana. Water hickory grows from Texas to South Carolina. Nutmeg hickory occurs principally in Texas and Louisiana.

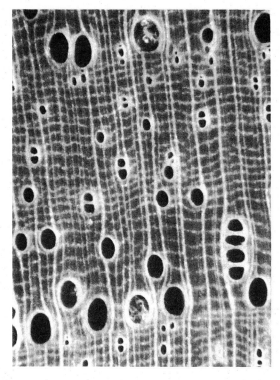

Fig. 27-33. A pecan hickory; *Carya* spp.

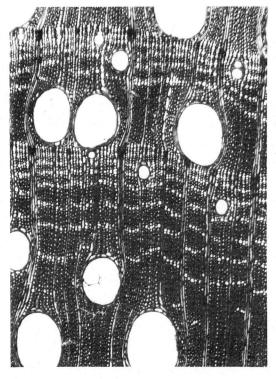
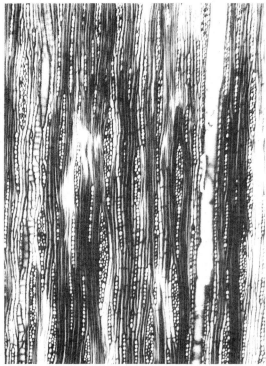

Fig. 27-34. Shagbark hickory, a true hickory (*Carya ovata*) (50×).

The wood of pecan hickory resembles that of true hickory but tends to be semi—ring—porous. It has nearly white sapwood, which is relatively wide, and somewhat darker heartwood. The wood is heavy and sometimes has very large shrinkage. Heavy pecan hickory finds use in tool and implement handles and flooring. The lower grades are used in pallets. Many higher grade logs are sliced to provide veneer for furniture and paneling.

Water hickory differs from the other hickories by having less clearly defined annual growth rings and large pores being more scattered (approaching diffuse—porous woods in structure). The fine, light—colored, lines of parenchyma are particularly distinct in this species. The wood is somewhat lighter in weight than most hickories.

True hickories

True hickories are found throughout most of the eastern half of the United States. The species most important commercially are shagbark (*Carya ovata*), pignut (*C. glabra*), shell bark (*C. Laciniosa*), and mockernut (*C. tomentosa*).

The greatest commercial production of the true hickories is in the Middle Atlantic and Central States. The Southern and South Atlantic States produce nearly half of all hickory lumber. The wood of true hickory is exceptionally tough, heavy, hard, strong, and shrinks considerably in drying. Both rings per inch and weight are limiting factors where strength is important.

The major use for hickory is for tool handles, which require high shock resistance. It also used for ladder rungs, athletic goods, agricultural implements, dowels, gymnasium apparatus, poles, and furniture. A considerable quantity of lower grade hickory is not suitable, because of knottiness or other growth features and low density, for the special uses of high—quality hickory. It appears particularly useful for pallets, blocking, and similar items. Hickory sawdust and chips and some solid wood are used by the major packing companies to flavor meat by smoking.

True hickories are ring—porous woods. The latewood vessels are solitary and in short, radial multiples sparsely distributed. Rays are normally spaced and uniform. Longitudinal parenchyma are essentially apotracheal, forming short tangential bands at right angles to the rays (parallel to the growth rings).

American holly (Fig. 27-35)

American holly (*Ilex opaca*) is sometimes called white holly, evergreen holly, and boxwood. The natural range of holly extends along the Atlantic coast, Gulf Coast, and Mississippi Valley.

Uses. The wood has a uniform and compact texture; it is moderately low in strength when used as a beam or column and low in stiffness, but it is heavy and hard, and ranks high in shock resistance. It is readily penetrable to liquids and can be satisfactorily dyed. It works well, cuts smoothly, and is used principally for scientific and musical instruments, furniture inlays, and athletic goods as a specialty wood.

Color. Both the heartwood and sapwood are exceptionally white, occasionally with a bluish tinge. The heartwood has an ivory cast.

Macroscopic structure. The wood is diffuse—porous. The margin of the growth ring is made up of dense, fibrous cells. The latewood vessels are quite small and arranged in long radial chains. The scarce longitudinal parenchyma are not visible to the unaided eye. The rays are of two widths, and the wider ones are barely distinct to the unaided eye.

Honey locust (Fig. 27-36)

The wood of honeylocust (*Gleditsia triacanthos*), alias thorny locust, honey shucks, and three—thorned acacia, possesses many desirable qualities such as attractive figure and color, hardness, and strength, but is little used because of its scarcity. Although the natural range of honeylocust has been extended by planting, it is found most commonly in the eastern United States, except for New England and the South Atlantic and Gulf Coastal Plains.

Uses. It is very heavy, very hard, strong in bending, stiff, resistant to shock, and is durable when in contact with the ground. When available, it is restricted primarily to local uses, such as fence posts and lumber for general construction. Occasionally it will show up with other species in lumber for pallets and crating.

Color. The sapwood is mostly 0.75 to 2 in. wide and yellowish, in contrast to the bright cherry red to reddish brown heartwood. The annual rings are distinct and irregular in width and outline.

Macroscopic structure. The wood is ring—porous. The earlywood pores are conspicuous without a lens and form a wide band (3—5 pores wide) that occupies about one—half of the total width of moderately wide growth rings. In the outer latewood, the pores are very small, with lumens that are not visible without a lens, and form short, wavy, light—colored tangential bands visible without a lens. Honey locust has gummy deposits in the vessels of its heartwood.

The rays are of two sizes; the larger size rays are distinct without a hand lens. The paratracheal and paratracheal—confluent parenchyma are not distinguished from the latewood pores without a lens.

Similar woods. Honey locust may be confused with Kentucky coffeetree. The gummy deposits of honey locust vessels are usually lacking in coffeetree, but this is not definitive. Coffeetree also has narrower sapwood and a much larger pith (often less than 0.15 in. in honey locust and greater than 0.20 in. in coffeetree). The pores in the outer latewood of coffeetree are somewhat larger and form more rounded groups than honey locust. The rays of coffeetree are less distinct to the unaided eye than those of honey locust and are all of the same width. Honey locust may have small pin knots from the thorns that grow on the truncks, which never occur in coffeetree.

Black locust (Fig. 27-37)

Black locust (*Robinia pseudoacacia*) is sometimes called yellow locust, white locust, green locust, or post locust. This species grows from Pennsylvania along the Appalachian Mountains to northern Georgia. The greatest production of black locust timber is in Tennessee, Kentucky, West Virginia, and Virginia.

Uses. Black locust is very heavy, very hard, very high in resistance to shock, and ranks very high in strength and stiffness. It has moderately small shrinkage. The heartwood has high decay resistance. Black locust is used extensively for round, hewed, or split mine timbers, and for fence—posts, poles, railroad ties, stakes, and fuel. Uses include rough construction, crating, ship treenails, and mine equipment.

Color. Locust has narrow, creamywhite sapwood. The heartwood, when freshly cut, varies from greenish yellow to dark brown.

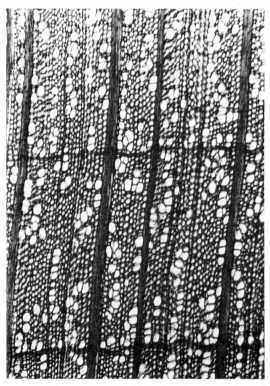

Fig. 27-35. American holly (*Illex opaca*) (50×).

Macroscopic features. The earlywood vessels are large but quite occluded with tyloses, which makes them appear as a band rather than as other ring—porous woods. A lens shows this band to consist of 2—3 rows of vessels. The latewood vessels are small and occur in wavy tangential bands toward the outer portion of the growth ring. The rays are not visible or barely visible without a lens. The annual rings are distinct and variable in width. Parenchyma are not easily distinguished from the vessels.

Similar woods. Black locust may be confused with Osage orange but black locust has vessels with distinct outlines in the earlywood and lacks the yellow—orange extractive of Osage orange wood. It is sometimes confused with mulberry.

Magnolia (Fig. 27-38)

Three species comprise commercial magnolia: southern magnolia (*Magnolia grandiflora*), sweetbay (*M. virginiana*), and cucumbertree (*M. acuminata*). Other names for southern magnolia are evergreen magnolia, magnolia, big laurel, bull bay, and laurel bay. Sweetbay is sometimes called swamp magnolia, or more often simply magnolia.

The natural range of sweetbay extends along the Atlantic and gulf coasts from Long Island to Texas, and that of southern magnolia from North Carolina to Texas. Cucumbertree grows from the Appalachians to the Ozarks northward to Ohio. Louisiana leads in production of magnolia lumber.

Uses. The wood, which has close, uniform texture and is generally straight grained, closely resembles yellow—poplar. It is moderately heavy, moderately low in shrinkage, moderately low in bending and compressive strength, moderately hard and stiff, and moderately high in shock resistance. Sweetbay is reported to be much like southern magnolia. The wood of cucumbertree is similar to that of yellow—poplar, and cucumbertree growing in the yellow—poplar range is not separated from that species on the market.

Magnolia lumber is used principally in the manufacture of furniture, boxes, pallet, venetian blinds, sash, doors, veneer, and millwork.

Color. The sapwood of southern magnolia is yellowish white, and the heartwood is light to dark brown with a tinge of yellow or green.

Macroscopic structure. Sweetbay is a diffuse porous wood. The vessels are in radial multiples of two to several. The rays are distinct, normally spaced, and uniform. The longitudinal parenchyma are marginal, marking the end of the growth ring as a whitish line.

Microscopic structure. The anatomy of this wood is very similar to yellow—poplar. Sweetbay, however, has sclariform intervessel pitting.

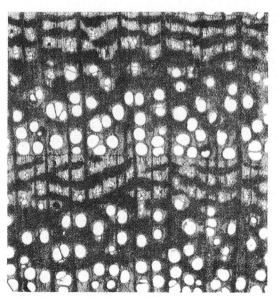

Fig. 27-36. *Gleditsia triacanthos* (Koehler) 15×.

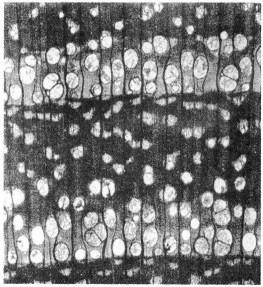

Fig. 27-37. *Robinia pseudoacacia* (Koehler) 15×.

Maple (Figs. 27-39 to 27-41)

Commercial species of maple in the U.S. include sugar maple (*Acer saccharum*), black maple (*A. nigrum*), silver maple (*A. saccharinum*), red maple (*A. rubrum*), boxelder (*A. negundo*), and bigleaf maple (*A. macrophyllum*). Sugar maple is also known as hard maple, rock maple, sugar tree, and black ample; black maple as hard maple, black sugar maple, and sugar maple; silver maple as white maple, river maple, water maple, and swamp maple; red maple as soft maple, water maple, scarlet maple, white maple, and swamp maple; boxelder as Manitoba maple, ash—leaved maple, three—leaved maple, and cut—leaved maple; and bigleaf maple as Oregon maple. The Middle Atlantic and Lake States account for about two—thirds of the maple production.

Uses. The wood of sugar maple and black maple is known as hard maple; that of silver maple, red maple, and boxelder as soft maple. Hard maple has a fine, uniform texture. It is heavy, strong, stiff, hard, resistant to shock, and has large shrinkage. Sugar maple is generally straight grained but also occurs as "bird's eye," "curly," and "fiddleback" grain. Soft maple,

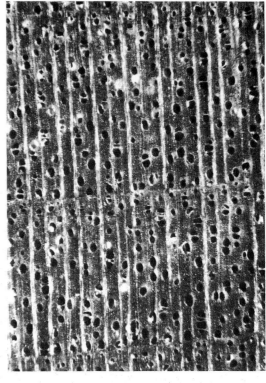

Fig. 27-39. *Acer macrophyllum.*

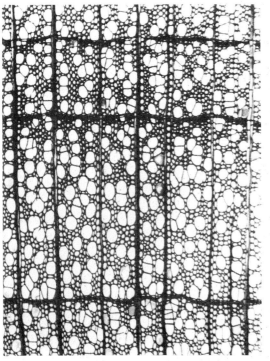

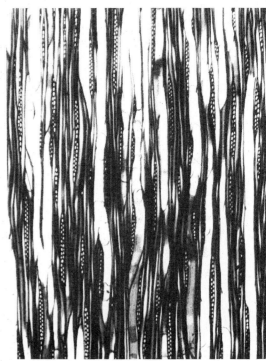

Fig. 27-38. Cucumbertree (*Magnolia acuminata*) (50×).

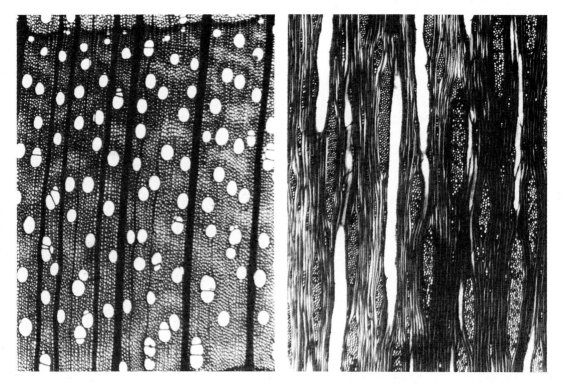

Fig. 27-40. Sugar maple (*Acer saccharum*) (50×).

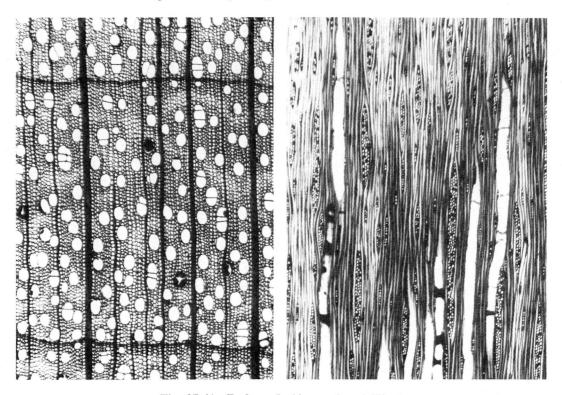

Fig. 27-41. Red maple (*Acer rubrum*) (50×).

while not as heavy, has been substituted for hard maple in the better grades, particularly for furniture.

Maple is used principally for lumber, veneer, crossties, and pulpwood. A large proportion is manufactured into flooring, furniture, boxes, pallets and crates, handles, woodenware, novelties, spools, and bobbins.

Color. The sapwood of the maples is commonly white with a slight reddishbrown tinge. It is from 3 to 5 or more inches thick. The heartwood is usually light reddish brown, but sometimes it is considerably darker. The annual rings are not very distinct; they are defined by an indistinct, darker line of marginal parenchyma intermixed with fibers.

Macroscopic features. Maples are diffuse porous woods, and the pores are not visible on any surface without magnification. The vessels are solitary (sometimes in radial multiples of 2 to 3) and evenly distributed within the growth ring. Tyloses are normally absent.

Rays are visible. Hard maple is distinguished from soft maple by the two distinct ray sizes of hard maple (which may be up to 8—seriate and as wide as the widest pores) and a lustrous appearance. The rays of soft maples vary in size, the widest of which is almost the same width as the widest vessels (about 5—seriate). Saturated $FeSO_4$ in water turns red maple blue—black but gives a green color with sugar maple. Parenchyma are not seen with a lens.

Microscopic features. Longitudinal parenchyma are marginal, defining growth ring boundaries and intermixed with fibers. Intervessel pitting of red maple is alternate, angular in outline, and crowded. The vessels of soft maple are longer than those of hard maple, whose lengths are seldom more than about three times their diameter.

Similar woods. Maples (simple vessel perforation plates) are distinguished from birches (sclariform perforation plates). The rays of birches are not distinct to the unaided eye, while the vessels are distinct. The rays of beech are much wider than the vessels (x).

Red mulberry (Fig. 27-42)

Red mulberry (*Morus rubra*), alias black mulberry, grows throughout the eastern U.S.,

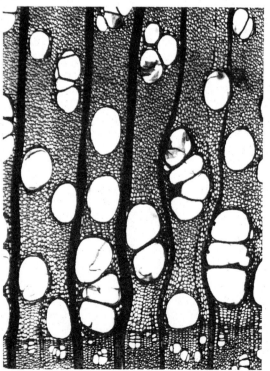

Fig. 27-42. Red mulberry (*Morus rubra*) (50×).

except in New England. The wood is heavy, straight grained. The growth rings are distinct and usually moderately wide.

Color. The sapwood is yellowish and about 0.5 in. wide. The heartwood is yellowish brown on freshly cut surfaces, turning to russet brown with exposure to air.

Macroscopic structure. The wood is ring—porous. The earlywood pores are large, distinct, and form a ring of 2—5 pores width. They are densely packed with tyloses that have a unique glistening appearance. The latewood pores form small irregular (nestlike) groups of 3—10 often forming wavy, tangential bands that are observed without a lens. The rays are evident without a lens and form flecks on the radial surfaces. The parenchyma are observed with a lens.

Red oaks (Figs. 27-43 and 27-44)

Most red oak lumber and other products come from the Southern States, the southern mountain regions, the Atlantic Coastal Plains, and the Central States. The principal species are: Northern red oak (*Quercus rubra*), scarlet oak (*Q. coccinea*), Shumard oak (*Q. shumardii*), pin oak

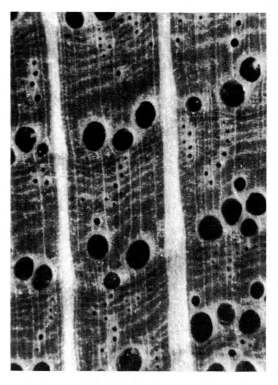

Fig. 27-43. *Quercus rubra.*

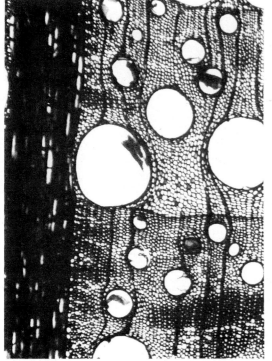
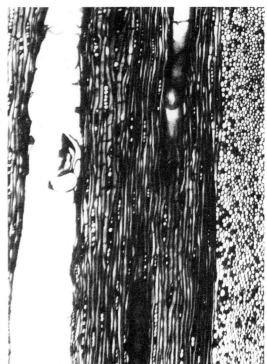

Fig. 27-44. Red oak group: pin oak (*Quercus palustris*) (50×).

(*Q. palustris*), Nuttall oak (*Q. nuttallii*), black oak (*Q. velutina*), southern red oak (*Q. falcata*), cherrybark oak (*Q. falcata* var. *pagodaefolia*), water oak (*Q. nigra*), laurel oak (*Q. laurifolia*), and willow oak (*Q. phellos*).

Uses. Wood of the red oaks is heavy. Rapidly grown second—growth oak is generally harder and tougher than finer textured old—growth timber. The red oaks have fairly large shrinkage in drying. The red oaks are largely cut into lumber, railroad ties, mine timbers, fence posts, veneer, pulpwood, and fuelwood. Ties, mine timbers, and fence posts require preservative treatment for satisfactory service. Quartersawn lumber is distinguished by the broad and conspicuous rays, which add to its attractiveness. Red oak lumber is remanufactured into flooring, furniture, general millwork, boxes, pallets and crates, caskets, woodenware, and handles.

Color. The sapwood is nearly white and usually 1 to 2 inches thick. The heartwood is brown with a tinge of red.

Macroscopic features. The red oaks are ring—porous, but may be more semi—ring—porous than the white oaks. The earlywood vessels form two or three rows (or four in wide growth rings). Latewood vessels have a dendritic arrangement (groups that are oblique or in a group that widens toward the outer limit of the growth ring) and are surrounded by paratracheal parenchyma. Individual vessels are solitary, rounded, sparsely distributed, and have thick walls.

Rays are either uniseriate or extremely broad of the type characteristic of oaks. Longitudinal parenchyma are abundant as paratracheal and apotracheal banded parenchyma. Tyloses are usually absent to sparse.

Similar woods. Sawed lumber of red oaks cannot be separated by species on the basis of the characteristics of the wood alone. Red oak lumber can be separated from white oak by the number of pores in the latewood and, as a rule, it lacks tyloses in the pores. The open pores of the red oaks make these species unsuitable for tight cooperage, unless the wood is sealed.

White oaks (Figs. 27-45 and 27-46)

White oak lumber comes chiefly from the South, South Atlantic, and Central States, including the Southern Appalachian area. Principal species are white oak (*Quercus alba*), chestnut oak (*Q. prinus*), post oak (*Q. stellata*), overcup oak (*Q. lyrata*), swamp chestnut oak (*Q. michauxii*), bur oak (*Q. macrocarpa*), chinkapin oak (*Q. muehlenbergii*), swamp white oak (*Q. bicolor*), and live oak (*Q. virginiana*)

Uses. The wood of white oak is heavy, averaging somewhat higher in weight than that of the red oaks. The heartwood has moderately good decay resistance. White oaks are used for lumber, railroad ties, cooperage, mine timbers, fenceposts, veneer, fuelwood, and many other products. High—quality white oak is especially sought for tight cooperage. Live oak is considerably heavier and stronger than the other oaks and was formerly used extensively for ship timbers. An important use of white oak is for planking and bent parts of ships and boats, heartwood often being specified because of its decay resistance. It is also used for flooring, pallets, agricultural implements, railroad cars, truck floors, furniture, doors, millwork, and many other items.

Color. The heartwood of the white oaks is generally grayish brown, and the sapwood, which is from 1 to 2 or more inches thick, is nearly white, although the sapwood is often stained by tannins that leach from the bark.

Macroscopic features. The white oak group are not separable as to species. White oaks are typically ring—porous, with the earlywood vessels forming one to two rows (or three in wide growth rings). The pores of the heartwood of white oaks are usually plugged with the membranous growth known as tyloses (although chestnut oak pores are more open). These tend to make the wood impenetrable by liquids, and for this reason most white oaks are suitable for tight cooperage. Latewood vessels have a dendritic arrangement (groups that are oblique or in a group that widens toward the outer limit of the growth ring) (*x*) and are surrounded by paratracheal parenchyma. Latewood vessels are solitary, angular, very small, closely spaced and thin walled. The red and white oaks are separable based on the appearance of the latewood vessels (see red oaks).

Rays are either uniseriate or extremely broad of the type characteristic of oaks. Ray height in red oaks is ¼ to 1 in., whereas those in white oak may be 0.5 to 5 in. high.   Parenchyma are paratracheal and apotracheal.

Osage orange (Fig. 27-47)

Osage orange (*Maclura pomifera*) is also called hedge—plant, yellowwood, and mock orange. It grows naturally from Texas to Mississippi and north to Oklahoma, but is widely planted outside this region for hedges.

Uses. The wood is very heavy, exceedingly hard, somewhat crossgrained, and without unique odor or taste. It is separated from black locust by its yellowish water—soluble extract of the wood.

Color. The sapwood is very narrow, usually less than 0.5 in. wide. The heartwood on freshly cut surfaces is golden yellow to brown with reddish brown streaks and turns darker with exposure to air.

Macroscopic features The wood is ring—porous. The large earlywood vessels form a light—colored ring, although the individual vessels are not distinguished due to the presence of tyloses (except near the bark). The latewood pores are very small and are grouped in short, wavy tangential bands. The rays are barely visible without a

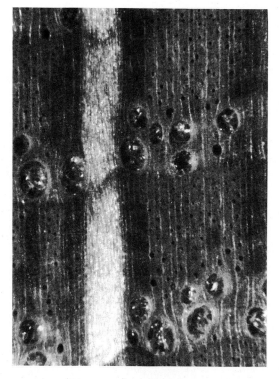

Fig. 27-45. *Quercus alba.*

Fig. 27-46. White oak (*Quercus alba*) (50×).

lens. The paratracheal and paratracheal— confluent parenchyma are difficult to distinguish from the latewood vessels with a lens. The annual rings are distinct and irregular in width and outline.

Similar woods. Black locust is very similar, but black locust lacks the water—soluble yellow dye and has more distinct pores.

Persimmon (Fig. 27-48)

Persimmon (*Diospyros virginiana*), alias date plum or possum wood, grows in the eastern half of the U.S., except in the Northeast. The wood is very heavy to extremely heavy, straight or wavy grained. It is easily turned on lathes.

Color. The sapwood is from 2 to 5 in. wide; it is white when freshly cut, but may darken slightly to grayish brown when exposed to air for some time. The heartwood is jet black or blackish brown, irregular in outline, and usually very small. The annual rings are marked by a row of larger pores at the beginning of the earlywood, but are not conspicuous because the pores are not numerous and more or less gradually decrease in size toward the outer portion of each ring.

Macroscopic features. The wood is semi—ring—porous with pores barely visible to the unaided eye. The thick—walled vessels are comparatively large, especially in the earlywood, decreasing in size either somewhat abruptly or gradually toward the latewood. Occasionally they are in rows of 2—5; tyloses are lacking.

The rays are very fine even under a lens. On the tangential surface they appear in tiers or stories producing very fine bands running across the grain and visible with the unaided eye. The apotracheal—banded parenchyma are visible with a lens as numerous very fine, light colored, irregular, tangential lines no wider than the rays; with the rays, they impart a net—like appearance.

Similar woods. The heartwood is characteristic among native U.S. woods. The sapwood is similar to that of hickory with the heavy weight, large pores, and fine lines of parenchyma, but can be distinguished by the fine horizontal bands on the tangential surface, which are due to the storied arrangement of the rays.

Sassafras (Fig. 27-49)

The range of sassafras (*Sassafras albidum*) covers most of the eastern half of the U.S. from southeastern Iowa and eastern Texas eastward.

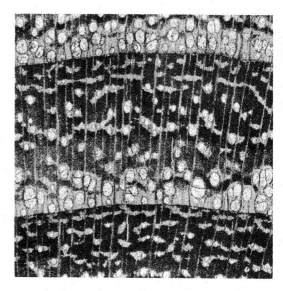

Fig. 27-47. *Maclura pomifera* (Koehler, 1917)15×.

Uses. Sassafras is moderately heavy, moderately hard, moderately weak in bending and endwise compression, quite high in shock resistance, and has high resistance to decay. It was highly prized by the Indians for dugout canoes, and some sassafras lumber is now used for small boats. It is used for fence posts and rails and general millwork, for foundation posts, and some wooden containers.

Color. The sapwood is light yellow and rarely over 0.55 in. wide; the heartwood varies

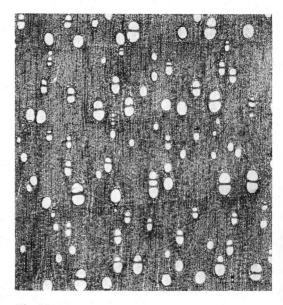

Fig. 27-48. *Diospyros virginiana* (Koehler), 15×.

from dull grayish brown to dark brown, sometimes with a reddish tinge. The annual rings are very distinct and of moderate width. The wood has an odor of sassafras on freshly cut surfaces. Sassafras has less luster than catalpa.

Macrostructure. The wood is ring—porous. The large pores of the earlywood from a band that is 3—5, or even 7, vessels wide; they are filled with glistening tyloses. The latewood vessels are very small (not visible without a lens), mostly isolated or in radial rows of two or three, and not crowded. The rays are fine, but distinct without a lens. The paratracheal—confluent and aliform parenchyma are noticeable as a narrow lighter—colored area in the latewood.

Similar woods. The wood of sassafras is easily confused with black ash, which it resembles in color, grain, and texture. See ash for distinguishing characteristics of other woods.

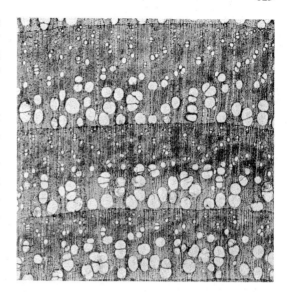

Fig. 27-49. *Sassafras albidum* **(Koehler), 15×.**

Sweetgum (Figs. 27-50 and 27-51)

Sweetgum (*Liquidambar styraciflua*), alias red gum and gum—tree, grows from southwestern Connecticut westward into Missouri and southward to the gulf. Lumber production is almost entirely from the Southern and South Atlantic States. The lumber from sweetgum is usually divided into two classes: sap gum, the light—colored (white with reddish tinge) wood from the sapwood, and red gum, the reddishbrown heartwood.

Uses. Sweetgum must be carefully dried due to its interlocked grain; it causes a ribbon stripe that is desirable for interior finish and furniture. The wood is rated as moderately heavy and hard. It is moderately strong, moderately stiff, and moderately high in shock resistance. Sweetgum is used principally for lumber, veneer, plywood, slack cooperage, railroad ties, fuel, and pulpwood. The lumber goes principally into boxes and crates, furniture, cabinets, interior trim, and millwork. Sweetgum veneer and plywood are used for boxes, pallets, crates, baskets, and interior woodwork.

Color. The sapwood is 1—5 in. thick and subject to blue stain. It is whitish with a pinkish tinge. The heartwood is reddish brown and variable; occasionally it streaked with darker color. Traumatic resin canals may be present.

Macroscopic structure. Sweetgum is a diffuse—porous wood. The pores tend to be solitary, are very numerous, closely spaced, and appear to occupy one—half to two—thirds of the area between rays on the transverse surface. They are visible only with magnification. The appearance of pore pairing occurs in the cross section where the perforation plate between two vessel elements is present. Rays are fine, closely spaced (numerous), and visible on the cross section and radial surface without magnification; they appear to occupy about half of the cross—sectional area. Longitudinal parenchyma are sparse; strands of apotracheal—diffuse and paratracheal parenchyma are occasionally present but difficult to see even with a lens. Growth—ring boundaries are indistinct and are marked by a few rows of thick—walled and somewhat flattened fibers.

Similar woods. The heartwood color is characteristic. Black tupelo has less vessel area, vessels in radial multiples, and irregularly distributed vessels. Cherry is usually darker, more lustrous, and has more distinct rays.

American sycamore (Fig. 27-52 to Fig. 27-54)

American sycamore (*Platanus occidentalis*) is also known as sycamore and sometimes as buttonwood, buttonball tree, water beech, and planetree. Sycamore grows from Maine to Nebraska, southward to Texas, and eastward to Florida. In the production of sycamore lumber, the Central and Southern States rank first.

Uses. The wood has a fine texture and interlocked grain; it shrinks moderately in drying and may warp; it is moderately heavy, moderately hard, moderately stiff, moderately strong, and has good resistance to shock. Sycamore is used principally for lumber, veneer, railroad ties, slack cooperage, fence posts, and fuel. Sycamore lumber is used for furniture, boxes (small food containers), pallets, flooring, handles, and butcher's blocks. The veneer is used for fruit and vegetable baskets, and decorative panels and door skins.

Color. The heartwood of sycamore is reddish brown and sometimes not sharply delineated from the sapwood; the sapwood is white to yellowish and 1.5—3 in. thick. The annual rings are distinct but are not differentiated into distinct earlywood and latewood.

Macroscopic features. The wood is diffuse—porous. The vessels are very small and not visible to the unaided eye. They are crowded in the earlywood but gradually decrease in number in the latewood. They have some tangential pairing in the latewood. Toward the end of the growth ring they forming a narrow, lighter colored band.

Rays are visible on all surfaces. They appear

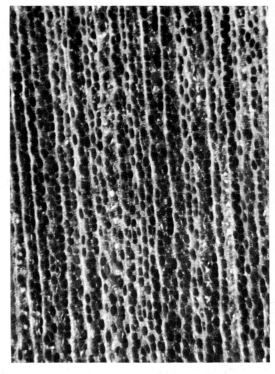

Fig. 27-50. *Liquidambar styraciflua.*

Fig. 27-51. Sweetgum (*Liquidambar styraciflua*) (50×).

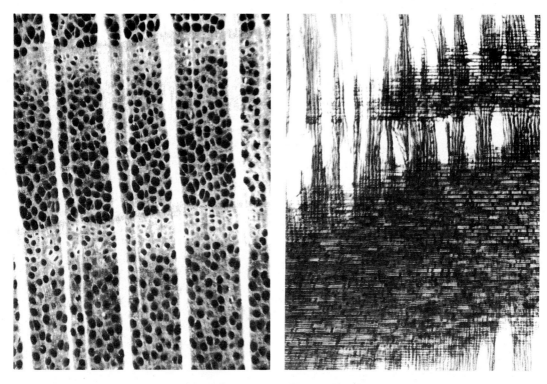

Fig. 27-52. *Platanus occidentalis.*   Fig. 27-53. *Platanus occidentalis, r* view, 50×.

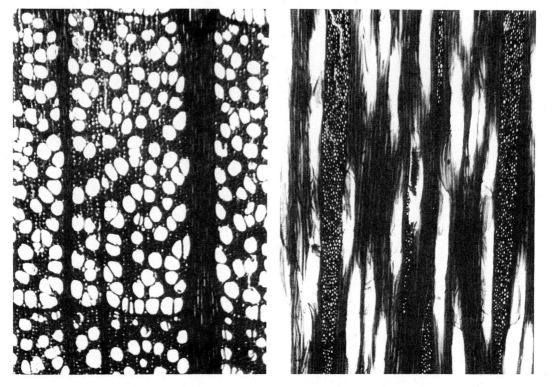

Fig. 27-54. American sycamore (*Platanus occidentalis*) (50×).

uniformly spaced on the cross section and of uniform height on the radial surface. These rays appear as "silver grain" on the radial surfaces. The parenchyma are not visible even with a lens.

Other species. Sycamore is not easily confused with other species. Its numerous conspicuous rays and interlocked grain make it easily recognizable. It is distinguished from beech by its rays, only a small proportion of which are broad in beech. Beech also has a distinct darker and denser band of latewood.

Tanoak (Figs. 27-55 and 27-56)

Tanoak (*Lithocarpus densiflorus*) has some importance commercially, primarily in California and Oregon. It is also known as tanbark—oak because at one time high—grade tannin in commercial quantities was obtained from the bark. This species is found in southwestern Oregon and south to Southern California, mostly near the coast but also in the Sierra Nevada range.

Uses. The wood is heavy, hard, and, except for compression perpendicular to the grain, has roughly the same strength properties as eastern white oak. Volumetric shrinkage during drying is

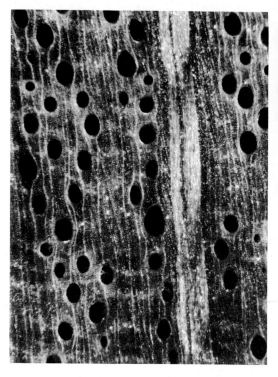

Fig. 27-55. *Lithocarpus densiflorus.*

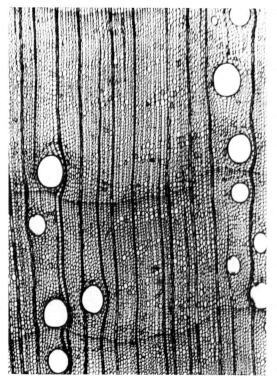

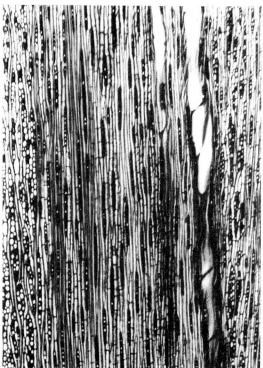

Fig. 27-56. Tanoak (*Lithocarpus densiflora*) (50×).

more than for white oak, and it has a tendency to collapse during drying. It is quite susceptible to decay, but the sapwood takes preservatives and stains easily. It has straight grain and machines and glues well. Because of tanoak's hardness and abrasion resistance, it is an excellent wood for flooring in homes or commercial buildings. It is suitable for industrial applications such as truck flooring. Tanoak treated with preservative has been used for railroad crossties. The wood has been manufactured into baseball bats with good results. It is also suitable for veneer, both decorative and industrial, and for high—quality furniture. It is used in corrugating medium.

Color. The sapwood of tanoak is light reddish brown when first cut and turns darker with age to resemble the heartwood, which also ages to dark reddish brown.

Macroscopic features. The wood is semi—ring—porous with a gradual transition in pore size. The latewood vessels form oblique groups or clusters. The apotracheal parenchyma are banded and abundant. They are visible with a lens as irregular wide lines. The rays are of two sizes; the large rays are irregularly distributed.

Tupelo (Fig. 27-57)

The tupelo group includes water tupelo (*Nyssa aquatica*), also known as tupelo gum, swamp tupelo, and gum; black tupelo (*N. sylvatica*), also known as black gum and sour gum; swamp tupelo (*N. sylvatica* var. *biflora*), also known as swamp blackgum, blackgum, tupelo gum, and sour gum; Ogeechee tupelo (*N. ogeche*), also known as sour tupelo, gopher plum, tupelo, and Ogeechee plum.

All except black tupelo grow principally in the southeastern U.S. Black tupelo grows in the eastern U.S. from Maine to Texas and Missouri. About two—thirds of the production of tupelo lumber is from the Southern States.

Uses. The wood has fine, uniform texture and interlocked grain that requires careful lumber drying. Tupelo wood is rated as moderately heavy. It is moderately strong, moderately hard and stiff, and moderately high in shock resistance.

Fig. 27-57. Water tupelo (*Nyssa sylvatica*) (50×).

Buttresses of trees growing in swamps or flooded areas contain wood that is much lighter in weight than that from upper portions of the same trees; this wood should be separated from the heavier wood to assure material of uniform strength. Tupelo is cut principally for lumber, veneer, pulpwood, and some railroad ties and slack cooperage. Lumber goes into boxes, pallets, crates, baskets, and furniture.

Color. Wood of the different tupelos is quite similar in appearance and properties. The heartwood is light brownish gray and merges gradually into the lighter colored sapwood, which is generally several inches wide.

Macroscopic structure. Black tupelo is diffuse—porous with small pores like sweetgum. The vessels are numerous and in radial multiples which may be irregularly distributed. Growth rings are inconspicuous to moderately distinct. Rays are slightly coarser than in sweetgum but are numerous and closely spaced; they are inconspicuous against the background color, but appear to occupy half of the surface (in the corss section, *x*). The longitudinal parenchyma occur in scattered paratracheal—scanty and apotracheal—diffuse arrangements, but are very difficult to see. See sweetgum for its distinction from black tupelo.

Microscopic structure. Intervessel pitting of black tupelo is opposite, pits are nearly square in outline, and there are up to 6 pits per row.

Black walnut (Fig. 27-58)

Black walnut (*Juglans nigra*) is also known as American black walnut. Its natural range extends from Vermont to the Great Plains and southward into Louisiana and Texas. About 75% of the timber is produced in the Central States.

Uses. Black walnut is normally straight grained, easily worked with tools, and stable in use. It has a faint characteristic odor that is manifested when it is worked. It is heavy, hard, strong, stiff, and has good resistance to shock. Black walnut wood is expensive and well suited for natural finishes. The outstanding uses of black walnut are for furniture, architectural woodwork, and decorative panels. Other important uses are gunstocks, cabinets, and interior finish. It is used either as solid wood or as plywood.

Color. The characteristic heartwood of black walnut varies from light to dark brown; the

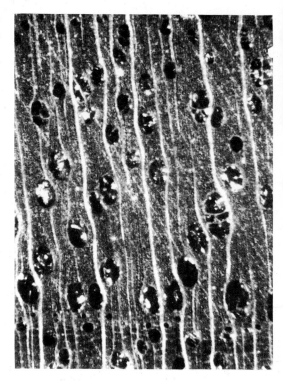

Fig. 27-58. *Juglans nigra.*

sapwood is nearly white and up to 3 in. wide in open—grown trees, but typically less than 1 in. wide. Annual rings are distinct and marked by the change in pore size.

Macroscopic features. Black walnut is not easily confused with any other native species. The wood is semi—ring—porous. Pores are barely visible in the cross section (with tyloses) to the unaided eye but are easily seen as darker streaks or grooves on the longitudinal sections. The pore arrangement is similar to that of hickories and persimmon, but the pores are smaller in size. The rays are very fine, inconspicuous, and not distinct without a lens. The apotracheal parenchyma are readily visible in the sapwood as several lighter colored, irregular, tangential bands; they are indistinct in the heartwood without a lens.

Microscopic features. The netlike (*reticulate*) thickening that occurs sporadically in the latewood vessels (*gash pits*) distinguishes it from butternut and English walnut of the same genus. Black walnut has crystals in strand parenchyma (that tend to dull woodworking tools) but butternut does not.

Black willow (Fig. 27-59)

Black willow (*Salix nigra*) is the most important of the many willows that grow in the United States. It is the only one to supply lumber to the market under its own name. Black willow is most heavily produced in the Mississippi Valley from Louisiana to southern Missouri and Illinois.

Uses. The wood of black willow is uniform in texture, with somewhat interlocked grain. The wood is light in weight. It has exceedingly low strength as a beam or post and is moderately soft and moderately high in shock resistance. It has moderately large shrinkage. Willow is cut principally into lumber. Small amounts are used for slack cooperage, veneer, excelsior, charcoal, pulpwood, artificial limbs, and fenceposts. Black willow lumber is remanufactured principally into boxes, pallets, crates, caskets, and furniture. Willow lumber is suitable for roof and wall sheathing, subflooring, and studding.

Color. The heartwood of black willow is grayish brown or light reddish brown frequently containing darker streaks. The sapwood is whitish to creamy yellow.

Macroscopic structure. The wood is semi—ring—porous. The latewood vessels are small, numerous, and solitary or in radial multiples of 2 to 4. The marginal parenchyma are not conspicuous. The rays are narrow, normally spaced, and visible with a lens.

Yellow—poplar (Figs. 27-60 and 27-61)

Yellow—poplar (*Liriodendron tulipifera*) is also known as poplar, tulip poplar, tulipwood, and hickory poplar. Sapwood from yellow—poplar is sometimes called white poplar or whitewood. Yellow—poplar grows from Connecticut and New York southward to Florida and westward to Missouri. The greatest commercial production of yellow—poplar lumber is in the South.

Uses. The wood is generally straight grained and comparatively uniform in texture. Old—growth timber is moderately light in weight and is reported as being moderately low in bending strength, moderately soft, and moderately low in shock resistance. It has moderately large shrinkage when dried but is not difficult to season and stays in place well after seasoning.

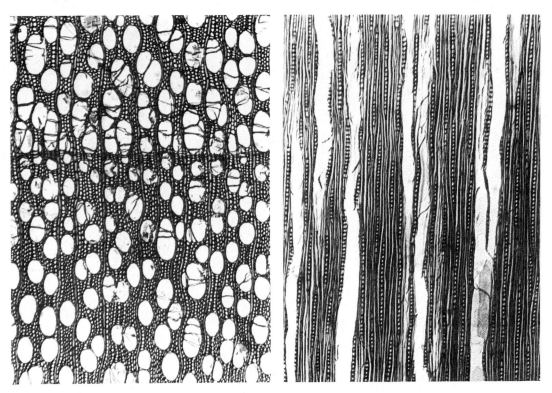

Fig. 27-59. Black willow (*Salix nigra*) (50×).

Much of the second—growth yellow—poplar is heavier, harder, and stronger than old growth. Selected trees produce wood heavy enough for gunstocks. Lumber is used in furniture, interior finish, siding, core stock for plywood, cabinets, and musical instruments, but use for core stock is decreasing as particleboard use increases. Yellow—poplar is frequently used for crossbands in plywood. Boxes, pallets, and crates are made from lower grade stock. Yellow—poplar plywood is used for finish, furniture, piano cases, and various other special products. Yellow—poplar is used for pulpwood and slack—cooperage staves.

Lumber from the cucumbertree (*Magnolia acuminata*) sometimes may be included in shipments of yellow—poplar because of its similarity.

Color. Yellow—poplar sapwood is white and often several inches thick; it merges gradually into the heartwood. The heartwood is yellowish brown, sometimes streaked with purple, green, black, blue, or red.

Macroscopic structure. Yellow—poplar is a diffuse—porous wood. The pores are numerous and in radial multiples that frequently are diago-

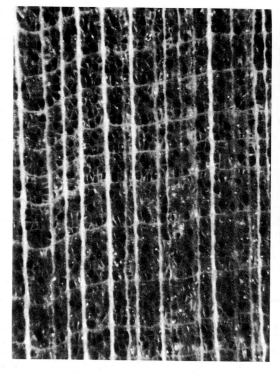

Fig. 27-60. *Liriodendron tulipifera.*

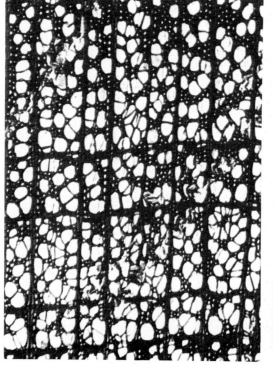

Fig. 27-61. Yellow-poplar (*Liriodendron tulipifera*) (50×).

nal. Rays are normally spaced, much fewer than in sweetgum, and visible to the unaided eye. Longitudinal parenchyma are marginal, forming a continuous white line of several rows in the outer portion of the latewood. The vessel distribution is variable, frequently being less numerous in the outer latewood.

Microscopic structure. Intervessel pitting is opposite, pits are rectangular in outline.

Similar woods. Yellow—poplar can usually be distinguished from other woods by its greenish tinge. Sweetbay (magnolia) is very similar, but can be separated by use of a microscope.

Non—North American Species
Eucalyptus (Fig. 27-62)

Eucalypts (*Eucalyptus* spp.) constitute a genus with over 500 species of greatly differing properties; some are suitable for pulping, while others are not. *E. marginata* is called jarrah and is native to the coastal belt of southwestern Australia. The heartwood is a uniform pinkish to dark red, often with a rich, dark red mahogany hue, turning to a deep brownish red with age and exposure. The sapwood is pale in color and usually very narrow in old trees. The texture is even and moderately coarse. The density is 44 lb/ft^3. The grain is often interlocked or wavy. Gum veins and pockets are a common defect. The heartwood is extremely resistant to preservatives.

E. diversicolor is called karri and is a large tree limited to Western Australia. It is 57 lb/ft^3 and paler in color than jarrah, but otherwise similar in appearance. *E. grandis* is the principal hardwood species used for pulping in Brazil; it is more difficult to pulp than *E. globulus*. Other species include (Tasmanian) bluegum (*E. globulus* Labill.), blackbutt (*E. pilularis* Sm.), and robusta (*E. robusta* Sm.).

Gmelina

Gmelina (*Gmelina arborea*) has some limited use in pulp and paper. It is distributed in India, Burma, and eastward to Vietnam. It is a common plantation species. It takes 20 years for it to grow to 100 feet high with a diameter of 2 feet.

The grain is interlocked and wavy. The texture is moderately coarse. The wood is pale straw yellow. The fibers are 0.8 to 1.3 mm long with diameters of 25—35 μm.

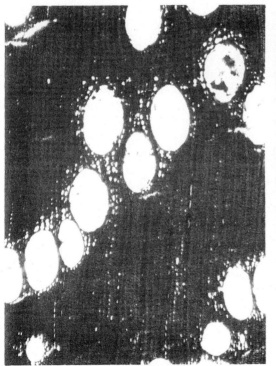 

Fig. 27-62. A eucalypt (*Eucalyptus grandis*) (50×).

28

WOOD FIBER ANATOMY AND IDENTIFICATION

28.1 FIBER ANALYSIS

Introduction

The references for this chapter are found on pages 537—540 of Chapter 25.

It is useful to look at wood in terms of its fiber anatomy and properties, since this is how it is used by the pulp and paper industry.

There are many excellent references available for fiber analysis (see the Annotated Bibliography in Chapter 25), and much of the information presented in earlier chapters can be used. This chapter gives an overview of fiber morphology from wood and shows what features to look for in fibers for their identification. Information is also given to help determine pulping and bleaching methods used to prepare the pulp.

It is fundamental that the properties on paper depend on the fiber properties and the method of fiber preparation (pulping, bleaching, and fiber processing). It is difficult to separate wood prior to pulping (latewood from earlywood, fast growth wood from slow growth wood, compression wood from normal wood, juvenile wood from mature wood, etc.), so variability of wood must be overcome by varying the processing of the wood and pulp. The availability of and price of wood are as important as the properties of the wood itself.

Fiber anatomy considerations

Identification of fibers is difficult since there is much less information available than in the wood; identification is often down to a few members within a genus, rather than to a particular species. Other species, like Douglas—fir with spiral thickening, are unique and can be identified as to the species. The situation is complicated by the fact that many papers can contain two, three, or more types of fibers, especially if the paper contains appreciable amounts of recycled fiber. Panshin and de Zeeuw (1980, pp. 657—666) include a key for fiber identification.

It is important to have authentic fiber samples of contributing species on hand for comparison with unknown samples. For example, if one is using a lot of "precommercial" thinnings with a high percentage of juvenile wood, fiber length will be shorter than published values.

It takes a large investment of time to become proficient at fiber identification, and many references containing micrographs should be used. This chapter allows one to visualize the major differences in pulps that contribute to the papermaking properties and is useful for simple identifications, such as determining when various places in a mill are affected by the changeover of a digester line from one species to another.

In mechanical pulps one can learn about the relative morphology of the contributing species by looking for fiber bundles.

Weight factors

When attempting quantitative analysis of fiber mixtures, the number of a particular type of fiber is multiplied by its *weight factors* since some types of fibers appear to be present in larger amounts than others, such as those with low surface areas. Weight factors are given in T 401 [based largely on Graff, J.H., *Paper Trade J.* 110(2):37(1940)]. The weight factors for coastal Douglas—fir and loblolly pine are 1.40, for most hardwoods 0.50, and for bleached straw 0.35. The higher the relative surface area (per given mass of pulp), the lower the weight factor. Parham and Gray (1990) give weight factors on page 35 [largely from D.W. Einspahr and J.D. Hankey, *Tappi* 61(12): 86-87(1978)]. One could determine one's own weight factors for a particular application. Weight factors are a crude method of taking number averages and converting them to weight averages.

Fiber pitting

Pitting, in general, is very useful in the determination of isolated fibers. However, one must keep in mind that pulping operations and refining are bound to affect the way a pit looks under the microscope. Intervessel pitting of hardwoods and ray cross—field pitting in softwoods are the principal means of identification. Ray cross—field pitting shows whether ray parenchyma and or tracheids are present; it also indicates the height of the rays.

Fiber and wood staining

Chemical stains can help with the analysis by identifying the type of pulping method used to generate the fibers. This might be more important in brown papers that might have some recycled newsprint rather than in white printing papers, but these might have some bleached CTMP pulp.

Hoadley (1990) references examples of species determination of wood using chemical stains including boxelder and other maples, red gum and black tupelo, elms, pond and sand pines, spruces and pines, and other groups. Separation of red and white oaks and red and sugar maples is discussed in Panshin and de Zeeuw (1980). Eastern spruce and balsam fir are separated by the method of Kutscha et al. in *Wood Sci. and Tech.* 12:293-308(1978). Separation of heartwood and sapwood is referenced (Kutscha and Sachs, 1962).

Iodine is used to detect the presence of starch, which might be present in sapwood but not heartwood or in paper such as in the detection of forged currency. It is used with iodide to form the soluble I_3^- species in water (0.3% I_2 and 1.5% KI in water). *Safranin O* is a general stain for cellulosic materials (red). It is used as a 0.5—1% solution in water or 50% ethanol.

Pulp stains are described in detail in Graff (1940), Isenberg (1967), and TAPPI Standard T401 om-82. Graff (1940) describes three types of stains. (1) Spot stains or groundwood reagents using chemicals such as aniline sulfate, phloroglucinol & hydrochloric acid, and *p*—nitroaniline for detection of mechanical pulps in paper. The first class is said here to be of limited value. (2) The iodine/iodide metallic salt stains including the Herzberg stain, the "A" stain, and the "C" stain; these give qualitative information about fibers. (3) Aniline dyes for the determination of the degree of cooking, bleaching, and purity of pulps.

Phloroglucinol is specific for lignin, causing it to turn red. It is used as a 2% solution in 18.5% HCl. Isenberg (1967) suggests mixing 1 g in 50 mL of EtOH to which is added 25 mL of concentrated HCl; this is made up immediately before use. One use is to show mechanical pulps in white papers.

Mechanical softwood and hardwoods are differentiated with the use of 2% aniline sulfate made acidic (1 drop concentrated H_2SO_4 per 50 mL) followed by 0.02% methylene blue after removal of the first dye by blotting. Softwoods are yellow and hardwoods are bluish green. The Mäule reaction can also be used.

Mechanical pulp bleached with dithionite no more than a few months old is determined by detecting traces of SO_2 or sulfite as H_2S liberated by stannous chloride. Several stains are available to determine the level of bleaching (or cooking) of pulps, such as Bright stain. These can be used to check the uniformity of bleaching (and pulping) in the mill (Isenberg, 1967). Stains even exist for the identification of dirt, although scanning electron microscopy (SEM-EDDAX) page 184) has made many of these methods somewhat obsolete.

Other techniques

Monosaccharide analysis (although tedious) will give a good indication of the ratio of hardwoods to softwoods since galactose and mannose come from softwoods. Dissolving pulps will have low levels of hemicelluloses, of course.

28.2 SOFTWOOD FIBER

Fibers over 2 mm long are most often softwood fibers. Be sure to use Table 26-3 (page 550) as a general guide to softwood fiber analysis. and Table 26-4 (page 552) as a summary of some important softwood fiber properties. Fig. 28-1 shows six types of softwood fibers.

Parham and Gray (1990) claim that Douglas—fir, redwood, baldcypress, podocarp, parana—pine, sugar pine, white pine, and hard pines should be distinguishable under most circumstances, although these separations require a high degree of skill. Spruce, larch, and hemlock may not be able to be distinguished from each other as with true—fir and western redcedar. If the source of the pulp is known then regional information can be used (i.e., European or American; eastern or western, etc.).

Ray cross—field pitting in softwoods is the principal means of identification. Examination of many ray cross—field pits shows whether ray tracheids are present, whether ray tracheids are marginal or interspersed, and may indicate the average height of the rays.

For example, Fig. 28-2 shows some fibers with ray cross—field pitting on the tracheids. The

Douglas—fir   Southern pine   Eastern white pine

Sitka spruce   Western hemlock   Western cedar

Fig. 28-1. Various softwood fibers isolated in the laboratory (60×).

western white pine shows the fenestriform ray cross—field pitting that is characteristic of the white pines and the small pits are from ray tracheids. Ponderosa pine has pinoid ray cross—field pitting with pits that are variable is size and up to 7 pits per cross—field. The Sitka spruce fiber shows piceoid ray cross—field pitting of uniform size with 3 or more pits per cross—field common.

28.3 HARDWOOD FIBER

Be sure to use Table 27-1 (page 590) as a general guide to hardwood pulp analysis. Whole fibers less than 2 mm long usually originate from hardwoods. Hardwood fibers are accompanied by vessels or vessel fragments depending on the pulping process. The fibers shown in this chapter were isolated by a laboratory pulping process that leaves the vessels largely intact. Furthermore, parenchyma cells have not been removed as occurs to various degrees in pulp processing. Fig. 28-3 shows fibers from hardwoods.

Hardwood libriform fibers or fiber tracheids are not helpful in species determination unless helical thickening is observed. Vascular or vasicentric tracheids [the shape of vasicentric tracheids is observed in red oak (upper right) in Fig. 28-3], indicate the presence of certain species. However, vessel elements offer much information based on their overall size, shape, intervessel pitting, and ray contact pitting. The intervessel pitting of *Populus* (observe black cottonwood in Fig. 28-3) and *Salix* genera are easy to recognize . Ring—porous woods may be difficult to determine since the large vessels are generally fragmented during processing.

Western white pine   Ponderosa pine   Sitka spruce

Fig. 28-2. Portions of fiber tracheids of three softwoods (600×).

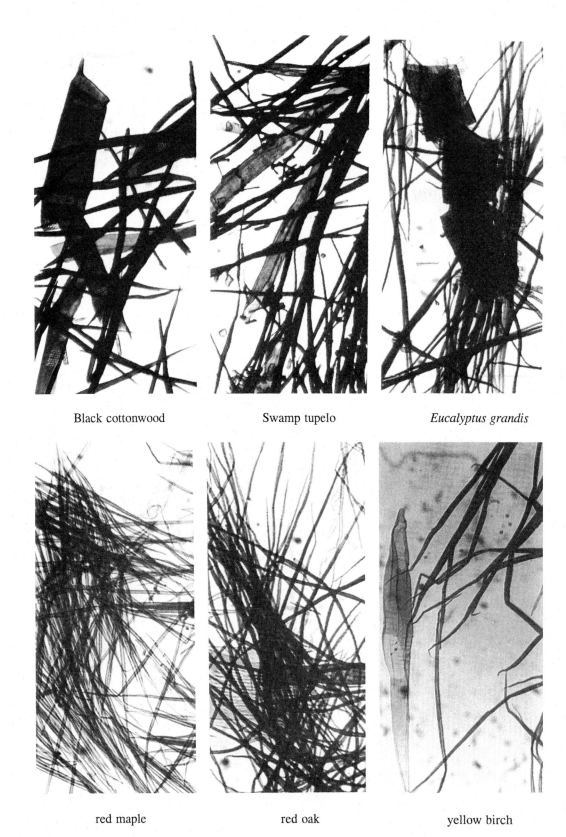

Black cottonwood   Swamp tupelo   *Eucalyptus grandis*

red maple   red oak   yellow birch

Fig. 28-3. Various hardwood fibers isolated in the laboratory (60×).

29
NONWOOD FIBER USE IN PULP AND PAPER

29.1 INTRODUCTION

Nonwood fiber is increasingly being used worldwide as a fiber source for papermaking. According to Atchison (1988), major nonwood pulp production includes (in millions of metric tons per year) straw, 5.4 (over 80% produced in China); sugar cane bagasse, 2.1; bamboo, 1.4; reeds, 1.6; cotton linters, 0.4; and other, 1.0. This chapter provides some background information on the topic. See pages 47-50 for more information.

Taxonomy

All seed plants (members of the phylum Spermaphyta) are divided into two subphyla: Gymnospermae (gymnosperms, those with naked seeds) and Angiospermae (angiosperms, those with seeds enclosed within the ovary of the flower).

Angiospermae are arranged into two groups: the *monocots* (*monocotyledons*) and *dicots* (*dicoty-ledons*). Monocotyledoneae have one leaf in the seed (a seed leaf is a cotyledon; a peanut consists of two cotyledons with a small embryo in the middle so peanuts are dicots) and include mostly herbaceous plants such as grasses, palms, lilies, and corn. A few plants with a wood—like stem such as bamboo that are used in pulp and paper are members of this class.

Dicotyledoneae have two leaves in the seed and include kenaf, hemp, flax, and the hardwoods. The numerous Dicotyledoneae are further divided into subclasses, orders, suborder, families, and some subfamilies.

Monocots have parallel veins in their leaves, and dicots have branching veins in the leaves.

Anatomy

The primary vascular tissue (xylem and phloem) exists in small bundles(Fig. 29-1). The

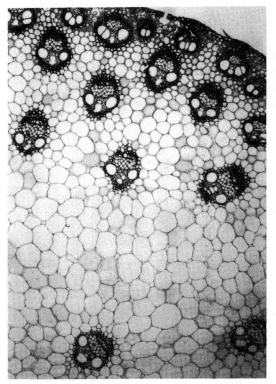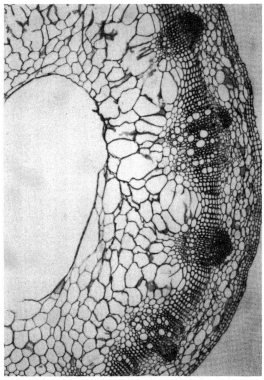

Fig. 29-1. Portions of a typical monocot stem (left) and herbaceous dicot stem (right) (50×).

phloem consists of fibers, sieve tube elements, companion cells, and parenchyma. In monocots these bundles are scattered in the stem throughout the ground tissue, whereas in dicots the bundles are all about the same distance from the center and form a ring of bundles around the pith. In dicots, the bundles unite as the stem grows outward; in many species the bundles are no longer distinct (Fig. 29-2). The scattering of the xylem remains in those monocots that undergo secondary growth.

The outer layer of cells is the epidermis. The layer just inside the epidermis is the cortex. The layer inside the cortex but outside the vascular tissue is the pericycle. Epidermal cells have sinuous or toothed margins and may have projections (in cells called trichomes) that are helpful in identification of samples.

Useful fibers

Useful fiber (that provides the strength for paper made from the pulp) is derived from the vascular tissue of monocots of barley, *Hordeum* spp.; rice, *Oryza* spp.; esparto, *Stipa tenacissima*; wheat, *Triticum* spp.; bamboo, *Phyllostachys* (Fig. 29-3); sugar cane, *Saccharum officinarum*; and others.

Bast, fibers from the phloem of dicots, is derived from hemp, *Cannabis sativa*; kenaf, *Hibiscus cannabinus*; flax, *Linum usitatissimum* (Fig. 29-3); and others. Fiber may obtained from the pericycle or cortex of some dicots. Cotton is the seed hair of the cotton plant (Fig. 29-3). Fiber from the vascular tissue of leafs is obtained from sisal (*Agave sisalina*) and manila hemp (*Musa textilis*).

Nonwood fiber identification

A variety of nonwood fibers can be identified by a series of specialized stains. For example, several are available for hemp and flax. Acidified potassium dichromate swells flax somewhat faster than hemp. Cyanine stains these materials differently. Other separations exist for cotton, linen, and wood; cotton, flax, jute, and hemp; animal and plant fibers; etc. (Standard T 401 and others).

Straw morphology considerations

Fig. 29-4 shows a cross section of the straw stem. The usable fibers are the dark fibers near the edge of the stem consisting of phloem and other tissue.

According to Huamin (1988), parenchyma cells occupy 38% of wheat straw based on the cross—sectional area of the stem, so the mass percentage may be lower. These cells are small and thin—walled, contribute to decreased pulp freeness, and do not add much to the strength of paper. They cook more slowly than the fiber cells and use more cooking chemicals. The lignin, cellulose, and xylan composition of the various cell types is very similar.

Petersen (1991) found the α—cellulose content of the internode of straw of four grains to be 37—42% while the leaves had only 28—30% α—cellulose. The fibers are about 15—20% longer in the nodes compared to the leaves. The author suggests that whole straw should not be pulped anymore than one would pulp trees with leaves and all. The internode is 35% of orchard grass (including the inflorescence) and 43 to 48% of many other grasses.

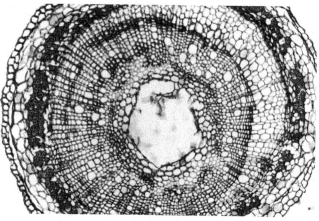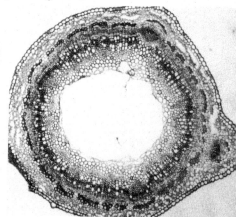

Fig. 29-2. Stem cross sections of *Cannabis* (left, 100×) and *Linum* (right, 40×).

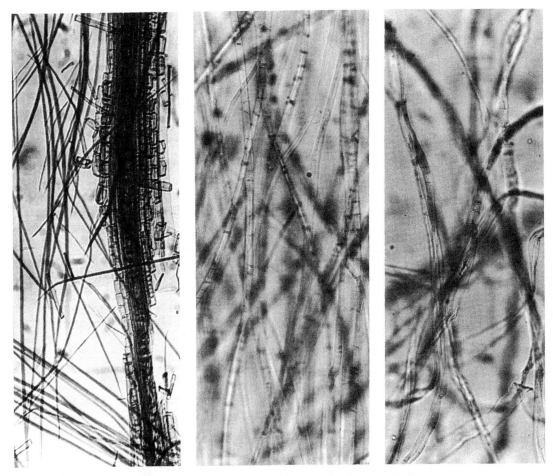

Fig. 29-3. Fibers of bamboo (left, with parenchyma, 60×), linen (center, 150×), and cotton seed hairs (right, 150×).

Depithing

Bagasse, and some other nonwood fibers, must be depithed before use. The pith contains small cells (parenchyma) that do not add to the pulp strength and reduce the freeness considerably.

29.2 PULPING METHODS FOR NONWOOD FIBERS

Introduction

Annual plants have been used since the early 1800s, except for cotton, which has been used much longer, of course. Straw has traditionally been pulped by boiling solutions of lime for board grades of paper; this led to a bright yellow pulp. Sodium hydroxide was used to make bleachable grades of pulp as well as board grades. The soda

anthraquinone (AQ) method is replacing the soda method and the results of the soda/AQ for straw are said to be similar to those for kraft pulping of straw. Other methods of pulping include neutral sulfite and chlorine systems.

Alkali, chlorine

The Pomilio process (Stephenson, 1951) was the most popular method at one time, being used in 20 mills. Developed around 1925, this process involved treating pulp with 8% NaOH (on dry pulp) in 60 ft cylindrical reaction towers where temperatures up to 130°C could be reached at the bottom. The digestion time was about 90 min. A chlorination step follows. The overall yield is 36—40%. The high use of chlorine and cellulose hydrolysis make this method obsolete.

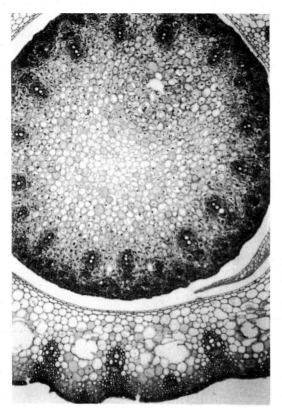

Fig. 29-4. Stem cross section of wheat (with part of a leaf) (50×).

Sulfite

Stephenson (1951) discussed a neutral sulfite (NSSC) method that enjoyed a lot of use in Europe. Sodium sulfite (10%) and 5% NaOH on pulp are combined with straw for a 6 hr cook at 160°C. The pulp yield is about 55%. Hypochlorite (5% on pulp) bleaching reduces the yield to 42%.

Aronovsky (1948) suggested an NSSC process for pulping wheat straw. This process uses 8% sodium sulfite and 2—3% sodium carbonate (on dry straw), with a liquor to straw ratio of 7:1, and a cooking time of two hours at 170°C. The pulp yield is 52—55% and can be brightened to 70% with 5—7% total chlorine. Bleaching above this brightness requires a three—stage process. The freeness was said to be 400—500 mL. Aronovsky (1949) found that the neutral sulfite process was unsuitable at atmospheric pressure, in agreement with recent unpublished results, but atmospheric pulping for 1 hour with 12% kraft chemicals in a hydrapuler at 90—98° was very effective. In other work (1947), he also suggested

2% lime and 2% sodium sulfite to give an exceptionally free pulp, although it would seem that calcium sulfite might precipitate.

Acid sulfite methods of pulping straw give poor results due to the relatively labile carbohydrates of straw. Relatively weak and brittle pulps result, and the process is unsuitable for materials with high silica contents.

Soda

England used the soda process to make much of its straw pulp at one time. In 1945, the United Kingdom pulped nearly 350,000 tons of straw for paper (FAO, 1952). Esparto uses less alkali and gave a higher yield than the cereal straws. Typically, esparto straw was pulped with 12% NaOH (on straw) for 5 hr at 160°C. Hypochlorite bleaching gave a yield of 36—40%.

Anon. (1948) discussed the French process developed by Huguenot for pulping straw. The continuous process involves pulping chopped straw with NaOH at 90—100°C for 30—50 minutes. The pulp then goes to a digester pulper at 80°C, where it is agitated for 4—7 hours. This process is also said to be suitable for waste paper. More information on soda pulping on nonwood plants is found on pages 47—50.

Alkali—oxygen, NACO process

Ceragioli (1975) describes alkali—oxygen pulping of straw. The author claims that sodium carbonate does not reduce the strength of pulp as does sodium hydroxide in the presence of oxygen. With carbonate, magnesium addition to protect carbohydrates from degradation appears to be unnecessary. A sodium carbonate—based pulping system has a simplified chemical recovery system.

The NACO process (U.S. patent 4,612,088 issued in 1986) uses oxygen and alkali (sodium carbonate and some sodium hydroxide) to pulp nonwood fibers (and upgrade secondary fiber) on a relatively small scale (Anon., 1984). A continuous, pressurized reactor (Turbo—Pulper™) was developed as part of the process. The process has been used commercially since 1986 in Italy at a 100 ton per day mill (IPZP Foggia) for nonwood pulp and 50 tons per day for upgrading OCC.

Unbleached straw pulp has a brightness of 50—52 (ISO) and kappa number of 15—16 with a yield of 48%. The brightness can be increased to 72% with one hypochlorite (H) stage or 82% with two H stages (6—7% as active chlorine). Ozone bleaching gives a brightness of 75—78%. Pulping is carried out at 7—8% consistency, 90 psig, and 130—135°C for a minimum of one hour.

29.3 CONSIDERATIONS FOR NONWOOD FIBER USE

Ash and silica both interfere with the pulping (and chemical recovery of pulping chemicals) of many nonwood fibers. The cellulose, hemicellulose, lignin, and extractive contents are fundamental to pulping. Silica is present at elevated levels; while wood has less than 0.1%, many annual crops such as straws have 0.5%—5% or more.

The chemistry of silica

An introduction to the chemistry of silica will serve as a basis for the behavior of silica during pulping operations. Silicon has an atomic weight of 28.09 and has a valence of 4 in the oxide forms. Silicon dioxide (SiO_2, also called silica or silicic anhydride) is 46.75% silicon and occurs in nature as a variety of minerals such as the quartz minerals and cristobalite. The Si—O bond is partially ionic (about 50%). SiO_2 occurs in crystalline and amorphous forms. The crystallized form is said to be inert to alkali.

SiO_2 is insoluble in acids and water. It is attacked by HF and ultimately converted to the volatile gas SiF_4. (This is the basis of some straw pretreatments with HF for silica removal.) The amorphous form is solubilized in alkali as the salts of silicic acid. Silicic acid has the formula of H_2SiO_3 and is the basis of silica gel desiccant and opal. The soluble ion under alkaline conditions is SiO_3^{2-} or $[SiO_2(OH)_2]^{2-}$.

Dean (1985) lists the two acid ionization constants of silicic acid and the solubility product of $CaSiO_3$ at 25°C. The pK_1 for silicic acid is 9.77 and the pK_2 is 11.80. (Methods that precipitate silica from black liquor using CO_2 must reduce the pH to about 9.0 to obtain near—complete precipitation of silica.) The pK_{sp} for $CaSiO_3$ is 7.60; $K_{sp} = 2.5 \times 10^{-8}$, corresponding to a solubility of 0.0184 g/L.

The chemical composition of straw

Aronovsky et al. (1943) analyzed a variety of straw materials as well (Table 29-1). The moisture contents were 6.6—8.4%. Barley is known to

have a relatively high content of pectins and gums, making a high water extractives level, although others have reported much lower results for water—soluble materials. Among the straws, the low ash content, high cellulose content, and long fibers of rye grain make it particularly useful for pulp.

Mineral composition, especially silica

Delga (1947) investigated the mineral (and elemental) composition of cereal stalks. They are high in soluble ash (1.54—4.7%). Potassium was 0.77—3.44% and silicon (not silica) was 0.35 to 0.73%. The highest level of sodium occurred in oat straw (0.85%), while the highest level of calcium was 0.26% in rye grain straw. Sulfur and phosphorus contents were low. Oat straws grown in two different soils did not vary in composition.

Fahmy and Fadl (1958) found Egyptian wheat straw to contain 8.34% ash and 3.44% silica. The leaf (sheath) is more concentrated with 14.9% ash and 6.90% silica while the stem is 5.90% ash and 2.24% silica. Wet or dry sorting of the prehydrolyzed raw material was said to remove the silica—rich epidermis with a 4% loss of material. Some work indicates that ethanol—benzene extraction allows silica to be removed with a second extraction step.

Silica is about 2.5—3.5% in bamboo; it is more concentrated in the nodes, where it may occur as nearly pure silica, than in the internode region. Bagasse is about 1.5% silica. The silica content of straw leaves is higher than that of the stems.

Table 29-1. Chemical composition (%) of cereal straws after Aronovsky (1943).

	Rice	Barley	Wheat	Rye	Oat
Extractives					
EtOH—benz.	4.6	4.7	3.7	3.2	4.4
Cold water	10.6	16.0	5.8	8.4	13.2
Hot water	13.3	16.1	7.4	9.4	15.3
1% NaOH	49.1	47.0	41.0	37.4	41.8
Lignin	11.9	14.5	16.7	19.0	17.5
Pentosans	24.5	24.7	28.2	30.5	27.1
α-cellulose	36.2	33.8	39.9	37.4	39.4
Ash	16.1	6.4	6.6	4.3	7.2
Nitrogen	0.6	1.1	0.4	0.7	0.5

These facts point out that if straw can be preprocessed to remove leaves and nodes, then many advantages may be realized.

Pulping and silica content

Fahmy and Fadl (1959) determined that the duration of alkaline pulping of wheat and rice straws was the most important parameter of the ash and silica content of the final pulp. Other variables have very little influence. For example, rice straw cooked at 150°C for 0.5 hour was lower in ash than that cooked for 4 hours, with the effect more pronounced for leaves than for stalks. This may partly be due to a lower pulp yield with increased cooking time. Bleached pulps from rice stalks cooked at 120°C for 0.5 hour had 0.064% silica and 0.15% ash.

Sodium sulfite pulping of bamboo with kraft green liquor removes less than 10% of the silica from straw compared to the 60—70% removed during kraft pulping. The sulfite process leads to high—strength and high—yield pulps.

Alkali chemical recovery and silica

Black liquor from straw pulping has about 5500 Btu/lb (on solids) compare to 6600 Btu/lb for black liquor from wood. A lower residual alkali (about 3 to 4 g/L) causes lignin and silica to precipitate during liquor concentration, making scaling and high viscosities big problems. Liquor viscosity is much higher than with wood pulp. A long time ago, Rinman developed a technique of adding some $Ca(OH)_2$ to the pulping liquor so that calcium silicate would precipitate onto fibers during the cook.

Grubshein (1961) pointed out that during causticization some of the sodium silicate is converted to insoluble calcium silicate. This decreases the causticizing efficiency and increases the lime mud volume. He concluded that the answer is removal of silica from the black liquor by one of two methods: 1) treat the black liquor with lime or 2) treat the black liquor with flue gases to lower the pH and precipitate silicic acid. Method 1 is covered in the patent by Gruen (1953), who suggested the precipitation with CaO occur near or above the boiling point of the black liquor for a short period of time (5—10 min) to decrease the amount of organic material precipitated. The process was patented in Germany by Schwalbe

(1929). (One mill in South Africa uses ferric oxide and alumina to precipitate silica.) The CO_2 method is more common since costly lime is not used. Also, scaling will be lower in subsequent evaporation. The use of CO_2 precipitation with kraft liquors would probably increase the TRS emissions in the recovery boiler since this process is akin to direct contact evaporation.

The experimental findings of Lengyel (1960) indicate that 6 to 8 g/L silica in black liquor can be evaporated without corrosion provided that addition of excess NaOH is used to dissolve incrustations that form. Silica at this level does not overly interfere with causticization but does increase the quantity of lime mud by about 100% and begins to retard the sedimentation rate. Corrosion and scaling are aggravated by allowing the black liquor to stand motionless for prolonged periods of time. Black liquor with more than 8 g/L silica should be treated with CO_2 (favored) or CaO (less favored). Lime mud can be purged if calcium silicate builds up in the system.

Sawheny (1988) reports that silica precipitates from soda black liquor from pH 10.2 to 9.1. At the lower pH large amounts of lignin also precipitate. The solubilities are also temperature dependent. In pilot plant tests involving several species of nonwood plants, careful carbonation (with a bubble reactor) of black liquor containing 6 g/L silica followed by filtering in a filter press resulted in 90% of the silica being removed as a solid containing 70% silica. The author claims rapid precipitation/sedimentation of large silica particles.

Ibrahim (1988) indicates that precipitation of silica from black liquor is most effective if the black liquor is preconcentrated to at least 8% solids. Either CaO or CO_2 would precipitate over 95% of the silica under these conditions. (The concentration is often about 4% solids off the brown stock washers.) Long settling times (6 hours) were needed. The long time period often means that the pH drops and lignin may continue to precipitate. The precipitate should be washed to recover useful alkali. Centrifugal separation of the silica precipitate was the most efficient precipitation method. The best precipitation of silica was achieved at pH 9—10 (at 50°C) with a flue gas (with CO_2 concentration of 6—8%) application of 50—150 m^3 per m^3 of black liquor.

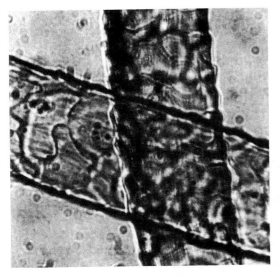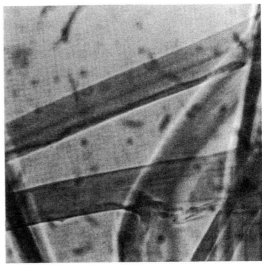

Fig. 29-5. Wool (left) and silk (right) fibers (600×).

29.4 OTHER FIBER TYPES

Animal fibers (Fig. 29-5)

Wool and human hair fibers have scales on the surface. Silk is formed by the secretion of proteins from two main glands of the silkworm caterpillar (*Bombyx mori*). Silk often can be seen as two separate component strands and it is usually of uneven diameter.

Polymer fibers

Synthetic fibers (Orlon and Dacron are shown in Fig. 29-6) can be quite long.

Fig. 29-6. Orlon and Dacron fibers (150×).

29.5 ANNOTATED BIBLIOGRAPHY

Nonwood fibers

1. Anon., Continuous manufacture of pulps from annual plants, *Papeterie* 70(9):271, 273-276 (Sept., 1948).

2. Anon., Italian mill tires new pulp process, *PPI* (Jan.):48, 49, 52(1984).

3. Anon., Tappi Standard T 259 om-83, Species identification of nonwood plant fibers, 11 p., 5 references. This reference has numerous micrographs. It has information on the fiber types obtained from species of commercial importance.

4. Anon., Tappi Standard T 401 om-88, Fiber analysis of paper and paperboard, 12 p., 32 references. This method describes preparation of slides with the use of stains for the identification of fibers from wood and nonwood plants.

5. Aronovsky, S.I., G.H. Nelson, and E.C. Lathrop, *Paper Trade J.* 117(25):38-48(1943).

6. Aronovsky, S.I., A.J. Ernst, and H.M. Sutcliffe, Pulping with sodium sulfite to produce strawboard, *Tech. Assoc. Papers* 30:321-323 (June, 1947).

7. Aronovsky, S.I., G.H. Nelson, A.J. Ernst, H.M. Sutcliffe and E.C. Lathrop, *Paper Trade J.* 127(11):154-162(1948) and *Tech. Assoc. Papers* 31:291-299(1948).

8. Aronovsky, S.I. and E.C. Lathrop, A new mechano-chemical process for pulping agricultural residues, *Tappi* 32(4):145-149(April, 1949).

9. Atchison, J.E., World capacities for nonwood plant fiber pulping increasing faster than wood pulping capacities, *Tappi Proceedings, 1988 Pulping Conference*, pp 25-45.

10. Casey, J.P., *Pulp and Paper: Chemistry and Chemical Technology*, 2nd, Ed., Vol. 1, Interscience, New York, 1960. Pages 398-425 discuss the pulping of various nonwood fibers.

11. Ceragioli, G., Wheat straw pulping by alkali-oxygen processes—cooking variables and pulp processes, pp. 227-252 (1975).

12. Dean, J.A., *Lange's Handbook of Chemistry*, 13th ed., McGraw-Hill, New York, 1985.

13. Delga, J., Chemical composition of the stalks of cereals, *Mém. services chim. etat* (Paris) 33:7-73 (1947).

14. Fahmy, Y. and M. Fadl, Digestion of wheat straw for the economical production of purified pulps of low silica content, *Textil-Rundschau* 13(12):709-719(1958).

15. Fahmy, Y. and M. Fadl, Solubility of ash and silicic acid during alkaline pulping of cereal straws, *Das Papier* 13(13/14):311-314(1959).

16. FAO, *Tropical Woods and Agricultural Residues as Sources of Pulp*, Rome, Italy 1952, 190 p. Sulfite pulping of Paraná-pine (23 p.), chemical pulps from Australian eucalyptus (6 p.), and agricultural residues (47 p.) are some highlights of this work.

17. Grubshein, B.D., Recovery of lime in the alkaline pulping of straw and reed, *Bumazh. Prom.* 36(8):12-13(1961). Russian.

18. Gruen, B. H. E., U.S. patent 2,628,155 (Feb. 10, 1953).

19. Huamin, Z., Separation of fibrous cells and parenchymatous cells from wheat straw and the characteristics in soda-AQ pulping, *1988 International Non-wood Fiber Pulping and Papermaking Conference*, China, pp 469-478.

20. Misra, D.K., Pulping and bleaching of nonwood fibers, in *Pulp and Paper: Chemistry and Chemical Technology*, Vol. 1, 3rd ed., Casey, J.P., Ed., Wiley, New York, 1980, pp. 504-568.

21. Petersen, P.B., Industrial application of straw, *1991 Wood and Pulping Chemistry*, Tappi Press, Atlanta, pp. 179-183.

22. Stephenson, J.N., Ed. *Preparation of Stock for Paper Making*, Vol. 2, McGraw-Hill, New York, 1951. Pages 1-91 discuss pulping of nonwood fibers.

Silica and kraft liquor recovery
23. Grubshein, B.D., Recovery of lime in the alkaline pulping of straw and reed, *Bumazh. Prom.* 36(8):12-13(1961). Russian.

24. Ibrahim, H., Silica is no longer a problem in the recovery of heat and chemicals from nonwood plant fibers black liquor, *1988 International Non-wood Fiber Pulping and Papermaking Conference*, China, pp 877-889.

25. Lengyel, P., Silica removal from black liquors, *Zellstoff u. Papier* 9(3):89-94(1960).

26. Sawheny, R.S., Desilicanization of black liquor-a glance, *1988 International Non-wood Fiber Pulping and Papermaking Conference*, China, pp 933-942.

27. Schwalbe, C.G., Process for desalination of alkaline silicious spent liquors, German Patent 522,730 (1929).

30
HYDRAULICS

30.1 FLOW OF LIQUIDS

Introduction

The material in this section is examined in more detail in many introductory works such as McGill (1980). Related concepts have been presented elsewhere, such as the static head of a headbox (page 213).

Reynolds number

The flow of liquids in pipes can be difficult to describe mathematically. In order to simplify the mathematical development, flow is often considered in circular—cross—section pipe (as is done here). The *Reynolds number* is an important parameter used to predict the flow pattern in a pipe. The Reynolds number is defined as:

$$Re = \frac{D\,u_b\,\rho}{\mu}$$

where

> Re = Reynolds number
> D = inside diameter of pipe
> u_b = bulk fluid velocity
> μ = fluid viscosity

If the Reynolds number is below 2000, the flow tends to be *laminar* or *streamline*; that is, the pressure and flow velocity at any point remains constant with time. Each point on a concentric circle has the same pressure and flow velocity. The flow velocity is highest at the center point.

If the Reynolds number is above 2000, the flow tends to be unstable or *turbulent*; that is, the flow velocity at a given point is not constant with time, although there may be a layer of laminar flow near the pipe wall. At high Reynolds numbers the flow becomes quite *turbulent*.

The addition of a small amount of fiber to water (on the order of 0.3% consistency although it can be higher at higher flow velocities) may actually reduce friction compared to water alone. When this is true, the plug of fibers tends to flow in the center of the pipe, while a clear water annulus flows near the edge of the pipe.

Flow in pipes

For laminar flow in smooth—walled, straight, circular pipes, the Hagen—Poiseuille law gives the flow rate Q as a function of the pressure gradient across the length of the pipe (ΔP), the pipe radius (r), the pipe length (L), and the fluid viscosity (μ):

$$Q = \pi\Delta P r^4/(8\mu L)$$

The Darcey equation gives frictional losses (h) as a function of frictional factor (f), pipe length (L), average velocity (v), and acceleration due to gravity (g) for flow in circular pipes of diameter D as:

$$h = fLv^2/(2\,Dg)$$

For Reynolds number below 2000, f has been empirically shown to be equal to 64/Re.

Water hammer

Water hammer is the sudden change in pressure caused by a column of water that suddenly changes velocity (as when a valve is closed).

30.2 PUMP BASICS

Introduction

Pumps may be considered as positive—displacement or dynamic types. Dynamic pumps usually rely on centrifugal force. Centrifugal pumps are the most common type of dynamic pump and the most common type of pump used in pulp and paper mills.

Cavitation and related problems

The inlet of any pump operates at a lower pressure than the source of the liquid; it is called the suction side of the pump. This is what causes the liquid to flow into the pump. If the pressure at the suction side of the pump drops below the vapor pressure of the liquid, gas bubbles (cavities) will form that interfere with the proper operation of the pump. The gas bubbles collapse on the output side of the pump where the pressure is high, making much noise and causing undue wear. This condition is known as *cavitation*.

Cavitation is aggravated by using liquids at high temperatures where their vapor pressure is high, by using filters with high pressure drops, or by using inlet lines that have high pressure drops at the operating flow rates.

Cavitation can be alleviated by having a generous geodetic height for the liquid entering the pump; this is accomplished by lowering the pump or raising the reservoir. Inlet pipes and filters with larger inside diameters will decrease the frictional and velocity pressure drops. If the flow rate is pulsating with a period less than a few seconds, the use of a dampening system can level the flow rate at the inlet. In some cases the liquid in the reservoir can be padded with a pressurized inert gas, although this can lead to bubble formation downstream (much like the *bends* that may occur with scuba divers). Booster pumps may be used to solve difficult cavitation problems.

The use of an adequate NPSH will avoid cavitation. **Much difficulty and premature wear encountered in the use of pumps are the result of an inadequate NPSH**.

NPSH, NPSHA, NPSHR

NPSH, *the net positive suction head*, is the absolute suction head minus the vapor pressure of the liquid being pumped. NPSHA is the net positive suction head available and must be greater than NPSHR, the net positive suction head required. NPSH increases as the square of the pump flow rate near and above the design flow rate of a pump. It does not increase this fast at flow rates much below the rated flow rate.

Good design practices on the inlet piping help insure an adequate NPSH. Suction piping should be at least twice the diameter of the pump inlet nozzle. Pump inlet velocities should be below 10 ft/s. Place pipe elbows at least 5—10 pipe diameters from the pump inlet, with the lower ratio applicable to large pipe diameters. When this is impossible, the use of rotating vanes before the elbow can solve problems. Inlet screening devices, if used, must not have high pressure drops.

High elevation reduces atmospheric pressure and NPSHA. Every 1000 feet of elevation reduces NPSHA the equivalent of 1.1 feet of water head.

Pump selection

One must consider the properties of the fluid to be controlled (abrasiveness, viscosity, corrosiveness, and specific gravity), the flow rates involved, the pressure of the system, the temperature range and gradient to which the pump will be exposed, and the length and frequency of cycling. The suction conditions are also very important. If a pump is being replaced, consider the reason the old pump is no longer suited to the job.

30.3 POSITIVE DISPLACEMENT PUMPS

Positive displacement pumps actually enclose the fluid to be moved through the system. The pressure and volume at which the liquid can be discharged depend only upon the mechanical strength and size of the properly operating pump and the energy supplied to the pump (neglecting leakage). Pressure relief valves are used to prevent damage to the system or buildup of high, dangerous pressures. Priming of positive displacement pumps may or may not be required.

Reciprocating and diaphragm pumps

Reciprocating pumps use a piston moving back and forth in a cylinder (see Fig. 30-16). Two check valves for the input and output at one end of the cylinder allow one—way flow; this gives a pulsed output. Because rotational motions of engines are not easily converted to reciprocating motions, this type of pulp is seldom used on mobile equipment. Two cylinders can be used to give a continuous output as in pumps for high—performance liquid chromatography, or one cylinder can be made to deliver with either stroke. The *diaphragm* pump (Fig. 30-1) is similar in principle, except that the piston assembly is replaced by a diaphragm and housing; diaphragm pumps are often powered by compressed air and occur as double—diaphragm pumps.

Rotary pumps (gear and lobe types)

Rotary pumps use a variety of means to generate flow. A pair of gears with many teeth (external to the gear and called an external gear pump) in *gear pumps* or a few lobes in *lobe pumps* is demonstrated in Fig. 30-2. Internal gear pumps use a gear within a gear design; *internal* means the teeth of one gear are on the inside, projecting inward. Gear pumps are useful for pumping

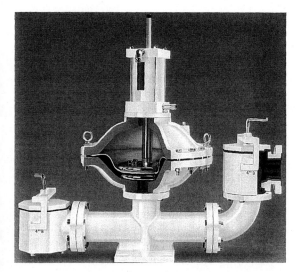

Fig. 30-1. Cutaway of a diaphragm pump with two check valves. Courtesy of Gorman—Rupp.

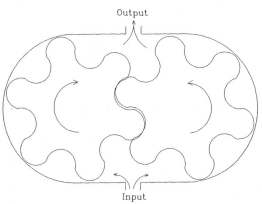

Fig. 30-2. Operation of a gear or lobe pump.

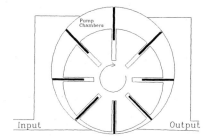

Fig. 30-3. Operation of a vane pump.

viscous liquids or slurries, but highly viscous liquids will require relatively large, slow—moving pumps to keep NPSHA greater than NPSHR. Helical screws (like a screw press shown on page 293) or sets of screws are also used. Piston and progressing cavity pumps have rotary forms.

A series of vanes in *vane pumps* is shown in Fig. 30-3. Vane pumps are commonly used on mobile hydraulic equipment. Fig. 30-3 shows an unbalanced vane pump; by use of an elliptical—shaped housing with a central rotor each pair of vanes can have two inputs alternated by two outputs for each rotation. Rotary vane pumps are also used for vacuum pumps.

Progressing cavity

The progressing cavity pump (page 717) provides axial pumping in a tube—shaped pumping chamber. The rotor (the inside portion that turns) is externally threaded or lobed, and the stator (the outside, stationary portion) is internally threaded or lobed. Cavities between the rotor and stator progress from the inlet to the outlet. While rotors are usually constructed from metal, stators are often replaceable and made of elastomers.

Peristaltic pump

The peristaltic (or tubing) pump (Fig. 30-4) uses a tube in the shape of a semicircle with two or more pinching rollers on the outer edge of a wheel that rotates on the inside of the semicircle. These rollers pinch the tubing and enclose small

amounts of fluid and push them forward in the tubing. The pump acts as a check valve. Only the tubing is contacted by the material to be pumped so seals are not required. These pumps are usually fairly small and have flow rates on the order of less than 1 mL/min at low pressures, although large units using tubing of 65 mm inside diameter can pump 600 L/min at 220 psi. The pumps are useful for metering materials.

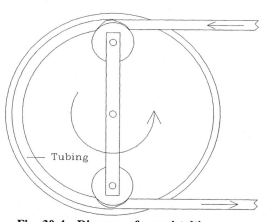

Fig. 30-4. Diagram of a peristaltic pump.

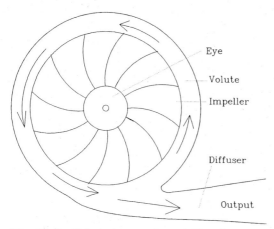

Fig. 30-5. Volute type of centrifugal pump. After McGill, 1980.

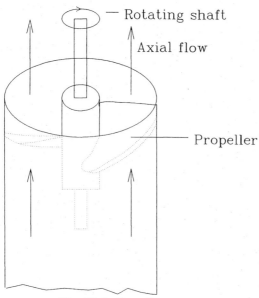

Fig. 30-6. Impeller pump.

Metering pumps

Positive displacement pumps are often used as metering pumps because they accurately control flow rates by changing the speed at which the pump is driven without regard to the back pressure. They are also used to pump relatively small volumes of liquids against high pressures.

Possible metering pumps include leakproof diaphragm pump, piston pump for precision, rotary designs for simplicity and low maintenance, or solenoid for proportional impulse control. The corrosive nature of the fluid to be pumped (which

is usually a function of temperature) is a factor in the selection of a pump. A fluid that is listed as compatible with a pump material at room temperature may not be compatible at high temperatures.

30.4 CENTRIFUGAL (DYNAMIC) PUMPS

Introduction

The *impeller* is the piece that rotates inside the casing to effect pumping (Fig. 30-5). The input of the pump is at the eye and the liquid enters perpendicular to the plane of the impeller in the *volute* or spiral design. Since the impeller is never in contact with the case, the input and output are connected hydraulically. One important characteristic of centrifugal pumps is that they must be primed (filled with water) before they can generate suction to overcome any lift required for the liquid (i.e, if the source of water is below the pump). Also, buildup of air in the system may cause the pump to stop pumping if lift is required.

Another type of dynamic pump, the *impeller* pump (Fig. 30-6), uses an inline input and output inside a tube—shaped housing. Fluid is conducted axially instead of radially.

Because centrifugal pumps do not enclose the fluid to be pumped, the flow rate (Q), for a given pump speed, will be a function of the back pressure, the pressure at which the pump discharges the liquid (H). H denotes the equivalent *head* (height) of water or other liquid required to achieve that pressure; a column of water 2.3067 feet high exerts a pressure of 1 psi at the bottom. The efficiency of a centrifugal pump depends on the pumping head and flow rate; the level of highest efficiency is called the best efficiency point (BEP), which is also the ideal condition to run pumps for low maintenance.

Centrifugal pump action is characterized by the pump performance curve, as demonstrated in Fig. 30-7 (upper curve of the graph on the right). The maximum pressure that can be achieved is usually sufficiently low such that a pressure relief valve is not required.

The system response versus flow rate is shown in Fig. 30-7 (lower curve in the graph on the right). The operating flow rate and pressure are the balance of system performance (flow and pressure) and pump (flow and pressure) performance. The system response is a function of the

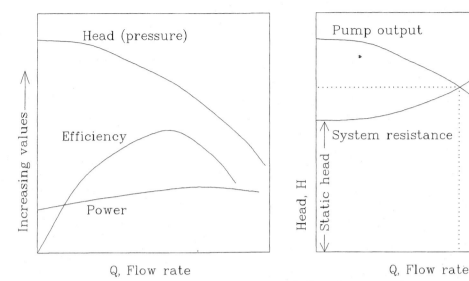

Fig. 30-7. Centrifugal pump efficiency curve (left) and pump performance curve (right).

height to which the liquid is pumped (the *static head*, which is constant with flow rate);   the frictional resistance of the pipes, valves, and bends; and the velocity head, the energy required to accelerate the fluid (the kinetic energy is equal to 0.5 mv^2). The velocity head is often small and usually ignored.

Cavitation in centrifugal pumps

Problems of centrifugal pumps related to cavitation include entrained air, suction recirculation, and discharge recirculation (Table 30-1). These three conditions cause pitting of the impeller due to the formation and collapse of gas bubbles. *Air entrainment*, where air is already in the liquid or is drawn into the system, causes pitting at the eye of the impeller, as does cavitation. *Suction recirculation*, due to unstable flow patterns between the impeller vanes at low flow rates (and high pressures), causes pitting halfway along the vanes. Many manufacturers specify a minimum pump flow rate to avoid pitting, vibration, and noise associated with suction recirculation. *Discharge recirculation*, which also usually occurs at low flow rates, causes pitting at the tips of the vanes and is often caused by a hydraulic design flaw known as vane passing frequency.

NPSH requirements can be reduced by using an inducer, a low head, axial flow impeller used in front of the impeller of a centrifugal pump.

Pump size

Selection of the proper size pump is very important.   In pulp and paper, flow rates are traditionally controlled by use of a pump operating at full power and a throttling valve to impede the flow as necessary.   This has been likened to driving a car with the accelerator pressed all the way down at all times and using the brake to control the speed!

In this type of operation it is very important to choose the proper size pump. Often pumps are oversized for lack of proper information, to allow for increased production in the future, or to counteract the effect of scale that may build up in pipes. A control valve should account for 30% or less of the pressure drop in piping at the maximum flow rate. The wasted power absorbed by a valve (which also causes the valve to wear out more quickly) is:

$$P = q \, \Delta p/(1714\eta)$$
where
P = power, in horsepower
q = flow rate, gal/min
Δp = pressure drop of the valve, psi
η = efficiency of the pump
At \$0.10/kWh, each one horsepower loss over the course of a year corresponds to \$653 (for a mill that operates continuously). So wasting 100 hp costs \$65,350 each year.

Table 30-1. Check chart for centrifugal pump problems. Reprinted from Karassik (1989) by courtesy of Marcel Dekker, Inc.

Symptoms	Possible cause of trouble
1. Pump does not deliver liquid	1-3, 5, 10, 12-14, 16, 21, 22, 25, 30, 32, 38, 40
2. Insufficient capacity delivered	2-7, 10-18, 21-25, 31, 32, 40, 44, 63, 64
3. Insufficient pressure developed	4, 6, 7, 10-16, 18, 21-25, 34, 39-41, 44, 63, 64
4. Pump loses prime after starting	2, 4, 6-11
5. Pump requires excessive power	20, 22-24, 26, 32-35, 39-41, 44, 45, 61, 69-71
6. Pump vibrates or is noisy at all flows	2, 16, 37, 43-61, 67, 78-81, 83-85
7. Pump vibrates or is noisy at low flows	2, 3, 17, 19, 27-29, 35, 38, 77
8. Pump vibrates or is noisy at high flows	2, 3, 10-18, 33, 34, 41
9. Shaft oscillates axially	17-19, 27, 29, 35, 38
10. Impeller vanes are eroded on visible side	3, 12-15, 17, 41
11. Impeller vanes are eroded on invisible side	12, 17, 19, 29
12. Impeller vanes are eroded at discharge near center	37
13. Impeller vanes are eroded at discharge near shroud or vane fillets	27, 29
14. Impeller shrouds bowed out or fractured	27, 29
15. Pump overheats and seizes	1, 3, 12, 28, 29, 38, 42, 43, 45, 50-55, 57-62, 77, 78, 82
16. Internal parts are corroded prematurely	66
17. Internal clearances wear too rapidly	3, 28, 29, 45, 50-55, 57, 59, 61, 62, 66, 67
18. Axially-split casing is cut through wire-drawing	63-65
19. Internal stationary joints are cut through wire-drawing	53, 63-65
20. Packed box leaks excessively or packing has short life	8, 9, 45, 54, 55, 57, 68-74
21. Packed box: sleeve scored	8, 9
22. Mechanical seal leaks excessively	45, 54, 55, 57, 58, 62, 75, 76
23. Mechanical seal: a) faces damaged; b) sleeve damaged; c) metal bellows fails	45, 54, 55, 57, 58, 62, 75, 76
24. Bearings have short life	3, 29, 41, 42, 45, 50, 51, 54, 55, 58, 77-85
25. Coupling fails	45, 50, 51, 54, 67

Table 30-1. Check chart for centrifugal pump problems. Reprinted from Karassik (1989) by courtesy of Marcel Dekker, Inc., Continued.

Legend for possible cause of trouble

A. Suction troubles

1. Pump not primed
2. Pump suction pipe not completely filled with liquid
3. Insufficient available NPSH
4. Excessive amount of air or gas in liquid
5. Air pocket in suction line
6. Air leaks into suction line
7. Air leaks into pump through stuffing boxes or through mechanical seal; Source of sealing liquid has air in it
8. Water seal pipe plugged
9. Seal cage improperly mounted in stuffing box
10. Inlet of suction pipe insufficiently submerged
11. Vortex formation at suction
12. Pump operated with closed or partially close suction valve
13. Clogged suction strainer
14. Obstruction in suction line
15. Excessive friction losses in suction line
16. Clogged impeller
17. Suction elbow in plane parallel to the shaft (for double-suction pumps)
18. Two elbows in suction piping at 90° to each other, creating swirl and prerotation
19. Selection of pump with too high a suction specific speed

B. Other hydraulic problems

20. Speed of pump too high
21. Speed of pump too low
22. Wrong direction of rotation
23. Reverse mounting of double-suction impeller
24. Uncalibrated instruments
25. Impeller diameter smaller than specified
26. Impeller diameter larger than specified
27. Impeller selection with abnormally high head coefficient
28. Running the pump against a closed discharge valve without opening a bypass
29. Operating pump below recommended minimum flow
30. Static head higher than shut off head
31. Friction losses in discharge higher than calculated
32. Total head of system higher than design of pump
33. Total head of system lower than design of pump
34. Running pump at too high a flow (for low specific speed pumps)
35. Running pump at too low a flow (for high specific speed pumps)
36. Leaky or stuck check valve
37. Too close a gap between impeller vanes and volute tongue or diffuser vanes
38. Parallel operation of pumps unsuitable for this purpose
39. Specific gravity of liquid differs from design conditions
40. Viscosity of liquid differs from design condition
41. Excessive wear at internal running clearances
42. Obstruction in balancing device leak-off line
43. Transients at suction source (imbalance between pressure at surface of liquid and vapor pressure at suction flange)

**C. Mechanical troubles:
a. General**

44. Foreign matter in impellers
45. Misalignment
46. Foundations insufficiently rigid
47. Loose foundation bolts
48. Loose pump or motor bolts
49. Inadequate grouting of baseplate
50. Excessive forces and moments from piping on pump nozzles
51. Improperly mounted expansion joints
52. Starting the pump without proper warm-up
53. Mounting surfaces of internal fits (at wearing rings, impellers, shaft sleeves, shaft nuts, bearing housings, (and so on) not 100% perpendicular to shaft axis
54. Bent shaft
55. Rotor out of balance
56. Parts loose on the shaft
57. Shaft running off center because of worn bearings
58. Pump running at or near critical speed
59. Use of too long a shaft span or too small a shaft diameter
60. Resonance between operating speed and natural frequency of foundation, baseplate, or piping
61. Rotating part rubbing on stationary part

62. Incursion of hard solid particles into running clearances
63. Use of improper casing gasket material
64. Inadequate installation of gasket
65. Inadequate tightening of casing bolts
66. Pump materials not suitable for liquid handled
67. Lack of lubrication of certain couplings

**C Mechanical troubles
b. Sealing area**

68. Shaft or shaft sleeves worn or scored at packing
69. Incorrect type of packing for operating conditions
70. Packing improperly installed
71. Gland too tight, resulting in no flow of liquid to lubricate packing
72. Excessive clearance at bottom of stuffing box, causing packing to be forced into pump interior
73. Dirt or grit in sealing liquid
74. Failure to provide adequate cooling liquid to water-cooled stuffing boxes
75. Incorrect type of mechanical seal for prevailing conditions
76. Mechanical seal improperly installed

**C. Mechanical troubles
c. At bearings**

77. Excessive radial thrust in single-volute pumps
78. Excessive axial thrust caused by excessive wear at internal clearances or by failure or excessive wear of balancing device (if such is used)
79. Wrong grade of grease or oil
80. Excessive grease or oil in antifriction bearing housings
81. Lack of lubrication
82. Improper installation of antifriction bearings (damage during installation, incorrect assembly of stacked bearings, use of unmatched bearings as a pair, (and so on)
83. Dirt getting into the bearings
84. Moisture contamination of lubricant
85. Excessive cooling of water-cooled bearings

Controlling the process flow

The process flow can be controlled by three methods. 1) With *flow splitting* a portion of the flow can be recirculated to the supply vessel or suction side of the pump, as in the stuff box of a paper machine. 2) The pump head may be changed by varying the speed of the pump (either by using a variable speed motor or some type of transmission). This method offers the potential to be energy efficient where flow rate differences are high. 3) The system curve can be altered, for example, by using a throttling valve to increase the system flow losses.

Variable speed motors

Variable speed motors (especially NEMA B squirrel cage AC induction motors) may be operated by alternating current (AC) variable frequency drive (VFD) or by varying the voltage and current to DC motors (Brink and Van Lieshout, 1993). With VFD normal alternating current is converted to direct current (DC) and from DC, using electronic circuits, to variable frequency and voltage AC. The speed of AC motors is dependent upon the frequency of the power source, if an adequate voltage is supplied. Variable torque motors are often used with centrifugal pumps. While the initial cost of these systems is higher than using a pump with a throttling valve, the energy savings can be substantial in many applications.

Impeller design and centrifugal pump performance

Over relatively small ranges for a given centrifugal pump (because of the effect on the volute, pumps losses, and other factors), the pump head is proportional to the peripheral speed of the impeller (and, therefore, proportional to the impeller diameter for a given rotational speed). The pump capacity is proportional to the square of the peripheral speed or propeller diameter for a given rotational speed (angular velocity), and the power consumption varies as the cube of the peripheral speed or propeller diameter for a given rotational speed. These relationships can be applied to a pump with a fixed propeller diameter operating at varying rotational speeds; i.e., pump head is proportional to rotational speed, and pump capacity is proportional to rotation speed squared.

Impellers are sometimes cut or trimmed in size to reduce output or motor overloading. When impellers are cut, both the vanes and the shroud should be cut to decreases losses from disk losses, which are proportional to the fifth power of the impeller diameter.

The capacity of a pump is proportional to the propeller width over a narrow range, but the static head is not usually affected to a large degree.

Vibrational analysis and troubleshooting

Vibrational analysis of pumps indicates some possible sources of trouble. The ratio of the frequency of the major vibrations to the motor speed gives important information. A ratio below 0.5 indicates suction recirculation; a ratio near 0.5 indicates oil whirl or whip in the bearings; a ratio of 1 indicates rotor unbalance or coupling problems; a ratio of 2 indicates axial misalignment of couplings; a ratio equal to the number of impeller vanes indicates vane passing or, less likely, suction recirculation; a ratio above 2 may indicate problems with antifrictional bearings.

Table 30-1 is an extensive guide to the troubleshooting of centrifugal pump by a leading authority in the field.

Sealless pumps

Diaphragm and peristaltic pumps are by their nature sealless. Centrifugal pumps can be made sealless by using magnetic drives or canned motors. Magnetic drives are more suited to pumping hot liquids (on the order of 750°F or 400°C) unless canned motor pumps are water cooled. Canned motor pumps can tolerate high pressures.

30.5 VALVES

Flow in valves (Anderson, 1980, page 670)

Valves are used as control elements for numerous processes with the pulp and paper industry. They may be used in an on or off mode only, but more often they are used to regulate the flow in conjunction with instrumental process control. In this mode they become a variable orifice to regulate the flow where an *actuator* is used to control the position of the valve.

This section is applicable to fluids of relatively low viscosity. Bernoulli's theorem predicts the flow through an orifice to be:

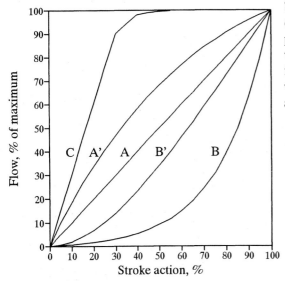

Fig. 30-8. Valve flow rates for linear, A, equal percentage, B, and quick opening, C types.

$$Q = C A (\Delta P)^{1/2}$$

where

Q = flow rate
C = constant for flow conditions
A = open area of valve
ΔP = pressure drop across the valve

The flow behavior of valves will certainly influence the behavior of the overall process control loop. Valve behavior (Fig. 30-8) is usually reported based on a constant pressure drop across the valve. It is described as linear (curve A), equal percentage (curve B), or quick opening (curve C). In fact, the pressure drop of a valve that is barely open is usually much higher than for a fully open valve. Curve A' shows the flow properties of a linear valve where the pressure drop varies by a factor of 5 as the valve is opened, and curve B' uses this pressure drop behavior with an equal percentage valve.

The flow capacity of a valve is measured by the *hydraulic valve coefficient* C_v, the flow of water through a wide—open valve in U.S. gal/min for a pressure gradient of 1 psi. For a given nominal valve size the highest C_v is achieved with ball valves followed by (in decreasing order) gate, plug, and globe valves.

The proper size valve must be selected so that it is large enough to allow the required flow,

but small enough to provide adequate control. (Likewise, one should pick the appropriate size pump.) A large valve allowing only a small flow will be almost shut during service, creating high flows near the seat, which may lead to undue wear. Flow rates can be given for liquids, gases, and steam based on the published C_v values:

$$C_v = Q \sqrt{\frac{G}{\Delta P}} \qquad \text{liquids} \quad (30\text{-}1)$$

$$C_v = \frac{Q}{1360} \sqrt{\frac{T_f G}{\Delta P\, P_2}} \qquad \text{gases} \quad (30\text{-}2)$$

$$C_v = \frac{W}{63.3} \sqrt{\frac{v}{\Delta P}} \qquad \text{steam} \quad (30\text{-}3)$$

where

Q = flow rate gal/min (for liquids)
     scfh (for gases)
W = lb/h (for steam or vapor)
G = specific gravity
T_f = temperature in Rankine (°F + 491.67)
P_2 = valve discharge pressure
v = downstream volume

For gas and steam flow ΔP must be less than $\frac{1}{2}P_1$, where P_1 is the pressure at the valve inlet. If ΔP is greater than $\frac{1}{2}P_1$, set ΔP and P_2 as equal to $\frac{1}{2}P_1$. Pulp slurries may deviate from the formula for a liquid.

The pressure within a valve is lower than either P_1 or P_2 because of the venturi effect (as a fluid moves faster, its pressure decreases). The lowest pressure (and cross sectional area) experienced at the orifice is called the *vena contracta*. If the *vena contracta* is less than the vapor pressure of the liquid while the pressure at the output of the valve is above the vapor pressure of the liquid, vapor bubbles will contract as they exit the valve, a condition known as *cavitation*. Cavitation causes undue noise and valve wear and should be avoided; it occurs when hot liquids are throttled. *Flashing* occurs if the pressure at the input of the valve is above the vapor pressure of the liquid while output of the valve is below the vapor pressure of the liquid.

Valve *rangeability* is the ratio of the maximum controlled flow to the minimum controlled

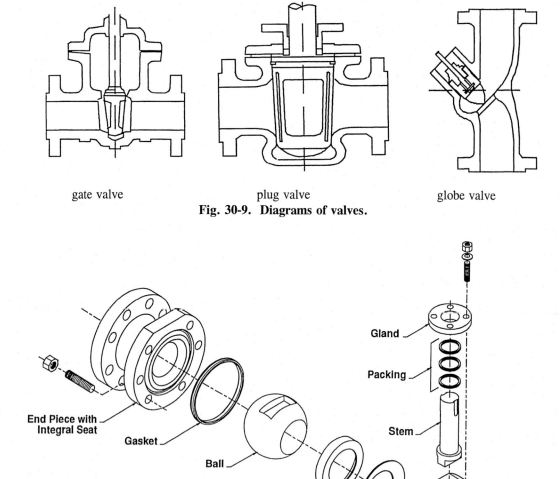

gate valve   plug valve   globe valve
Fig. 30-9. Diagrams of valves.

Fig. 30-10. Exploded diagram of a ball valve. Courtesy of Keystone International, Inc.

flow.   A rangeability of 100:1 is very high. Valves are often selected so that the maximum operating flow rate is 70% of the maximum possible flow rate so that control can be maintained in unusual situations.

Valve types (Figs. 30-9 and 30-10)
    Valves are used to control the flow of fluids. Shutoff valves are used in an open or closed state. Check valves (nonreturn valves) allow flow in one direction but not the other direction.   Control valves are used to vary the flow as needed.   We will consider control valves in detail in this section, although some of these designs are used for shut—off valves.

    Double—ported valves have balanced construction allowing for the use of small actuators, but these types are not suitable in applications requiring tight shutoff.

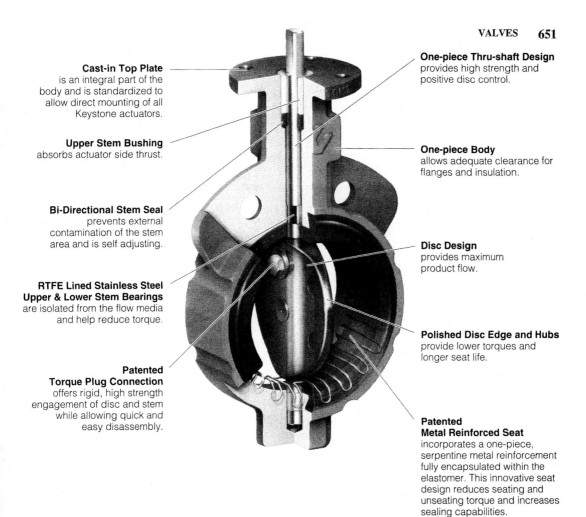

Cast-in Top Plate
is an integral part of the
body and is standardized to
allow direct mounting of all
Keystone actuators.

Upper Stem Bushing
absorbs actuator side thrust.

Bi-Directional Stem Seal
prevents external
contamination of the stem
area and is self adjusting.

**RTFE Lined Stainless Steel
Upper & Lower Stem Bearings**
are isolated from the flow media
and help reduce torque.

**Patented
Torque Plug Connection**
offers rigid, high strength
engagement of disc and stem
while allowing quick and
easy disassembly.

One-piece Thru-shaft Design
provides high strength and
positive disc control.

One-piece Body
allows adequate clearance for
flanges and insulation.

Disc Design
provides maximum
product flow.

Polished Disc Edge and Hubs
provide lower torques and
longer seat life.

**Patented
Metal Reinforced Seat**
incorporates a one-piece,
serpentine metal reinforcement
fully encapsulated within the
elastomer. This innovative seat
design reduces seating and
unseating torque and increases
sealing capabilities.

Fig. 30-11. Diagram of a butterfly valve. Courtesy of Keystone International, Inc.

Pinch valves use a flexible tube which is constricted against two members, at least one of which is movable. *Diaphragm* valves use a flexible diaphragm (disk) against a fixed seat. Both of these allow the control mechanism to be completely sealed from the fluid without the use of seals or glands, although they are limited to relatively low temperatures and pressures.

Common types of rigid valves include ball plug (Fig. 30-10) [which include the segmented (or V-notch) types], butterfly (Fig. 30-11), lubricated plug or taper (eccentric) plug (like those used on glass laboratory burettes), knife gate valve (Fig. 30-12), and globe valves. *Rotary valves* (rotating or *quarter—turn*), including ball, butterfly, and plug valves, are position—seated devices. Gate (or slide) valves are *linear—action* valves with a disk or wedge moving up and down to give

seating. Globe (or *obturating*) valves use a plug (such as a sphere or needle) that is forced directly into the seat; technically globe valves are one particular type of obturating valve, based on the shape of the stopper or plug, but the term is used for almost all obturating valves. The properties of valves are compared in Fig. 30-13 and Table 30-2.

Globe valves have high friction loss due to the irregular flow path, but have high rangeability. Globe valves have cage—guided, top—guided, split—body, and top—and—bottom—guided configurations of body type and plug guiding. Cage—guided globe valves are fairly versatile and common since their introduction in the 1960s, but this design is not suited for use with abrasive or highly viscous liquids. The post—guided globe valve is useful with harsh chemicals. If used with bellows stem seals, fugitive emissions are decreased.

Fig. 30-12. Gate valve. Courtesy of Keystone International, Inc.

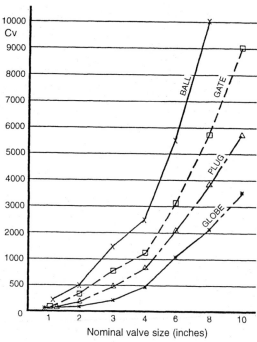

Fig. 30-13. C_v values of valves.

Ball valves for use with abrasive fluids should use a integral seat since there are no behind—the—seat cavities to fill with particles.

Eccentric plug valves are suitable when solids are present or crystallization may occur; they offer stick resistance and erosion resistance due to the camming action of the valve plug and the geometry of the body. The flow path is streamlined, and the tapered seating surfaces are away from the flow path. Erosion (abrasion) resistance is enhanced by the selection of the proper material for the trim materials. These include chrome plated stainless steels, solid Alloy 6 parts, and various ceramics. Reverse flow direction also protects against erosion. Lubrication, in those valves that use it, may be a source of difficulty.

Butterfly valves are fairly compact, but they generally are not used with high fluid velocities (above 30 ft/s) since they then require expensive, high—torque actuators for their control. Butterfly valves offer good protection against erosion only in situations where shutoff is not required.

Gate valves give full port design and are resistant to erosion at the seat. They are often used for shutoff control. When large gate valves are used with steam, they may jam when the valve is closed and then cooled since thermal contraction puts a high pressure on the gate; the valve may have to be heated before it can be opened. Gate

valves are subject to buildup of solids and have high maintenance requirements.

Check valves come in several designs including ball and seat (usually with a spring pushing the ball against the seat as in Fig. 30-16), wafer swing (Fig. 30-14), or guided disk.

Selection of valves (Anon., 1992)

Like most items, the choice of valves maximizes performance, cost, and service life; the least expensive suitable valve is usually the best choice. Corrosion is a major consideration for

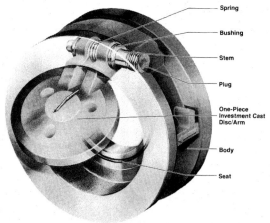

Fig. 30-14. Wafer check valve. Courtesy of Keystone International, Inc.

Table 30-2. Comparison of the properties of valves. Courtesy of Keystone International, Inc.

Valve Type	Process Isolation	Flow Path	Maintenance	Operation
Valvtron Ball Valve	• Seats several times harder than stellite • Zero leakage • Constant wiping • All metal-to-metal • Seats protected in open position • Integral seats	• Highest C_v • Seat faces not in flow path • Non interrupted flow path • Straight through flow path • Full bore	• Spring loaded, hard faced stem and gland minimizes packing leaks • Very hard seat faces prevent seat damage • Simple and easy to maintain • No matched parts required	• 1/4 Turn • Easily automated • No lock up • Continuous seat wiping action • No displacement of internal parts
Globe	• Seats susceptible to damage • Seating action traps particles • All metal-to-metal • Integral • Flow erosion in open position	• Tortuous flow path • Very low C_v • Seat faces in flow path • Reduced bore	• Seat damage due to crushing between seat faces • Galling common • Easy to maintain • Small stems • Maintained in line	• Multi-turn • Difficult to automate • Displaces volume
Wedge Gate	• Wedge action damages seat faces • Seating action traps particles damaging seat faces • All metal-to-metal • Integral seats • Flow erosion in open position	• Interrupted flow path • High C_v • Straight through flow path • Reduced bore	• Seat damage due to wedging action • Difficult to achieve tight shut off after rework • Cavity packing can result in inoperability • Small stems	• Multi-turn • Difficult to automate • Displaces volume
Plug	• Difficult to match seats • Not tight shut off • Non-wiping • All metal-to-metal	• Irregular port • Medium C_v value • Seats not in flow path • Straight through flow passage	• Build up can prevent operation • Difficult to maintain • Hardened seats • Weak stems • Lubrication required	• 1/4 Turn • Difficult to automate • Requires balancing • Displaces volume
Butterfly	• Zero leakage • All metal-to-metal • Seats exposed to the flow • Thin section will distort under thermal conditions • Non-wiping	• Seat and disc in flow • Low C_v • Reduced flow passage	• Weak stems • Flow erosion • Body must be removed from the line • Seat easily damaged	• Cams into seat • 1/4 Turn • Easily automated • Displaces volume

control valves used by the pulp and paper industry. Stainless steel is used widely in valves, although some valves will require other materials to prevent excessive corrosion. Cavitation and erosion caused by abrasive slurries are two other considerations.

One must consider the properties of fluid to be controlled (abrasiveness, viscosity, corrosiveness, and specific gravity), the flow rates involved where control may be necessary, the pressure of the system and the pressure drop across the valve, the temperature range (the range of -50° to 400°F does not usually require special consideration) and gradient to which the valve will be exposed, and the length and frequency of cycling. If a valve is being replaced, one may also wish to consider the reason the old valve is no longer suited to the job.

The coefficient of flow C_v required determines the size of the required valve. Globe valves are much more expensive than other valves for a given coefficient of flow, followed by eccentric rotating plug valves, V—notch ball valves, and butterfly valves in that order.

Water is usually controlled by standard butterfly valves. Pulp stock is often controlled with stainless steel segmented (or V—notch) ball valves. When precise basis weight control is required, precision electric actuators give immediate response to computer signals. When the stock consistency is below 2%, high—end butterfly valves of stainless steel may be used. Offset disks will lengthen the seal life.

Liquors for chemical pulping are corrosive, although they are used under moderate temperatures and pressures in most cases. Stainless steel butterfly valves are used. High—performance varieties are used when tight shutoff is required. Dual seal types are required when the pressure may originate on either side of the valve, such as those used around digesters. When solids are present or crystallization may occur (as in green liquor lines) eccentric plug valves offer stick resistance and erosion resistance.

Boiler feedwater must be pumped at very high pressures. This requires high—pressure globe valves with cage—guided trims to balance internal pressures so that the actuators will be of a manageable size. High—temperature steam control valves use this same type, although low—

pressure steam may use eccentric plug or high—performance butterfly valves.

Another consideration may be whether the valve meets the requirements of regulating agencies such as the U.S. EPA if regulations apply.

Material selection (Anon., 1992)

One should consider the pipe specification when selecting a material for a valve. Metallic seats will not soften and swell, unlike many elastomer seats subjected to steam.

Wet chlorine requires the use of Hastelloy C. Stainless steels can usually be used with dry chlorine. Titanium is recommended for calcium hypochlorite, chlorine dioxide, and sodium chlorate. Nickel—based alloys are useful with sodium hydroxide, hydrogen peroxide, and sulfur dioxide. The use of alloys is considered on Section 33.6. Elastomer—lined globe valves, especially those of Teflon (PTFE), are suitable for may corrosive chemicals, although they do not last as long as metallic valves in noncorrosive environments.

Abrasive slurries, high temperatures, or corrosive materials may require the use of surface modification or coatings. Surface modifications include the diffusion process or nitriding process. In the nitriding process the metal surface (0.003 to 0.015 in.) is transformed to a hard iron—nitride material suitable for mild abrasion and steam resistance. The diffusion process is very expensive and recommended only for high temperatures (to 2000°F).

Coatings include cobalt alloy (metal) or silicone carbide (ceramic). Coatings may be applied by electroplating for metals, thermal plasma spraying for ceramics (chromium oxide, silicone dioxide), or high—velocity oxygen fuel (HVOF) for many metal alloy powders (chromium carbide, tungsten carbide). For ball valves, a common materials configuration with good corrosion resistance for general pulp and paper use is a 316 SS base with a chromium—plated ball and a chromium carbide—plated seat. Ceramics offer high corrosion resistance.

HVOF was patented by J. Browning in 1983 (U.S. Patent 4,460,421) and has become an important process (Buchanan, 1993). HVOF may be followed by a post heat—treatment for some metal alloys (nickel, chromium, cobalt). HVOF uses speeds of 0.75 mi/s (1.2 km/s) to produce a highly abrasive resistance surface. Coatings are usually less than 0.012 in. (0.025 in. for HVOF spray followed by heat treatment) thick.

Fugitive emissions and valve stem packing

Fugitive emissions (think leaks) are becoming less and less tolerated by regulating agencies (either direct leaks or organic and other vapors above ~500 ppm) (Dresch, 1994). Often, fugitive emissions occur at the stem. Quarter—turn valves do not have much motion in this area and often provide good protection against fugitive emissions. Diaphragm and pinch valves isolate the fluid from the control mechanism in a manner that avoids leaks around the stem. Bellows stem seals do not have stem leakage either.

The valve stem packing, a type of gasket, is a resilient material packed under pressure so that it fills the space between the valve stem and the stuffing box and presses against both of these surfaces. Packings such as graphite [for fire—proofing the seal and temperatures above 450°F (232°C), but not commonly used in pulp and paper] or PTFE (Teflon) or are impermeable. By controlling leaks between the packing and the stem, leaks between the stuffing box and the packing do not matter if the packing itself is impermeable (which is usually the case).

External springs (live loads) such as belleville washers are sometimes used to insure that proper pressure exists in the stem packing over a long period of time. This may be necessary to offset the effects of thermal expansion, consolidation, or cold flow of Teflon packing. On the other hand, this method increases the stem friction that must be overcome when the valve is adjusted.

The packing itself is often a series of U— or V—shaped layers. Often two layers of packing are separated by a lantern ring.

Actuators

Actuators are used to control the position of the valve. The actuator may be controlled by air (pneumatic), electricity, or hydraulics.

Electric actuators are generally more compact and easier to interface into computer—controlled systems than pneumatic actuators. They, of course, do not require an air supply.

Pneumatic actuators may be considered in high—humidity environments or in actuators that must be explosion proof (i.e., do not have any uncontained electric sparks). Types include cylinder, vane, piston, spring—diaphragm, rack and pinion, and scotch yoke. A spring may be used to return the valve to a normally open or normally closed position.

30.6 HYDRAULIC POWER SYSTEMS

Introduction

Hydraulics is the study of fluids whether in motion or at rest. *Hydrodynamics* is the study of fluids in motion, and *hydrostatics* considers the properties of fluids in static equilibrium (motionless). Concepts from these fields will be used as necessary to explain the operation of hydraulic devices. A water turbine or wheel operates with a large change in the kinetic energy of water; they are hydrodynamic devices.

Specifically, the use of hydraulics (or pressure hydraulics) for power transmission will be considered here. Power transmission is the result of the force of a confined liquid. *The confined liquid merely transmits the force generated by the power supply; the flow contributes to the other component of work, i.e., displacement.* The amount of work accomplished depends on the overall force and the overall distance to which it is applied. The power supply may be an electric motor, gasoline engine, or hand power. Although the liquid must flow to cause motion, its velocity is usually sufficiently low so as to have only a small kinetic energy component relative to the overall work accomplished (making the hydrodynamic component a trivial consideration). Some common systems that use hydraulics are hand—operated hydraulic jacks and presses, power steering and brakes on many vehicles, backhoes, and hitch controls of agricultural tractors.

Hydraulic systems offer many advantages including a high level of flexibility due to their compact size per given level of power, the use of small forces to control large forces, their relatively simple and economical design and operation, and self—lubricating components. Energy is easily transferred by fluid under pressure instead of cumbersome systems of gears and chains or pulleys and belts. Vibration is usually minimal in hydraulic systems.

Safe operation of hydraulic systems is important since the high pressure involved is potentially dangerous. A failure of the system such as the accidental release of the system's oil may lead to catastrophe, such as when loaded booms or weights fall suddenly.

Pascal's law

When fluid is in static equilibrium (motionless) it obeys Pascal's law. That is, it is only under compressive forces (measured in units of pressure, force per area) and those forces act with equal intensity (pressure) in all directions at any point in the fluid. The pressure is normal (90°) to any surface on which it acts.

We can demonstrate the operation of the hydraulic jack to multiply an overall force with the use of this concept. Fig. 30-15 shows how forces can be multiplied with hydraulics. A force is applied to the pump piston by the handle. This force provides a pressure to the hydraulic fluid that is inversely proportional to the piston area. The pressure of the fluid against the piston provides a force that is proportional to the area (perpendicular to its direction of motion) of the piston.

EXAMPLE 1. Calculate the mechanical advantage of the system in Fig. 30-15 by calculating the pressure generated on the fluid by the hand pump when a force of 10 lb is exerted on the pump piston and the force generated by the fluid on the piston carrying the load.

SOLUTION: The surfaces of both pistons are normal to the direction of movement. The radius of the pump piston is 0.5 in. so that its area $= \pi r^2 = 0.7854$ in.². The pressure generated by the piston $=$ force/area $=$ 12.73 lb/in.². The same pressure acts upon the piston with the load, but its area is 7.069 in.². The overall force is pressure $\times$ area $=$ 90 lb/in.². A ninefold increase in force is realized (if frictional losses are ignored); since the radius of the load piston is three times that of the pump piston, and since area is proportional to radius², this is logical.

PROBLEM: Suppose the pump piston was only 0.1 in. in diameter. What would be the mechanical advantage of the system?

EXAMPLE 2. Show that the work (force times distance) is the same for both pistons of Example 1.

SOLUTION: The following five equations or relationships are all fundamental:
1) force $\times$ distance $=$ work;
2) area $\times$ distance $=$ volume;
3) pressure $=$ force/area;
4) the change in volume of fluid is constant for both pistons;
5) the pressure is constant for both pistons.
In 1) force will be substituted with pressure $\times$ area from 3) and distance will be substituted with volume/area from 2) with areas cancelling to give work $=$ pressure $\times$ volume. Since pressure and volume are equal for each piston, work must also be equal.

Flow through an orifice and other obstacles

A restriction in a line (such as an orifice or valve) causes a pressure gradient across the restriction when the fluid is in motion. When the fluid is not in motion, the pressure gradient across the restriction decreases to zero since the pressure

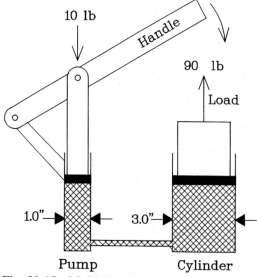

Fig. 30-15. Multiplication of forces with hydraulics.

through the fluid is constant according to Pascal's law. When there is a flow across a restriction useful mechanical energy is converted to (useless) heat energy. The production of heat energy is proportional to both the pressure gradient and the flow rate. Of course, there is no flow through a valve that is completely closed.

In effect any line is really an orifice so there is a pressure drop across it. Consequently there is always some energy loss through a line. However, the energy losses can minimized by using a line of proper diameter to avoid high flow velocities, avoiding sharp turns in flow patterns, and other proper design practices.

Hydraulic jack

The system in Fig. 30-15 is not very practical since there is a very limited range of motion. It takes many operations of the hand pump to lift the object a suitable distance. A *check valve* allows flow in one direction only so that several operations of the handle may be combined to lift the load. It works by means of a ball that seats against a seal. Fig. 30-16 shows the addition of check valves, a control valve, and a reservoir to the system to make a complete hydraulic jack.

Bernoulli's principle

Bernoulli's principle is valid for any fluid (liquid or gas); it is especially important to fluids moving at a high velocity. Its principle is the basis of venturi scrubbers, thermo—compressors, aspirators, and other devices where fluids are moving at high velocities. It also explains cavitation in fluids (such as in valves and pumps). The sum of pressure (potential energy) and kinetic energy in any system is constant (i.e., energy is conserved if frictional losses are ignored). Thus when a fluid flows through areas of different diameters, there is a change in velocity. The change in velocity leads to a change in kinetic energy and so the pressure changes as well. A decreased pipe diameter means an increase in velocity and kinetic energy and a decrease in pressure. This is demonstrated in Fig. 30-17.

Hydraulic fluid

The hydraulic fluid has many purposes including the principal one of power transmission by transferring forces from one location to another. It provides lubrication of the internal surfaces; its cleanliness is important since abrasive particles in the oil can lead to undue wear of the system. The hydraulic fluid also acts as a sealant and coolant. Hydraulic fluids can be used up to about 82°C (180°F) although cooling is recommended for temperatures above 70°C (160°F). Air coolers consist of tubing with metal fins to increase the surface area. Water coolers use cold water to conduct heat from the hydraulic fluid.

Two important criteria for hydraulic fluid are its viscosity and the amount and type of additives in it. Additives include antiwear chemicals, corrosion inhibitors, and antioxidants. The viscosity is important to maintain a thin layer of fluid on surfaces to avoid wear. When fluid is used at higher temperatures, it must be more viscous. The viscosity index is a measure of how the viscosity changes with temperature; the higher the index, the smaller is the change in viscosity with temperature. The *pour point* is the lowest temperature at which a fluid will flow (rapidly enough to pour) under its own weight.

Reservoirs

Reservoirs (sump or tank) are used to maintain a reserve of fluid; they have secondary

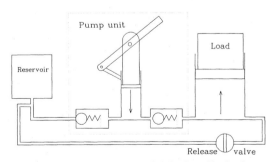

Fig. 30-16. Diagram of a hydraulic jack.

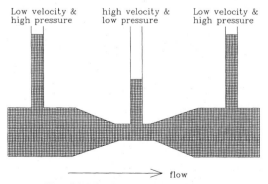

Low velocity & high pressure   high velocity & low pressure   Low velocity & high pressure

flow

Fig. 30-17. Bernoulli's theorem.

purposes of oil cooling and separation of air and contaminants from the oil. Small amounts of fluid may leak from systems, and cylinders cause system volume changes according to their positions. Reservoirs may be pressurized (an air pressure supply maintains the pressure) or vented to the atmosphere. Obviously, pressurized systems must be sealed; therefore, they are more resistant to contamination and condensation formation. A relatively large reservoir can also help keep the hydraulic fluid cool. Components of reservoirs include filler cap, fluid level gauge, outlet (to pump), inlet with filter to remove particles that would otherwise damage the pump, drain plug (which may be magnetic to trap metal particles), inspection plate, and pressure regulator. A baffle may be used to separate inlet and outlet flows for cooling and contaminant separation.

Reservoirs should be relatively high and narrow to give a column of fluid above the pump inlet. This decreases the chance of air entering the system by formation of a vortex. Any air will greatly reduce the efficiency of the system since the fluid then becomes compressible.

Filters and strainers

Strainers are coarse filters used to remove relatively large pieces of debris. Filters have a porous medium that is used to remove insoluble solids from hydraulic fluid. Cotton, cloth, and resin—reinforced paper are often the porous media. A good filter will remove solids of 10 μm diameter and larger. The pressure drop across a filter with fine pores may be as much as 25 psi, so they should not be used at pump inlets since cavitation may result. Strainers may be used with caution at pump inlets if the pressure drop is small enough to avoid problems. Filters are sometimes used with automatic bypass valves that open as the filter becomes plugged so that flow in the system can continue. A color indicator may be used to indicate the flow bypass condition.

Transfer of hydraulic fluid

Hydraulic fluid travels through tubes that are depicted as single lines in schematic diagrams. A solid line designates a working line that carries the bulk of the fluid. Pilot lines are used to control components and are represented by lines with long dashes. A drain line is represented by short dashes and is used to carry leakage oil back to the reservoir.

Hydraulic fluid may be transferred with pipes, tubing, or hose. Pipes are often of (nonstainless) steel (do not use galvanized steel since zinc is fairly reactive with some fluid additives). They are rigid and not meant to be bent. The *schedule number* refers to the wall thickness; the actual thickness and pressure rating depend on the overall nominal pipe diameter. Schedules are 10, 20, 30, 40, 60, 80, 100, 120, 140, and 160 with 40, 80, and 160 as common designations. Schedule 40 is used for low—pressure lines, schedule 80 is used for high—pressure lines, and schedule 160 is used for very high pressure lines. Pipes of larger inside diameter generally operate at lower pressures. For example, a nominal pipe diameter of ½ in. for steel pipe has an outside diameter of 0.84 in. The inside diameters are 0.622, 0.564, and 0.466 in. for schedules 40, 80, and 160, respectively. The operating pressure limits are 2300, 4100, and 7300 psi and the burst pressures are 15,500 psi, 21,000 psi, and 26,700 psi, respectively. (Notice the *safety factors* of 6.74, 5.12, and 3.66; minimum safety factors are 8 below 1000 psi, 6 in the pressure range of 1000—2500 psi, and 4 above 2500 psi.) For steel pipe of 2 in. nominal diameter, the outside diameter is 2.375 in., and the inside diameters are 2.067, 1.939, and 1.689 in., for schedule 40, 80, and 160, respectively. The working pressures are 1500, 2500, and 4600 psi, respectively.

Tubes are often of stainless or nonstainless steel (do not use galvanized steel) that is seamless or welded. They are usually of relatively small diameter and relatively thin—walled. Tubes may be bent into shape, but generally should not bend during routine operation of a hydraulic system. Tubes are used with a variety of fittings. *Flare* fittings are usually 37°, although 45° may be used in certain applications. *Swage* fittings work by compressing the tubing so that its outside diameter increases to seal around a retaining ring. *Bite* fittings use a retaining ring which is compressed into (bites) the tube. Some fittings may be *brazed* on to form a seal with the tubing. Avoid the use of straight—line connections that do not allow for expansion and contraction.

Hydraulic hoses have an inner tube, reinforcement layer(s), and a cover. The inner tube is often of synthetic rubber. The reinforcement layers are constructed of steel or fabric braided mesh, although several of spiral plies wrapped in opposite directions are also used. Skive is the process of removing the outer cover of a hydraulic hose. Avoid loops, twists, rubbing, heat, kinks, and strained connections with hydraulic hoses.

Hydraulic pumps and motors

The energy required to drive components of a hydraulic system is derived by the pump, which has been discussed in detail. Usually, hydraulic pumps are *positive displacement*.

The theoretical flow rate (Q) is equal to the displacement (d) per revolution times the number of revolutions per minute (n). If Q is reported in gallons per hour and d is in in.^{3} then $Q = d \times n/(231$ in.^{3}/gal). The power generated (in horsepower) is equal to the outlet pressure in psi times the flow rate in gallons per minute divided by 1714 (gal·lb/in.^{2}·min·hp).

Hydraulic motors operate very similarly to hydraulic pumps (positive displacement pumps), except that the forced flow is converted to rotary motion. In fact, most hydraulic pumps can be used as hydraulic motors with little or no modification. A hydrostatic transmission uses a pump at the input and a motor at the output.

Accumulators

In systems where the hydraulic power is used intermittently, accumulators can be used to store hydraulic fluid under pressure so that when it is required a fluid delivery rate higher than that of the pump may be achieved. Accumulators are pressurized chambers that hold the fluid. Pressure is maintained with a gas (contained in a bladder), a weight, or a spring. A weight does provide a constant pressure, although this system cannot be used practically on mobile hydraulic systems.

Valves

The path of the fluid in motion is directed by valves. Valves also control the pressure or flow rate of the fluid. Relief valves limit the maximum pressure in the system. The *poppet* prevents flow when it closes against the seat. An envelope is the valve symbol and as many envelopes are used as there are distinct valve positions.

Pressure control valves operate in a balanced condition; that is, a force is used to counteract the pressure of the hydraulic fluid. The opposing force may be generated by a spring with a valve that moves into a balanced position for easy operation. The detailed presentation of valves and groups of valves is beyond the scope of this work.

Cylinders

Cylinders are linear actuators and may be single or double acting. They consist of the barrel, the piston, the piston rod, and suitable end caps and seals. Cylinder cushions may be used at one or both ends to prevent hammering. Fig. 30-15 depicts a single—acting *ram*; it requires the load to return it to the unextended position. Double—acting cylinders have two hydraulic lines, one on each side of the piston, so that the cylinder can be made to operate with power in either direction without depending on an external force. Telescoping cylinders have four or five concentric cylinders (*sleeves*) when collapsed that can each extend separately so that the extended length can be much more than the collapsed length.

Motors

Hydraulic motors are analogous to hydraulic pumps but operate in "reverse" with a high—energy fluid being converted into rotary motion with a certain angular velocity and torque. Gear, vane, and piston motors are common.

30.7 TROUBLESHOOTING HYDRAULIC POWER SYSTEMS

Safety during repair of hydraulic systems

Normal safety rules should be observed; notify all people involved of the who, what, where, when, and whys of the repair and how it will affect them. When working with hydraulic systems it is preferable to shut the system down completely. Electrical switches must be tagged and locked out. The fluid pressure must be relieved, and all gauges should indicate no pressure in the system. Accumulators must be bled. Heavy weights supported by the hydraulics must be locked into position. When working on these systems always assume there is pressure in the system and allow fluid to drain before completely breaking a connection. Fluid should be collected in suitable vessels to prevent floors from becoming slick with it. Always read the safety instructions for a given system.

Troubleshooting hydraulic systems

When troubleshooting hydraulic systems it is important to have the drawings available. Keep in mind that modifications to the system may not be transcribed to the original drawings. The operator should always be consulted to determine how the system is operating differently from normal. If the system is safe to operate, one may wish to operate it to determine the extent of the problem.

Contaminated hydraulic fluid is the most common cause of failure in hydraulic systems. The fluid should be kept clean to avoid problems.

Pump cavitation

Pump cavitation occurs when voids (gas bubbles) form in the fluid within the pump, usually near the inlet. As the voids collapse, noise (like thunder) is emitted. (Cavitation is really boiling of the liquid at reduced pressure.) Pump cavitation often occurs because fluid is not entering the pump fast enough at the pump intake. Many causes can contribute to this. If the fluid viscosity is too low (such as in a cold system) then it flows slowly. If the fluid reservoir is far below the intake, then a partial vacuum is already formed. Dirty intake filters, intake lines with small internal diameters, and air in the system all decrease the rate of fluid entering the pump. Problems with the pump itself may also cause pump cavitation.

Low or no system pressure

Low system pressure can be caused by a variety of conditions. Dirty fluid will quickly make check valves plug into their open positions. Relief valves may stick in an open position or, worse yet, in a closed position where fluid pressure builds up until some component of the system fails catastrophically. Faulty or improperly set relief valves can lead to low system pressure. Reservoirs must have adequate fluid. The reservoirs should be large enough to allow air bubbles to escape before the fluid reenters the system. Unexpected contributions include improperly installed shaft seals.

If there is a complete lack of system pressure, make sure the pump is rotating and rotating in the correct direction.

The energy of a hydraulic system is:

h.p. = [gal/min × pressure (psig)]/ 1714

Hydraulic symbols

Table 30-3 lists many hydraulic symbols. The energy triangle of a pump points outward showing it is a source of hydraulic energy; the energy triangle of a motor points inward showing it uses hydraulic energy. Two energy triangles indicate the pump or motor is reversible.

30.8 ANNOTATED BIBLIOGRAPHY

Pumps

1. The Hydraulic Institute (Parsippany, New Jersey, USA) is a nonprofit industry association working in the engineering, manufacturing, and application of pumping equipment since 1917. They publish *Pump Standards*.

2. Buchanan, E.R., High velocity spray coatings, *Pumps and Systems Mag.* 1(3):57-59 (1993). This is a good summary of hard coatings use.

3. Karassik, I.J., *Centrifugal Pump Clinic*, 2nd ed., Marcel Dekker, Inc., New York, 1989.

Valves

4. Anonymous, Valves, *Pulp & Paper Buyers Guide* :25-29(1992).

5. Dresch, C., The selection and maintenance of valves for the control of fugitive emissions, *Tappi J.* 77(8):46-50(1994); *ibid.*, *Valve Mag.* Spring, 1994, from Valve Manufacturing Association of America, 1050 17th St. NW, Washington, D.C. 20036.

Miscellaneous

6. Brink, K.L. and R.F. Van Lieshout, Application and selection guidelines for ac system drives: part 1, *Tappi J.* 76(4):143-155(1993) and part 2, *ibid*; 76(5):173-180(1993).

7. Anders, J.E., Sr., *Industrial Hydraulics Troubleshooting*, McGraw-Hill, 1982, 176 p. This is a practical, but not comprehensive, overview of troubleshooting already installed systems and starting up new systems. A number of pages (with photographs) describe catastrophic failures caused by dirt and contaminants in systems. Other sections describe

Table 30-3. Hydraulic symbols.

accumulator bottle	flow control -pressure and temperature	lubricator -pneumatic
accumulator bottle -gas charged	filter	line to vented manifold
accumulator bottle -spring loaded	flexible line	pneumatic flow arrow
adjustment bars, continuous	flow divider rotary	pneumatic pressure regulator
air motor (pneumatic)		pressure compensator
air motor -bi-directional	flow meter	pressure switch
air motor -oscillating	gauge -pressure	pneumatic air shutoff
flow arrow	gauge -temperature	pump -combination
check valve	hydraulic connection	
compressor	heater	pump -double
heat exchanger, cooler	hydraulic motor	pump -single fixed- displacement
cylinder, cushions at both ends	hydraulic motor -bi-directional	pump -two stage
cylinder, double acting	hydraulic motor -oscillating	quick disconnect
cylinder, double ended rod	intensifier	quick disconnect with check valves
cylinder, large differential rod	line passover -no connection	rotation arrow -one direction
cylinder, single acting rod	line, plugged	rotation arrow -both directions
electric motor	line to reservoir -above fluid	restriction -fixed
exhaust, pneumatic	line to reservoir -below fluid	restriction -variable

Table 30-3. Hydraulic symbols (Continued).

separator -pneumatic system	valve -counter balance	valve, -pedal operation
spring -fixed	valve -deceleration normally open	valve, -pilot pressure oper.
spring -adjustable		valve -pushbutton oper.
temperature controller	valve -manual shutoff	valve, -servo controlled
test connection		valve, -solenoid pilot pressure operation
thermometer	valve, -air pressure oper.	valve, -servo controlled
valve -normally closed	valve, -detent operated	valve, -external temperature compensation
valve -normally open	valve -lever operated	valve, -internal temperature compensation
variable adjustment		valve, -pressure reducing
valve -bidirectional	valve, -manually operated	valve, -pressure relief
valve -cross-flow	valve mechanical operation	valve -sequencing
valve flow down and left	valve, -electric motor oper.	

methods for cleaning hydraulic systems and keeping them clean with filter systems.

8. McGill, R.J., *Measurement and Control in Papermaking*, Adam Hilger Ltd., Bristol, 1980, 424 p. This book includes basic hydraulics, flow properties of pulp slurries, pumping, valves, viscosity, and other properties of water suspensions in addition to the title subject. Some aspects are dated.

Tappi Technical Information Sheets
9. TIS 0406-17, Selection of motors for centrifugal pump drives, 1 p.

10. TIS 0410-01, Fan pump calculations, 2 p.; -12, Pipe friction pressure loss of pulp suspensions: literature review and evaluation of data and design methods, 5 p.; -14, Generalized method for determining the pipe friction loss of flowing pulp suspensions, 4 p.; -15, Optimum consistency for pumping pulp, 3 p.

11. TIS 0420-04, Guidelines for stainless steel, bonnetless, flanged, wafer and knife gate valves, 2 p.; -07, Guidelines for stainless steel, bonneted, flanged and buttweld end wedge gate valves, 3 p.; -10, Horizontal end suction centrifugal stock pumps, 3p.

31
PROCESS CONTROL

31.1 INTRODUCTION

Introduction

The objective of process control is to keep key process operating parameters within narrow bounds of the *reference value* or *setpoint*. The importance of this and some aspects of this have been discussed in Chapter 20. This section will describe the theory behind control circuits to maintain automatic control over a process.

Automatic control

The basis of automatic control (where a machine or electronic circuit is the control as opposed to a human being) is the control loop. A control loop for a given process must have at least one *sensor*, the *controller* (that decides what to do with the information that is collected), and a *control element* to which the results are applied.

A good example of automatic control is electric heating of a house in a cold environment, a process where a single variable is controlled. A thermostat (sensor) monitors the temperature. When the temperature drops below the setpoint a switch (controller) is closed and the heater (control element) comes on. When the temperature rises above the setpoint, the switch opens and the heater is turned off.

The difference between the actual value and the setpoint is called the *error*. The controller reads the error and makes a decision. The action taken can be very simple to very complicated, depending on the system and the size of the acceptable error. When heating a house the actuator turns the furnace on or off if the error is below or above the setpoint, respectively. It is a very simple control, but the error ranges to 2°C, not very precise by many standards.

This system is called a *process control loop*. This is a *feedback loop* because the desired result is determined and the information (the error) is fed back to the controller for appropriate action.

In a feedforward loop it is generally not convenient to use the desired result to control the process so another variable must be substituted.

For example, we refine pulp to influence the properties of the final sheet; however, it is not convenient to use any paper properties to control refining level because the *dead time* is too high. The dead time is a measure of the length of time before a change in the actuator (refining level) causes a change at the sensor (paper properties). Changes in refining level may not be seen in the paper for several minutes to several days if stock storage is used. In this case the refining level is controlled by monitoring the energy input to the refiners for a given amount of pulp. Two sensors (stock flow rate and energy consumption) are combined into one variable (refining intensity). One must know the relationship between refining intensity and paper properties.

31.2 SENSORS

Definition

The sensor is a device which converts a change in a parameter (such as temperature, position, or pressure) to a change in a electric or pneumatic signal (information). (A sensor is one type of *transducer*, a device that changes a signal from one form to another form, such as a change in voltage to a change in current or a change in light to a change in voltage.)

Transfer function

A transfer function relates the value of the parameter to be measured with the output of the sensor. For example, with a reference point of 0°C, type J thermocouples (iron—constatan) give a signal of 0.0515 mV per degree Celsius (the *sensitivity*) over a narrow range. The relationship is not linear over a wide range of temperatures, so one might want to use a more sophisticated transfer function; in fact, the National Bureau of Standards specifies a 5th—order polynomial equation for the temperature range of 0° to 760°C with a accuracy of 0.1°C for the type J thermocouple.

This signal undergoes some processing in most (except the most basic) systems. For example, electrical information is reliably transmitted in the form of a current. One standard specifies that the signal from a sensor be conditioned such that a signal in the range of 4—20 mA be reported for the range of the variable. The controller usually converts this to a voltage by taking the signal across a resistor (i.e., the $I \cdot R$ drop). In pneumatic systems, one common standard calls for a signal output of 3—15 psi (20—100 kPa).

Response times of sensors

The *time response* of a sensor is a measure of how the sensor responds to an actual change as a function of time. For example, all thermocouples have mass. This mass takes a certain period of time to change temperature. Thus, when the thermocouple at one temperature is immersed in a liquid of a different temperature, the output of the thermocouple will change with time until it finally gives a reading that is within its normal error. This might take a fraction of a second to many seconds depending on the geometry of the thermocouple. The time constant of the sensor is usually much lower than that of the system, so it should not be a factor. The time constant, however, may depend upon how the sensor is used; for example, temperature measuring devices respond much more quickly in a water bath (where heat transfer is high) than in air (where heat transfer is low).

A first—order time response (such as the charging of a capacitor, or, approximately, most temperature sensing devices) follows the equation:

$$b(t) = b_i + (b_f - b_i) [1 - e^{-t/\tau}]$$

when $b_i = 0$, then: $b(t) = b_f [1 - e^{-t/\tau}]$

where b is the output value as a function of time, t, initially, i, or finally, f. The symbol τ is called the time constant. If the initial output is zero then after one time constant of time has passed the sensor reaches only 63.2% of its final output. After 2, 3, 4, 5, and 6 time constants, the sensor reaches 86.5%, 95.0%, 98.2%, 99.3%, and 99.75% of its final value, respectively.

Second—order responses will cause an oscillation in the output (as will be shown in Fig. 31-4), that is impossible to solve precisely, but can be characterized mathematically.

Optical sensors

The principles of light have been presented in Chapter 24, which (implicitly) provides some details for measurement of light. *Photoconductors* use a thin band (in a serpentine pattern to increase the effective length) of a material that becomes conductive when exposed to light of a minimum frequency (energy level). Two common materials are CdS (which requires light of wavelength 515 nm or less and has a time constant of about 100 ms) and CdSe (which requires light of wavelength 716 nm or less and has a time constant of about 10 ms). Photovoltaic cells produce a voltage and current that are a function of the light level incident to them. These have time constants on the order of 1 to 100 μs. For very low levels of light (used in liquid scintillation counters to determine radioactivity) photomultiplier tubes are used.

31.3 TEMPERATURE SENSORS

Resistance—temperature detectors (RTDs)

The resistance of metals is a function of temperature and increases with increasing temperature. The relationship is fairly linear over a wide temperature range, although second—order transfer functions are often used. Resistance—temperature detectors (RTDs) often use platinum or nickel. The sensitivity of platinum is a function of temperature, but is about 0.00385 $\Omega/\Omega/°C$ (calibrated to the internationally accepted DIN 43760 standard.) (One common U.S. practice is to configure RTDs at 0.00392 $\Omega/\Omega/°C$.) Nickel is 0.005/°C at room temperature. Thus, a 100.0Ω platinum RTD at 0°C will have a resistance of 100.385 Ω at 1°C.

This small signal difference is conditioned by using a Wheatstone bridge circuit which may contain a *compensating line* to nullify the effect of long leads to the RTD. The current used in RTD devices must be low enough so that the RTD does not alter the environment through *self—heating*. RTD time constants are about 0.2 to 0.5 seconds in flowing water to more than 10 times that time in air. RTD devices can be used from −100 to 600°C (or 900°C with ceramic materials).

Thermistors

The resistance of semiconductors is also a function of temperature, but semiconductor materials (oxides of Ni, Mn, Fe, Co, Cu, Ti, etc.) can be made whose resistance increases drastically with increasing temperature. This fact is used to

Fig. 31-1. Diagram of a thermocouple where *V* is the voltage generated by the circuit.

make thermistors. High sensitivities of 0.1 $\Omega/\Omega/°C$ (or 30 mV/°C) are possible. Typically, an uncoated thermistor has a time constant of 10 seconds in air and 1 second in air. Coating with Teflon will increase these values by 2.5×. They are used in the range of −50° to 150°C. Thermistors are subject to self—heating and are often used in Wheatstone bridge circuits.

Thermocouples

Thermocouples consist of a pair of wires made of dissimilar metals joined together at one or more *junctions* that generate a voltage when subjected to a temperature gradient (Fig. 31-1). (This phenomenon is known as the *Seebeck effect*, first reported by Thomas Seebeck in 1821.) The sensitivity of type J thermocouples is about 50 $\mu V/°C$.

Table 31-1 shows some characteristics of thermocouples. Since low voltages are used, signal amplification with differential amplifiers is used. A reference junction must be used, but this is easily accomplished in one of a number of convenient methods. Noise reduction techniques should be used to avoid interfering voltages that may be otherwise generated in the system.

Type T and J thermocouples are inert and can be used in oxidation, reduction, or vacuum environments. Type J should never be exposed to temperatures above 760°C because a magnetic transformation can cause an irreversible change in its calibration. Type K should not be used in sulfurous or reducing environments, nor in vacuums for very long due to a calibration change. Type E should not be used in reducing or vacuum environments. Tungsten—rhenium thermocouples can be used at temperatures up to 2760°C (but not in oxidation environments). Thermocouples can be made with time constants of 0.05 seconds (0.001 in. wire) and 1 second (0.005 in. wire) in still air.

Other temperature sensors

Integrated circuit temperature transducers, which appeared in the 1980s, have similar advantages and disadvantages of thermistors, but their output is linear such as 1 $\mu A/K$ or 10 mV/K.

Bimetallic strips (where strips of two different metals of differing coefficients of thermal expansion are stacked and bonded together) have been used in thermostats for years. They are mechanically connected to a switch, such as a mercury switch, to control heaters.

Optical pyrometry is used to measure the temperature of objects. This relies on blackbody radiation and is discussed in Chapter 24.

31.4 MECHANICAL SENSORS

Introduction

Mechanical sensors are used to measure variables such as position, velocity, acceleration,

Table 31-1. Properties of the common types of thermocouples.

Thermocouple type	Metal type 1 (positive)	Metal type 2 (negative)	$\mu V/°C$ at 20°C	Color code	Suggested temp. range
J	Iron	Constatan™	51	Black	0—760°C
T	Copper	Constatan	40	Blue	−200—370°C
K	Chromel™	Alumel™	40	Yellow	−200—1260°C
E	Chromel	Constatan	62	Purple	−200—900°C
S	90%Pt/10%Rh	Platinum	7		0—1480°C
R	87%Pt/13%Rh	Platinum	7		0—1480°C

force, pressure, levels (such as a liquid in a tank) and flow. This section describes some sensors used by the pulp and paper industry; more information on these and other sensors is found in Johnson (1993), which is the source of much of the information in this section.

Electrical position sensors

Position sensors rely on a change of capacitance, resistance, inductance, or reluctance . The capacitance of a capacitor depends on the distance between two plates or on how two plates overlap

A potentiometric displacement sensor uses a wire of relatively high resistance with a wiper in electrical contact with the wire. As the object moves along the wire, the resistance of the circuit changes. By measuring the resistance, the position of the object along the wire is known.

Variable inductance is achieved by the use of a ferromagnetic *core* in the shape of a rod that is surrounded by a hollow *coil* of wire. As the core enters the coil the inductance of the coil is changed. With reluctance sensors, the magnetic flux coupling between two or more coils is varied by a ferromagnetic core.

Reluctance sensors, a type of transformer, are the basis of the prevalent LVDT (*linear variable differential transformer*) shown in Fig. 31-2. The signal is easily conditioned to give a DC voltage that in linearly proportional to position over part of the range of motion of the core; the sensitivity can be on the order of 0.001 mm movement. It is possible to alter the design to measure the angular displacement as the core is rotated.

Other position sensors

Ultrasound can be used as a *position sensor* by measuring the time it takes a pulse of ultra-sound to "echo" back to the transmitter. It is similar in principle to a radar. This technique has been used on early "self—focusing" cameras, some consumer measuring devices used to measure room sizes in houses, and other ranging devices.

Strain gauges and load cells

Strain is the change in length of a material caused by a stress (force) on the object. For example, tensile loads on a metal rod cause the metal rod to elongate, while at the same time the cross—sectional area decreases slightly. If the force is limited to the elastic range of the metal rod, the metal rod will return to its original shape when the tensile force is released. A compression force causes the metal bar to contract with a slight increase in cross—sectional area.

Strain gauges use this phenomenon to measure strain by the change in resistance of metal. As metal increases in length and decreases in cross—sectional area, its resistance will increase; however, the change in resistance is largely due to the change in length. A metal strain gauge is made of fine wire or foil that zigzags (Fig. 31-3) to give a useful resistance (often 120 Ω, but common values are 60—1000 Ω). This design makes the gauge sensitive to strain in only one direction.

The *gauge factor* is the relative change in resistance divided by relative change in length (i.e., the definition of strain). Metal strain gauges have a gauge factor of about 2; thus a strain of 0.1% will cause a 0.24 Ω change in a 120 Ω strain gauge. The sensitivity of metal strain gauges can be 10^{-6} strain. Specialized semiconductor strain gauges based on silicon have gauge factors of —50 to —200, but these devices are nonlinear, making them more difficult to use.

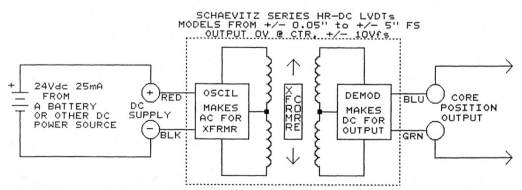

Fig. 31-2. Diagram of a linear variable differential transformer (LVDT). Courtesy of H.R. Holbo.

Of course, if the temperature of the metal changes, this will have a confounding effect. A second strain gauge is set perpendicular to the direction of strain and counteracts the change in temperature when they each make up separate legs in a Wheatstone bridge (Fig. 31-3). Many other configurations are possible to measure twist or other parameters.

To make a *load cell*, one attaches one or more sets (one to measure, one for temperature compensation) to a material. For example, if the load cells are attached to a support of a hopper, then the weight in the hopper can be measured as a corresponding strain. The ratio of the change in weight to the change in strain depends on the size of the support, with a large support giving a small change in strain per change in weight. Thus, the same strain gauge can be used to make highly sensitive load cells or load cells capable of measuring very large weights.

Pressure sensors

Pressure is force per area. If a fluid is at rest, one refers to static pressure. If a fluid is in motion, one refers to dynamic pressure, which is a function of the motion of the fluid. The relationship between *gauge pressure* is the difference between *absolute pressure* and *atmospheric pressure*. For example, a flat tire on a automobile is said to have a (gauge) pressure of zero, but it actually has an absolute pressure of 14.7 psi. The pressure (p) exerted by a liquid is a function of the density of the liquid (ρ), the acceleration due to gravity (g), and the height of the liquid (h).

$$p = \rho g h$$

Pressure can be measuring by monitoring the movement of a bellows with a LVDT. Strain gauges can be mounted on diaphragms to measure pressure; this method has been miniaturized onto integrated circuits.

Pressure can be measured mechanically by using a *Bourdon tube*, a coiled tube that is partially flattened (the pressure version of the coiled bimetallic strip used to measure temperature).

Very low pressures, i.e., very high vacuums, which indicates a gas phase, requiring specialized techniques. The temperature of a heated metal filament (or its resistance) depends on the gas pressure, since the gas molecules can conduct heat from the filament. Pressures below 10^{-3} can be measured with ionization gauges.

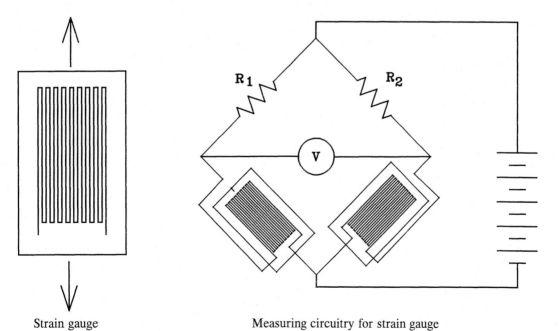

Strain gauge   Measuring circuitry for strain gauge

Fig. 31-3. Diagram of a strain gauge and its circuitry.

Flow sensors

Flow rates of materials moved on conveyors can be determined by weighing a portion of the conveyor (dividing by the length of the section that is weighed) and multiplying by the speed of the conveyor to give units of weight (or mass) per unit of time. This method can be approximated by using an LVDT to measure the sag of the belt.

Flow rates in pipes are often measured indirectly by measuring the pressure (or head) that causes flow in a pipe. Flow in pipes can be measured by placing a restriction in the pipe and measuring the pressure drop (Bernoulli's principle) in the restriction. *Rotameters* use an obstruction in the flow path (a ball that rises in a tapered, vertical column); the flow provides the force to raise the ball against gravity. A vane that is pushed upward by the fluid is an indicator of fluid velocity if the vane's position is monitored.

Magnetic flow meters use an insulated portion of pipe where two coils (on the outside top and bottom of the pipe) generate a magnetic field. The movement of charged particles in the fluid generates a voltage perpendicular to the magnetic field. The voltage generated between two electrodes placed outside the two sides of the pipe is an indication of the velocity. Knowing the cross—sectional area of the pipe section gives the volumetric flow rate. The fluid must be at least a partial conductor (i.e., water with electrolytes).

Level sensors

The level of liquid, slurry, or solid in a tank or hopper can be measured in several fashions. A float can be used in conjunction with an LVDT core or other displacement sensor. An ultrasonic position sensor at the top of the vessel can be used to detect the top of the material. A pressure sensor can measure the pressure the material exerts at the bottom of the vessel. Similarly, load cells can be used. Occasionally the electrical properties of the material (resistance or capacitance) can be used to measure the level.

31.5 CONTROLLERS

Introduction

Controllers respond only to errors; therefore, some error will be present in the system for control to occur; without error, a controller would not be required. Sophisticated controllers

will minimize this error by *tuning* the controller to the system, a process to be described later. *Regulation* is a process of keeping the error within acceptable limits. In the case of our house it might be 22°C ±1°C.

A *transient* is an upset caused by a change in the setpoint or some other factor that influences the response. In our heated house, what happens if we change the temperature (adjust the thermostat) to 26°C? The *transient response* is the error as a function of time, which is to be minimized.

Fig. 31-4 shows an example of a transient response curve. One control strategy is to minimize the area under the error time curve relative to the new (or continuing if there is an upset in the system) setpoint. A second control criterion is to have the amplitude of each peak one—quarter of that of the preceding peak in the *quarter amplitude dampening system*, which was the industry standard for many years since its description by J.P Ziegler and N.B. Nichols in 1942.

Let us consider the four individual control strategies available. The simplest form is the on or off control, which has been modeled above in the case of heating a house. In this case, there will always be cycling with the temperature going appreciably above the setpoint and then below the setpoint. In the remaining control strategies, one has the choice of varying the actuator from off to various levels of on to fully on.

(An on—off system can be used to approximate various intensities. In microwave ovens, the megatron is always on or always off, but various cooking levels and defrosting are achieved by turning the megatron on for an increasing percentage of a short time cycle. The time cycle is usually on the order of about 2 seconds. Industrial electric heater controls often employ this technique so that devices are not required to partially turn on huge loads. These times are usually unnoticeable relative to the time constant of the system.)

Controller action for a continuous actuator may be *proportional*, *integral*, or *derivative*. Many controllers use a combination of all three techniques and are called *PID controllers*. We will consider a benchtop stirred hot water bath with a heating element whose output (heat) can be controlled continuously from 0 watts to 1000 watts as an example of a device to be controlled.

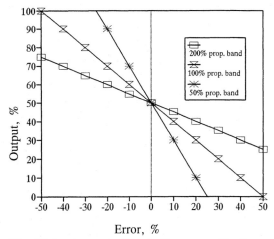

Fig. 31-5. A proportional diagram with three proportional bands.

Proportional controllers

Proportional controllers give an output to the actuator that is a multiple of (proportional) to the error; they respond to the size of the error. The multiple is the *gain* (= Δ output/Δ input). When the error is zero (the measurement equals the setpoint), the output is 50%. Fig. 31-5 shows some examples with various proportional band values. When the proportional band is too high, a large error is required to make the necessary change. When the proportional band is too low, a small error causes a large change in the output

that overcompensates; this can send the system oscillating out of control, much like feedback from a public address system. Gain is related to the proportional band expressed as a percentage:

$$\text{Gain} = 100/(\text{Proportional Band, \%})$$

Using our water bath as an example demonstrates one problem with this methodology. Suppose we want to heat our water at 30°C and room temperature is 25°C. It is quite likely that our 1000 watt heater operating at 50% output (when the error is 0) will tend to heat the water much above out setpoint of 30°C. Proportional controllers operate with an *offset*; that is, they try to maintain the measurement at a value that is different from the setpoint. A high gain will minimize the offset (but has the problems mentioned above); an infinite gain is essentially an off/on control. If one adds the integral controller this problem is solved.

Integral controllers

(An obsolete term for integral controllers is reset.) When an error occurs, the integral mode causes a continuous change in the output until the error disappears. This mode responds to the size of the error and the length of time the error occurs. In the case of our water heater, when trying to control the temperature to 30°C with a proportional controller, the temperature will

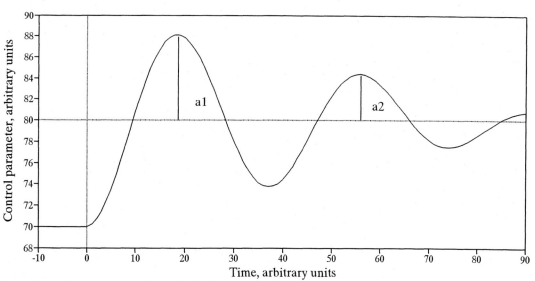

Fig. 31-4. Response of a (hypothetical) controlled system to a change in setpoint from 70 to 80 units at time 0.

consistently overshoot the mark. The integral mode would step in and continuously lower the offset until the temperature goes below 30°C, which might be a power level of 20%. The curves in Fig. 31-5 would be in effect, except that they would intersect at 20% output for this device operating at this temperature.

The effect of integral controllers is measured in units of minutes per repeat or its reciprocal. A high number of minutes per repeat indicates a low action by the integral controller.

Derivative controllers

Derivative controllers respond to the change in error with time. If the absolute value of error is rapidly decreasing and the setpoint is near, it is likely that the setpoint will be reached and the error will change sign and increase in the other direction. The derivative mode is useful in preventing this situation, especially in systems with large time constants. Derivative response is usually measured in minutes; the time is a measure of how far ahead the derivative + proportional action is ahead of the proportional action only in an open loop; the higher this value, the greater the derivative action..

Since a spike (a large, but brief response in the measured value) would cause a large action by the derivative mode, where no action is appropriate, the derivative mode is not always appropriate. This is true with signals that are inherently noisy (such as flow rate measurements) or systems with long dead times. Sometimes the derivative mode is only used when a change in setpoint occurs. The derivative mode us usually well suited for temperature measurements and control.

PID controllers

Proportional—plus—integral—plus—derivative controllers are described (Anderson, 1980) as:

$$m = g\left(e + \frac{1}{R}\int e\, dt + D\frac{de}{dt}\right)$$

where:

D = derivative time
e = error
g = gain
m = controller output (0—100%)
R = integral time
t = time

31.6 ACTUATORS

Once the sensors determine the status and once the controller determines what action is required, *actuators* carry out the appropriate response. Actuators take the result of control signals and cause a change on the control element. They are not the control elements. The term actuators seems to imply movement, but that is not always the case. An actuator may be an electronic relay (such as a silicon—controlled rectifier, SCR) that switches power to the control element.

Solenoids convert an electrical current into a relatively small back and forth motion. When a coil is energized, a metal plunger is drawn into it. A spring may be used to return the plunger to its original position. Motors convert electric signals into rotary motion. Motors are, in turn, used for movement, pumping of fluids, valve positions, etc. Motors can also be driven by air (pneumatic) or hydraulic fluid. Pneumatic actuators convert air pressure into movement. Actuators designed for valves are discussed on page 655.

31.7 ANNOTATED BIBLIOGRAPHY

1.   Anderson, N.A., *Instrumentation for Process Measurement and Control*, 3rd ed., Chilton Co., Radnor, Penn., 1980, 498 p. This work is useful for controller theory.

2.   Johnson, C., *Process Control Instrumentation Technology*, 4th ed., Regents/Prentice—Hall Englewood Cliffs, New Jersey, 1993, 592 p. The work is up to date and a good overall reference on the topic. It includes analog and digital signal conditioning, thermal, mechanical, and optical sensors, process control, actuators, and controllers. A later edition is available.

3.   Omega, *The Temperature Handbook*, *The Data Acquisition Systems*, *pH and Conductivity*, *Pressure and Strain*, *Heaters*, and *Flow and Level*. This is a series of catalogs by the Omega Technology Co. (Stamford, Connecticut) of products they sell; however, these substantial catalogs (available at no charge) contain invaluable practical information and tutorials. **Every** mill engineering library should have a set of these books.

32
CORRUGATED CONTAINERS

32.1 INTRODUCTION

Information on printing, paper used in corrugated boxes, paper production, paper testing, and other topics related to the manufacture of corrugated boxes can be found in the appropriate chapters of this book. Many of the same techniques described in other sections of the book apply to the manufacture of corrugated boxes including slitting, scoring, and crowning of press rolls.

Brief history

The production of corrugated paper was patented by Albert L. Jones in 1871 as a packaging material. It was used for protecting glass containers from each other during shipping. In 1874, Albert Long patented the use of corrugated paper glued to another piece of paper (single—faced corrugated board). The Shefton machine of 1895 could make single—face or double—face corrugated board and operated below 10 ft/min.

In the early part of this century, railroads had a virtual monopoly on long—distance transportation. Many of the railroads owned large areas of timber and had ownership in companies that manufactured wooden crates. The railroads, therefore, charged a tariff for products transported with fiber (solid paperboard) boxes and corrugated boxes. In 1914, the Pridham decision forced railroads to rescind this tariff. This marked the beginning of the large—scale use of fiber and corrugated boxes in transportation. The amount used (in billions of square feet) was 6.6 in 1923, 44.3 in 1943, and 251 in 1979.

Box types

Fig. 32-1 shows diagrams of several types of corrugated materials. Corrugated paper glued to one sheet on linerboard is called *single—face board*. It is used primarily as a spacer inside packages. If a second sheet of linerboard is glued to the other side the product is known as *double—face* board, and this product accounts for the largest percentage of corrugated board products.

Double—wall corrugated boxes are made with three layers of linerboard separated by two layers of corrugating medium. (A small amount of triple—wall corrugated boxes is produced.) Double—wall corrugated boxes usually use two different sizes of flutes, and the smaller flute is usually toward the outside of the box for several reasons. The smaller flute gives an even surface for the outer liner which provides a good printing surface; it also provides higher puncture resistance. (A common configuration for double—wall corrugated boxes is a B flute on top of a C flute.)

Flute types

The size of the flutes controls the properties of the final board. Small flutes increase the puncture resistance of boxes, while large flutes enhance the stacking strength of boxes. Page 162 gives the basic flute types, but the situation has increased in complexity. Flute C was originally 42 flutes per foot, but is now 39 flutes per foot. The C flute has replaced the A flute for most purposes since it requires less paper and gives similar properties to the A flute.

The E flute increased in use in the late 1980s in corrugated boxes designed to compete in the folding carton industry. The F flute was intro-

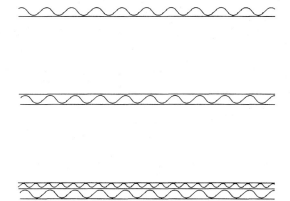

Fig. 32-1. Some examples of corrugated board including single—face (top), double—faced (middle), and double—walled.

duced in 1991 (about 0.03 in. high and about 150 flutes per inch) as a smaller alternative to the E flute and folding carton stock. Larger flutes are recent developments to allow double—face boxes to replace double—wall boxes. The K flute is 0.25 in. thick. The S flute is a jumbo flute.

32.2 MANUFACTURE

Introduction to the corrugator

The first part of the operation is accomplished by the *single—facer* section (Fig. 32-2). First the corrugating medium is unwound and conditioned to the proper moisture content and temperature with a preheater and steam shower. Preconditioning rolls, if used, are usually 12 to 18 in. in diameter. The corrugating medium must have enough water to make it flexible for proper flute formation. The adhesive usually has enough water to gelatinize the starch.

Splicers for the corrugating medium and linerboards attach the end of one roll of paper to the beginning of the next roll so production can be continued without having to stop the machine. Automatic splicers are activated by the (small)

diameter of the roll in use or by detection of the paper end. Modern splicers have an overlap of only a few inches, which causes little problem in the production process or the final product.

The corrugating medium is sent between a set of corrugating rolls (12 to 24 in. in diameter) to flute the medium. After the corrugating nip, adhesive is applied to the tips of the flutes on one side of the sheet. Linerboard (which is preheated, but not usually premoistened) is pressed against the corrugating medium by the pressure roll (with pressures of 300—500 pounds per linear inch) while it is still supported by one of the corrugating rolls; the pressure roll, or single—facer, is 16—18 in. in diameter and heated to 177°C (350°F). The initial bond that is formed is called the green bond. Control of the moisture content of the linerboard prevents curl of the board product during its production. In some (older) corrugators, thin metal plates called *fingers* hold the medium against the corrugating roll to maintain its shape while the liner is applied to the fluted corrugating medium. Most designs are now fingerless and use a vacuum or air positive pressure.

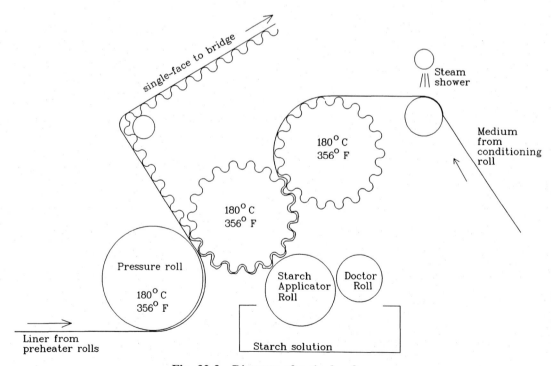

Fig. 32-2. Diagram of a single—facer.

The single—face product then goes to the *double—facer* (alias *double—backer*) section to apply a second layer of linerboard on the opposite side of the fluted corrugating medium. Starch adhesive is applied to the exposed tips of the fluted corrugating medium, the liner is put in contact with the medium, and the combination is placed under a belt that travels over steam chests; the nip at this point is extremely dangerous as it can drag a person's arm into it with severe consequences. (From here on out the board must be handled in a flat condition since wrapping it around rolls will cause permanent deformation.) The pressure applied in the *hot plate section* (or *double-backer*) cannot be so high as to crush the flutes. The product goes through the hot plate section for several seconds to allow the starch adhesive to set. It goes through a cooling section where it is pulled by the addition of a lower belt. The later section is often called the *traction section* because it pulls the web through as in the Langstron.

The adhesive application system had used tapered speed drives at one time, but developments in drive technology have changed this. For example, the speed of the adhesive applicator surface might be 20 ft/min slower than the paper speed regardless of the paper's speed in order to cause some wiping action; the more wiping action, the more adhesive transferred. When the machine is not moving, the applicator roll is separated from the paper. The distance between the starch applicator roll and the doctor blade is about 0.15 mm (0.006 in.).

Miscellaneous aspects of corrugators

Recent corrugators are 2500 mm (99 in.) wide and operate at 6 m/s (1000 ft/min). The corrugating medium may initially travel at 8 m/s if the take—up ratio is 1.33 (or 8 m/s divided by 6 m/s) as for a B flute. An 85" corrugator will use 4—5 tons of steam per hour requiring a minimal pressure of 175 psig.

Corrugating adhesive

Initially starch (one part in about 15 parts of water) was used as the adhesive, but it took a long time to set, which caused numerous problems for box makers. By 1920, sodium silicates (water-glass) were the principal adhesives used in box-making. This caustic material was strong but brittle and had some major problems in its use.

In 1936 J.V. Bauer patented the use of starch adhesive that was cooked on the box during assembly (which, of course, requires that the box be heated to set the starch adhesive). By 1960 starch regained its position as the principal adhesive for boxmaking. Corn is the most common source of the starch. The use of starch is just over 2 lb/1000 ft^2 of double—faced board.

In the Stein—Hall process, cooked starch (about 15—17% of the total starch) is used to make a suitable formulation for the uncooked starch. Sodium hydroxide is used in the formulation to lower the temperature at which the starch will gelatinize on the box; corn starch will gelatinize around 70°C, but NaOH lowers this to about 65°C for making corrugated boxes. Borax is used to increase the tackiness of the adhesive and act as a buffer for the NaOH. A preservative must be used to prevent microorganisms from attacking the starch because they can quickly alter the viscosity of the formulation. Other materials in addition to cooked starch can be used to make a suitable formulation with the uncooked starch, especially in water—resistant adhesives.

There must be adequate water in the starch formulation and sufficient heating of the formulation on the corrugator to get a suitable bond. The cooking of the starch and the formation of the bond take place within a few one—hundredths of a second on fast—moving single—facer machines (where high pressure between the corrugated sheet supported by the corrugator and the linerboard are possible). On double—backer machines, the time for cooking must be several seconds since much lower pressures are used, so a higher viscosity starch solution can be used.

A recent practice at some box plants is to use flexographic wash water as the water in the starch adhesive. One should know exactly what ink components are in this wash water and how they will affect the system. DiDominicis and Klein (1983) give much information on the starch—based adhesive used in corrugated boxes.

White boxes

White boxes can be made using a thin layer of white pulp on the top liner (applied with a second headbox) which gives a blotchy appearance. A second method is to use an entirely white linerboard, which is expensive. A third method is to coat the corrugated box.

Corrugator finishing operations

The continuous board must then be cut into *blanks* (individual sheets) that will each make an entire box. The continuous board first goes through the *slitter/scorer section*. Here, the board is cut and scored in the machine direction. A series of sheeter knives make cuts in the cross machine direction to produce the blanks.

Blanks to containers

The blanks go through a series of operations to become containers. They go through a printing operation. They must be scored in the cross machine direction. Slots must be formed by stamping out portions of the box to define the tabs for assembly.

Overall considerations

The cost of paper delivered to the box plant is usually 60—80% of the final cost of a box. Paper use (especially trim and other waste) must be minimized to stay competitive with other box plants. Boxes are constructed to meet requirements of the rail (especially Rule 41) or trucking industries (Rule 222 for common carriers). It takes about 60—70 tons of paper to produce 1 million square feet of corrugated board. A corrugator with a width of 97" and operating at 600 ft/min will use over 3500 tons of paper per month.

About 5—7% of the incoming paper will become scrap material. This is usually handled by a pneumatic system that goes to a cyclone separator. From the separator the waste paper goes to a baler, where it is apt to be sold to a pulp mill using waste paper.

Scheduling orders

A wide variety and number of orders may be processed in a relatively short time. These must be scheduled to meet deadlines, to obtain the maximum capacity of machines, to make the most efficient use of the machine width, and other factors. Also, the proper type and suitable amount of paper must be on hand (a 5—6 week inventory of paper is generally suitable). Scheduling orders efficiently, as is any scheduling operation, can be an extremely complex task using linear programming and other tools. Computer modeling of the scheduling process is extraordinarily complicated; thus, this position is of key importance to the smooth operation of the box plant.

Partial rolls of paper that remain after a job should be measured, recorded in inventory, and used in the next suitable job so one is not inundated with partial rolls (called butts) in inventory.

32.3 TEST METHODS

Testing of corrugated board

A variety of tests are used to test corrugated board, the two paper components of board, or assembled boxes. Some of these, depicted in Fig. 32-3, are performed between two flat platens. Various types of compression (or crush) tests are critical since most corrugated boxes fail in compression when stacked.

For routine quality control, sealed, empty boxes are tested on relatively large test devices with parallel platens (moving at 10—13 mm/min during the test) in the *box compression test*, BCT, (Tappi Standard T 804). A number of different box designs (flute types, liner types and weights, etc.) can be compared directly provided the boxes are all the same size. Inner packing may be used for testing design loads. This test best represents the performance of most corrugated boxes in service (i.e., where they are stacked on top of each other). In reality, however, loading will not be uniform, humidity conditions will vary, loading times will be much longer; loads substantially lower that of the test value can be used in service.

Testing of board components

The remaining tests described here are done on small samples on a test device such as that shown in Fig. 7-16 on page 179. The *pin adhesion test* (T 821) tests the strength of the bond of the flute tips of the corrugating medium to one of the liner boards in a sample of corrugated board. The sample size is 2 in. by 6 in. for boards with A or C flutes and 1.25 in. by 4 in. for boards with B flutes. Low values for this test are caused by poor adhesive or poor adhesive application (if failure is at the glue line) or containerboard with a low internal bond strength (if failure occurs in the paper).

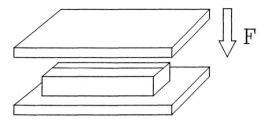

The box compression test (BCT) measures the resistance of an assembled box to compressive force.

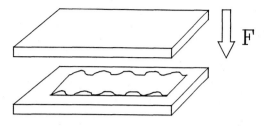

The concora medium test, CMT, measures the crushing resistance of a fluted sample of medium.

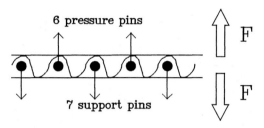

6 pressure pins

7 support pins

The pin adhesion test, PAT, measures the resistance of the flute tips to separation from the liner.

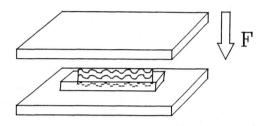

The corrugated crush test, CCT, is performed on medium (in a jig) fluted in the laboratory.

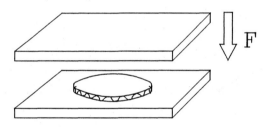

The flat crush test, FCT, measures the resistance of the flutes to a crushing force.

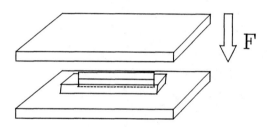

The concora liner test, CLT, measures the edgewise compression in a supported sample of liner.

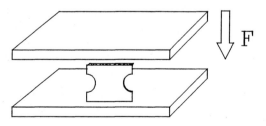

The edge crush test, ECT, measures the compression strength as in boxes that are stacked.

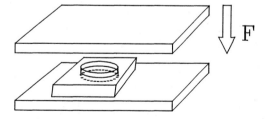

The ring crush test, RCT, measures edgewise compression in a ring of liner supported by a jig.

Fig. 32-3. Some common tests performed on corrugating medium, linerboard, corrugated board, and completed boxes. One test machine is shown in Fig. 7-16 on page 179.

The *flat crush test* (T 808) measures the resistance of the flutes to crushing force from a sample cut from manufactured corrugated board. T 808 specifies a circular sample of 5 or 10 in.[2] in area. Low values for this test might be due to medium of low strength, crushed flutes, or flutes that have not been well formed (such as leaning flutes). This test is not suitable for double— or triple—wall boards since the inner liner(s) are subject to lateral motion during the test, but would be much less so in products in use. This test does not simulate forces applied to boxes in use, so the correlation between this test and how a box will perform in service is not always high. The bending stiffness of the corrugated board should complement the FCT (Markström, 1992).

The *edge crush test* or *edgewise compression test*, ECT (T 811) also uses a sample cut from manufactured board. The shape may be rectangular or as that shown in the Fig. 32-3; the latter shape is designed to give a failure away from the edge. The force is applied parallel to the flutes, just as in a container in service. This test is a good indicator of the general performance of a box in transportation. ECT can be related to BCT by use of the McKee (1963) equation (Markström, 1992). There are many modifications to this test which are discussed and quantitatively compared by Markström (1992).

The remaining tests described here are designed to test the properties of the medium or liner, i.e., they are quality control tests of the raw materials. The sum of the compression strengths of the liner and medium layers correlates well with the compression strength of the corrugated board.

The *short—span compression test*, SCT, has been described on page 177. T 826 specifies a sample 15 mm wide. This test provides similar information to other compression tests except sample geometry does not enter into the results. Geometry is an issue for at least two reasons: slender samples, such as a straw in the lengthwise direction, fail by buckling long before the material fails by compression; samples that fail at the edge give lower strength results than are truly representative. For this reason it is replacing other, older compression tests such as those described below (Markström, 1992).

The *concora medium test*, CMT, or *flat crush test* [T 809 specifies a sample 0.5 × 6.0 in. (12.7 × 152.4 mm)] measures the crushing resistance of a single—face sample prepared in the laboratory. It tests the medium's resistance to crushing.

The *corrugated crush test*, CCT, or *fluted edge crush* measures the edgewise compression strength of a piece of corrugated medium. This provides a measure of the corrugating medium's contribution to the compression strength of a corrugated box. It is an alternative to the ring crush test (T 818). T 824 calls for a sample 0.5 in. by 6 in. with the longer dimension in the machine direction. The sample is held in a specimen holder during the test. The *concora liner test*, CLT, is very similar to the CCT, except the sample is not corrugated.

The *ring crush test*, RCT, measures the edgewise compression in a ring of liner or medium that is supported in the groove of a jig. Two variations of the test are T 818 (flexing beam) and T 822 (rigid platen).

32.4 ANNOTATED BIBLIOGRAPHY

Corrugated containers—thermal resistance, misc.

1. Bormett, D.W., Overall effective thermal resistance of corrugated fiberboard containers, USDA For. Ser. Res. Pap. FPL 406 (1981), 12 p. The effective thermal resistance of corrugated fiberboard containers was determined versus air velocity (0—5.4 m/s) and board thickness. Differences were noted for heating versus cooling for boards less than 20 mm thick. This information is useful for consideration of how produce and frozen foods will behave during shipping. The effective thermal resistance of a corrugated fiberboard package was found to be about 0.18 K·m²/W (although the outer boundary air and box—product interface may effectively double or triple this value) for a kraft linerboard package (0.33 mm thick) to 1.10 for a corrugated fiberboard box having walls 51 mm thick as shown in Table 32-1.

2. TAPPI Technical Information Sheets (TIS) with numbers 0303 to 0305, and several 0306

to 0309 cover various aspects of corrugating medium and starch adhesive preparation.

3. DiDominicis, A.J. and G.H. Klein, Corrugating, in *Pulp and Paper Chemistry and Chemical Technology*, 3rd ed., J.R. Casey, Ed., 1983, Wiley, New York, pp 2373-2398.

4. Markström, H., *Testing Methods and Instruments for Corrugated Board*, 3rd ed., Lorentzen & Wettre, Stockholm, 1992, 77p. While promoting the publisher's test instruments, it is a useful, relatively unbiased source for the theory of the test methods. It is usually available on a complimentary basis.

5. McKee, R.C., J.W. Gander, and J.R. Wachuta, Compressive strength formula for corrugated boxes, *Paperboard Packaging* 48:19, 149-159(1963).

Table 32-1. Overall effective thermal resistance (values listed are R or K·m²/W) of a cubical corrugated container. From USDA For. Ser. Res. Paper FPL 406 (1981).

Board caliper, mm	Air velocity, m/s		
	0.0	1.8	5.4
0.00	0.33	0.14	0.13
0.33	0.42	0.19	0.18
4.23	0.45	0.31	0.28
12.7	0.61	0.50	0.50
25.4	0.80	0.73	0.70
50.8	1.10	1.04	1.04

33
MISCELLANEOUS TOPICS

33.1 VISCOSITY AND SURFACE TENSION OF WATER FROM 0 TO 100°C

CRC Handbook of Chemistry and Physics (67th ed.) gives the viscosity of water from 0° to 100°C based on a contribution from the National Bureau of Standards that is not subject to copyright. The viscosity of water is given in Fig. 33-1 and is calculated from 0°—20°C [Hardy, R. C., and R. L. Cottington, *J. Res. NBS* 42:573(1949)] as:

The viscosity of water from 20° to 100°C is calculated according to Swindells, J.F., NBS, unpublished data as:

$$\eta_T = 1.002 \times 10^{\frac{1.3272(20-T) - 0.001053(T-20)^2}{T+105}}$$

The surface tension of water against air is and the dielectric constant of water (ϵ) (*Lange's Handbook of Chemistry and Physics*, 13th ed., page 10-99) are also given in Fig. 33-1.

$$\eta_T = 10^{\left(\frac{1301}{998.333 + 8.1855(T-20) + 0.00585(T-20)^2} - 1.30233\right)}$$

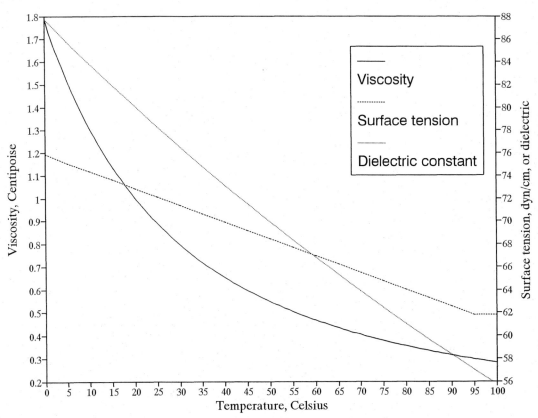

Fig. 33-1. Viscosity, surface tension, and dielectric constant of water from 0 to 100°C.

33.2 VAPOR PRESSURE OF WATER FROM 0 TO 280°C

The absolute vapor pressure of water (psia) is shown from 0—100°C (Fig. 33-2) and 100—180°C (Fig. 33-3). The following equation is useful for calculating the pressure to within 0.05 psi of published data for 100—180°C (*Handbook of Chemistry and Physics*, 40th ed., pp. 2328—2330, CRC Co.).

$$\text{psia} = -6.487 + 0.1229\,T + 0.000599566\,T^2 - 0.0000171538\,T^3 + 0.0000002004133\,T^4$$

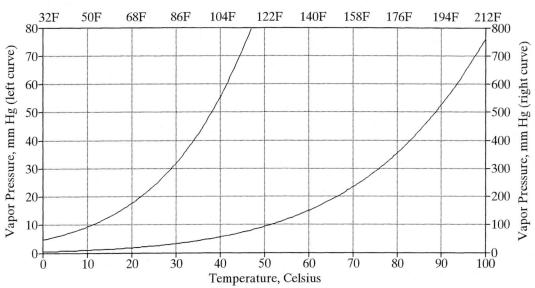

Fig. 33-2. Vapor pressure of water from 0 to 100°C.

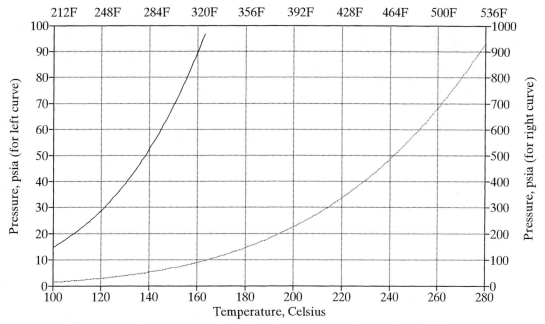

Fig. 33-3. Water vapor pressure as a function of temperature from 100 to 280°C with expanded scale for pressures below 100 psia.

33.3 PROPERTIES OF AIR AND WATER

The properties of air are particularly important for combustion processes. The composition of air is given in Table 33-1. General properties of air are given in Table 33-2. The importance of water to the industry also goes without saying; the properties on water are summarized in Table 33-3.

33.4 WATER CONDITIONING

BETZ® (Trevose, Pennsylvania) has published *BETZ Handbook of Industrial Water Conditioning*, which is a very good source of information on this topic. Much of the information in the sections on water treatment is a condensation of this extensive work.

Table 33-1. Content of dry, atmospheric air.

| Component | ------Content, %------ | |
	Volume	Mass
N_2	78.08	75.52
O_2	20.95	23.14
Ar	0.94	1.29
CO_2	0.033	0.05
Others	0.003	--

Table 33-2. Properties of dry air at 25°C.

Property	Value
Average molecular weight	28.964 g/mol
Speed of sound	340 m/s
Specific heat	0.240 cal/g/K

Table 33-3. Properties of water at 25°C.

Property	Value
Specific heat	0.998 cal/g/K
Heat of fusion (freezing, 0°C)	79.72 cal/g
Heat of vaporization	540 cal/g
Density	0.997 g/cm³
Index of refraction (sodium light)	1.3329

Water treatment methods discussed in this section are often termed *external treatments* since they are designed for large quantities of water not designated for use inside boilers or for use in other specific applications.

Water must be properly conditioned to prevent dangerous corrosion and scale formation in boiler tubes that carry water, to be used in cooling systems, and if it will enter or be reused within the mill. Wastewater treatment has already been addressed in Chapter 11.

Contaminants in water

The theoretical basis of corrosion was presented in Section 13.11. A review of this section will help explain why certain impurities of water cause corrosion. The solubility products of some compounds were presented in Section 13.8 and help explain why certain contaminants tend to cause scale formation. Section 14.7, a discussion on coordinate chemistry, is relevant to this section.

Contaminants in water may be classified as dissolved solids, dissolved gases, and suspended solids (which make water appear turbid or murky). In some cases the distinction is not clear since hydrogen sulfide may occur as the dissolved solid of the metallic salt. Suspended solids, which can easily cause deposits, are removed by filtration or settling, often after clarification treatments.

Any ionic material in water increases corrosion by increasing the conductivity of water. Insoluble ions, especially most calcium salts that decrease in solubility with increasing temperature, will contribute to scale. Water hardness is a measure of the calcium and magnesium content. Other culprits or synergistic elements include aluminum (carried over from poor clarifier operations), magnesium, silicates, and sulfates. In the Southern U.S., barium introduced through the water and wood supply contributes to scale formation. Acids and bases are ionic in nature and contribute to corrosion. These materials can be removed by ion exchange or other demineralization treatments. Chelating agents such as EDTA bind with and solubilize the worst scale-forming metal cations.

Dissolved gases such as oxygen and carbon dioxide increase corrosion. These can be removed by deaeration. Corrosion inhibitors decrease their damage. Oxygen increases pitting and other forms of corrosion. Carbon dioxide causes corrosion in steam condensate systems.

A variety of materials cause fairly specific problems. Ferrous or ferric salts will interfere with wet end chemistry and decrease the brightness of paper. Ammonia complexes with copper and zinc alloys, making them soluble. Work in my laboratory shows that relatively small amounts of fluoride can decrease rosin sizing with alum.

Aeration

Aeration of water is used to remove dissolved gases by "dilution" or flushing them away. Since air contains large amounts of nitrogen and oxygen, these gases will remain in the water after treatment. Aeration decreases the amount of free CO_2 and NH_3. Carbonates and ammonium salts, however, will remain depending upon the pH of the water. Aeration also precipitates iron as ferric hydroxide and manganese as manganese dioxide.

Chlorination

Chlorination of water is used to kill microorganisms with adverse health effects. It is used to produce drinking water. Some mills chlorinate their water supply in order to decrease the growth of slime fungi and bacteria. The active species is hypochlorous acid. The chemistry of chlorine is covered in Section 17.2. Chlorine is occasionally used as an oxidant to remove iron and manganese from water. Chloride, however, causes scale and corrosion in boilers.

Clarification

Clarification of fresh water is used to prepare water for domestic and industrial use. Effluent may also be treated by this method. This method is used to remove suspended solids by coagulation. Inorganic salts of aluminum, iron, magnesium, or silicates can be used. Bentonite (aluminum silicate) is also used. The mechanism involves colloidal chemistry, which is discussed in detail in Chapter 21.

Polyelectrolytes, water—soluble polymers with ionized groups for charge interactions, are also used. The charged groups may be anionic, cationic, both cationic and anionic, or only slightly ionic. The sludge obtained with polyelectrolytes tends to be of much lower volume than those obtained with alum treatment. Low molecular

weight cationic polyelectrolytes (such as polyamines below 500,000) are used for primary coagulation. High molecular weight ones (above 20,000,000) are used for flocculation by bridging particles together. After flocculation, the flocs must be separated from the water by conventional clarification units.

Filtration

Filtration is used after clarification of water for drinking or boiler purposes. Water passes downward through a sand and/or anthracite bed through a total depth of 40 to 80 cm (15 to 30 in.) A multilayer gravel bed supports the filtering media and allows for collection of the filtered water. These may be gravity or pressure filters.

Do not use cold water to backwash a hot process filter since thermal compression and expansion will cause leaks to develop at the flanges and oxygen in the cold water will cause corrosion.

Precipitation softening

Water hardness is a measure of the amount of calcium and magnesium present and is reported as parts per million (as calcium carbonate). Moderate to high hardness water of 150—500 ppm can be partially softened by precipitation of the offending species. Calcium and magnesium associated with bicarbonate anions are termed *carbonate* or *temporary* hardness since they can be removed by heating. Calcium and magnesium associated with chloride or sulfate are termed *permanent* or *noncarbonate* hardness. Lime or sodium aluminate may be added to raw water to precipitate noncarbonate salts of calcium and magnesium. This can be done under a variety of conditions depending upon the quality of the raw water and the purpose of the treatment.

Cold lime softening uses calcium hydroxide to precipitate magnesium ions as magnesium hydroxide. Sodium carbonate is used to precipitate calcium ions as calcium hydroxide. Residual calcium is about 35 ppm. Since calcium carbonate and magnesium hydroxide have lower solubility at high temperatures, hot lime softening can decrease magnesium to 2 ppm and calcium to 25 ppm.

Ion exchange

Ion exchange resins are used to *soften* water by replacing the cations with sodium ions (and possibly the anions with chloride ions) of sodium chloride. They may also be used to *demineralize* water where the cations are replaced by H^+ ions and the anions are replaced by OH^- ions.

In 1905 the German chemist Gans used sodium aluminosilicate materials (*zeolites*) to soften water. These materials can absorb calcium or magnesium cations and liberate sodium ions, the basis of water softening. The zeolite could be regenerated by running concentrated NaCl solution through it to liberate $CaCl_2$ and $MgCl_2$. *While aluminosilicates are seldom used anymore, the term zeolite softener is occasionally used for any cationic exchange resin.*

In 1944, cationic exchange resins were developed based on polystyrene crosslinked with about 6–8% divinylbenzene. This resin had a much higher *capacity* than previous resins, meaning that the same volume of resin could exchange many times more ions than previous materials. In 1948, the anion exchange analog was developed. Using

these two resins in sequence allows the complete demineralization of water, instead of merely exchanging cations for sodium ions. The macroreticular anion resins that have discrete pores are used first in the sequence to remove organic materials from the raw water that would otherwise cause fouling of the standard gelular resins. Many modifications of the original polystyrene divinylbenzene have been made to accomplish a wide variety of separations (such as the determination of molecular weight distributions of polymers described in Section 18.4).

Figure 33-4 shows the structure of polystyrene divinylbenzene resin. The functional groups used in the major types of ion exchange resins are also showed. In cationic exchange resins, the counter anion is tied to the resin and not free to migrate. Cations in solution can exchange with the resin cations until all the resin cations have been exchanged. At this point, a solution containing the desired cation in relatively

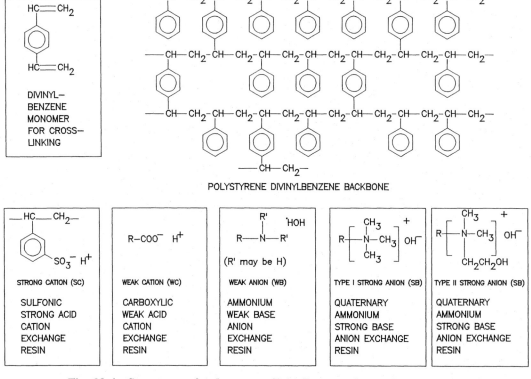

Fig. 33-4. Structures of polystyrene divinylbenzene ion exchange resins.

Table 33-4. Boiler feedwater guidelines.

| Drum pressure psig | ----------Feedwater---------- -Concentration, parts/billion- | | | Boiler ppm |
	Iron	Cu	Hardness total as $CaCO_3$	SiO_2
0-300	100	50	300	150
<450	50	25	300	90
<600	30	20	200	40
<750	25	20	200	30
<900	20	15	100	20
<1000	20	15	50	8
<1500	10	10	0 as OH^-	2
<2000	10	10	0 as OH^-	1

high concentration must be used to regenerate the column by backwashing. For water softening the column can be regenerated with about 10% sodium chloride. For demineralization, an acid such as hydrochloric acid must be used. Since not all of the hydrochloric acid will be absorbed by the column, an acid waste stream is created.

33.5 BOILER FEEDWATER TREATMENT

The treatment of water for boilers began with the invention of the first practical steam engine by James Watt in 1763 (patented in 1769). Although Watt alluded to the possibility of high pressure engines in his patent, he usually used steam pressures on the order of 50 kPa (7 psig). Engineers soon found that adding a few potatoes to the boiler reduced the amount of scale formed in the boiler.

With the development of higher temperature and pressure boilers, water treatment became more sophisticated. From 1857 to 1900 numerous patents described the use of tannins and other materials for treating boiler water. The use of disodium phosphate and trisodium phosphate came during this time period. Still, boiler explosions killed hundreds of people each year during this period of time due to corrosion and alkaline embrittlement of metal. The buildup of scale inside boiler tubes decreases heat transfer, so that the metal tubes will be much hotter, leading to higher rates of corrosion and reduced strength.

Boiler feedwater is purified to remove contaminants that would interfere with the safe operation of boilers. Methods used to purify boiler feedwater include those already discussed and additional methods. Additives are used in the feedwater to counteract those impurities that remain in the feedwater.

Specifications

Table 33-4 gives some guidelines of the ASME Research Committee on Water in Thermal Power Systems in terms of the tolerable amounts of impurities in boiler waters.

Deaeration

Deaeration of water is accomplished by decreasing the concentration of the offending gas species (O_2, CO_2, NH_3, H_2S) over the water sometimes with heating of the water, which decreases the solubility of most gases (e.g., Fig. 11-1 for oxygen). This can be accomplished by applying a vacuum over the water, a method suitable for water distribution systems. However, the gases over boiler water are usually purged (flushed away) with the use of steam. Steam does not contaminate the water and raises the temperature of the water (which must be done anyway), thereby decreasing the solubility of the gases.

Oxygen scavengers

Oxygen may be introduced into the system inadvertently even with proper preparation of boiler feedwater. Trace amounts of oxygen can be removed by oxygen scavengers. Sodium sulfite is the most commonly used additive. It reacts with oxygen to form sodium sulfate: $2Na_2SO_3 + O_2 \rightarrow 2NaSO_4$. It is added continuously at about 10 lb per lb of oxygen in the feedwater. The reaction is quick at temperatures above 100°C (212°F). The reaction is most rapid at a pH of 9—10. Catalysts such as the divalent cations of the transition elements are often used.

Hydrazine (N_2H_4, which is converted to nitrogen and water) is also used as an oxygen scavenger. Because it reacts slowly, it is useful only in high—temperature applications (900 psi pressure corresponding to 278°C or 532°F) where sodium sulfite may decompose and sodium sulfate is undesirable. Hydrazine also reacts with iron to give a protective coating of magnetite (Fe_3O_4).

Other boiler feedwater additives

Maintaining the pH of boiler water in the range of 9 to 11 decreases the rate of corrosion;

9.2 to 9.8 is ideal. In low—pressure steam or hot water heating systems, nitrite or chromate provides a protective film on metal surfaces.

Filming amines provide a physical barrier to protect metal from corrosion. They are effective in the pH range of 5.5 to 8.5. Amines that are commonly used include those with long—chain hydrocarbons that orient away from the metal toward the water, such as octadecylamine.

Cyclohexylamine, diethylaminoethanol, dimethylpropanolamine, and morpholine are sufficiently volatile to travel with the steam. The amine in the vapor helps keeps the pH sufficiently high to avoid corrosion.

Scale control

Scale tends to build up in boiler water (called water side deposits) for two reasons. First, bicarbonate ions are converted to carbonate ions and carbonate salts of calcium and some other cations have low solubility. Second, the solubility of calcium carbonate and similar compounds decreases with increasing temperature. Heat transfer surfaces, where the temperature change is most dramatic and fast, tend to be where scale forms.

The buildup of scale is controlled by the use of dispersants with chelants and/or phosphate. Chelants like EDTA (Section 14.7) keep cations like calcium from precipitating by forming strong, soluble complexes with them. Phosphate tends to precipitate these materials early on so they do not later precipitate on heat transfer surfaces. Compare the solubility products of the compounds in Table 13-5 to those of $CaHPO_4$, 1×10^{-9}; $Ca_3(PO_4)_2$, 2×10^{-29}; $Ba_3(PO_4)_2$, 3.4×10^{-23}; $Mg_3(PO_4)_2$, $\approx 10^{-23}$ to 10^{-27}; and $Zn_3(PO_4)_2$, 9×10^{-33}. Remember, however, that the phosphate salts of divalent cations have five species in the solubility product.

EXAMPLE 1.   Compare the concentration of calcium ion in calcium carbonate and calcium phosphate at 25°C.

SOLUTION: $[Ca^{2+}][CO_3] = 4 \times 10^{-9}$
$[Ca^{2+}] = 6.3 \times 10^{-5}$

$[Ca^{2+}]^3 \, [(PO_4^{3-})]^2 = 2 \times 10^{-29}$;
$[Ca^{2+}] = (2 \times 10^{-29})^{1/5} = 1.8 \times 10^{-6}$

This applies only at high pH (11—12) where carbonate and phosphate ions occur. At moderate pH (5—7) calcium bicarbonate is fairly soluble (apparently too soluble to find K_{sp} values readily in tables), but calcium hydrogen phosphate is still fairly insoluble.

PROBLEM: What is the solubility of calcium over calcium sulfate and calcium hydrogen phosphate? This is applicable at pH values found in boiler water.

NTA (nitrilo triacetic acid) is a weaker chelating agent than EDTA but is used for boiler water treatment.

Excessive boiler water carryover with the steam will cause corrosion and scale formation on turbines and other critical equipment.

Demands of high—temperature boilers

New boilers for the pulp and paper industry operate at 900—1500 psig. While water hardness is always of concern, the formation of metal oxide scales is equally important. Iron (and possibly copper) oxide is often the principal contaminant of scale formation in high—temperature boilers.

Polymeric dispersants have been used for scale control at lower temperatures since the 1950s. These include negatively charged polymers, such as polyacrylate, polymethacrylate, and polyacrylate—acrylamide copolymer, apparently bind to the positively charged metal oxide particles, keep them in solution, and prevent them from growing, i.e., steric stabilization. (One might hypothesize that maleic acid copolymers may chelate better than these other polymers.) Unlike chelants or precipitants, dispersants do not react stoichiometrically.

In the mid—1980s Betz introduced a polyethylene polymer (with every other carbon containing a methyl group and a phosphonate group) for use at high temperature.

Boiler water blowdown

As steam is removed from the boiler, salts and other materials remain behind, allowing these materials to concentrate in the boiler water. The concentration of salts and metal oxides are typically 10—50 times higher in the boiler water than in the feedwater. Some boiler water is continuously removed from the system to prevent undue concentration of these materials that would eventually precipitate and cause scale.

33.6 CORROSION

Corrosion

Corrosion, while always an important consideration, has become of particular concern to the pulp and paper industry since the 1970s, when mills have been forced to reuse water in various processes. As pressure continues to increase water closure in mills, the problems of corrosion will increase. Indeed, corrosion may be one of the limiting factor in determining how much closure can be obtained. Some basic aspects of corrosion have been considered in Section 13.11 on page 330.

Types of metals for corrosion resistance

Table 13-8 (page 331) shows some metals with their relative resistance to corrosion in moving sea water. A few additional aspects of stainless steels will be considered here. In addition to the compositions shown in Table 13-8, most stainless steels have 1—2% Mn, although 201 and 202 have 5.5—7.5% and 7.5—10%, respectively. P is limited to less than 0.06 or 0.045% in most stainless steels, although type 303 allows up to 0.2%. S is limited to 0.15%, 0.06%, or 0.03% in many types. Si is limited to 1.0% in many stainless steels; however, type 302B specifies 2—3%, type 314 specifies 1.5—3%, and type 403 limits it to 0.5%. Stainless steels with an L suffix are low carbon. For example, types 304 and 316 have maximums of 0.08% carbon while 304L and 316L have maximums of 0.03% carbon.

Type 317 is designed for maximum corrosion resistance of standard stainless steels; its composition is 18—20% Cr, 11—15% Ni, 2% Mn, and 3—4% Mo. Type 347 is 17—19% Cr, 9—13% Ni, 2% Mn, 1% Si, 0.045% P, 0.03% S, and at least 10 times the amount of carbon as Cb—Ta.

In addition to the stainless steels, which have matrixes of iron, alloys which are mixtures of chromium and molybdenum in a nickel matrix are very important.

Duplex stainless steels have been used in selected corrosive environments since the 1960s. New duplex stainless steels that are more easily welded have been available since the mid—1980s. These materials offer the potential to replace other stainless steels in a wide range of applications.

Acid cleaning and corrosion (Crowe, 1992)

Many components, such as kraft digesters, are washed with acid to remove scale. $CaCO_3$ and other carbonates are major species of scale; acid solubilizes this material by converting the CO_3^{2-} into CO_2 and H_2O by the addition of $2H^+$. When HCl is used as the source of acid the remaining $CaCl_2$ is water soluble. Formic (HCOOH), sulfamic, and nitric acids are also used for acid washing. The use of additives (corrosion inhibitors) with these acids may decrease the amount of corrosion caused by acid washing. These inhibitors are often organic sulfides or amines. One inhibitor is p—aminophenol, which has trade names of Activol and Rodinal. Another agent is diethylhydroxylamine and is commonly used in steam cycles. Propargyl alcohol (2—propyn—1—ol) is used in strong acid baths to decrease hydrogen embrittlement and steel corrosion. [The addition of copper sulfate to hot, dilute, sulfuric acid often limits its corrosion to stainless steel.]

Often paper mill workers allow suppliers to give proprietary chemicals as corrosion inhibitors. This gives the mill no way to effectively test it against other manufacturers, products without lengthy studies. Even research papers report results with unknown inhibitors (Crowe, 1992). It is also difficult to find information on corrosion inhibitors that may be particularly effective to the industry. A mill could even do its own corrosion tests with small metal coupons in various solutions to determine what might be most suitable (Crowe, 1992). The coupon test is appropriate since acid washing is a short—term corrosion.

Silicates are often other major components of scale. But, as indicated on page 637, silicates are soluble under highly alkaline conditions; therefore, they are not present in many components of the kraft process such as digesters, liquor evaporators, etc. On the other hand, aluminum, calcium, and silicate together can form a troublesome scale.

The removal of scale by acid washing is a tradeoff of many factors. Some acid washing solution circulation is needed to help dissolve the scale, but too much circulation leads to increased rates of corrosion. The time of exposure to the acid cleaning solution must be such that most of the scale is removed, but not so long that the amount of corrosion is unnecessarily high.

Ferric ion and iron sulfide in HCl increases corrosion because as the ferric ion is reduced iron metal is oxidized. Iron oxides are removed by HCl treatment.

Nitric acid is often used instead of HCl for cleaning stainless steel surfaces of digesters because HCl causes pitting or stress corrosion. Nitric acid rapidly attacks carbon steel.

Formic acid vapors cause corrosion since the inhibitors are not necessarily volatile. Formic acid can decompose to CO and H_2O, so it must be used with ventilation and containers with relief valves.

Sulfamic acid (NH_2SO_3H) is used below $50°—60°C$ since higher temperatures (above about $70°C$) leads to some formation of sulfate, which causes scale formation by formation of insoluble metal sulfates (such as $CaSO_4$).

Corrosion in the wash and bleach plants

As late as 1970, most brown stock pulp washers were constructed of mild carbon steel, but now stainless steel is specified.

Chlorine dioxide is surprisingly harsh on nickel—based alloys. Even type 317L stainless steel (3—4% Mo) is now considered inadequate for many bleach plants. The stainless steels with high levels (as much as 6%) of molybdenum are fairly resistant to attack by ClO_2 and have been incorporated into many bleach plant washers.

Titanium apparently reacts very quickly with dry chlorine so its use must be limited to areas where exposure to dry chlorine is avoided. However, it seems to have very good corrosion resistance in most bleach plants.

Corrosion in the recovery boiler

Corrosion in the recovery boiler is the most difficult challenge to the pulp and paper mill corrosion engineer. The potential for explosions makes corrosion monitoring of the recovery boiler very important. High temperatures, molten salts, water, oxygen, and condensation of materials all contribute to the possibility of rapid corrosion rates. This area is discussed in detail by Sharp (1992).

Prevention of corrosion in boilers during shutdown

When power boilers are shut down water may condense on the fireside surfaces if the temperature of these surfaces falls below the dew point. Sulfur deposits may react with this water to form very acidic solutions that can quickly corrode the tubes. The corrosion is often localized, causing severe pitting.

This type of damage can be minimized during shutdowns by taking the following steps (in the order presented): removing deposits as quickly as possible, washing the deposit—free surfaces with water, air drying the surfaces, applying a thin layer of lightweight oil, and using a desiccant, such as unslaked lime, which is replaced as necessary, in the ash pits.

To prevent problems in the future, one might take several samples of deposits, dissolve them separately in small amounts of water, and measure the pH of the water. A pH below 4 may signal potential problems with the sulfur content of the fuel supply.

Corrosion in the paper machine

Corrosion of bronze rolls on paper machines was studied by Magar (1992). Five corrosion inhibitors were tested that all contained mercapto-benzothiazole as the active ingredient. The inhibitors were all more effective if added prior to exposure of the test liquid to the bronze.

Stress corrosion crack detection

The detection of cracks caused by stress corrosion is often accomplished nondestructively by the use of liquid penetrant or magnetic particle tests. One study (Reid and Reid, 1993) indicates that some surface metal must be removed by light surface grinding for these methods to be effective; sandblasting and power wire brushing are inadequate preparation methods. The paper refutes the idea that such grinding forms cracks or enlarges existing cracks for commonly used alloys.

33.7 SAFETY PRACTICES FOR HAZARDOUS CHEMICALS

Table 33-5 summarizes some steps to take when handling potentially hazardous chemicals. Material Safety Data Sheets are available from the manufacturers and suppliers of chemicals.

Table 33-5. Procedures for handling hazardous chemicals.

1. Read the Material Safety Data Sheet (MSDS).

2. Follow the manufacturer's directions for handling the material.

3. Label all vessels and plumping carrying the material in a clear fashion with indelible ink.

4. Be sure to use appropriate protective equipment, clothing, gloves, and safety glasses or face shields. Have plenty of these materials on hand, near the point of use, so they are used.

5. Do not handle these materials or feed them into the process by hand; instead use high—quality metering pumps equipped with power interlocks to prevent their addition during shutdowns.

6. Locate showers and eye wash stations near the point of use of toxic chemicals.

7. Do not smoke or eat when handling these materials.

8. Do not wipe or scratch your eyes, mouth, head, or other body parts with your gloved hands or clothing that has been in contact with poisonous or corrosive chemicals.

33.8 TRANSPORTATION SAFETY

Introduction

This section was prepared with the help of Mankui (David) Chen who collected most of the references and wrote some of the first draft of the text. The author gratefully acknowledges his help.

The chemical industry occupies an important place in the U.S. economy; its total output accounts for about 10% of the GNP in manufacturing. Hauling chemicals is a major source of revenue for railroads and motor carriers. The share of freight handled by rail has decreased from about 70% in 1946 to less than 40% in 1990. Trucking's share increased from 5 to 25% in the same time. The railroad industry was deregulated with the Staggers Rail Act in 1980. In 1989 there

was a record year of nearly 1.5 million rail carloads of chemicals and related materials.

Intermodal transportation involves shipping a material with several forms of transportation (a combination of truck, ship, rail, and even air). For example, a motor carrier might deliver and pick up a bulk handling container at a plant. The container would be transported by rail over a large distance and then transferred to a motor carrier for transportation to the final destination. This trend has not been widely practiced in the past since the railroad and trucking industries have been major competitors rather than cooperators. The ability of a shipper to deal with one carrier is a driving force for the partnerships that are forming between the various types of carriers.

Chemicals can be classified into hazardous materials and nonhazardous materials. Hazardous materials may be corrosive, explosive, flammable, and/or toxic. Any major accident in their transportation can directly endanger people's lives and the environment. Hazardous material handling regulations include substance classification (corrosive, explosive, etc.), packaging, handling, labeling and placarding, accident reports, information hotlines, filing route plans, etc.

Some incidents involving "nonhazardous" chemicals may not threaten people, but can destroy an entire ecosystem. For example (Elmer—DeWitt, 1991), a Southern Pacific tanker car derailed on a treacherous canyon bridge six miles north of Dunsumuir on July 14, 1991. Its contents of 19,500 gal. of metam sodium spilled into the Sacramento River. Metam sodium was classified as a nonhazardous material at the time of the accident. It wiped out aquatic plants, nymphs, caddis flies, mayflies, and at least 100,000 trout through a 45 mile stretch.

Every year there are about four billion tons of hazardous materials and radioactive waste shipped in the U.S. Accidental chemical releases seem inevitable—every day there are about 20 accidents where hazardous materials are released in interstate transportation. These accidents result in hundreds of deaths, thousands of injuries, great loss of resources, and damage to the environment. The public grows increasingly concerned about this environmental issue.

Government agencies and Congress react to this situation by proposing new laws. Chemical companies are taking many steps to ensure the safe handling of their products as the liability is shifting from the carrier to the shipper.

Safety regulations and codes

There are many regulations, rules, and codes concerning the transportation of hazardous materials issued by the federal government, the states, associations, and manufacturers. Three very important and influential regulations from the federal government, the shipper, and the carrier, respectively, are considered here as examples.

The *Hazardous Materials Transportation Act* of 1974 was the first complete federal law about hazardous materials transportation; it was originally passed by Congress as the result (in part) of public concern about the air accident at Logan Airport in 1974 (Gerhart, 1992). Three crewmen aboard a Boeing 707 died in a crash at Logan Airport in Boston, Mass., in 1974. Investigators found the dead men had been incapacitated by fumes from 10 quarts of improperly packaged nitric acid. Although correctly marked and labeled, the acid was stored on its side and leaked. The regulation covers substance classification, packaging, handling, incident report, placarding, and other topics.

In November 1990, Congress approved the *Uniform Safety Amendments* for this act (Stitt, 1991). It introduced a long list of new regulations covering emergency response information systems, specific highway routes, training programs, registration requirements, safe permit system, inspection and certification, and in place financial responsibility (insurance).

This bill allows federal preemption of state and local hazardous materials laws, calls for state and local participation in setting routes and standards, and sets up a petition process for objections to routing of hazardous materials. It expanded federal authority to prosecute violators and increased civil and criminal penalties.

Responsible Care (McConville, 1992; Ainsworth, 1993) is considered the standard for the chemical makers' total approach to carrying out business. It was set up in 1988 by the Chemical Manufacturers Association (CMA), launched in 1989, and parts of it continue to be ratified. It provides a framework for good safety practice to guide manufacturers, distributors, and carriers. It consists of six operating codes covering every facet of the industry:

·Community awareness and responsible care
·Pollution prevention
·Process safety
·Distribution
·Employee health and safety
·Product stewardship

Product stewardship is a catch—all for making "health, safety, and environmental protection an integral part of designing, manufacturing, marketing, distributing, using, recycling, and disposing of products."

The most stringent code for logistics is the Distribution Code of Management Practice, which includes the following five elements:

A. Risk management
B. Compliance review and training
C. Carrier safety
D. Handling and storage
E. Emergency procedures

The purpose of this code is to reduce the risk of chemical exposure to the public, carriers, distributors, chemical industry employees, and the environment. To meet the requirements of this code, CMA's 178 member companies had to make drastic changes to their policies and procedures and document their compliance.

Many of the goals of the program are borrowed from TQM such as "constantly upgrading manufacturing procedures and practices."

The *Key Train and Key Route Program* is a program from chemical carriers designed to improve safety and assure compliance with regulations. One of the most advanced cross—industry efforts has brought together representatives of the railroad, chemical, and rail equipment industries to develop standards to improve safety. The group, the Inter—Industry Task Force on the Safe Transportation of Hazardous Materials by Rail, called for defining trains and routes of highest concern as "key trains " and "key routes" for limiting train speeds (50 mph) and for enhancing track maintenance and derailment safeguards. The regulation detailed practices for yard operations, storage, and employee training (Anon., 1990).

In addition to the regulations listed above, there is a new suggestion that carriers obtain ISO 9000 certification to compete in the international market. TRANSCARE (The Transportation Community Awareness and Emergency program) calls for the combination of community education and emergency response training with public relation efforts.

Carriers

According to the Interstate Commerce Act, 54, Statute 889, carriers are classified as public or private. Private carriers are carriers that are (an incidental) part of the shipping company through ownership and/or leasing; they are not subject to government price regulations.

Public carriers can be *common*, *contract*, or *exempt*. Common carriers agree to serve all who wish to do business, and their prices are regulated by the government. Contract carriers are bound by a contract between the carrier and the shipper, are limited in the number of firms they can serve, and are not subject to government price control. Exempt carriers are not subject to government price control but are limited to certain products, geographical regions, and organizations formed. All carriers are subject to government safety regulations.

To comply with government regulations and meet public expectations, logistics departments of chemical companies are putting a great deal of effort in the selection of carriers, establishment of partnerships with carriers, and improvement of communication with carriers and communities.

Partnerships

Partnership relations between shippers and carriers are now becoming standard for logistics management (Rotman and Gibson, 1989; Morris, 1990). Companies like Du Pont, ICI, and Monsanto are giving this much concern. They believe that if service can be improved and if safety regulations can be completely carried out, a carrier should be a partner from the beginning. By reducing their carrier base and by formalizing that relationship with contract carriers, many shippers achieve high standards of safety and performance.

Chemical shippers are reducing their carrier bases to a few trusted companies; at the same time, they are attempting to establish more durable relationships and monitor more closely the safety and service of the selected carriers. For example, Monsanto cut its carrier base from 1500 to 100 within three years; FMC's chemical products group slashed its base of bulk carriers from 40 to 12 and packaged carriers from 500 to 60 in 1992. With fewer carriers, the existing carriers obtain more business and gain incentive to improve service and safety.

One approach used by some companies to ensure safety and service standards by truck carriers involves the use of contract rather than common carriers. For example, Du Pont uses about 95% of its bulk truckers on a long—term contract basis. As a major step in beginning its comprehensive safety program, Ethyl Co. decided that all shipments of hazardous materials would be by contract carriers (Cooke, 1992) using Ethyl owned or leased trailer and tank equipment. Dow Chemical Co. contracts most of its truck transportation with motor carriers to ensure that they adhere to its standards for safe handling. These companies got away from the traditional, pre-regulation relationship with carriers and built up close relationships with contract carriers, sharing information and even objectives, and in some case contract carriers could paint their trucks to match the companies' logos.

Selection of carriers and drivers

Carefully scrutinizing transportation providers is the first and critical step for hazardous material shippers to achieve safe transportation. The screening processes used by some companies gives us a good model. Auditors from the Department of Transportation (DOT) are responsible for checking every aspect of a carrier's safety techniques, from drivers' logs to equipment maintenance records. If a DOT auditor finds violations, he or she will rate the carrier accordingly with three options: satisfactory, conditional, and unsatisfactory. Most companies require that the hazardous material carriers have the satisfactory rating. Some shippers also carry out on—site surveys of all their contract carriers. Du Pont and Rohm and Haas perform thorough evaluations and on—site surveys. They go into their multiple facilities and review maintenance programs, logs and training. They want to see how safe the carriers are with their products (Gordon, 1990).

Besides examining the DOT audit and on—site survey, some shipping companies think it is necessary to train the drivers themselves; Du Pont trains carrier employees extensively. Its hazard material training program called RHYTHM (Remember How You Treat Hazardous Materials) has become an industry standard (Gordon, 1990). Ethyl Co. (Cooke, 1992) has developed its own instructional and certification program for the drivers of the contract motor carriers who have excellent safety records at the time of their selection. Ethyl wants to be sure that the carriers' drivers are fully knowledgeable about the chemicals they would be hauling and to be sure that the drivers are taught appropriate emergency—response procedures including proper use of personal protective equipment. The principal benefit of this program is the improved safety in transporting their hazardous materials.

Chemical distribution is growing increasingly sophisticated and requires improved *communication*. Traditional paper documentation and the hotline system called CEMTRCE (an emergency response communication standard) seems unsuitable. Electronic data interchange (EDI) into a nationwide computer system has to be established. The system would provide fire fighters with the exact details on all U.S. shipments of hazardous materials and provide emergency personnel with dial—up access to a mainframe database of electronic manifests listing details regarding the content of every hazardous materials' shipment. Since the present EDI technology is not advanced enough to process this vital information, the proposal to the Congress to mandate the use of a national computer tracking system has not been passed yet (Hamilton, 1993).

33.9 OPERATIONS MANAGEMENT

Operations are used to convert resources (inputs include people, capital, raw materials) into products (output). Products are usually either manufactured goods (tangible products like a new car) or services (intangible goods such as banking services). The output should have more value than the input to "create wealth."

Operations management is the decision—making process whereby all aspects of the process are selected based on the criteria selected for the process. Processes must be flexible to respond to changing market conditions and customer needs, efficient in order to capture market share from other competitors, and produce high—quality products. Requirements for an operation include research and development to design a product; engineering to scale the process to industrial size; capital for equipment, facilities, inventory, and raw materials; marketing; accounting; human resources; and information systems.

Other criteria include product safety and limited environmental impact. Consumers and society have placed higher values on these criteria in recent years.

A system consists of a group of related activities, items, and events. A change in one item will cause a change in other part(s) of the system. Any system is a part of a larger system and is usually made up of subsystems.

Flow charts are used to show overall processes. Process chart symbols include an "o" for operation, arrow for transportation, triangle for storage, D for delay, and square for inspection. Sometimes symbols are combined if operations are combined; for example, a triangle inside a square indicates storage and inspection.

Flow charts are often used in conjunction with time and motion studies to reduce bottlenecks and improve the overall process.

33.10 PROCESS SIMULATION

Introduction

Process simulation in pulp and paper usually implies the mathematical solution of mass and energy balances facilitated by the use of computers. [There are many other types of simulation as well, such as simulating the steps in an industrial process or a grocery store. In the latter, the computer would "create" customers that enter the store at times based on a probability distribution. These customers would spend various lengths of time in the store and purchase various items. The average length of time (and variability) they would be required to spend in a checkout line (queue) and other parameters could be determined. These simulations allow one to predict where bottlenecks will occur and how scheduling of employees might change the situation.]

Steady—state simulators work on the assumption that variables are constant with time or reach a constant level. The heart of simulation is the selection of mathematical equations that represent the relationships between variables. These mathematical equations are called *models*. In turn, many of the mathematical equations are based on certain assumptions.

The results of simulation are as good as the assumptions of the models, the models, and the degree of precision of the actual values of a variable. The *sensitivity* of a variable is the amount that a small change in the value of that variable causes the final result to change. If a small change in the value of a variable causes a large change in the result, it is important to know the value of that variable with a high level of precision. The term garbage—in—garbage—out is used to describe simulation processes where there is a high level of uncertainty in assumptions, models, and the actual values of variables, so the results are essentially meaningless.

Simulation does not have to be elaborate to be useful and effective. Some of the most useful simulations are quite simple. The degree of cook in the kraft process is modeled by the H—factor which combines the variables time and temperature. The mathematical relationship is sound and the variables of time and temperature can both be measured with a high degree of precision. The effects of time and temperature of the pulping process on the yield can be predicted accurately for an infinite number of combinations ahead of time. That is why this has enjoyed widespread use. The model does not describe the influence of pulping chemicals on the pulp yield but other models with these refinements have been described (page 371). Section 17.6 describes the simulation of fiber cleaning systems.

Process simulation, however, usually implies the use of a specially designed program to keep track of large numbers of variables. The first process simulation programs of large scope were developed in the late 1960s. Several popular ones (roughly in order of decreasing popularity) for pulp and paper include GEMS, which was developed by L.L. Edwards at the University of Idaho, MAPPS, which was developed by the Institute of Paper Chemistry and is a modification of GEMCS, MASSBAL, developed by SACDA at the University of Western Ontario, and FlowCalc, which is written in Basic and is designed specifically for microcomputer use; all of these have versions that run on IBM/PC compatible computers. Many people are performing sophisticated simulations using spreadsheet programs such as QuattroPro® or Lotus 1—2—3®.

Models should be checked for agreement with reality by *validation*. The model should be tested under several sets of conditions (where the real—world results are known) that were not used to develop the model. If the model consistently gives results that are in agreement with real—world results, then it is probably very useful. However, one should be careful to not used the model under conditions where its underlying assumptions are not valid or it has not been validated.

Once a process has been adequately modelled and validated, variables can be changed in the computer and the result will be determined. This allows a large number of conditions to be studied without having to perform numerous, expensive mill trials. Processes can be made more efficient or economically favorable. It may also be a useful tool for operator training.

Simulation a particular process

As always, one must clearly define the objectives of the simulation project. The simulation of a process requires much background information on the process. What variables are involved? What are the relationships between the variables? What equipment will be used in the process? What are the typical operating conditions?

This information is used to formulate mathematical models. These models are then converted into a form that the process simulation program can use. The model is then tested and validated. The model can then be used under to predict real—world behavior.

Processes may be dynamic or steady—state. The grocery store is an example of dynamic simulation. Sometimes there may be many people in the store, and other times there may be very few. The values of variables of dynamic processes change with time. Most process simulation is approximated by assuming the process reaches a steady state where the values of the variables do not change with time after steady state is reached; in this case the system can be modeled as a series of algebraic equations. The two principal solution algorithms of steady state simulations are *sequential* and *simultaneous*. In sequential simulation the variables are solved sequentially in a defined order, solved again as better values for each variables are determined, and resolved until the convergence (change in the value of the variables from a solution to the next solution) is within acceptable limits. With simultaneous solutions each equation is put into a linear form and all of the equations are solved simultaneously.

33.11 SUPERABSORBENCY

Superabsorbent polymers (SAPs) have been used in hygiene products since the late 1960s (Masuda, 1993). About 250,000 tons are used globally each year, most of which are used in disposable diapers for babies. Most SAPs for hygiene products are the sodium salts of moderately crosslinked polyacrylic acid (not all of the acid groups are in the salt form) in a dry powder of 300 μm particles. The crosslinking agents are typically glycol diacrylate or N,N'—methylenebisacrylamide. Residual monomer (which may be a skin irritant) levels are now below 100 ppm.

The famous polymer scientist P. J. Flory had theoretically determined the important parameters for water absorption by polymers. The important factors for *absorption capacity* include a polymer with a high affinity for water (i.e., a polar polymer), a low crosslinking density in the polymer (just enough to keep the polymer insoluble), and a high ion density on the polymer.

Electrolytes dissolved in the water to be absorbed decrease the amount of water ultimately absorbed by the crosslinked polymer, and here the type of electrolyte is important, with polyvalent cations decreasing absorption to a higher degree than monovalent cations (as expected from colloid chemistry principles); these factors become important when SAP systems are extended to non-hygiene product uses such as those pressed between two papers and used in sheet form.

The *rate of absorption* is determined by physical properties, especially the surface area of the SAP. Finer particles generally absorb faster, although if the particle size is too small, a gummy material is obtained that does not absorb water quickly. Masuda (1993) states that diaper dryness is correlated to absorbency under load (AUL, $r = 0.81$) and diaper leakage is inversely correlated to gel—stability ($r = 0.93$). The trend has been to use lower amounts of fluff fiber and higher levels of SAP in hygiene products; these products may be as much as 30% SAP.

33.12 BEARINGS

Introduction

This section will provide a brief overview of the different types of bearings and their use. This information comes largely from the SKF USA Inc. (King of Prussia, Pennsylvania) *Product Service Guide* (April, 1992). Fig. 33-5 shows the basic types of bearings that are discussed here.

Bearings range in size from the precision miniature bearings used in medical and aerospace applications to deep grove ball bearings with bore sizes exceeding 55 in.

Single row deep groove ball bearings—Conrad type

This popular type of bearing is a relatively simple, nonseparable (since elastic deflection of the rings is used to introduce the balls into the bearing during the Conrad assembly method) design. It can carry significant radial loads, and, due to the uninterrupted raceway grooves and the high degree of conformity between balls and raceways, substantial thrust loads in either direction, even at very high speeds. Single row deep groove ball bearings are available in open form or with various types of seals or snap rings.

Single row deep groove ball bearings of the cartridge—type

Cartridge—type bearings have the same standard bore and outside diameter as single row deep groove ball bearings but are as wide as double row bearings. They are supplied with two seals or shields.

Max type (filling slot)/single row ball bearings

Max type single row ball bearings (also called filling slot type) employ a filling slot (rather

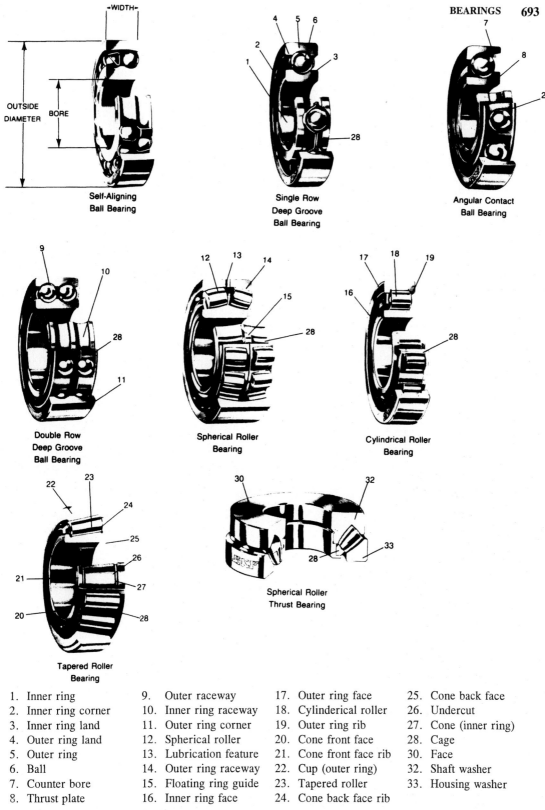

Self-Aligning
Ball Bearing

Single Row
Deep Groove
Ball Bearing

Angular Contact
Ball Bearing

Double Row
Deep Groove
Ball Bearing

Spherical Roller
Bearing

Cylindrical Roller
Bearing

Tapered Roller
Bearing

Spherical Roller
Thrust Bearing

1. Inner ring	9. Outer raceway	17. Outer ring face	25. Cone back face
2. Inner ring corner	10. Inner ring raceway	18. Cylinderical roller	26. Undercut
3. Inner ring land	11. Outer ring corner	19. Outer ring rib	27. Cone (inner ring)
4. Outer ring land	12. Spherical roller	20. Cone front face	28. Cage
5. Outer ring	13. Lubrication feature	21. Cone front face rib	30. Face
6. Ball	14. Outer ring raceway	22. Cup (outer ring)	32. Shaft washer
7. Counter bore	15. Floating ring guide	23. Tapered roller	33. Housing washer
8. Thrust plate	16. Inner ring face	24. Cone back face rib	

Fig. 33-5. Eight basic bearing types and their terminology. Courtesy of SKF USA, Inc.

than ring deflection as in the Conrad type) to introduce the balls into the bearing. This permits the use of the greatest possible number and size of balls. This type is capable of carrying very high radial loads as well as some thrust loads.

Angular contact ball bearings

The most popular type are the single row bearings; these can carry appreciable thrust load in one direction combined with a radial load or without a radial load.

Paired mounting of two or more single row angular contact ball bearings is commonly used to accommodate heavier or two—directional thrust loads and/or radial loads in various combinations. Tandem mounting is used for heavy unidirectional thrust loads and duplex mounting (either face—to—face or back—to—back) for two directional thrust loading. Special flush ground bearings are needed for paired mounting to assure either a predetermined amount of axial internal clearance or an even load sharing of bearings. Contact angles of 40° are common except for large bearings (above 110 mm bore), which often use 37°.

Double row ball bearings

This design is similar to single row ball bearings. Since they have two rows of balls (and are wider) they can carry appreciably more radial load. They are designed with one—piece inner and outer rings. Most designs have contact angles that converge outside the bearing to provide overall application rigidity.

The use of larger and more balls allows for increased load carrying capacity. Both Conrad type, with uninterrupted raceways, and Max type, with a filling slot, bearings are used. Contact angles of 0°, 25°, and 30° are available.

Self—aligning ball bearings

Self—aligning bearings have two rows of balls and a common spherical raceway in the outer ring, which provides then with the self—aligning feature. Therefore, they are well suited for applications with misalignment from mounting errors or shaft deflections. This bearing cannot exert a bending influence on the shaft, which is important for applications requiring extreme accuracy at high speeds. It is recommended for radial loads combined with moderate dual—direction thrust loads.

Bearings in papermaking machines (SKF, 1989)

Large paper machines may use as many as 3000 bearings, the majority of which are spherical roller bearings. Spherical roller bearings are able to support substantial axial and radial load combinations of the rolls and cylinders used. They tolerate misalignment of between the shaft and housing. Bearings should have a rated life of at least 200,000 hours in the dryer section and 120,000 hours (14 years) in the other sections.

Well—lubricated bearings will show little signs of wear and corrosion and have lives as long as new bearings. For sealing, a well greased multi—stage labyrinth gives good protection when the housing is exposed to moisture. Bearings in the wet section are often grease lubricated, and bearings in the other sections are oil lubricated. High—speed paper machines usually use oil lubrication throughout with a circulating oil system.

33.13 ELECTRIC CIRCUIT SYMBOLS

Table 33-6 gives the meaning of electrical power symbols.

33.14 ANNOTATED BIBLIOGRAPHY

Water treatment

1. Aono., *BETZ Handbook of Industrial Water Conditioning*, 8th ed., BETZ®, Trevose, Pennsylvania, 1980, 437 p. with numerous color plates. This is a good source of information on this topic and is often available to mill personnel on a complimentary basis.

Boiler water treatment and control

2. A series of six articles on boiler water appeared in *Pulp and Paper* starting in 1979. They appeared in consecutive issues from 53(9) to 54(1). These articles were redone starting in 1990 as follows: 64(7):97-100; 64(8):130-135; 64(9):237-241; 64(10):138-142; 64(11):114-119; and 65(1):87-91. This is useful and concise.

3. Anon., Troubleshooting low or high pH in boilers that use high purity water, *Tappi J.* 75(9):275-277(1992). This describes how to maintain control of pH in boiler water.

4. Korbas, G. and J.K. Kasper, Water treatment considerations for pulp and paper mill boil-

Table 33-6. Electric power symbols.

Symbol	Name
	alarm
	ammeter
	antenna
	autotransformer
	battery
	broken delta
	breaker combination starter
	broken delta with protective relay
	bus
	capacitor
	chassis ground
	contact set
	current transformer, single phase
	control ground
	delta-delta connection
	delta winding
49	thermal overload relay
50	instant. overcurrent ground sensor relay
51	overcurrent protective relay
86	locking out relay
87	differential protective relay
	diesel engine
	diode
	connection disk, round (crossover)
	connection dot, square (crossover)
	delta-wye connection
	fan
	fused combination starter
	fused disconnect
	fuse
	fused switch
	generator
	ground sensor current transformer
	induction motor
	inductor
	lamp
	low voltage power circuit breaker
	meter
	meter switch
	molded case breaker
	moderate voltage circuit breaker
	moderate voltage starter
	normally closed contact
	switch, normally closed
	switch, normally closed, momentary contact
	normally open contactor
	switch, normally open
	switch, normally open momentary contact
	open delta
	potential transformer, single phase
	potential transformer, 3-phase
	power ground
	full wave rectifier
	resistor
	variable resistor box
	rheostat

Table 33-6. Electric power symbols (continued).

silicon controlled rectifier	thermal overload switch	wye winding
speaker	turbine	single winding transformer
disconnect stabilizers	volts amps reactive meter	two winding transformer
substation ground	variable resistor	three winding transformer
surge arrester	volt meter	wyw-wye connection
synchronous function box	watt hour meter	zig-zag connection
synchronous motor	watt meter	zigzag winding

ers, in *TAPPI 1989 Environmental Conference Proceedings*, pp 231-237. Feedwater for steam generation from power boilers must be very pure so that much less than one ppm dissolved solids will be found in the steam. Tolerable levels of contaminants depend on the nature of the system. Water treatment methods and water conditioners must be used so that corrosion can be minimized.

Corrosion

5. Garner, A., Pulp and Paper Industry, Alloy Selection, in J.J. McKetta, *Encyclopedia of Chemical Processing and Design*, Vol. 33, Marcel Dekker, New York, 1990. "The corrosion resistance of recently developed stainless steels is reviewed with reference to their performance in pulp mill bleach plants." This is a listing of many stainless steels available and their properties.

6. *Proceedings of the 7th International Symposium on Corrosion in the Pulp & Paper Industry*, Orlando, Nov. 16-20, 1992, Tappi

Press. This is a good resource on general aspects of corrosion in pulp and paper processes. Areas receiving high coverage in this resource are corrosion aspects of kraft recovery (especially the recovery boiler) and, to a lesser extent, pulp bleaching. Some particularly useful articles in this resource follow.

Crowe, D.C., Corrosion in acid cleaning solutions for kraft digesters, pp. 33-39. Kraft digester material was tested for corrosion with the four commonly used acids and various corrosion inhibitors.

Magar, I.J., Protective coatings for paper machine wet end service: an update, pp. 173-180. Of 42 systems from 14 manufacturers evaluated in salt fog exposures, the best system performance was obtained with an organic zinc primer with an amine adduct—cured, epoxy topcoat.

Sharp, W.B.A., Overview of recovery boiler corrosion, pp. 23-31, with 93 references. The author divides the recovery boiler into eight different zones and discusses each zone.

7. Reid, J.C. and D.C. Reid, Influence of non-destructive test procedures on detection of stress—corrosion cracks, *Tappi J.* 76(8):70-85(1993).

Transportation safety
8. Ainsworth, S.J., Chemical transportation & distribution, *Chem. & Eng. News*, 71(March 8):8-28(1993).

9. Anon., Railroads, chemical industry link up for safety operations, *Purchasing* (Mar. 22):29(1990).

10. Cooke, J.A., Taking safety seriously, *Traffic Management*, 31(April):50, 51, 53(1992).

11. Elmer-Dewitt, P., Death of a river, *Time Mag.* 138(July 29):24(1991).

12. Gerhart, C., Heed new rules for hazardous materials shipping, *Alaska Business Monthly* 8(June):42-47(1992).

13. Gordon, J., In search of perfection, *Distribution* 89(Sept.):41-43(1990).

14. Hamilton, D., Hazardous materials haulers hail rejection of EDI plan, *Journal of Commerce and Commercial* (May 26):11B(1993).

15. McConville, D.J., Responsible care gains respect, *Distribution* 91(1):52-55(1992).

16. Morris, G.D.L. Getting it there, global transportation links the chemical marketplace, *Chemicalweek* 146(6, Feb. 14):26-48(1990).

17. Rotman, D. and Gibson, W.D., The quickening pace of chemical distribution, *Chemical Week*, 144(12, Mar. 22):22-36(1989).

28. Stitt, W., Keeping highways safe, *The Business Journal* 8(April 8):12(1991).

Process simulation
19. Syberg, O., N.W. Wild, and H.A. Simons, Ed., *Introduction to Process Simulation*, 2nd ed., Tappi Press, Atlanta, 1992. This is the best available introduction to process simulation in pulp and paper, although shallow.

Superabsorbency
20. Masuda, F., Trends in the development of suberabsorbent polymer for diapers, *Polym. Mat. Sci. Eng.*, 69:484(1993).

Bearings
21. SKF, *SKF Bearings in Papermaking Machines*, SKF, King of Prussia, Pennsylvania, 1989, 44p. This very useful, detailed information is published by a major bearing manufacturing company and includes many formulas and calculations.

22. TIS 420-05, Design guide for the selection of rolling contact bearings for paper mill service, 5p.

23. TIS 0420-08, Mounting, sealing, lubrication, and troubleshooting guide for paper mill rolling contact bearings, 13 p.

APPENDIX

Transportation Acronyms

DOT	U.S. Dept. of Transportation
RSPA	Research and Special Programs Administration of the DOT
ANPRM	Advance Notice of Proposed Rulemaking
CMA	Chemical Manufacturers Association
NPRM	Notice of Proposed Rulemaking
RPI	Railway Progress Institute
hazmat	hazardous material

34

UPDATES AND BIBLIOGRAPHY

This bibliography is not intended to be comprehensive. It includes selected articles such as classic studies, comprehensive treatises on specific topics, and additions and recent updates of the first 20 chapters from the first edition (which remain unaltered in the second edition). Articles, though recent, which are not comprehensive and do not add new knowledge are not listed. Specific analytical techniques are listed in the last section.

In some areas where comprehensive sources are not available, recent articles are listed to point one in the right direction. Often literature cited within these recent sources provides additional useful information. Also, there are numerous conference proceedings, monographs, bibliographies, and other sources not included here. The author simply cannot obtain all of these materials for review. Some trends and predictions not suited for other formal sections of this volume are presented here with references.

Additionally, this chapter provides a space for recent advances in the industry, brief discussions on topics that do not fit in other chapters, and other housekeeping activities.

34.1 GENERAL PULP AND PAPER

Industry statistics

In 1989 U.S. exports reached a record 17,200,000 t. U.S. imports were 16,000,000 t, much of which was newsprint from Canada.

McKeever, D.B., The United States Wood—pulp Industry, Resour. Bull. FPL-RB-18, USDA For. Ser. (1987) 29 p. This is a useful resource on the U.S. pulp and paper industry from 1961 to 1983. In 1983, 315 woodpulp mills produced 174,000 ton/day (552 ton/day/mill). In 1961, 350 wood pulp mills produced 90,000 ton/day (256 ton/day/mill). In 1983 the wood source was 38% softwood roundwood, 22% hardwood roundwood, and 40% plant by—products. Sulfate pulp accounted for 70% of production. The south produced 66% of all U.S. pulp. Foreign trade in

woodpulp had never exceeded 10% of domestic production through 1983. A breakdown of each U.S. mill is given.

Knocel, J.A., *Am. Papermak.* 55(4):18-19(1992); 59(4):24-25(1996) and McReady, M. *PaperAge* 108 (3):27,34(March, 1992) list sales and earnings for U.S. and Canadian pulp and paper companies. The U.S. (1995) and Canadian (1991) sales are presented in Table 34-1 for some of the larger companies.

Table 34-1. Sales of major U.S. and Canadian (1991) paper companies.

Company		1995 Sales $Billions
Abitibi-Price	(CAN$)	2.80
Boise Cascade		5.06
Bowater		2.00
Can. Pacific For. Prod.		1.98
Champion International Corp.		6.79
Domtar	(CAN$)	1.80
Federal Paper Board		1.91
Fletcher Challenge Canada	(CAN$)	1.04
Georgia—Pacific		14.29
International Paper		19.80
James River		6.80
Kimberly—Clark		13.79
Louisiana—Pacific		2.84
MacMillan Bloedel	(CAN$)	2.73
Mead		5.18
Noranda Forest	(CAN$)	4.12
Pentair		1.65
Potlatch Corp.		1.61
Repap Enterprises	(CAN$)	1.07
Rayonier		1.60
Sonoco Products		2.71
Stone Container		7.35
Temple—Inland		3.46
Union Camp		4.21
Westvaco		3.27
Weyerhaeuser		11.79
Willamette Industries		3.87

Pulp manufacturing costs

Fromson, D.A., One and only chance, *Paper-maker* 57(11):39, 41, 42(1994). This article gives a breakdown (fiber, chemicals, energy, labor, materials) of costs for pulp manufacture in six countries. Softwood pulp manufacturing costs totaled $276 (USD) in Chile, $328 in the south-eastern U.S., $361 in Canada, $422 in New Zealand and Nordic, $444 in the western U.S., and $598 in Japan. Hardwood pulp manufacturing costs ranged from $217—488; newsprint manufac-turing ranged from $269—$577.

Chemical usage

Ferguson, K.H., Chemical markets remain stable as industry activity levels off, *Pulp & Paper* 70(3):79, 80, 85(Mar., 1996). In 1996, the annual U.S. consumption of chemicals in the paper industry is estimated at 557,000 ton of alum at $120 per ton, 1.2 million tons of calcium carbon-ate at $100—$200 per ton, 3.5 to 3.8 million tons of kaolin clay at $32—$65 per ton for filler grades to more than $160 per ton for the premium coating grades, 4 to 4.5 million pounds (seems very low) of modified starch at $0.13—$0.29 per pound (up to $0.50 per pound for some grades), 1.4 billion pounds of styrene—butadiene latex for coating binders at $0.60—$0.70 per pound, and 250,000 tons of titanium dioxide at $0.85—$1.00 per pound. Other wet—end additives are discussed.

Piccone, J., Prices to increase as demand catches up with supplies, *Pulp Paper Can.* 96(1):18, 19, 22(1995). This is a good analysis of the availabil-ity and use of chemical raw materials by the Canadian pulp and paper industry. It shows some trends: in 1993 this industry used 195,000 t of Cl_2, but in 1994 it was only 157,000 t.

34.2 WOOD

Wood procurement

Slinn, R.J., The impact of industry restructuring on fiber procurement, *J. Forestry* (2)17-20(1989). "The past and present roles of the pulp and paper sector in wood procurement."

Wood handling

Piggott, H.R. and R.A. Thompson, Drum bark-ing: key factors for design and performance, *Tappi J.* 70(8):37-41(1987). No references.

Wood physics and transport properties

Woods like redwood that have a high level of water—soluble extractives have a low apparent value for the FSP; it is believed that these extractives use some of the sites that water nor-mally uses. Perhaps this effect is observed in paper with certain water—soluble materials (like polymers). On the other hand, hemicellulose has a high FSP so starch might increase the FSP of paper. Pulps generally have a slightly lower FSP than the woods from which they were derived.

Cellulose molecules have an index of refrac-tion of 1.599 in the axial direction and 1.532 in the transverse direction. The dielectric constant is 5.7 and the breakdown voltage is 500 kV/cm.

Dunlop, F., U. S. Dept. Ag. Bull. 110 (1912). The specific heat of wood as 0.266 at 0°C and 0.382 at 100°C for wood with 0% moisture.

A correction for specific heat at 0% mois-ture, c_o, is calculated from the fractional moisture content of wood, m, and the specific heat due to wood—water bound energy, A, as follows:

$$\text{Specific heat} = A + (m + c_o)/(1 + m)$$

where at 0.10 water content A varies from 0.02 at 85°F to 0.04 at 140°F and at 0.30 water content A varies from 0.04 to 0.09 for 85—140°F.

Huang, C.—L., Revealing the fibril angle in wood sections by ultrasonic treatment, *Wood and Fiber Sci.* 27(1):49-54(1995).

James, W.L, Y.—H.Yen, and R.J. King, A microwave method for measuring moisture con-tent, density, and grain angle of wood, USDA, For. Ser. Res. Note FPL—0250 (1985), 9 p. The attenuation, phase shift, and depolarization of a polarized 4.81 GHz wave transmitted through a wood specimen can provide estimates of the moisture content, density, and grain angle of the specimen.

Jiménez, G, W.T. McKean, and R.R. Gustafson, Using a kraft pulping model to improve pulp uniformity, *Tappi J.* 73(7)173-176(1990). This article covers diffusion of solutions at high pH (12—14), such as in kraft pulping, where swelling of wood increases the diffusion in the radial and tangential directions to that approaching the longi-tudinal direction.

Siau, J.F., *Transport Processes in Wood*, Springer—Verlag, New York, 1984. 245 p. This is not written for pulp and paper specifically, but for those needing information on the wood—moisture relationship, transport of fluids in wood (of use in pulping), capillary and water potential, thermal conductivity, and moisture movement under steady—state or unsteady—state conditions, this is a good starting point.

Stamm, A.J., *Wood and Cellulose Science*, Ronald Press Co., New York, 1964. 549 p. This book might be entitled "The Physical Chemistry of Wood." It covers microscopy; X—ray structure; solution properties of lignin, cellulose, and cellulose derivatives; sorption behavior of wood with water vapor and other gases; surface area and accessibility; shrinking and swelling of wood and cellulose in various liquids; mechanical, electrical, and thermal properties of wood; diffusion through wood; wood drying and preservation; adhesion; and fiber bonding. Stamm pioneered many aspects of the physical chemistry of wood.

Steinhagen, H.P., Thermal conductivity properties of wood, green or dry, from —40° to +100°C: a literature review, USDA For. Ser. Rep. general tech. report FPL-9, (1977), 10 p. This work investigating heat transfer in frozen logs and includes information on the specific heat of wood from 0—30% moisture. Values are given in the USDA reference below.

Stone, J.E., The effective capillary cross—sectional area of wood as a function of pH, *Tappi*, 40(7):539-541(1957). This classic work shows (see page 88) that the diffusion of wood in the radial and tangential direction, while normally about 20% of that of the longitudinal direction, begins to increase significantly above pH 12.5 and increases to 80% of that of the longitudinal direction at pH above 13.5, typical of kraft pulping. This is why kraft chips are screened by thickness, and not size, as the thickness is the limiting factor.

USDA, Agriculture Handbook No. 72, *Wood Handbook*, 1986 gives many properties of wood. The thermal conductivity (k) of softwood for construction is about 0.75 Btu·in./(ft^2·hr·°F) compared to aluminum at 1500, steel at 310, glass at 7, concrete at 6, plaster at 5, and glass wool at 0.25. k in Btu·in./(ft^2·hr·°F) perpendicular to the grain (its about 2 to 2.8 times higher parallel to the grain) as a function of percent moisture content ovendry basis, M, and specific gravity, S, is:

$$k \approx S(1.39 + 0.028M) + 0.165 \qquad M < 30\%$$

$$k \approx S(1.39 + 0.038M) + 0.165 \qquad M > 30\%$$

The coefficient of thermal expansion of ovendry wood parallel to the grain (α_l) is 0.0000017 to 0.0000025 per °F over the temperature range of —60 to 130°F. For ovendry wood of specific gravity 0.1 to 0.8:

$$\alpha_r = [(\text{sp. gr}) \times 18 + 5.5] \times 10^{-6} \text{ per °F}$$

$$\alpha_t = [(\text{sp. gr}) \times 18 + 10.2] \times 10^{-6} \text{ per °F}$$

The electrical resistivity of wood varies from 10^{14}—10^{16} ohm—meters when ovendry to 10^{3}—10^{4} at the fiber saturation point. Wood is about twice as conductive in the longitudinal direction as in the other two directions.

Wahren, D. (Bonano, E.J., Ed.), *Paper Technology, Part 1: Fundamentals,* Institute of Paper Chemistry, Appleton, Wisconsin, 1980. pp. 60-96. The wood—water relationship, especially in regard to drying paper, capillary action and other forces that hold paper together during pressing and drying, and principles of drying are well covered from the pulp and paper point of view.

Wenzl, H.F.J., *The Chemical Technology of Wood*, Academic Press, New York, 1970, pp. 410-416. The permeability of wood to gases and liquids is discussed, especially in regard to sulfite pulping.

34.3 ROUNDWOOD PROPERTIES

Sapwood thickness
Lassen, L.E. and E.A. Okkonen, Sapwood thickness of Douglas—fir and five other western softwoods, USDA For. Ser. Res. Pap. FPL 124 (Oct. 1969). The sapwood thickness was measured on increment cores at breast height (4.5 ft above the ground) on the title species. Sapwood thickness, like any wood property, is highly variable and depends on many factors. Generally, fast—growing trees have a wider band of sapwood than slow—

growing trees. Sapwood thickness decreases with increasing trunk height in some species. Table 34-2 summarizes the results of this work, but does not indicate the high variability encountered.

Wood density

The U.S. Forest Products Laboratory has numerous studies on wood density of trees. For example, report FPL 176-177 (rev. 1975) is a survey of wood density and structural properties of the southern pines. At least several hundred samples from each species were taken. From these samples, regression equations have been developed to predict the average specific gravity of trees based on the results of increment cores.

34.4 WOOD CHEMISTRY

General wood chemistry
Browning, B.L., Ed., *The Chemistry of Wood*, R.E. Krieger Pub. Co., Huntington, New York, 1975, 689 p. While first published in 1963 by Interscience, the addendums of each chapter of the later edition add little. This volume is particularly useful for detailed chemical compositions of hemicellulose, lignin, bark, and extractives. The chemical and physical properties of fibers, cellulose, and cellulose derivatives are well covered.

Browning, B.L., *Methods of Wood Chemistry*, Interscience, New York, 1967. This is an excellent resource for laboratory procedures of wood chemistry. It is required reading for workers in this area. Volume I, 384 p.; Volume II, 384 p.

Hon, D. N.—S. and N. Shiraishi, Ed., *Wood and Cellulosic Chemistry*, Marcel Dekker, Inc., New York, 1991, 1020 p. This is an extensive look at many aspects of wood chemistry. The work is divided into three sections of approximately equal length: I. Structure and Chemistry, II. Degradation, and III. Modification and Utilization. The areas of fiber chemistry and morphology and the chemical structure of wood components are particularly useful.

Wenzl, H.F.J., *The Chemical Technology of Wood* (translated from the German by Brauns, F.E. and Brauns, D.A.) Academic Press, New York, 1970, 692 p. This is a fairly practical volume on wood chemistry including world distributions of forests, anatomical and physical properties of wood, wood chemistry, and chemical wood processing including pulping, acid hydrolysis, and pyrolysis. There are about 150 pages each on sulfite and alkaline pulping chemistry. Each chapter contains 150—500 references covering the time period of 1945—1967 and many from the nineteenth century.

Lignin chemistry
Adler, E., Lignin chemistry—past, present and future. *Wood Sci. Technol.* 11:169-218(1977). The article summary is "some pertinent results and views from the earlier history of lignin chemistry, pointing to the importance of the arylpropane skeleton, are outlined. Later development, beginning with the dehydrogenation theory and experimental studies on the dehydrogenation polymerization of *p*—hydroxycinnamyl alcohols, is then reviewed. Finally, recent degradative work resulting in a detailed picture of lignin structure is discussed."

Table 34-2. Average sapwood thickness (in.) at 4.5 ft verses species and tree diameter.

Species	---------- Tree diameter inside bark (inches) ----------					
	5	10	15	20	25	30
Coast Douglas—fir	0.75	1.5	1.8	1.9	2.0	2.1
Interior Douglas—fir	0.55	1.2	1.4	1.5	1.6	1.7
Ponderosa pine	2.0	3.0	4.5	5.0	5.0	5.0
Lodgepole pine	1.3	1.9	2.2	2.5	3.0	3.5
Englemann spruce	0.9	1.4	1.7	2.0	2.0	2.1
Western redcedar	0.7	0.8	0.9	1.0	1.1	1.1
Western larch	0.7	0.7	0.7			

Sarkanen, K.V. and C. H. Ludwig, *Lignins: Occurrence, Formation, Structure and Reactions*, Wiley—Interscience, New York, 1971, 916 p. This is a useful work on lignin chemistry.

Carbohydrate analysis
Biermann, C.J. and G.D. McGinnis, Eds., *Analysis of Carbohydrates by GLC and MS*, CRC Press, Boca Raton, Florida, 1988, 292 p. This book covers hydrolysis of wood, pulp, and other polysaccharides and analysis of carbohydrates by gas—liquid chromatography and mass spectrometry.

Mineral composition of wood and wood ash
The composition of wood ash is relevant to impurities that might build up in pulping systems with increasing closure (such as Al, Si, and Ca in the kraft chemical recovery system) and as a potential feedstock. Bark is considered to have a 10—fold amount of minerals such as Ca, Si, and Al compared to wood.

Bailey, J.H.E. and D.W. Reeve, Spatial distribution of trace elements in black spruce by imaging microprobe secondary ion mass spectrometry, *J. Pulp Paper Sci.* 20(3):J83-J86(1994).

Campbell, A.G., Recycling and disposing of wood ash, *Tappi J.* 73(9):141-146(1990). In preference to being put in landfills, wood ash may be used as a mineral source for agriculture or even as a source of potash and calcium. The macro— and microelemental analyses of six wood ashes are included. The macro elements are Ca (7.4—33.1%), K (1.7—4.2%), Al (1.59—3.2%), Fe (0.33—2.10), Mg (0.7—2.2%), P (0.3—1.4%), Mn (0.3—1.3%), and Na (0.2—0.5%). The composition in wood ash depends on the ashing procedure as some elements are relatively volatile.

Guyette, R.P., Cutter, B.E., and G.S. Henderson, Inorganic concentrations in the wood of western redcedar grown on different sites, *Wood and Fiber Sci.* 24(2):133-140(1992). This includes the analysis of 30 elements in wood. The mean and (standard deviation) in ppm for several elements in sapwood are Ca, 1347 (167); Fe, 2.27 (1.19); K, 478 (135); Mg, 103 (27.6); Mn, 25.7 (28.6); Na, 16 (13.2); P, 71 (19.5); Si, 23.9 (28.9). These results with those of Campbell would indicate that most iron in wood ash comes from the processing equipment or additives.

Okada, N., et al., Trace elements in the stem of trees, V, *Mokuzai Gakkaishi* 39(10):1111-1118(1993); ibid VI 39(10):1119-1127(1993). These papers give radial distributions in Japanese softwoods and hardwoods, respectively. In softwoods, alkali metal (group I) concentrations are generally higher in heartwood than sapwood, but the opposite is true for Mn and Cl. Alkaline earth (group II) concentrations did not change abruptly except for Mg. In hardwoods, the elements were generally of higher concentration in the sapwood than the heartwood. Concentrations of metals are typically Ca, 500—2000 ppm; K, 500—2000; Mg, 50—400 ppm; Cl, highly variable by species from 10—1000 ppm; Mn, highly variable by species, 0.5—500 ppm; Al, 10—100 ppm.

34.5 RECYCLING AND SECONDARY FIBER

Progress in Paper Recycling
Progress in Paper Recycling, Doshi & Associates, Appleton, Wisconsin, published quarterly since November, 1991, is an excellent resource in this area. It is up—to—date and practical.

Flotation and washing for deinking
A comparison of flotation and washing methods for deinking is presented in Table 34-3. Paper containing water—based flexographic inks often uses an acid bath for its removal; therefore, the overall process may be an acid flotation—base flotation method. In 1994, the global capacity for flotation deinking was 20 million tons per year, most of which is for newspapers and magazines.

Newsprint alone (which is essentially unfilled in the U.S.) must be deinked by washing (and water—based inks must be absent), but mixtures of newsprint and magazines (or magazines alone) are usually deinked by flotation.

Ferguson, L.D., Deinking chemistry (two parts), *Tappi J.* 75(7):75-83 and 75(8):49-58(1992). This is a detailed look at deinking chemistry.

Agglomeration deinking
Olson, C.R., Richmann, S.K., Sutman, F.J., Letscher, M.B., Deinking of laser—printed stock using chemical densification and forward cleaning, *Tappi J.* 76(1):136-144(1993).

Table 34-3.   Comparison of washing and flotation methods of deinking paper.

Washing (like washing machine for clothes)

NaOH: fiber swelling, breaks up esters (e.g., soy bean oil); colloidally separates ink from fiber

sodium silicate: "buffer" and sequestering agent to protect peroxide

hydrogen peroxide: prevents mechanical pulps from yellowing above pH of 10; bleaching

dispersant: sodium salt of stearic acid (i.e., soap; don't use calcium, just as one avoids hard water to wash one's body or clothes) or low—foaming (low number of C atoms in alkyl chain), nonionic alkylphenol ethoxylates or linear alcohol ethoxylates

fillers and fines: largely removed in the process

general notes: emulsion particles of about 1 μm are formed; hydrophilic particles more easily removed;   process from 1800s;   only small particles (less than 5 μm) removed; limited to only a few types of paper; low consistency with high reject volume using 10,000 or more gallons of water per ton pulp; long pulping times; high brightness; low yield

Chemicals used in one or both processes:
Alkylphenol ethoxylate
$CH_3(CH_2)_n(C_6H_4)(OCH_2CH_2)_mOH$

Linear alcohol ethoxylate
$CH_3(CH_2)_n(OCH_2CH_2)_mOH$

Fatty acid soap
$CH_3(CH_2)_nCOO^-\ Na^+$

Fatty acid ethoxylate
$CH_3(CH_2)_n(CO)(OCH_2CH_2)_mOH$

Flotation

NaOH: fiber swelling, breaks up esters (e.g., soy bean oil); colloidally separates ink from fiber

sodium silicate: "buffer" and sequestering agent to protect peroxide

hydrogen peroxide: prevents mechanical pulps from yellowing above a pH of 10; bleaching

collector system: fatty acid (0.5—5%, usually 1%) and calcium (200 ppm) to make fatty acid water—insoluble, forming large particles; nonpolar surfactant, esp. C—9 alkyl phenol; or large primary alcohol linked to 6—16 monomers of ethylene oxide; the large nonpolar particles 30—60 μm adhere to air bubbles

fillers: 8—14% (clay) help process, but only 25—30% is removed, so often OMG is added when ONP is treated by flotation but this has problems (heavy metals that are a disposal concern, inks that are harder to remove, bale—to—bale variability in ash composition)

general notes: new method (1960s) from metal ore technology;   required for modern polymeric inks that do not disperse well; many choices of paper grade; 1—2% consistency; 3000 gallons of water per ton pulp; low brightness; high yield

Finely divided solids (fillers, alumina, fines, and other colloids) contribute to foam problems by surrounding air bubbles to decrease the surface tension.

Miscellaneous notes:

Postdeinking fiber treatment may use dispersion to break large particles into pieces the eye cannot detect individually, but pulp brightness decreases. Carryover of calcium can cause scaling and corrosion problems, carryover of surfactants can interfere with the wet end chemistry and surface sizing operations. Combined flotation/washing systems provide some difficulties because washing requires a good dispersant that would prevent large particles from forming during flotation. Dispersant collectors called *displectors* have been developed. These are often ethoxylated derivatives of fatty acids. Mixtures of various surfactants are also used commercially. If one uses paper machine whitewater in the secondary fiber plant, alum may help the ink bind to the fiber.

Effluent must be clarified of the ink so the water can be reused. Two chemical treatments may be used including a primary flocculent which is a low molecular weight cationic polymer or alum and a secondary flocculent of high molecular weight anionic polymer. Water soluble flexo inks have a large amount of dispersant in their formulation and typically high levels of flocculent to clarify the water.

Agglomeration deinking is a relatively new method of deinking that is applied to office papers printed with laser printing inks. Betz and PPG offer chemistry systems in this area. Surfactants are used to agglomerate small particles of ink into large particles that can be removed by screens or forward hydrocleaners.

Snyder, B.A. and J.C. Berg, Liquid bridge agglomeration: a fundamental approach to toner deinking, *Tappi J.* 77(5):79-84(1994).

Enzymatic deinking, Biodeinking

Welt, T. and R.J. Dinus, Enzymatic deinking—a review, *Progress Paper Recycling* 4(2):36-47(Feb., 1995), with 81 references. Enzymatic deinking uses cellulases (to, hypothetically, remove the fibrils from the surface of fibers where the ink anchors itself to the fiber), hemicellulases, pectinases, or, less studied, esterases (lipases), which would saponify inks with triglycerides (soy bean oil). The process was not commercial at the time of this article.

Changes in ink formulations

Trends in ink formulations to decrease their impact on the environment and make printed papers easier to recycle include the decrease in the use of volatile organic compounds (VOC, i.e., hydrocarbon and other solvents). Corrugated flexo inks recently were 10—20% VOC, but now range from 0—2%. The use of appreciable amounts of heavy metals in inks is under attack. Reductions in lead, chromium, copper, and, to a lesser extent, barium have begun to occur.

Use of ONP in OCC

The use of ONP in OCC allows a higher level of postconsumer waste to be included in OCC, which is important for some classifications of paper. The strength of ONP can be increased by chemical treatment (such as with NaOH), refining, the use of starch adhesives, and fiber fractionation.

General aspects

When using batch pulping, high consistency is usually preferable; however, newsprint is sometimes pulped in continuous pulpers at low consistency. The hole size in pulpers is often 1/4, 1/2, or 3/4 in. in diameter.

Three inch forward cleaners are used at 0.7 to 0.8% consistency unless the fibers are fairly short, when up to 1% consistency is used. Four inch cleaners are used at 1% consistency.

One method to improve the quality of recycled fiber is to fractionate it by size; OCC can be fractionated into small fibers for use in corrugating medium and large fibers to be used in linerboard.

Bobalek, J.F. and M. Chaturvedi, The effects of recycling on the physical properties of handsheets with respect to specific wood species, *Tappi J.* 72(6):123-125(1989). Ibid., *Tappi Proceedings, 1988 Pulping Conference*, pp. 183—187. Commercial bleached kraft loblolly pine, eucalyptus, quaking aspen, and northern pine were subjected to 0—3 cycles of handsheet formation without wet pressing (with light refining) with concomitant strength and other properties reported. It is unclear whether these handsheets were air dried or oven dried, but these conditions do not seem to duplicate commercial production practices.

Cox, J., Wastepaper growth highlighted in API's capacity survey, *Am. Papermaker* 54(2):37-40(1990).

Iannazzi, F.D. and R. Strauss, Municipal solid waste and the paper industry: the next five years, *Pulp & Paper* 64(3):222-225(March, 1990).

Olkinuora, M., U.S. leads the world in the production and consumption of wastepaper, *Pulp & Paper* 64(3):130-132(March, 1989). This paper gives figures for 1986 when the U.S. was recovering 27.4% of its waste paper. By 1990 this figure was over 30% and will most likely continue to rise as pressure mounts to reduce the amount of solid waste going into landfills, of which waste paper is approximately 50% of domestic waste. Japan recovers about 50% of its paper, which is near the sustainable maximum as fibers lose strength with each use.

Sorenson, D., Environmental concerns, economics drive paper recycling technology, *Pulp & Paper* 64(3)56-57:(March, 1990).

Gyrocleaner

Marson, M., New lightweight cleaner units solve mill's plastic problems, *Pulp & Paper* 64(6):93-96(1990). The development and operation of the

Gyrocleaner mentioned and shown on page 271 of this book are described.

Flotation deinking vs. flotation clarification
Doshi, M.R., Flotation deinking vs. flotation clarification, *Progress Paper Recycling* 1(2): 83(Feb., 1992).

Dispersed—air flotation deinking is a separation process where ink is separated from fiber (a selective process). Large bubbles are used, 0.1 to 1.0 mm in diameter, that quickly rise to the surface, leaving fibers behind; particles must be relatively large and hydrophobic to be removed.

Dissolved—air flotation clarification to clean process water, however, is designed to remove all particles (a collective process). Small air bubbles are used, 0.01 to 0.1 mm, under conditions of relatively low turbulence.

Stickies
Several mills report that the use of 0.006 in. slotted screens has appreciably decreased the problems associated with stickies in recycled pulp.

Doshi, M.R., Properties and control of stickies, *Progress Paper Recycling* 1(1):54-63(Nov., 1991).

Doshi, M.R., Quantification, control, and retention of depositable stickies, *Progress Paper Recycling* 2(1):83(Nov., 1992). These two articles are the basis of the discussion in this section.

Stickies are designated as either micro or macro because these two classes behave differently during the papermaking process. Micro stickies are smaller than 150 μm and will pass through a 0.006 in. slotted screen. A second classification is passing through a 200 mesh screen (< 76 μm).

Microstickies are determined by tests such as those of Buckman (plastic bottle), Berol (polypropylene film), Doshi (microfoam), and PIRA (paper machine wire). Talc and zirconium are used for the control of microstickies.

Macrostickies are determined by making handsheets that are then hotpressed and dried. The stickies are determined by counting with the needle nest, counting specks with the Tappi method, or dying the fibers with a water—soluble black dye. Another method that requires standards dyes the stickies with a water—insoluble dye.

Some manufacturers say talc and related materials remove macro stickies by increasing their density and allowing their removal in forward cleaners, but this is probably not true.

Twin—wire machines seem to have more difficulty with stickies (wire plugging) than fourdrinier machines. In addition to talc, zirconium compounds are also used for stickies control.

Sticky control agents should be added after the refiners, but as far upstream as possible to allow them ample time to work. Control agents include talc (used at 0.6—2% on pulp), dispersants, zirconium compounds (0.1% on pulp), and alum sequestering agents (to prevent agglomeration of stickies). Synthetic fibers were used at one time (0.1% on pulp), but this method is no longer used.

Dirt counting
Glowacki, J.J., Ponderosa boosts dirt count accuracy with PC/scanner, *Pulp & Paper* 69(2):93-94(1995). Dirt counting has come a long way over the last several years. This system uses a standard computer image scanner with software to give a dirt size distribution analysis. These systems are easy to use, highly repeatable, and quick.

34.6 PROPERTIES OF PULPS

Comparison of pulping methods
There are several papers in the literature that compare the strength and other properties of paper made with various pulping methods, cooking times, and refining levels. These studies are usually conducted on laboratory-made pulps and handsheets.

Liebergott, N. and T. Joachimides, Choosing the best brightening process, *Pulp Paper Can.* 80(12):T391-T395 (1979).

Wegner, T.H., G.C. Myers and G.F. Leatham, Biological treatments as an alternative to chemical pretreatments in high yield wood pulping, *Tappi J.* 74(3):189-193(1991). This paper includes fiber classification by Bauer—McNett analysis.

34.7 MECHANICAL PULPING

Thermomechanical methods
Prusas, Z.C., M.J. Rourke, and L.O. Uhrig, Variables in chemi—thermomechanical pulping of

northern hardwoods, *Tappi J.* 70(10):91-95(1987). The concentrations of NaOH in the first stage and sulfite in the second stage of Mead's CTMP process were investigated with refining energy and other pulp and paper properties reported. Some good basic data for these parameters are included.

Biomechanical pulping
Biomechanical pulping is an experimental technique where wood chips are inoculated with specific types of decay fungi (especially white rot) or lignin—degrading enzymes to break some lignin bonds prior to mechanical pulping; this offers the advantages of reducing power requirements in the refiners and giving higher strength pulps.

Akhtar, M., M.C. Attridge, G.C. Myers, T.K. Kirk, and R.A. Blanchette, Biomechanical pulping of loblolly pine with different strains of the white—rot fungus *Ceriporiopsis subvermispora*, *Tappi J.* 75(2):105-109(1992), with 24 references.

34.8 CHEMICAL PULPING

Digesters
Horng, A.J., D.M. Mackie, and J. Tichy, Factors affecting pulp quality from continuous digesters, *Tappi J.* 70(12):75-79(1987). One important factor that influences pulp quality from a continuous digester is the rate of liquor circulation.

Kraft process, general aspects
Pulp and Paper Manufacture, Volume 1, 2nd ed., MacDonald, R.G., Ed., McGraw—Hill, New York, 1969, pp. 347-627 (Chap. 8—10). Especially noteworthy are the history of the process, the liquor impregnation discussion, the demonstration of carbohydrate removal by type of carbohydrate and pulping process, the effect of chip quality on kraft pulping, the effect of effective alkali charge, effective alkali concentration, and sulfidity on the pulping of Douglas—fir sawdust, sweet gum, and other species, and a table with H—factor as a function of temperature in Chapter 8. Chapter 10 is an expansion of the article by Swartz directly below with (in addition to most of the above) information on process control, continuous pulping, causticizing curves, chemical compositions at various points in causticizing, performance of rotary kilns as a function of size, liquor titration calculations, and material specifica-

tions for construction of equipment. Chapter 11 discusses the thermodynamics of the kraft recovery cycle with an energy balance around the recovery process based on one ton of pulp. Much of the last chapter was from the first edition of 1950.

Swartz, J.N. and R.C. MacDonald, Alkaline pulping, in *Pulp and Paper Science and Technology*, Volume 1, Libby, C.E., Ed. McGraw-Hill, New York, 1962, pp. 160—239. This is a discourse on alkaline pulping from an engineering viewpoint including information on chemical recovery, the development of the modern recovery boiler, mass balances of cooking and washing, heat transfer reactions, heat balance of multiple effect evaporators, and tall oil and turpentine recovery. A table relates Baumé to solids content (from 13.2—54.7% solids) to the specific gravity of black liquor from a southern kraft mill. (Specific gravity is a very useful method for determining solids content of black liquor if calibrated by gravimetric analysis for a given mill.)

Schwartz, S.L. and M.W. Bray, Chemistry of the alkaline wood pulp processes, V. Effect of chemical ratio at constant initial concentration and the effect of initial concentration of the rate of delignification and hydrolysis of Douglas—fir by the sulphate process, *Techn. Assoc. Papers* 22(600)1939. The comprehensive study of the effect of effective alkali charge and effective alkali concentration on the pulping (as yield and lignin content) of Douglas—fir sawdust mentioned in the reference immediately above is from this article.

Kraft pulping chemistry
Chiang, V.L., et al., Alkali consumption during kraft pulping of Douglas—fir, western hemlock, and red alder, *Tappi J.* 70(2):101-104(1987). This study shows that 22% of the EA was used for initial delignification of the softwoods while alder required about 40%. Only about 2—4% Na_2O on wood was required for bulk delignification of the three species, although this accounts for about 70% of lignin removal. (Studies like this indicate why RDH pulping is so effective.)

Milanova, E. and G.M. Dorris, On the determination of residual alkali in black liquors, *Nor. Pulp Paper Res. J.* 4(1):4-9, 15(1994). The concept of residual alkali is important, but not easily defined. This is a good starting point on the topic.

Tormund, D. and A. Teder, New findings on sulfide chemistry in kraft pulping, *TAPPI 1989 International Symposium on Wood and Pulping Chemistry Proc.* pp. 247-254. The older literature gives the second ionization constant of H_2S as 10^{-12} to 10^{-15}. According to more recent publications cited in this work [such as Giggenbach, W. *Inorganic Chem.* 10(7):1333(1971) and Licht S. and J. Manassen, *J. Electrochem. Soc.* 134:918 (1987)] it is closer to 10^{-17} in alkaline aqueous solutions. This means one should always assume that Na_2S is completely hydrolyzed and exists in solution as NaHS and NaOH in <u>any</u> part of the kraft process; there is negligible S^{2-}. This article also studied the selective absorption of sulfide to woody materials.

Kraft polysulfide pulping
Hara, S.—i., New polysulfide pulping process at Shirakawa and Hachinoche, *TAPPI Proceedings, 1991 Pulping Conference*, pp. 107—113. Polysulfide white liquor, produced using a granular activated carbon catalyst, gives improved pulp yield and lowers the recovery boiler load while maintaining pulp qualities.

Jiang, J.E., Extended modified cooking of Southern pine with polysulfide: effects on pulp yield and physical properties, *Tappi J.* 77(2):120-124(1994). Results as immediately above with an additional claim of lowered refining energy.

Mao, B. and N. Hartler, Improved modified kraft cooking, Part 3: Modified vapor—phase polysulfide pulping, *Tappi J.* 77(11):149-153.

Extended delignification
A variety of pretreatment processes (many of which are experimental) applied to pulp lower the amount of bleaching chemicals required, leading to lower levels of chlorinated organic materials. Chemicals involved in various processes include oxygen (enjoying widespread commercial use, and gaining), nitrogen dioxide, ozone, sulfur dioxide, hydrogen peroxide, and sodium sulfite.

Fossum, G. and A. Marklund, Pretreatment of kraft pulp is the key to easy final bleaching, *Tappi J.* 71(11):79-84(1988).

34.9 KRAFT RECOVERY

Brownstock washing
Pelton, R., Defoamers can improve brownstock washer operation, *Pulp Paper Can.* 90(2):61-68(1989). This is a well—referenced source of information.

Multiple effect evaporation
Bergstrom, R.E., Evaporation of alkaline black liquor, in Stephenson, J.N., *Pulp and Paper Manufacture, Vol. 1, Preparation & Treatment of Wood Pulp*, McGraw—Hill, New York, 1950, pp. 535-568. Detailed calculations of multiple—effect evaporators are given.

Venkatesh, V and X.N. Nguyen, 3. Evaporation and concentration of black liquor, in Hough, G, Ed., *Chemical Recovery in the Alkaline Pulping Process*, TAPPI Press, Atlanta, 1985, pp. 15-85. A detailed discussion on black liquor evaporation including types of evaporators, operating conditions of a sextuple—effect evaporator, and strong black liquor concentration is given here. Their operating conditions of the sextuple—effect evaporator is identical to that presented by Smook, *Handbook for Pulp & Paper Technologists*, pp. 126-127, which is identical to that presented by Swartz et al. in MacDonald's *Pulp and Paper Manufacture*, Vol. 1, 1969, pp. 512-513. A similar sextuple—effect evaporator example is given by Swartz and MacDonald in Libby, C.E. *Pulp and Paper Science and Technology*, Vol. 1, pp. 194-195 (1962).

Zaman, A.A. and A.L. Fricke, Viscosity of softwood kraft black liquors at low solids concentrations: effect of solids content, degree of delignification and liquor composition, *J. Pulp Paper Sci.* 21(4):J119-J126(1995).

Recovery boiler
Fujisaki, A., H. Takatsuka, and M. Yamamura, World's largest high pressure and temperature recovery boiler, *Pulp Paper Can.* 95(11):T452-T459(46-53)(1994). This unit uses 2400 tons per day of black liquor solids producing steam at 500°C. The furnace is 12 m by 12 m by 55 m high. Automation is a key feature.

Ibach, S., Conversion to high solids firing, *Pulp Paper Can.* 97(3):T88-94(March 1996). This is an excellent article on the succesful operation of kraft recovery boilers using high solids firing. It is required reading on the topic.

Jones, A.K. and P.J. Chapman, Computational fluid dynamics combustion and modelling—a comparison of secondary air systems design, *Tappi J.* 76(7):195-202(1993). The use of CFD to model recovery boilers is increasing.

Thorslund, G.L., Visual recovery boiler inspection, *Tappi J.* 77(12):65-74(1994). An inspection plan should include the water, steam, and fire sides with the latter including pressure parts, attachments, casings, and wall openings.

Recovery boiler and nitrogen oxide emissions
Nichols, K.M., L.M. Thompson, and H.J. Empie, A review of NO_x formation mechanisms in recovery furnaces, *Tappi J.* 76(1):119-124(1993). This study claims that most of the NO_x arises from nitrogen in the fuel.

Prouty, A.L., R.C. Stuart, and A.L. Caron, Nitrogen oxide emissions from a kraft recovery furnace, *Tappi J.* 76(1):115-118(1993).

Effect of dead load on recovery & delignification
Ahlers, P.—E., H. Norrström, and B. Warnqvist, Chlorides in kraft recovery: effects on process and equipment, *Pulp Paper Can.* 79(5):T169-T173(63-66, 68)(1978). Any mill with over 10 g/L NaCl in the white liquor should certainly consult these two articles.

Mortimer, R.D. and B.I. Fleming, An empirical relationship to describe the effect of deadload on delignification, *Tappi J.* 66(1):98-99(1983). Deadload, inactive materials in kraft white liquor, such as NaCl, Na_2CO_3, and Na_2SO_4, decreases the rate of delignification. Normal levels of these compounds in commercial kraft pulping have a negligible effect, but some mills with high levels of NaCl will experience a decrease in delignification. For example, a soda—AQ black spruce pulp with no deadload in the white liquor has a kappa number of 53. The kappa number is increased with deadload. At 37.2 g/L as Na_2O (of

the chemical specified): NaCl gives 68.8, Na_2CO_3 gives 60.3, and Na_2SO_4 gives 64.5. The log (kappa) is proportional to the concentration of deadload contaminant. The slope of the relationship is similar for kraft, soda, and soda—AQ pulping methods.

Novel black liquor recovery technologies
A variety of techniques are being investigated to recover kraft liquor, especially methods that are amenable to small scales for incremental capacity increases. Another goal is to eliminate the possibility of smelt—water explosions.

Empie, H.J, Alternative kraft recovery processes, *Tappi J.* 74(5):272-276(1991). Several of these techniques involve methods of gasification or fluidized beds. The Chemrec plasma gasifier when operated at 1250°C (2280°F) followed by rapid cooling offers the potential of direct recovery of alkali and sulfur reduction approaching 100%. Although there is still much development needed on this process, Kamyr plans to build a 6000 lb/h demonstration unit.

Paleologou, M., P.—Y. Wong, and R.M. Berry, A solution to caustic/chlorine imbalance: bipolar membrane electrodialysis, *J. Pulp Paper Sci.* 18(4):J138-J145(1992). In this article sodium sulfate was recovered as sodium hydroxide and sulfuric acid with the application of electricity across two ion—selective membranes.

Although not covered in this article, the same technique can be used to recover NaCl as NaOH and HCl. This technique can be used to recover alkali from the alkali extraction of pulp bleaching with the added advantage of decolorization.

Dorr—Oliver, Inc., The fluid bed calciner, in Hough, G, Ed., *Chemical Recovery in the Alkaline Pulping Process*, TAPPI Press, Atlanta, 1985, pp. 299-303.

Novel black liquor recovery, auto—causticizing
A variety of alternate recovery processes have been proposed. One class of these processes are the auto—causticizing recovery methods where an amphoteric oxide (or related material) is present in the black liquor prior to recovery. Upon dissolution of the smelt after combustion of the

black liquor, the additive and sodium hydroxide are recovered. (Actually Na_2S behaves in this fashion as it generates NaOH directly in the dissolving tank.) For example, use of sodium borate in the white liquor allows NaOH to be regenerated directly when the smelt from the recovery boiler is dissolved in water; this eliminates the lime cycle and kiln.

In the DARS (direct alkali recovery system) developed by G.H. Covey, iron oxide serves the purpose. The DARS process has been used at the Fredericia, Denmark mill, which pulps straw. The chemistry of generating NaOH by these methods is well known and has been used in industry. Titanium—based recovery was also proposed in the late 1960s by K. Kato. The ferric oxide—sodium ferrite cycle, which has had some use to produce NaOH since it was patented in 1883 by Lowig, is summarized as:

$$Na_2CO_3 + Fe_2O_3 \rightarrow 2NaFeO_2 + CO_2(\uparrow)$$

$$2NaFeO_2 + H_2O \rightarrow 2NaOH + Fe_2O_3(\downarrow)$$

Janson, J., Pulping processes based on autocausticizable borate, *Svensk Papperstidning* 83:392-395(1985).

Ring formation in lime kilns
Backman, R, M. Hupa, and H. Tran, Modelling the sodium enrichment in lime kilns, *Pulp Paper Can.* 94(11):78-83(1993). Sodium and sulfur compounds can become molten in the lime kiln and increase the tendency for ring formation.

Notidis, E. and H. Tran, Survey of lime kiln operation and ringing problem, *Tappi J.* 76(5):125-131(1993). A survey of 134 kilns showed that ringing caused 20% of kiln downtime, although over 80% of kilns had this problem within the last two years. Frequent scheduled shutdowns decreased problems due to ringing. Nearly 80% of 61 kilns had trouble with ringing before they started burning NCG in the kiln, indicating that NCG may not be the primary cause of ringing. Most people felt sodium content was the major factor.

Older kilns and kilns that switch between oil and natural gas fuels seem to be particularly susceptible to this problem. Two—thirds of these burn noncondensible gases. The lime mud solids

content was 67—73%, with an average of 70%. The sodium content was 2% (on weight) as Na_2O on dry solids before the precoat filter and 0.5% after the mud washer. $CaCO_3$ was 90—95% of the solids with major impurities of Na, Mg, S, P, Si, Al, K, and Fe present.

The ring deposit consists largely of $CaCO_3$ and $CaSO_4$. While CaO may deposit initially, work indicates the colder temperature of the ring allows CaO to react with CO_2 to form $CaCO_3$, in the process of recarbonization.

Other operating parameters for mills are available in this study. While U.S. mills average 29% (range is 18—39%) sulfidity, Scandinavian mills average 34%. U.S. active alkali averages 100 g/L, while Scandinavian average 121 g/L (as Na_2O). The causticizing efficiency averages 80%.

Tran, H., X. Mao, and D. Barham, Mechanisms of ring formation in lime kilns, *J. Pulp Paper Sci.* 19(4):J167-J175(1993). High sodium and unstable kiln operation are major factors in ring formation.

34.10 BLEACHING

The area of bleaching is one of the most rapidly changing in the industry. Numerous publications appear. I recommend using the annual index of *Tappi J.* to find recent papers on the many aspects of this topic.

Oxygen supply
Baeuerlin, C.R., M.H. Kirby, and G. Berndt, Vacuum pressure swing adsorption oxygen for oxygen delignification, *Tappi J.*, 74(5):85-92(1991). Oxygen for use at rates above 130 tons/day is usually produced by cryogenic distillation of air on—site. Oxygen for use at rates below 15 tons per day is best supplied by having oxygen delivered by truck or rail. For intermediate levels of oxygen supply, an on—site vacuum pressure swing adsorption unit may be economical. The system works by adsorbing N_2, CO_2, and H_2O from air with synthetic zeolite sieves.

Oxygen bleaching
While high consistency oxygen bleaching was more common than medium consistency at one time, the trend has now switched. Most people now use medium consistency oxygen bleaching;

although the steam and alkali costs are slightly higher, the capital costs are lower.

Ozone bleaching

Ozone is a bleaching agent that has been studied much in the laboratory and in pilot plant operations. It now emerging as a commercial process. Ozone is used at high consistency (25—45%), at pH 2—3, below 60°C to replace chlorine bleaching. Several recent studies have studied ozone bleaching after xylanase treatment.

Byrd, Jr., M.V., J.S. Grazel, and R.P. Singh, Delignification and bleaching of chemical pulps with ozone: a literature review, *Tappi J.* 75(3):207-213(1992).

Byrd, Jr., M.V. and K.J. Knoernschild, Design considerations for ozone bleaching, A guide to ozone use, generation, and handling, *Tappi J.* 75(5):101-106(1992).

Gottlieb, P.M., P.E. Nutt, S.R. Miller, and T.S. Macas, Mill experience in high—consistency ozone bleaching of southern pine pulp, *Tappi J.* 77(6):117-124(1994).

Liebergott, N., B. van Lierop, and A. Skothos, The use of ozone in bleaching pulps, *TAPPI Proceedings, 1991 Pulping Conference*, pp. 1-23 (81 references). The article was reprinted in two parts in *Tappi J.* 75(1):145-152(1992) and 75(2):117-124.

O'Brien, J., PaperAge tours Union Camp's Franklin, Va., ozone bleaching facility, *PaperAge* 109(7):9-12(July, 1993).

Hypochlorous acid bleaching, MONOX—L™
Klykken, P.G. and M.M. Hurt, Molecular chlorine free bleaching with MONOX—L, *1991 TAPPI Pulping Conference Proceedings*, pp. 523-532.

Hurst, M.M., R Korhonen, K. Vakeva, and R. Yant, High brightness pulp bleaching with Monoox—L, *Pulp Paper Can.* 93(11):41-44(1992). Hypochlorous acid is purported to be an effective bleaching agent when used with a proprietary additive (said to be a C—2 monoalkyl amine) that protects cellulose from acid hydrolysis. (One wonders if this additive might be useful during chlorine bleaching.) A mill trial using this process is described here. This method helps keep chlo-

rine and caustic use balanced. The production of dioxins is said to be low. Use of a weak hypo by-product from the ClO_2 generators is also helpful.

Chlorine monoxide bleaching
Bolker, H.I. and N. Liebergott, Chlorine monoxide gas: a novel reagent for bleaching pulp, *Pulp Paper Mag. Can.* 73(11)82-88(1972). Chlorine monoxide, Cl_2O, was found to be an effective bleaching agent at 15—30% consistency and could replace chlorine dioxide in later bleaching stages. For example, chlorinated, extracted sulfite pulp could be brought to 94 brightness with a single stage using 0.6% Cl_2O on pulp. Chlorinated, extracted kraft pulp was brought to 82 brightness when treated with 1% Cl_2O for one minute; a second stage brought the brightness to 94%. Strength properties of the pulps were as good as those pulps bleached with chlorine dioxide. With unbleached kraft pulp, 2% Cl_2O was more effective than bleaching with 5.9% Cl_2, with 94% less color in the effluent.

Xylanase treatment
Daneault, C., C. Leduc, and J.L. Valade, The use of xylanases in kraft pulp bleaching: a review, *Tappi J.* 77(6):125-131(1994). Enzymatic bleaching pretreatments (hypothetically) rely on breaking lignin—carbohydrate bonds in pulp.

34.11 FIBER PHYSICS AND PREPARATION

Fiber bonding revisited

An increased level of fiber—to—fiber bonding gives a sheet with higher density, lower opacity, and increased shrinkage and swelling with changes in moisture content. In the case of the latter observation, the change in length of fibers (which does not normally occur with a change in moisture content in individual fibers) in well—bonded papers is attributed to a change in width of fibers laying across fibers, as demonstrated in Fig. 34-1.

Consider a fiber that lays parallel to the force applied in a sample of paper (Fig. 34-2). This fiber is held by a series of fibers perpendicular to it. The force on the bonds is highest at the ends of the fiber and lowest in the middle of the fiber (Fig. 34-2, top graph); the tensile load on the fiber is highest at the center of the fiber (second graph);

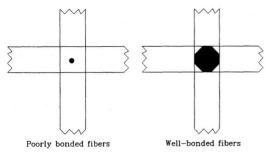

Poorly bonded fibers   Well—bonded fibers

Fig. 34-1. In poorly bonded fibers, shrinkage of width of one fiber does not cause the length of another fiber to decrease very much, but this is not true in well—bonded fibers.

therefore, the probability of fiber failure is highest in the middle, especially in well—bonded papers (third graph), and the probability of bond failure is highest at the ends of the fiber (bottom graph).

Drying under restraint

It is a fact that much of increased strength in the machine direction of paper over the cross machine direction is due to drying under restrain, and not due to fiber alignment. Two studies in this area demonstrate this with handsheets.

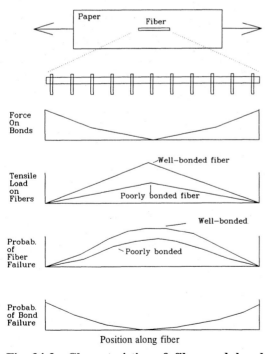

Fig. 34-2. Characteristics of fiber and bond forces along the length of a fiber.

Setterholm, V.A., W.A. Chilson, and A.T. Luey, Effect of temperature and restraint during drying on the tensile properties of handsheets, USDA FPL rep. no. 2265 (1962), 6 p.

Handsheets prepared from a bleached southern pine kraft pulp were dried at 73°, 150°, and 220°F under four degrees of restraint. Analysis of stress—strain data from tension tests revealed that increasing either drying temperature or restraint during drying increased the tensile properties, and that restraint had the greater effect.

Setterholm, V.A. and W.A. Chilson, Effect of restraint during drying on the tensile properties of handsheets, US For. Ser., Res. Pap. FPL 11 (April, 1964), 25 p. This study shows, among many other things, that drying under restraint increases the density of sheets (and presumably the degree of fiber bonding) and their tensile strength values. Much data is given in this reference.

Fiber physics

Chemical pulp fibers have a higher fiber saturation point, a lower wet strength, and a higher dry strength than wood fibers.

Bolam, F., Ed., *The Formation and Structure of Paper*, Tech. Sect. British Paper and Board Makers. Assoc., London, 1962, 2 vol., 910 p. and 49 p. index. This is a highly technical look at fiber and paper physics. Many micrographs are included that show well on the books' coated paper. A must for anyone serious about paper physics.

Refining

Gentle action in refining (more impacts of lower energy each, for example, by reducing the bar or groove width thereby increasing the number of bars) leads to higher strength properties for fiber refined to a given freeness. For this reason, series refining has some advantages over parallel refining. However, with series refining, the maximum pressure of the last refiner must be below 100 psi.

If there is too much depth between the bars then the pressure across the refiner increases, the idle load increases, and there is more pumping action. If the depth is too small, then the plates wear out quickly.

The peripheral speed of disk refiners is generally in the range of 4700—5700 ft/min. If it is too low, the refiner may plug, but low speeds are conducive to fiber cutting. If the peripheral speed is unnecessarily high, energy is wasted.

The level of refining can be monitored by specific freeness, temperature change, pulp freeness, couch vacuum, or dry end properties. Notice the feedback time increases with these methods. Specific energy is a very common and easy way of monitoring the refining.

Kraft pulping (unbleached spruce) leaves very little uronic acid in the pulp whereas sulfite pulping leaves much higher amounts behind. This is one reason sulfite pulps are easier to refine than kraft pulps. These pulps differ in fiber swelling and response to alkali and metal ions.

Knutsen, D.P. and W.J. Bublitz, Unpublished. Energy use of the PFI mill in terms of hp—day/ton can be calculated by the following equation, based on the fact that the idle power of the PFI mill (which should be determined on each machine) is about 0.0034 watt·hr/revolution. The amount of pulp is usually 30 grams (oven—dry basis).

$$\text{hp–days/ton} = \frac{\text{PFI watt·hr} - 0.0034 \times \text{PFI rev.}}{\text{grams pulp}}$$

$$\times 50.67 \, \frac{\text{g·hp·day}}{\text{watt·hr·ton}}$$

Screening
McCarthy, C., Various factors affect pressure screen operation and capacity, *Pulp & Paper* 62(9):233-237(Sept., 1988). This is a good, general article on screening and has a useful figure on contour types of screen cylinders.

34.12 PAPERMAKING

Papermaking chemistry
Scott, W.E., *Principles of Wet End Chemistry*, Tappi Press, 1996. This is a revision of the 1992 work *Wet End Chemistry: An Introduction*. This book is very helpful, although the technical quality of the computer—generated figures is fairly poor.

Papers from mechanical pulps
Santkuyl, R.J., SC grades without the supercalender: hot soft—calendering can broaden the grade offering, *Pulp Paper Can.* 96(1):49-54,

T22-T27(1995). This is a very good overview of printing papers made from mechanical pulps.

A wide variety of printing papers are made from mechanical pulps, and the range continues to increase. Some common acronyms include SC for supercalendered (which is further broken down to three commercial grades SCA, SCB, and SCC); MF for machine finished; MFC for machine finished coated, a relatively new process; news for newsprint; and LWC for light weight coated. Fig. 34-3 shows the approximate relative brightness and gloss of these grades. Keep in mind that calendering operations improve gloss, but do not affect brightness, which improves largely by the use of fillers and coatings.

Santkuyl points out that the names we choose are based on the process to make them and not on the properties of the papers themselves, which would be much more appropriate for the end users (i.e., the customer). Confusion easily arises when SC grades can be produced without supercalendering (for example, by hot soft—calendering). (We are accustomed to calling paper made from thermomechanical pulp as groundwood grades when, of course, the stone groundwood pulping process is much different from the thermomechanical pulping process.)

Supercalendering must be done off—machine because it cannot be threaded at machine speeds. The traditional rolls for the supercalender were cotton—filled rolls. Recently, synthetic composite coverings are being used instead of the cotton—filled rolls, allowing for better calendering action under conditions not suited for the cotton-filed rolls. Santkuyl points out a variety of reasons why on—machine calendering is preferred to supercalendering, including lower equipment requirements and lower operating costs. Depending on the equipment used, hot soft—calendering is carried out from 200°F (93°C) with a pressure of 750 lb/in. (131 kN/m) (as low in Fig. 34-3) to 400°F (204°C) with 2000 lb/in. (350 kN/m) (as high in Fig. 34-3).

Pressing and drying
A useful general reference on this is *1989 Practical Aspects of Pressing and Drying Seminar*, TAPPI Press, Atlanta, Georgia.

Fillers and paper strength

Fillers reduce the strength of paper. Finding numbers for strength loss is not easy, but Libby, vol. 2 (1962) gives some values on page 76.

34.13 PAPER

Surface properties of paper

Conners, T.E. and S. Banerjee, *Surface Analysis of Paper*, CRC Press, Boca Raton, Florida, 1995, 368 p. This work provides information on the application of modern surface analysis techniques to paper.

Handmade paper

Hunter, D., Handmade papers, in *Pulp and Paper Manufacture, Vol. 3*, MacDonald, R.G., Ed., McGraw—Hill, New York, 1969, pp. 1-18.

Mechanical properties

Seth, R.S., Implications of the single—ply Elmendorf tear strength test for characterizing pulp, *Tappi J.* 74(8):109-113(1991). This work points out that the tear strength of paper when measuring a single sheet can vary by as much as 60% compared to that of tests using multiple sheets. The magnitude of the difference depends on the degree of bonding in the sheet. This fact is particularly important since the performance of paper in usage may correlate better with single—ply results rather than multiply results.

TAPPI TIS 0304-24 (1989), Calibration and maintenance of the burst tester. This helps insure a tester is giving correct readings.

Paper abrasiveness

Laufmann, M., G Bräutigam, N. Gerteiser, and H. Rapp, Natural ground $CaCO_3$ as copy paper filler, *TAPPI Notes, 1985 Alkaline Papermaking*, pp. 61-72. This paper discusses measurement of abrasiveness of fillers. There is some concern that the use of calcium carbonate fillers in printing papers will cause increased wear on the paper machine or in service such as in photocopy machines.

Pitt, R., Fillers and fabric life in alkaline papermaking, *TAPPI Notes, 1985 Alkaline Papermaking*, 47-50. This is of related interest.

34.14 COATING AND CONVERTING

Converting

Kline, J.E., *Paper and Paperboard, Manufacturing and Converting Fundamentals*, Miller Freeman Pub. Inc., San Francisco, 1982. Section III (pp. 152-228) contains five useful chapters corrugating operations and raw materials; packaging; tissue and related grades; and cut size, bond, copy paper, and computer paper.

Coating

Coating has been discussed on pages 250, 254-258, and 261-262. Some additional details are considered here. Coating operations are usually carried out with pigments (clay, calcium carbonate, titanium dioxide, or others) to improve the gloss and printing characteristics of paper. Other coatings are used to make carbonless papers and other specialty papers. The optical properties of coated and filled papers are considered separately in Chapter 24.

Coatings can be applied at the size press or after the paper is formed and dried. The coating is applied to the paper, the coating is smoothed,

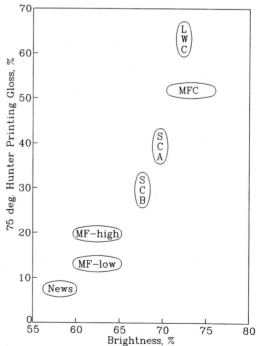

Fig. 34-3. Spectrum of mechanical printing grades after Santkuyl (1995).

and the coating is dried. Air knife coating is limited to one side of the sheet at a time. It is useful for carbonless papers where physical contact with the paper is not possible. Some high—end papers use two— or even three—pass coating.

Coating pigments

The general properties of the most common coating pigments (kaolin clay, calcium carbonate, and titanium dioxide) were described earlier (pages 194-195 and 439-440). (Titanium dioxide absorbs light in the near UV range and should not be used with fluorescent whitening agents.) Particles for coating formulations must have small size distributions, generally below 1—2 μm. Delaminated clay of ultrafine particle sizes are used in coatings to give high gloss. Fine particle sizes give lower gloss. Coarse particles are used for matte finished sheets. Of the additional pigments listed here, aluminum trihydrate has appreciable use, but the others enjoy relatively little use.

Aluminum trihydrate, ATH, is a high—brightness pigment with a bluish color; the preferred name is aluminum hydroxide, $Al(OH)_3$. It is often used as an extender for TiO_2 and acts as a spacer between the TiO_2 particles. About 35,000 tons were used by the world's paper industry in 1993.

Satin white, first used in the last quarter of the 1800s, has appreciable use in premium grades of paper in Europe, but has little use in the U.S. It is made by reacting $CaCO_3$ with alum to give a precipitate of needle—shaped particles consisting of calcium sulfate. In paints it is called satin white. Pearl filler is also calcium sulfate but is prepared in a different manner.

Silica and *silicate* pigments have high brightness, small particle size, and low abrasion. The composition is H_2SiO_3. They do not absorb UV light and, therefore, work well with optical brighteners. They are used on premium grades of paper often with titanium dioxide.

Barium sulfate is an expensive filler with good properties. Zinc oxide is occasionally used as a coating material, but is more often used in electrophotography.

Plastic pigments are usually spherical particles of polymers. Polystyrene is a hard plastic and is the only material presently used for plastic pigments. The size distribution is monodisperse, although two (or more) different—sized particles may be mixed together to give a high packing density. Particle sizes are available from 0.2 to 1.0 μm. The cost is high, but high sheet gloss can be obtained if the polystyrene is heated above its glass transition temperature of about 100°C. The density is about 1.05 (much lower than other fillers and fibers, making it good for lightweight papers) and the index of refraction is 1.60.

The rheology of the coating solutions is very important. It is desired to have a relatively low viscosity at high solids content so that drying costs can be decreased. The solids content of coating formulations is usually 50—75%. Low solids content coatings with relatively high levels of binders are used in the size press.

Coating binders and adhesives

The use of coating binders must be limited so that it does not fill in the voids between pigment particles and, therefore, decrease the opacity of the coating. Typical levels are 5 to 20% of the dry solids of the coating formulation. Binder formulations often consist of a soluble binder and an emulsion binder. Binders must be particularly strong for paper destined for offset printing.

Information on coating binder migration, which causes ink mottling and poor adherence of adhesive in many instances, is discussed on page 477. Materials may be added to the coating formulation to crosslink or precipitate the binder.

Starch (page 199) accounts for most of the *natural binders* used in coating formulations. Corn starch is most commonly used, but other starches can be used. Starch is often modified to improve its properties. Hydroxyethylated starch is commonly used on premium grades of paper; oxidized starch is also used and is less expensive. Starch viscosity is decreased by enzymes (amylases), acid hydrolysis, or oxidative cleavage of glycosidic linkages (called concersion or scission in the paper industry). Proper cooking gives a well—dispersed starch; cooking is accomplished by the batch or continuous (jet) process. The cooking process is sometimes monitored by observing the granules under a microscope with polarized light; the granules are not polarized after they are well dispersed. The starch is stabilized by crosslinking on papers designed for offset (lithographic) printing so that is does not dissolve in the fountain solutions and picking is limited.

Soy protein (commercially available since 1937) is used in some paper coatings and in relatively large amounts in paperboard coatings. The soy protein is received as a powder and is often carboxylated, which is solubilized with the addition of some alkali. Other formulations can be dissolved in cold water. A wide variety of viscosities are available. Soy protein replaced casein in the 1950s. *Casein* (milk protein) has been used widely in the past.

Synthetic binders are used as latexes (emulsions). Latex originally was the term for emulsions of natural rubber, but the term now applies to many emulsions of synthetic polymers. *Styrene—butadiene rubber (SBR)* is the most commonly used *synthetic binder*. Polystyrene is a tough, strong, brittle material, and polybutadiene is an elastic, rubbery material. The butadiene is usually carboxylated to improve its bonding ability. The relative amounts of these components alters the properties of this copolymer. Styrene content of 50% gives a tacky material with good binding properties; styrene content of 70% gives a material that is more brittle but gives good gloss. The latex particles are 0.1 to 0.3 μm in diameter. Synthetic binders often have poorer ink receptivity than starch or protein binders, but give enhanced ink gloss due to improved ink holdout.

Polyvinyl acetate (PVA) is the major component used in white glues for wood. PVA may be used with other natural or synthetic binders. It is somewhat water sensitive and gives a hard surface. It is useful for web offset coatings because it is somewhat porous and imparts stiffness in the paper (useful for improving the stiffness in the cross machine direction); these properties decrease paper blistering during printing. Vinyl acetate is used with acrylic monomers to make copolymers. Polyvinyl alcohol has use in specialty coatings.

Acrylic latexes are high—quality binders with many advantages, but the SBR binders have limited their use until recently. They provide good stability against aging by light, heat, and chemicals and fold strength. Polymethyl methacrylate is one acrylic polymer (Lucite® or Plexiglas®).

Coating additives

A wide variety of materials are added to improve and control coating formulations (see Chapter 21, Colloid and Surface Chemistry). Many are similar to the papermaking additives, although retention and drainage are not considerations. One must consider additives that are already included with the pigments, such as dispersants, and adhesives, such as surfactants.

Dispersants are used to keep pigment particles from agglomerating, which would decrease their opacity. They are used at about 1% of the pigment weight. Polyphosphates are used with clay calcium carbonate, titanium dioxide, and other mineral pigments. Polyphosphates, such as tetrasodium pyrophosphate ($Na_4P_2O_7$, also called sodium diphosphate) are formed by dehydrating sodium salts of phosphoric acid (such as disodium phosphate, Na_2HPO_4). Polyphosphates may hydrolyze to low molecular forms in water with heating or low pH; this process liberates acid, which further catalyzes the reaction. Polyphosphate anions are effective at sequestering multivalent cations. Aqueous solutions of sodium silicates (water glass) are also used as dispersants.

Polyelectrolytes, such as low molecular weight sulfonated materials or polyacrylates, may be used as a second dispersant in case the polyphosphate degrades or is used improperly. Ethoxylated fatty alcohols are used as nonionic dispersants, which prevent the pigments from getting close enough to agglomerate (steric hindrance).

Defoaming agents or foam control agents are used to prevent entrained air from forming in the coating mixture. Entrained air increases the viscosity of the coating formulation and leads to defects in the coated product.

Biocides are used especially when starch, a favorite food of microorganisms everywhere, is used as the binder. Good housekeeping, proper machine design, and occasional sterilization help control microorganisms. If problems are encountered, one should isolate the offending microorganism and use a suitable biocide.

Viscosity control agents allow the rheological (flow) characteristics to be regulated. The use of most binders, and the desire to have a high solids contents, usually means that most coating formulations will have a high viscosity. Some chemicals actually reduce the viscosity of coating formulations, including urea and polyvinyl pyrrolidone with molecular weight below 40,000. Urea, used at 5—20% of the weight of starch, stabilize the viscosity behavior of modified starch.

It is relatively easy to increase viscosity by using polymers with high molecular weight (starch, alginates, or CMC).

Antistatic (electricity) agents work by absorbing water from the atmosphere (glycerides or glycols), by adding conductivity to the coating (potassium chloride or quaternary ammonium compounds and other ionic surfactants), or by putting a charge on the surface that is opposite to the charge that otherwise tends to build.

Other additives include lubricants or plasticizers, such as calcium or ammonium stearate, PEG, wax, or polyethylene, which are used to prevent dusting. Specialized coatings contain silicone or fluorocarbon release agents for paper to which removable labels are attached, encapsulated pigments for carbonless papers, etc.

Additives to control the appearance of the final product include dyes and fluorescent whiteners. Additives that absorb harmful ultraviolet radiation (salicylates, benzophenones, and other aromatic compounds), which would otherwise eventually cause fading or other photochemical degradation, may be used.

Miscellaneous

The National Cash Register Company (NCR) obtained the original patents for carbonless paper in 1954. Commercial coated papers range from grade 1 to 5. Grades 1 to 3 are "free sheets"; they contain no mechanical pulp. All number 5 grade contains some mechanical pulp.

Walter, J.C., *The Coating Process*, Tappi Press, 1993, 260 p. This is a useful, general reference on paper coating including materials, the preparation of coating formulations, and the various coating processes.

34.15 FLEXIBLE PACKAGING

Hammond, F.M. and M.W. Potts, An overview of flexible—packaging markets and structures, *Tappi J.* 72(1):81-88(1989).

Thompson, K.I., A look at 20th century film development, *Tappi J.* 73(6)137-140(1990). This

is an interesting history of the development of plastic films for packaging. Cellophane was the material that started it in 1924 and was king until the early 1960s, when polyethylene surpassed it. Equipment technology and film markets are also discussed. It is too bad more detail is not given.

34.16 ENVIRONMENTAL

Cluster rules
Swan, C.E., Cluster rules update, 58(11):23-25(1995). In late 1993, the U.S. Environmental Protection Agency (EPA) proposed that water and air regulations would be combines for the U.S. pulp and paper industry. The proposal was nicknamed the "Cluster Rules". The industry has said these regulations are unnecessarily burdensome. As of this writing, the situation is still unresolved, and future reguations are still uncertain.

Sludge disposal: incineration, energy recovery
Solid waste disposal by landfilling is becoming increasingly expensive (with a tipping fee on the order of $75/ton) and of concern for many reasons (such as the possibility of a particular landfill being targeted as a hazardous waste site in the future for reasons presently unidentified). Sludge disposal by incineration is one area where landfilling of solid wastes can be decreased. For example, sludge from the primary clarifiers typically contains over 90% wood fiber, some calcium salts, etc. Most mills do not use this material as fuel because it is difficult to dewater and they do not require the fuel.

Jacobson, W., Fluid—bed combustion: a solution to increasing paper sludge problems, *PaperAge* 106(Recycling Annual):21-22(1990). Another major source of sludge is that generated from secondary fiber reclamation. Although quite variable, paper sludges have similar heating values to wood and about 8—10% ash content; deinking sludges have about 11—18 MJ/kg (4600—7500 Btu/lb) (dry basis) and ash contents of 15—40%. Fluid—bed combustion of these sludges is said to be quite efficient when they are mixed with other fuels such as wood.

Rude, J., Sludge disposal: a problem or opportunity, *PaperAge* 106(7):24(1990). The author considers drying of sludge for hog fuel by rotary driers.

Fig. 34-4. A progressing cavity pump. Courtesy of Roper Pumps.

Water
Upchurch, J., G. Hodge, and P. Speir, Where does all the water go? *TAPPI 1989 Environmental Conference Proceedings* pp. 225-229. The paper discusses the importance of a mill water balance and how to obtain one for a mill. Examples of methods to save or reuse water are given. The process water in thousands of gallons per dry metric ton of a state—of—the—art kraft mill are as follows: pulping, 3—4; conventional bleaching, 8—10;   bleaching system with oxygen delignification, 4—7; power—recovery—recausticizing, 3—4; paper machine, 3—5; and other uses, 2—3.

34.17 MISCELLANEOUS TOPICS

Pumps
    Figure 34-4 shows a progressing cavity pump, which is described on page 643

Millwide Information Systems
Miklovic, D.T., and H.J. Dammeyer, Manufacturing automation protocol and its implications to the forest products industry, *Tappi J.* 70(3):67-74(1987). This has a four page glossary.

34.18 TESTING METHODS

Automated (on—line) black liquor titrations
    Many manufactures make automated titrators that are applicable to automated liquor analysis. This is an area where much operator time could be saved while reliability and precision could be increased in the control of the kraft recovery (or other) process, especially in these days of tighter process/quality control. Management does not mind buying expensive production equipment, but is less amenable to buying relatively inexpensive testing equipment to insure the processing equipment is used to its maximum potential.

Cleveland, C., On—line analysis of pulping liquor with a process titrator, *Tappi J.* 73(11):157-161(1990). This is used for sodium hydroxide—sodium carbonate liquor for corrugating media.

Black liquor analysis
Busayasakul, N., J.M. Genco, and J.C. Hassler, Carbon analysis in kraft digester control, *Tappi J.* 70(4):149-153(1987). Use of total organic carbon (TOC) content of black liquor as a means of controlling the cooking process, just as one uses lignin content (kappa number) of pulp, is de-

scribed. This method is not likely to be used routinely in pulp and paper mills.

Lignin structural analysis

Gardner, D.J., T.P Schultz, and G.D. McGinnis, The pyrolytic behavior of selected lignin preparations, *J. Wood Chem. Tech.* 5(1):85-110(1985). Chemical analysis of lignin by a variety of methods such as nitrobenzene oxidation, thermal analysis, pyrolysis, and IR were investigated. This is a good introductory article describing the tools that tell us what is known about lignin.

Hatfield, G.R., G.E. Maciel, O. Erbatur, and G. Erbatur, Qualitative and quantitative analysis of solid lignin samples by carbon—13 nuclear magnetic resonance spectrometry, *Anal. Chem.* 59:172-179(1987). Maciel's laboratory has studied numerous aspects of NMR of lignins.

Lignin content by infrared or UV spectroscopy

The traditional method of lignin measurement using permanganate ion can be supplanted by optical spectroscopy methods like near—infrared (NIR) spectroscopy or visible—ultraviolet (UV) spectroscopy. Unlike permanganate titration of pulp, where only a single piece of data is obtained, spectroscopy methods offer reflectance curves that offer the potential for quantifying several components by simultaneous equations at several wavelengths. These methods are faster than titrations and lend themselves to continuous, on—line measurements. In the case of IR, information about the polysaccharides is also available, which, in turn, allows yield determinations with a high degree of precision.

Angeus, L. and S.—A. Damlin, Optimal control with on—line kappa number analysis, *Pulp Paper Can.* 93(2):T32-T36(1992). These workers measured the kappa number using UV absorbance at 280 nm with success.

Backa, S. and A. Brolin, Determination of pulp characteristics by diffuse reflectance FTIR, *Tappi J.* 74(5):218-226(1991).

Easty, D.B., S.A. Berben, F.A. DeThomas, and P.J. Brimmer, Near—infrared spectroscopy for the analysis of wood pulp: quantifying hardwood—softwood mixtures and estimating lignin content, *Tappi J.* 73(10):257-261 (1990).

Michell, A.J., Infra—red spectroscopy transformed—new applications in wood and pulping chemistry, *Appita* 41(5):375-380(1988).

Schultz, T.P. and D.A. Burns, Rapid secondary analysis of lignocellulose: comparison of near infrared (NIR) and Fourier transform infrared (FTIR), *Tappi J.* 73(5):209-212 (1990).

Ion—selective electrodes

Ion—selective electrodes (ISEs) are very similar in use to pH electrodes. They are used for chloride, potassium, calcium, carbon dioxide/carbonate, oxygen, and a variety of other ions. These methods are particularly suited for field analysis and on—line measurements.

Lenz, B.L. and J.R. Mold, Ion—selective electrode method compared to standard methods for sodium determination in mill liquors, *Tappi J.* 54(12): 2051 2055(1971). Sodium ion can be quickly measured directly with ion specific electrodes, with about 1% error, over a wide range of concentrations. A small sample is diluted with a small amount of ionic strength adjuster (ISA, an ammonium buffer solution) and the solution measured as if for pH, but in this case for $p(Na^+)$. This method would probably be useful for measuring sodium loss in pulp by incubating the pulp with ISA until ammonium ion has exchanged sodium ion. It could also measure sodium concentrations at various stages of the recovery process.

Cooper, Jr., H.B.H., Continuous measurement of sodium sulfide in black liquor, *Tappi* 58(6):59-62(1975). This is a general article on the subject.

Schwartz, J.L. and T.S. Light, Analysis of alkaline pulping liquor with sulfide ion—selective electrode, *Tappi J.* 53(1):90-95(1970). New electrodes often use a salicylate buffer with 1:3 sample:buffer dilution. This article has much experimental detail.

Aluminum ion concentrations

Avery, L.P., Evaluation of retention aids, the quantitative alum analysis of a papermaking furnish and the effect of alum on retention, *Tappi* 62(2)43-46 (1979). The concentration of alum was measured in an elegant fashion using a F⁻ ion-specific electrode. Fluoride complexes with Al^{3+} and a titration with NaF using the ISE allows the determination of aluminum concentration (complex formation values for F are available in *Lange*). This method can be used with poly-aluminum chlorides (PAC), but low aluminum values may be obtained as the polymer forms of aluminum may react slowly. Perhaps a method could be developed where excess fluoride is added (if the approximate concentration of PAC is known), the sample aged, and the fluoride back titrated with alum (or measured directly).

Varriano—Marston, E. and H. Cheeseman, 8—Hydroxyquinoline as a stain for determining the distribution of alum in paper, *Tappi J.* 69(6):116-117(1986).

Ion chromatography

Ion chromatography is a liquid chromatography technique that separates ions for quantitative and qualitative analysis. The procedure requires a high—performance liquid chromatograph (HPLC), usually available at corporate research centers or university laboratories. The method is applicable to pulping liquors, bleaching liquors, and other process streams.

Cox, D., P. Jandik, and W. Jones, Applications of single—column ion chromatography in the pulp and paper industry, *Pulp Paper Can.* 88(9):T318-T321(1987).

Easty, D.B. and J.E. Johnson, Recent progress in ion chromatographic analysis of pulping liquors: determination of sulfide and sulfate, *Tappi J.* 70(3):109-111(1987).

Murarka, S.K. and N.S. Fairchild, Ion chromatography: Is it the 'now' technique for monitoring process streams in a paper mill? Part I: Principles and instrumentation, *Pulp Paper Can.* 88(12):T438-T440(1987); ibid, Part II: Selected applications, *Pulp Paper Can.* 88(12):T441-T446(1987).

Bleaching chemicals

Fisher, R.P., M.D. Marks, and S.W. Jett, Measurement methods for chlorine and chlorine dioxide emissions from bleach plants, *Tappi J.* 70(4):97-102(1987). Chlorine and chlorine dioxide can be captured from air by neutral potassium iodide solution. A two—point titration is used to give contents of both species.

Starch in paper

Biermann, C.J., Quantitative analysis of starch in paper, *Tappi J.* 76(5):12(1993). A useful improvement over present TAPPI Standard methods for starch analysis in paper would be a hybrid method of enzymatic hydrolysis of starch followed by specific detection of glucose. Specific detection of glucose could be done with glucose analyzers available at drug stores or more sophisticated ones. Such a method would have the advantage of taking very little operator time, being portable, and being specific for glucose.

Imaging techniques

Paumi, J.D., Laser vs. camera inspection in the paper industry, *Tappi J.* 71(11):129-135(1988). The title technique is used to scan paper webs on—line at machine speeds up to 6000 ft/min for detection of any defects.

Tomimasu, H., D. Kim, M. Suk, and P. Luner, Comparison of four paper imaging techniques: β—radiography, electrography, light transmission, and soft X—radiography, *Tappi J.* 74(7):165-176(1991). These techniques are used for measuring the uniformity of paper, i.e., its formation.

Pitch and contaminant analysis, "stickies"

Buildup of pitch and polymer contaminants on paper machine wires and clothing is a big problem. The stock temperature often determines where the stickies will deposit. Often the source is presumed to be the pulp or contaminants from secondary fiber operations. These assumptions are not always warranted; analysis of the residue is instrumental to problem solving. Analytical techniques include sample purification by extraction, gas chromatography, NMR, FTIR, size—exclusion (gel—filtration) chromatography, etc.

Biermann, C.J. and M.—K. Lee, Analytical techniques for analyzing white pitch deposits, *Tappi J.* 73(1):127-131(1990). This is another general paper on the subject.

Ekman, R. and B. Holmbom, Analysis by gas chromatography of the wood extractives in pulp and water samples from mechanical pulping of spruce, *Nordic Pulp Paper J.* (1):16-24(1989).

Sithole, B.B., Modern methods for the analysis of extractives from wood and pulp: a review, *Appita*

45(4):260-246(1992). This is a concise review with a lot of information and 69 references.

Sjöström, J. and B. Holmbom, Size—exclusion chromatography of deposits in pulp and paper mills, *J. Chromatogr.* 411:363-370(1987).

Sjöström, J. and B. Holmbom, A scheme for chemical characterization of deposits in pulp and paper production, *Paperi ja Puu* (2):151-156(1988). This is a useful, general paper on the subject.

Sjöström, J., B. Holmbom, and L. Wiklund, Chemical characteristics of paper machine deposits from impurities in deinked pulp, *Nordic Pulp Paper Res. J.* (4):123-131(1987).

Suckling, I.D. and R.M. Ede, A quantitative ^{13}C nuclear magnetic resonance method for the analysis of wood extractives and pitch samples, *Appita*, 43(1):77-80(1990). Stickies can be measured qualitatively by hot press tests and other tests.

Gaseous pollutant measurement
Barker, N.J and J.L. Siqueira, Continuous, remote monitoring of specific reduced sulfur compounds in ambient air by gas chromatography—photoionization detector, *1988 Environment Conference Proceedings*, CPPA, pp. 11—15 (1988). Ambient TRS measurements 2 and 10 km from a mill are described in this work.

Dirt in wood, pulp, and paper
Clark, J. d'A., R.S. Von Hazemburg, and R.J. Knoll, *Paper Trade Journal* 96(5):40(1933). The authors published a chart of dirt sizes. They included 20 particles shapes for each of 17 sizes from 0.01 mm^2 to 5 mm^2. This is the basis of the TAPPI dirt estimation chart used in the tests described below.

TAPPI Standards are available for quantifying dirt by light reflectance in pulp (Standard T 213) and paper and paperboard (Standard T 437). In fact, both tests are very similar. With the widespread availability of optical character recognition (OCR) systems, these TAPPI Standards undoubtedly will become computerized or obsolete.

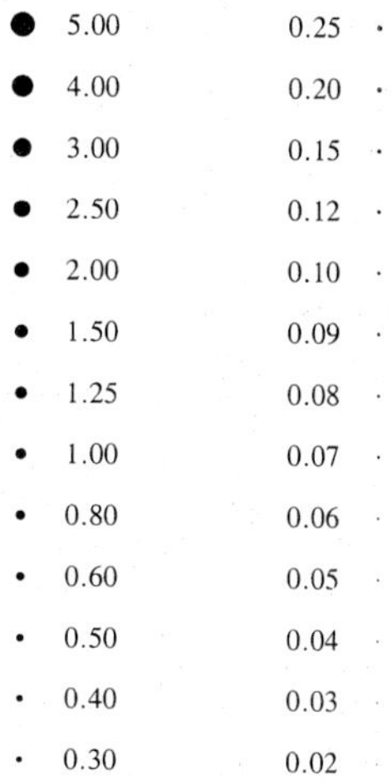

● 5.00	0.25 ·
● 4.00	0.20 ·
● 3.00	0.15 ·
● 2.50	0.12 ·
● 2.00	0.10 ·
● 1.50	0.09 ·
● 1.25	0.08 ·
● 1.00	0.07 ·
· 0.80	0.06 ·
· 0.60	0.05 ·
· 0.50	0.04 ·
· 0.40	0.03 ·
· 0.30	0.02

Fig. 34-5. A series of circles with areas in mm^2.

[Standard T 537 is similar, but merely counts dirt particles larger than a set threshold such as 0.02 mm^2 eba (explained below). It is said to be designed for OCR purposes of paper [OCR is used to determine dirt, only that dirt is determined for an (unspecified) OCR application.]

Dirt estimation is made with the TAPPI dirt estimation chart. The chart consists of a series of 23 pairs of black circles and rectangles (each rectangle has a length 20 times the width) comprising areas from 0.02 mm^2 to 5 mm^2. The rectangles are not used but are included as a leftover of Clark et al. (1933). The *equivalent black area* (eba) of a dirt particle is defined as the area of the dot from the dirt estimation chart that most closely resembles the visual appearance (not the actual size) of the dirt particle. A dark dirt particle on a brown piece of paper would give a smaller eba than the same particle on white paper. The limiting eba (minimum size to be counted) of pulp is 0.08 mm^2, and in paper is 0.04 mm^2. Dirt is usually reported in mm^2/m^2, which is ppm. The dirt estimation chart is not exact, and correction values must be used for many of the spots. Corrections are included in Standard T 213 only. Fig. 34-4 shows a series of circles of various approximate areas. A group of these are included at the end of the book for practice use.

ANSWERS TO SELECTED PROBLEMS

CHAPTER 1

13. The *Abstract Bulletin of IPST* for articles on dioxins in pulp mill effluents. The *Chemical Abstracts* (or medical abstracts) for the long—term toxicity of dioxin since this is independent of the source of dioxin.

CHAPTER 2

1. False.

2. Circle—brown, softwood, hardwood, soft-wood, mechanical.

3. a) Loss of extractives   b) Wood decay

4. 1. Chip size distribution
   2. Chip rot content
   3. Extractives content
   4. Age (from pith) of wood in chip
   5. Wood species
   6. Wood density
   7. Moisture content

5. Fines—overcook to give a low yield and they tend to plug the digester's liquor circulation system. Fines tend to contain high percent-ages of dirt and grit.

   Oversize—undercook, give a large amount of screen rejects and shives; also, larger amounts of bleaching chemicals will be used.

6. Typically 20-30% depending on the process, chip price, etc. Sometimes as high as 50%.

7. Cellulose, hemicellulose, and lignin.

8. Circle cellulose then hemicelluloses.

9. Lignin.

13. Briefly, dissolving pulp (relatively pure cellu-lose) is swelled in alkali and then treated with monochloroacetic acid. The cellulosate ion ($R-O^-$) is a nucleophile and attacks the pri-mary chloride of monochloroacetic acid. The result is:

$$Cell-O^- + Cl-CH_2-COO^- \rightarrow$$

$$Cell-O-CH_2COO^- \ Na^+ + Cl^-$$

14. The fibril angle of the S—2 cell wall layer.

15. They adsorb and desorb water from the air around them depending upon the temperature and relative humidity.

16. Hydrogen bonding.

18. The maximum level of recycling paper on a <u>sustained</u> basis is considered to be 50—60%. At rates higher than this a significant portion of fibers have been through one, two, or several more uses and have lost most of their strength.

19. $20—40/ton. $400, or more than half the cost of virgin pulp.

CHAPTER 3

1. Mechanical separation by softening and fatigue of the compound middle lamella or chemical dissolution of the lignin that binds the fibers together.

3. Chemical pulp has higher tensile strength and sheet density; mechanical pulp has higher yield, unbleached brightness, and opacity.

4. The lignin softens to the point that fiber sepa-ration occurs at the middle lamella (instead of at the primary cell wall) and results in fibers that are "coated" with lignin. Since hydro-gen bonding does not occur with lignin, strength is greatly reduced in the final sheet. The higher temperature darkens the pulp.

5. TMP, or especially CTMP, is much stronger than SGW since fiber liberation is much more specific in the TMP process, leading to longer fibers. This increase in strength means lower amounts of chemical pulp are required.

7. Batch and continuous digesters.

8. The kappa number is a measure of residual lignin; consequently, this is a measure of lignin content as a function of yield. The lower the lignin content at a given yield, the more selective (good), the pulping process.

Other factors such as cellulose viscosity, pulp strength, and bleachability must ultimately be determined to make a final judgment regarding the efficacy of a particular pulping process.

9. No. If the resultant pulp has a low cellulose viscosity and strength, it will be unsuitable for making paper.

10. Anthraquinone is a pulping additive used in some kraft and alkaline sulfite pulping methods. It undergoes a cyclic redox process. Lignin is reduced while cellulose and hemicelluloses are oxidized. The oxidation of the carbohydrates is of particular interest since the reducing end groups (the free anomeric carbon atoms) are oxidized from aldehyde to carboxylic acids that prevents the alkaline peeling reaction. (This is the same end from which the peeling reaction occurs.)

11. NaOH and Na_2S.

12. The pulping time and temperature are combined into the H-factor. Only those two variables are considered in the H-factor; other variables such as sulfidity and active alkali will also influence the degree of cook.

13. The spent liquor cannot be recovered for reuse in the pulping process. Since there is only a limited market for the spent liquor for drilling muds, dispersants, and other uses, the spent liquor is a large disposal problem for large—scale pulp production based on this method.

14. Free (sulfurous acid, H_2SO_3) and combined (sulfite ion, SO_3^{2-}) SO_2.

CHAPTER 4

1. The kraft recovery process allows 1) recovery of inorganic chemicals for reuse, 2) recovery of the intrinsic energy of the waste organic compounds to supply energy for mill processes, 3) the disposal of the organic materials, which would otherwise be an environmental problem.

2. a) $2NaOH + CO_2 \rightarrow Na_2CO_3 + H_2O$

   b) $NaCO_3 + Ca(OH)_2 \rightleftarrows 2NaOH + CaCO_3$

c) $CaCO_3 \xrightarrow{1000°C} CaO + CO_2$

d) $H_2O\ (l) \xrightarrow{heat} H_2O\ (g)$

3. Turpentine is recovered from the digester relief gases during digester heating. Tall oil is collected by skimming the surface of the partially concentrated black liquor.

4. Spent NSSC cooking liquor from the semi-chemical pulping process is added to the black liquor of the kraft pulping cycle to the extent it can be used as make—up chemical.

5. High dilution will result in less sodium loss through the pulp, increase the evaporation costs as more water must be removed, and decrease the bleaching costs as more of the soluble, residual lignin is removed.

6. One extra ton of water per ton of pulp is produced, and each ton of water requires 4.24^{-1} ton of steam, 472 additional pounds of steam will be required per ton of pulp.

$$\frac{2000\ \text{lb water}}{1\ \text{ton pulp}} \times \frac{1\ \text{lb steam}}{4.24\ \text{lb water}} = 472\ \text{lb steam/ton}$$

7. Long tube vertical evaporators and falling film evaporators.

8. Concentration of black liquor by direct contact with the hot flue gases of the recovery boiler strips significant quantities of sulfur compounds from the black liquor, causing extremely odorous emissions, wasting sulfur, and contributing to acid rain. Preoxidation of black liquor prior to combustion decreases the problem but decreases the fuel value of the black liquor. Indirect evaporators (concentrators) are now used, as they decrease sulfur emissions without the need for black liquor preoxidation.

9. As the solid content increases beyond 65% the viscosity increases exponentially, making the liquor difficult to pump. On the other hand, removal of water prior to combustion of the black liquor increases the theoretical amount of recoverable energy.

10. Reduction occurs in the bottom of the recovery boiler and these conditions alow sulfur chemicals to be recovered in their reduced form (Na_2S). Oxidation occurs in the top of the recovery boiler and limits CO emissions.

11. The ESP is used to capture particulate matter, especially Na_2SO_4 and soot particles, for reuse in the recovery boiler thereby decreasing pollution.

12. The sodium sulfide would be partly converted to sodium sulfite, sulfate, and similar compounds. This would precipitate as calcium sulfite or sulfate during causticizing.

13. The lime travels from the top of the kiln to the bottom. Unless the lime mud is dried separately, the lime mud first dries, is then heated to temperature, and then undergoes calcining as it travels through the kiln.

14. The dregs accumulate in the lime mud, thereby decreasing the lime availability and causing problems in lime mud settling.

15. It is a measure of the extent of conversion of sodium carbonate (inactive chemical) to sodium hydroxide (active pulping chemical) in the causticizing process.

CHAPTER 5

1. Mechanical pulps are high in lignin content. During their bleaching the lignin is preserved, but the color-bearing groups (chromophores) are altered to decrease their light absorbance. Bleaching of chemical pulps involves lignin removal as the primary goal—no lignin, no lignin color.

2. Color reversion is a consideration in mechanical pulps since large amounts of lignin are still present.

3. The lignin content can be measured by permanganate ion titration (kappa number or k number) or by light absorption in the UV or IR range. The latter may be converted to an equivalent kappa or k number as if the pulp were titrated with permanganate. This is done since titration with permanganate is still the most common method.

4. $30 \times 0.147 = 4.4\%$ lignin (approximately).

5. Cellulose viscosity is the viscosity of cellulose in solution. It is an indicator of the degree of polymerization of the cellulose and the strength of chemical pulps. Overly harsh pulping and bleaching treatments decrease the DP of cellulose and pulp strength.

6. Treatment of pulps with chlorine does not remove lignin since this is necessarily done under acidic conditions where the phenolic groups are not in the salt form that contributes to water solubility. Alkali extraction produces more phenolic groups and converts the phenolic groups to phenolate groups that are water soluble. In a nutshell, chlorine "predisposes" the lignin to removal by alkali extraction. In fact, about 75% of the lignin present before chlorination is removed by the subsequent alkali extraction.

If additional bleaching chemicals were used prior to alkali extraction, most of them would be consumed by the lignin that would otherwise have been removed by the inexpensive extraction stage.

7. Chlorine (followed by alkali extraction) is not overly specific for lignin removal. This treatment would not be done in the later stages of bleaching where the lignin content is very low.

8. If the pH is too low the cellulose degrades by acid hydrolysis. If the pH is too high, the cellulose degrades by oxidation. Overchlorination results in cellulose degradation and the production of large amounts of chlorinated organic compounds including dioxins.

10. Bleaching in several stages allows different parts of the lignin molecule to be attacked. It is more efficient in that lower amounts of chemical are used overall.

11. If the pH drops to low then hypochlorite is in equilibrium with hypochlorous acid which degrades the pulp by oxidation.

12. The high NaCl content increase the dead load and corrosion in the recovery boiler.

13. Effluents from the later stages, such as a D stage, can be used to wash pulp at the earlier stages. Chlorination water can be reused within the chlorination stage since this stage does not remove large quantities of lignin (most of it is removed in the subsequent alkali extraction stage). Oxygen delignification effluent can be sent to the brown stock washers.

14. By allowing the oxygen bleaching effluent to be used in the brown stock washers, lignin, alkali, and water are reused rather than being sent with the mill effluent discharge.

CHAPTER 6

2. Refining increases the surface area of fibers to improve fiber—to—fiber bonding. Refining causes lamination of the fibers to improve fiber flexibility (and therefore fiber—to—fiber bonding). Refining also hydrates the fibers to make them more flexible (e.g., compare dry spaghetti to wet spaghetti).

3. Generally, refining at high consistency increases fiber—to—fiber brushing. This means less fiber cutting due to refiner bar—fiber interactions.

4. Laboratory refining does not duplicate the conditions of commercial refining; therefore, it is not a perfect comparison of how fibers will actually behave in the commercial process.

5. The level of refining is usually determined by the freeness of the pulp. Other methods such as the size distribution of the fibers can be used as well. Handsheet testing is a good indicator of the effectiveness of refining.

7. At elevated pH the carboxylic acids are no longer in the acid form, but in the salt form. In this form refining is more effective because the fibers are hydrated more easily. The paper strength properties are improved because the fibers are more flexible. This is especially true when the salt form is with sodium ions. (Hard water in a mill would hinder the process at pH above 7 or so.)

8. Fiber size distributions can be determined with a series of screens, by optical scanning techniques, or by projection.

9. Fiber size distributions indicate what is happening during mechanical pulping or refining. Low freeness values are usually caused by fines generation, but also by fibrillation.

10. The formation and testing of laboratory handsheets indicate much about the quality of the fiber for papermaking. They provide the papermaker with useful information for troubleshooting if problems develop.

CHAPTER 7

2. The moisture content of paper (and its strength properties) under standard conditions depends upon whether the paper loses or gains water. Therefore, TAPPI standards state that paper should be placed in a hot, dry room before conditioning to standard conditions.

3. The paper may be over dried or under dried relative to the moisture content it will achieve under standard atmospheric conditions. It may take well over an hour for paper to achieve its equilibrium moisture content. Also, see question 2.

4. The fold test is extremely sensitive to the moisture content of the paper. For example, consider how the moisture content of spaghetti influences how easily it is bent without breaking. The analogy is fair since both spaghetti and cellulose fibers have polysaccharides of similar structures.

7. Fiber alignment is usually credited with this phenomenon, but the fact that paper is dried under tension in the machine direction, but not in the cross machine direction, accounts for much of this effect. Individual fibers dried under tension are stronger and stiffer than fibers not dried under tension.

8. In this case, the basis weight is an average over an area of 144 in.^{2} In the case of the fold endurance test, an area of about 15 mm by 1-2 mm is tested, or up to 30 mm^2 is

tested. Small imperfections in formation would not be noticed in the first case, but could be most of the test area in the second case. (This is related to the concept of statistical sample size where variability is inversely proportional to the square root of sample numbers.)

9. Scanning electron microscopy with X—ray analysis and atomic absorption spectroscopy.

10. Paper brightness is the relative amount of reflected light at 457 nm.

11. The two commonly used tests are the G.E. brightness and Elrepho brightness, which differ in the geometry of the light source and light detection.

12. All of these, except increased refining, will increase the opacity.

CHAPTER 8

1. Pressure screens and vortex cleaners.

2. Control additives are not designed to become an integral part of the paper. They are used to improve the papermaking process. Functional additives are used to become an integral part of the paper and influence the properties of the paper itself.

3. PCC generally costs less than wood fiber and can save money. PCC also improves the opacity and brightness of the resulting paper.

4. Titanium dioxide is used since it has a high index of refraction and brightness.

5. 1. Allows use of $CaCO_3$ as a filler, which would otherwise decompose at pH below 7 by the reaction:
$$CaCO_3^{2-} + 2\,H^+ \rightarrow CO_2 \uparrow + H_2O$$

2. Less corrosion of the paper machine.

3. Increased longevity of paper due to decreased acid-catalyzed hydrolysis of cellulose and hemicellulose.

4. Paper has higher strength since the fibers are more flexible at higher pH.

5. Allows the use of secondary fiber containing $CaCO_3$ filler.

6. Clay fillers.

7. Since strength is the most important feature of this paper, fillers are not used.

8. 1. The viscosity and polarity (surface tension) of the liquid. In the case of rosin sizing, the coordinating potential of any ligand in the liquid is important.

2. Base sheet properties (bulky, porous sheets pull in more water by capillary action than dense sheets).

3. The contact angle of the liquid on the surface. (If the angle is less than 90°, we say it is unsized.) This is a function of (2) and the polarity of the surface, such as whether or not hydrophobic groups (from internal sizing) are on the surface of the paper.

9. One does not normally size these papers since they are designed to be absorbent.

10. Both of these are similar in that they have a hydrophobic (water-hating or nonpolar) portion which resists water and a functional group that ultimately bonds with the fiber surface. The main difference is the mechanism of bonding. With rosin, aluminum ions complex with rosin and the fiber surface to achieve bonding, whereas with the synthetic size agents, covalent ester linkages are formed. The latter group, while expensive, are much more effective in that much lower quantities are required.

11. Rosin size is most effectively used under acidic to neutral conditions. PCC has a pH of about 8, which makes it difficult to use with rosin sizing.

13. Starch; the bond between glucose units is α instead of the β linkage occurring in cellulose. The mechanism of adhesion is hydrogen bonding. In this regard, starch is similar to hemicellulose in increasing the strength of paper.

CHAPTER 9

1. An object in motion tends to remain in motion in the same direction. When the web

goes through a curve the water tends to go straight (an example of centrifugal force), away from the web. An S—shaped curve allows this effect to occur on both sides of the web consecutively.

2. The large fibers make an excellent filter mat next to the screen; this allows smaller particles to be trapped that might otherwise escape through the screen. The sweetener stock is recovered along with the fines.

3. Small—scale turbulence (microturbulence) is desired and necessary to keep fibers suspended in the pulp slurry and well dispersed. Large—scale turbulence (macroturbulence and splashing) is not desired since this will lead to large amounts of fiber in some areas and little fibers in other areas. This would have a disruptive effect on paper formation.

4. Paper leaves the press section at higher consistencies using less energy in the dryer section. The higher consolidation of the web gives better fiber-fiber bonding leading to a stronger paper.

5. To prevent rewetting—the tendency of the water from the felt to rewet the web when the pressure is released.

6. Pressure is applied at the ends of press rolls. The press rolls are made slightly larger in diameter in the middle (perhaps 0.05 in.) so that the pressure is evenly dispersed across the width of the paper machine.

7. No, rather they tend to get plugged with fillers, pitch, and other debris. Calcium and aluminum ions may help precipitate materials, so felt washing solutions often have a chelating agent added to them.

8. Pitch from wood, polymers from secondary fiber, and additives used in the mill such as defoamers and sizing agents. Again aluminum and calcium ions may help precipitate these materials. Also, the felts collapse and the void volume decreases.

9. This increases pressing efficiency. Heating water lowers its viscosity and increases its rate of removal while pressing.

10. Tissue paper is too weak to withstand the numerous felt transfers of conventional dryers.

11. Tissue paper, see the answer immediately above, and writing papers with a machine glaze (the surface of the paper in contact with the dryer gets a smooth sheen on it).

12. The earlier dryer sections are relatively cool to prevent a large amount of steaming that would cause blisters in the paper, while the later sections are much hotter to drive off the chemisorbed water.

13. A coarse dryer felt will allow faster water removal, but a fine dryer felt leaves fewer impressions on the sheet.

14. Typically it takes 1.2-1.6 lb of steam to remove 1 lb of water from the paper in the dryer section. In multiple effect evaporators with a steam economy of 5, 0.2 lb of steam removes 1 lb of water from the black liquor.

15. Pigments and binders.

16. The refining stage offers the most control over the final caliper of paper. Glassine, a very dense paper, is made with much refining. Calendering is used to control the caliper across the sheet for uniform winding. Pressing has an appreciable influence on the final sheet caliper. Refining (most important), pressing (moderately important), and calendaring (relatively unimportant) all increase the density of paper.

17. Small particles such as fines and fillers are washed out of the wire side, hence the ash content (fillers contribute to ash since they are inert and not combustible) is higher on the felt side. This contributes to two—sidedness of paper.

18. Two—sided paper tends to curl since one side might absorb water slower than the other side or water might have more of an effect on one side than on the other.

A thermostat coil is a bimetallic strip of metal. One side of the strip is one type of metal and the other side of the strip is a metal with a much different coefficient of thermal expansion; therefore, a change in temperature causes a curling effect. (With paper, a change in moisture content is more important than a change in temperature.)

19. Strength is desired for obvious reasons, such as to prevent breaks in the web in the printing press. Bulk is desired because it contributes to opacity. (The more air in the sheet, the more fiber-air interfaces that scatter light.) Unfortunately, the bulkier the sheet the lower is its strength (that is, there is an inverse relationship). As in life, most things in pulp and paper are a compromise.

20. Wet strength adhesives often form covalent linkages. Most papers, however, are held together by hydrogen bonding without the use of added adhesives. Starch is often added to paper with a concomitant increase in the dry strength due to hydrogen bonding. Hemi-celluloses have a similar effect.

21. A corrugated box is like an I—beam. The center of the beam only separates the top and bottom, where the forces are highest. Consider a 2 by 4 piece of wood laying flat supported on each end by a cement block. When one jumps in the middle, it flexes quite a bit. When put on the narrow side, flexing decreases substantially because this beam is now "thicker".

22. Starch is a hydrophilic molecule; its mechanism for imparting sizing to paper is much different from that of rosin or the alkaline sizing agents. Starch acts to "plug" the capillaries on the paper surface, thereby decreasing the rate of water penetration.

CHAPTER 10

3. A high efficiency can be achieved with high fiber reject rates, but much useful fiber is lost.

4. Flotation and washing. Some processes use both methods.

5. Sodium hypochlorite would tend to react with all of the lignin (of the large amount present) and not just the chromophores.

6. Originally these were centrifugal cleaners "operated in reverse" that removed light-weight materials through the top, where fibers usually exit. Both the contaminants and the fibers leave through the bottom in through—flow cleaner. Through—flow cleaners are used in secondary fiber plants to remove plastics, styrofoam, and other materials less dense than wood fibers

CHAPTER 11

1. Material that is suspended (not dissolved) in the water.

2. BOD, color, suspended solids, and other materials such as acids or bases.

3. Color.

4. Both oxygen and carbon dioxide have lower solubilities in water at high temperatures. The limited solubility of oxygen in the warm summer months (along with lower water flow rates in rivers) often means that the allowable BOD discharge rate is lower.

5. Lower oxygen solubility in the water and decreased flow rates.

6. This is not easily removed and requires the tertiary water treatments. It is seldom practiced commercially.

7. The biological oxygen demand test must be carried out with microorganisms in order to simulate the fate of organic materials in rivers or lakes. Microorganisms take several days to metabolize the available food, especially mill effluents, which contain many materials that are difficult to metabolize.

8. One gallon is about 3.78 kg (or liters) so 75.6×10^6 kg effluent is produced daily and 27.6×10^9 kg is produced annually. This is equal to 27.6×10^{12} g annually or 27.6×10^{15} mg effluent annually. Since 10 parts dioxin per quadrillion part effluent are produced, less than 300 mg of dioxin is produced each year.

9. H_2S and CH_3SH can both be removed by alkali scrubbing as they have an ionizable

proton, that is, they are weak acids. By reaction with a metal hydroxide, a salt of low volatility is produced.

$$RSH + NaOH \rightleftarrows (RS^- + Na^+) + H_2O$$

10. The reactions of sulfides with the methoxy groups of lignin. It is an undesirable product since it is highly odorous and not easily trapped.

11. Nitrogen oxides form during any high—temperature combustion process where nitrogen from air can react with excess oxygen. Excess oxygen must be present or carbon monoxide emissions would be very high.

12. Ambient sampling is done in air in the vicinity of a mill. Point sampling is done directly at the source (such as the smokestack) of gaseous effluents.

13. Some pollutants are formed under reduction conditions (CO and TRS) while others are formed under oxidation conditions (SO_2 and NO_x), so there is always a tradeoff between levels of these pollutants.

CHAPTER 12

4. The following shows that a ream which weighs 52.3 pounds means the paper has a basis weight of 85 g/m^2.

$$\frac{1 \text{ lb}}{\text{ream} \times 24 \text{ in.} \times 36 \text{ in.}} \times \frac{1 \text{ ream}}{500 \text{ sheets}}$$

$$\times \frac{454 \text{ g}}{1 \text{ lb}} \times \frac{(39.37 \text{ in.})^2}{(1 \text{ m})^2} = 1.62 \text{ g/m}^2$$

5. This cannot be done since the units are incompatible.

CHAPTER 13

2. d) All of the above.

3. c) Na_2S.

4. a, c, f, g, i, j.

6. c, since hydrogen bonding occurs.

7. a) 0.546 mole.

   b) 0.0537 mole.

   c) 2.69 mole.

8. d) 50%.

9. 30.1 g CO_2.

10. 0.74 M.

11. 35 g of NaOH; 437.5 mL as (875 mL of 0.5 N H_2SO_4).

12. a) 1.33 is the pH.

    b) 1.70 is the pH since there are two ionizable protons in sulfuric acid.

    c) 11.1 is the pH.

15. Cr is +6.

16. Na loses an electron, is itself oxidized, and is the reducing agent since it causes water to be reduced.

17. One lb of NaOH is 11.3 moles of Na. Since each sodium atom requires one electron, this is 96,500 amp seconds (the number of electrons in a mole) times 11.3 or 303 amp hours. At 4 volts this is 1214 watt hours or 1.214 kWh, but this is the ideal minimum.

18. Corrosion occurs at a small point made the anode with a high surface area cathode, so the reaction is intense at a small point.

19. If the magnesium strip is used up, then corrosion can occur to the water heater itself. The high temperature and hot water can make corrosion very quick. This is a perfect example of "an ounce of prevention being worth a pound of cure."

21. C_6H_{12}.

CHAPTER 14

1. The normality of the diluted NaOH = (21.37 mL $\times$ 0.1027 N)/20.0 mL = 0.1097 N. The normality of the concentrated NaOH = 0.1097 N $\times$ (1000/25) = 4.39 N NaOH. The concentration of the NaOH = 4.39 N NaOH $\times$ 40 g/eq = 175.6 g/L.

2. The amount of water present during the titra-

tion is not critical as long as there is not undue dilution; it does not enter into the calculations above; however, it is important to know the precise dilution factor as this does enter into the calculations.

3. Use of Eq. 14-2 gives with pK_a of 4.76 for acetic acid gives:

    $pH = 4.76 + \log (0.2/0.5) = 4.36$

4. a, c, d.

5. No, not without large dilution factors.

6. Just like pH indicators such as phenolphthalein, the bleach effluent contains compounds whose colors are very pH dependent. Measuring the color at a low pH would underestimate its true color in the environment.

8. The following reaction shows the liberation of acid that could cause acid hydrolysis of cellulose and hemicellulose over a long period of time.

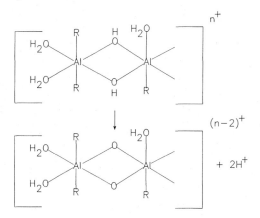

CHAPTER 15

1. Lumber: $150,000 \times (1.5 \times 3.5)/(2.0 \times 4.0) = 98,437$ true board feet
    98,437 true board feet $\times (1 \text{ ft}^3)/(12 \text{ bd ft}) = 8203 \text{ ft}^3$ lumber

    8203 ft^3 lumber $\times (62.4 \text{ lb/ft}^3 \times 0.44) = 225,225$ lb oven—dry wood as lumber

    225,225 lb lumber $\times (2000 \text{ lb/ton}) = 112.6$ tons oven—dry wood; 112.6 tons dry lumber $\times (1.20 + 1.0) = 247.7$ tons wet lumber

Chips: 112.6 tons dry wood $\times$ (33.3 chips/50 lumber) = 75.1 tons dry chips (62.5 BDU)

$$75.1 \text{ ton} \times \frac{2000 \text{ lb}}{1 \text{ ton}} \times \frac{\text{ft}^3}{10 \text{ lb chips}} \times \frac{1 \text{ RR car}}{18 \times 200 \text{ ft}^3} = 4.17$$

75.1 tons dry chips $\times$ (1.2 + 1.0) = 165.2 tons wet chips

Weight/car: 165.2 wet tons/4.17 cars = 39.6 tons

Sawdust: 112.6 tons dry $\times$ (16.6/50) = 37.53 tons dry; (31.3 BDU or 82.57 wet tons)

Rail cars: 2.61 RR cars; 39.6 tons/car

Revenue: 62.5 BDU $\times$ \$160/BDU + 31.3 BDU $\times$ \$60/BDU = \$11,878

2. 1 m^3 of water weighs 1000 kg (1 metric ton) and 1 ft^3 of water weighs 62.4 lb, oven—dry weight: 1000 kg/m$^3 \times 0.42 = 420$ kg/m^3; 62.4 lb/ft$^3 \times 0.42 = 26.2$ lb/ft^3

    green weight: 420 wood $\times$ (100 kg green/55 kg dry) = 763 kg/m^3; 26.2 wood $\times$ (100 lb green/55 lb dry) = 47.6 lb/ft^3

    weight of water: 763 green — 420 oven—dry = 343 kg water/m^3; 53.7 green — 26.2 oven—dry = 27.5 lb water/ft^3

3. Weight of dry wood/ft^3: $62.4 \times 0.45 = 28.1$ lb/ft^3

    Green weight: 28.1 lb/ft$^3 \times (1/(1 - 0.45)) = 51.1$ lb/ft^3

    Oven—dry weight/m^3: $1000 \times 0.45 = 450$ kg/m^3

    Green weight: 450 kg/m$^3 \times (1/(1 - 0.45)) = 818$ kg/m^3

    Vol. of chips: (28.1 lb/ft^3 wood)/(10 lb/ft^3) = 2.81 ft^3 chips/ft^3 wood; and the same applies to give: 2.81 m^3 chips/m^3 wood

5. Use of Eq. 15-8 shows the breaking length is 2.48 km, which is weaker than most types of paper.

$$L = \frac{3000 \text{ psi}}{0.85 \times 1422 \text{ psi}\cdot\text{km}^{-1}} = 2.48 \text{ km}$$

CHAPTER 16

3a. liquor sulfidity = 37.5% = Na_2S/AA; since ½Na_2S = (AA — EA), 37.5% = 2 × (AA — EA)/AA × 100%; AA = 115.3 g/L as Na_2O.

NaOH = (2 × EA) — AA = 72.1 g/L.

Na_2CO_3 = 20.45 g/L as Na_2O; TA = 135.8 g/L.

1.017 N HCl = 31.53 mg Na_2O/mL titrant. 10 mL liquor = 10 mL × 93.7 mg/mL EA = 937 mg EA/aliquot; therefore, Titration "A" is 937 mg/(31.53 mg Na_2O/mL titrant) = 29.72 mL titrant; B = 36.57 mL; C = 43.07 mL.

3b. Mol/L of NaOH = (2A — B) mL × 1.017 N/10 mL = 2.32 M.

Na_2S = NaHS = (B — A) mL × 1.017 N/10 mL = 0.697 M. Na_2CO_3 = 0.66 M.

3c. There is 100 g dry wood. EA = 93.7 g/L × 0.2 L/(100 g wood) × 100% = 18.7% EA on wood.

4. 600/hr × 1.43 hr = 858

858/(910/hr) = 0.94 hr

5. It is a molar comparison.

6. Combined is 3%. Sulfurous acid is 1%, bisulfite is 6%, and sulfite is 0%. (Appreciable amounts of both sulfurous acid and sulfite ion cannot exist in solution simultaneously.)

7. By difference, the free SO_2 is 6%. To get the concentration of each of the species we realize by the strong equilibrium shift to the right in the equation below that both sulfurous acid and sulfite ion do not exist in solution together (in any significant concentration). Therefore we have to "remove" at least one of the species. Thus 2% sulfite ion (combined form, leaving 0% sulfite ion) and 2% sulfurous acid (free form, leaving 4% sulfurous acid) react to give 4% bisulfite ion. According to Fig. 16-8, the pH will be about 2 since we have a ratio of 50% of the sulfurous acid species and 50% of the bisulfite species.

$$SO_3{}^{2-} + H_2SO_3 \rightarrow 2 \; HSO_3{}^{-}$$
$$2\% \; + \; 2\% \; \rightarrow \; 4\%$$

9. Iodometric titration gives free SO_2 (10%), and titration with NaOH gives free SO_2 (6.5%), leaving 3.5% SO_2 as combined. 3.5% free reacts with 3.5% combined to give 7% bisulfite and 3% sulfurous acid as the actual chemical present on a SO_2 basis.

CHAPTER 17

1. ClO_2 has five electrons transferred per Cl atom, while Cl_2 has one electron transferred per Cl atom. Therefore, ClO_2 gives much more bleaching power for the chlorine content. Furthermore, ClO_2 reacts principally by oxidation with most of the chlorine ending up as harmless, inorganic Cl^-. On the other hand, Cl_2 reacts principally by substitution of chlorine for hydrogen in the organic compounds. The resultant chlorinated organic compounds increase the toxicity of effluents. (Chlorinated organic chemicals include DDT, dioxins, PCBs, herbicides such as 2,4-D and 2,4,5-T.)

2. $NaOCl + H_2O + 2I^- \rightarrow I_2 + NaCl + 2OH^-$

3. $[ClO_2]$ = 0.0606 N by Eq. 14-1. Table 17-2 shows this is 0.819 g/L.

5. 100% — (0.16% lignin/kappa × 34 kappa + 2%) = 92.6%. If the original pulp yield is 46%, the overall yield of bleached pulp is 0.926 × 46% = 42.6%.

6. Five gallons is about 24 L. (When using large portions of water like this, it is more accurate and faster to weigh it rather than try to measure its volume. Also, start with 23.5 L so that it can be diluted to the exact volume that corresponds to 0.1000 N.) The equivalent weight of potassium permanganate

is one—fifth the formula weight or 158/5 = 31.6. Weigh 31.6 g/eq × 24 L × 0.1 eq/L = 75.84 g of $KMnO_4$ and add it to the water. (Freshly distilled or boiled water should be used.) Add 100 mL of water, 25 mL of 2 M H_2SO_4, and 5 mL of 2 M KI (using graduated cylinders) to a 250 mL Erlenmeyer flask. Pipet 25.00 mL of the $KMnO_4$ solution into the flask. Titrate the liberated iodine with thiosulfate until the blue color of the starch indicator disappears. Repeat the titration. Calculate the concentration of $KMnO_4$ and dilute to give 0.1000 N. Check the concentration with additional titrations.

8. The height of water is 20.4 meters; volumetric flow = 2.0 m^3/sec.

   The slice height is 0.01 m = 1 cm. The production is: 20 m/sec × 10 m × 50 g/m^2 × 1 kg/1000 g = 10 kg/sec.

   10 kg/sec × 60 sec/min × 60 min/hr × 24 hr/day × 365 day/year × 1 MT/1000 kg = 315,360 MT/year = 347,000 tons/year.

CHAPTER 18

1. There is a series of small void spaces or tunnels that small compounds can enter but large ones cannot. The apparent volume for small molecules is much higher than that for large molecules; hence, for a series of similar molecules, the small molecules are eluted (come off the column) last—a characteristic unique to gel permeation chromatography.

2. $\overline{DP}_n = 300$; $\overline{DP}_w = 333.3$. Since the styrene monomer has a mass of 104, the number average molecular weight is 104 × 300 = 31,200.

5. Anionic polymerization.

CHAPTER 19

1. Sodium borohydride reduces the end group of cellulose (or hemicellulose) that contains the free anomeric carbon (that is, the free aldehyde group, the reducing end). This is the same end from which the peeling reaction occurs. The reducing end is converted from an aldehyde to an alcohol which does not undergo peeling in alkaline solution.

5. The cellulose viscosity of mechanical pulps is generally very high and not a limiting factor in the strength properties of these pulps. There has not been any severe treatment during the processing of these pulps that would lower the degree of polymerization of the cellulose.

CHAPTER 20

4. 95.45% lie within ±2 standard deviations. 99.73% lie within ±3 standard deviations. 99.994% lie within ±4 standard deviations.

CHAPTER 21

1. Soap forms micelles which have non polar centers that can dissolve oil or grease so that an emulsion is formed.

4. No.

7. Blue light is scattered to a much higher degree than other colors. We see the scattered blue light in the sky.

CHAPTER 22

1. CTMP pulp produced with sulfite is probably sulfonated; these groups will affect the surface charge of the pulp.

3. Quaternary ammonium groups.

4. They have a higher surface area.

5. One of the polymers is of high molecular weight and, therefore, shear sensitive.

CHAPTER 23

2. Offset printing uses an intermediate blanket to transfer the ink from the plate to the paper. It refers to lithographic printing.

3. Halftone printing allows shades of gray to be reproduced using a grid of dots so that any point on the paper is either black or white. It is used for reproduction of photographs.

4. Stochastic printing uses a computer—generated "random" dot placement. It gives high-

er resolution and avoids moiré patterns. The latter advantage is especially important in multicolor printing.

5. Four—color printing uses four colors of ink, which allows reproduction of a wide variety of hues at various levels of saturation. It uses subtractive mixing of color inks, where each ink (ideally) selectively removes one of the additive primaries. In other words, each ink transmits (reflects) two of the additive primaries and absorbs one of the additive primaries. The additive primaries are the three colors that respectively selectively stimulate the three types of cones in the human eye.

6. Proper color reproduction is possible only if each of the four inks is printing at the proper intensity (as checked by a densitometer).

7. Gravure and letterpress put much less stress on the coating adhesive than offset, so paper made for gravure will probably give picking problems when used for offset.

CHAPTER 24

1. Optical brighteners absorb light in the ultraviolet range and give off light in the visible range. The effect is identical to that used in black light paints to produce posters and, in a sense, makes the object "brighter than bright," by giving off more visible light than shines on them, if the light source has significant amounts of UV light.

2. Yellowish paper absorbs more blue light than yellow light and does not appear to the human eye to be bright. By adding blue dye the paper, the amount of yellow light absorbed is increased to "balance" the light coming off. While such paper actually reflects less light, it appears to be brighter.

3. The way the eye and brain perceive the color of an object depends very much on the source of light. This is exemplified by observing objects under sodium-vapor lights such as yellow street lights, or, less directly, by taking pictures with the same type of film with no filters, using different sources of light such as incandescent, fluorescent, or sunlight.

4. Red, orange, and perhaps some yellow depending on the source of the color; blue alone if a high—quality filter is used or if it light from a laser; black; green, since this color is not absorbed efficiently, but largely reflected, there is no way for the plant to trap its energy.

6. Blackbody radiation at higher temperature tends to flatten the spectral distribution of the light source and is a more efficient light source.

7. The new paint had a slightly higher gloss. This is one problem with "computer paint matches," but I feel the match will improve as the new paint weathers and loses some of its gloss.

8. Opacity is improved since light is scattered to a higher degree.

9. Light scattering decreases with particle size below 50% of the wavelength of the light involved. Is this true of light absorption? Try to think of an example in order to come to a conclusion.

10. Using Eq. 24-8 or a table gives k/s as 0.0100 and 0.00132 for 0.87 and 0.95, respectively. Therefore, k is 0.0005 and 0.000066 for 0.87 and 0.95 absolute brightness, respectively, a very large change in k!

11. Since sW is 2.0 and R_∞ is 0.87, the Tappi opacity, $C_{0.89}$, is read as 0.74 from Fig. 24-16.

12. Refining leads to a sheet of higher density (better fiber—fiber conformation) so that there is less air—fiber interface to scatter light. This effect is more important than the relatively minor increase in surface area of the pulp.

CHAPTER 25

2. No.

4. Hardwoods.

AUTHOR INDEX[1]

[1]Most proper names are included here, including equipment and process names.

ABC titration of white liquor, **363**, *366*

Abietic acid, *38*, 198

Absorbance vs transmittance, *342*

Abstract indexes

Abstract Bulletin, Institute of Paper Science Technology, **6**

Chemical Abstract Index, **6**

Accept fraction, 82

Accumulator, blow heat, 79

Acid hydrolysis of carbohydrates, **405**, *406*

Acid ionization constant (K_a), *319*, **320**

Acid sulfite pulping, **95**

Acrylic polymers, *398*

Activated sludge process, **288**, *289*

Active alkali, kraft liquor, **89**, 359

Activity coefficient of ions, **320**, 323

Additives, for papermaking, **9**, **193-208**

alum chemistry, **205, 206**, *325*, **344-346**

antifoamer, **203, 435**

biocide, *204*, **205**

brightener, fluorescent, 195, **196**

chelating agents, **203, 204**

control additive, **201-205**

defoamers, **203, 435**

deposit control agents, **203, 204**

diatomaceous earth, pitch control, 195

drainage aid, **202, 447**, 449

dry strength additives, **200, 452**

polyacrylamide, 200, 441, **452**

dyes, **195-197**, *197*, 458

fillers, *194*, **195, 206**

calcium carbonate, **195**, 197, 199, 207

clay (kaolin), **195**, 422, **439-440**

precipitated calcium carbonate (PCC), **195, 196**

talc, 195

titanium dioxide, **195**

formation aid, **202, 447**

functional additive, **194-201**

internal sizing, **197-199**, 206-208, **429-431, 449-451**

reverse sizing, 198

rosin, 34, **197**, *198*, *344-346*, **449-450**

synthetic agents, AKD and ASA, *198*, 199, **207**, 450

metering and pumping of, **194**

retention aid, **201, 202**, 443-447, 451

starch, 199, 442

wet strength additives, **200, 201, 452**

epoxidized polyamine-polyamide resin, **200**, 452

glyoxal-polyacrylamide copolymer, *200*, **201**

melamine-formaldehyde, *200*, 452

urea-formaldehyde, **200**, 452

Aerated stabilization basins (ASB), **288**, 40

Air

composition of, *680*

properties of, *680*

Air flotation, *See* Deinking

Air knife coater, **254**, *259*

Air pollution, **290-293**

ambient sampling, 293

black liquor oxidation, **290**, *291*

gas analysis, **292, 293**, 720

nitrogen oxides, *291*, 292, 295

point sampling, 292

sulfide, reaction with lignin, *290*

sulfur dioxide, 291, 292

total reduced sulfur, **290-293**

Air quality tests, **290-293**

nitrogen oxides, *291*, 292

sulfur dioxide (SO_2), **291**

total reduced sulfur (TRS), **290**

AKD, *198*, **207, 450-451**

Alkaline pulping, *See* Kraft pulping

peeling reaction, **407-410**, *409*

Alkaline sulfite pulping, **95**, 99

Anilox rolls, 486

Alum chemistry, **205, 206**, *325*, **344-346**

Alum measurement, **718-719**

American Men and Women of Science, **11**

American Papermaker, **7**

Amphoteric elements, **318**

Anaerobic effluent treatment, **295**

Angiosperms, **16**, 18

Anode, 329-331

Anthraquinone (AQ), **83**, *86*, 97, **99**

APPITA, ix

Appita Journal, **7**

Arabinoglycans, **34**, *35*

ASA, *198*, **207, 450-451**

Ash, 14, 29, 32, **39**, 44, 46

wood, composition, **702**

Asplund process, 69, **70, 71**

ASTM, ix

Atactic polymer, **397**

Autocausticizing of kraft liquor, **708-709**

Bagasse, 47, **50**

Bamboo, 47, **50**

Bark, *13*, **14**, 525

composition, *13*, 14

fuel or heating value, **14**

pulp color, **21**

tolerances in wood, **13**

waxy materials of, **14**

Barkers, *24, 25*

cambial shear barker, **24**, 7

cutterhead barker, **24**, 7

drum barker, *24*, 699, 6

flail, *25*, 8

hydraulic barker, **25**

Basic specific gravity, wood, *42-46*, 349, 699

silviculture and, *536-537*

[1]**Boldface** numbers indicate a definition, discussion, or source of information, *italic* numbers indicate a figure or table, and <u>underlined</u> numbers indicate a color plate number.